BIOPHYSICO-CHEMICAL PROCESSES OF HEAVY METALS AND METALLOIDS IN SOIL ENVIRONMENTS

THE WILEY BICENTENNIAL–KNOWLEDGE FOR GENERATIONS

Each generation has its unique needs and aspirations. When Charles Wiley first opened his small printing shop in lower Manhattan in 1807, it was a generation of boundless potential searching for an identity. And we were there, helping to define a new American literary tradition. Over half a century later, in the midst of the Second Industrial Revolution, it was a generation focused on building the future. Once again, we were there, supplying the critical scientific, technical, and engineering knowledge that helped frame the world. Throughout the 20th Century, and into the new millennium, nations began to reach out beyond their own borders and a new international community was born. Wiley was there, expanding its operations around the world to enable a global exchange of ideas, opinions, and know-how.

For 200 years, Wiley has been an integral part of each generation's journey, enabling the flow of information and understanding necessary to meet their needs and fulfill their aspirations. Today, bold new technologies are changing the way we live and learn. Wiley will be there, providing you the must-have knowledge you need to imagine new worlds, new possibilities, and new opportunities.

Generations come and go, but you can always count on Wiley to provide you the knowledge you need, when and where you need it!

WILLIAM J. PESCE
PRESIDENT AND CHIEF EXECUTIVE OFFICER

PETER BOOTH WILEY
CHAIRMAN OF THE BOARD

BIOPHYSICO-CHEMICAL PROCESSES OF HEAVY METALS AND METALLOIDS IN SOIL ENVIRONMENTS

Edited by

ANTONIO VIOLANTE
PAN MING HUANG
GEOFFREY MICHAEL GADD

BICENTENNIAL 1807 WILEY 2007 BICENTENNIAL

WILEY-INTERSCIENCE

A JOHN WILEY & SONS, INC., PUBLICATION

Published by John Wiley & Sons, Inc., Hoboken, New Jersey.
Published simultaneously in Canada.

For general information on our other products and services or for technical support, please contact our Customer Care Department within the United States at (800) 762-2974, outside the United States at (317) 572-3993 or fax (317) 572-4002.

Wiley also publishes its books in a variety of electronic formats. Some content that appears in print may not be available in electronic formats. For more information about Wiley products, visit our web site at www.wiley.com.

Wiley Bicentennial Logo: Richard J. Pacifico

Library of Congress Cataloging-in-Publication Data:

Biophysico-chemical processes of heavy metals and metalloids in soil environments / edited by Antonio Violante, Pan Ming Huang, Geoffrey Michael Gadd.

p. cm.— (Wiley-IUPAC series. Biophysico-chemical processes in environmental systems)

Includes index.

ISBN 978-0-471-73778-0 (cloth)

1. Heavy metals—Environmental aspects. 2. Soil pollution. 3. Soils–Heavy metal content. I. Violante, A. (Antonio) II. Huang, P. M. III. Gadd, Geoffrey M.

TD879.H4B55 2007
628.5'5—dc22

2007019078

Printed in the United States of America

10 9 8 7 6 5 4 3 2 1

CONTENTS

CONTRIBUTORS

Michael J. Borda
Golder Associates, Inc.
200 Century Parkway, Suite C
Mt. Laurel, NJ 08054
USA
Tel.: 856-793-2005, ext. 34323
Fax: 856-793-2006
E-mail: m.j.borda@gmail.com

François Courchesne
Département de Géographie
Université de Montréal
C.P. 6128, succ. Centre-Ville
Montréal, Québec H3C 3J7
CANADA
Tel.: 514-343-8027
Fax: 514-343-8008
E-mail:
francois.courchesne@umontreal.ca

Louise J. Criscenti
Geochemistry Department
Sandia National Laboratories
P.O. Box 5800, MS 0754
Albuquerque, NM 87185-0750
USA
Tel.: 505-284-4357
Fax: 505-844-7354
E-mail: ljcrisc@sandia.gov

David Emerson
American Type Culture Collection
10801 University Boulevard
Manassas, VA 20110
USA
Tel.: 703-365-2804
Fax: 703-365-2790
E-mail: demerson@gmu.edu

Michael E. Essington
Department of Biosystems
Engineering and Soil Science
The University of Tennessee
2506 E.J. Chapman Drive
Knoxville, TN 37996-4531
USA
Tel.: 865-974-7266
Fax: 865-974-4514
E-mail: messington@utk.edu

Petr S. Fedotov
Vernadsky Institute of Geochemistry
and Analytical Chemistry
Russian Academy of Sciences
19 Kosygin Street
119991 Moscow
RUSSIA
Tel.: +7 (495) 1378608
Fax: +7 (495) 9382054
E-mail: fedotov_ps@mail.ru

Scott Fendorf
Department of Geological and Environmental Sciences
Stanford University
Stanford, CA 94305
USA
Tel.: 650-723-5238
Fax: 650-725-2199
E-mail: fendorf@stanford.edu

Geoffrey Michael Gadd
Division of Molecular and Environmental Microbiology
College of Life Sciences
University of Dundee
Dundee DD1 5EH
Scotland
UK
Tel.: +44 (0)1382 384765
Fax: +44 (0)1382 388216
E-mail: g.m.gadd@dundee.ac.uk

Sabine Goldberg
George E. Brown, Jr., Salinity Laboratory
USDA-ARS
450 West Big Springs Road
Riverside, CA 92507
USA
Tel.: 951-369-4820
Fax: 951-342-4962
E-mail: sgoldberg@ussl.ars.usda.gov

Markus Gräfe
Ross Street Building, Room 322
Faculty of Agriculture, Food and Natural Resources
The University of Sydney
Sydney, NSW 2006
AUSTRALIA
Tel.: +61(0)2-9036-6359
Fax: +61(0)2-9351-5108
E-mail: m.grafe@usyd.edu.au

Claire-Sophie Haudin
Andra, 1-7 rue Jean Monnet
92298 Châtenay-Malabry
FRANCE
E-mail: cshaudin@wanadoo.fr

Mitchell J. Herbel
BioScreen Testing Services
3904 Del Amo Boulevard
Torrance, CA 90503
USA
Tel.: 310-214-0043, ext. 113
Fax: 650-725-2199
E-mail: mherbel@bioscreen.net

Philippe Hinsinger
INRA–SupAgro–UMR 1222 Biogéochimie du Sol et de la Rhizosphère
Place Viala
F-34060 Montpellier cedex 1
FRANCE
Tel.: +33 (0)4 99 61 22 49
Fax: +33 (0)4 99 61 30 88
E-mail: philippe.hinsinger@supagro.inra.fr

Pan Ming Huang
Department of Soil Science
University of Saskatchewan
51 Campus Drive
Saskatoon, SK S7N 5A8
CANADA
Tel.: 001-306-966-6838
Fax: 001-306-966-6881
E-mail: pmh936@mail.usask.ca

Benjamin D. Kocar
Department of Geological and Environmental Sciences
Stanford University
Stanford, CA 94305
USA
Tel.: 650-723-4152
Fax: 650-725-2199
E-mail: Kocar@stanford.edu

Gummuluru S. R. Krishnamurti
313-855 West 16th Street
North Vancouver, BC V7P 1R2
CANADA

Tel.: 604-904-9168
E-mail: kgummuluru@yahoo.ca

Elisabetta Loffredo
Dipartimento di Biologia e Chimica Agroforestale ed Ambientale
Università di Bari
Via Amendola 165/A
70126 Bari
ITALY
Tel.: +39-080-5442282
Fax: +39-080-5442850
E-mail: loffredo@agr.uniba.it

Lena Q. Ma
Soil and Water Science Department
University of Florida
Gainesville, FL 32611-0290
USA
Tel.: 352-392-9063, ext. 208
Fax: 352-392-3902
E-mail: Lqma@ufl.edu

J. Patrick Megonigal
Smithsonian Environmental Research Center
P.O. Box 28
647 Contees Wharf Road
Edgewater, MD 21037
USA
Tel.: 443-482-2346
Fax: 443-482-2380
E-mail: megonigalp@si.edu

Ricardo Melamed
CETEM/MCT–Center for Mineral Technology
Ministry of Science and Technology
Av. Ipe 900
Ilha da Cidade Universitaria
Rio de Janeiro 21941-590
BRAZIL
Tel.: 55 (21) 865-7222 or 865-7303
Fax: 55 (21) 260-2837
E-mail: RMELAMED@cetem.gov.br

Manuel Miró
Department of Chemistry
University of the Balearic Islands
Faculty of Sciences
Carretera de Valldemossa Km 7.5
E-07122 Palma de Mallorca
Illes Balears
SPAIN
Tel.: +34 971 259576
Fax: +34 971 173426
E-mail: manuel.miro@uib.es

Ravendra Naidu
Cooperative Research Centre for Contamination Assessment and Remediation of the Environment
SPRI Building, Room W2-34
Mawson Lakes, South Australia 5095
AUSTRALIA
Tel.: 61 8 83025041
Fax: 61 8 83025041
E-mail: rava.naidu@unisa.edu.au

Scott C. Neubauer
Baruch Marine Field Laboratory
University of South Carolina
P.O. Box 1630
Georgetown, SC 29442
USA
Tel.: 843-546-3623
Fax: 843-546-1632
E-mail: scott@belle.baruch.sc.edu

Massimo Pigna
Dipartimento di Scienze del Suolo, della Pianta e dell'Ambiente
Università di Napoli Federico II
Via Università 100
80055 Portici (Napoli)
ITALY
Tel.: +39 081 2539176
Fax: +39 081 2539186
E-mail: massimo.pigna@unina.it

Nicola Senesi
Dipartimento di Biologia e Chimica Agroforestale e Ambientale
Università di Bari
Via Amendola 165/A

70126 Bari
ITALY
Tel.: +39 080 5442853
Fax: +39 080 5442850
E-mail: senesi@agr.uniba.it

George Shaw
Division of Agricultural and Environmental Sciences
University of Nottingham
University Park
Nottingham NG7 2RD
UK
Tel.: 44 (0)115 95 13206
Fax: 44 (0)115 95 16267
E-mail: george.shaw@nottingham.ac.uk

Donald L. Sparks
Department of Plant and Soil Sciences
University of Delaware
Newark, DE 19717-1303
USA
Tel.: 302-831-8153
Fax: 302-831-0605
E-mail: dlsparks@udel.edu

Siobhan Staunton
UMR Biogéochemie du Sol et de la Rhizosphère
INRA–Place Viala
F-34060 Montpellier Cedex
FRANCE
Tel.: 00 33 4 99 61 23 31
Fax: 00 33 499 61 30 88
E-mail: staunton@montpellier.inra.fr

Katharine J. Tufano
Department of Geological and Environmental Sciences
Braun Hall, Building 320, Room 118
Stanford University
Stanford, CA 94305
USA
Tel.: 650-723-4152
Fax: 650-725-2199
E-mail: ktufano@stanford.edu

Antonio Violante
Dipartimento di Scienze del Suolo, della Pianta e dell'Ambiente
Università di Napoli Federico II
Via Università 100
80055 Portici (Napoli)
ITALY
Tel.: +39 081 2539176
Fax: +39 081 2539186
E-mail: violante@unina.it

Guo Wang
Department of Resources and Environmental Sciences
Fujian Agriculture and Forestry University
Fuzhou, Fujian 350002
PEOPLE'S REPUBLIC OF CHINA
Tel.: 0591-83789357
Fax: 0591-83776849
E-mail: gwang572003@yahoo.com.cn

PREFACE

Pollution induced by heavy metals and metalloids in soils is a serious environmental problem because, in comparison with the atmosphere and water, the soil environment has a much lower ability to recover from toxic effects. In soil, trace elements potentially toxic to plants and other living organisms are involved in chemical and biological reactions such as solution and surface complexation, precipitation, sorption–desorption, and oxidation–reduction. These elements interact with a series of clay minerals, humic substances, metal oxides, microorganisms, extracellular enzymes, biopolymers, and other organic and inorganic ligands. Their behavior depends on chemical and physicochemical as well as biological processes and their interactions with microbial activities. These physicochemical–biological interactions would influence the transfer of these elements from the inorganic and organic soil constituents to the soil solution and to plants and contaminate the terrestrial food chain, thus endangering human and animal health. Biogeochemical processes operating in soil environments that affect the fate, behavior, and bioavailability of metals and metalloids are currently an area of active research.

As a recognition of the importance of biophysico-chemical processes of metals and metalloids in soil environments, we initiated the first volume of the IUPAC–Wiley book series "Biophysico-Chemical Processes in Environmental Systems." This volume, which consists of 15 chapters, is organized into three parts: Fundamentals of Biotic and Abiotic Interactions of Trace Metals and Metalloids with Soil Components; Transformations and Dynamics of Metals and Metalloids as Influenced by Soil–Root–Microbe Interactions; and Speciation, Mobility, and Bioavailability of Metals and Metalloids and Restoration of Contaminated Soils.

This book, contributed by a multidisciplinary group of soil and environmental scientists, provides the scientific community with a critical evaluation of the state of the art on the fundamentals of reactions and processes of these elements in soil environments, their speciation, mobility, bioavailability, and toxicity and their impact on the development of innovative restoration strategies. The book is an important guide to scientists interested in environmental sciences, soil chemistry and mineralogy, soil biochemistry, soil microbiology, and plant nutrition and physiology as related to understanding the biophysico-chemical processes of these pollutants in soil environments. The latest advances in spectroscopy to study various aspects of heavy metal and metalloid interactions with soil inorganic and organic components are also reviewed critically and addressed clearly. The authors are scientists who are experts in their fields and the chapters have been reviewed critically by at least two external referees. We wish to thank all of the referees, who generously contributed their time and talent to maintain the high quality of this volume. We also express our thanks to the staff of IUPAC and John Wiley & Sons for their invaluable support and cooperation in the publication of the book.

A. Violante
P. M. Huang
G. M. Gadd

SERIES PREFACE

Scientific progress is based ultimately on unification rather than fragmentation of knowledge. Environmental science is the fusion of physical and life sciences. Physical, chemical, and biological processes in the environment are not independent but rather, interactive processes. Therefore, it is essential to address physical, chemical, and biological interfacial interactions to understand the composition, complexity, and dynamics of ecosystems. Keeping separate these domains, no matter how fruitful, one cannot hope to deliver on the full promise of modern environmental science. The time is upon us to recognize that the new frontier in environmental science is the interface, wherever it remains unexplored.

The Division of Chemistry and the Environment of the International Union of Pure and Applied Chemistry (IUPAC) has recently approved the creation of an IUPAC-sponsored book series entitled *Biophysico-Chemical Processes in Environmental Systems* to be published by John Wiley & Sons, Hoboken, NJ. This series addresses the fundamentals of physical–chemical–biological interfacial interactions in the environment and the impacts on (1) the transformation, transport, and fate of nutrients and pollutants; (2) food chain contamination and food quality and safety; and (3) ecosystem health, including human health.

The book *Biophysico-Chemical Process of Heavy Metals and Metalloids in Soil Environments* is Volume I of this newly created Wiley–IUPAC series. The transformation of metals and metalloids is influenced by phsicochemical–biological interfacial interactions. The impacts of these interactive processes are especially important in the rhizoshpere, which is the bottleneck of the contamination of the terrestrial food chain, and in the location near the decomposition of biological residues. In contrast to classical books which focus largely on separate physical, chemical, and biological processes, this book is unique in integrating the frontiers of knowledge on both fundamentals and impacts of interfacial interactions of physical, chemical, and biological processes pertaining to the transformation, transport, bioavailability, and toxicity of metals and metalloids in soil environments.

This book can be used as an advanced reference book on biophysico-chemical processes of metals and metalloids in soils and related environments for senior undergraduate and graduate students in soil and environmental sciences. It is an essential reference for chemists and biologists studying environmental systems. It will serve as a useful resource book for instructors, research scientists, professional consultants, and others working on environmental and ecological systems.

SERIES EDITORS
P. M. HUANG
N. SENESI

ABOUT THE EDITORS

Antonio Violante is professor of soil chemistry at the Department of Soil, Plant and Environmental Sciences, University of Naples Federico II, Italy, where he has been actively involved in research and teaching since 1970. He received his academic degree in chemistry from the University of Naples in 1969. He spent two sabbatical years at the University of Wisconsin, Madison (1976–1977) and the University of Saskatchewan, Saskatoon, Canada (1981–1982) and was invited to serve as a visiting professor in the Department of Soil Science at the University of Saskatchewan in 1985, 1992, and 2003. He was Head of the Dipartimento di Scienze Chimico-Agrarie from 1994 to 1999 and is coordinator of the Doctoral School in Agrobiology and Agrochemistry.

Dr. Violante has taught courses in agricultural chemistry, soil chemistry, and soil fertility for undergraduate students and in environmental biogeochemistry and soil mineralogy for Ph.D. students. He has trained many M.S. and Ph.D. students and postdoctoral fellows and has received visiting scientists worldwide. Three of his students have received awards for their theses in soil chemistry.

He has contributed to research on the interface between soil chemistry and mineralogy and soil biology. His special areas of research include the formation mechanisms of aluminum hydroxides and oxyhydroxides, the surface chemistry and reactivities of short-range-ordered precipitation products of Al and Fe, the influence of biomolecules on the sorption and desorption of nutrients and xenobiotics on and from variable charge minerals and soils, the factors that influence the sorption and residual activity of enzymes on phyllosilicates, variable charge minerals, organomineral complexes, and soils and the chemistry of arsenic in soil environments.

Dr. Violante has served on many committees of the Italian Society of Soil Science (president of the session on soil chemistry) and the Italian Society of Agricultural Chemistry. He is vice-president and liaison officer of Gruppo Italiano AIPEA. He has been the scientific chairman and chief organizer of international and national congresses.

Dr. Violante is the author or coauthor of 167 refereed research articles and book chapters and invited reviews, and has coedited six books. He has also presented papers at many scientific congresses and symposia and has given invited lectures at universities and research institutes worldwide. He has international research and teaching experience in Canada, the United States, Europe, China, and Chile. He is a Fellow of the Soil Science Society of America and American Society of Agronomy.

Pan Ming Huang received his Ph.D. degree in soil science at the University of Wisconsin, Madison, in 1966. He is professor emeritus of soil science at the University of Saskatchewan, Saskatoon, Canada. His research work has significantly advanced the frontiers of knowledge on the formation chemistry and nature and surface reactivity of mineral colloids, organic matter, and organomineral complexes of soils and sediments and their role in the dynamics, transformations, and fate of nutrients, toxic metals, and xenobiotics in terrestrial and aquatic environments. His research findings, embodied in over 300 refereed scientific publications, including nine research papers published in *Nature* and others in leading journals, book chapters, and books, are fundamental to the development of sound strategies for managing land and water resources.

He has developed and taught courses in soil physical chemistry and mineralogy, soil analytical chemistry, and ecological toxicology. He has trained and inspired M.Sc. and Ph.D. students and postdoctoral fellows, and has received visiting scientists from all over the world. He has served on numerous national and international scientific and academic committees. He has also served as a member of many editorial boards, including those of the *Soil Science Society of America Journal; Geoderma; Chemosphere; Water, Air and Soil Pollution; Soil Science and Plant Nutrition;* and *Pedosphere*. He has served as a titular member of the Division of Chemistry and the Environment of the International Union of Pure and Applied Chemistry and is the founding chairman of the Working Group MO, "Interactions of Soil Minerals with Organic Components and Microorganisms," and the founding chair of Commission 2.5, "Soil Physical/Chemical/Biological Interfacial Reactions," of the International Union of Soil Sciences. He received the Distinguished Researcher Award from the University of Saskatchewan and the Soil Science Research Award from the Soil Science Society of America. He is a Fellow of the Canadian Society of Soil Science, the Soil Science Society of America, the American Society of Agronomy, the American Association for the Advancement of Science, and the World Innovation Foundation.

Geoffrey Michael Gadd is an international leader in the field of metals and microorganisms and is well known for research that has advanced understanding of cellular and physicochemical mechanisms underlying microbial–metal transformations, and their environmental and biotechnological significance, especially in relation to metal and radionuclide geomicrobiology. He holds a personal chair in microbiology at the University of Dundee and has served as Head of the Department of Biological Sciences, Head of the Division of Environmental and

Applied Biology, and as deputy research director of the School of Life Sciences; currently, he is Head of the Division of Molecular and Environmental Microbiology. Several scientific societies have recognized Professor Gadd's research through specific awards (e.g., the Berkeley Award of the British Mycological Society, the Charles Thom Award of the Society for Industrial Microbiology), and he is a Fellow of the Linnean Society, the Institute of Biology, the American Academy of Microbiology, and the Royal Society of Edinburgh. From 2004 to 2006 he served as president of the British Mycological Society (the first to serve a three-year term of office) and is currently convener of the Environmental Microbiology Group of the Society for General Microbiology. He has contributed to the vitality of his field via significant contributions to many professional societies and national and international editorial boards, advisory groups, and committees. Professor Gadd has published over 190 refereed scientific papers, over 40 edited books (two as sole editor), and over 35 invited book chapters.

PART I

FUNDAMENTALS OF BIOTIC AND ABIOTIC INTERACTIONS OF METALS AND METALLOIDS WITH SOIL COMPONENTS

1

IMPACTS OF PHYSICOCHEMICAL–BIOLOGICAL INTERACTIONS ON METAL AND METALLOID TRANSFORMATIONS IN SOILS: AN OVERVIEW

P. M. Huang

University of Saskatchewan, Saskatoon, Saskatchewan, Canada

Biophysico-Chemical Processes of Heavy Metals and Metalloids in Soil Environments,
Edited by Antonio Violante, Pan Ming Huang, and Geoffrey Michael Gadd

1.1. INTRODUCTION

Soil, geoderma, is the "skin" of planet Earth and the central organizer of the terrestrial ecosystem. It is a life-sustaining material which is a structurally porous and biologically active medium that has evolved over time on continental land surfaces. This material is formed and continues to develop through weathering processes driven by geological, biogeochemical, topological, climatic, chronological, and anthropogenic influences.

Soils are complex interactive biogeochemical reactors, reservoirs of organisms (microorganisms, protozoans, and nematodes), and major compartments of the terrestrial ecosystem. Minerals, organic components, and organisms are among major solid components of soils. These components are not separate entities but rather, form a unified system whose members are constantly in association with each other in the environment (Huang and Schnitzer, 1986; Huang et al., 1995; Banfield and Nealson, 1997). The association of microorganisms with soil mineral and organic colloids is depicted in Figure 1.1. Interactions among the components have an enormous impact on the physics, chemistry, and biology of soil and surrounding ecosystems (Huang, 2002). However, physical, chemical, and biological processes are not independent but rather, interactive processes in soil environments. A fundamental understanding of physical, chemical, and biological interfacial interactions in soil is essential in restoring, enhancing, and sustaining ecosystem productivity, services, and integrity and for establishing soil's impact on human welfare. In view of the mineral–organic component–organism interactions and interactive physical, chemical, and biological processes in the terrestrial ecosystem, a commission on Soil Physical/Chemical/Biological Interfacial Reactions was created within the structure of the International Union of Soil Sciences in 2004.

The transformation of metals and metalloids in the environment is influenced by physicochemical–biological interfacial interactions. The impact of these interactive interfacial processes on metal and metalloid transformations is especially important in the rhizosphere (Huang and Germida, 2002; Huang and Gobran, 2005) and near the decomposition of plant residues (Huang and Schnitzer, 1986), where the types and concentrations of substrates are different from those of the bulk soil because of enhanced biological activity. In the rhizosphere, root exudation leads to colonization by various populations of bacteria, fungi, protozoans, and nematodes. The distribution of microorganisms with distance from the root is illustrated in Figure 1.2. Plant–microbe interactions result in intense biological processes in the rhizosphere. These intense biological processes, in turn, affect physicochemical reactions in the rhizosphere. Physicochemical properties that can be different in the rhizosphere include acidity, concentration of complexing biomolecules, redox potential, ionic strength, moisture, and nutrient status. In the rhizosphere, it is difficult to separate the effects of microbial activity on the chemistry of metals and metalloids from the effects of plant root activity. Microorganisms act in a fashion similar to that of plant roots in that they can accumulate metals and metalloids through uptake and adsorption as well as mobilizing them through the action of microbial exudates, diffusates, and excretions.

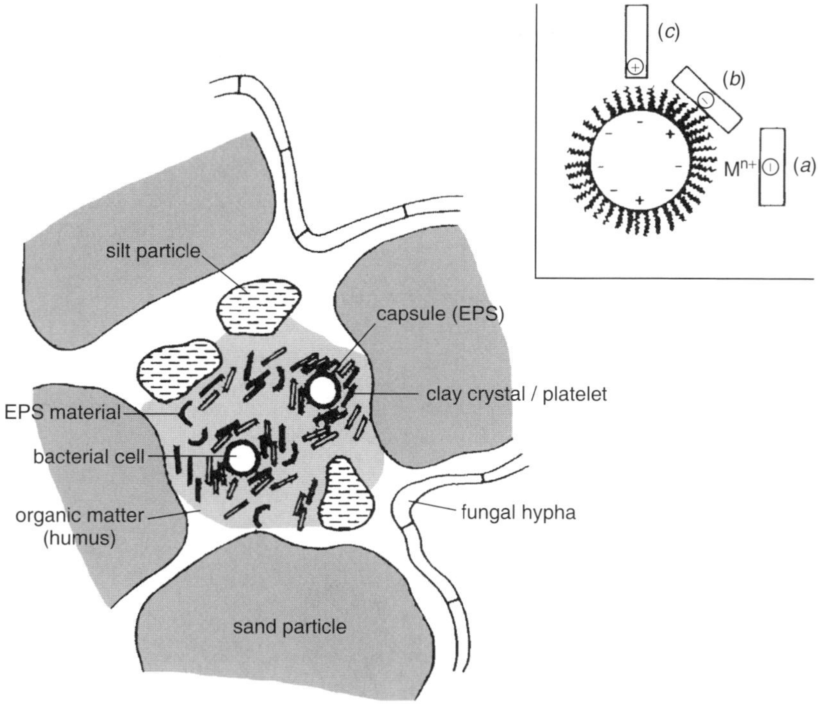

Figure 1.1. Interaction of bacteria and fungi with mineral particles in a soil aggregate. Bacterial cells with a coat of extracellular polysaccharides (EPSs) are enveloped by clay particles. The pore space where clays and bacteria interact, bounded by silt- and sand-size particles, is relatively enriched in organic matter, including EPS residues. Fungal hyphae are attached to the outside surface of an aggregate. Inset shows an enlarged view of a bacterial cell with its complement of EPS. At normal soil pH conditions, the cell has a net negative surface charge. Most clay particles adhere to the cell surface by bridging through polyvalent cations, represented by M^{n+} (*a*), although some may be attached directly by electrostatic interactions, either in face-to-face (*b*) or edge-to-face (*c*) association. (From Theng and Orchard, 1995.)

The total rhizosphere environment is governed by an interacting trinity of the soil, the plant, and the organisms associated with the root (Lynch, 1990a,b), as illustrated in Figure 1.3.

Therefore, the role of physicochemical–biological interfacial interactions in controlling the transformation, transport, fate, and toxicity of metals and metalloids in soil and surrounding environments, especially the rhizosphere, which is the bottleneck of contamination of the terrestrial food chain, deserves increasing attention. In this chapter we present an overview of this emerging and extremely important area of science, to advance our knowledge of the interface between physicochemical and biological reactions and processes in the environment.

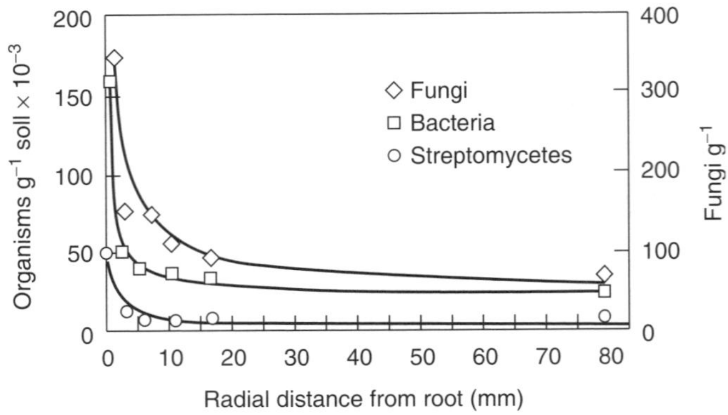

Figure 1.2. Distribution of organisms with distance from the roots of 18-day-old lupin (*Lupinus angustifolius* L.). (From Rovira and Davey, 1974.)

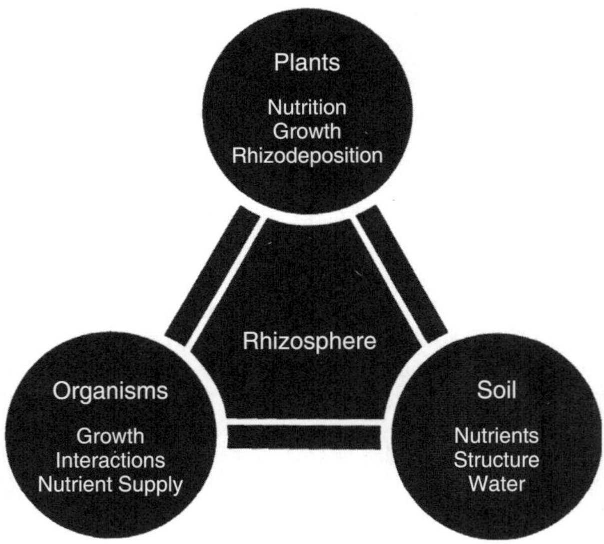

Figure 1.3. Rhizosphere trinity. The total rhizospere environment is determined by an interacting trinity of the soil, the plant, and the organism associated with the roots. (From Lynch, 1990a.)

1.2. MINERAL–ORGANIC MATTER–MICROORGANISM INTERACTIONS IN SOIL ENVIRONMENTS

Interactions of soil minerals with organic matter and microorganisms exert a vital role in the formation of short-range-ordered metal oxides, humic substances, mineral colloid–humic complexes, aggregates, and the structure and porosity of soils. These nonliving and living reactive components and the structural stability and associated nanoporosity of soils control the transformation, transport, fate, and toxicity of metals and metalloids in the environment.

1.2.1. Formation of Short-Range-Ordered Metal Oxides

Short-range-ordered Al and Fe oxides, especially nanoparticles, are indeed among the most reactive components of acidic and neutral soils (Bigham et al., 2002; Huang et al., 2002). In addition to humic substances, biomolecules, which are present in root exudates, microbial metabolites, and decomposition products of biological tissues play a very significant role in influencing the formation and transformation of these metal oxides and the resulting alteration of their surface properties pertaining to the speciation, transport, fate, and toxicity of metals and metalloids in soil environments (Huang and Violante, 1986; Schwertmann et al., 1986; Sposito, 1996, 2004; Huang and Wang, 1997; Huang et al., 2002; Violante et al., 2002; Huang, 2004). Details of this issue are treated by Violante et al., in Chapter 5.

As the third most abundant element on Earth's surface (after O and Si), Al is a major element in all mineral soils. There are only a few crystalline Al oxides, hydroxides, or oxyhydroxides, and only one, gibbsite, occurs to any great extent in soils. Aluminum, however, also forms a series of highly reactive soluble species and poorly crystalline to noncrystalline mineral colloids (Figure 1.4). The mode depicted in the figure incorporates multiple reaction paths. Reaction path I is believed to be common when systems are neutralized rapidly within seconds or hours. Under these conditions, little or no Al_{13} polynuclear species $[AlO_4Al_{12}(OH)_{24}(H_2O)^{12}_{7+}]$ forms; the formation of $Al(OH)_3$ nuclei is rapid, and gibbsite appears within days or weeks of aging. Reaction paths II and III both involve the formation of Al_{13}. Reaction path II is applicable when systems are neutralized relatively slowly compared with reaction path I. The individual Al_{13} ions can remain in solution for months to years, but eventually they transfer by one of the three pathways. If no Al_{13} nuclei are present, the soluble Al_{13} ions can slowly dissociate into Al^{3+} ions and deposit onto $Al(OH)_3$ nuclei (path IIa). These soluble Al_{13} may also aggregate and form Al_{13} nuclei (path IIb). If some Al_{13} nuclei are already present, the remaining Al_{13} ions can be deposited on them within weeks and months (path IIc). Path III is applicable when Al_{13} forms under conditions that promote the rapid formation of Al_{13} aggregates via an anion bridging mechanism (i.e., outer-sphere associations). These aggregates rearrange themselves rapidly into Al_{13} nuclei. The Al_{13} nuclei in paths IIb, IIc, and III then transform into a poorly ordered phase, microcrystalline boehmite (pseudoboehmite), which then transforms into gibbsite. The relative importance of each pathway depends greatly on reaction conditions. These soluble Al species and poorly ordered to noncrystalline Al species formed in the reaction pathways are highly reactive and thus important in influencing reactions and processes in the environment. The transformation of Al via the various pathways described above are strongly influenced by the nature and concentration of solution inorganic and organic ions and solid-state ions such as clay minerals and humic substances

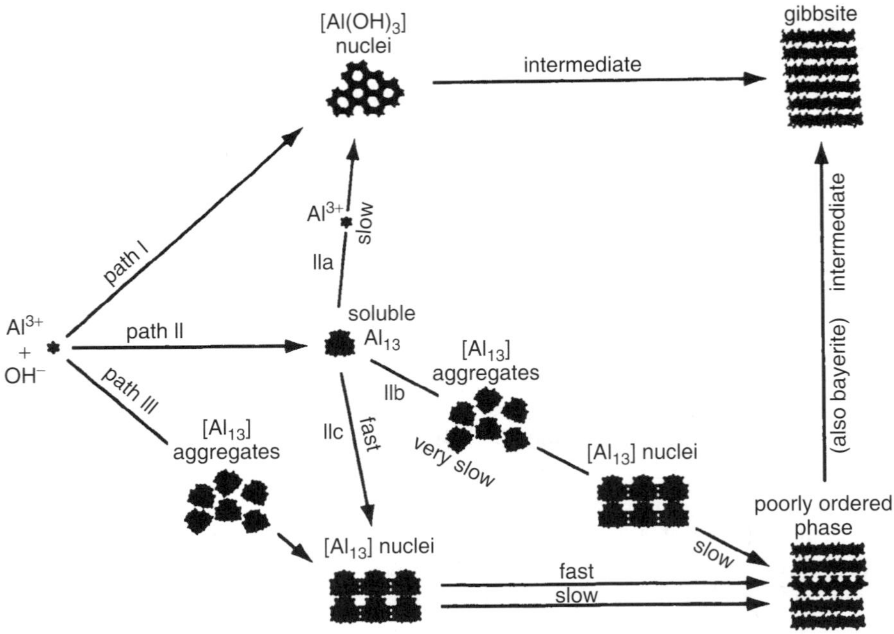

Figure 1.4. Reaction scheme proposed for Al hydroxide mineral formation from hydrolyzed Al solution. (Modified from Bertsch and Parker, 1996; Huang et al., 2002.)

(Huang and Violante, 1986, Huang, 1988; Bertsch and Parker, 1996; Krishnamurti et al., 1999, 2004; Huang et al., 2002).

Organic substances such as fulvic acid (FA) and humic acid (HA) and many low-molecular-mass organic acids (LMMOAs) form stable complexes with Al and disrupt the crystallization of Al hydroxides and thus promote the formation of short-range-ordered (SRO) Al hydroxides (Kwong and Huang, 1975, 1979a,b, 1981; Lind and Hem, 1975; Kodama and Schnitzer, 1980; Violante and Huang, 1985; Singer and Huang, 1990; Violante et al., 2002; Colombo et al., 2004). Organics greatly influence the surface properties of Al transformation products. For instance, the presence of organic acids during aging of Al hydroxide gels for 40 days increases the specific surface of the precipitation products up to 30-fold over that of the control and higher acid concentrations results in higher specific surfaces (Table 1.1). The surface charge characteristics of the precipitation products are also altered dramatically (Figure 1.5). These SRO transformation products of Al should thus exert substantial influence on the adsorption–desorption, transformation, and dynamics of metals and metalloids in the environment.

Interactions of soil minerals with organic substances and microorganisms have a great impact on the formation and transformation of SRO Fe oxides (Schwertmann et al., 1986; Cornell and Schwertmann, 2003). Microorganisms may influence Fe transformation through reduction and oxidation reactions. Soil

TABLE 1.1. Specific Surface Area of Al Hydroxide Precipitation Products[a]

Organic Acid	Organic Acid Concentration (mol L^{-1}) 1.0×10^{-6}	1.0×10^{-4}
	$m^2\ g^{-1}$	
None	20	20
p-Hydroxybenzoic	22	28
Aspartic	27	587
Tannic	95	195
Malic	36	635
Citric	117	295

Source: Kwong and Huang (1981).

[a]Formed in the presence of tannic acid and selected low-molecular-weight organic acids in systems at an initial Al concentration of 1.1×10^{-3} mol L^{-1} and OH/Al molar ratios of 3.0 and aged for 40 days at room temperatures.

organic matter and microorganisms play a vital role in influencing redox cycling of Fe, hydrolytic reactions of Fe, crystallization of Fe precipitation products, and the subsequent formation of pedogenic Fe oxides (Huang and Wang, 1997). Furthermore, the fine-scale morphology, mean surface roughness, fractal dimension, specific surface, microporosity, and surface charge properties of Fe oxides are strongly influenced by biomolecules (Liu and Huang, 1999, 2003). The surface of Fe oxides is the region of their interactions with the soil solution, organic and inorganic particles, plant roots, microorganisms, and other soil biota. Surface properties of Fe oxides should have a profound impact on microaggregate formation, soil biota habitat, water flux, metal and metalloid flux, and the ability of soils to respond to management. Therefore, surface properties of Fe oxides formed under the influence of biomolecules deserve close attention in advancing our understanding of the surface chemistry pertaining to biogeochemical transformations of metals and metalloids in the terrestrial ecosystem.

1.2.2. Formation of Humic Substances and Organomineral Complexes

Humic substances are formed through biotic and abiotic processes (Stevenson, 1994; Huang, 2004). A variety of biomolecules, such as carbohydrates, phenolic compounds, and amino acids, can participate as raw materials. Soil mineral surface chemistry plays a vital role in the catalysis of abiotic formation of humic substances. The Maillard reaction (Maillard, 1913) is perceived to be a major pathway in humification because of significant similarities between humic substances and melanoidins formed through this pathway involving sugar–amino acid condensations (Ikan et al., 1996). The presence of characteristic products of the Maillard reaction (alkyl pyrazines) was detected in archaeological plant remains up to 1500 years in age (Evershed et al., 1997). The great appeal of the Maillard reaction in humification processes lies in the two types of precursors, sugars and amino acids, which are among the most abundant constituents of

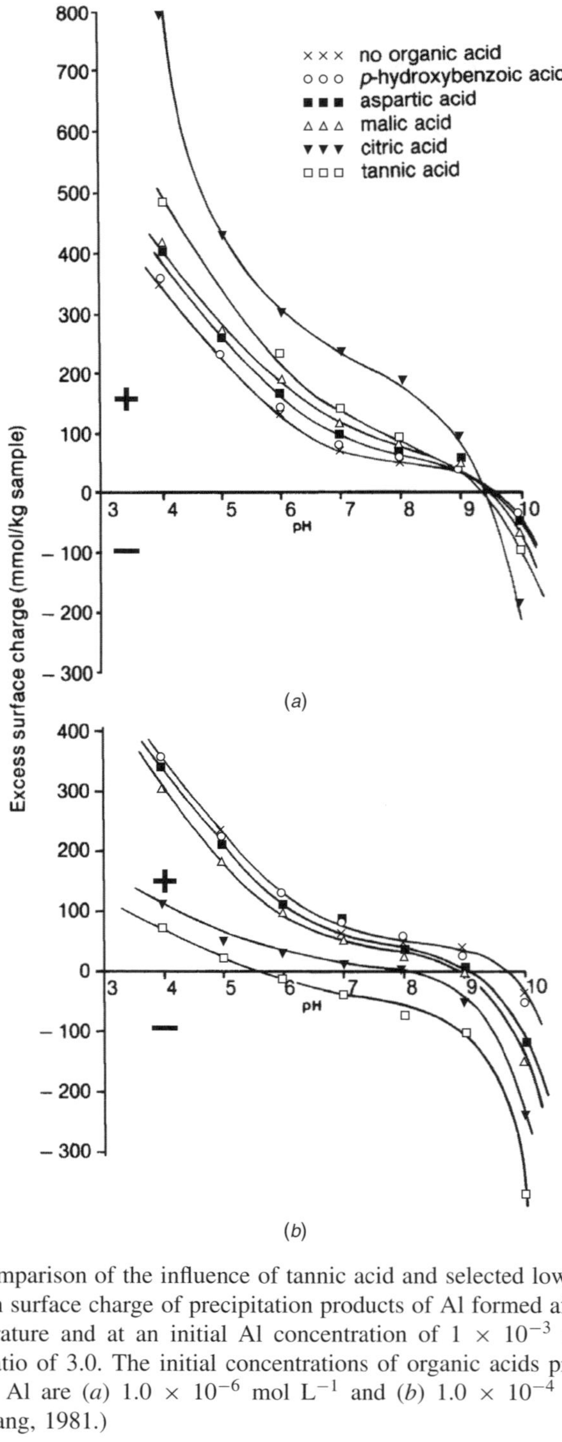

Figure 1.5. Comparison of the influence of tannic acid and selected low-molecular-mass organic acids on surface charge of precipitation products of Al formed after 40-day aging at room temperature and at an initial Al concentration of 1×10^{-3} mol L^{-1} and an OH/Al molar ratio of 3.0. The initial concentrations of organic acids present during the precipitation of Al are (*a*) 1.0×10^{-6} mol L^{-1} and (*b*) 1.0×10^{-4} mol L^{-1}. (From Kwong and Huang, 1981.)

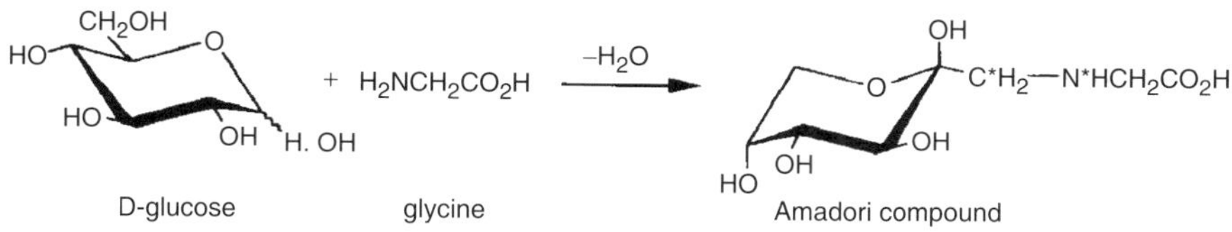

Figure 1.6. Formation of the Amadori compound from D-glucopyranose (D-glucose) and glycine. Note that a molecule of water is split off. The Amadori compound is an intermediate product in the Maillard reaction, which is perceived as one of the pathways in humification. (From Mossine et al., 1994.)

terrestrial and aquatic environments (Anderson et al., 1989). A major criticism of the Maillard reaction has been that it is very slow under ambient conditions (Hedges, 1988). To elucidate some details of the process, Jokic et al. (2001a) applied molecular topological analysis to investigate the initial reaction between D-glucose and glycine to form the Amadori compound fructosylglycine, which is an intermediate product in the Maillard reaction (Figure 1.6). Their calculations show that fructosylglycine and water and D-glucose and glycine as separate entities are very close to each other in terms of their ground-state energy. Therefore, the potential energy barrier is high and the reaction between D-glucose and glycine alone to form fructosylglycine is thus very slow at room temperature. Jokic et al. (2001b) demonstrated that the action of δ-MnO_2 under ambient environmental conditions accelerates the Maillard reaction by about two orders of magnitude, lending credence as an important abiotic pathway for the formation of humic substances in nature.

Soil minerals also play an important role in accelerating abiotic polymerization of phenolic compounds, polycondensation of phenolic compounds and amino acids, and subsequent formation of humic substances (Wang et al., 1986; Huang, 1990, 2000). Kumada and Kato (1970), Filip et al. (1977), and Wang and Li (1977) are among the pioneers in the study of browning of polyphenols catalyzed by clay-size phyllosilicates. Since the early 1980s, Huang and co-workers have investigated the sequence of catalytic ability of phyllosilicates and their reaction sites in the polymerization of phenolic compounds and subsequent formation of humic substances (Shindo and Huang, 1985a,b; Wang and Huang, 1986, 1988, 1994). Among Al, Fe, and Mn oxides, hydroxides, and oxyhydroxides, Mn oxides are the most reactive catalysts in the transformation of phenolic compounds by virtue of their high oxidation potentials, high specific surfaces, and high surface reactivity (Shindo and Huang, 1982, 1984; Wang and Huang, 2000a,b). Manganese oxides (birnessite, cryptomelane, and pyrolusite), which are common in soils, act as Lewis acids that accept electrons from phenolic compounds, resulting in the formation of semiquinones, oxidative polymerization, and the genesis of humic substances. Therefore, the catalytic power of these oxides in the formation of humic substances via abiotic processes in soil and surrounding environments merits close scrutiny.

TABLE 1.2. Visible Absorbances of Supernatants in a Series of Reaction Systems

	Absorbance	
Reaction System	400 nm	600 nm
25°C, 60 days		
Glucose–glycine	0.077 ± 0.004	0.021 ± 0.001
Glucose–glycine–δ-MnO_2	0.367 ± 0.006	0.067 ± 0.002
Glucose–glycine–catechol	10.8 ± 0.6	3.8 ± 0.2
Glucose–glycine–catechol–δ-MnO_2	90.4 ± 3.8	37.6 ± 1.7
45°C, 15 days		
Glucose–glycine	0.342 ± 0.018	0.103 ± 0.006
Glucose–glycine–δ-MnO_2	14.7 ± 0.68	1.50 ± 0.10
Glucose–glycine–catechol	5.60 ± 0.35	2.13 ± 0.08
Glucose–glycine–catechol–δ-MnO_2	73.7 ± 1.3	24.7 ± 1.5

Source: Jokic et al. (2004).

The Maillard reaction and the polyphenol transformation model are considered as separate significant pathways for the formation of humic substances. In nature, however, it is likely that these two pathways do not occur separately, but rather, interact closely with each other, since sugar, amino acids, and polyphenols coexist in soil solutions and natural waters. Jokic et al. (2004) reported that soil mineral colloids such as δ-MnO_2 significantly accelerate humification processes in systems containing glucose, glycine, and catechol under ambient conditions (Table 1.2). The promoting effect of δ-MnO_2 on a system consisting of carbohydrate, amino acid, and polyphenol is a complex process involving mineral surface sorption and condensation. This points to a linking of the polyphenol and Maillard reactions in humification pathways, which is a significant advance in our understanding of the chemistry of humification and its impact on biogeochemical reactions and processes in the environment.

Humic substances are the principal organic components of soils and waters, in which they interact with metal ions, metal oxides, metal oxyhydroxides, metal hydroxides, and clay minerals to form complexes of widely differing physicochemical and biological stabilities (Senesi and Loffredo, 2005). These interaction products affect the moisture and aeration regime; the exchange capacity; the transformation, transport, and fate of metals and metalloids; and other physical, chemical, and biological reactions and processes. In soils and waters, large amounts of humic substances are adsorbed on the surfaces of hydrous metal oxides and clay minerals. The importance of metal–humic and clay–humic interactions in predominantly mineral soils, with the exception of extremely sandy ones, is illustrated by the data presented by Greenland (1965), which show that between 52 and 98% of the organic C in a wide range of soils is associated with the clay fraction. It is likely that most of the remaining organic C is linked to metal oxides, oxyhydroxides, and hydroxides (Schnitzer and Kodama, 1977). Short-range-ordered mineral colloids are especially reactive in binding and stabilizing organic matter, due to their high specific surface area and high charge density (Wada, 1995; Torn et al., 1997; Guggenberger and Haider, 2002).

1.2.3. Formation of Aggregates and Porosity

Interactions of mineral particles with microorganisms and organic matter in the formation of aggregates and porosity of soils are depicted in Figure 1.1. Microbial cells with a coat of extracellular polysaccharides (EPSs) are enveloped by clay particles. At normal soil pH conditions, the microbial cell has a net negative surface charge. Most clay particles adhere to the microbial cell surface with a coat of EPS by bridging through polyvalent cations. Some may be attached directly by electrostatic interactions, either in face-to-face or edge-to-face association. The pore space where clays and bacteria interact, bounded by silt- and sand-size particles, is relatively enriched in organic matter, including EPS residues.

Polyvalent cation bridging may be direct [scheme (1.1)] or effected by hydrogen bonding to water molecules in the primary hydration shell of a polyvalent cation [scheme (1.2)]:

$$[\text{clay}]^{-} \cdots \text{M}^{n+} \cdots {}^{-}[\text{EPS} - \text{B}] \tag{1.1}$$

or

$$[\text{clay}]^{-} \cdots \text{M}^{n+} \cdots \underset{}{\overset{\text{H}\atop|}{\text{O}}}\text{H} \cdots {}^{-}[\text{EPS} - \text{B}] \tag{1.2}$$

where M^{n+} denotes a polyvalent cation of valency n and B is a bacterial cell (Theng and Orchard, 1995). The predominant mechanism would depend on the nature of the polyvalent cation but even more on the hydration status of the soil. Dehydration conditions would clearly favor the mechanism shown in scheme (1.1). In addition, short-range interactions (van der Waals, hydrogen bonding) may come into play because the EPS chains are induced to make close contact with the surface, increasing the net energy of interactions.

Mineral colloids are coated with hydroxy Al (or Fe) polymers in both interlayers (Brown, 1953; Rich and Obenshain, 1955; Rich, 1960, 1968, Jackson, 1962; Barnhisel and Bertsch, 1989; Huang et al., 2002) and on the edges and external planar surfaces (Huang and Kozak, 1970; Huang et al., 2002). Furthermore, hydroxyaluminosilicate ions, which are common in soil solutions, especially in temperate and subtropical regions, can be adsorbed on the surfaces of mineral colloids (Lou and Huang, 1988). The coated mineral colloids thus behave as a positively charged species or display amphoteric characteristics. Therefore, mineral colloids can interact strongly with negatively charged microbial cells coated with EPS in soil environments. This type of bonding, which is stronger than cation bridging, is also expected to occur with discrete hydrous Al and Fe oxides in the pH range of soils. The attachment of microorganisms to SRO mineral colloids and the crystal edges of phyllosilicates through electrostatic interactions and subsequent binding to mineral surfaces would also be predicted to occur when the soil pH falls below 6 because all of these surfaces would then be largely positively charged.

In the majority of cases, minerals in top soils are partially covered with organic materials, especially humic substances (HSs), which are microbially resistant and most prevalent. The most common mode of mineral colloid–organic matter–microbe interactions may be represented by scheme (1.3). In humic-rich calcareous Mollisols, Ca would be the predominant bridging cation:

$$[\text{mineral colloid} - \text{HS}]^{-} \cdots \text{M}^{n+} \cdots {}^{-}[\text{EPS} - \text{B}] \quad (1.3)$$

In Andisols, Oxisols, Ultisols, and the B horizons of Spodosols, HSs occur largely as complexes with Al and Fe or their respective poorly crystalline and noncrystalline oxides (Oades et al., 1989; Theng et al., 1989). In soils with little organic matter and in subsoils, mineral colloid–microbe interactions are governed largely by the mineralogical composition and pH of the system.

Interactions between soil mineral particles, organic substances, and organisms can occur at many different size scales, as these materials have a large size range in soils (Figure 1.7). Therefore, it is important to indicate the size scale being considered when discussing soil aggregates and soil structure and the mechanisms of their stabilization, because the potential mechanisms for stabilization vary with aggregate size. The adsorption of organic molecules such as microbially derived polysaccharides and other unaltered and altered biomolecules onto mineral surfaces is important in binding together the clay microstructures and silt particles into small microaggregates with 2- to 50-μm diameters and density greater than 2.0 Mg m^{-3} (Baldock, 2002). Many microaggregates contain pieces of fungal hyphae, bacteria, or bacterial colonies coated with EPSs and clay minerals (Oades and Waters, 1991). The polysaccharides are present throughout the matrix but concentrated in pores between clay microstructure.

Particulate organic matter (POM) is an important stabilizing agent at larger size scales: large microaggregates and small macroaggregates (Jastrow and Miller, 1998). Soil structure can be stabilized by POM through two mechanisms related to its physical properties and its susceptibility to biological decomposition (Baldock, 2002). The bridging can result from a combination of binding to aggregate surfaces, penetration through aggregates, and the formation of a network capable of holding groups of aggregates together. POM also enhances the stability of soil structure by providing a substrate for microorganisms to enhance the production of fungal hyphae and microbial metabolites such as polysaccharides. Mechanisms of stabilization of soil structure can operate over larger distances to bind microaggregates together to form macroaggregates (Figure 1.7). In view of the distances involved, the stabilization of macroaggregates is related to the presence of nonliving POM capable of spanning distances greater than 100 μm or the existence of a network of fungal hyphae and plant roots that physically enmeshes microaggregates.

The formation of aggregates alters the spatial arrangement of soil particles within soil aggregates, and the pore-size distribution even more (Theng and

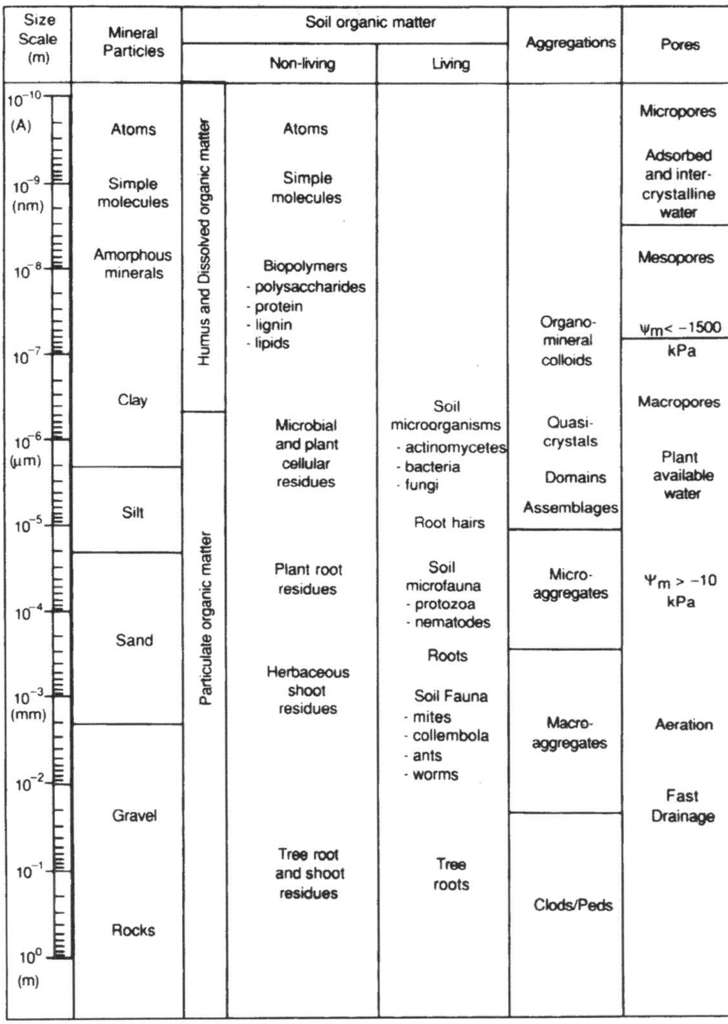

Figure 1.7. Size scales associated with soil mineral particles, organic components, pores, and aggregations of mineral and organic components. The definitions of pore size have used those developed by IUPAC (micropores, <2 nm; mesopores, 2 to 50 nm; and macropores, > 50 nm). Alternatively, the pore sizes corresponding to the lower ($\psi_m = -1500$ kPa) and upper ($\psi_m = -100$ kPa) limits of water availability to plants may be used to define the boundaries between the various classes of pore size. (From Baldock, 2002.)

Orchard, 1995). Such a modification of aggregate "architecture" results in a complex and discontinuous pattern in pore spaces of various sizes and shapes (Haider and Guggenberger, 2005) and would thus greatly influence the physical, chemical, and biological properties of soils. The dimension of pore sizes ranges from <2 nm (micropores) to 2 to 50 nm (mesopores) to >50 nm (macropores). The pore neck determines the accessibility by microbes. Microbial access to metals and metalloids adsorbed in small pores may be limited by steric factors and thus prevent microbial and biochemical transformations of these elements. Most recently, research findings show that organic substances such as tannate ligands induce the development of microporosity through structural perturbation of Al hydroxides (Yu et al., 2007). Furthermore, the kinetics of adsorption of metals such as Cd are greatly influenced by structural perturbation-induced microporosity development (Yu et al., 2006). Therefore, the impact of the formation of aggregates and porosity of various sizes and shapes through mineral–organic matter–microbe interaction on the transformation of metals and metalloids in soil environments should be an important and exciting area of research for years to come.

1.3. SOLUTION COMPLEXATION REACTIONS OF METALS

Complexation reactions in the soil solution affect metal transformation and transport in soils, especially the rhizosphere (Huang and Gobran, 2005). Metal complexation reactions with ligands in the soil solution play a significant role in controlling the chemical reactivity, bioavailability, and toxicity of metals in soil environments. The most important complexing functional groups of soil organic matter can be classified based on their affinity for hard, borderline, and soft metals (Buffle, 1988). For soft metals (e.g., Cd^{2+}, Pb^{2+}, Hg^{2+}), the order of donor atom affinity is S > N > O; the reverse order is true for hard cations (e.g., alkaline-earth metals Ca^{2+} and Mg^{2+}). Borderline metals, including most transition metals (e.g., Fe^{3+}, Cu^{2+}, Zn^{2+}, Mn^{2+}), have properties intermediate between hard and soft metals and thus demonstrate appreciable affinity for both hard and soft ligands. In general, the competitive reactions for a given ligand essentially involve hard and borderline metals for O sites, and borderline and soft metas for N and S sites, with competition between hard and soft metals being weak (Buffle, 1988; Senesi and Loffredo, 2005). The typical affinity sequence of soil organic matter for divalent metal ions (at pH 5) generally parallels the metal electronegativity values of Linus Pauling; however, the type, source, and concentration of soil organic matter, the amount of metal bound, and the method used to measure metal bonding and pH can affect metal binding affinity (Stevenson and Ardakani, 1972; Davies et al., 1969). Details of metal complexation reactions by humic substances are discussed by Senesi and Loffredo in Chapter 4.

Biomolecules are present in soils, especially in the rhizosphere, because of root exudates and microbial metabolites (Lynch, 1990a,b; Marschner, 1998). Furthermore, in view of the stability constants of these biomolecular ligands with metals (NIST, 2004), a substantial fraction of the soluble metal ions in soil solution may be complexed with a series of biomolecules commonly present in microbial metabolites and root exudates. The concentrations in the rhizosphere of such metals as Cu, Mn, Zn, and Co vary with the seasons and are related to the presence of complexing biomolecules (Nielson, 1976; Linehan et al., 1989). The pH of the rhizosphere soil is generally lower than that of the corresponding

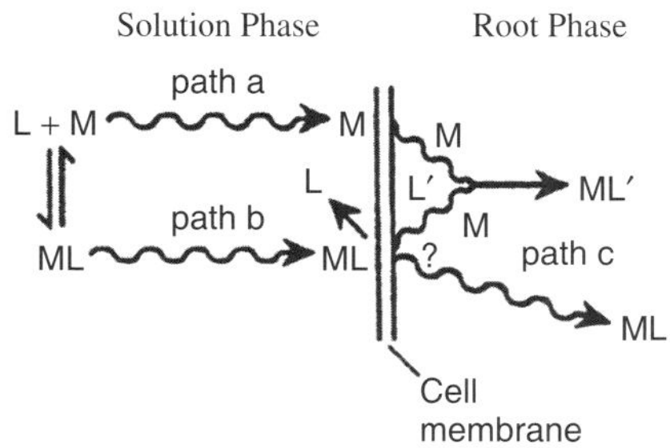

Figure 1.8. Conceptual model for potential uptake by plants of metal ions (M) complexed by organic ligands (L) (Laurie et al., 1991). (After McLaughlin et al., 1998.)

bulk soil. The cadmium availability index (CAI) values in the rhizosphere are higher, indicating that more Cd is complexed with biomolecules at the soil–root interface (Krishnamurti et al., 1996). The plant root and prolific microbial activity should result in increased amounts of biomolecules. A larger fraction of the metal contaminant will thus be in a complexed and usually soluble form in the soil–root interface than in the bulk soil.

Research on the metal speciation of the soil solution has been encouraged by the free metal ion hypothesis in environmental toxicology (Lund, 1990). This hypothesis states that the toxicity or bioavailability of a metal is related to the activity of the free aquo ion. This hypothesis is gaining popularity in studies of soil–plant relations (Parker et al., 1995). However, some evidence is now emerging that free metal ion hypothesis may not be valid in all situations (Tessier and Turner, 1995). Plant uptake of metals varies with the types of chelators present in solution at the same free metal activity. Furthermore, given the same chelate, total metal concentration in solution affects metal uptake by plants. Either kinetic limitations to dissociation of the complex or uptake of the intact complex could explain these observations (Laurie et al., 1991). The possible reactions of complexed metals at the soil–root interface and the potential uptake by plants of metal–organic complexes are depicted in Figure 1.8.

1.4. REDUCTION–OXIDATION REACTIONS OF METALS AND METALLOIDS

Changes in the oxidation state of trace metals can occur depending on the redox condition of the environment. Redox reactions are thus important in influencing the chemical speciation of a number of metals and metalloids, notably Hg, As, Se, Cr, Pu, Co, Pb, Ni, and Cu (Oscarson et al., 1981; Bartlett and James, 1993; Alloway, 1995; Myneni et al., 1997; Huang, 2000; James and Bartlett, 2000; Adriano, 2001; Sparks, 2003). Redox reactions also exert a great influence in the transformation and reactivity of Fe and Mn oxides in soils that have an enormous capacity to adsorb metals and metalloids (Huang and Germida, 2002). Furthermore, reduction of sulfate to sulfide in an anerobic environment also affects the transformation, solubility, and bioavailability of these pollutants through the formation of highly insoluble metal sulfides.

Masschelyn and Patrick (1994) have summarized the critical redox potentials for the transformation of some metal contaminants in soil environments. There has been little information on how changes in soil redox potential in the rhizosphere could affect the transformation of metal and metalloid contaminants. The generation of biomolecules through root exudation and microbial metabolism in the rhizosphere influences the redox potential. From a thermodynamic point of view, complexation of ligands with metals both on solid and solute phases has a dramatic effect on the redox potentials (Stumm and Morgan, 1996). For exemplification, Fe(II) and Fe(III) are used to illustrate the consequences of complexation on the redox potentials because (1) more data are available with this redox pair than with others, and (2) the transformations of Fe are especially important in the redox cycling of electrons in natural environments. The Fe(III)/Fe(II) redox couple can be adjusted with appropriate ligands to any redox potential within the entire range of the stability of water. As illustrated in Figure 1.9, the redox potential at pH 7, E^{o}_{H} (pH 7), decreases in the presence of most complexing ligands, especially chelates with oxygen donor atoms, such as citrate, EDTA, and salicylate, because these ligands form stronger complexes with Fe(III) than with Fe(II). Phenanthroline, which stabilizes Fe(II) more than Fe(III), is an exception. But Fe(II) complexes are usually stronger reductants than Fe^{2+}. The range of redox potentials for heme derivatives given on the right in Figure 1.9 illustrates the possibilities involved in bioinorganic systems. The principles exemplified here are applicable to other redox systems. Furthermore, when metals are complexed with ligands, the kinetics of their oxidation are substantially retarded. This complexation effect is clearly illustrated in the kinetics of Fe(II) oxidation as influenced by a series of organic ligands (Table 1.3). Therefore, the consequences of complexation on the redox potentials of soils and related environments and the impact on the transformations of metals and metalloids warrant in-depth research.

Although complexation with most complexing ligands, such as biomolecules in the rhizosphere, should decrease the redox potentials, the creation of an oxidized zone adjacent to the plant root in wetland soils has been identified as one process affecting the chemistry of Zn, Cu, and As in soils. In wetland soils, steep gradients in redox potentials develop around plant roots. This process is reflected in the precipitation of FeOOH (iron plaque) on the roots (Otte et al., 1989; Kirk and Bajita, 1995). Compared with the surrounding soil, these Fe-rich plaques on the roots of the saltmarsh plant *Aster tripolium* are enriched in Zn and Cu (Otte et al., 1989). Zinc also accumulates in the rhizosphere (*Oryza sativa* L.), which is the result of the formation of a zone of oxidation of Fe^{2+} to Fe^{3+} adjacent to the roots (Kirk and Bajita, 1995). Zinc concentration is higher in red roots (with iron plaque) than in white roots; a positive effect of the Fe concentration on the root surface, up to a certain level, on Zn uptake into the xylem fluid has been demonstrated (Otte et al., 1989). Above this level of Fe coating, Zn uptake by the plant is reduced, which is attributed to complete coating of the root surface by

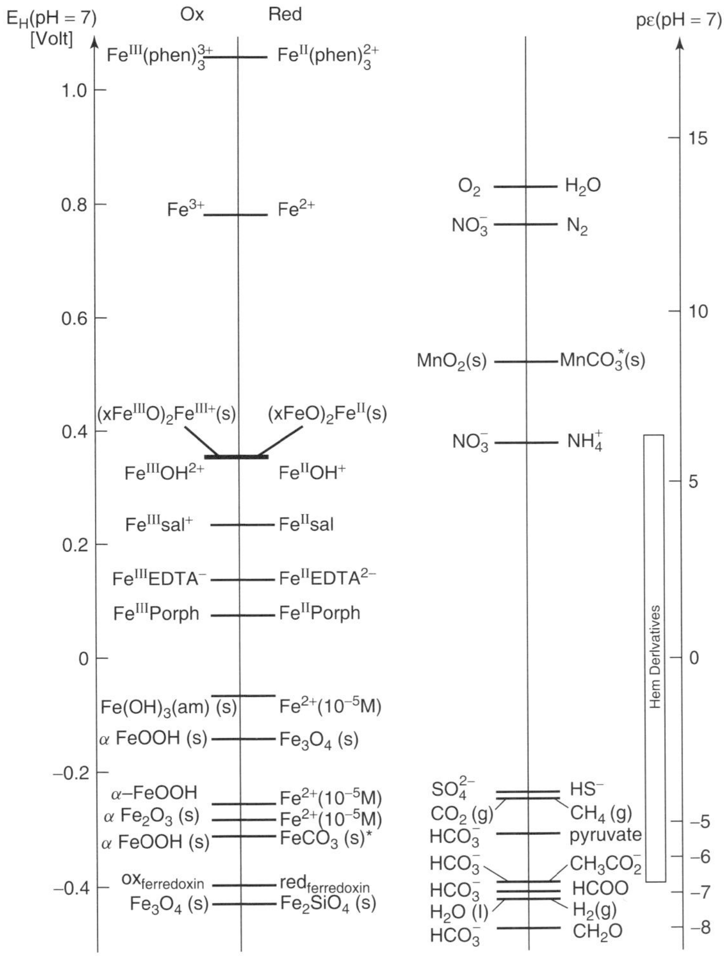

Figure 1.9. Representative Fe(II)/Fe(III) redox couples at pH 7 (phen, phenanthroline; sal, salicylate; porph, porphyrin; *, valid for $[HCO_3^-] = 10^{-3}$ mol L^{-1}). Complex formation with Fe(II) and Fe(III) both on solid and solute phases has a dramatic effect on the redox potentials; thus electron transfer by the Fe(II), Fe(III) system can occur at pH 7 over the entire range of the stability of water; E_H (−0.5 to 1.1 V). (= $Fe^{III}O)_2Fe^{III}$ refers to Fe^{3+} adsorbed inner spherically to a surface of a hydrous ferric oxide. The range of redox potentials for heme derivatives given on the right illustrates the possibilities involved in bioinorganic systems. (From Stumm and Morgan, 1996.)

FeOOH and blocking of absorption sites. In reduced conditions, As is mobilized as a result of reduction of Fe and Mn oxides and reduction of As(V) to As(III). However, in the rhizosphere in wetlands, As is immobilized due to oxidation of As and its adsorption onto FeOOH (Otte et al., 1991). Most of the As in the rhizosphere is probably retained on the root surface (Otte et al., 1991, 1995).

TABLE 1.3. Kinetics of Oxygenation of Fe(II)[a]

Ligand	Stability Constant (log *K*) of Fe–Ligand Complex[b] Fe(II)	Fe(III)	Rate Constant[c] ($min^{-1} \times 10^4$)
Oxalate	2.52	8.63	26.7 ± 1.2
Acetate	1.22	3.93	24.5 ± 1.1
Tartrate	4.85	9.00	8.6 ± 0.8
Citrate	3.86	10.69	7.6 ± 0.9

Source: Krishnamurti and Huang (1990).
[a]As 0.01 mol L^{-1} ferrous perchlorate in the absence and presence of organic complexing ligands.
[b]The log *K* values are for $\mu = 0.01$ and at 25°C.
[c]The rate constant was obtained at a ligand/Fe molar ratio of 0.1 at pH 6.00 and 23.5°C.

1.5. SORPTION–DESORPTION REACTIONS OF METALS AND METALLOIDS

Metals and metalloids undergo a series of reactions involving both the aqueous and solid phases, which vary in space and time. Sorption may be considered as a major process controlling the partitioning of metals and metalloids between the aqueous and solid phases in soils. It may represent the combined effects of ion exchange, nonspecific and specific adsorption, surface complexation, and (co)precipitation on the transfer of ions from the aqueous phase to the solid phase. Adsorption refers to ion exchange, nonspecific and specific adsorption, and surface complexation. When no molecular-scale data on which to base a decision as to mechanism are available, the prima facie loss of an adsorptive from aqueous solution to the solid phase can be termed sorption to avoid the implication that either adsorption or precipitation has taken place (Sposito, 2004). As a general rule, a surface precipitation mechanism is favored by high adsorptive concentrations and long reaction times in sorption processes. The reactions between adsorptive metal and metalloid ions and soil particles can be portrayed as a web of sorption processes mediated by two parameters, time scale and surface coverage (Charlet, 1994; Hayes and Katz, 1996; Sposito, 2004). Surface complexes (i.e., adsorption complexes) are the expected products of these reactions when time scales are sufficiently short and surface coverage is sufficiently low, *sufficiently* always being defined operationally in terms of conditions attendant to the sorption process (Figure 1.10). As time scales are lengthened and surface coverage increases, or as chemical conditions are altered (e.g., pH changes) for a fixed reaction time, adsorbate "islands" comprising a small number of adsorptive ions bound closely together will form. These reaction products are termed *multinuclear surface complexes* (Figure 1.10) by analogy with their counterpart in aqueous chemistry. They are more likely for adsorptive ions that readily form polymeric structures in aqueous solution. Multinuclear surface complexes may,

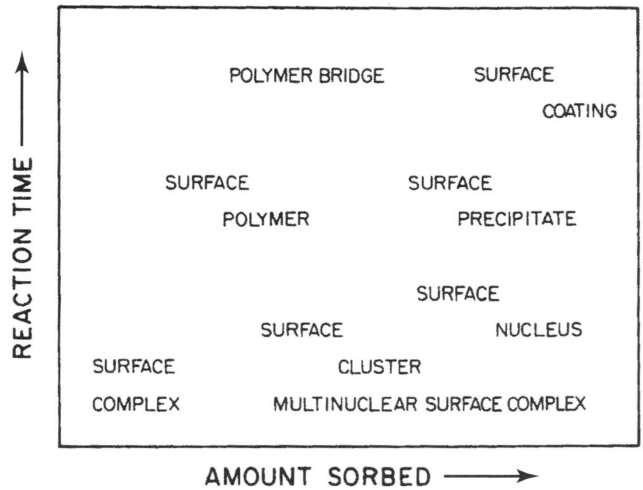

Figure 1.10. Regions of stability in a two-dimensional field defined by the quantity of sorbate and the sorption time, for the products of sorption reactions with natural particles. (From Schindler and Stumm, 1987.)

in turn, grow with time to become colloidal structures that are precursors of surface polymers, or if they are well organized on a three-dimensional lattice, of surface precipitates (Charlet, 1994).

Humic substances contain a large number of complexing sites per molecule, and thus behave as other natural "multiligand" complexing agents, such as metal oxides (Buffle, 1988). The principal molecular characteristics that govern the complexing ability of humic substances are polyfunctionality, polyelectrolyte character, hydrophobicity, and the capacity to form intermolecular associations and change molecular conformation. The major functional groups in humic substances that can bind metal ions are O-bearing groups, including carboxylic, phenolic, alcohoholic, and enolic hydroxyl groups as well as carbonyl functionalities of various types (Stevenson and Fitch, 1986; Senesi, 1992; Senesi and Loffredo, 2005). Amino groups and S- and P-bearing groups are also involved in metal binding. Two main types of complexes may be formed between metal ions and humic substances: (1) inner-sphere complexes, which result in the formation of bonds with some covalent character between the ligand atom(s) and the metal ion, both completely or partially dehydrated; and (2) outer-sphere complexes, which result in electrostatic attraction between the ligand(s) and the metal ion that remains completely hydrated. The electronic and steric environment of the ligand site, such as its chemical network, geometry, and conformation, and physical and chemical characteristics of the surrounding medium, including pH, ionic strength, and metal concentration, can exert a marked influence on the overall interaction process (Buffle, 1988). The binding of metal ions by humic substances is treated in detail by Senesi and Loffredo in Chapter 4.

The transformation of metals and metalloids is influenced significantly by adsorption–desorption reactions in soil environments. These reactions are affected by physicochemical and biological interfacial interactions, which should be especially important in the rhizosphere. Few studies have investigated the adsorption–desorption reactions as influenced by physicochemical and biological interfacial interactions. Microbial metabolites, decay products of biomass, and root exudates include a series of biomolecules which are capable of forming complexes with metal ions (Robert and Berthelin, 1986; Stevenson, 1994; Huang, 2004). There is increasing evidence showing that biomolecules may modify the transformation and dynamics of metals and metalloids (McLaughlin et al., 1998; Huang and Germida, 2002; Violante et al., Chapter 5, this volume). For example, the increase in Cd release in the presence of biomolecules is evidently due to the surface complexation of the particulate-bound Cd in soil with biomolecules, which is reflected in an increase in the release of Cd from soils with the increase in the stability constant of Cd–biomolecule complexes (Krishnamurti et al., 1997). Furthermore, the rate coefficients of the Cd release from the soils, calculated from the parabolic diffusion equation, are influenced substantially by the type of biomolecules (Table 1.4). The rate coefficients of Cd release within each ligand vary from soil to soil, which is attributable to the reactivity of the particulate-bound Cd of the soils toward biomolecules. The activity of Cd species in the soil solution of the soil–root interface governs the labile pool of soil Cd. The importance of the metal/organic complex–bound particulate Cd species in determining the bioavailability of soil Cd has been demonstrated by Krishnamurti et al. (1995a). The rate coefficients of Cd release from the soils by biomolecules (Table 1.4), which is a measure of the rate of the release of soil Cd to soil solution through complexation of soil Cd with biomolecules, follow the same sequence as that of Cd availability index (CAI) values of the soils (Krishnamurti et al., 1995b). Furthermore, the amounts of Cd released from the soils by renewal of biomolecules (Krishnamurti et al., 1997), which is an indication of the Cd sustaining power of the soils, follow the same order as the CAI values of the soils. More research should be conducted to understand the dynamics of adsorption–desorption of metals and metalloids under the influence of physicochemical and biological interactive processes.

TABLE 1.4. Overall Diffusion Coefficients of Cd Release from Soils by 10^{-2} mol L^{-1} Low-Molecular-Mass Organic Acids During a 0.25⁻ to 1-h Reaction Period

	Overall Diffusion Coefficient (μmol kg^{-1} h$^{-0.5}$)				
Soil Site	Acetic Acid	Citric Acid	Fumaric Acid	Oxalic Acid	Succinic Acid
Luseland	0.112 ± 0.010	0.200 ± 0.015	0.199 ± 0.012	0.079 ± 0.006	0.090 ± 0.005
Waitville	0.046 ± 0.004	0.049 ± 0.003	0.050 ± 0.005	0.036 ± 0.004	0.019 ± 0.003
Jedbergh	0.036 ± 0.005	0.196 ± 0.009	0.041 ± 0.003	0.026 ± 0.004	0.009 ± 0.003

Source: Krishnamurti et al. (1997).

1.6. PRECIPITATION–DISSOLUTION REACTIONS OF METALS AND METALLOIDS

Dissolution or precipitation reactions are generally slower than reactions among dissolved species, but it is quite difficult to generalize about rates of precipitation and dissolution (Stumm and Morgan, 1996). There is a lack of data concerning many geochemically important solid–solution reactions. Furthermore, there is not only a continuum between surface complexation (adsorption) and precipitation, but there is also obviously a continuum between heterogeneous nucleation and surface precipitation.

For many of the more abundant elements, such as Al, Fe, and Mn, precipitation of mineral forms is common and may greatly influence or even control their solubility. For most trace elements, direct precipitation from solution through homogeneous nucleation appears to be less likely than adsorption–desorption, by virtue of the low concentration of these metals and metalloids in soil solutions in well-aerated dryland soils. When soils become heavily polluted, metal solubility may reach a level to satisfy the solubility product to cause precipitation. Precipitation may also occur in the immediate vicinity of the phosphate fertilizer zone, where the concentration of heavy metals and metalloids present as impurities may be sufficiently high. Precipitation of trace metals as sulfides may have a significant role in metal transformation in reduced environments where the solution sulfide concentration is sufficiently high to satisfy the solubility product constants of metal sulfides (Robert and Berthelin, 1986).

In aerobic soils, although precipitation of trace metals through homogeneous nucleation is not likely, heterogeneous nucleation may play a significant role in metal transformation because of the presence of mineral, organic, and microbial surfaces that catalyze the nucleation set of crystallization (Huang and Germida, 2002). The energy barrier to nucleation is reduced or removed by surfaces. This is especially true in cases where there are crystallographic similarities between the surface and the precipitation phase. This catalytic process reduces the extent of supersaturation necessary for precipitation to occur. However, precipitation reactions are often slower than adsorption–desorption reactions in soil environments. Figure 1.11 illustrates the growth process from surface complex to surface nucleus to surface precipitate for Cr^{3+} sorbed by poorly crystalline goethite at pH 4 (Manceau et al., 1992).

Besides physicochemical reactions, metals have easy access to bacterial surfaces through diffusion. Metal sorption and precipitation on bacterial surfaces are interfacial effects. Surface metal concentrations frequently exceed the stoichiometry expected per reactive chemical sites within cell walls (Beveridge, 1989; McLean et al., 2002). The sorption of metals can be so great that precipitates can be formed, and distinct minerals are eventually formed through microbial biomineralization (i.e., the formation of minerals by microbes).

In the rhizosphere, activities of free metal ions may be decreased through uptake by plants and microbes. Metal contaminants are complexed substantially with biomolecules in the rhizosphere due to higher concentrations of complexing

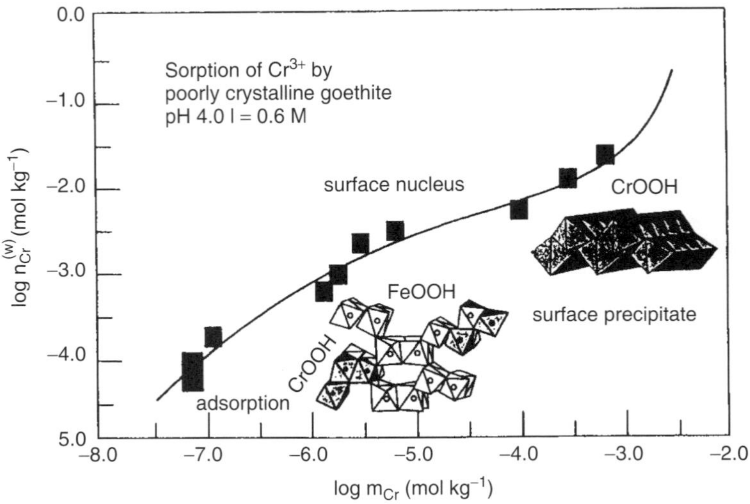

Figure 1.11. Sorption isotherm for Cr^{3+} reacted with poorly crystalline goethite (Cs = 10 kg m^{-3}), illustrating movement from lower left to upper right (Manceau et al., 1992) in the stability field of Figure 1.10 (Schindler and Stumm, 1987). (After Sposito, 2004.)

biomolecules than in bulk soils. Activities of free metal ions should thus be decreased further. Therefore, compared with bulk soils, the activities of free trace metal ions in the soil solution of the rhizosphere in aerobic dryland soils should be even less controlled by precipitation through homogeneous nucleation. In the rhizosphere, precipitation of metals through heterogeneous nucleation on microbial surfaces on the one hand and metal mobilization by biomolecules on the other hand, as a result of root exudation and intense microbial activity, warrants in-depth research for years to come.

1.7. MICROBE-MEDIATED METAL AND METALLOID RELEASE FROM SOIL PARTICLES

Soil biomass–organic matter is present as live plant (roots), animals (e.g., invertebrates), and microbial tissues. One kilogram of surface soil may contain as many as 10 billion bacteria, 2 billion actinomycetes, 120 million fungi, 25 million algae, and a faunal population in the millions (Table 1.5). Added to this biomass is the kilometer or so total length of roots from a single plant in the top meter of soils. Therefore, surface soils are teeming with dense and diverse populations of microorganisms and invertebrates. They play a myriad of important functions in the transformation of metals and metalloids in the environment (Gadd, 2001; Gadd et al., 2001; Glasauer et al., 2005; Gadd, Chapter 2, and Violante et al., Chapter 5, this volume).

TABLE 1.5. Vertical Distribution of Different Groups of Microorganisms in Soil ($\times 10^3$ g^{-1} Soil)

Depth (cm)	Aerobic Bacteria	Anaerobic Bacteria	Actinomycetes	Fungi	Algae
3–8	7800	1950	2080	119	25
20–25	1800	379	245	50	5
35–40	472	98	49	14	0.5
65–75	10	1	5	6	0.1
135–145	1	0.4	—	3	—

Source: Eijsackers (1994).

Chemical weathering of minerals as a part of the soil formation process can be enhanced by microbial activity by a factor as high as 10^6 (Kurek, 2002). Microorganisms can dissolve minerals by direct and indirect action under aerobic and anaerobic conditions (Ehrlich, 2002; Kurek, 2002). The modes of attack of minerals by microorganisms include (1) direct enzymatic oxidation or reduction of a reduced or oxidized mineral component; (2) indirect attack with a metabolically produced redox agent or inorganic and organic acids; (3) indirect attack by metabolically produced alkali, usually in the form of ammonia; (4) indirect attack by a metabolically produced ligand that forms a highly soluble product with a mineral component; and (5) indirect attack by a biopolymer. The mode of microbial attack of a mineral may involve a combination of these mechanisms. The microorganisms may be dispersed in soil solutions in some cases of attack; in others, they may grow in biofilms on the surface of susceptible minerals.

Oxidation of sulfur entities of metal sulfides to obtain energy is an example of direct dissolving action under aerobic conditions (Kurek, 2002). When oxidized metal compounds [e.g., Fe(III), Mn(IV), As(V)] act as electron acceptors, anaerobic respiration becomes an example of direct dissolving action under anaerobic conditions (Ahmann et al., 1994; Ehrlich, 2002). Volatilization of metals and metalloids or biomethylated metals and metalloid compounds from the soil into the atmosphere can be a mechanism of detoxification of toxic elements such as Hg, As, and Se for microorganisms (Gadd, 1993).

1.8. MICROBIAL SORPTION AND UPTAKE OF METALS AND METALLOIDS

All microorganisms contain biopolymers such as proteins, nucleic acids, and polysaccharides, which provide reactive sites for binding metal ions (Hughes and Poole, 1989). These binding sites include negatively charged groups such as carboxylate, thiolate, and phosphate, and groups such as amines, which coordinate to the metal center through lone pairs of electrons. Large concentrations of metals are frequently associated not only with living microbial biomass but also with dead cells (Berthelin et al., 1995), because microbial biopolymers have the ability to bind metals. Metals bind with various degrees of tenacity to the largely anionic outer surface layers of microbial cells. Metal binding by microbial cells alters cell wall composition and induces morphological, ultrastructural, and surface charge changes (Venkateswerlu et al., 1989; Collins and Stotzky, 1996). Bacterial cell walls and membranes may act as foci for the accumulation of metals in soils, since some metals are bound by cell walls and envelopes to a greater extent than by clay minerals such as kaolinite and smectite (Table 1.6).

Microorganisms have a range of metal transport systems, which are often highly specific for certain metals and capable of accumulating metals against large concentration gradients (Kurek, 2002). Certain microorganisms synthesize compounds that bind specific metals with high affinity (Lynch, 1990a,b). For example, some microorganisms make Fe-binding siderophores, which are biomolecules, usually a phenolate or hydroxamate ligand (Nielands, 1981). Some microorganisms produce biomolecules termed *metallothioneins* that are small cystine-rich proteins that strongly bind Cd, Cu, and Zn. Biomolecules of this type or related compounds are significant in influencing the transformation, transport, bioavailability, and toxicity of metals in soil environments, especially in the rhizosphere.

Bacteria and fungi have the ability to uptake and concentrate elements such as Ca, Fe, and K. For example, Weed et al. (1969) reported that fungi can adsorb K from solution and thus shift K equilibrium in suspensions of trioclahedral and dioclahedral micas and transfer them into vermiculites. Some microorganisms can promote the transformation of one mineral into another by a process termed *mineral diagenesis*, which can be an indirect effect of aerobic and anaerobic microbial metabolisms (Ehrlich, 2002; Kurek, 2002).

Most plants in natural habitats form associations with mycorrhizae. Arbuscular mycorrhizal fungi (AMF) are obligate symbionts, and infection of plant roots exerts a metabolic load on the host plant (Reid, 1990). During infection and

TABLE 1.6. Metals Bound by Native *Baccilus subtilis* Walls, *Escherichia coli* Envelopes, Kaolinite, and Smectite

	Amount of Metal Bound (μmol g^{-1}) (Oven-Dry Weight)[a]			
Metal	Walls	Envelopes	Kaolinite	Smectite
Ag	423 ± 15	176 ± 3	0.46 ± 0.02	43 ± 0.3
Cu	530 ± 13	172 ± 9	5 ± 0.03	197 ± 4
Ni	654 ± 25	190 ± 3	4 ± 0.2	173 ± 10
Cd	683 ± 19	221 ± 6	6 ± 0.2	1 ± 0.02
Pb	543 ± 11	254 ± 5	3 ± 0.2	118 ± 6
Zn	973 ± 13	529 ± 32	37 ± 1	65 ± 2
Cr	435 ± 37	102 ± 2	8 ± 0.5	39 ± 5

Source: Walker et al. (1989).

[a]The data represent the average of three to five determinations for each sample from duplicate experiments and the standard error.

colonization of host plant roots, mycorrhizal fungi produce mycelia inside root cortical cells. They may form storage structures termed *vesicles*, may also form structures called *arbuscules*, and serve as the site of ion exchange between the host plant and the mycorrhizal fungus. The fungi also form extracellular hyphae that penetrate out of the root and explore the soil in search of nutrients, including metals. Mycorrhizal infection thus expands the volume of soil that a root can penetrate. Therefore, plant uptake of metals and metalloids may be facilitated by mycorrhizae, since these elements are diffusion limited in soils. However, the effect of mycorrhizal infection on metal and metalloid uptake depends on the ability of the fungal symbiont to absorb metals and metalloids and to transfer them to the symbiotic roots through extensive vegetative mycelium. Mycorrhizal fungi release low-molecular-mass organic acids into soils to enhance the solubilization of particulate-bound metals and metalloids and thus might facilitate their uptake by plant roots. To date, there are considerable gaps in our knowledge of the mechanisms of the effect of fungi on the transformation and uptake of metals and metalloids.

1.9. BIOMINERALIZATION OF METALS AND METALLOIDS AND FINE-GRAINED MINERAL DEVELOPMENT

Microbial *biomineralization*, the formation of minerals by microorganisms, is another important activity of microorganisms (Beveridge, 1989). It involves metal and metalloid transformations and the development of fine-grained minerals of tremendous range and type (McLean et al., 2002). Most biominerals generated by microorganisms are nanoparticles (Banfield and Zhang, 2001). Research on biomineralization indicates that specific molecular interactions at inorganic–organic interfaces can result in the controlled nucleation and growth of inorganic crystals (Mann et al., 1993). A central tenet of biomineralization is that the nucleation, growth, morphology, and aggregation (assembly) of the inorganic crystals are regulated by organized assemblies of organic macromolecules, the *organic matrix*. Control over the crystallochemical properties of biominerals is achieved by specific processes involving molecular recognition at inorganic–organic interfaces. Electrostatic binding or association, geometric matching (epitaxis), and stereochemical correspondence are important in these recognitions. The subtle differences in the kinetics of these recognition processes on different crystal faces lead to specific changes in crystal morphology. Biomineralization should have global consequences in dynamics, transformation, toxicity, and the fate of metals and metalloids in the environment.

Geochemical modeling of metal speciation is only beginning to include bacteria as geochemically active surfaces. The physical and chemical characteristics of bacteria, such as their large surface area-to-volume ratio, serve to increase the metal-binding capacity of their charged surfaces, leading to precipitation and formation of mineral phases on their cell walls or surface biopolymers (McLean et al., 2002). The mechanisms by which bacteria initiate the formation of minerals in the bulk solution vary widely among species. There may be a combination of biochemical and surface-mediated reactions during the process. Bacteria surface layers may positively adsorb and indirectly serve as a nucleation template. Bacteria can also more directly initiate mineral precipitation by producing reactive compounds such as enzymes, siderophores, metallothioneins, which bind metals or catalyze their transformations. Furthermore, bacteria can instigate the spontaneous precipitation of metals by altering the chemistry of their microenvironments (Beveridge et al., 1997; Douglas and Beveridge, 1998).

Metal complexing ligands such as metallothioneins and siderophores serve to sequester metals from the environment for incorporation into cellular components (McLean et al., 2002). Reactive inorganic ligands such as sulfide and phosphate may also be produced as cellular metabolic by-products. Sulfide reacts with metals to form metal sulfides, a common reaction in anoxic environments with sulfate-reducing bacteria. Phosphate pumped out of cells can also react with metal ions to form cell-associated and extracellular precipitates. Another example of microbially mediated fine-grained mineral development is the formation of Mn oxides. Microbial oxidation of Mn(II) is a major process that can produce Mn oxide coatings on soil particles 10^5 times faster than can abiotic oxidation (Tebo et al., 1997). The microbially mediated formation of Mn oxide is illustrated in Figure 1.12. Manganese oxides are highly reactive minerals that help to restrict

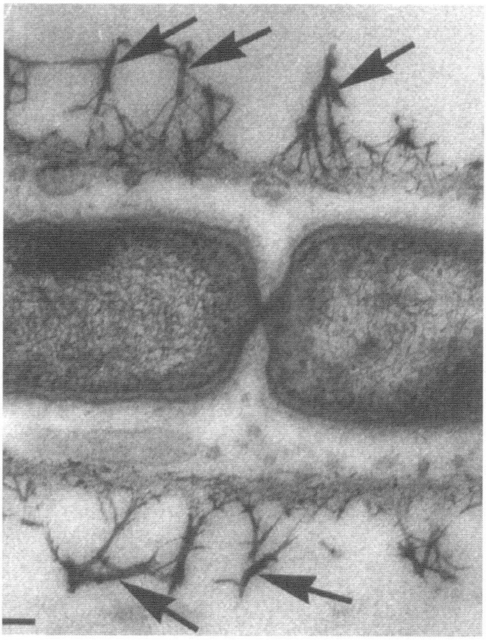

Figure 1.12. Thin section of *Leptothrix* sp., which is precipitating manganese oxide on its outermost structure, called a sheath. The arrows point to the manganese mineral phase, identified by EDS. Scale bar = 150 nm. (From McLean et al., 2002.)

the mobility of metals in soils and related environments through adsorption on their surfaces. Biogenic Mn oxides formed by *Leptothrix discophora* ss-1 have a significantly larger specific surface and higher Pb adsorption capacity than those of abiotically precipitated Mn oxides (Nelson et al., 1999). Bioformation of minerals should thus have a significant bearing on remediation of metal contamination of the environment.

1.10. IMPACTS ON THE TERRESTRIAL ECOSYSTEM

1.10.1. Global Ion Cycling

Transformations and transport of metals and metalloids are part of natural biogeochemical cycling. One of the characteristics of the cycle of metal and metalloid mobilization and deposition is that the form of the metal and metalloid is changed. This change of metal or metalloid speciation has a great effect on its fate and impact on ecosystem health (Benjamin and Honeyman, 1992; Hayes and Traina, 1998; Adriano, 2001). These trace elements are found in the environment in solid, solution, and gaseous phases, associated with thousands of different compounds. The critical processes controlling global metal and metalloid cycling are adsorption–desorption, precipitation–dissolution, complexation, oxidation–reduction, and volatilization (Benjamin and Honeyman, 1992; Hayes and Traina, 1998; Adriano, 2001; Kurek, 2002; McLean et al., 2002; Lollar, 2004). Transport in solution and aqueous suspensions is a major mechanism for metal and metalloid movement in the ecosystem. This transport process is influenced profoundly by adsorption–desorption on surfaces of minerals, organic matter, organomineral complexes and microbes; precipitation–dissolution, especially in reduced environments when sulfide concentration is sufficiently high; and a series of inorganic and organic complexation reactions in solution and particulate phases. Microorganisms can help dissolve, precipitate, complex, oxidize, or reduce metals and metalloids and can initiate the formation of fine-grained minerals, including nanoparticles.

Natural particles suspended in the air can be transported to regions far from their sources. This is important for transporting many metals and metalloids in the ecosystem. A few metals and metalloids, most notably Hg, As, and Se, can exist not only in the solid and liquid phases but also as gases in ambient environments. The loss of Hg from the aqueous phase can result from reduction of Hg^{2+} to Hg^0 and alkylation to form methyl- or dimethylmercury. Through microbial activity, the methylated forms can be converted to Hg^0, which is more volatile and less toxic. Microbial mediation can also transform several other trace elements (e.g., As, Se) to organometallic compounds (Gadd, 1993). These volatile organometallic compounds can dominate the transport of these trace elements in local environments. However, bacterial mediation of alkylation of metals such as Hg is influenced substantially by Hg speciation. Mineral colloids vary in their ability to affect the bioavailability and methylation of Hg(II) in aqueous systems

(Farrell et al., 1998). In particular, MnO_2 (birnessite) is a very effective scavenger of Hg(II) even in the presence of competing ligands and also an inhibitor of the methylation of Hg(II), due to its ability to bind Hg(II) tenaciously . Therefore, soil physicochemical–biological interfacial interactions play important roles in influencing the speciation and cycling of metals and metalloids in the environment. The role of microorganisms in influencing metal transformation and mobilization in the environment is addressed in detail by Gadd in Chapter 2.

1.10.2. Geomedical Problems

Geomedicine is the science dealing with the environmental factors that influence the geographical distribution of pathological and nutritional problems relating to human and animal health (Låg, 1980). Hunger and malnutrition are of concern for large groups of populations, especially in developing countries. In addition to prevention of starvation, promotion of better nourishment and better quality of food and feed is essential. Pollution of the environment and related health problems have increased rapidly with industrialization. Knowledge of soil science is indispensable for solutions to many geomedical problems (Låg, 1994). The effect of soil physical, chemical, and biological reactions and processes on the quality of vegetation and the food and feed produced and related geomedical problems should be investigated in depth.

The transformation and bioavailability of trace elements are profoundly influenced by soil physical, chemical, and biological interactions (Huang and Gobran, 2005). Many trace elements are of concern to animal nutrition and human health (National Academy of Sciences, 1974, 1977; Adriano, 2001). These include Se, Fe, I, Zn, Cu, Mn, Mo, Cr, F, Co, Si, V, Ni, As, and Mg. One trace element may serve in one, several, or dozens of different metalloenzymes or tissue constituents.

To facilitate fundamental understanding of the linkage of trace elements in soils with plant–animal–human–environment systems and related geomedical problems and to provide practical solutions to their deficiency and toxicity problems, it is essential to promote research on the relationship between soil physicochemical–biological interactions and the impacts on the transformation, transport, bioavailability, toxicity, and fate of trace elements in the terrestrial environment.

1.10.3. Ecotoxicological Problems

Ecotoxicology is a fairly new field of science developed from the traditional field of toxicology and environmental chemistry that gave rise to environmental toxicology. When applied to the ecosystem setting, environmental toxicology virtually becomes ecotoxicology. *Ecotoxicology* is defined as "the study of fate and effect of toxic agents in ecosystems" (Cairns and Mount, 1990). Ecotoxicology integrates several disciplines, including environmental biogeochemistry, toxicology, and ecology, dealing with mechanisms, processes, and responses. Therefore, ecotoxicological research deals with the interactions among organisms, the toxic agents (in this case metals and metalloids), and the environment.

There are several levels of biological organization in ecotoxicology research (Eijsackers, 1994; Newman, 1995): molecular level (e.g., RNA/DNA), cellular level (e.g., enzymes, hormones, ATP content, phytochelatins, metallothioneins), individual level (e.g., growth rates, organ functioning, behavioral responses), population level (e.g., lethality, reproduction, genetic drift), community level (e.g., species diversity, community structure), and ecosystem level (e.g., nutrient cycling, energy fluxes). Our present understanding of effects along the ecological spectrum of organization is depicted in Figure 1.13, which becomes poorer as the level of organization increases. The lower level of ecological organization is generally more sensitive and responds more rapidly to toxicant effects than are higher levels of organization.

Soil plays the central role as organizer of the terrestrial ecosystem (Coleman et al., 1998). It may be perceived as the center of the ecosystem, which evolves because of interactions of the lithosphere, hydrosphere, atmosphere, and biosphere. A factor of central importance of soil to ecological studies is that soils on a global scale have a range of characteristics, which enable an enormous array of microorganisms, plants, animals, and humans to coexist and thrive. Among the environmental compartments, about 90% of environmental pollutants are bound with soil particles and 9% of the pollutants are bound with aquatic sediments (Table 1.7). These soil- and sediment-bound pollutants are in dynamic equilibrium with the hydrosphere, atmosphere, and biosphere. Soil physicochemical and

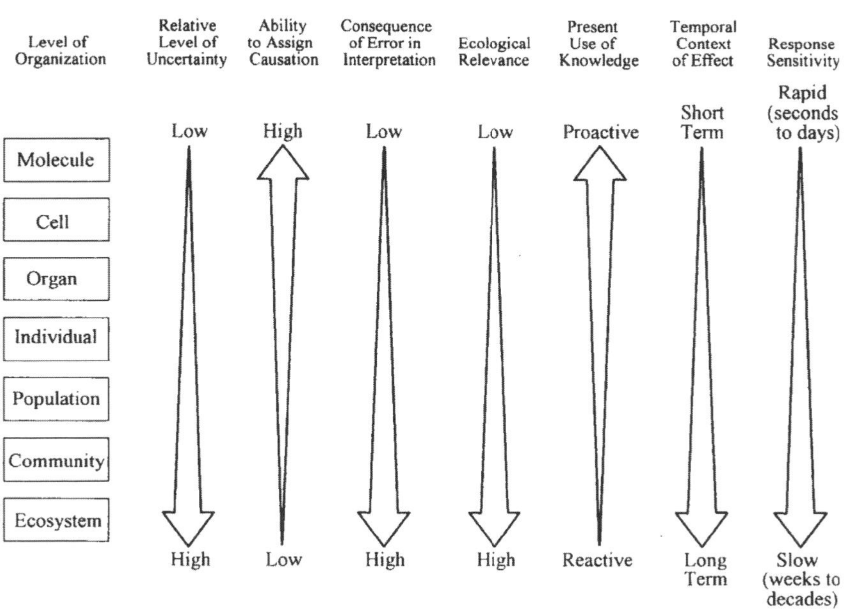

Figure 1.13. Features of ecotoxicological effects based on the level of ecological organization. (From Newman, 1995.)

TABLE 1.7. Theoretical Pollutant Distribution in the Environment at Equilibium

Compartment	Distribution in Compartment[a] (%)
Air	0.35
Water	0.01
Sediment	9.10
Soil	90.50
Aquatic biota	0.01

Source: Crosby (1982).
[a]Assumes approximately 100 kg of pollutant (MW 100) introduced into 10 km^3 of the environment.

biological interfacial interactions have vital roles in governing the transformation, speciation, transport, bioavailability, and toxicity of metals and metalloids of ecological concerns. These interactions may enhance the release of metals and metalloids from soils and sediments and thus pose a threat to ecosystem integrity, including human health.

The long-term ecological effects of metals and metalloids introduced to soils are very difficult to assess, as only a few such experiments exist. Even less is known about the adverse long-term effects of metals and metalloids on soil microorganisms than on crop growth and metal uptake. McGrath et al. (1995) summarized the evidence for the impact of metals on the growth of selected plants and on the effects of metals on soil microbial activity and soil fertility in the long term. They integrated the information from long-term controlled field experiments in the United States, the United Kingdom, Sweden, and Germany. Adverse effects of metals on soil microbial parameters may indeed be not only additive (McGrath et al., 1995) but also interactive. Furthermore, the influence of soil type, organic matter, mineral colloids, microorganisms, pH, and other relevant soil properties on the lowest observed adverse effect concentrations (LOAECs) of a metal for each parameter may be interactive. For instance, heavier soils (i.e., more clayed), near-neutral soils (i.e., higher pH), or soils having a high content of organic matter and short-range-ordered mineral colloids may be expected to reduce the detrimental effects of metals to soil microorganisms by binding the contaminants and rendering them less bioavailable, thus giving high LOAECs.

1.10.4. Biodiversity

The functioning and stability of the terrestrial ecosystem are determined by plant biodiversity and species composition (Schulze and Mooney, 1993; Tilman et al., 1996; Hooper and Vitousek, 1997). However, the ecological mechanisms by which they are regulated and maintained are not well understood. It is essential to identify these mechanisms to ensure successful management for conservation and restoration of diverse natural ecosystems.

All terrestrial ecosystems consist of above- and belowground components that interact to influence community- and ecosystem-level processes and properties (Wardle, 2002; Wardle et al., 2004). These components are closely interlinked at the community level, reinforced by a greater degree of specificity between plants and soil organisms than had previously been supposed. As such, above- and belowground communities can be powerful mutual drivers, with both positive and negative feedback.

Soil is a focal point of the terrestrial ecosystem (Odum, 1989; Coleman et al., 1998). It is the most complicated biomaterial on the planet. As with any material, physical habitat is of prime importance in determining and regulating biological activity (Young and Crawford, 2004). Therefore, soil is Earth's most important resource in sustaining biological activity in both above- and belowground ecosystems. Belowground biodiversity is a major factor contributing to the maintenance of plant biodiversity and to ecosystem functioning and stability. Floristically rich systems may display greater stability under stress (Tilman and Downing, 1994), are more likely to mitigate global problems posed by atmospheric CO_2 enrichment (Naeem et al., 1994), and are more productive (Tilman et al., 1996). However, the diversity of terrestrial vegetation systems is under stress everywhere. The present reduction in biodiversity on Earth and its potential threat to ecosystem stability and productivity are of concern (Tilman et al., 1996; van der Heijden et al., 1998). There is a need to determine how biotic relationships interact with abiotic agents to drive community and ecosystem properties and processes.

Enzymatic activity in soil has been regarded as an indication of soil microbial activity. Certain enzymatic activity (e.g., dehydrogenase enzyme) has decreased on several occasions in direct relationship with soil metal concentration (Reddy et al., 1987; Adriano, 2001). Reduced populations of actinomycetes, bacteria, and fungi in ridgetop soils contaminated with Zn, Cd, and Pb from zinc smelters in Pennsylvania were observed (Jordan and Lechevalier, 1975). Studies of litter–soil fauna indicate marked differences in the taxa of invertebrates at the contaminated sites (Table 1.8). In particular, millipedes (Diplopoda), earthworms (Annelida), and woodlice (Isopoda) were greatly decreased in population at the contaminated site. However, the influence of the speciation of metals and metalloids, which should be governed by soil physicochemical and biological interfacial interactions, on soil microbial and enzymatic activity and the taxa and activity of soil fauna remain obscure. Furthermore, subsequent impact of the reduction of belowground biodiversity on aboveground biodiversity remains to be uncovered.

1.10.5. Risk Assessment, Management, and Restoration of Soil Environments

Metal contamination of soils from anthropogenic chemicals and their transformations has become a major concern because of the critical role of soil resources in sustaining ecosystem integrity and human health and prosperity. Metal contaminants in soils may not only affect their production potential and biodiversity

TABLE 1.8. Number of Invertebrates in Litter Collected from Hallen Wood, Avonmouth, England Compared with a Relatively Uncontaminated Woodland (Wetmoor Wood) (no. kg^{-1})

	Hallen Wood (Contaminated)	Wetmoor Wood (Control)
Isopoda		
Oniscus asellus	3.9	14.8
Trichoniscus pusillus	0	112
Diplopoda		
Polydesmidae	0.6	58.5
Julidae	0	8
Glomeridae	0	15.6
Chilopoda		
Lithobiidae	7.8	86
Geophilomorpha	23	195
Arachnida		
Acari	9,034	14,370
Aranae	17.4	60
Pseudoscorpionidae	14	50
Insecta		
Collembola	1,457	6,436
Coleoptera	63	36
Coleoptera (larvae)	8.4	3
Diptera (larvae)	321.4	215.6
Annelida		
Lumbricus rubellus	2	2.2
L. terrestris	0	21.5
Aporrectodea longa	0	3
A. caliginosa	0	22
Octoclasium cyaneum	0	6.7

Source: Martin and Bullock (1994).

adversely but also compromise the quality of the food chain and the underlying groundwater. Therefore, regulatory-driven risk assessment and management are essential in restoration of soil environments.

Soil quality may be defined in relation to the functions that soils perform in ecosystems (Table 1.9). *Soil quality*, which is used interchangeably with *soil health*, can be deduced or estimated by measuring key indicators, including physical, chemical, and biological properties. Therefore, soil quality reflects the composite picture of soil physical, chemical, and biological properties and processes that interact with each other to determine its conditions. The chemical speciation, concentration, and bioavailability of metals and metalloids are important indicators for soil quality, which, in turn, is related to soil functions. Assessing exposure to metals and metalloids in soil environments includes determination of the pathways to exposure (inhalation, drinking water, food, or

TABLE 1.9. Indices for Soil Quality and Their Relationship to Soil Functions

	Soil Function			
Indicator	Promote Plant Growth	Regulate Water Flow	Buffer Environmental Changes	Promote Biodiversity
Nutrient availability	Direct	Indirect	Direct	Direct
Organic matter	Indirect	Direct	Direct	Direct
Infiltration	Direct	Direct	Indirect	Indirect
Aggregation	Direct	Direct	Indirect	Direct
pH	Direct	Direct	Indirect	Direct
Soil fauna	Indirect	Indirect	Indirect	Direct
Bulk density	Direct	Direct	Indirect	Indirect
Topsoil depth	Direct	Indirect	Indirect	Direct
Salinity	Direct	Direct	Indirect	Direct
Cation exchange capacity	Indirect	Indirect	Indirect	Indirect
Water-holding capacity	Direct	Direct	Indirect	Direct
Soil enzymes	Indirect	Indirect	Indirect	Indirect
Soil flora	Indirect	Indirect	Indirect	Direct
Heavy metal bioavailability	Direct	Indirect	Direct	Direct

Source: National Research Council (1993).

TABLE 1.10. Pathways for Human Exposure to Soilborne Contaminants[a]

Pathway	MEI
Soil → plant → human	Human lifetime plant ingestion; general population
	Human lifetime plant ingestion; home gardener
Soil → human	Child
Soil → plant → animal → human	Human lifetime ingestion of animals products; animals raised on forage
Soil → animals → human	Human lifetime ingestion of animals products; animals ingest soil
Soil → dust → human	Human lifetime dust inhalation
Soil → surface water → human	Human lifetime ingestion of surface water and fish
Soil → groundwater → human	Human lifetime ingestion of groundwater
Soil → air → human	Human lifetime inhalation of volatilized contaminants

Source: Logan (1998).

[a]The most exposed individual (MEI) of the population is also identified. Based on U.S. Environmental Protection Agency exposure assessment for land-applied contaminants in sewage sludge.

direct soil ingestion) (Table 1.10) and the extent of contaminant transfer from soil to the various pathway components (Logan, 1998). In environmental exposure assessment, a *transfer coefficient* must be determined for each component of the pathway: namely, the fraction of the component that is transferred from one component to another (e.g., from soil to plant, soil to air, soil to surface water, soil to groundwater). The transfer coefficient can be viewed as an *availability index*, although other terms (*extractability, solubility, accessibility, volatility*) have also been used. This will depend on the mechanisms of transfer, which should be governed by physicochemical and biological interfacial interactions in soil environments.

The cleanup of soils polluted by hazardous metal and metalloid contaminants has become a matter of urgent public concern. Nature has evolved ways of remediating problems of many contaminants without human intervention. Over a very long period of time, natural degradation might remove many of the organic contaminants in soil, but may result in an accumulation of metals and metalloids. Affordable novel technologies are needed to enhance natural remediation processes and restore healthy ecosystems. Natural remediation usually influences the extent of mobility and bioavailability of contaminants in the soil complex (Adriano et al., 2004; Bollag et al., 2005). The various natural remediation–bioavailability processes, which influence contaminant dynamics, are depicted in Figure 1.14. In the initial phase (A), the dynamics of contaminants between the aqueous and solid phases are governed by various biogeochemical processes (e.g., adsorption–desorption, precipitation–dissolution, redox reactions, and complexation). The nature of these processes determines the kinetics of natural remediation that can be parameterized by bioavailability. The role of the rhizosphere in contaminant transformation and mobility can be very important. The next phase (B, B′) involves the transport of contaminants to organisms. The contaminant can be transported in soluble, colloidal, and/or particulate form. The particulate form can play a significant role in humans and animals. For

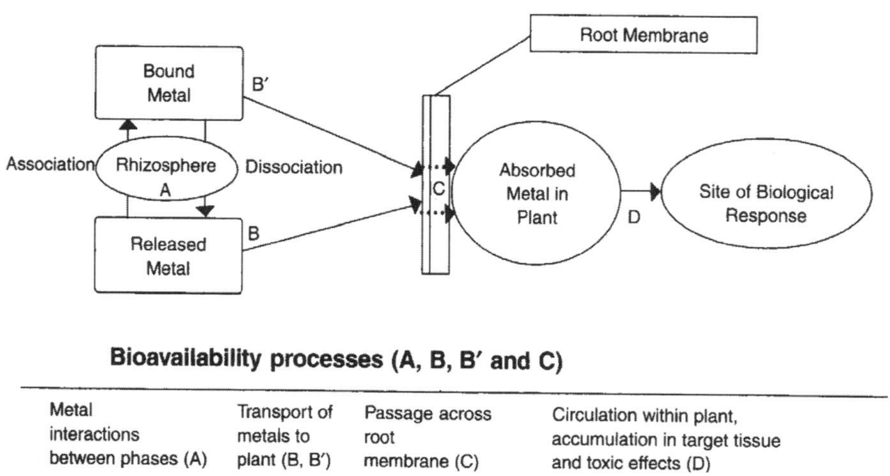

Figure 1.14. Bioavailability processes in soils. Note the delineation of the entire scheme into "natural remediation" (left) and "bioavailability" (right). (Modified from the National Research Council, 2003; Adriano et al., 2004.)

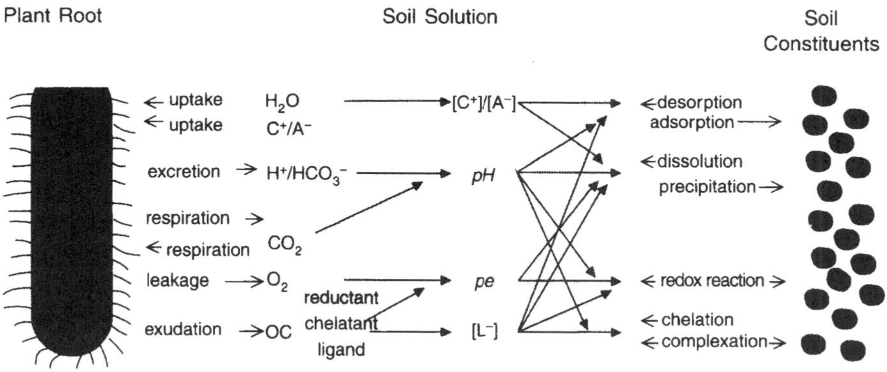

Figure 1.15. Schematic rhizosphere, showing the various exudates and how they can influence abiotic factors and mechanisms at the soil–solution interface (OC, organic carbon; C^+, cation; A^-, anion; L^-, ligand; re, redox potential.) (Modified from Hinsinger, 2001; Adriano et al., 2004.)

example, ingestion of soil particles by children is an important exposure pathway for lead; ingestion of soil particles by grazing livestock is an important exposure pathway for animals feeding on contaminated pastures. From the aqueous phase, metals and metalloids can be mobilized in either soluble or colloidal form, the latter being viewed as facilitated transport. The next phase (C) involves passing through a biological membrane, which in many instances can serve as a biofilter for contaminants. The last phase (D) involves circulation and assimilation in the metabolic machinery of the organisms, culminating in some form of biological response which refers to the endpoint of interest.

The rhizosphere, which is characterized by distinct physical, chemical, and biological conditions (Figures 1.14 and 1.15), is the focal point of the potential role of higher plants in accelerating natural remediation in contaminated soils. Rhizosphere conditions are created by the plant roots and their microbial associations. Rhizospheric biogeochemical processes are influenced substantially by soil physical, mineralogical, chemical, and biological features and also by edaphic and climatic conditions. Special tools and techniques are required to study the characteristics and processes of the rhizosphere because of its limited areal extension (Wenzel et al., 2001).

The physicochemical and biological reactions occurring in the rhizosphere play a significant role in the natural remediation and subsequent bioavailability of metals and metalloids to plants. Plant roots may change the physical, chemical, and biological conditions of the soil in the rhizosphere, which compared to bulk soil is enriched with biomolecules of plant and microbial origin, including low-molecular-mass organic acids (LMMOAs), sugars, amino acids, lipids, coumarins, flavonoids, proteins, enzymes, other aliphatics, aromatics, and carbohydrates, among others (Pomilio et al., 2000). Among these biomolecules, the LMMOAs are the most abundant and chemically very reactive with soil particles. The commonly found LMMOAs are acetic, butyric, citric, fumaric, lactic, malic, malonic, oxalic, propionic, tartaric, and succinic acids (Szmigielska et al., 1996; Koo, 2001). These LMMOAs in the rhizosphere affect the dynamics and mechanisms of the transformation of metals and metalloids in soils by their effects on acidification, complexation, precipitation–dissolution, redox reactions, microbial activity, rhizosphere physical properties, and root morphology (Huang and Germida, 2002).

The use of bioremediation technology for metals and metalloids is just becoming recognized as an efficient and effective strategy. Cadmium, lead, zinc, nickel, cobalt, and other base metals generally exist as cations in the aqueous phase, and adsorption to negatively charged sites of the surface layers of bacteria can remove those metals very effectively (Mullen et al., 1989; Volesky, 1990). Highly toxic and mobile oxyanions of uranium, chromium, arsenic, and selenium are of particular concern in many soils and groundwaters throughout the world. Their speciation in the environment and hence their reactivity, mobility, bioavailability, and toxicity are influenced significantly by microbially mediated reactions. Oxyanions such as arsenates, selenates, and chromates do not bind to the bacterial surface to any significant extent. Bioremediation of these oxyanions is based predominantly on microbially catalyzed redox conversions to insoluble forms (McLean et al., 2002). Application of bioremediation technology to decontaminate polluted sites is still a developing science. The mechanisms driving microbial activity and degradation pathways of specific pollutants should be further elucidated before successful and better controlled site-specific treatments can be applied (Bollag et al., 2005). It is of great importance to consider the fact that the growth, metabolism, and activity of soil microorganisms is strongly influenced by their interactions with mineral and organic colloids that bind organic chemicals, inorganic ions, and water films to their surfaces (Stotzky, 1986; Theng and Orchard, 1995; Huang, 2004). Recent advances in biotechnology have enabled modification of organisms at the molecular level for improved degradative performance (Bollag et al., 2005).

Besides biotic processes, abiotic processes play a very important role in remediation of metal- and metalloid-contaminated sites. Many soil ameliorants have been used to remediate contaminated soils (Table 1.11). These soil ameliorants are rather inexpensive and readily available on a global scale. Application of limestone to elevate soil pH to about 6.5 is recommended when agricultural soils are amended with sewage sludge high in metals, especially cadmium (Page et al., 1983; Pierzynski et al., 2000). The addition of certain clay minerals, such as vermiculite, to contaminated soils may be effective in immobilizing metals and radionuclides such as ^{137}Cs and ^{90}Sr (Adriano et al., 1997; Mench et al., 1997). Zeolites, which serve as molecular sieves, have been used as soil ameliorants (Gworek, 1992a,b). Application of phosphate and potash fertilizers may have some beneficial effects on certain contaminants. When mixed with hydroxyapatite, $Ca_{10}(PO_4)_6(OH)_2$, lead of soil particles will dissolve and precipitate as hydroxypyromorphite, $Pb_{10}(PO_4)_6(OH)_2$ (Ma et al., 1993; Melamed and Ma, Chapter 15, this volume). Use of potash fertilizer is based on the similar physiological behavior of K toward Cs. As such, they compete for the same sites

TABLE 1.11. Selected Ameliorants Adapted to Metal-Contaminated Soils

Technique	Target Contaminants	Soil Processes Involved	Constraints
Limestone	Metals, radionuclides	Precipitation, sorption	Ineffective for oxyanions; certain crops (lettuce, spinach, tuber, and others); short term
Zeolite	Metals, radionuclides	Ion exchange, sorption, fixation	Insufficient data; short term
Apatite	Metals	Sorption, precipitation, complexation	Selective, insufficient data
Clay mineral	Metals, radionuclides	Ion exchange, sorption, fixation	Type of clay; short term

Source: Adriano et al. (1998).

on plant roots, causing antagonistic effects. The ameliorants listed in Table 1.11 have been applied in agricultural soils but may provide only an interim solution in stabilizing contaminants. Their efficacy over the long term remains to be studied.

Furthermore, abiotic and biotic reactions are not independent but rather, interactive processes in soil environments. Interactions of abiotic and biotic processes are thus very important in governing the dynamics and fate of metals and metalloids in soils, especially at the soil–root interface. Abiotic and biotic interactions in the rhizosphere in influencing the stabilization of contaminants and the efficacy of ameliorants need to be investigated. The impact of physical, chemical, and biological interfacial interactions on risk assessment and management of metal and metalloid contamination and restoration of ecosystem health merits close attention.

1.11. CONCLUSIONS AND FUTURE DIRECTIONS

Soil is the most diverse ecosystem and Earth's most important resource in sustaining all life in the terrestrial environment. Physical, chemical, and biological processes are not independent but rather, interactive with each other. The interactions at physical–chemical–biological interfaces govern the mechanisms of transformations, speciation, dynamics, bioavailability, toxicity, and fate of metals and metalloids in soil and related environments.

Soil particles, be they minerals, organic matter, microorganisms, protozoans, or nematodes, are constantly in close association and interaction. They play key roles in affecting physical, chemical, and biological processes in soil environments. The interactions of these soil components, mediated by soil solution and atmosphere, govern mineral weathering transformations; formation of short-range-ordered mineral colloids, humic substances, and organomineral complexes; microbial and enzymatic activity; formation and stability of soil structure and porosity of various sizes and shapes; dynamics of aggregate turnover; and subsequent impact on the biogeochemical cycling of metals and metalloids and related geomedical, ecotoxicological, and biodiversity problems.

Fundamental understanding of soil physical, chemical, and biological interfacial interactions at the molecular level is essential to understanding the behavior of metals and metalloids in the pedosphere and to restoring terrestrial ecosystem health on the global scale. Future research on this extremely important and exciting area of science should be stimulated to sustain and enhance ecosystem productivity, services, and integrity and the impact on human health and prosperity.

Acknowledgments

This study was supported by Discovery Grant G2383 of the Natural Sciences and Engineering Research Council of Canada.

REFERENCES

Adriano, D. C. (2001). *Trace Elements in the Terrestrial Environments: Biogeochemistry, Bioavailability, and Risks of Metals*, Springer-Verlag, New York.

Adriano, D. C., Albright, J., Whicker, F. W., and Iskandar, I. K. (1997). Remediation of metal- and radionuclide-contaminated soil. In *Remediation of Metal-Contaminated Soils*, ed. Iskandar, I. K., and Adriano, D. C., Science Reviews, Northwood, Middlesex, England, 27–45.

Adriano, D. C., Chlopecka, A., and Kaplan, D. I. (1998). Role of soil chemistry in soil remediation and ecosystem conservation. In *Soil Chemistry and Ecosystem Health,* ed. Huang, P. M., Adriano, D. C., Logan, T. J., and Checkai, R. T., Soil Science Society of America, Madison, WI, 361–386.

Adriano, D. C., Wenzel, W. W., Vangronsveld, J., and Bolan, N. S. (2004). Role of assisted natural remediation in environmental cleanup. *Geoderma* **122**, 121–142.

Ahmann, D., Robert, A. L., Krumholz, L. R., and Morel, F. M. M. (1994). Microbe grows by reducing arsenic. *Nature* (*London*) **371**, 750.

Alloway, B. J. (1995). Soil processes and the behavior of metals. In *Heavy Metals in Soils*, ed. Alloway, B. J., Blackie Academic and Professional, London, 11–37.

Anderson, H. A., Bick, W., Hepburn, A., and Steward, M. (1989). Nitrogen in humic substances. In *Humic Substances II: In Search of Structure*, ed. Hayes, M. H. B., MacCarthy, P., Malcolm, R. L., and Swift, R. S., Wiley-Interscience, Chichester, West Sussex, England, 223–253.

Baldock, J. A. (2002). Interactions of organic materials and microorganisms with minerals in the stabilization of structure. In *Interactions Between Soil Particles and Microorganisms: Impact on the Terrestrial Ecosystem*, ed. Huang, P. M., Bollag, J.-M., and Senesi,

N., IUPAC Series on Analytical and Physical Chemistry of Environmental Systems, Vol. 8, Wiley, Chichester, West Sussex, England, 85–131.

Banfield, J. F., and Nealson, K. H. (1997). *Geomicrobiology: Interactions Between Microbes and Minerals*, Reviews in Mineralogy and Geochemistry, Vol. 35, Mineralogical Society of America, Washington, DC.

Banfield, J. F., and Zhang, H. (2001). Nanoparticles in the environment. In *Nanoparticles and the Environment*, ed. Banfield, J. F., and Navrotsky, A., Reviews in Mineralogy and Geochemistry, Vol. 44, Mineralogical Society of America, Washington, DC, 1–58.

Barnhisel, R. I., and Bertsch, P. M. (1989). Chlorites and hydroxy-interlayered vermiculite and smectite. In *Minerals in Soil Environments*, ed. Dixon, J. B., and Weed, S. B., Soil Science Society of America, Madison, WI, 729–788.

Bartlett, R. J., and James, B. R. (1993). Redox chemistry of soils. *Adv. Agron.* **50**, 151–208.

Benjamin, M. M., and Honeyman, B. D. (1992). Trace metals. In *Global Biogeochemical Cycles*, ed. Butcher, S. S., Charlson, R. J., Orians, G. H., and Wolfe, G. V., Academic Press, London, 317–352.

Berthelin, J., Munier-Lamy, C., and Leyval, C. (1995). Effect of microorganisms on mobility of heavy metals in soils. In *Environmental Impact of Soil Component Interactions, Vol. II, Metals, Other Inorganics, and Microbial Activities*, ed. Huang, P. M., Berthelin, J., Bollag, J.-M., McGill, W. B., and Page, A. L., CRC Press/Lewis Publishers, Boca Raton, FL, 3–17.

Bertsch, P. M., and Parker, D. R. (1996). Aqueous polynuclear aluminum species. In *The Environmental Chemistry of Aluminum*, ed. Sposito, G., CRC Press/Lewis Publishers, Boca Raton, FL, 117–168.

Beveridge, T. J. (1989). Role of cellular design in bacterial metal accumulation and mineralization. *Annu. Rev. Microbiol.* **43**, 147–171.

Beveridge, T. J., Makin, S. A., Kadurugamuwa, J. L., and Li, Z. (1997). Interactions between biofilms and the environment. *FEMS Microbiol. Rev.* **20**, 291–303.

Bigham, J. M., Fritzpatrick, R. W., and Schulze, D. G. (2002). Iron oxides. In *Soil Mineralogy with Environmental Applications*, ed. Dixon, J. B., and Schulze, D. G., Soil Science Society of America, Madison, WI, 323–366.

Bollag, J. -M., Berthelin, J., Adriano, D. C., and Huang, P. M. (2005). Impact of soil mineral–organic component–microorganism interactions on restoration of terrestrial ecosystems. In *Soil Abiotic and Biotic Interactions and Impact on the Ecosystem and Human Welfare*, ed. Huang, P. M., Violante, A., Bollag, J.-M., and Vityakon, P., Science Publishers, Enfield, NH, 51–70.

Brown, G. (1953). The dioctahedral analogue of vermiculite. *Clay Miner. Bull.* **2**, 64–70.

Buffle, J. (1988). *Complexation Reactions in Aquatic Systems: An Analytical Approach*. Ellis Horwood, Chichester, West Sussex, England.

Cairns, J., and Mount, D. I. (1990). Aquatic toxicology. *Environ. Sci. Technol.* **24**, 154–161.

Charlet, L. (1994). In *Chemistry of Aquatic Systems*, ed. Bidoglio, G., and Stumm, W., Kluwer Academic, Boston, MA, 273–305.

Coleman, D. C., Hendrix, P. F., and Odum, E. P. (1998). Ecosystem health: an overview. In *Soil Chemistry and Ecosystem Health*, ed. Huang, P. M., Adriano, D. C., Logan, T. L., and Checkai, R. T., Soil Science Society of America, Madison, WI, 1–20.

Collins, Y. E., and Stotzky, G. (1996). Changes in the surface charge of bacteria caused by heavy metals does not affect survival. *Can. J. Microbiol* **42**, 621–627.

Colombo, C., Ricciardella, M., Di Cerce, A., Maiuro, L., and Violante, A. (2004). Effect of tannate, pH, sample preparation, ageing and temperature on the formation and nature of Al oxyhydroxides. *Clays Clay Miner.* **52**, 721–733.

Cornell, R. M., and Schwertmann, U. (2003). *The Iron Oxides*, VCH, Weinheim, Germany.

Crosby, D. G. (1982). Environmental chemistry: an overview. *Environ. Toxicol. Chem.***1**, 1–8.

Davies, R. I., Cheshire, M. V., and Graham-Bryce, I. J. (1969). Retention of low level copper by humic acids. *J. Soil Sci.* **20**, 65–71.

Douglas, S., and Beveridge, T. J. (1998). Mineral formation by bacteria in natural microbial communities. *FEMS Microbiol. Ecol.* **26**, 79–88.

Ehrlich, H. L. (2002). Interactions between microorganisms and minerals under anaerobic conditions. In *Interactions Between Soil Particles and Microorganisms: Impact on the Terrestrial Ecosystem*, ed. Huang, P. M., Bollag, J.-M., and Senesi, N., IUPAC Series on Analytical and Physical Chemistry of Environmental Systems, Vol. 8, Wiley, Chichester, West Sussex, England, 459–494.

Eijsackers, H. (1994). Ecotoxicology of soil organisms: seeking the way in a pitch-dark labyrinth. In *Ecotoxicology of Soil Organisms*, ed. Donker, M. H., Eijsackers, H., and Heimbach, F., CRC Press/Lewis Publishers, Boca Raton, FL, 3–32.

Evershed, R. P., Bland, H. A., van Bergen, P. F., Carter, J. F., Horton, M. C., and Rowley-Conwy, P. A. (1997). Volatile compounds in archaeological plant remains and the Maillard reaction during decay of organic matter. *Science* **278**, 432–433.

Farrell, R. E., Huang, P. M., and Germida, J. J. (1998). Biomethylation of mercury(II) adsorbed on mineral colloids common in freshwater sediments. *Appl. Organomet. Chem.* **12**, 613–620.

Filip, Z., Flaig, W., and Rietz, E. (1977). Oxidation of some phenolic substances as influenced by clay minerals. *International Isot. Radiat. Soil Organic Matter Studies II*, Bulletin of International Atomic Energy Agency, Vienna, Austria, 91–96.

Gadd, G. M. (1993). Microbial formation and transformation of organometallic and organometalloid compounds. *FEMS Microbial. Rev.* **11**, 297–310.

Gadd, G. M. (2001). Accumulation and transformation of metals by microorganisms. In *Biotechnology: A Multi-volume Comprehensive Treatise,* Vol. 10, *Special Processes*, ed. Rehm, H.-J., Reed, G., Puhler, A., and Stadler, P., Wiley-VCH, Weinheim, Germany, 225–264.

Gadd, G. M., Bridge, T. A. M., Gharieb, M. M., Sayer, J. A., and White, C. (2001). Microbial processes for solubilization or immobilization of metals and metalloids and their potential for environmental bioremediation. In *Industrial and Environmental Biotechnology*, ed. Ahmed, N., Quereshi, F. M., and Khan, O. Y., Horizon Scientific Press, Wymondham, Norfolk, England 55–80.

Glasauer, S., Burford, E. P., Harper, F. A., Gadd, G. M., and Beveridge, T. J. (2005). Metals and metalloids, transformation by microorganisms. In *Encyclopedia of Soils in the Environment*, ed. Hillel, D., Rosenzweig, C. Powlson, D., Scow, K., Singer, M., and Sparks, D. L., Elsevier, Oxford, 438–447.

Greenland, D. J. (1965). Interaction between clays and organic compounds in soils, II: Adsorption of soil organic compounds and its effect on soil properties. *Soil Fert.* **28**, 521–532.

Guggenberger, G., and Haider, K. M. (2002). Effect of mineral colloids on biogeochemical cycling of C, N, P, and S in soil. In *Interactions Between Soil Particles and Microorganisms: Impact on the Terrestrial Ecosystems*, ed. Huang, P. M., Bollag, J.-M., and Senesi, N., IUPAC Series on Analytical and Physical Chemistry of Environmental Systems, Vol. 8, Wiley, Chichester, West Sussex, England, 267–322.

Gworek, B. (1992a). Lead inactivation in soil by zeolites. *Plant Soil* **143**, 71–74.

Gworek, B. (1992b). Inactivation of cadmium in soils by synthetic zeolites. *Environ. Pollut.* **75**, 269–271.

Haider, K., and Guggenberger, G. (2005). Soil mineral and organic components: impact on biological processes, human welfare, and nutrition. In *Soil Abiotic and Biotic Interactions and Impact on the Ecosystem and Human Welfare*, ed. Huang, P. M., Violante, A., Bollag, J.-M., and Vityakon, P., Science Publishers, Enfield, NH, 3–16.

Hayes, K. F., and Katz, L. E. (1996). Application of x-ray absorption spectroscopy for surface complexation modeling of metal ion sorption. In *Physics and Chemistry of Mineral Surfaces*, ed. Brady, P. V., CRC Press, Boca Raton, FL, 147–223.

Hayes, K. F., and Traina, S. J. (1998). Metal ion speciation and its significance in ecosystem health. In *Soil Chemistry and Ecosystem Health*, ed. Huang, P. M., Adriano, D. C., Logan, T. J., and Checkai, R. T., Soil Science Society of America, Madison, WI, 45–84.

Hedges, J. I. (1988). Polymerization of humic substances in natural environments. In *Humic Substances and Their Role in the Environment*, ed. Frimmel, F. H., and Christman, R. F., Wiley, Chichester, West Sussex, Engalnd, 45–58.

Hinsinger, P. (2001). Bioavailability of metals as related to root-induced chemical changes in the rhizosphere. In *Trace Elements in the Rhizosphere*, ed. Gobran, G. R., Wenzel, W. W., and Lombi, E. CRC Press, Boca Raton, FL, 25–41.

Hooper, D. U., and Vitousek, P. M. (1997). The effect of plant composition and diversity on ecosystem processes. *Science* **227**, 1302–1305.

Huang, P. M. (1988). Ionic factors affecting aluminum transformation and the impact on soil and environmental sciences. *Adv. Soil Sci.* **1**, 1–78.

Huang, P. M. (1990). Role of soil minerals in transformations of natural organics and xenobiotics. In *Soil Biochemistry*, Vol. 6, ed. Bollag, J.-M., and Stotzky, G., Marcel Dekker, New York, 29–115.

Huang, P. M. (2000). Abiotic catalysis. In *Handbook of Soil Science*, ed.-in-chief Summer, M. E., CRC Press, Boca Raton, FL, B303–B332.

Huang, P. M. (2002). Foreseeable impacts of soil mineral–organic matter–microorganism interactions on society: ecosystem health. In *Soil Mineral–Organic Matter– Microorganism Interactions and Ecosystem Health*, Vol. 28A, *Dynamics, Mobility, and Transformation of Pollutants and Nutrients*, ed. Violante, A., Huang, P. M., Bollag, J.-M., and Gianfreda, L., Elsevier, Amsterdam, The Netherlands, 1–36.

Huang, P. M. (2004). Soil mineral–organic matter–microorganism interactions: fundamentals and impacts. *Adv. Agron.* **82**, 391–472.

Huang, P. M., and Germida, J. J. (2002). Chemical and biological process in the rhizosphere: metal pollutants. In *Interactions Between Soil Particles and Microorganisms: Impact on Terrestrial Ecosystem*, ed. Huang, P. M. Bollag, J.-M., and Senesi, N., IUPAC Series on Analytical and Physical Chemistry of Environmental Systems, Vol. 8, Wiley, Chichester, West Sussex, England, 381–438.

Huang, P. M., and Gobran, G. R. (eds.) (2005). *Biogeochemistry of Trace Elements in the Rhizosphere*, Elsevier, Amsterdam, The Netherlands.

Huang, P. M., and Kozak, L. M. (1970). Adsorption of hydroxy-aluminum polymers by muscovite and biotite. *Nature (London)* **228**, 1084–1085.

Huang, P. M., and Schnitzer, M. (eds.) (1986). *Interactions of Soil Minerals with Natural Organics and Microbes*, Soil Science Society of America, Madison, WI.

Huang, P. M., and Violante, A. (1986). Influence of organic acids on crystallization and surface properties of precipitation products of aluminum. In *Interactions of Soil Minerals with Natural Organics and Microbes*, ed. Huang, P. M., and Schnitzer, M., Soil Science Society of America, Madison, WI, 159–221.

Huang, P. M., and Wang, M. K. (1997). Formation chemistry and selected surface properties of iron oxides. In *Advances in GeoEcology: Soil and Environments*, ed. Auerswald, K., Stanjek, H., and Bigham, J., Catena, Reiskirchen, Germany, 241–270.

Huang, P. M., Berthelin, J., Bollag, J.-M., McGill, W. B., and Page, A. L. (eds.) (1995). *Environmental Impact of Soil Component Interactions*, Vol. I, *Natural and Anthropogenic Organics*, and Vol. II , *Metals, Other Inorganics, and Microbial Activities*, CRC Press/Lewis Publishers, Boca Raton, FL.

Huang, P. M., Wang, M. K., Kämpf, N., and Schulze, D. G. (2002). Aluminum hydroxides. In *Soil Mineralogy with Environmental Applications*, ed. Dixon, J. B., and Schulze, D. G., Soil Science Society of America, Madison, WI, 261–289.

Hughes, M. N., and Poole, R. K. (1989). *Metals and Microorganisms*, Chapman & Hall, New York.

Ikan, R., Rubinsztain, Y., Nissenbaum, A., and Kaplan, I. R. (1996). Geochemical aspects of the Maillard reaction. In *The Maillard Reaction: Consequences for the Chemical and Life Sciences*, ed. Ikan, R., Wiley, Chichester, West Sussex, England, 1–25.

Jackson, M. L. (1962). Interlayering of expansible layer silicates in soils by chemical weathering. *Clays Clay Miner.* **11**, 29–46.

James, B. R., and Bartlett, R. J. (2000). Redox phenomena. In *Handbook of Soil Science*, ed.-in-chief, Sumner, M. E., CRC Press, Boca Raton, FL, B169–B194.

Jastrow, J. D., and Miller, R. M. (1998). Soil aggregate stabilization and carbon sequestration: feedbacks through organomineral associations. In *Soil Processes and the Carbon Cycle*, ed. Lal, R., et al., CRC Press, Boca Raton, FL, 207–223.

Jokic, A., Zimpel, Z., Huang, P. M., and Mezey, P. G. (2001a). Molecular shape analysis of a Maillard reaction intermediate. *SAR QSAR Environ. Res.* **12**, 297–307.

Jokic, A., Frenkel, A. I., Vairavamurthy, M. A., and Huang, P. M. (2001b). Birnessite catalysis of the Maillard reaction: its significance in natural humification. *Geophys. Res. Lett.* **28**, 3899–3902.

Jokic, A., Wang, M. C., Liu, C. Frenkel, A. I., and Huang, P. M. (2004). Integration of the polyphenol and Maillard reactions into a unified abiotic pathway for humification in nature: the role of δ-MnO_2. *Org. Geochem.* **35**, 747–762.

Jordan, M. J., and Lechevalier, M. P. (1975). Effects of zinc-smelter emissions on forest soil microflora. *Can. J. Microbiol.* **21**, 1855–1865.

Kirk, G. J. D., and Bajita, J. B. (1995). Root-induced iron oxidation, pH changes and zinc solubilization in the rhizosphere of lowland rice. *New Phytol.* **131**, 129–137.

Kodama, H., and Schnitzer, M. (1980). Effect of fulvic acid on the crystallization of aluminum hydroxides. *Geoderma* **24**, 195–205.

Koo, B. J. (2001). Assessing bioavailability of metals in biosolid-treated soils: root exudates and their effects on solubility of metals. Ph.D. dissertation, University of California, Riverside, CA.

Krishnamurti, G. S. R., and Huang, P. M. (1990). Kinetics of Fe(II) oxygenation and the nature of hydrolytic products as influenced by ligands. In *Proc. 9th International Clay Conferance*, Strasbourg, France, ed. Farmer, V. C., and Tardy, Y., Publié avec le concours l'Université Louis Pasteur et du Center National de la Recherche Scientifique, Strasburg, France, 195–204.

Krishnamurti, G. S. R., Huang, P. M., Van Rees, K. C. J., Kozak, L. M., and Rostad, H. P. W. (1995a). Speciation of particulate-bound Cd in soils and its bioavailability. *Analyst* **120**, 659–665.

Krishnamurti, G. S. R., Huang, P. M., Van Rees, K. C. J., Kozak, L. M., and Rostad, H. P. W. (1995b). A new soil test method for the determination of plant-available cadmium in soils. *Commun. Soil Sci. Plant Anal.* **26**, 2857–2867.

Krishnamurti, G. S. R., Huang, P. M., and Van Rees, K. C. J. (1996). Studies on soil rhizosphere: speciation and availability of Cd. *Chem. Speciat. Bioavailab.* **8**, 23–28.

Krishnamurti, G. S. R., Cieśliński, G., Huang, P. M., and Van Rees, K. C. J. (1997). Kinetics of cadmium release from soils as influenced by organic acids: implication in cadmium availability. *J. Environ. Qual.* **26**, 271–277.

Krishnamurti, G. S. R., Wang, M. K., and Huang, P. M. (1999). Role of tartaric acid in the inhibition of the formation of Al_{13} tridecamer using sulfate precipitation. *Clays Clay Miner.* **47**, 658–663.

Krishnamurti, G. S. R., Wang, M. K., and Huang, P. M. (2004). Effects of pyrogallol on Al_{13} tridecamer formation and humification. *Clays Clay Miner.* **52**, 734–741.

Kumada, K., and Kato, H. (1970). Browning of pyrogallol as affected by clay minerals. *Soil Sci. Plant Nutr.* **16**, 195–200.

Kurek, E. (2002). Microbial mobilization of metals from soil minerals under aerobic conditions. In *Interactions Between Soil Particles and Microorganisms: Impact on the Terrestrial Ecosystem*, ed. Huang, P. M., Bollag, J.-M., and Senesi, N., IUPAC Series on Analytical and Physical Chemistry of Environmental Systems, Vol. 8, Wiley, Chichester, West Sussex, England, 189–225.

Kwong, K. F. Ng Kee, and Huang, P. M. (1975). Influence of citric acid on the crystallization of aluminum hydroxides. *Clays Clay Miner.* **23**, 164–165.

Kwong, K. F. Ng Kee, and Huang, P. M. (1979a). The relative influence of low-molecular-weight complexing organic acids on the hydrolysis and precipitation products of aluminum. *Soil Sci.* **128**, 337–342.

Kwong, K. F. Ng Kee, and Huang, P. M. (1979b). Nature of hydrolytic precipitation products of aluminum as influenced by low molecular weight complexing organic acids. In *Proc. 6th International Clay Conference*, Oxford, ed. Farmer, V. C., and Mortland, M. M., Elsevier, Oxford, UK, 527–536.

Kwong, K. F. Ng Kee, and Huang, P. M. (1981). Comparison of the influence of tannic acid and selected low-molecular-weight organic acids on precipitation products of aluminum. *Geoderma* **26**, 179–193.

Låg, J. (1980). *Geomedical Aspects in Present and Future Research*, Universitatsforlaget, Oslo, Norway.

Låg, J. (1994). Geomedicine, an expanded application of soil science. In *Interactions of Soil Components, Agricultural Ecosystems, and Health, Trans. 15th World Congress of Soil Science* Vol. 3a, ISSS Commission II Symposium, Acapulco, Mexico, 557–567.

Laurie, S. H., Tancock, N., McGrath, S. P., and Sanders, J. R. (1991). Influence of complexation on the uptake by plants of iron, manganese, copper and zinc: effect of DTPA in a multi-metal and computer simulation study. *J. Exp. Bot.* **42**, 509–513.

Lind, C. J., and Hem, J. D. (1975). *Effects of Organic Solutes on Chemical Reactions of Aluminum*, Water Supply Pap. 1827-G, U.S, Geological Survey, Washington, DC.

Linehan, D. J., Sinclair, A. H., and Mitchell, M. C. (1989). Seasonal changes in Cu, Mn, Zn, and Co concentrations in soil in the root zone of barley (*Hordeum vulgare* L.). *J. Soil Sci.* **40**, 103–115.

Liu, C., and Huang P. M. (1999). Atomic force microscopy and surface characteristics of iron oxides formed in citrate solutions. *Soil Sci. Soc. Am. J.* **63**, 65–72.

Liu, C., and Huang, P. M. (2003). Kinetics of lead adsorption by iron oxides formed under the influence of citrate. *Geochim. Cosmochim. Acta* **67**, 1045–1054.

Logan, T. J. (1998). Impact of soil chemical reactions on food chain contamination and environmental quality. In *Future Prospects for Soil Chemistry*, ed. Huang, P. M., Sparks, D. L., and Boyd, S. A., Soil Science Society of America, Madison, WI, 191–204.

Lollar, B. S. (ed.) (2004). *Environmental Geochemistry: Treatise on Geochemistry*, Vol. 9, Elsevier, Amsterdam, The Netherlands.

Lou, G., and Huang, P. M. (1988). Hydroxy-aluminosilicate interlayers in montmorillonite: implications for acidic environments. *Nature* (*London*) **335**, 625–627.

Lund, W. (1990). Speciation analysis: Why and how? *Fresenius'J. Anal. Chem.* **337**, 557–564.

Lynch, J. M. (1990a). Introduction: some consequences of microbial rhizosphere competence for plant and soil. In *The Rhizosphere*, ed. Lynch, J. M., Wiley, Chichester, West Sussex, England, 1–10.

Lynch, J. M. (1990b). Microbial metabolites. In *The Rhizosphere*, ed. Lynch, J. M., Wiley, Chichester, West Sussex, England, 177–206.

Ma, Q. Y., Logan, T. J., Ryan, J. A., and Traina, S. J. (1993). In situ lead immobilization by apatite. *Environ. Sci. Technol.* **27**, 1803–1810.

Maillard, L. C. (1913). Formation de matières humiques par action de polypeptides sur sucres. *C. R. Acad. Sci.* **156**, 148–149.

Manceau, A., Charlet, L. Boisset, M. C., Didier, B., and Spadini, L. (1992). Sorption and speciation of heavy metals on hydrous Fe and Mn oxides: from microscopic to macroscopic. *Appl. Clay Sci.* **7**, 201–223.

Mann, S., Archibald, D. A., Didymus, J. M., Douglas, T., Heywood, R. R., Meldrum, F. C., and Reeves, N. J. (1993). Crystallization at inorganic–organic interfaces: biominerals and biomimetic synthesis. *Science* **261**, 1286–1292.

Marschner, H. (1998). Soil–root interface: biological and biochemical processes. In *Soil Chemistry and Ecosystem Health*, ed. Huang, P. M., Adriano, D. C., Logan, T. J., and Checkai, R. T., Soil Science Society of America, Madison, WI, 191–231.

Martin, M. H., and Bullock, R. J. (1994). The impact and fate of heavy metals in an oak woodland ecosystem. In *Toxic Metals in Soil–Plant Systems*, ed. Ross, S. M., Wiley, Chichester, West Sussex, England, 327–365.

Masschelyn, P. H., and Patrick, W. H., Jr. (1994). Selenium, arsenic, and chromium redox chemistry in wetland soils and sediments. In *Biogeochemistry of Trace Elements*, ed. Adriano, D. C., Science and Technology Letters, Northwood, Middlesex, England, 615–625.

McGrath, S. P., Chaudri, A. M., and Giller, K. E. (1995). Long-term effects of metals in sewage sludge on soils, microorganisms and plants. *J. Ind. Microbiol.* **14**, 94–104.

McLaughlin, M. J., Smolders, E., and Merckx, R. (1998). Soil–root interface: physicochemical processes. In *Soil Chemistry and Ecosystem Health*, ed. Huang, P. M., Adriano, D. C., Logan, T. L., and Checkai, R. T., Soil Science Society of America, Madison, WI, 233–277.

McLean, J. S., Lee, J.-U., and Beveridge, T. L. (2002). Interactions of bacteria and environmental metals, fine-grained mineral development and bioremediation strategies. In *Interactions Between Soil Particles and Microorganisms: Impact on the Terrestrial Ecosystem*, ed. Huang, P. M., Bollag, J.-M., and Senesi, N., IUPAC Series on Analytical and Physical Chemistry of Environmental Systems, Vol. 8, Wiley, Chichester, West Sussex, England, 227–261.

Mench, M., Amans, V., Sappin-Didier, V., Farques, S., Gomez, A., Loffler, M., Masson, P., and Arrouays, D. (1997). A study of additives to reduce availability of Pb in soils to plants. In *Remediation of Metal-Contaminated Soils*, ed. Iskandar, I. K., and Adriano, D. C., Science Reviews, Northwood, Middlesex, England, 185–202.

Mossine, V. V., Glinsky, G. V., and Feather, M. S. (1994). The preparation and characterization of some Amadori compounds (1-amino-1-deoxy-D-fructose derivatives) derived from a series of aliphatic ω-amino acids. *Carbohydr. Res.* **262**, 257–270.

Mullen, M. D., Wolf, D. C., Ferris, F. G., Beveridge, T. J., Flemming, C. A., and Bailey, G. W. (1989). Bacterial sorption of heavy metals. *Appl. Environ. Microbiol.* **55**, 3143–3149.

Myneni, S. C. B., Tokunaga, T. K., and Brown, G. E., Jr. (1997). Abiotic selenium redox transformation in the presence of Fe(II,III) oxides. *Science* **278**, 1106–1109.

Naeem, S., Thompson, L. J., Lawler, S. P., Lawton, J. H., and Woodfin, R. M. (1994). Declining biodiversity can alter the performance of ecosystems. *Nature (London)* **368**, 734–737.

National Academy of Sciences (1974). Geochemistry and the environment. In *The Relation of Selected Trace Elements to Health and Disease*, Vol. I, NAS, Washington, DC.

National Academy of Sciences (1977). Geochemistry and the environment. In *The Relation of Other Selected Trace Elements to Health and Disease*, Vol. II, NAS, Washington, DC.

National Research Council (1993). *Soil and Water Quality: An Agenda for Agriculture*, National Academies Press, Washington, DC.

National Research Council (2003). Bioavailability of contaminants in soils and sediments: processes, tools, and applications. In *Report of Committee on Bioavailability of Contaminants in Soils and Sediments*, Water Science and Technology Board, NRC, Washington, DC, 432 pp.

Nelson, Y. M., Lion, L. W., Ghiorse, W. C., and Shuler, M. L. (1999). Production of biogenic Mn oxides by *Leptothrix discophora* ss-1 in a chemically defined growth medium and evaluation of their Pb adsorption characteristics. *Appl. Environ. Microbiol.* **65**, 175–180.

Newman, M. C. (1995). *Quantitative Method of Aquatic Ecotoxicology*, CRC Press, Boca Raton, FL.

Nielands, J. B. (1981). Microbial iron compounds. *Annu. Rev. Biochem.* **50**, 715–731.

Nielson, N. E. (1976). The effect of plants on the copper concentration in the soil solution. *Plant Soil* **45**, 679–687.

NIST (2004). *Critically Selected Stability Constants of Metal Complexes*, Version 8.0, Standard Reference Database 46, National Institute of Standards and Technology, U.S. Department of Commerce, Gaithersburg, MD.

Oades, J. M., and Waters, A. G. (1991). Aggregate hierarchy in soils. *Aust. J. Soil Res.* **29**, 815–828.

Oades, J. M., Gillman, G. P., and Uehara, G. (1989). Interactions of soil organic matter and variable-charge soils. In *Dynamics of Soil Organic Matter in Tropical Ecosystems*, ed. Coleman, D. C., Oades, J. M., and Uehara, G., Nif TAL Project, University of Hawaii, Honolulu, HI, 69–95.

Odum, E. P. (1989). *Ecology and Our Endangered Life-Support Systems*, Sinauer Associates., Sunderland, MA.

Oscarson, D. W., Huang, P. M., Defosse, C., and Herbillon, A. (1981). Oxidative power of Mn(IV) and Fe(III) oxides with respect to As(III) in terrestrial and aquatic environments. *Nature* (*London*) **291**, 50–51.

Otte, M. L., Rozem, J., Koster, L., Haarsma, M. S., and Broekman, R. A. (1989). Iron plaque on roots of *Aster tripolium* L.: interaction with Zn uptake. *New Phytol.* **111**, 309–317.

Otte, M. L., Dekkers, M. J., Rozema, J., and Broekman, R. A. (1991). Uptake of arsenic by *Aster tripolium* in relation to rhizosphere oxidation. *Can. J. Bot.* **69**, 2670–2677.

Otte, M. L., Kearns, C. C., and Doyle, M. O. (1995). Accumulation of arsenic and zinc in the rhizosphere of wetland plants. *Bull. Environ. Contam. Toxicol.* **55**, 154–161.

Page, A. L., Gleason, T. L., Smith, J. E., Iskandar, I. K., and Summers, L. E. (eds.) (1983). *Utilization of Municipal Wastewater and Sludges on Land*, University of California, Riverside, CA.

Parker, D. R., Chaney, R. L., and Norvell, W. A. (1995). Chemical equilibrium models: applications to plant nutrition. In *Soil Chemical Equilibrium and Reaction Models*, ed. Loeppert, R. H., et al., America Society of Agronomy and Soil Science Society of America, Madison, WI, 163–200.

Pierzynski, G. M., Sims, J. T., and Vance, G. F. (2000). *Soil and Environmental Quality*, Lewis Publishers, Boca Raton, FL.

Pomilio, A. B., Leicach, S. R., Grass, M. Y., Ghersa, C. M., Santoro, M., and Vitale, A. A. (2000). Constituents of the root exudates of *Avena fauta* grown under far-infrared-enriched light. *Phytochem. Anal.* **11**, 304–308.

Reddy, G. B., Faza, A., and Bennett, R. C., Jr. (1987). Activity of enzymes in rhizosphere and non-rhizosphere soils amended with sludge. *Soil Biol. Biochem.* **19**, 203–205.

Reid, C. P. P. (1990). Mycorrhizas. In *The Rhizosphere*, ed. Lynch, J. M., Wiley, New York, 281–315.

Rich, C. I. (1960). Aluminum in interlayers of vermiculite. *Soil Sci. Soc. Am. Proc.* **24**, 26–32.

Rich, C. I. (1968). Hydroxy interlayers in expansible layer silicates. *Clays Clay Miner.* **16**, 15–30.

Rich, C. I., and Obenshain, S. S. (1955). Chemical and clay mineral properties of a Red-Yellow Podzolic soil derived from muscovite schist. *Soil Sci. Soc. Am. Proc.* **19**, 334–339.

Robert, M., and Berthelin, J. (1986). Role of biological and biochemical factors in soil mineral weathering. In *Interactions of Soil Minerals with Natural Organics and Microbes*, ed. Huang, P. M., and Schnitzer, M., Soil Science Society of America, Madison, WI, 453–495.

Rovira, A. D., and Davey, C. B. (1974). Biology of the rhizosphere. In *The Plant Root and Its Environment*, ed. Cursova, E. W., University of Virginia Press, Charlottesville, VA, 153–204.

Schindler, P. W., and Stumm, W. (1987). The surface chemistry of oxides, hydroxides, and oxide minerals. In *Aquatic Surface Chemistry: Chemical Process at the Particle–Water Interface*, ed. Stumm, W., Wiley, New York, 83–110.

Schnitzer, M., and Kodama, H. (1977). Reactions of minerals with soil humic substances. In *Minerals in Soil Environments*, ed. Dixon, J. B., and Weed, S. B., Soil Science Society of America, Madison, WI, 741–770.

Schulze, E. D., and Mooney, H. A. (1993). *Biodiversity and Ecosystem Function*, Springer-Verlag, Berlin.

Schwertmann, U., Kodama, H., and Fisher, W. R. (1986). Mutual interactions between organics and iron oxides. In *Interactions of Soil Minerals with Natural Organics and Microbes*, ed. Huang, P. M., and Schnitzer, M., Soil Science Society of America, Madison, WI, 223–250.

Senesi, N. (1992). Metal–humic substance complexes in the environment: molecular and mechanistic aspects by multiple spectroscopic approach. In *Biogeochemistry of Trace Elements*, ed. Adriano, D. C., Lewis, Publishers, Boca Raton, FL, 429–496.

Senesi, N., and Loffredo, E. (2005). Metal ion complexation by soil humic substances. In *Chemical Processes in Soils*, ed. Tabatabai, M. A., and Sparks, D. L., Soil Science Society of America, Madison, WI, 563–617.

Shindo, H., and Huang, P. M. (1982). Role of Mn(IV) oxide in abiotic formation of humic substances in the environment. *Nature* (*London*) **298**, 363–365.

Shindo, H., and Huang, P. M. (1984). Catalytic effects of manganese(IV), iron(III), aluminum, and silicon oxides on the formation of phenolic polymers. *Soil Sci. Soc. Am. J.* **48**, 927–934.

Shindo, H., and Huang, P. M. (1985a). The catalytic power of inorganic components in the abiotic synthesis of hydroquinone-derived humic polymers. *Appl. Clay Sci.* **1**, 71–81.

Shindo, H., and Huang, P. M. (1985b). Catalytic polymerization of hydroquinone by primary minerals. *Soil Sci.* **139**, 505–511.

Singer, A., and Huang, P. M. (1990). Effect of humic acid on the crystallization of aluminum hydroxides. *Clays Clay Miner.* **38**, 47–52.

Sparks, D. L. (2003). *Environmental Soil Chemistry*, 2nd ed., Academic Press, San Diego, CA.

Sposito, G. (ed.) (1996). *The Environmental Chemistry of Aluminum*, CRC Press/Lewis Publishers, Boca Raton, FL.

Sposito, G. (2004). *The Surface Chemistry of Natural Particles*, Oxford University Press, New York.

Stevenson, F. J. (1994). *Humus Chemistry: Genesis, Composition, Reactions*, 2nd ed., Wiley, New York.

Stevenson, F. J., and Ardakani, M. S. (1972). Organic matter reactions involving micronutrients in soils. In *Micronutrients in Agriculture*, ed. Mortvedt, J. J., et al., Soil Science Society of America, Madison, WI, 79–114.

Stevenson, F. J., and Fitch, A. (1986). Chemistry of complexation of metal ions with soil solution organics. In *Interactions of Soil Minerals with Natural Organics and Microbes*, ed. Huang, P. M., and Schnitzer, M., Soil Science Society of America, Madison, WI, 29–58.

Stotzky, G. (1986). Influence of soil mineral colloids on metabolic processes, growth, adhesion, and ecology of microbes and viruses. In *Interactions of Soil Minerals with Natural Organics and Microbes*, ed. Huang, P. M., and Schnitzer, M., Soil Science Society of America, Madison, WI, 305–428.

Stumm, W., and Morgan, J. J. (1996). *Aquatic Chemistry: Chemical Equilibria and Rates in Natural Waters*, 3rd ed. Wiley, New York.

Szmigielska, A. M., Van Rees, K. C. J., Cieslinski, G., and Huang, P. M. (1996). Low molecular weight dicarboxylic acids in rhizosphere soils of durum wheat. *J. Agric. Food Chem.* **44**, 1036–1040.

Tebo, B. M., Ghiorse, W. C., van Waasbergen, L. G., Siering, P. L., and Caspi, R. (1997). Bacterially mediated mineral formation: insight into manganese(II) oxidation from molecular genetic and biochemical studies. In *Geomicrobiology: Interactions Between Microbes and Minerals*, ed. Banfield, J. F., and Nealson, K. H., Reviews in Mineralogy and Geochemistry, Vol. 35, Mineralogical Society of America, Washington, DC, 225–266.

Tessier, A., and Turner, D. R. (1995). *Metal Speciation and Bioavailability in Aquatic Systems*, IUPAC Series on Analytical and Physical Chemistry of Environmental Systems, Vol. 3, Wiley, Chichester, West Sussex, England.

Theng, B. K. G. and Orchard, V. A. (1995). Interactions of clays with microorganisms and bacterial survival in soil: a physicochemical perspective. In *Environmental Impact of Soil Component Interactions*, Vol. II, *Metals and Other Inorganics and Microbial Activities*, ed. Huang, P. M., Berthelin, J., Bollag, J.-M., McGill, W. G., and Page, A. L., CRC Press/Lewis Publishers, Boca Raton, FL, 123–143.

Theng, B. K. G., Tate, K. R., and Sollins, P. (1989). Constituents of organic matter in temperate and tropical soils. In *Dynamics of Soil Organic Matter in Tropical Ecosystems*, ed. Coleman, D. C., Oades, J. M., and Uehara, G., Nif TAL Project, University of Hawaii, Honolulu, HI, 5–31.

Tilman, D., and Downing, J. A. (1994). Biodiversity and stability in grassland. N*ature* (*London*) **367**, 363–365.

Tilman, D., Wedlin, D., and Knops, J. (1996). Productivity and sustainability influenced by biodiversity in grassland ecosystems. *Nature* (*London*) **379**, 718–720.

Torn, M. S., Trumbore, S. E., Chadwick, O. A., Vitousek, P. M., and Hendricks, D. M., (1997). Mineral control of soil organic carbon storage and turnover. *Nature* (*London*) **389**, 170–173.

van der Heijden, M. G. A., Klironomos, J. N., Ursic, M., Moutoglis, P., Streitwolf-Engel, R., Boller, T., Wiemken, A., and Sanders, I. R. (1998). Mycorrhizal fungal diversity determines plant biodiversity, ecosysem variability and productivity. *Nature* (*London*) **396**, 67–72.

Venkateswerlu, G., Yorder, M. J., and Stotzky, G. (1989). Morphological ultra-structural, and chemical changes induced in *Cunninghamella blakesleeana* by copper and cobalt. *Appl. Microb. Biotechnol.* **31**, 204–210.

Violante, A., and Huang, P. M. (1985). Influence of inorganic and organic ligands on the formation of aluminum hydroxides and oxyhydroxides. *Clays Clay Miner.* **33**, 181–192.

Violante, A., Krishnamurti, G. S. R., and Huang, P. M. (2002). Impact of organic substances on the formation and transformation of metal oxides in soil environments. In *Interactions Between Soil Particles and Microorganism: Impact on the Terrestrial Ecosystem*, ed. Huang, P. M., Bollag, J.-M., and Senesi, N., IUPAC Series on Analytical and Physical Chemistry of Environmental Systems, Vol. 8, Wiley, Chichester, West Sussex, England, 133–188.

Volesky, B. (1990). Removal and recovery of heavy metals by biosorption. In *Biosorption of Heavy Metals*, ed. Volesky, B., CRC Press, Boca Raton, FL, 7–43.

Wada, K. (1995). Role of aluminum and iron in the accumulation of organic matter in soils with variable charge. In *Environmental Impact of Soil Component Interactions*, Vol. I, *Natural and Anthropogenic Organics*, ed. Huang, P. M., Berthelin, J., Bollag, J.-M., McGill, W. B., and Page, A. L., CRC Press/Lewis Publishers, Boca Raton, FL, 47–58.

Walker, S. G., Flemming, C. A., Ferris, F. G., Beveridge, T. J., and Bailey, G. W. (1989). Physicochemical interaction of *Escherichia coli* cell envelopes and *Bacillus subtilis* cell walls with two clays and ability of the composite to immobilize heavy metals from solution. *Appl. Environ. Microbiol.* **55**, 2976–2984.

Wang, M. C., and Huang, P. M. (1986). Humic macromolecular interlayering in nontronite through interaction with phenol monomers. *Nature (London)* **323**, 529–531.

Wang, M. C., and Huang, P. M. (1988). Catalytic power of nontronite, kaolinite, and quartz and their reaction sites in the formation of hydroquinone-derived polymers. *Appl. Clay Sci.* **4**, 43–57.

Wang, M. C., and Huang, P. M. (1994). Structural role of polyphenols in influencing the ring cleavage and related chemical reactions as catalyzed by nontronite. In *Humic Substances in the Global Environment and Implications in Human Health*, ed. Senesi, N., and Miano, T. M., Elsevier, Amsterdam, The Netherlands, 173–180.

Wang, M. C., and Huang, P. M. (2000a). Characteristics of pyrogallol-derived polymers formed by catalysis of oxides. *Soil Sci.* **165**, 737–747.

Wang, M. C., and Huang, P. M. (2000b). Ring cleavage and oxidative transformation of pyrogallol catalyzed by Mn, Fe, Al, and Si oxides. *Soil Sci.* **165**, 934–942.

Wang, T. S. C., and Li, S. W. (1977). Clay minerals as heterogeneous catalysts in preparation of model humic substances. *Z. Pflanzenernaehr. Bodenkd.* **140**, 669–676.

Wang, T. S. C., Huang, P. M., Chou, C.-H., and Chen, J.-H. (1986). The role of soil minerals in the abiotic polymerization of phenolic compounds and formation of humic substances. In *Interactions of Soil Minerals with Natural Organics and Microbes*, ed. Huang, P. M., and Schnitzer, M., Soil Science Society of America, Madison, WI, 251–281.

Wardle, D. A. (2002). *Communities and Ecosystems: Linking the Aboveground and Belowground Components*, Princeton University Press, Princeton, NJ.

Wardle, D. A., Bardgett, R. D., Klironomos, J. N., Setälä, H., van der Putten, W. H., and Wall, D. H. (2004). Ecological linkages between aboveground and belowground biota. *Science* **304**, 1629–1633.

Weed, S. B., Davey, C. B., and Cook, M. G. (1969). Weathering of micas by fungi. *Soil Sci. Soc. Am. Proc.* **33**, 702–706.

Wenzel, W. W., Wieshammer, G., Fitz, W. J., and Puschenreiter, M. (2001). Novel rhizobox design to assess rhizosphere characteristics at high spatial resolution. *Plant Soil* **237**, 37–45.

Young, I. M., and Crawford, J. W. (2004). Interactions and self-organization in the soil–microbe complex. *Science* **304**, 1634–1637.

Yu, G., Saha, U. K., Kozak, L. M., and P. M. Huang (2006). Kinetics of cadmium adsorption on aluminum precipitation products formed under the influence of tannate. *Geochim. Cosmochim. Acta* **70**, 5134–5145.

Yu, G., Saha, U. K., Kozak, L. M., and P. M. Huang (2007). Combined effects of tannate and aging on structural and surface properties of aluminum precipitates. *Clays Clay Miner.* **55**, 369–379.

2

TRANSFORMATION AND MOBILIZATION OF METALS, METALLOIDS, AND RADIONUCLIDES BY MICROORGANISMS

G. M. Gadd
University of Dundee, Dundee, Scotland

2.1. INTRODUCTION

In the soil environment, microorganisms are intimately associated with the biogeochemical cycling of all important elements, including metals, and these processes are of global and microscale importance (Gadd, 2006; Gadd et al., 2005, 2007).

Biophysico-Chemical Processes of Heavy Metals and Metalloids in Soil Environments,
Edited by Antonio Violante, Pan Ming Huang, and Geoffrey Michael Gadd

However, microbial influence on metal geochemistry has probably not received as much scientific attention as have physical and chemical processes. For metals (and related substances such as metalloids, metal radioisotopes, actinides, and organometallic compounds) in their various forms, microbial activities can result in mobilization or immobilization, depending on the processes involved and the (micro)environment where the organism is located (Figure 2.1). Given organisms may carry out both processes, with repeated stages of mobilization and immobilization, and this may result in a variety of changes in metal speciation and biogeochemical effects (Ehrlich, 1997, 2002). Some changes are an indirect consequence of microbial morphology, structure, and metabolism; others are components of specific metabolic systems for the acquisition of important nutrients such as essential metals and macronutrients (e.g., phosphorus). These phenomena are of wider consequence to other members of the soil biota, especially plants. For example, root-inhabiting free-living and symbiotic rhizosphere microorganisms, especially bacteria and fungi, have a major influence on plant

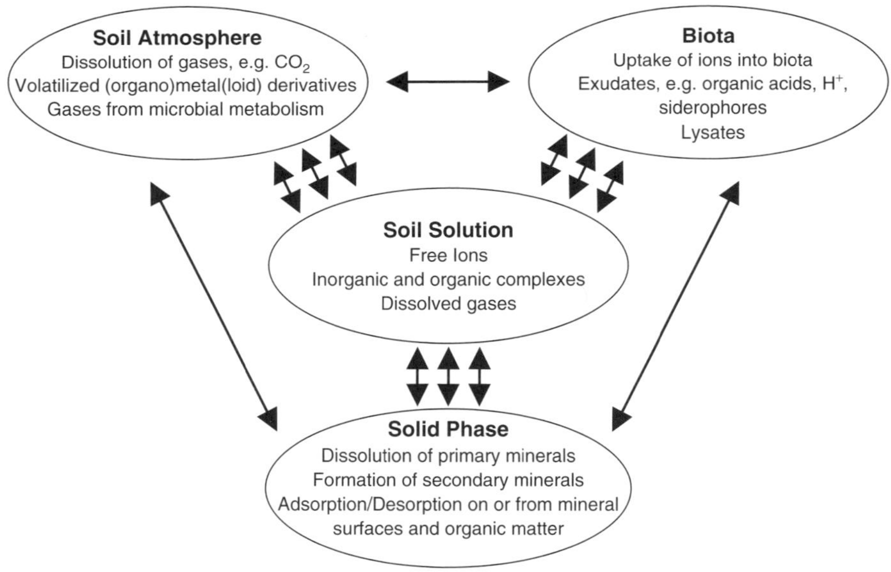

Figure 2.1. Soil-solution reactions involved in metal(loid) speciation and partitioning between solid, liquid, and gaseous phases. The soil solution (shown as the central compartment in the diagram) has a pivotal role in most reactions of metabolic and geochemical significance. The arrows linking the outer components of the diagram that represent the biota, soil atmosphere, and solid phase indicate those direct reactions that may occur with little or no significant function of the soil solution. These could include utilization or production of gases by the biota (e.g., O_2 and CO_2, respectively), direct attack of solid phases by biota, and gas adsorption onto solid phase surfaces. However, in the absence of water, such reactions may be limited, especially those dependent on microbial metabolism. (Derived from Cotter-Howells and Paterson, 2000.)

nutrition by means of effects on phosphate availability but also on concomitant metal circulation. Further, microbial populations are intimately involved in the actual formation of mineral soil: the contribution of microbial activities to rock weathering, mineral dissolution, and element cycling is significant during the early phases of soil formation and is related intimately to metal cycling and microbial mineral and metal transformations (Ehrlich, 2002; Gadd, 2007).

Metals are significant natural components of all soils where their presence in the mineral fraction comprises a store of potentially mobile metal species as important components of, for example, clays, minerals, and iron and manganese oxides, which in turn have a dramatic influence on soil geochemistry (Ehrlich, 2002). Metals are also present in the organic fraction, frequently as bound forms, with some metal recycling occurring as a result of organic matter decomposition. The aqueous phase provides a mobile medium for chemical reactions, metal transfer, and circulation through the soil to organisms and to the aquatic environment (Figure 2.1).

Metals exhibit a range of toxicities toward microorganisms, depending on physicochemical and biotic factors, and although toxic effects can arise from natural activities, toxic effects on soil microbial communities are more commonly associated with anthropogenic contamination or redistribution of toxic metals. This can arise from aerial and aquatic sources, as well as agricultural practices, industrial activity, and domestic and industrial wastes. In some cases, microbial activity can result in remobilization of metals from other wastes and transfer into aquatic systems. It is commonly accepted that toxic metals (and their chemical derivatives and related substances) can have significant effects on microbial populations, and almost every index of microbial activity can be affected. However, metal toxicity is greatly affected by physicochemical factors and the chemical behavior of the particular metal species in question. Further, a link between bioavailability and toxicity is not always clear, a good example being the amelioration of the toxicity of several metals by acidic pH conditions (Gadd, 1993a). Despite potential toxicity, many microorganisms survive, grow, and flourish in apparently metal-polluted locations, and a variety of mechanisms, both active and incidental, contribute to resistance and tolerance (Gadd, 1992a). Virtually all mechanisms depend on some change in metal speciation leading to decreased or increased mobility. Such transformations between soluble and insoluble phases are central to metal biogeochemistry and therefore provide a direct link between microbial responses and elemental cycles. Thus, metal–microbe interactions are extremely important and underpin many aspects of soil biology and function. The objective of this chapter is to outline the major mechanisms by which microorganisms can affect metal transformations and to draw attention to their significance in the soil, especially regarding metal mobility, biogeochemical cycling, and plant nutrition and remediation technologies for contaminated habitats.

2.2. METAL–MICROBE INTERACTIONS

Toxic metals may have a considerable impact on populations of soil microorganisms and their activities (Brookes and McGrath, 1984; Gadd, 2005) and effects may be observed as a reduction in biomass as well as activity: Almost every microbial process has the potential to be affected by toxic metals (Brookes and McGrath, 1984; Aoyama and Nagumo, 1997; Kuperman and Carreiro, 1997). However, experimentation based around metal additions to soils may have little relation to field results (Giller et al., 1998). It seems obvious that metal effects on natural soil communities are complicated and difficult to characterize because of the many complex contributing factors. Temporal and spatial variations in soil composition result in accompanying changes in metal speciation, so prediction of the mobility and availability of metals in soils is extremely difficult (Staunton, 2004). It follows that such effects will also be mirrored in difficulties in assessing responses and changes in microbial populations toward toxic metals. Most knowledge to date is derived from data relating to only a few toxic metals (e.g., Cu and Zn) or from studies on, for example, sewage sludge applications that contain metals at relatively low concentrations (Shi et al., 2002). Many contaminated sites contain mixtures of metals as well as organic pollutants: each may have reciprocal physical and chemical effects on the other. One study found that the soil microbial community was affected predominantly by hydrocarbons rather than by the associated Pb and Cr contamination (Shi et al., 2002). Baath (1989) found that metal concentrations that affected microbial communities in soil differed more than 100-fold between studies. These discrepancies may be caused by differences in the bioavailability of metals as a consequence of different physical and chemical properties of the soils investigated (Giller et al., 1998; see also Huang, chapter 1, Krishnamurti and Naidu, chapter 11, and Violante et al., chapter 5, this volume). Therefore, although some gross generalizations are possible regarding toxic metal influence on microbial communities, individual cases are likely to be site specific and extremely complex.

A variety of experimental systems have been investigated to study the stress exerted by toxic metals on soil microbial communities. These have included short- versus long-term microcosm incubations, and dosing with a single metal or a combination of metals with or without the simultaneous addition of organic matter. Differences in soil pH, organic matter content, cation exchange capacity, temperature, and moisture content introduce a high level of variability (Baath, 1989; Giller et al., 1998). Giller et al. (1998) caution against an automatic assumption that a simple, negative relationship exists between stress and diversity in microbial communities (Del Val et al., 1999, Degens et al., 2001). However, studies of the impact of metals on bacterial diversity in soil have generally shown a negative influence (Hirsch et al., 1993; Smit et al., 1997; Sandaa et al., 1999; 2001; Muller et al., 2001). Toxic metals contained in sewage sludge dry solids may permanently degrade the microbial decomposer communities of agricultural soil. For example, there were 120 operational taxonomic units in a control soil but only 90 in a zinc-treated soil, a decrease of 25% (Moffett et al., 2003).

Numerous field studies have also demonstrated the adverse effect of different toxic metals on soil microbial communities (Baath, 1989; Said and Lewis, 1991; Roane and Kellogg, 1996; Khan and Scullion, 2000). In both short and long-term exposure to Zn, Cu, and Ni, microbial community structure was altered as assessed from soil total phospholipid fatty acid (PLFA) profiles (Frostegaard et al., 1993; Pennanen et al., 1996; Baath et al., 1998; Kelly et al., 1999). Several studies have also found that metals decrease soil microbial biomass (Frostegaard et al., 1993; Fliessbach et al., 1994; Roane and Kellogg, 1996; Konopka *et al.*, 1999) and viable population size (plate counts) (Kelly and Tate, 1998). Microbial adaptation may be an important response to the presence of soil contaminants and may result in increased activity of the remaining microbiota (Duxbury and Bicknell, 1983; Doelman, 1986).

The nature of the soil environment can also affect metal speciation and microbial distribution and therefore both individual and community responses. Recent research has demonstrated that bacterial communities are not distributed randomly in soil, and spatial patterns have been identified in the distribution of bacteria and bacterial function at scales from several millimeters to several meters (Foster, 1988; Robertson et al., 1997; Nunan et al., 2002, 2003). Evidence for bacterial patchiness at scales below 1 mm is now available (Grundmann et al., 2001; Nunan et al., 2002, 2003). It has been suggested that microscale spatial patterns may have a regulatory effect on bacterial activity as a result of the diffusional limitations imposed on nutrient uptake and on metabolite excretion and dispersion. These may have rate-limiting or stimulatory effects on several microbially mediated processes (Darrah et al., 1987; Grundmann et al., 2001).

Metal-contaminated soils contain a number of organic ligands, particularly in the rhizosphere, so a significant fraction of the bioavailable soil metals probably exists in complexed form (see also Krishnamurti and Naidu, chapter 11, Senesi and Loffredo, chapter 4, and Essington, chapter 10, this volume). Oxalate and citrate complexed with Cd, Cu, Mg, Pb, and Zn have been used as model metal–organic complexes. Using soil respiration as an index of biodegradation, it was found that citrate–metal complexes were generally more degradable than oxalate–metal complexes. Biodegradability of the metal–oxalate or metal–citrate complexes varied for the different metals and followed the order Mg $>$ Zn $>$ Cu $\equiv$ Pb $>$ Cd, with the addition of Cd complexes decreasing soil respiration. Such observations emphasize that the formation of complexes between toxic metals and low-molecular-mass organic ligands may affect soil functionality, especially in the rhizosphere of contaminated soils (Renella et al., 2004).

Several studies have shown that microbial population responses to toxic metals are characterized by shift from bacteria, including streptomycetes, to fungi (Mineev et al., 1999; Chander et al., 2001a,b; Kostov and Van Cleemput, 2001; Olayinka and Babalola, 2001; Khan and Scullion, 2002). However, other studies have shown a higher metal sensitivity of the fungal component of the microbial biomass (Pennanen et al., 1996). The time needed for development of community tolerance may be highly variable (Doelman and Hanstra, 1979; Diaz-Ravina and Baath, 1996). An increase in metal tolerance of a bacterial community after metal addition may be attributed to the immediate death of sensitive species followed by adaptation of bacterial survivors (Diaz-Ravina and Baath, 1996). Using phospholipid fatty acid to reveal changes in species composition, it was found that bacterial community tolerance increased in all metal treatments compared to unpolluted control (Baath et al., 1998). In anaerobic bacterial consortia, dehalogenation, aromatic degradation, and methanogenesis all showed differential sensitivities to a variety of toxic metals (Kuo and Genthner, 1996).

Field studies on the microbial communities of boreal coniferous forest humus exposed to environmental stress showed that the structure of the microbial community was influenced by changes in humus pH and metal concentrations at levels where few or no effects were evident on microbial biomass or metabolic activity. Changes in the relative proportions of gram-negative and gram-positive bacteria, including actinomycetes, occurred as well as adaptation to the environmental disturbance in question. Increased metal tolerance of the humus bacterial community resulted partly from a change in microbial species composition (Pennanen, 2001).

Microbial community structure, potential microbial activity, and arsenic (As) resistance were affected by As, chromium (Cr), and copper (Cu) contamination in soils of abandoned wood-impregnating plants (Turpeinen et al., 2004). Phospholipid fatty acid and 16S rRNA gene terminal restriction fragment length polymorphism (t-RFLP) profiles indicated that exposure to high metal contamination or subsequent effects of this exposure changed the microbial community structure permanently. The total number of colony-forming units (CFU) was not affected by metal contamination, and the As(V)-resistant bacterial ratio to total heterotrophic plate counts was high (0.5 : 1.1) and relatively independent of the concentration of As. In contrast, the proportion of As(III)-resistant bacteria was dependent on the concentration of As in the soils, and a significant positive relationship was found between the bioavailability of As and the proportion of As(III)-resistant bacteria. Dominant As-resistant isolates from contaminated soils were *Acinetobacter*, *Edwardsiella*, *Enterobacter*, *Pseudomonas*, *Salmonella*, and *Serratia* species. No differences were noted in glucose utilization between contaminated and control soil samples within sites. This study, combining complementary culture-dependent and culture-independent methods, suggests that microbes are able to respond to soil metal contamination and maintain metabolic activity apparently through changes in microbial community structure and selection for resistance (Turpeinen et al., 2004).

The impact of copper amendment on the bacterial community in agricultural soil was investigated by a two-year field experiment complemented by short-term microcosm studies. Bioavailable copper was determined with a copper-specific bioluminescent *Pseudomonas* reporter strain. Amounts of total Cu as well as of bioavailable Cu in the field declined throughout the experimental period. Bacterial community structure was examined by t-RFLP analysis of community DNA amplified with primers specific for 16S rDNA from the Bacteria domain, the *Rhizobium–Agrobacterium* group, and the *Cytophaga* group. Similarity analysis of t-RFLP profiles from field samples demonstrated an impact of copper at the domain level and within the *Rhizobium–Agrobacterium* group. Comparable Cu effects were observed for microcosms, but in addition an impact on community structure within the *Cytophaga* group was observed (Tom-Petersen et al., 2003).

The combined use of rDNA and rRNA libraries was used to indicate which bacterial taxa might be metabolically active in metal-contaminated bulk and rhizosphere soil (Gremion et al., 2003). Most dominant taxa were relatively less represented in the rRNA libraries than in the rDNA libraries. However, sequences belonging to the *Actinobacteria* dominated both bulk and rhizosphere soil libraries derived from rRNA (50 and 60% of all phylotypes, respectively) providing evidence that this group of bacteria was probably metabolically active in metal-contaminated soil. Identification and tracking of these potentially active, yet largely uncultured bacteria might help us to understand how phytoremediation techniques influence dominant bacterial populations and their functions and whether such soil bacteria play an important role in metal uptake by plants. Cultivation of members of these groups will allow further study of their metal resistance, metal-transforming capacities, and potential use as inocula in (phyto)remediation processes (Gremion et al., 2003).

Picard and Bosco (2003) found that the percentage of bacterial isolates capable of auxin production (*aux*+) in the rhizosphere of *Achillea ageratum* increased with soil antimony concentration and that this *aux*+ population was genetically very diverse. These data appear to contradict numerous other studies that have observed a lower diversity index in between bacteria isolated from high-metal-contaminated soils than in those isolated from low-metal-contaminated soils (Kelly et al., 1999; Ellis et al., 2001; Fritze et al., 2000; Sandaa et al., 2001). However, comparisons may be difficult since many studies are concerned with short-term metal effects in experimentally contaminated soils (Kelly et al., 1999; Fritze et al., 2000) or long-term contaminated soils, but without consideration of plant relationships (Ellis et al., 2001). Few investigations examine the effects of toxic metals on microorganisms in the rhizosphere of plants naturally colonizing long-term polluted soils. For nickel-rich soils, it has been shown that bacteria isolated from the rhizosphere of the hyperaccumulator plant *Alyssum bertolonii* were genetically greatly diverse from those isolated from nonrhizospheric soil (Mengoni et al., 2001). Another study showed that zinc had no or only weak effects on the total number of bacteria isolated from the rhizosphere of *Thlaspi caerulescens* or *Trifolium pratense* (Delorme et al., 2001), which is in accord with the data of Picard and Bosco (2003). An investigation on *Rhizobium leguminosarum* biovar *trifolii* in a range of metal-contaminated soils is also partly in agreement with the latter study, since it observed that metals had a quantitative effect on the rhizobia population but not on genetic diversity (Obbard et al., 1993).

Serpentine soils have high levels of heavy metals (Ni, Co, Cr), and low levels of important plant nutrients (P, Ca, N). Because of these inhospitable conditions, serpentine soils support a specialized flora, including such plant species as the nickel hyperaccumulator *Alyssum bertolonii*. In a study that examined bacterial communities at various distances from *A. bertolonii* roots in different serpentine areas, it was found that the bacterial communities possessed high genetic diversity and heterogeneity (Mengoni et al., 2004). Other studies have shown that serpentine bacterial communities can tolerate additions of nickel better than can isolates from unpolluted soil and that there is an increase in metal-resistant bacteria in soil near Ni-hyperaccumulating plants (Delorme et. al., 2001; Gremion et al., 2003; Hery et al., 2003). Such bacteria may play a role in regulating metal availability to plants (de Souza et al., 1999; Whiting et al., 2001; Abou-Shanab et al., 2003). Adaptation to nickel of bacterial communities of two extreme neocaledonian soils (an ultramafic soil and an acidic soil) was investigated by nickel spiking and compared with adaptation in a non-neocaledonian soil. Bacterial community structure was analyzed with the ribosomal intergenic spacer analysis (RISA) technique, and bacterial populations that responded to nickel stress were identified by cloning and sequencing. In the ultramafic soil, a shift occurred consisting of the emergence of a bacterial group closely related to the *Ralstonia–Oxalobacter–Burkholderia* group followed later by a shift in the Actinomycete group. Only a few changes occurred in the bacterial community structure of the neocaledonian soils compared with those of the reference soil, which was more affected by nickel spiking, indicating that neocaledonian soil bacteria are particularly well adapted to nickel (Hery et al., 2003). In another study, heterotrophic nickel-resistant bacteria were isolated and characterized from three different serpentine outcrops (Mengoni et al., 2001). The proportion of nickel-resistant CFU was higher in proximity to the nickel-hyperaccumulating *A. bertolonii* than in free soil. The most represented genera in all three different serpentine outcrops were *Pseudomonas* and *Streptomyces*. *Pseudomonas* strains were found to be predominant in the plant rhizosphere, whereas *Streptomyces* strains were present mainly in the soil (Mengoni et al., 2001).

In recent years, culture-independent methods have been used in preference to traditional isolation techniques for microbial community analysis. However, it is questionable whether uncultured organisms from a given sample are important for determining the impact of anthropogenic stress on indigenous communities. In an attempt to clarify this, soil samples were taken from a site with patchy metal contamination, and the bacterial community structure was assessed with a variety of approaches. There were small differences in microscopic epifluorescence bacterial counts; denaturing gradient gel electrophoresis (DGGE) profiles of 16S rRNA gene fragments (16S-DGGE) amplified directly from soil samples were very similar. However, the proportion of bacteria from the soil samples that were culturable on standard plate-counting media varied between 0.08 and 2.2%, and these values correlated negatively with metal concentrations (Ellis et al., 2003). The primary conclusion from this and other data were that metal contamination did not have a significant effect on the total genetic diversity present but affected the physiological status, so that the number of bacteria capable of responding to laboratory culture and their taxonomic distribution were altered. Thus, it appears that plate counts may be a more appropriate method of determining the effect of heavy metals on soil bacteria than culture-independent approaches (Ellis et al., 2003). Further advances in the design of solid media and culture methods for soil bacteria may make plate counting even more useful for assessing the impact of metal pollution. In fact, because some culturing techniques have shown that functionally important bacterial groups (e.g., rhizobia) decrease in abundance in metal-contaminated soil, this has been used as an indicator for metal toxicity (Giller et al., 1993; Dahlin et al., 1997; Chaudri et al., 2000).

The individual bases for metal resistance in bacteria have received considerable attention. Bacterial plasmids have resistance genes to many toxic metal and metalloid species (e.g., Ag^{+}, AsO_2^{-}, AsO_4^{3-}, Cd^{2+}, Co^{2+}, CrO_4^{2-}, Cu^{2+}, Hg^{2+}, Ni^{2+}, Sb^{3+}, TeO_3^{2-}, Tl^{+}, and Zn^{2+}). Related systems are also frequently located on bacterial chromosomes (e.g., Hg^{2+} resistance in *Bacillus*, Cd^{2+} efflux in *Bacillus*, and arsenic efflux in *Escherichia coli*) (Silver and Phung, 1996). Resistance mechanisms generally involve efflux from the cells or enzymatic detoxification (Silver and Phung, 1996; Silver, 1998; Nies, 1992a, 1995, 1999, 2003), although less-specific interactions (e.g., sorption) may contribute to the overall response. Efflux pumps, determined by plasmid and chromosomal systems, are either ATPases or chemiosmotic systems, with mechanisms often showing similarity in different types of bacteria (see Silver and Phung, 1996; Ji and Silver, 1992a,b; Nies, 1992b; Ji et al., 1994; Silver, 1998). It appears that mercury selects for higher frequencies of Hg^{2+}-resistant bacteria in polluted habitats (Silver and Phung, 1996). There are still many unknown interactions with a variety of metals to be fully characterized, and even for the well-known bacterial Hg, As, Cd, and Cu resistance mechanisms, fundamental mechanisms of resistance or gene regulation at the molecular level are perhaps not yet fully understood (Silver and Phung, 1996). However, as more information becomes available, genomic sequences can be used to survey mechanisms of uptake, resistance to, and homeostatic regulation of several metals and metalloids in bacteria and other microorganisms (Cánovas et al., 2003). Analysis of the genome of *Pseudomonas putida* KT2440 revealed a very high capacity to tolerate heavy metals and metalloids. The complete chromosomal sequence allowed the categorization of 61 open reading frames likely to be involved in metal tolerance or homeostasis, plus seven more possibly involved in metal resistance mechanisms, with some systems appearing to be duplicated. *P. putida* possessed two systems for arsenic (*arsRBCH*), one for chromate (*chrA*), four to six systems for divalent cations (two *cadA* and two to four *czc* chemiosmotic antiporters), two systems for monovalent cations (*pacS* and *cusCBA*, plus one cryptic *silP* gene containing a frameshift mutation), two operons for Cu chelation (*copAB*), one metallothionein for metal(loid) binding, one system for Te/Se methylation (*tpmT*), and four ABC transporters for the uptake of essential Zn, Mn, Mo, and Ni (one *nikABCDE*, two *znuACB*and, and one *mobABC*) (Cánovas et al., 2003).

Polyphosphate has also been implicated in metal tolerance and removal in several microorganisms (Aiking et al., 1984; Pettersson et al., 1985). Polyphosphate is reversibly synthesized by polyphosphate kinase (PPK) and irreversibly hydrolyzed by exopolyphosphatase (PPX) (Kornberg et al., 1999). Some work has indicated that the ability to turn over polyphosphate reserves is more important for metal resistance than just the ability to accumulate polyphosphate: *E. coli* harboring multiple copies of the genes for PPK (*ppk*) and PPX (*ppx*) was able to tolerate increased concentrations of, for example, cadmium, whereas *E. coli* engineered with only *ppk* (so that it accumulated significant amounts of polyphosphate) had little more metal tolerance than polyphosphate-free strains (Keasling and Hupf, 1996). It was proposed that the phosphate secreted, generated from polyphosphate hydrolysis, precipitated metals outside the cell, thereby reducing toxicity. For *Acidithiobacillus ferrooxidans*, polyphosphate levels were determined when the bacterium was grown in or shifted to a high copper concentration (100 mmol L^{-1}). Under these conditions, cells showed a rapid decrease in polyphosphate levels with a concomitant increase in exopolyphosphatase activity and a stimulation of phosphate efflux. This study supported a model for metal detoxification in which heavy metals stimulate polyphosphate hydrolysis, and the metal–phosphate complexes formed are transported out of the cell as part of a metal tolerance mechanism in *A. ferrooxidans* (Alvarez and Jerez, 2004).

All nutritional groups of fungi can be affected by toxic metals (Gadd, 2005, 2007). A relative decrease in arbuscular mycorrhizal (AM) fungi and an increase in other fungi has been reported for zinc-polluted soil (Kelly et al., 1999). Toxic metals (Cd, Cr, Cu, Ni, Pb, and Zn) led to a decrease in the number of AM fungi and low colonization of plant roots, and as a result, changes in mycorrhizal species diversity (Del Val et al., 1999; Moynahan et al., 2002; Mozafar et al., 2002). Toxic metals also reduce plant root colonization by ectomycorrhizal fungi and ectomycorrhizal species composition (Fay and Mitchell, 1999; Hartley et al., 1999; Markkola et al., 2002). Investigations on metal toxicity toward mycorrhizal fungi have revealed extensive inter- and intraspecific variation in sensitivity (Jones and Muehlchen, 1994; Hartley et al., 1997; Vodnik et al., 1998; Blaudez et al., 2000; Meharg and Cairney, 2000). Some studies have suggested that selection for resistant ecotypes occurs where the degree of toxic metal contamination and selection pressure is high (Colpaert et al., 2000; Lacourt et al., 2000; Sharples et al., 2000, 2001).

Plant symbiotic mycorrhizal fungi can accumulate metals from soil components, and this may have consequences for metal nutrition of the symbiosis as well as increased or decreased toxicity (Fomina et al., 2005a, 2006). Since plants growing on metalliferous soils are generally mycorrhizal, an important ecological role for the fungus has frequently been postulated, although such a role (e.g., phytoprotection) is often difficult to confirm (Meharg and Cairney, 2000). Ericaceous plants appear to be entirely dependent on the presence of ericoid mycorrhizas for protection against Cu, the fungus preventing metal translocation to plant shoots (Bradley et al., 1981, 1982). Arbuscular mycorrhizas (AMs) from metal-contaminated sites are often more metal tolerant than other isolates to, for example, Cd and Zn, suggesting a possible benefit to the plant via increased metal resistance, nutrient uptake, and so on, although in some instances, AM plants do not necessarily require fungal colonization for survival (Griffioen, 1994). It is often postulated that mycorrhizas provide a barrier to the uptake of potentially toxic metals (Wilkins, 1991; Hetrick et al., 1994; Wilkinson and Dickinson, 1995; Leyval et al., 1997; Meharg and Cairney, 2000), although this has not been confirmed in every case. Further, in some instances, AM may mediate enhanced accumulation of essential metals, which unless regulated, may lead to phytotoxicity (Killham and Firestone, 1983). It is generally concluded that local conditions in metal-contaminated sites may determine the cost–benefit relationship between the plant and the AM fungus, since detrimental, neutral, or beneficial interactions have all been documented (Meharg and Cairney, 2000). For ericaceous mycorrhizas, clear host protection is observed for ericaceous plants (e.g., *Calluna* sp., *Erica* sp. *Vaccinium* sp.) growing on polluted and/or naturally metalliferous soils (Bradley et al., 1981, 1982). Further, ericaceous plants are generally found on nutrient-deficient soils and it is likely that the mycorrhiza additionally benefit the plants by enhanced nutrient uptake (Smith and Read, 1997). A protective metal-binding effect of ectomycorrhizal fungi (EcM) has been postulated frequently (e.g., Leyval et al., 1997), although other workers point out the lack of clear evidence (Dixon and Buschena, 1988; Colpaert and Van Assche, 1993).

The most frequent soil saprotrophic microfungi isolated from heavily metal-polluted habitats in Argentina, the Czech Republic, and Ukraine were reported to be species of *Penicillium, Aspergillus, Trichoderma, Fusarium, Rhizopus, and Mucor,* as well as *Paecilomyces lilacinus, Nectria invertum, Cladosporium cladosporioides, Alternaria alternata,* and *Phoma fimeti* (Kubatova et al., 2002; Massaccesi et al., 2002; Fomina, Manichev, Kadoshnikov, and Nakonechnaya, unpublished). Melanized fungi such as *Cladosporium* sp., *Alternaria alternata*, and *Aureobasidium pullulans* were often isolated from soil containing high concentrations of copper and mercury (Zhdanova et al., 1986) and can be dominant members of the mycoflora of metal-contaminated phylloplanes (Mowll and Gadd, 1985). Dark septate endophytes were found to be dominant fungi among isolates from roots of *Erica herbacea* L. in Pb-, Cd-, and Zn-polluted soil (Cevnik et al., 2000).

As with bacteria, intracellular metal concentrations may be regulated by transport, including efflux mechanisms (Gadd, 1993b; Macreadie et al., 1994; Blaudez et al., 2000). Such mechanisms are involved in normal metal homeostasis but also have a role in metal detoxification. The fungal vacuole also has an important role in the regulation of cytosolic metal ion concentrations and metal(loid) detoxification (White and Gadd, 1986; Gadd, 1993b, 1995b; Gharieb and Gadd, 1998; Liu and Culotta, 1999). The absence of a vacuole or a functional vacuolar H^+-ATPase in *Saccharomyces cerevisiae* is associated with increased sensitivity and a largely decreased capacity of the cells to accumulate Zn, Mn, Co, and Ni (Ramsay and Gadd, 1997), metals known to be detoxified primarily in the vacuole (Gadd, 1993b; Joho et al., 1995). For Cu and Cd, intracellular detoxification in fungi appears to depend predominantly on sequestration in the cytosol by induced metal-binding molecules (Hayashi and Mutoh, 1994; Macreadie et al., 1994; Ow et al., 1994; Rauser, 1995). These include low-molecular-mass cysteine-rich proteins (metallothioneins) and peptides derived from glutathione (phytochelatins) (Mehra and Winge, 1991; Macreadie et al., 1994; Ow et al., 1994; Rauser, 1995; Wu et al., 1995; Inouhe et al., 1996; Gharieb and Gadd, 2004). However, it should be noted that other determinants of tolerance can occur in these and other organisms (e.g., transport phenomena and extracellular precipitation) (Gadd et al., 1984; Gadd and White, 1985, 1989; Inouhe et al., 1996; Yu et al., 1996).

2.3. METAL AND MINERAL TRANSFORMATIONS

The range of interactions between microorganisms, metals, and soil components can be extremely complex, and because of this, many inferences are necessarily taken from simplified laboratory experiments, which probably do not adequately reflect the range of natural situations that may be encountered. However, it is possible to make some general observations and conclusions regarding the processes involved, their relative importance, and their involvement in metal speciation in the soil. The relative role of fungi and bacteria depends on associated biota (other microorganisms, plants, and animals) as well as on physicochemical properties of the environmental matrix (e.g., pH, water, inorganic and organic ions, molecules, compounds, colloids, and particulates).

The ability of microorganisms to influence metal speciation in the soil results from indirect and direct properties which influence the balance of metal species between soluble and insoluble phases (Figures 2.1 to 2.3). Mobilization of metals can be achieved by various mechanisms: protonation, chelation, and chemical transformations. Metal immobilization can occur by precipitation or crystallization of insoluble organic or inorganic compounds (e.g., oxalates and sulfides, respectively), or by sorption, uptake, and intracellular sequestration (Glasauer et al., 2004a). In addition, redox reactions can mobilize or immobilize metals, depending on the particular metal species and microenvironment (Figure 2.2).

2.3.1. Mobilization

Fungi and bacteria promote metal speciation into soluble forms via the dissolution of metal compounds, including oxides, phosphates, sulfides, and more complex minerals, and by desorption from exchange sites on clay minerals or organic matter (Figures 2.2 and 2.3). A number of mechanisms appear to be significant. Fungi and bacteria can acidify their environment in a number of ways: for example, proton efflux via the plasma membrane H^+-ATPase, maintenance of charge balance, or as a result of respiratory carbon dioxide accumulation leading to carbonic acid formation. Acidification can result in metal desorption by protonation of binding sites. Microorganisms produce a number of extracellular metabolites that can complex metals in solution, including polysaccharides, pigments, organic acids, and siderophores.

Siderophores are highly specific Fe(III) ligands (formation constants often $> 10^{30}$) and are excreted by a wide variety of fungi and bacteria to aid iron assimilation because of the low solubility of Fe^{3+} at pH values where most life exists. Siderophores are the most common means of acquisition of iron by bacteria and fungi (Crichton, 1991), and in soil, many different species, including plants, will compete for Fe. Although produced primarily as a means of obtaining iron,

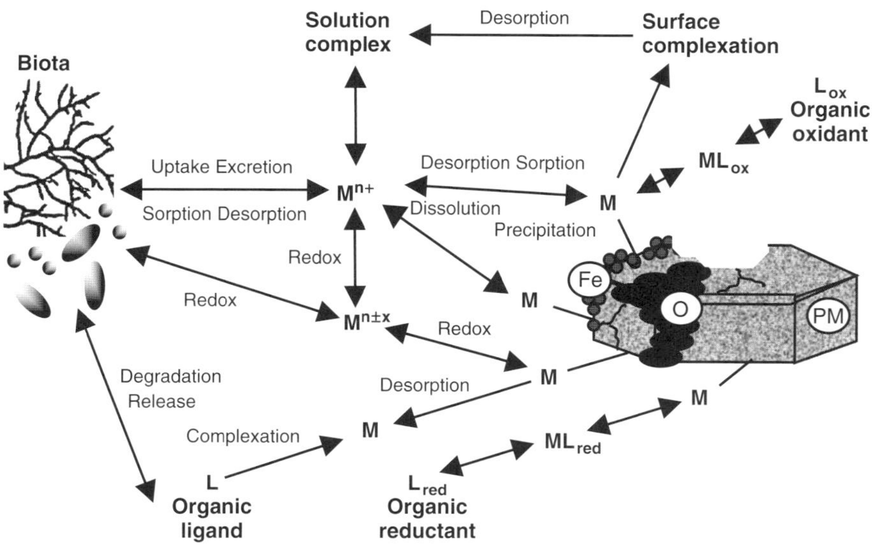

Figure 2.2. Some molecular-scale processes involved in metal speciation, transformation, and fate in the soil environment. Only the involvement of microbiota is shown, although other organisms will also have roles in the overall process by means of exudates and excretion, decomposition products, and physical activities (e.g., tunneling and predation). The diagram highlights filamentous fungi and bacteria as the prime groups of microorganisms involved in soil processes (Fe, iron oxide coating; O, organic coating; PM, primary mineral). Organic ligands may arise from a number of sources, including direct excretion by biota or from degradation. (Adapted from Brown et al., 1999.)

siderophores are also able to bind other metals, such as magnesium, manganese, chromium(III), gallium(III), and radionuclides such as plutonium(IV) (Birch and Bachofen, 1990) (see also Hinsinger and Courchesne, chapter 7, and Neubauer et al., chapter 9, this volume).

Citric and oxalic acids are common organic acids in soils and can be released into the soil by free-living and mycorrhizal fungi, lichens, and plant roots. The metal-solubilizing ability among natural soil fungal communities appears to be high (Sayer et al., 1995; Martino et al., 2003). Organic acids have important functions as a result of their acidification and metal-binding properties. Phosphate can be released from common soil compounds such as $CaHPO_4$, $Ca_3(PO_4)_2$, $FePO_4{\cdot}4H_2O$, and $AlPO_4$ by direct ligand exchange between phosphate and the organic anion or by binding of the metal to the anion. The bioavailability of solubilized metal–ligand complexes depends on its environmental fate, including sorption onto components of the soil matrix or potential utilization by biota. Metal citrates can be highly resistant to microbial degradation, which can result in leaching of metals from soil. In addition, organic acid production by microbes is an important agent of mineral deterioration (Adeyemi and Gadd, 2005), and these substances play a role in both biogenic chemical weathering and soil formation

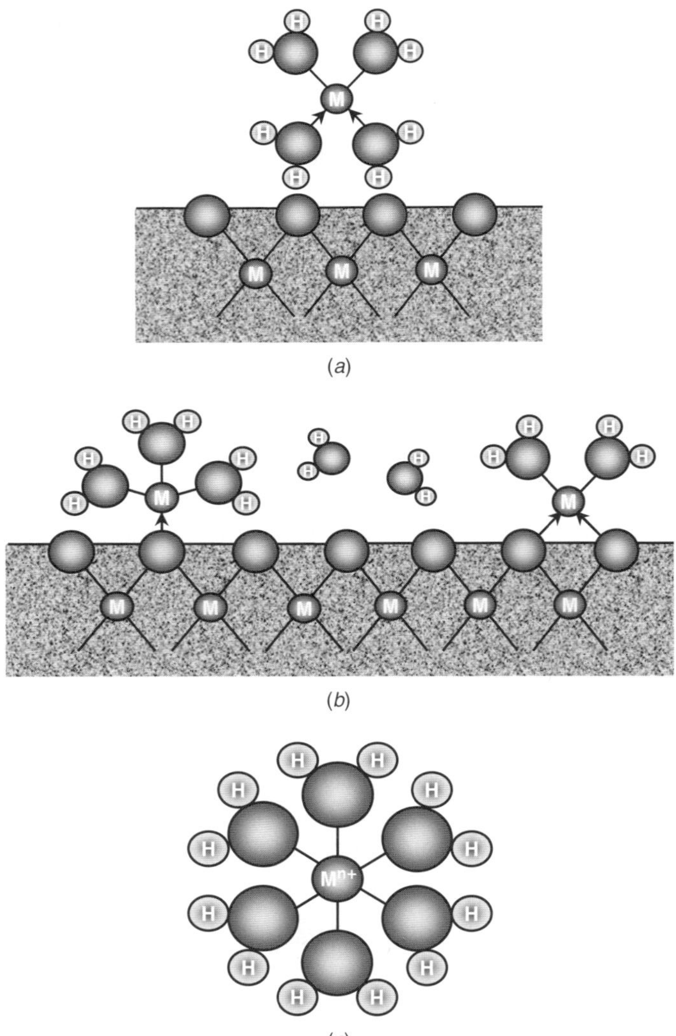

Figure 2.3. Surface complexation phenomena in the retention or desorption of metals from mineral surfaces. Nonspecific (exchangeable) adsorption consists of electrostatic bonds only and the ions retain their hydration sphere (outer-sphere complexes): specific (nonexchangeable) adsorption requires removal of the hydration sphere (inner-sphere complexes). Alkali and alkaline earth metals tend to form outer-sphere complexes, hence their tendency to be loosely bound and readily exchangeable with other ions in solution. Transition metals tend to form inner-sphere complexes, which are more strongly bound and less exchangeable (Cotter-Howells and Paterson, 2000). Representation of (*a*) an outer-sphere complex, (*b*) an inner-sphere complex, and (*c*) a solution complex (see also Figure 2.2). The solid substrate is textured with the solution above this. Unlabeled spheres represent oxygen atoms, and the spheres labeled M represent metals in the substrate or in solution. Smaller shaded spheres labeled H are hydrogen atoms. (Adapted from Brown et al., 1999; Cotter-Howells and Paterson, 2000.)

(Gadd et al., 2007; Gadd, 2007). Fungi can also physically weather minerals by etching and fracturing mineral grains during hyphal growth.

Bacterial production of short-chain organic acids and element-specific ligands (siderophores), which in combination are able to change pH and enhance chelation, results in increased mobilization of many metals and other elements, including lanthanides and actinides. *Pseudomonas fluorescens* could leach uranium and other trace elements from shale mine tailings, and this was attributed to the production of pyoverdine chelators. Mobilization of U from ore could occur aerobically at neutral to alkaline conditions, which may be an important process determining the high metal contents observed in the surroundings of U mines (Kalinowski et al., 2004) (see also Staunton et al., chapter 13, this volume).

Microorganisms also mobilize certain metals, metalloids, and organometallic compounds by reduction, oxidation, and methylation (Gadd, 1993b, 2000; Gharieb et al., 1999; Lovley, 2000). The solubilities of Fe and Mn increase on reduction of Fe(III) to Fe(II) and Mn(IV) to Mn(II) (Lovley and Coates, 1997; Lovley, 2000; McLean et al., 2002). Most iron reduction is carried out by specialized anaerobic bacteria that use Fe(III) as a terminal electron acceptor. Dissimilatory metal-reducing bacteria can use a variety of metal(loid)s with an appropriate redox couple, including Fe(III), Mn(IV), Se(IV), Cr(VI), and U(VI) (Oremland et al., 1991; Stolz and Oremland, 1999). Whereas reduction increases Fe and Mn solubility, the solubility of other metals, such as U(VI) to U(IV) and Cr(VI) to Cr(III), decreases, resulting in immobilization (Phillips et al., 1995; Smith and Gadd, 2000). Reduction of Hg(II) to Hg(0) by bacteria and fungi, mediated by mercuric reductase, results in diffusion of elemental Hg out of cells (Silver, 1996, 1998; Hobman et al., 2000). In contrast, certain *Bacillus* and *Streptomyces* sp. can oxidize Hg(0) to Hg(II) (Smith et al., 1998). Some bacteria can methylate certain metals, including Hg and Sn, as well as several metalloids (see Gadd, 1993b, 2000).

Oxides of Fe(III) and Mn(IV) absorb metals strongly, and this may hinder metal mobilization. Microbial reduction of Fe(III) and Mn(IV) may provide a means of releasing such metals, and this process may be enhanced with the addition of humic materials or related compounds (Lovley and Coates, 1997). Bacterial Fe(III) reduction resulted in release of Mn and Co from goethite, where 5% of the iron was substituted by these metals (Bousserrhine et al., 1999). Iron-reducing bacterial strains solubilized 40% of the Pu present in contaminated soils within 6 to 7 days (Rusin et al., 1994), and both iron- and sulfate-reducing bacteria solubilized Ra from uranium mine tailings, largely by disruption of reducible host minerals (Landa and Gray, 1995).

Most chemolithotrophic (autotrophic) leaching is carried out by acidophilic bacteria, which fix CO_2 and obtain energy from the oxidation of Fe(II), or by reduced sulfur compounds, which cause the solubilization of metals because of the resulting production of Fe(III) and H_2SO_4 (Rawlings, 1997; Schippers and Sand, 1999). The microorganisms involved include sulfur-oxidizing bacteria (e.g., *Acidithiobacillus thiooxidans*), iron- and sulfur-oxidizing bacteria (e.g., *Acidithiobacillus ferrooxidans*), and iron-oxidizing bacteria (e.g., *Leptospirillum ferrooxidans*) (Ewart and Hughes, 1991; Bosecker, 1997). As a result of sulfur and iron oxidation, metal sulfides are solubilized concomitant with the pH of their immediate environment being decreased, therefore resulting in solubilization of other metal compounds, including metals sorbed to soil and mineral constituents (Ewart and Hughes, 1991; Rawlings and Silver, 1995; Bosecker, 1997; Rawlings, 1997).

2.3.2. Immobilization

A number of mechanisms lead to metal immobilization. Although these can reduce the external free metal concentration, solubilization may be promoted in some circumstances because of the shift in equilibrium, which will result in the release of more metal species into solution.

Biosorption can be defined as the microbial uptake of soluble and insoluble organic and inorganic metal species by physicochemical mechanisms such as adsorption, and in living cells, metabolic activity may influence this process because of changes in pH, Eh, organic and inorganic nutrients, and metabolite excretion (White et al., 1995). Biosorption can also be a prelude for the formation of stable minerals (Beveridge and Doyle, 1989; Southam, 2000; McLean et al., 2002). Cationic metal species can also be accumulated within cells via membrane transport systems of varying affinity and specificity. Once inside cells, metal species may be bound (e.g., to metallothioneins, phytochelatins), precipitated (e.g., as reduced forms), localized within intracellular structures or organelles (e.g., fungal vacuoles), or translocated to specific structures (e.g., fungal fruiting bodies) (Gadd, 1996, 1997; White et al., 1997; Gadd and Sayer, 2000).

Peptidoglycan carboxyl groups are major binding sites for cations in gram-positive bacterial cell walls with phosphate groups contributing significantly in gram-negative species (Beveridge and Doyle, 1989; McLean et al., 2002). Chitin is an important component of fungal cell walls, and this is an effective biosorbent, especially for radionuclides (Tsezos and Volesky, 1982; Tobin et al., 1994). Fungal phenolic polymers and melanins also possess many potential metal-binding sites (Gadd, 1993a; Fomina and Gadd, 2002). Fungi can be efficient sorbents of metal ions over a wide range of pH values, and although they may take up less metal per unit dry weight than clay minerals, they are more efficient sorbents per unit surface area (Morley and Gadd, 1995). It seems that microbial binding and biomineralization (mineral formation) processes have a more significant role in metal speciation and mobility in the soil than has been considered previously (Krantz-Rulcker et al., 1993, 1996; Ledin et al., 1996; Marley et al., 1996; McLean et al., 2002). It should also be appreciated that fungal morphology allows interconnection of hyphae and the ability of fungi to translocate nutrients (and organelles, oxygen, and metabolites) may be important in the translocation of metal species and concentration in fruiting bodies (Haselwandter and Berreck, 1994; Dighton and Terry, 1996; Gray et al., 1996; Connolly and Jellison, 1997). Elevated metal and radionuclide concentrations, particularly radiocesium, commonly occur in fruiting bodies of basidiomycetes during growth in polluted environments (Gadd, 1993a; Haselwandter and Berreck, 1994; Anderson et al., 1997; Gadd, 1997; Wainwright and Gadd, 1997). In fact, it has been concluded that the fungal component of soil could immobilize the total Chernobyl radiocesium fallout received in upland grasslands (Dighton et al., 1991).

A range of metal-binding compounds are produced by microorganisms. Nonspecific metal-binding compounds range from simple organic acids and alcohols to polysaccharides and humic and fulvic acids (Birch and Bachofen, 1990; Beech and Cheung, 1995; Bridge et al., 1999; Sayer and Gadd, 2001). Extracellular polymeric substances (EPSs), a mixture of polysaccharides, mucopolysaccarides, and proteins (Zinkevich et al., 1996), are produced by bacteria and fungi (Beech and Cheung, 1995; White and Gadd, 1998a). Extracellular polysaccharides can also adsorb or entrap particulate matter such as metal sulfides and oxides (Flemming, 1995; Vieira and Melo, 1995), and these processes may be particularly important in biofilms (White and Gadd, 1998a, 2000). The genetics and biochemistry of metal resistance mechanisms have been studied primarily in free-swimming organisms; however, many bacteria in the environment exist in surface-attached biofilms. Biofilm bacteria are usually embedded in an extracellular polymeric substance (EPS) matrix composed of polysaccharides, proteins, and nucleic acids. This may confer increased resistance to external stresses compared to free-swimming organisms (Teitzel and Parsek, 2003). It has been found that bacterial biofilms may be from 2 to 600 times more resistant than free-swimming cells to toxic metal stress (Teitzel and Parsek, 2003). For planktonic cells, logarithmically growing cells were more resistant to copper and lead stress than stationary-phase cells. However, biofilms were observed to be more resistant to heavy metals than either stationary-phase or logarithmically growing planktonic cells. Extracellular polymeric substances that enclose a biofilm may be responsible for protecting cells from metal stress by binding and retarding their diffusion within the biofilm (Teitzel and Parsek, 2003).

Th^{4+} or Al^{3+} induced a precipitate to form in all *Bradyrhizobium* and *Sinorhizobium* cultures tested, which suggested a defense mechanism based on metal precipitation by extracellular polymers (Santamaría et al., 2003). Among the metals tested, only Fe^{3+}, Al^{3+}, and Th^{4+} were able to induce the formation of precipitate. Al^{3+} is probably the natural soil component against which this defence mechanism could be directed, and a different defence mechanism based on extracellular aluminium precipitation within a gelatinous residue has been described for *P. fluorescens* (Appanna and St. Pierre, 1996). However, this polymer was composed mainly of phosphatidylethanolamine. While metal binding to extracellular polymers and bacterial surfaces have been proposed as the reason for increased metal resistance of biofilm-growing bacteria, this proposed defense mechanism involved the physical removal of the capsule after metal binding (Santamaría et al., 2003).

Microbial reduction of certain metals to a lower redox state may result in reduced mobility and toxicity (Lovley, 2001; Finneran et al., 2002a). Such processes may accompany other metal precipitation mechanisms [e.g., in sulfate-reducing bacterial systems where reduction of Cr(VI) can be a result of indirect reduction by Fe^{2+} and sulfide produced]. Aerobic or anaerobic reduction of Cr(VI) to Cr(III) is widespread in microorganisms, (Smith and Gadd, 2000; McLean and Beveridge, 2001). U(VI) can be reduced to U(IV) by certain Fe(III)-dissimilatory microorganisms, and this reduction in solubility can be the basis of U removal from contaminated waters and leachates (Lovley and Coates, 1997; Lovley, 2001; Finneran et al., 2002b). Nitrate can inhibit U(VI) reduction, but once depleted, both U(VI) and Fe(III) can be reduced concurrently. When nitrate was added to sediments in which U(VI) had been reduced, U(VI) reappeared in solution (Finneran et al., 2002b). When *Pseudomonas aeruginosa* biofilms were exposed to relatively high $AuCl_3$ (i.e., 0.5 to 5 mmol L^{-1}), reduction of the auric ion resulted in the formation of both extracellular and intracellular metallic gold colloids. Most mineralization occurred on cell surfaces with lesser amounts within cells and little throughout the EPS (Karthikeyan and Beveridge, 2002). Both dissimilatory Fe(III)-reducing bacteria and archaea have the ability to reduce and precipitate gold (Kashefi et al., 2001) and silver (Beveridge and Murray, 1976; Korenevskii et al., 1999). In glass columns with sandy soil from a former antiknocking agents factory, hydrophobic tetraalkyllead (up to 324 mg L^{-1}) was transformed in oxygen-saturated water to inorganic lead, and after 740 days, nearly 50% of the organic lead was converted (Gallert and Winter, 2004).

Sulfur and sulfate-reducing bacteria are geochemically important in reductive precipitation of toxic metals [e.g., U(VI), Cr(VI), Tc(VII), Pd(II)] (Aubert et al., 1998; Lloyd and Macaskie, 1998; Lloyd et al., 1999a,b). Some sulfate-reducing bacteria such as *Desulfotomaculum reducens* share physiological properties of both sulfate- and metal-reducing groups of bacteria and can use Cr(VI), Mn(IV), Fe(III), and U(IV) as sole electron acceptors (Tebo and Obraztsova, 1998).

Shewanella spp. are facultatively anaerobic, metal-reducing bacteria found in aquatic and sedimentary environments (Nealson et al., 1991; DiChristina and DeLong, 1993). They can reduce soluble or mineral-bound metals, including Mn(IV) and Fe(III), by a dissimilatory mechanism in which the oxidized metal serves as an electron acceptor for the membrane-bound electron transport chain during respiration (Nealson and Saffarini, 1994). Contact is necessary for reduction of solid-bound metal, and *S. putrefaciens* attaches closely to minerals under aerobic and anaerobic conditions (Glasauer et al., 2001, 2003). Survival by metal respiration must preclude blockage of important membrane sites by, for example, sorption or precipitation, yet bacteria can retain high amounts of metals at the cell surface (Beveridge, 1989). Intracellular sequestration of metals may minimize metal-related problems, and *S. putrefaciens* can form internal Fe oxide particles during active metabolism (Glasauer et al., 2002). Mineral-bound Mn(IV) is also a known electron acceptor during dissimilatory metal reduction and Mn-rich granules result in the cytoplasm of *S. putrefaciens* when either birnessite or pyrolusite (β-MnO_2) serves as the electron acceptor during growth (Glasauer et al., 2004b).

Sulfate-reducing bacteria (SRB) oxidize organic compounds or hydrogen coupled with the reduction of sulfate, producing sulfide (White and Gadd, 1996a,b; 1997). The solubility products of most heavy metal sulfides are very low, so that even a moderate output of sulfide can remove metals from solution (White and Gadd, 1998a,b).

Bacterial Fe oxidation is ubiquitous in environments with sufficient Fe^{2+} and conditions to support bacterial growth such as drainage waters and tailings piles in mined areas, pyritic and hydric soils (bogs and sediments), drain pipes and irrigation ditches, and plant rhizospheres. Iron oxidizers found in acidic soil environments are acidophilic chemolithotrophs such as *A. ferrooxidans*, significant for its role in generating acid mine drainage (Ewart and Hughes, 1991). Fungi also oxidize metals in their environment. Desert varnish is an oxidized metal layer (patina) a few millimeters thick found on rocks and in soils of arid and semiarid regions, and is believed to be of fungal and bacterial origin (Burford et al., 2003).

2.3.3. Metalloid Transformations

Main microbial metalloid transformations in the soil are reduction and methylation. For selenium, some bacteria can use oxyanions of arsenic and selenium as terminal e^- acceptors in dissimilatory reduction and also reduce and incorporate Se into organic components (e.g., selenoproteins) (assimilatory reduction). The detection of arsenate- and selenate-respiring bacteria in numerous pristine and contaminated environments and their rapid appearance in enrichment culture suggest that they are widespread and metabolically active in nature. Although the bacterial species that have been isolated and characterized are still few in number, they are scattered throughout the bacterial domain and include gram-positive bacteria, β-, γ-, and ε-Proteobacteria, and the sole member of a deeply branching lineage of the bacteriaum *Chrysiogenes arsenatus*. The oxidation of a number of organic substrates (i.e., acetate, lactate, pyruvate, glycerol, ethanol) or hydrogen can be coupled to the reduction of arsenate and selenate, but the actual donor used varies from species to species. Both periplasmic and membrane-associated arsenate and selenate reductases have been characterized (Stolz and Oremland, 1999).

Methylation and volatilization of methylated selenium derivatives are also common properties of soil bacteria and fungi and may be important in Se transport from soil to aquatic environments (Dungan and Frankenberger, 1999). Selenate (SeO_4^{2-}) and selenite (SeO_3^{2-}) can be reduced to Se^0, with SeO_3^{2-} reduction appearing more common than SeO_4^{2-} reduction. However, only SeO_4^{2-} can support bacterial growth under anaerobic conditions (Oremland et al., 1989; Stolz and Oremland, 1999; Oremland and Stolz, 2000).

The opposite process of Se^0 oxidation can occur in soils (Dowdle and Oremland, 1998; Losi and Frankenberger, 1998). It is possible that Se^0 oxidation is similar to S oxidation and may be mediated by heterotrophs and autotrophs (Losi and Frankenberger, 1998). In aerobic soil slurries, Se^{4+} was the main product, with lower amounts of Se^{6+} being produced as a result of the activities of eterotrophic and autotrophic thiobacilli (Dowdle and Oremland, 1998).

Selenium methylation is a ubiquitous property of microorganisms and can occur in soils, sediments, and water (Gadd, 1993b). Bacteria and fungi are the most important Se methylaters in soil (Karlson and Frankenberger, 1988) with the most frequently produced compound being dimethyl selenide (DMSe) (Karlson and Frankenberger, 1988, 1989; Thompson-Eagle et al., 1989). Environmental factors that affect microbial activity can markedly affect Se methylation, (e.g., pH, temperature, organic amendments, Se speciation): Addition of organic amendments can stimulate methylation (Dungan and Frankenberger, 1999). The opposite process of demethylation can also occur in soil and water systems: anaerobic demethylation may be mediated by methylotrophic bacteria (Oremland et al., 1989).

Tellurium may also be transformed by reduction and methylation (Chasteen and Bentley, 2003). Reduction of tellurite to Te^0 results in a gray to black coloration of microbial colonies, with extracellular and intracellular precipitation being observed (Gharieb et al., 1999). Dimethyl telluride (DMTe) is the main product of Te methylation (Chasteen and Bentley, 2003).

Arsenic methylation can be mediated by many organisms, with compounds having the general structure $(CH_3)_nAsH_{3-n}$ and mono-, di-, and trimethylarsine ($n = 1, 2, 3$, respectively) being major volatile compounds (Bentley and Chasteen, 2002). Reduction of arsenic oxyanions by reductase enzymes is also frequent and a determinant of As resistance. However, there does not appear to be involvement of such reductases in biomethylation (Bentley and Chasteen, 2002). Microbes have evolved mechanisms for arsenic resistance and enzymes that oxidize As(III) to As(V) or reduce As(V) to As(III). Degradation of produced organoarsenicals (e.g., methylarsenic compounds) also occurs. As with other metals and metalloids, there is a global arsenic geocycle, where microbial metabolism and mobilization (or immobilization) are important processes (Mukhopadhyay et al., 2002).

2.4. MICROBIAL METAL AND MINERAL TRANSFORMATIONS IN SOIL GEOCHEMISTRY

Biologically induced mineralization (biomineralization) refers to biotic modification of the local microenvironment, creating conditions that promote chemical precipitation of extracellular mineral phases (Hamilton, 2003). This can result from microbial oxidation or reduction of a metal species as well as metabolism-dependent microbial transformations of metal species. Microbial surfaces provide chemically reactive sites for adsorption and complexation (≡ biosorption), as described previously (Figure 2.2). The latter processes can lead to the nucleation and formation of mineral precipitates around biomass (Beveridge, 1989; Fortin et al., 1997; McLean et al., 2002; Burford et al., 2006). Highly reactive bacterial cell surfaces bind metal ions (Beveridge, 1989), inorganic anions (e.g., silicates; Urrutia and Beveridge, 1994), and even preformed nanominerals (Glasauer et al., 2001) from the environment, leading to fine-grained mineral phases on cell surfaces. This reactivity arises from the presence of a wide array of ionizable groups, such as carboxylates and phosphates, present in the lipopolysaccharide (LPS) of gram-negative cell walls and the peptidoglycan, teichuronic acids, and teichoic acids of gram-positive bacteria (Beveridge, 1989). Sometimes nanomineral phases can even form directly in the cytoplasmic components of bacteria, such as magnetite, greigite (Bazylinski and Moskowitz, 1997), or amorphous iron hydroxide (Glasauer et al., 2002). External metal precipitation and biomineralization may also result from indirect effects of bacterial metabolism on microenvironmental geochemistry (Lee and Beveridge, 2001). Precipitation can also be due to the production of reactive components which either bind metals or catalyze precipitation (Macaskie et al., 1994; Mossman et al., 1999; Wang et al., 2001). The nature of the resulting mineral(s) may depend on the nature of the cell surface, the cellular microenvironment, and the presence of reactive anions (e.g., sulfide, carbonate, and phosphate). However, such biomineralization can also occur independent of microbial activity and on dead biomass. Mineral phases may undergo further changes in, for example, crystallinity, with time, and some minerals may incorporate other metals into their structure (Watson et al., 1995, 2000; Brown et al., 1999; McLean et al., 2002). Microbial modifications of their microenvironment may also result in physicochemical conditions that promote spontaneous metal precipitation (McLean et al., 2002).

The formation of solid mineral phases by bacterial processes may result in mineral deposition over geological time scales (Beveridge et al., 1983). Although much work on microbial biomineralization has involved bacteria, it should be stressed that all major microbial groups have roles in metal immobilization and mineral formation (e.g., cyanobacteria, microalgae, and fungi) (Gadd, 2004a, b; Gadd et al., 2005). Whereas cyanobacterial and microalgal processes are of greater significance in aquatic environments and in the early stages of soil formation (e.g., rock colonization), fungi have highly important roles in soil regarding mineral dissolution and mineral formation (Gadd, 2006).

Calcium oxalate crystals are commonly associated with free-living, pathogenic, and plant symbiotic fungi and are formed by the precipitation of solubilized calcium as the oxalate (Gharieb et al., 1998; Gadd, 1999, 2007). Calcium oxalate is the most common form of oxalate encountered in the environment, occurring mostly as the dihydrate (weddellite) or the more stable monohydrate (whewellite) (Gadd, 1999). Calcium oxalate has an important influence on biogeochemical processes in soils, acting as a reservoir for calcium and influencing phosphate availability (Figure 2.4*Á*). Fungi can produce other metal oxalates with a variety of different metals and metal-bearing minerals (e.g., Cd, Co, Cu, Mn, Sr, and Zn) (White et al., 1997; Gadd, 1999; Sayer et al., 1999; Jarosz-Wilkołazka and Gadd, 2003; Fomina et al., 2005b, 2007; Gadd et al., 2007; Gadd, 2007) (Figure 2.4*b*).

In many arid and semiarid regions, calcareous soils and near-surface limestones (calcretes) are often cemented secondarily with calcite ($CaCO_3$). Although this phenomenon has been partly attributed to physicochemical processes, the abundance of calcified fungal filaments in weathered profiles of chalky limestone and Quaternary calcretes indicates fungal activity (Verrecchia and Dumont, 1996; Gadd, 1999; Verrecchia, 2000) (Figure 2.5). Calcite has been proposed to form indirectly through fungal excretion of oxalic acid and precipitation of calcium oxalate, which results in dissolution of the internal pore walls of the limestone matrix and enriches the matrix solution in free carbonate. During passage of the solution through the pore walls, calcium carbonate recrystallizes as a result of a decrease in CO_2, thus hardening the limestone. Calcite crystals act as sites of further secondary calcite precipitation when the hyphae decompose. Through these processes, fungal activity is important in lithification (hardening), alteration, and

(a)

(b)

Figure 2.4. Scanning electron micrographs showing crystalline deposition of (*a*) whewellite ($CaC_2O_4{\cdot}H_2O$) and weddelite ($CaC_2O_4{\cdot}2H_2O$) and (*b*) moolooite ($CuC_2O_4{\cdot}xH_2O$) on the filamentous mycelium of *Serpula himantioides*. (From E. P. Burford and G.M. Gadd, unpublished; Burford et al., 2006).

diagenesis of subsurface limestone formations. Mineralized carbonate precipitates are also found in association with bacterial biofilms (Glasauer et al., 2002).

2.5. MICROBIAL METAL AND MINERAL TRANSFORMATIONS IN ENVIRONMENTAL BIOTECHNOLOGY

Several microbial metal and mineral transformations have potential for the treatment of contaminated land (Gadd, 2004a, b, 2005; Fomina and Gadd, 2007). Autotrophic leaching has been used to remediate metal-contaminated solids, including sewage sludge (Sreekrishnan and Tyagi, 1994), soil, and red mud, the latter being the main waste product of Al extraction from bauxite (Vachon et al., 1994). Although some processes could be used in situ (e.g., leaching using

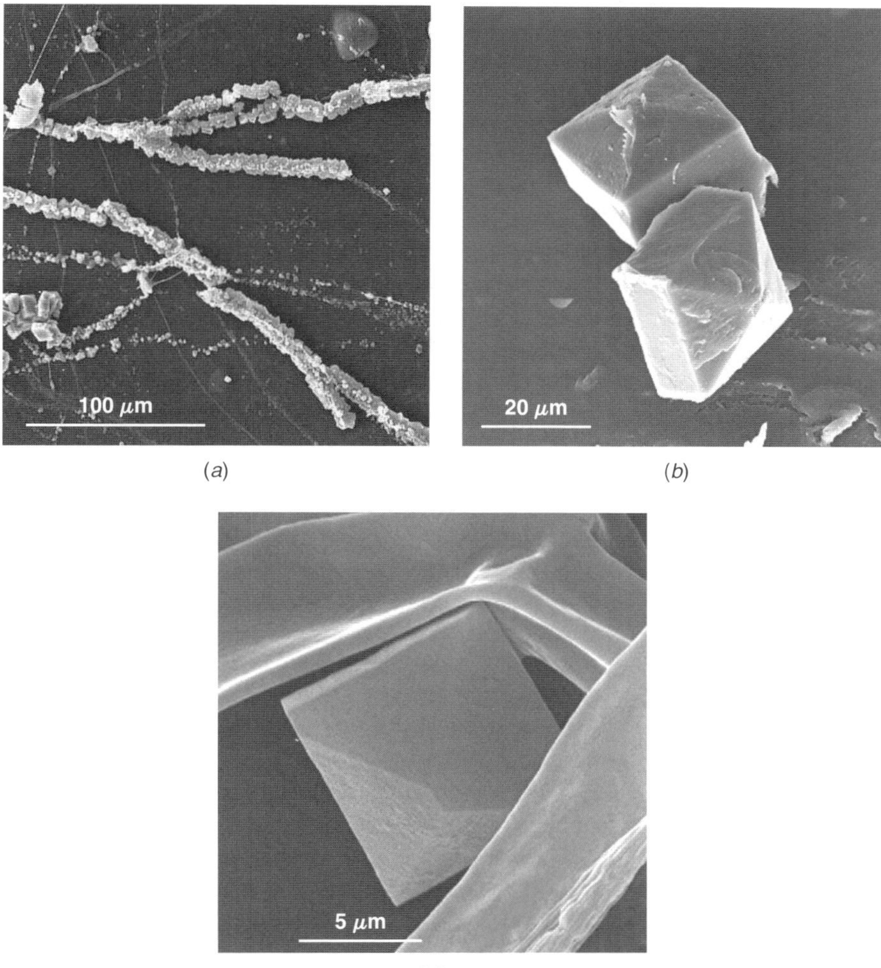

Figure 2.5. Scanning electron micrographs of (*a*) hyphae of *Serpula himantioides* mineralized with calcite ($CaCO_3$) and whewellite (calcium oxalate monohydrate, $CaC_2O_4{\cdot}H_2O$), (*b*) calcite precipitated in proximity to and (*c*) on biomass of a *Doratomyces* sp. (From E. P. Burford and G.M. Gadd, unpublished; Burford et al., 2006.)

S-oxidizing bacteria), many are probably most appropriate for ex situ use in bioreactors, where the mobilized or immobilized metal can be separated from soil components (White et al., 1998). Living or dead fungal and bacterial biomass and metabolites have been used to remove metals, radionuclides, and metalloids from solution by biosorption or chelation (Macaskie, 1991; Gadd, 2001). The metalloregulatory protein MerR, which exhibits high affinity and selectivity toward mercury, was exploited for the construction of microbial biosorbents specific for mercury removal (Bae et al., 2001, 2002, 2003). Whole-cell sorbents were constructed with MerR genetically engineered onto the surface of *E. coli* cells by using an ice nucleation protein anchor. The presence of surface-exposed MerR on the engineered strains enabled sixfold-higher Hg^{2+} biosorption than that found in the wild-type cells. Hg^{2+} binding via MerR was very specific, with no observable decline even in the presence of 100-fold excess Cd^{2+} and Zn^{2+}(Bae et al., 2003). Since metalloregulatory proteins are currently available for a wide variety of toxic metals, these results suggest that microbial biosorbents overexpressing metalloregulatory proteins may be used similarly for the cleanup of other important metals (Bae et al., 2000, 2003).

Phytochelatins (PCs) are metal-binding cysteine-rich peptides, enzymatically synthesized in plants and yeasts from glutathione in response to heavy metal stress by PC synthase (EC 2.3.2.15). In an attempt to increase the ability of bacterial cells to accumulate heavy metals, the *Arabidopsis thaliana* gene encoding PC synthase (*AtPCS*) was expressed in *E. coli*. When bacterial cells expressing AtPCS were placed in the presence of, for example, cadmium or the metalloid arsenic, cellular metal contents were increased 20- and 50-fold, respectively. Overexpression of PC synthase in bacterial strains appears to be a promising way to improve the heavy metal (such as Cd) or metalloid (such as As) content of organisms for use in bioremediation processes (Valls et al., 2000; Sauge-Merle et al., 2003).

Microbial activities in anaerobic subsurface environments also offer possibilities for metal and radionuclide bioremediation. Metal(loid)s that form insoluble precipitates when reduced include Se(0), Cr(III), Tc(IV), and U(IV) (Thompson-Eagle and Frankenberger, 1992; Lovley and Coates, 1997; Stolz and Oremland, 1999; Hockin and Gadd, 2003, 2006). Microbial reduction of U(VI) to U(IV) has been proposed as a bioremediation strategy for uranium-contaminated groundwater (Lovley et al., 1991; Lovley, 1995) as reduction of U(VI) under anaerobic conditions produces U(IV), which typically precipitates from groundwater as the insoluble mineral uraninite. The results of Finneran et al. (2002b) demonstrate that there is no net U(VI) reduction until nitrate has been reduced, and studies with sediments as well as pure cultures suggest that both the preferential reduction of nitrate over U(VI), as well as microbial oxidation of U(IV) with nitrate, contribute to the persistence of soluble U(VI) in the presence of nitrate. However, it has also been suggested that anaerobic oxidation of U(IV) to U(VI) with nitrate serving as the electron acceptor may provide a novel strategy for solubilizing and extracting microbial U(IV) precipitates from the subsurface (Finneran et al., 2002a, b).

In respect to bioremediation, reduction is generally limited to anaerobic processes and is ineffective against single-oxidation-state metals. Precipitation by other means has the advantage of producing chemically stable forms of metal, and its use is not limited to reducible metals. With respect to biological metal precipitation, both metal sulfides and metal phosphates have been investigated due to their low solubilities. Sulfide production is generally an anaerobic process.

The release of phosphate via the hydrolysis of an organic phosphate has been shown to be an effective method for metal precipitation of metals that precipitate on the surface of cells as insoluble metal phosphates (Macaskie, 1991). In addition, metal precipitation by secreted phosphate generated from polyphosphate hydrolysis has been suggested as a mechanism to remove metals and actinides from aqueous waste streams. *Acinetobacter johnsonii* was effective in removing lanthanum from solution (Boswell et al., 1999, 2001). In the bioreactor used, biomass was cycled between an aerobic phase, in which phosphate was accumulated and stored as polyphosphate, and an anaerobic phase, in which phosphate was released from microorganisms. Another mixed bacterial consortium, enriched for its ability to cycle phosphate, was able to remove over 98% of 1.5 mmol L^{-1} uranyl nitrate to loadings of over 0.5 g of uranium per gram of dry-cell weight (Renninger et al., 2001). Another two-stage system utilized *P. aeruginosa*, chosen for its metal tolerance and ability to accumulate large amounts of polyphosphates. Metal-binding occurred after degradation of the polyphosphate and concomitant release of phosphate from the cells. This system removed uranyl from solution as uranyl phosphate, but therefore in a manner that did not requires an organic phosphate source, a chelation system, or living cells (Renninger et al., 2004).

The sulfide produced from sulfate reduction plays a major role in metal sulfide immobilization in sediments but has also been applied to bioremediation of metals in waters and soil leachates. One process used sulfur- and iron-oxidizing bacteria to liberate metals from soils in the form of an acidic sulfate solution that enabled almost all the metals to be removed by bacterial sulfate reduction (White et al., 1998). Large-scale bioreactors have in fact been developed using bacterial sulfate reduction for treating metal-contaminated waters (Barnes et al., 1992; Gadd, 1992b).

Phytoremediation is the use of plants to remove or detoxify environmental pollutants (Baker and Brooks, 1989; Salt et al., 1998). Although free-living and symbiotic microorganisms influence plant productivity, metal bioavailability, and interactions, there are few integrated studies, and many phytoremediation studies are carried out without reference to contributory microbial processes. Phytoremediation can be divided into phytoextraction (pollutant removal from soil into shoots and leaves), phytodegradation (pollutant degradation by plant–microbe systems), rhizofiltration (absorption of pollutants by plant roots), phytostabilisation (plant-mediated reduction of pollutant bioavailability), phytovolatilization (plant-mediated volatilization of pollutants), and phytoscrubbing (plant removal of atmospheric pollutants). Most attention has focused on metals with two remediation strategies: chelate-assisted and continuous phytoextraction. Application of various synthetic chelates can enhance plant metal accumulation (Huang et al., 1997; Salt et al., 1998). Chelate-assisted phytoextraction has been developed because plants do not naturally accumulate important toxic elements (e.g., Pb, Cd, As, and many radionuclides) to levels that would be significant in a remediative context. Continuous phytoextraction of metals relies on properties of plants that lead to accumulation in aerial plant tissues. However, many natural "hyperaccumulators" often exhibit low biomass and slow growth rates, and none are known for such important elements as Pb, Cd, As, and U (Salt et al., 1998). Ni, Zn, and Se appear to be the elements accumulated to the highest levels (Salt et al., 1998). Plants possess metal resistance mechanisms analogous to those of microorganisms (i.e., chelation, intracellular compartmentation, transformations, etc.), although plants may be relatively metal-sensitive compared to microorganisms. Manipulation of metal tolerance may provide a means for phytoremediation: bacterial Hg^{2+}-reductase has been expressed in *Arabidopsis thaliana* (Rugh et al., 1996). In addition, endophytic bacteria have also been investigated in this context. In an organic context, engineered endophytes have been shown to increase plant tolerance to toluene (Barac et al., 2004; Newman and Reynolds, 2005), while for metals, recombinant heavy metal–resistant endophytic bacteria have been examined for effects on metal uptake by their host plants (Lodewyckx et al., 2001). Field exploitation of genetically transformed plant–bacterial systems for phytoremediation seems to be a distant possibility in the current sociopolitical climate.

Application to soils of certain amendments that immobilize metals (e.g., lime or phosphate treatment) have demonstrated, under field conditions, enhanced natural remediation, resulting in improved vegetation growth, increased microbial activity and diversity, and reduced off-site metal transport. However, whereas long-term stability of certain metal complexes such as metal pyromorphites has been shown in model systems (Adriano et al., 2004), the influence of plant roots and their microbial and mycorrhizal association on such stability has often been neglected. Pyromorphite [$Pb_5(PO_4)_3Cl$], which can form in urban and industrially contaminated soils, can be solubilized by phosphate-solubilizing fungi, with concomitant production of lead oxalate (Sayer et al., 1999; Fomina et al., 2004).

Some of the properties described above already present in certain strains have or can be combined or improved through genetic manipulation. In other cases, knowledge of microbe/metal-related reactions may allow further optimization of the process desired by altering the physicochemical conditions of the contaminated area. A combination of genetic engineering with appropriate ecoengineering of polluted sites may be of importance in future bioremediation strategies (Valls and de Lorenzo, 2002), although subject to significant legal and political barriers.

2.6. CONCLUSIONS AND FUTURE DIRECTIONS

In this chapter, some aspects of the transformation and mobilization of metals by microorganisms in relation to the soil environment have been outlined. It seems evident that metal–microbe interactions are important in several soil contexts, not least in the biogeochemical cycling of metals, associated elements, and nutrients, and in plant productivity. The application and potential of many natural microbial and microbe–plant processes are also growing topics in the area of bioremediation. However, analysis and understanding of the effects of toxic metals on microbial communities is relatively limited despite extensive research, and many studies are site specific. Clearly, the complexity of interactions between metal species and soil components, as well as between metal species and microorganisms, are major factors, as are the limitations of chemical and biological techniques that can be meaningfully applied. Laboratory and short-term field experimentation may provide extensive information, although this often has limited relevance to natural and contaminated natural environments. The lack of meaningful baseline data and ignorance of changes in metal speciation, microbial communities, and soil structure and composition can also hamper studies of long-term and historically contaminated sites. Pure culture studies may be open to the usual and predictable criticisms, although in the area of metal–microbe interactions, fundamental data obtained have been of paramount importance, having wide relevance to our understanding of prokaryotic and eukaryotic biology, as well as providing insights into natural roles and potential in environmental biotechnology. Although novel and developing molecular and genomic techniques for characterizing microbial communities and their functions will catalyze important steps forward in our understanding of metals and soil microbial communities, it seems clear that culture-based techniques are still of value, and indeed some work indicates that these may be more relevant in several contexts. Some doubt has been thrown on the idea that the unculturable majority of soil microorganisms are having a significant role. To probe metabolic diversity, improved means of culturing the "unculturable" population would be very helpful. Genomic information on environmentally relevant microorganisms is increasingly available, and along with reliable techniques for, for example, DNA extraction, phylogenetic analysis, cell imaging, biolabeling, and the use of marker molecules, improvements in understanding should progress and serve to reconcile important areas of laboratory and field-based research. Future studies are also dependent on an interdisciplinary approach involving all aspects of biology, chemistry, mathematical modeling, mineralogy, and geochemistry. Whether such "systems biology" increases or decreases our understanding of soil as a complex environment is a matter for conjecture.

Acknowledgments

The author gratefully acknowledges research support from the Biotechnology and Biological Sciences Research Council, the Natural Environment Research Council, and British Nuclear Fuels plc.

REFERENCES

Abou-Shanab, R. A., Angle, J. S., Delorme, T. A., Chaney, R. L., van Berkum, P., Moawad, H., Ghanem, K., and Ghozlan, H. A. (2003). Rhizobacterial effects on nickel extraction from soil and uptake by *Alyssum murale*. *New Phytol.* **158**, 219–224.

Adeyemi, A. O., and Gadd, G. M. (2005). Fungal degradation of calcium-, lead- and silicon-bearing minerals. *Biometals* **18**, 269–281.

Adriano, D. C., Wenzel, W. W., Vangronsveld, J., and Bolan, N. S. (2004). Role of assisted natural remediation in environmental cleanup. *Geoderma* **122**, 121–142.

Aiking, H., Stijnman, A., van Garderen, C., van Heerikhuizena, H., and van't Riet, J. (1984). Inorganic phosphate accumulation and cadmium detoxification in *Klebsiella aerogenes* NCTC 418 growing in continuous culture. *Appl. Environ. Microbiol.* **47**, 374–377.

Alvarez, S., and Jerez, C. A. (2004). Copper ions stimulate polyphosphate degradation and phosphate efflux in *Acidithiobacillus ferrooxidans. Appl. Environ. Microbiol.* **70**, 5177–5182.

Anderson, P., Davidson, C. M., Littlejohn, D., Ure, A. M., Shand, C. A. and Cheshire, M. V. (1997). The translocation of caesium and silver by fungi in some Scottish soils. *Commun. Soil Sci. Plant Anal.* **28**, 635–650.

Aoyama, M., and Nagumo, T. (1997). Effects of heavy metal accumulation in apple orchard soils on microbial biomass and microbial activities. *Soil Sci. Plant Nutr.* **43**, 601–612.

Appanna, V. D., and St Pierre, M. (1996). Aluminum elicits exocellular phosphatidylethanolamine production in *Pseudomonas fluorescens. Appl. Environ. Microbiol.* **62**, 2778–2782.

Aubert, C., Lojou, E., Bianco, P., Rousset, M., Durand, M.-C., Bruschi, M., and Dolla, A. (1998). The *Desulfuromonas acetoxidans* triheme cytochrome c_7 produced in *Desulfovibrio desulfuricans* retains its metal reductase activity. *Appl. Environ. Microbiol.* **64**, 1308–1312.

Baath, E. (1989). Effects of heavy metals in soil on microbial processes and populations (a review). *Water Air Soil Pollut.* **47**, 335–379.

Baath, E., Diaz-Ravina, M., Frostegaard, A., and Campbell, C. D. (1998). Effect of metal-rich sludge amendments on the soil microbial community. *Appl. Environ. Microbiol.* **64**, 238–245.

Bae, W., Chen, W., Mulchandani, A., and Mehra, R. (2000). Enhanced bioaccumulation of heavy metals by bacterial cells displaying synthetic phytochelatins. *Biotechnol. Bioeng.* **70**, 518–523.

Bae, W., Mehra, R. K., Mulchandani, A., and Chen, W. (2001). Genetic engineering of *Escherichia coli* for enhanced uptake and bioaccumulation of mercury. *Appl. Environ. Microbiol.* **67**, 5335–5338.

Bae, W., Mulchandani, A., and Chen, W. (2002). Cell surface display of synthetic phytochelatins using ice nucleation protein for enhanced heavy metal bioaccumulation. *J. Inorg. Biochem.* **88** ,223–227.

Bae, W., Wu, C. H., Kostal, J., Mulchandani, A., and Chen, W. (2003). Enhanced mercury biosorption by bacterial cells with surface-displayed MerR. *Appl. Environ. Microbiol.* **69**, 3176–3180.

Baker, A. J. M., and Brooks, R. R. (1989). Terrestrial higher plants which accumulate metallic elements: a review of their distribution, ecology and phytochemistry. *Biorecovery* **1**, 81–126.

Barac, T., Taghavi, S., Borremans, B., Provoost, A., Oeyen, L., Colpaert, J. V., Vangronsveld, J. and van der Lelie, D. (2004). Engineered endophytic bacteria improve phytoremediation of water-soluble, volatile, organic pollutants. *Nature Biotechnol.* **22**, 583–588.

Barnes, L. J., Janssen, F. J., Sherren, J., Versteegh, J. H., Koch, R. O. and Scheeren, P. J. H. (1992). Simultaneous removal of microbial sulphate and heavy metals from wastewater. *Trans. Inst. Mining Metall.* **101**, 183–190.

Bazylinski, D. A., and Moskowitz, B. M. (1997). Microbial biomineralization of magnetic iron minerals: microbiology, magnetism, and environmental significance. *Rev. Mineral.* **35**, 181–223.

Beech, I. B., and Cheung, C. W. S. (1995). Interactions of exopolymers produced by sulphate-reducing bacteria with metal ions. *Int. Biodeterior. Biodegrad.* **35**, 59–72.

Bentley, R., and Chasteen, T. G. (2002). Microbial methylation of metalloids: arsenic, antimony and bismuth. *Microbiol. Mol. Biol. Rev.* **66**, 250–271.

Beveridge, T. J. (1989). Role of cellular design in bacterial metal accumulation and mineralization. *Annu., Rev. Microbiol.* **43**, 147–171.

Beveridge, T. J. and Doyle, R. J. (1989). *Metal Ions and Bacteria*, Wiley, New York.

Beveridge, T. J., and Murray, R. G. E. (1976). Uptake and retention of metals by cell walls of *Bacillus subtilis*. *J. Bacteriol.* **127**, 1502–1518.

Beveridge, T. J., Meloche, J. D., Fyfe, W. S., and Murray, R. G. E. (1983). Diagenesis of metals chemically complexed to bacteria: laboratory formation of metal phosphates, sulfides and organic condensates in artificial sediments. *Appl. Environ. Microbiol.* **45**, 1094–1108.

Birch, L., and Bachofen, R. (1990). Complexing agents from microorganisms. *Experientia* **46**, 827–834.

Blaudez, D., Jacob, C., Turnau, K., Colpaert, J. V., Ahonen-Jonnarth, U., Finlay, R., Botton, B., and Chalot, M. (2000). Differential responses of ectomycorrizal fungi to heavy metals in vitro. *Mycol. Res.* **104**, 1366–1371.

Bosecker, K. (1997). Bioleaching: metal solubilization by microorganisms. *FEMS Microbiol. Rev.* **20**, 591–604.

Boswell, C. D., Dick, R. E., and Macaskie, L. E. (1999). The effect of heavy metals and other environmental conditions on the anaerobic phosphate metabolism of *Acinetobacter johnsonii*. *Microbiology*. **145**, 1711–1720.

Boswell, C. D., Dick, R. E., Eccles, H., and Macaskie, L. E. (2001). Phosphate uptake and release by *Acinetobacter johnsonii* in continuous culture and coupling of phosphate release to heavy metal accumulation. *J. Ind. Microbiol. Biotechnol.* **26**, 333–340.

Bousserrhine, N., Gasser, U. G., Jeanroy, E., and Berthelin, J. (1999). Bacterial and chemical reductive dissolution of Mn-, Co- Cr-, and Al-substituted goethites. *Geomicrobiol. J.* **16**, 245–258.

Bradley, R., Burt, A. J., and Read, D. J. (1981). Mycorrhizal infection and resistance to heavy metals. *Nature* **292**, 335–337.

Bradley, B., Burt, A. J., and Read, D. J. (1982). The biology of mycorrhiza in the Ericaceae, VIII: The role of mycorrhizal infection in heavy metal resistance. *New Phytol.* **91**, 197–209.

Bridge, T. A. M., White, C., and Gadd, G. M. (1999). Extracellular metal-binding activity of the sulphate-reducing bacterium *Desulfococcus multivorans*. *Microbiology* **145**, 2987–2995.

Brookes, P. C., and McGrath, S. P. (1984). Effects of metal toxicity on the size of the soil microbial biomass. *J. Soil Sci.* **35**, 341–346.

Brown, G. E., Foster, A. L., and Ostergren, J. D. (1999). Mineral surfaces and bioavailability of heavy metals: a molecular-scale perspective. *Proc. Natl. Acad. Sci. USA* **96**, 3388–3395.

Burford, E. P., Fomina, M., and Gadd, G. M. (2003). Fungal involvement in bioweathering and biotransformation of rocks and minerals. *Mineral. Mag.* **67**, 1127–1155.

Burford, E.P., Hillier, S., and Gadd, G.M. (2006). Biomineralization of fungal hyphae with calcite ($CaCO_3$) and calcium oxalate mono- and dihydrate in carboniferous limestone microcosms. *Gemicrobiology J.* **23**, 599–611.

Cánovas, D., Cases, I., and de Lorenzo, V. (2003). Heavy metal tolerance and metal homeostasis in *Pseudomonas putida* as revealed by complete genome analysis. *Environ. Microbiol.* **5**, 1242–1256.

Cevnik, M., Jurc, M., and Vodnik, D. (2000). Filamentous fungi associated with the fine roots of *Erica herbacea* L. from the area influenced by the Zerjav lead smelter (Slovenia). *Phyton Ann. Rei. Bot.* **40**, 61–64.

Chander, K., Dyckmans, J., Hoeper, H., Joergensen, R. G., and Raubuch, M. (2001a). Long-term effects on soil microbial properties of heavy metals from industrial exhaust deposition. *J. Plant Nutr. Soil Sci.* **164**, 657–663.

Chander, K., Dyckmans, J., Joergensen, R. G., Meyer, B., and Raubuch, M. (2001b). Different sources of heavy metals and their long-term effects on soil microbial properties. *Biol. Fertil. Soils* **34**, 241–247.

Chasteen, T. G., and Bentley, R. (2003). Biomethylation of selenium and tellurium: microorganisms and plants. *Chem. Rev.* **103**, 1–26.

Chaudri, A. M., Allain, C. M. G., Barbosa-Jefferson, V. L., Nicholson, F. A., Chambers, B. J., and McGrath, S. P. (2000). A study of the impacts of Zn and Cu on two rhizobial species in soils of a long-term field experiment. *Plant Soil* **221**, 167–179.

Colpaert, J. V., and Van Assche, J. A. (1993). The effect of cadmium on ectomycorrhizal *Pinus sylvestris* L. *New Phytol.* **123**, 325–333.

Colpaert, J. V., Vandenkoornhuyse, P., Adriaensen, K., and Vangronsveld, J. (2000). Genetic variation and heavy metal tolerance in the ectomycorrhizal basidiomycete *Suillus luteus*. *New Phytol.* **147**, 367–379.

Connolly, J. H., and Jellison, J. (1997). Two-way translocation of cations by the brown rot fungus *Gloeophyllum trabeum*. *Int. J. Biodeterior. Biodegrad.* **39**, 181–188.

Cotter-Howells, J. D., and Paterson, E. (2000). Minerals and soil development. In *Environmental Mineralogy*, ed. Vaughan, D. J., and Wogelius, R. A., Eotvos University Press, Budapest, Hungary, 1–124.

Crichton, R. R. (1991). *Inorganic Biochemistry of Iron Metabolism*, Ellis Horwood, Chichester West Susset, England.

Dahlin, S., Witter, E., Maartensson, A., Turner, A., and Baath, E. (1997). Where's the limit? Changes in the microbiological properties of agricultural soils at low levels of metal contamination. *Soil Biol. Biochem.* **29**, 1405–1415.

Darrah, P. R., White, R. E., and Nye, P. H. (1987). A theoretical consideration of the implications of cell clustering for the prediction of nitrification in soil. *Plant Soil* **99**, 387–400.

Degens, B. P., Schipper, L. A., Sparling, G. P., and Duncan, L. C. (2001). Is the microbial community in a soil with reduced catabolic diversity less resistant to stress or disturbance? *Soil Biol. Biochem.* **33**, 1143–1153.

Delorme, T., Gagliardi, J., Angle, J., and Chaney, R. (2001). Influence of zinc hyperaccumulator *Thlaspi caerulescens* J. and C. Presl. and the non metal accumulator *Trifolium pratense* L. on soil microbial populations. *Can. J. Microbiol.* **47**, 773–776.

Del Val, C., Barea, J. M., and Azcon-Aguilar, C. (1999). Diversity of arbuscular mycorrhizal fungus populations in heavy-metal-contaminated soils. *Appl. Environ. Microbiol.* **65**, 718–723.

de Souza, M. P., Chu, D., Zhao, M., Zayed, A. M., Ruzin, S. E., Schichnes, D., and Terry, N. (1999). Rhizosphere bacteria enhance selenium accumulation and volatilization by Indian mustard. *Plant Physiol.* **119**, 565–573.

Diaz-Ravina, M., and Baath, E. (1996). Development of metal tolerance in soil bacterial communities exposed to experimentally increased metal levels. *Appl. Environ. Microbiol.* **62**, 2970–2977.

DiChristina, T. J., and DeLong, E. F. (1993). Design and application of rRNA-targeted oligonucleotide probes for the dissimilatory iron- and manganese-reducing bacterium *Shewanella putrefaciens*. *Appl. Environ. Microbiol.* **59**, 4152–4160.

Dighton, J., and Terry, G. (1996). Uptake and immbilization of caesium in UK grassland and forest soils by fungi, following the Chernobyl accident. In *Fungi and Environmental Change,* ed. Frankland, J. C., Magan, N., and Gadd, G. M., Cambridge University Press, Cambridge, 184–200.

Dighton, J., Clint, G. M., and Poskitt, J. (1991). Uptake and accumulation of ^{137}Cs by upland grassland soil fungi: a potential pool of Cs immobilization. *Mycol. Res.* **95**, 1052–1056.

Dixon, R. K., and Buschena, C. A. (1988). Response of ectomycorrhizal *Pinus banksia* and *Picea glauka* to heavy metals in soil. *Plant Soil* **105**, 265–271.

Doelman, P. (1986). Resistance of soil microbial communities to heavy metals. In *Microbial Communities in Soil*, ed. Jensen, V., Kjoller, A., and Sorensen, C. H., Elsevier Applied Science, London, 369–384.

Doelman, P., and Hanstra, L. (1979). Effects of lead on the soil bacterial microflora. *Soil Biol. Biochem.* **11**, 487–491.

Dowdle, P. R., and Oremland, R. S. (1998). Microbial oxidation of elemental selenium in soil slurries and bacterial cultures. *Environ. Sci. Technol.* **32**, 3749–3755.

Dungan, R. S., and Frankenberger, W. T. (1999). Microbial transformations of selenium and the bioremediation of seleniferous environments. *Bioremediation* **3**, 171–188.

Duxbury, T., and Bicknell, B. (1983). Metal-tolerant bacterial populations from natural and metal-polluted soils. *Soil Biol. Biochem.* **15**, 243–250.

Ehrlich, H. L. (1997). Microbes and metals. *Appl. Microbiol. Biotechnol.* **48**, 687–692.

Ehrlich, H. L. (2002). *Geomicrobiology*, Marcel Dekker, New York.

Ellis, R., Neish, B., Trett, M., Best, G., Weightman, A., Morgan, P., and Fry, J. C. (2001). Comparison of microbial and meiofaunal community analyses for determining impact of heavy metal contamination. *J. Microbiol. Meth.* **45**, 171–185.

Ellis, R. J., Morgan, P., Weightman, A. J., and Fry, J. C. (2003). Cultivation-dependent and-independent approaches for determining bacterial diversity in heavy-metal-contaminated soil. *Appl. Environ. Microbiol.* **69**, 3223–3230.

Ewart, D. K., and Hughes, M. N. (1991). The extraction of metals from ores using bacteria. *Adv. Inorg. Chem.* **36**, 103–135.

Fay, D. A., and Mitchell, D. T. (1999). A preliminary study of the mycorrhizal associations of the tree seedlings growing on mine spoil at Avoca, Co. Wicklow. *Biol. and Environ. Proc. R. Ir. Acad.* **99B**, 19–26.

Finneran, K. T., Anderson, R. T., Nevin, K. P., and Lovley, D. R. (2002a). Bioremediation of uranium-contaminated aquifers with microbial U(VI) reduction. *Soil Sediment. Contam.* **11**, 339–357.

Finneran, K. T., Housewright, M. E., and Lovley, D. R. (2002b). Multiple influences of nitrate on uranium solubility during bioremediation of uranium-contaminated subsurface sediments. *Environ. Microbiol.* **4,** 510–516.

Flemming, H.-K. (1995). Sorption sites in biofilms. *Water Sci. Technol.* **32**, 27–33.

Fliessbach, A., Martens, R., and Reber, H. H. (1994). Soil microbial biomass and microbial activity in soil treated with heavy metal contaminated sewage sludge. *Soil Biol. Biochem.* **26**, 1201–1205.

Fomina, M., and Gadd, G. M. (2002). Metal sorption by biomass of melanin-producing fungi grown in clay-containing medium. *J. Chem. Technol. Biotechnol.* **78**, 23–34.

Fomina, M., and Gadd, G. M. (2007). Metal and mineral transformations: a mycoremediation perspective. In *Exploitation of Fungi*, ed. Robson, G. D., Van West, P., and Gadd., G. M., Cambridge University Press, Cambridge, 236–254.

Fomina, M. A., Alexander, I. J., Hillier, S., and Gadd, G. M. (2004). Zinc phosphate and pyromorphite solubilization by soil plant–symbiotic fungi. *Geomicrobiol. J.* **21**, 351–366.

Fomina, M. A., Alexander, I. J., Colpaert, J. V., and Gadd, G. M. (2005a). Solubilization of toxic metal minerals and metal tolerance of mycorrhizal fungi. *Soil Biol. Biochem.* **37**, 851–866.

Fomina, M., Hillier, S., Charnock, J. M., Melville, K., Alexander, I. J., and Gadd, G. M. (2005b). Role of oxalic acid over-excretion in toxic metal mineral transformations by *Beauveria caledonica. Appl. Environ. Microbiol.* **71**, 371–381.

Fomina, M., Charnock, J. M., Hillier, S., Alexander, I. J., and Gadd, G. M. (2006a). Zinc phosphate transformations by the *Paxillus involutus*/pine ectomycorrhizal association. *Microb. Ecol.* **52**, 322–333.

Fomina, M., Charnock, J., Bowen, A. D., and Gadd, G. M. (2007). X-ray absorption spectroscopy (XAS) of toxic metal mineral transformations by fungi. *Environ. Microbiol.* **9** 308–321.

Fortin, D., Ferris, F. G. and Beveridge, T. J. (1997). Surface-mediated mineral development by bacteria. In *Reviews in Mineralogy*, Vol. 35, ed. Banfield, J., and Nealson, K. H., Mineralogical Society of America, Washington, DC, 161–180.

Foster, R. C. (1988). Microenvironments of soil-microorganisms. *Biol. Fertil. Soils* **6**, 189–203.

Fritze, H., Perkiomaki, J., Saarela, U., Katainen, R., Tikka, P., Yrjala, K., Karp, M., Haimi, J., and Romantschuk, M. (2000). Effect of Cd-containing wood ash on the microflora of coniferous forest humus. *FEMS Microbiol. Ecol.* **32**, 43–51.

Frostegaard, A., Tunlid, A., and Baath, E. (1993). Phospholipid fatty acid composition, biomass, and activity of microbial communities from two soil types experimentally exposed to different heavy metals. *Appl. Environ. Microbiol.* **11**, 3605–3617.

Gadd, G. M. (1992a). Metals and microorganisms: a problem of definition. *FEMS Microbiol. Lett.* **100**, 197–204.

Gadd, G. M. (1992b). Microbial control of heavy metal pollution. In *Microbial Control of Pollution,* ed. Fry, J. C., Gadd, G. M., Herbert, R. A., Jones, C. W., and Watson-Craik, I., Cambridge University Press, Cambridge, 59–88.

Gadd, G. M. (1993a). Interactions of fungi with toxic metals. *New Phytol.* **124**, 25–60.

Gadd, G. M. (1993b). Microbial formation and transformation of organometallic and organometalloid compounds. *FEMS Microbiol. Rev.* **11**, 297–316.

Gadd, G. M. (1995). Signal transduction in fungi. In *The Growing fungus*, ed. Gow, N. A. R., and Gadd, G. M., Chapman and Hall, London, 183–210.

Gadd, G. M. (1996). Influence of microorganisms on the environmental fate of radionuclides. *Endeavour* **20**, 150–156.

Gadd, G. M. (1997). Roles of microorganisms in the environmental fate of radionuclides. In *Health Impacts of Large Releases of Radionuclides*, ed. Lake, J. V., Bock, G. R., and Cardew, G., CIBA Found. Symp. 203, Wiley, Chichester, West Susset, England, 94–108.

Gadd, G. M. (1999). Fungal production of citric and oxalic acid: importance in metal speciation, physiology and biogeochemical processes. *Adv. Microb. Physiol.* **41**, 47–92.

Gadd, G. M (2000). Microbial interactions with tributyltin compounds: detoxification, accumulation, and environmental fate. *Sci. Total Environ.* **258**, 119–127.

Gadd, G.M. (2001). Metal transformations. In *Fungi in Bioremediation*, ed. Gadd, G.M., Cambridge University Press, Cambridge, 359–382.

Gadd, G. M. (2004a). Mycotransformation of organic and inorganic substrates. *Mycologist* **18**, 60–70.

Gadd, G. M. (2004b). Microbial influence on metal mobility and application for bioremediation. *Geoderma* **122**, 109–119.

Gadd, G. M. (2005). Microorganisms in toxic metal polluted soils. In *Microorganisms in Soils: Roles in Genesis and Functions,* ed. Buscot, F., and Varma, A., Springer-Verlag, Berlin, 325–356.

Gadd, G. M. (ed.) (2006). *Fungi in Biogeochemical Cycles*, Cambridge University Press, Cambridge.

Gadd, G. M., and Sayer, J. A. (2000). Fungal transformations of metals and metalloids. In *Environmental Microbe–Metal Interactions*, ed. Lovley, D. R., American Society of Microbiology, Washington, DC, 237–256.

Gadd, G. M. and White, C. (1985). Copper uptake by *Penicillium ochro-chloron:* influence of pH on toxicity and demonstration of energy-dependent copper influx using protoplasts. *J. Gen. Microbiol.* **131**, 1875–1879.

Gadd, G. M., and White, C. (1989). Heavy metal and radionuclide accumulation and toxicity in fungi and yeasts. In *Metal–Microbe Interactions*, ed. Poole, R. K., and Gadd, G. M., IRL Press, Oxford, 19–38.

Gadd, G. M., Chudek, J. A., Foster, R., and Reed, R. H. (1984). The osmotic responses of *Penicillium ochro-chloron*: changes in internal solute levels in response to copper and salt stress. *J. Gen. Microbiol.* **130**, 1969–1975.

Gadd, G. M., Semple, K., and Lappin-Scott, H. (eds.) (2005). *Microorganisms in Earth Systems: Advances in Geomicrobiology*, Cambridge University Press, Cambridge.

Gadd, G. M., Burford, E. P., Fomina, M., and Melville, K. (2007). Mineral transformation and biogeochemical cycles: a geomycological perspective. In *Fungi in the Environment,* ed. Gadd, G. M., Dyer, P. S., and Watkinson, S. C., Cambridge University Press, Cambridge, 78–111.

Gallert, C., and Winter, J. (2004). Degradation of alkyllead compounds to inorganic lead in contaminated soil. *Water Res.* **38**, 4204–4212.

Gharieb, M. M., and Gadd, G. M. (1998). Evidence for the involvement of vacuolar activity in metal(loid) tolerance: vacuolar-lacking and -defective mutants of *Saccharomyces cerevisiae* display higher sensitivity to chromate, tellurite and selenite. *Biometals* **11**, 101–106.

Gharieb, M. M., and Gadd, G. M. (2004). Role of glutathione in detoxification of metal(loid)s by *Saccharomyces cerevisiae. Biometals* **17**, 183–188.

Gharieb, M. M., Sayer, J. A., and Gadd, G. M. (1998). Solubilization of natural gypsum ($CaSO_4{\cdot}2H_2O$) and the formation of calcium oxalate by *Aspergillus niger* and *Serpula himantiodes. Mycol. Res.* **102**, 825–830.

Gharieb, M. M., Kierans, M., and Gadd, G. M. (1999). Transformation and tolerance of tellurite by filamentous fungi: accumulation, reduction and volatilization. *Mycol. Res.* **103**, 299–305.

Giller, K. E., Nussbaum, R., Chaudri, A. M., and McGrath, S. P. (1993). *Rhizobium meliloti* is less sensitive to heavy metal contamination in soil than *R. leguminosarum* bv. *trifolii* or *R. loti. Soil Biol. Biochem.* **25**, 273–278.

Giller, K., Witter, E., and McGrath, S. P. (1998). Toxicity of heavy metals to microorganisms and microbial processes in agricultural soils: a review. *Soil Biol. Biochem.* **30**, 1389–1414.

Glasauer, S., Langley, S., and Beveridge, T. J. (2001). Sorption of Fe (hydr)oxides to the surface of *Shewanella putrefaciens*: cell-bound fine-grained minerals are not always formed de novo. *Appl. Environ. Microbiol.* **67**, 5544–5550.

Glasauer, S., Langley, S., and Beveridge, T. J. (2002). Intracellular iron minerals in a dissimilatory iron-reducing bacterium. *Science* **295**, 117–119.

Glasauer, S., Weidler, P. G., Langley, S., and Beveridge, T. J. (2003). Controls on Fe reduction and mineral formation by a subsurface bacterium. *Geochim. Cosmochim. Acta* **67**, 1277–1288.

Glasauer, S., Beveridge, T. J., Burford, E. P., Harper, F. A., and Gadd, G. M. (2004a). Metals and metalloids, transformations by microorganisms. In *Encyclopedia of Soils in the Environment*, ed. Hillel, D., Rosenzweig, C., Powlson, D. S., Scow, K. M, Singer, M. J., Sparks, D. L., and Hatfield, J., Elsevier, Amsterdam, The Netherlands 438–447.

Glasauer, S., Langley, S., and Beveridge, T. J. (2004b). Intracellular manganese granules formed by a subsurface bacterium. *Environ. Microbiol.* **6**, 1042–1048.

Gray, S. N., Dighton, J., and Jennings, D. H. (1996). The physiology of basidiomycete linear organs, 3: Uptake and translocation of radiocaesium within differentiated mycelia of *Armillaria* spp. growing in microcosms and in the field. *New Phytol.* **132**, 471–482.

Gremion, F., Chatzinotas, A., and Harms, H. (2003). Comparative 16S rDNA and 16S rRNA sequence analysis indicates that Actinobacteria might be a dominant part of the metabolically active bacteria in heavy metal-contaminated bulk and rhizosphere soil. *Environ. Microbiol.* **5**, 896–907.

Griffioen, W. A. J. (1994). Characterization of a heavy metal-tolerant endomycorrhizal fungus from the surroundings of a zinc refinery. *Mycorrhiza* **4**, 197–200.

Grundmann, G. L., Dechesne, A., Bartoli, F., Flandrois, J. P., Chasse, J. L., and Kizungu, R. (2001). Spatial modeling of nitrifier microhabitats in soil. *Soil Sci. Soc. Am. J.* **65**, 1709–1716.

Hamilton, W. A. (2003). Microbially influenced corrosion as a model system for the study of metal microbe interactions: a unifying electron transfer hypothesis. *Biofouling* **19**, 65–76.

Hartley, C., Cairney, J. W. G., Sanders, F. E., and Meharg, A. A. (1997). Toxic interactions of metal ions (Cd^{2+}, Pb^{2+}, Zn^{2+} and Sb^{3-}) on in vitro biomass production of ectomycorrhizal fungi. *New Phytol.* **137**, 551–562.

Hartley, J., Cairney, J. W. G., Freestone, P., Woods, C., and Meharg, A. A. (1999). The effects of multiple metal contamination on ectomycorrhizal Scots pine (*Pinus sylvestris*) seedlings. *Environ. Pollut.* **106**, 413–424.

Haselwandter, K., and Berreck, M. (1994). Accumulation of radionuclides in fungi. In *Metal Ions in Fungi,* ed. Winkelmann, G., and Winge, D. R., Marcel Dekker, Inc, New York, 259–277.

Hayashi, Y., and Mutoh, N. (1994). Cadystin (phytochelatin) in fungi. In *Metal Ions in Fungi,*ed. Winkelmann, G., and Winge, D. R., Marcel Dekker, New York, 311–337.

Hery, M., Nazaret, S., Jaffre, T., Normand, P., and Navarro, E. (2003). Adaptation to nickel spiking of bacterial communities in neocaledonian soils. *Environ. Microbiol.* **5**, 3–12.

Hetrick, B. A. D., Wilson, G. W. T. and Figge, D. A. H. (1994). The influence of mycorrhizal symbiosis and fertilizer amendments on establishment of vegetation in heavy metal mine spoil. *Environ. Pollut.* **86**, 171–179.

Hirsch, P. R., Jones, M. J., McGrath, S. P., and Giller, K. E. (1993). Heavy metals from past applications of sewage sludge decrease the genetic diversity of *Rhizobium leguminosarum* biovar *trifolii* populations. *Soil Biol. Biochem.* **25**, 1485–1490.

Hobman, J. L., Wilson, J. R., and Brown, N. L. (2000). Microbial mercury reduction. In *Environmental Microbe–Metal Interactions,* ed. Lovley, D. R., ASM Press, Washington, DC, 177–197.

Hockin, S., and Gadd, G. M. (2003). Linked redox-precipitation of sulfur and selenium under anaerobic conditions by sulfate-reducing bacterial biofilms. *Appl. Environ. Microbiol.* **69**, 7063–7072.

Hockin, S., and Gadd, G. M. (2006). Removal of selenate from sulphate-containing media by sulphate-reducing bacterial biofilms. *Environ. Microbiol.* **8**, 816–826.

Huang, J. W. W., Chen, J. J., Berti, W. R., and Cunningham, S. D. (1997). Phytoremediation of lead-contaminated soils: role of synthetic chelates in lead phytoextraction. *Environ. Sci. Technol.* **31**, 800–805.

Inouhe, M., Sumiyoshi, M., Tohoyama, H., and Joho, M. (1996). Resistance to cadmium ions and formation of a cadmium-binding complex in various wild-type yeasts. *Plant Cell Physiol.* **37**, 341–346.

Jarosz-Wilkołazka, A., and Gadd, G. M. (2003). Oxalate production by wood-rotting fungi growing in toxic metal-amended medium. *Chemosphere* **52**, 541–547.

Ji, G., and Silver, S. (1992a). Regulation and expression of the arsenic resistance operon from *Staphylococcus aureus* plasmid pI258. *J. Bacteriol.* **174**, 3684–3694.

Ji, G., and Silver, S. (1992b). Reduction of arsenate to arsenite by the arsC protein of the arsenic resistance operon of *Staphylococcus aureus* plasmid pI258. *Proc. Nat. Acad. Sci. USA* **89**, 9474–9478.

Ji, G., Garber, E. A., Armes, L. G., Chen, C.-M., Fuchs, J. A. and Silver, S. (1994). Arsenate reductase of *Staphylococcus aureus* plasmid p1258. *Biochemistry.* **33**, 7294–7299.

Joho, M., Inouhe, M., Tohoyama, H., and Murayama, T. (1995). Nickel resistance mechanisms in yeasts and other fungi. *J. Ind. Microbiol.* **14**, 164–168.

Jones, D., and Muehlchen, A. (1994). Effects of the potentially toxic metals, aluminium, zinc and copper on ectomycorrhizal fungi. *J. Environ. Sci. Health A Environ. Sci. Eng.* **29**, 949–966.

Kalinowski, B. E., Oskarsson, A., Albinsson, Y., Arlinger, J., Odegaard-Jensen, A., Andlid, T., and Pedersen, K. (2004). Microbial leaching of uranium and other trace elements from shale mine tailings at Ranstad. *Geoderma* **122**, 177–194.

Karlson, U., and Frankenberger, W. T. (1988). Effects of carbon and trace element addition on alkylselenide production by soil. *Soil Sci. Soc. Am J.* **52**, 1640–1644.

Karlson, U., and Frankenberger, W. T. (1989). Accelerated rates of selenium volatilization from California soils. *Soil Sci. Soc. Am J.* **53**, 749–753.

Karthikeyan, S., and Beveridge, T. J. (2002). *Pseudomonas aeruginosa* biofilms react with and precipitate toxic soluble gold. *Environ. Microbiol.* **4**, 667–675.

Kashefi, K., Tor, J. M., Nevin, K. P., and Lovley, D. R. (2001). Reductive precipitation of gold by dissimilatory Fe(III)-reducing Bacteria and Archaea. *Appl. Environ. Microbiol.* **67**, 3275–3279.

Keasling, J. D., and Hupf, G. A. (1996). Genetic manipulation of polyphosphate metabolism affects cadmium tolerance in *Escherichia coli*. *Appl. Environ. Microbiol.* **62**, 743–746.

Kelly, J. J. and Tate, R. L. (1998). Effects of heavy metal contamination and remediation on soil microbial communities in the vicinity of a zinc smelter. *J. Environ. Qual.* **27**, 609–617.

Kelly, J. J., Haggblom, M., and Tate, R. L. (1999). Changes in soil microbial communities over time resulting from one time application of zinc: a laboratory microcosm study. *Soil Biol. Biochem.* **31**, 1455–1465.

Khan, M., and Scullion, J. (2000). Effect of soil on microbial responses to metal contamination. *Environ. Pollut.* **110**, 115–125.

Khan, M., and Scullion, J. (2002). Effects of metal (Cd, Cu, Ni, Pb or Zn) enrichment of sewage-sludge on soil micro-organisms and their activities. *Appl. Soil Ecol.* **20**, 145–155.

Killham, K., and Firestone, M. K. (1983). Vesicular arbuscular mycorrhizal mediation of grass response to acidic and heavy metal depositions. *Plant Soil* **72**, 39–48.

Konopka, A., Zakharova, T., Bischoff, M., Oliver, L., Nakatsu, C., and Turco, R. F. (1999). Microbial biomass and activity in lead-contaminated soil. *Appl. Environ. Microbiol.* **65**, 2256–2259.

Korenevskii, A. A., Khamidova, K., Avakyan, Z. A., and Karavaiko, G. I. (1999). Silver biosorption by micromycetes. *Mikrobiologiya* **68**, 172–178.

Kornberg, A., Rao, N. N., and Ault-Riche, D. (1999). Inorganic polyphosphate: a molecule of many functions. *Annu. Rev. Biochem.* **68**, 89–125.

Kostov, O., and Van Cleemput, O. (2001). Microbial activity of Cu contaminated soils and effect of lime and compost on soil resiliency. *Compost Sci. Util.* **9**, 336–351.

Krantz-Rulcker, C., Allard, B., and Schnurer, J. (1993). Interactions between a soil fungus, *Trichoderma harzianum* and IIB metals: adsorption to mycelium and production of complexing metabolites. *Biometals*, **6**, 223–230.

Krantz-Rulcker, C., Allard, B., and Schnurer, J. (1996). Adsorption of IIB metals by 3 common soil fungi: comparison and assessment of importance for metal distribution in natural soil systems. *Soil Biol. Biochem.* **28**, 967–975.

Kubatova, A., Prasil, K., and Vanova, M. (2002). Diversity of soil microscopic fungi on abandoned industrial deposits. *Cryptogamic. Mycol.* **23**, 205–219.

Kuo, C.-W., and Genthner, B. R. S. (1996). Effect of added heavy metal ions on biotransformation and biodegradation of 2-chlorophenol and 3-chlorobenzoate in anaerobic bacterial consortia. *Appl. Environ. Microbiol.* **62**, 2317–2323.

Kuperman, R. G., and Carreiro, M. M. (1997). Soil heavy metal concentrations, microbial biomass and enzyme activities in a contaminated grassland ecosystem. *Soil Biol. Biochem.* **29**, 179–190.

Lacourt, I., D'Angelo, S., Girlanda, M., Turnau, K., Bonfante, P., and Perotto, S. (2000). Genetic polymorphism and metal sensitivity of *Oidiodendron maius* strains isolated from polluted soil. *Ann. Microbiol.* **50**, 157–166.

Landa, E. R., and Gray, J. R. (1995). US Geological Survey: results on the environmental fate of uranium mining and milling wastes. *J. Ind. Microbiol.* **26**, 19–31.

Ledin, M., Krantz-Rulcker, C., and Allard B. (1996). Zn, Cd and Hg accumulation by microorganisms, organic and inorganic soil components in multicompartment systems. *Soil Biol. Biochem.* **28**, 791–799.

Lee, J.-U., and Beveridge, T. J. (2001). Interaction between iron and *Pseudomonas aeruginosa* biofilms attached to Sepharose surfaces. *Chem. Geol.* **180**, 67–80.

Leyval, C., Turnau, K., and Haselwandter, K. (1997). Effect of heavy metal pollution on mycorrhizal colonization and function: physiological, ecological and applied aspects. *Mycorrhiza* **7**, 139–153.

Liu, X. F., and Culotta, V. C. (1999). Mutational analysis of *Saccharomyces cerevisiae* Smf1p, a member of the Nramp family of metal transporters. *J. Mol. Biol.* **289**, 885–891.

Lloyd, J. R., and Macaskie, L. E. (1998). Enzymatic recovery of elemental palladium using sulfate-reducing bacteria. *Appl. Environ. Microbiol.* **64**, 4607–4609.

Lloyd, J. R., Ridley, J., Khizniak, T., Lyalikova, N. N., and Macaskie, L. E. (1999a). Reduction of technetium by *Desulfovibrio desulfuricans*: biocatalyst characterization and use in a flow-through bioreactor. *Appl. Environ. Microbiol.* **65**, 2691–2696.

Lloyd, J. R., Thomas, G. H., Finlay, J. A., Cole, J. A., and Macaskie, L. E. (1999b). Microbial reduction of technetium by *Escherichia coli* and *Desulfovibrio desulfuricans*: enhancement via the use of high activity strains and effect of process parameters. *Biotechnol. Bioeng.* **66**, 122–130.

Lodewyckx, C., Taghavi, S., Mergeay, M., Vangronsveld, J., Clijsters, H., and van der Lelie, D. (2001). The effect of recombinant heavy metal resistant endophytic bacteria in heavy metal uptake by their host plant. *Int. J. Phytoremediat.* **3**, 173–187.

Losi, M. E., and Frankenberger, W. T. (1998). Microbial oxidation and solubilization of precipitated elemental selenium in soil. *J. Environ. Qual.* **27**, 836–843.

Lovley, D. R. (1995). Bioremediation of organic and metal contaminants with dissimilatory metal reduction. *J. Ind. Microbiol.* **14**, 85–93.

Lovley, D. R. (ed.) (2000). *Environmental Microbe–Metal Interactions*, ASM Press, Washington, DC.

Lovley, D. R. (2001). Anaerobes to the rescue. *Science* **293**, 1444–1446.

Lovley, D. R., and Coates, J. D. (1997). Bioremediation of metal contamination. *Curr. Opin. Biotechnol.* **8**, 285–289.

Lovley, D. R., Phillips, E. J. P., Gorby, Y. A., and Landa, E. R. (1991). Microbial reduction of uranium. *Nature* **350**, 413–416.

Macaskie, L. E. (1991). The application of biotechnology to the treatment of wastes produced by the nuclear fuel cycle: biodegradation and bioaccumulation as a means of treating radionuclide-containing streams. *Crit. Rev. Biotechnol.* **11**, 41–112.

Macaskie, L. E., Jeong, B. C., and Tolley, M. R. (1994). Enzymically accelerated biomineralization of heavy metals: application to the removal of americium and plutonium from aqueous flows. *FEMS Microbiol. Rev.* **14**, 351–367.

Macreadie, I. G., Sewell, A. K., and Winge, D. R. (1994). Metal ion resistance and the role of metallothionein in yeast. In *Metal Ions in Fungi*, ed. Winkelmann, G., and Winge, D. R., Marcel Dekker, New York, 279–310.

Markkola, A. M., Ahonen-Jonnarth, U., Roitto, M., Strommer, R., and Hyvarinen, M. (2002). Shift in ectomycorrhizal community composition in Scots pine (*Pinus sylvestris* L.) seedling roots as a response to nickel deposition and removal of lichen cover. *Environ. Pollut.* **120**, 797–803.

Martino, E., Perotto, S., Parsons, R., and Gadd, G. M. (2003). Solubilization of insoluble inorganic zinc compounds by ericoid mycorrhizal fungi derived from heavy metal polluted sites. *Soil Biol. Biochem.* **35**, 133–141.

Massaccesi, G., Romero, M. C., Cazau, M. C., and Bucsinszky, A. M. (2002). Cadmium removal capacities of filamentous soil fungi isolated from industrially polluted sediments, in La Plata (Argentina). *World J. Microbiol. Biotechnol.* **18**, 817–820.

McLean, J., and Beveridge, T. J. (2001). Chromate reduction by a pseudomonad isolated from a site contaminated with chromated copper arsenate. *Appl. Environ. Microbiol.* **67**, 1076–1084.

McLean, J. S., Lee, J.-U., and Beveridge, T. J. (2002). Interactions of bacteria and environmental metals, fine-grained mineral development, and bioremediation strategies. In *Interactions Between Soil Particles and Microorganisms*, ed. Huang, P. M., Bollag, J.-M., and Senesi, N., Wiley, New York, 228–261.

Meharg, A. A., and Cairney, J. W. G. (2000). Co-evolution of mycorrhizal symbionts and their hosts to metal-contaminated environments. *Adv. Ecol. Res.* **30**, 69–112.

Mehra, R. K., and Winge, D. R. (1991). Metal ion resistance in fungi: molecular mechanisms and their related expression. *J. Cell. Biochem.* **45**, 30–40.

Mengoni, A., Barzanti, R., Gonnelli, C., Gabbrielli, R., and Bazzicalupo, M. (2001). Characterization of nickel-resistant bacteria isolated from serpentine soil. *Environ. Microbiol.* **3**, 691–698.

Mengoni, A., Grassi, E., Barzanti, R., Biondi, E. G., Gonnelli, C., Kim, C. K., and Bazzicalupo, M. (2004). Genetic diversity of bacterial communities of serpentine soil and of rhizosphere of the nickel-hyperaccumulator plant *Alyssum bertolonii*. *Microb. Ecol.* **48**, 209–217.

Mineev, V. G., Gomonova, N. F., Zenova, G. M., and Skvortsova, I. N. (1999). Changes in the properties of soddy-podzolic soil and its microbocenosis under intensive anthropogenic impact. *Eurasian J. Soil Sci.* **32**, 413–417.

Moffett, B. F., Nicholson, F. A., Uwakwe, N. C., Chambers, B. J., Harris, J. A., and Hill, T. C. J. (2003). Zinc contamination decreases the bacterial diversity of agricultural soil. *FEMS Microbiol. Ecol.* **43**, 13–19.

Morley, G. F., and Gadd, G. M. (1995). Sorption of toxic metals by fungi and clay minerals. *Mycol. Res.* **99**, 1429–1438.

Morley, G. F., Sayer, J. A., Wilkinson, S. C., Gharieb, M. M., and Gadd, G. M. (1996). Fungal sequestration, solubilization and transformation of toxic metals. In *Fungi and Environmental Change*, ed. Frankland, J. C., Magan, N., and Gadd, G. M., Cambridge University Press, Cambridge, 235–256.

Mossman, D. J., Reimer, T., and Durstling, H. (1999). Microbial processes in gold migration and deposition: modern analogues to ancient deposits. *Geosci. Can.* **26**, 131–140.

Mowll, J. L., and Gadd, G. M. (1985). The effect of vehicular lead pollution on phylloplane mycoflora. *Trans. Br. Mycol. Soc.* **84**, 685–689.

Moynahan, O. S., Zabinski, C. A., and Gannon, J. E. (2002). Microbial community structure and carbon-utilization diversity in a mine tailings revegetation study. *Restor. Ecol.* **10**, 77–87.

Mozafar, A., Ruh, R., Klingel, P., Gamper, H., Egli, S., and Frossard, E. (2002). Effect of heavy metal contaminated shooting range soils on mycorrhizal colonization of roots and metal uptake by leek. *Environ. Monitor. Assess.* **79**, 177–191.

Mukhopadhyay, R., Rosen, B. P., Pung, L. T., and Silver, S. (2002). Microbial arsenic: from geocycles to genes and enzymes. *FEMS Microbiol. Rev.* **26**, 311–325.

Muller, A. K., Westergaard, K., Christensen, S., and Sorensen, S. J. (2001). The effect of long-term mercury pollution on the soil microbial community. *FEMS Microbiol. Ecol.* **36**, 11–19.

Nealson, K. H., and Saffarini, D. (1994). Iron and manganese in anaerobic respiration: environmental significance, physiology, and regulation. *Annu. Rev. Microbiol.* **48**, 311–343.

Nealson, K. H., Myers, C. R., and Wimpee, B. B. (1991). Isolation and identification of manganese-reducing bacteria and estimates of microbial Mn(IV)-reducing potential in the Black Sea. *Deep-Sea Res.* **38**, S907–S920.

Newman, L. A., and Reynolds, C. M. (2005). Bacteria and phytoremediation: new uses for endophytic bacteria in plants. *Trends Biotechnol.* **23**, 6–8.

Nies, D. H. (1992a). Resistance to cadmium, cobalt, zinc, and nickel in microbes. *Plasmid* **27**, 17–28.

Nies, D. H. (1992b). Czcr and Czcd, gene-products affecting regulation of resistance to cobalt, zinc, and cadmium (*czc* system) in *Alcaligenes eutrophus*. *J. Bacteriol.* **174**, 8102–8110.

Nies, D. H. (1995). The cobalt, zinc, and cadmium efflux system *czcabc* from *Alcaligenes eutrophus* functions as a cation-proton antiporter in *Escherichia coli*. *J. Bacteriol.* **177**, 2707–2712.

Nies, D. H. (1999). Microbial heavy-metal resistance. *Appl. Microbiol. Biotechnol.* **51**, 730–750.

Nies, D. H. (2003). Efflux-mediated heavy metal resistance in prokaryotes. *FEMS Microbiol. Rev.* **27**, 313–339.

Nunan, N., Wu, K. J., Young, I. M., Crawford, J. W., and Ritz, K. (2002). In situ spatial patterns of soil bacterial populations, mapped at multiple scales, in an arable soil. *Microb. Ecol.* **44**, 296–305.

Nunan, N., Wu, K. J., Young, I. M., Crawford, J. W., and Ritz, K. (2003). Spatial distribution of bacterial communities and their relationships with the micro-architecture of soil. *FEMS Microbiol. Ecol.* **44**, 203–215.

Obbard, J., Sauerbeck, D., and Jones, K. (1993). *Rhizobium leguminosarum* biovar. *trifolii* in soils amended with heavy metal contaminated sewage sludges. *Soil Biol. Biochem.* **25**, 227–231.

Olayinka, A., and Babalola, G. O. (2001). Effects of copper sulphate application on microbial numbers and respiration, nitrifier and urease activities, and nitrogen and phosphorus mineralization in an alfisol. *Biol. Agric. Hortic.* **19**, 1–8.

Oremland, R., and Stolz, J. (2000). Dissimilatory reduction of selenate and arsenate in nature. In *Environmental Microbe–Metal Interactions,*ed. Lovley, D. R., ASM Press, Washington, DC, 199–224.

Oremland, R. S., Hollibaugh, J. T., Maest, A. S., Presser, T. S., Miller, L. G., and Culbertson, C. W. (1989). Selenate reduction to elemental selenium by anaerobic bacteria in sediments and culture: biogeochemical significance of a novel sulfate-independent respiration. *Appl. Environ. Microbiol.* **55**, 2333–2343.

Oremland, R. S., Steinberg, N. A., Presser, T. S., and Miller, L. G. (1991). In situ bacterial selenate reduction in the agricultural drainage systems of western Nevada. *Appl. Environ. Microbiol.* **57**, 615–617.

Ow, D. W., Ortiz, D. F., Speiser, D. M., and McCue, K. F. (1994). Molecular genetic analysis of cadmium tolerance in *Schizosaccharomyces pombe*. In *Metal Ions in Fungi,*ed. Winkelmann, G., and Winge, D. R., Marcel Dekker, New York, 339–359.

Pennanen, T. (2001). Microbial communities in boreal coniferous forest humus exposed to heavy metals and changes in soil pH: a summary of the use of phospholipid fatty acids, Biolog (R) and H-3-thymidine incorporation methods in field studies. *Geoderma* **100**, 91–126.

Pennanen, T., Frostegaard, A., Fritze, H., and Baath, E. (1996). Phospholipid fatty acid composition and heavy metal tolerance of soil microbial communities along two heavy metal-polluted gradients in coniferous forests. *Appl. Environ. Microbiol.* **62**, 420–428.

Pettersson, A., Kunst, L., Bergman, B., and Roomans, G. M. (1985). Accumulation of aluminum by *Anabaena cylindrica*into polyphosphate granules and cell walls: an x-ray energy-dispersive microanalysis study. *J. Gen. Microbiol.* **131**, 2545–2548.

Phillips, E. J. P., Landa, E. R., and Lovley, D. R. (1995). Remediation of uranium contaminated soils with bicarbonate extraction and microbial U(VI) reduction. *J. Ind. Microbiol.* **14**, 203–207.

Picard, C., and Bosco, M. (2003). Soil antimony pollution and plant growth stage affect the biodiversity of auxin-producing bacteria isolated from the rhizosphere of *Achillea ageratum* L. *FEMS Microbiol. Ecol.* **46**, 73–80.

Ramsay, L. M., and Gadd, G. M. (1997). Mutants of *Saccharomyces cerevisiae* defective in vacuolar function confirm a role for the vacuole in toxic metal ion detoxification. *FEMS Microbiol. Lett.* **152**, 293–298.

Rauser, W. E. (1995). Phytochelatins and related peptides. *Plant Physiol.* **109**, 1141–1149.

Rawlings, D. E. (1997). Mesophilic, autotrophic bioleaching bacteria: description, physiology and role. In *Biomining: Theory, Microbes and Industrial Processes,*ed. Rawlings, D. E., Springer-Verlag, Berlin, 229–245.

Rawlings, D. E., and Silver, S. (1995). Mining with microbes. *Biotechnology* **13**, 773–338.

Renella, G., Landi, L., and Nannipieni, P. (2004). Degradation of low molecular weight organic acids complexed with heavy metals in soil. *Geoderma* **122**, 311–315.

Renninger, N., McMahon, K. D., Knopp, R., Nitsche, H., Clark, D. S., and Keasling, J. D. (2001). Uranyl precipitation by biomass from an enhanced biological phosphorus removal reactor. *Biodegradation* **12**, 401–410.

Renninger, N., Knopp, R., Nitsche, H., Clark, D. S., Jay, D., and Keasling, J. D. (2004). Uranyl precipitation by *Pseudomonas aeruginosa* via controlled polyphosphate metabolism. *Appl. Environ. Microbiol.* **70**, 7404–7412.

Roane, T. M., and Kellogg, S. T. (1996). Characterization of bacterial communities in heavy metal contaminated soils. *Can. J. Microbiol.* **42**, 593–603.

Robertson, G. P., Klingensmith, K. M., Klug, M. J., Paul, E. P., Crum, J. R., and Ellis, B. G. (1997). Soil resources, microbial activity, and primary production across an agricultural ecosystem. *Ecol. Appl.* **7**, 158–170.

Rugh, C. L., Wilde, H. D., Stack, N. M., Thompson, D. M., Summers, A. O., and Meagher, R. B. (1996). Mercuric ion reduction and resistance in transgenic *Arabidopsis thaliana* plants expressing a modified bacterial *merA* gene. *Proc. Natl. Acad. Sci. USA* **93**, 3182–3187.

Rusin, P. A., Quintana, L., Brainard, J. R., Strietelmeier, B. A., Tait, C. D., Ekberg, S. A., Palmer, P.D., Newton, T.W. and Clark, D.L. (1994). Solubilization of plutonium hydrous oxide by iron reducing bacteria. *Environ. Sci. Technol.* **28**, 1686–1690.

Said, W. A., and Lewis, D. L. (1991). Quantitative assessment of the effects of metals on microbial degradation of organic chemicals. *Appl. Environ. Microbiol.* **57**, 1498–1503.

Salt, D. E., Smith, R. D., and Raskin, I. (1998). Phytoremediation. *Annu. Rev. Plant Physiol. Plant Mol. Biol.* **49**, 643–668.

Sandaa, R.-A., Torsvik, V., Enger, O., Daae, F. L., Castberg, T., and Hahn, D. (1999). Analysis of bacterial communities in heavy metal contaminated soils at different levels of resolution. *FEMS Microbiol. Ecol.* **30**, 237–251.

Sandaa, R.-A., Torsvik, V., and Enger, O. (2001). Influence of long-term heavy-metal contamination on microbial communities in soil. *Soil Biol. Biochem.* **33**, 287–295.

Santamaría, M., Díaz-Marrero, A. R., Hernández, J., Gutiérrez-Navarro, A. M., and Corzo, J. (2003). Effect of thorium on the growth and capsule morphology of *Bradyrhizobium*. *Environ. Microbiol.* **5**, 916–924.

Sauge-Merle, S., Cuine, S., Carrier, P., Lecomte-Pradines, C., Luu, D.-T., and Peltier, G. (2003). Enhanced toxic metal accumulation in engineered bacterial cells expressing *Arabidopsis thaliana* phytochelatin synthase. *Appl. Environ. Microbiol.* **69**, 490–494.

Sayer, J. A., and Gadd, G. M. (2001). Binding of cobalt and zinc by organic acids and culture filtrates of *Aspergillus niger* grown in the absence or presence of insoluble cobalt or zinc phosphate. *Mycol. Res.* **105**, 1261–1267.

Sayer, J. A., Raggett, S. L., and Gadd, G. M. (1995). Solubilization of insoluble metal compounds by soil fungi: development of a screening method for solubilizing ability and metal tolerance. *Mycol. Res.* **99**, 987–993.

Sayer, J. A., Cotter-Howells, J. D., Watson, C., Hillier, S., and Gadd, G. M. (1999). Lead mineral transformation by fungi. *Curr. Biol.* **9**, 691–694.

Schippers, A., and Sand, W. (1999). Bacterial leaching of metal sulfides proceeds by two indirect mechanisms via thiosulfate or via polysulfides and sulphur. *Appl. Environ. Microbiol.* **65**, 319–321.

Sharples, J. M., Chambers, S. M., Meharg, A. A., and Cairney, J. W. G. (2000). Genetic diversity of root-associated fungal endophytes from *Calluna vulgaris* at contrasting field sites. *New Phytol.* **148**, 153–162.

Sharples, J. M., Meharg, A. A., Chambers, S. M., and Cairney, J. W. G. (2001). Arsenate resistance in the ericoid mycorrhizal fungus *Hymenoscyphus ericae*. *New Phytol.* **151**, 265–270.

Shi, W., Becker, J., Bischoff, M., Turco, R. F., and Konopka, A. E. (2002). Association of microbial community composition and activity with lead, chromium, and hydrocarbon contamination. *Appl. Environ. Microbiol.* **68**, 3859–3866.

Silver, S. (1996). Bacterial resistances to toxic metals: a review. *Gene* **179**, 9–19.

Silver, S. (1998). Genes for all metals: a bacterial view of the periodic table. *J. Ind. Microbiol. Biotechnol.* **20**, 1–12.

Silver, S., and Phung, LeT. (1996). Bacterial heavy metal resistance: new surprises. *Annu. Rev. Microbiol.* **50**, 753–789.

Smit, E., Leeflang, P., and Wernars, K. (1997). Detection of shifts in microbial community structure and diversity in soil caused by copper contamination using amplified ribosomal DNA restriction analysis. *FEMS Microbiol. Ecol.* **23**, 249–261.

Smith, W. L., and Gadd, G. M. (2000). Reduction and precipitation of chromate by mixed culture sulphate-reducing bacterial biofilms. *J. Appl. Microbiol.* **88**, 983–991.

Smith, S. E., and Read, D. J. (1997). *Mycorrhizal Symbiosis*, 2nd ed., Academic Press, San Diego, CA.

Smith, T., Pitts, K., McGarvey, J. A., and Summers, A. O. (1998). Bacterial oxidation of mercury metal vapour, Hg(0). *Appl. Environ. Microbiol.* **64**, 1328–1332.

Southam, G. (2000). Bacterial surface-mediated mineral formation. In *Environmental Microbe–Metal Interactions*, ed. Lovley, D. R., ASM Press, Washington, DC, 257–276.

Sreekrishnan, T. R., and Tyagi, R. D. (1994). Heavy metal leaching from sewage sludges: a technoeconomic evaluation of the process options. *Environ. Technol.* **15**, 531–543.

Staunton, S. (2004). Sensitivity analysis of the distribution coefficient, K_d, of nickel with changing soil chemical properties. *Geoderma* **122**, 281–290.

Stolz, J. F., and Oremland, R. S. (1999). Bacterial respiration of arsenic and selenium. *FEMS Microbiol. Rev.* **23**, 615–627.

Tebo, B. M., and Obraztsova, A. Y. (1998). Sulfate-reducing bacterium grows with Cr(VI), U(VI), Mn(IV), and Fe(III) as electron acceptors. *FEMS Microbiol. Lett.* **162**, 193–198.

Teitzel, G. M., and Parsek, M. R. (2003). Heavy metal resistance of biofilm and planktonic *Pseudomonas aeruginosa*. *Appl. Environ. Microbiol.* **69**, 2313–2320.

Thompson-Eagle, E. T., and Frankenberger, W. T. (1992). Bioremediation of soils contaminated with selenium. In *Advances in Soil Science,* ed. Lal, R., and Stewart, B. A., Springer-Verlag, New York, 261–309.

Thompson-Eagle, E. T., Frankenberger, W. T., and Karlson, U. (1989). Volatilization of selenium by *Alternaria alternata*. *Appl. Environ. Microbiol.* **55**, 1406–1413.

Tobin, J. M., White, C., and Gadd, G. M. (1994). Metal accumulation by fungi: applications in environmental biotechnology. *J. Ind. Microbiol.* **13**, 126–130.

Tom-Petersen, A., Leser, T. D., Marsh, T. L., and Nybroe, O. (2003). Effects of copper amendment on the bacterial community in agricultural soil analyzed by the T-RFLP technique. *FEMS Microbiol. Ecol.* **46**, 53–62.

Tsezos, M., and Volesky, B. (1982). The mechanism of uranium biosorption by *Rhizopus arrhizus*. *Biotechnol. Bioeng.* **24**, 385–401.

Turpeinen, R., Kairesalo, T., and Haggblom, M. M. (2004). Microbial community structure and activity in arsenic-, chromium- and copper-contaminated soils. *FEMS Microbiol. Ecol.* **47**, 39–50.

Urrutia, M. M., and Beveridge, T. J. (1994). Formation of fine-grained silicate minerals and metal precipitates by a bacterial cell surface (*Bacillus subtilis*) and implications on global cycling of silicon. *Chem. Geol.* **116**, 261–280.

Vachon, R. P. D., Tyagi, J., Auclair, C., and Wilkinson, K. J. (1994). Chemical and biological leaching of aluminium from red mud. *Environ. Sci. Technol.* **28**, 26–30.

Valls, M., and de Lorenzo, V. (2002). Exploiting the genetic and biochemical capacities of bacteria for the remediation of heavy metal pollution. *FEMS Microbiol. Rev.* **26**, 327–338.

Valls, M., Atrian, S., de Lorenzo, V., and Fernandez, L. A. (2000). Engineering a mouse metallothionein on the cell surface of *Ralstonia eutropha* CH34 for immobilization of heavy metals in soil. *Nat. Biotechnol.* **18**, 661–665.

Verrecchia, E. P. (2000). Fungi and sediments. In *Microbial Sediments*, ed. Riding, R. E., and Awramik, S. M., Springer-Verlag, Berlin, 69–75.

Verrecchia, E. P., and Dumont, J.-L. (1996). A biogeochemical model for chalk alteration by fungi in semiarid environments. *Biogeochemistry.* **35**, 447–470.

Vieira, M. J., and Melo, L. F. (1995). Effect of clay particles on the behaviour of biofilms formed by *Pseudomomonas fluorescens*. *Water Sci. Technol.* **32**, 45–52.

Vodnik, D., Byrne, A. R., and Gogala, N. (1998). The uptake and transport of lead in some ectomycorrhizal fungi in culture. *Mycol. Res.* **102**, 953–958.

Wainwright, M., and Gadd, G. M. (1997). Fungi and industrial pollutants. In *The Mycota, Vol. v, Environmental and Microbial Relationships,*ed. Wicklow, D. T., and Soderstrom, B. E., Springer-Verlag, Berlin, 85–97.

Wang, C. L., Lum, A. M., Ozuna, S. C., Clark, D. S., and Keasling, J. D. (2001). Aerobic sulfide production and cadmium precipitation by *Escherichia coli* expressing the *Treponema denticola* cysteine desulfhydrase gene. *Appl. Microbiol. Biotechnol.* **56**, 425–430.

Watson, J. H. P., Ellwood, D. C., Deng, Q. X., Mikhalovsky, S., Hayter, C. E., and Evans, J. (1995). Heavy metal adsorption on bacterially-produced FeS. *Miner. Eng.* **8**, 1097–1108.

Watson, J. H. P., Cressey, B. A., Roberts, A. P., Ellwood, D. C., Charnock, J. M., and Soper, A. K. (2000). Structural and magnetic studies on heavy-metal-adsorbing iron sulphide nanoparticles produced by sulphate-reducing bacteria. *J. Magn. Magn. Mater.* **214**, 13–30.

White, C., and Gadd, G. M. (1986). Uptake and cellular distribution of copper, cobalt and cadmium in strains of *Saccharomyces cerevisiae* cultured on elevated concentrations of these metals. *FEMS Microbiol. Ecol.* **38**, 277–283.

White, C., and Gadd, G. M. (1996a). Mixed sulphate-reducing bacterial cultures for bioprecipitation of toxic metals: factorial and response-surface analysis of the effects of dilution rate, sulphate and substrate concentration. *Microbiology.* **142**, 2197–2205.

White, C., and Gadd, G. M. (1996b). A comparison of carbon/energy and complex nitrogen sources for bacterial sulphate-reduction: potential applications to bioprecipitation of toxic metals as sulphides. *J. Ind. Microbiol.* **17**, 116–123.

White, C., and Gadd, G. M. (1997). An internal sedimentation bioreactor for laboratory-scale removal of toxic metals from soil leachates using biogenic sulphide precipitation. *J. Ind. Microbiol. Biotechnol.* **18**, 414–421.

White, C., and Gadd, G. M. (1998a). Accumulation and effects of cadmium on sulphate-reducing bacterial biofilms. *Microbiology.* **144**, 1407–1415.

White, C., and Gadd, G. M. (1998b). Reduction of metal cations and oxyanions by anaerobic and metal-resistant organisms: chemistry, physiology and potential for the control and bioremediation of toxic metal pollution. In *Extremophiles: Physiology and Biotechnology*, ed. Grant, W. D., and Horikoshi, T., Wiley, New York, 233–254.

White, C., and Gadd, G. M. (2000). Copper accumulation by sulphate-reducing bacterial biofilms and effects on growth. *FEMS Microbiol. Lett.* **183**, 313–318.

White, C., Sayer, J. A., and Gadd, G. M. (1997). Microbial solubilization and immobilization of toxic metals: key biogeochemical processes for treatment of contamination. *FEMS Microbiol. Rev.* **20**, 503–516.

White, C. A., Sharman, K., and Gadd, G. M. (1998). An integrated microbial process for the bioremediation of soil contaminated with toxic metals. *Nat. Biotechnol.* **16**, 572–575.

White, C., Wilkinson, S. C., and Gadd, G. M. (1995). The role of microorganisms in biosorption of toxic metals and radionuclides. *Int. Biodeter. Biodegrad.* **35**, 17–40.

Whiting, S. N., Leake, J. R., McGrath, S. P., and Baker, A. J. M. (2001). Zinc accumulation by *Thlaspi caerulescens* from soils with different Zn availability: a pot study. *Plant Soil* **236**, 11–18.

Wilkins, D. A. (1991). The influence of sheathing (ecto-)mycorrhizas of trees on the uptake and toxicity of metals. *Agric. Ecosyst. Environ.* **35**, 245–260.

Wilkinson, D. M., and Dickinson, N. M. (1995). Metal resistance in trees: the role of mycorrhizae. *Oikos* **72**, 298–300.

Wu, J. S., Sung, H. Y., and Juang, R. J. (1995). Transformation of cadmium-binding complexes during cadmium sequestration in fission yeast. *Biochem. Mol. Biol. Interact.* **36**, 1169–1175.

Yu, W., Farrell, R. A., Stillman, D. J., and Winge, D. R. (1996). Identification of SLF1 as a new copper homeostasis gene involved in copper sulfide mineralization in *Saccharomyces cerevisiae. Mol. Cell. Biol.* **16,** 2464–2472.

Zhdanova, N.N., Redchitz, T. I., and Vasilevskaya, A. I. (1986). Species composition and sorption properties of Deuteromycetes in soils polluted by industrial wastewater [in Russian]. *Mikrobiol Zhu.* **48**, 44–50.

Zinkevich, V., Kang, H., Bogdarina, I., Hill, M. A., and Beech, I. B. (1996). The characterisation of exopolymers produced by different species of marine sulphate-reducing bacteria. *Int. Biodeter. Biodegrad.* **37** 163–172.

3

KINETICS AND MECHANISMS OF SORPTION–DESORPTION IN SOILS: A MULTISCALE ASSESSMENT

M. J. Borda
Golder Associates, Inc., Mt. Laurel, New Jersey
D. L. Sparks
University of Delaware, Newark, Delaware

3.1. INTRODUCTION

In natural systems, the bioavailability of micro- and macronutrients, as well as contaminants, is controlled by interaction with soils and soil components. One of the most important interactions is via sorption. The overarching term *sorption* is used to describe any system where a sorbate (e.g., an atom, a molecule, a molecular ion) interacts with a sorbent (i.e., a solid surface) resulting in an accumulation at the sorbate–sorbent interface (Stumm and Morgan, 1981). Typically,

Biophysico-Chemical Processes of Heavy Metals and Metalloids in Soil Environments,
Edited by Antonio Violante, Pan Ming Huang, and Geoffrey Michael Gadd

the broad term *sorption* is used if the exact mechanism of attachment of the species to the surface is unknown.

Sorption occurs on many scales, both temporal and spatial, ranging from milliseconds to months and years and from the atomic to the global scale. The past decade has seen advances in understanding these phenomena and has been punctuated by the development of technologies capable of observing these systems with increased resolution. Our intent in this chapter is to assemble some of the results reported in the recent literature dealing with advances in the study of sorption–desorption to establish the state of the art and present the future experimental challenges that face scientists working in this field.

We begin by focusing on the limitations of equilibrium-based approaches in understanding systems that are kinetically controlled and the importance of using, and developing, techniques that are capable of assessing sorption–desorption reactions at high spatial, and temporal, resolution. With an understanding of limitations and experimental challenges, the current methods for assessing sorption–desorption are discussed, including macroscopic and molecular-scale approaches leading to a new description of soil chemical processes. This description will not be fragmented by traditional descriptions of end-member processes (e.g., adsorption vs. precipitation), but rather, will cover the continuum of processes that represent the range of pathways, from interaction to incorporation, that control the bioavailability of species. From this new description, future experimental challenges for scientists in this field are discussed.

3.1.1. Sorption Terminology

A starting point for any discussion of the kinetics and mechanisms of sorption–desorption processes is an understanding of the terminology that is central to the topic. There are well-developed terms associated with sorption–desorption processes, and a brief introduction is warranted.

Adsorption defines the accumulation of a substance, or material, at the interface between a solid surface and a bathing solution (Sparks, 2002). Within the adsorption framework, the individual components are referred to as the *adsorbate*, the accumulating material at the interface, and the *adsorbent*, or solid surface (Sparks, 2002). If adsorption occurs and results in the formation of a stable molecular phase at the interface, this entity can be described as a *surface complex*. Two general surface complexes exist and are described by the configuration geometry of the adsorbate at the adsorbent surface. These are the inner- and outer-sphere surface complexes, defined by the presence, or absence, of the hydration sphere of the adsorbate molecule upon interaction. When at least one water molecule of the hydration sphere is retained upon adsorption, the surface complex is referred to as an *outer-sphere complex* (Sposito, 1984); when an ion or molecule is bound directly to the adsorbent without the presence of the hydration sphere, an *inner-sphere complex* is formed.

In the event that adsorption occurs and continues through the formation of a three-dimensional surface species, the new species can be defined as a *surface precipitate*. Later, the concept of a continuum of sorption processes is developed to describe the "gray area" between adsorption and precipitation where a number of systems naturally exist.

3.2. TIME SCALES FOR SORPTION PROCESSES

As mentioned previously, sorption–desorption processes can occur over wide time scales. The type of soil component has a major effect on the sorption rate. For example, sorption reactions are often more rapid on clay minerals such as kaolinite and smectite than on vermiculitic and micaceous minerals. This is due in large part to the availability of sites for sorption. For example, kaolinite has readily available planar external sites, and smectites have predominantly internal sites that are quite available for retention of sorbates. Thus, sorption reactions on these soil constituents are often quite rapid, even occurring on time scales of seconds and milliseconds (Sparks, 1989, 2002). On the other hand, vermiculite and micas have multiple sites for retention of ions, including planar, edge, and interlayer sites, with some of the latter sites being partially to totally collapsed. Therefore, sorption and desorption reactions on these sites can be slow, tortuous, and mass transfer controlled. Often, an apparent equilibrium may not be attained even after several days or weeks. Thus, with vermiculite and mica, sorption can involve two to three different reaction rates: high rates on external sites, intermediate rates on edge sites, and low rates on interlayer sites (Sparks, 2002). Sorption reactions on oxides, hydroxides, and humic substances depend on the type of surface and type of sorbate, but the chemical reaction (CR) (devoid of mass transfer processes) appears to be rapid and not measurable using most traditional batch and flow techniques. For example, CR sorption rates of metals and oxyanions on goethite occur on millisecond time scales (Sparks and Zhang, 1991; Grossl et al., 1997). Half-times for divalent Pb, Cu, and Zn sorption on peat range from 5 to 15 seconds (Bunzl et al., 1976). A number of studies have shown that heavy metal sorption on oxides (Barrow, 1986; Bruemmer et al., 1988; Ainsworth et al., 1994; Scheidegger et al., 1997a, 1998) and clay minerals (Lövgren et al., 1990) increases with residence time (contact time between the metal and the sorbent). The mechanisms for these lower reaction rates are not well understood, but have been ascribed to diffusion phenomena, sites of lower reactivity, and surface nucleation–precipitation (Scheidegger et al., 1997a; Sparks, 2002). Sorption–desorption of metals, oxyanions, radionuclides, and organic chemicals on soils can be very slow and may demonstrate a residence time effect, which has been ascribed to diffusion into micropores of inorganic minerals such as some metal oxides and into humic substances, to retention on sites of varying reactivity, and to surface nucleation–precipitation (Scheidegger et al., 1997a; Strawn and Sparks, 1999; Alexander, 2000; Pignatello, 2000; Sparks, 2002; Trivedi et al., 2002).

When one studies sorption processes, be they with metals, oxyanions, radionuclides, or organic chemicals, a biphasic behavior is almost always observed

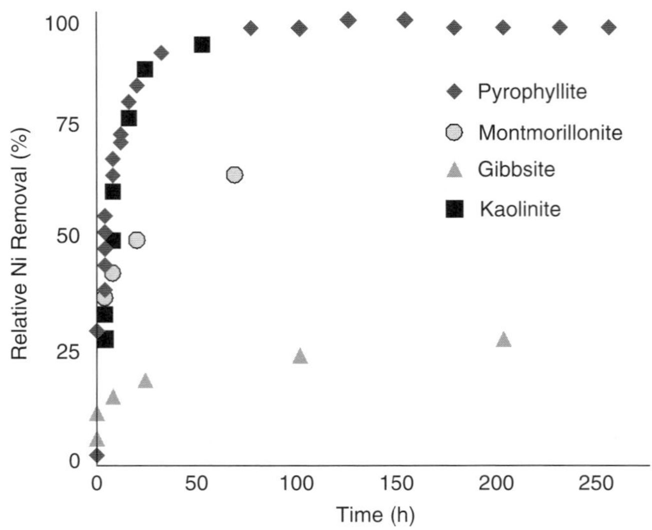

Figure 3.1. Kinetics of Ni sorption (%) on pyrophyllite, kaolinite, gibbsite, and montmorillonite from a 3 mM Ni solution, an ionic strength $I = 0.1$ $NaNO_3$, and a pH of 7.5. (From Scheidegger et al., 1997a.)

(Figures 3.1 to 3.3). Sorption is initially rapid, with a large amount of the total sorption occurring within a few hours and the remaining sorption often occurring over long periods of time. The rapid step, which occurs over milliseconds to hours, can be ascribed to chemical reaction and film diffusion processes. During this rapid reaction process, a large portion of the sorption may occur. For example, in Figure 3.1 one sees that about 90% of the total Ni sorbed on kaolinite and pyrophyllite occurred within the first 24 hours. For Pb sorption on a Matapeake soil, 78% of the total Pb sorption occurred in 8 minutes. Following the initial fast reaction, slow sorption continued, but only about 1% additional Pb was sorbed after 800 hours (Figure 3.2). Figure 3.3 shows a biphasic reaction for As(V) sorption on ferrihydrite. Within 5 minutes, a majority of the total sorption had occurred. Slow sorption continued for at least 192 hours.

3.2.1. Surface Precipitation of Metals

As the amount of metal cation or anion sorbed on a surface (surface coverage or loading, which is affected by the pH at which sorption occurs) increases, sorption can proceed from mononuclear adsorption to surface precipitation (a three-dimensional phase). There are several thermodynamic reasons for surface precipitate formation: (1) the solid surface may lower the energy of nucleation by providing sterically similar sites (McBride, 1991); (2) the activity of the surface precipitate is less than 1 (Sposito, 1986); and (3) the solubility of the surface precipitate is lowered because the dielectric constant of the solution near the surface is less than that of the bulk solution (O'Day et al., 1994). There are

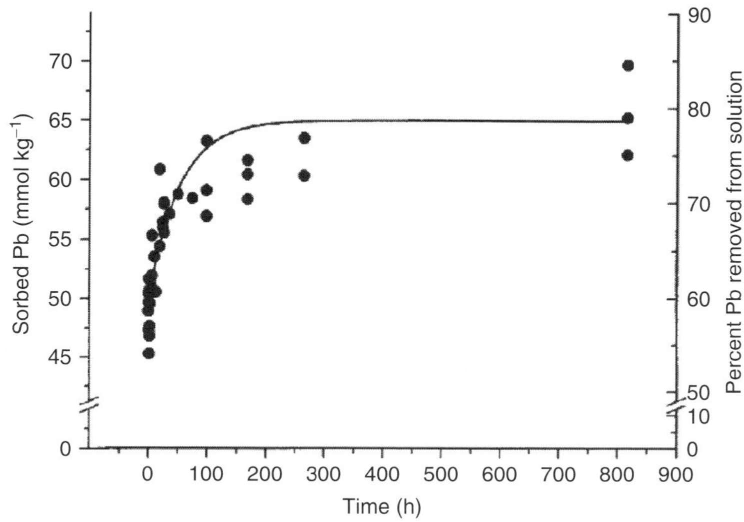

Figure 3.2. Kinetics of Pb sorption on a Matapeake soil from a 12.25 mM Pb solution, an ionic strength $I = 0.05$ M, and a pH of 5.5. (From Strawn and Sparks, 2000.)

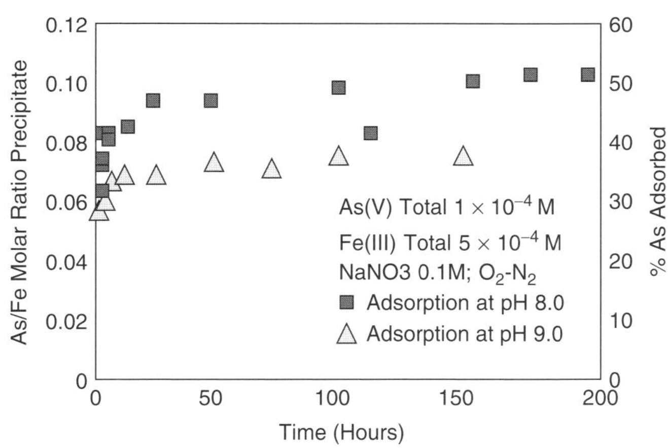

Figure 3.3. Kinetics of As(V) sorption on ferrihydrite at pH 8.0 and 9.0. (From Fuller et al., 1993.)

several types of surface precipitates. They can arise via polymeric metal complexes (dimers, trimers, etc.) that form on mineral surfaces and via the sorption of aqueous polymers (Chisholm-Brause et al., 1990a). Homogeneous precipitates can form on a surface when the solution becomes saturated and the surface acts as a nucleation site. When adsorption attains monolayer coverage, sorption continues on the newly created sites, causing a precipitate on the surface (Farley et al., 1985; Sposito, 1986; Chisholm-Brause et al., 1990a; McBride, 1991;

O'Day et al., 1994). When the precipitate consists of chemical species derived from both the aqueous solution and dissolution of the mineral, it is referred to as a *coprecipitate*. The composition of the coprecipitate varies between that of the original solid and a pure precipitate of the sorbing metal. The ionic radius of the sorbing metal and sorbent ions must be similar for coprecipitates to form. Thus, Co(II), Mn(II), Ni(II), and Zn(II) form coprecipitates on sorbents containing Al(III) and Si(IV) but not Pb(II), which is considerably larger (1.20 Å). Coprecipitate formation is most limited by the rate of mineral dissolution rather than by the lack of thermodynamic favorability (McBride, 1994; Scheidegger et al., 1998). If the formation of a precipitate occurs under solution conditions that would in the absence of a sorbent be undersaturated with respect to any known solid phase, this is referred to as *surface-induced precipitation* (Towle et al., 1997; Sparks, 2002).

Thus, there is often a continuum between surface complexation (adsorption) and surface precipitation. This continuum depends on several factors: (1) the ratio of the number of surface sites versus the number of metal ions in solution, (2) the strength of the metal–oxide bond, and (3) the degree to which the bulk solution is undersaturated with respect to the metal hydroxide precipitate (McBride, 1991). At low surface coverages, surface complexation (e.g., outer- and inner-sphere adsorption) tends to dominate. As surface loadings increase, nucleation occurs and results in the formation of distinct entities or aggregates on the surface. As noted by Bleam and McBride (1985, 1986), aggregates of Mn(II) and Mg(II), referred to as *metal clusters*, have been observed to form with a significant dependence on the material properties of the substrate. These clusters are isolated from one another and do not dramatically affect the overall surface properties of the adsorbate. As surface loadings increase further, surface precipitation becomes the dominant mechanism (Sparks, 2002, 2005). For example, Fendorf et al. (1994) and Fendorf and Sparks (1994) used x-ray adsorption fine structure (XAFS) spectroscopy, Fourier transform infrared (FTIR) spectroscopy, and high-resolution transmission electron microscopy (HRTEM) to study Cr(III) sorption on Si-oxide. At low Cr(III) surface coverage (<20%), adsorption was the dominant process, with an inner-sphere monodentate complex forming. As Cr(III) surface coverage increased (>20%), surface precipitation occurred and was the dominant process.

Using in situ XAFS, it has been shown by a number of scientists that multinuclear metal hydroxide complexes and surface precipitates Co(II), Cr(III), Cu(II), Ni(II), and Pb(II) can form on metal oxides, phyllosilicates, soil clays, and soils (Chisholm-Brause et al., 1990a,b; Roe et al., 1991; Charlet and Manceau, 1992; Fendorf and Sparks, 1994; O'Day et al., 1994; Bargar et al., 1995; Papelis et al., 1996; Scheidegger et al., 1996, 1997b, 1998; Scheidegger and Sparks, 1996; Towle et al., 1997; Elzinga and Sparks, 1999; Roberts et al., 1999; Thompson et al., 1999b; Ford and Sparks, 2000; Scheckel and Sparks, 2001). These metal hydroxide phases occur at metal loadings below a theoretical monolayer coverage and in a pH range well below the pH where the formation of metal hydroxide precipitates would be expected according to the thermodynamic solubility product (Scheidegger and Sparks, 1996; Sparks, 2002, 2005). Recently, mixed metal hydroxide precipitates have been found in field-contaminated soils (Juillot et al., 2003; Nachtegaal et al., 2005).

Scheidegger et al. (1997b) were the first to show that sorption of metals such as Ni on an array of phyllosilicates and Al oxide could result in formation of mixed metal–Al hydroxide surface precipitates which appear to be coprecipitates. The precipitate phase shares structural features common to the hydrotalcite group of minerals and the layered double hydroxides (LDH) observed in catalyst synthesis. The LDH structure is built of stacked sheets of edge-sharing metal octahedra containing divalent and trivalent metal ions separated by anions between the interlayer spaces (Figure 3.4). The general structural formula can be expressed as $[Me^{2+}_{1-x}\ Me^{3+}_{x}\ (OH)_2]^{x+} \cdot (x/n)A^{n}\text{-}mH_2O$, where, for example, Me^{2+} could be Mg(II), Ni(II), Co(II), Zn(II), Mn(II), or Fe(II), and Me^{3+} is Al(III), Fe(III), or Cr(III) (Towle et al., 1997). The LDH structure exhibits a net positive charge x per formula unit, which is balanced by an equal negative charge from interlayer anions A^{n-}, such as Cl^-, Br^-, I^-, NO_3^-, OH^-, ClO_4^-, and CO_3^{2-}; water molecules occupy the remaining interlayer space (Allmann, 1970; Taylor, 1984). The minerals takovite, $Ni_6Al_2(OH)_{16}CO_3 \cdot H_2O$, and hydrotalcite, $Mg_6Al_2(OH)_{16}CO_3 \cdot H_2O$, are among the most common natural mixed-cation hydroxide compounds containing Al (Taylor, 1984). Figure 3.4 shows a Ni–Al LDH phase (Sparks, 2002, 2005).

XAFS data, showing the formation of Ni–Al LDH phases on soil components, are shown in Figure 3.5 and Table 3.1. Radial structure functions (RSFs), collected from XAFS analyses, for Ni sorption on pyrophyllite, kaolinite, gibbsite, and montmorillonite were compared to the spectra of crystalline $Ni(OH)_2$ and takovite. All spectra showed a peak at $R \approx 0.18$ nm, which represents the first

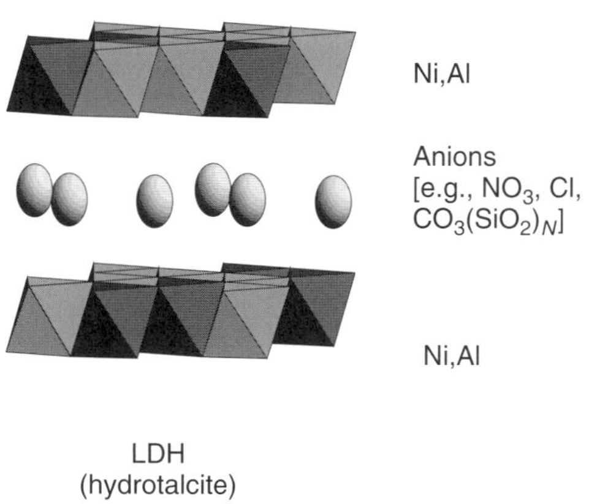

Figure 3.4. Structure of an Ni–Al LDH showing brucitelike octahedral layers in which Al^{3+} substitutes for Ni^{2+}, creating a net positive charge balanced by hydrated anions in the interlayer space. (From Scheinost et al., 1999.)

coordination shell of Ni. A second peak representing the second Ni shell was observed in the spectra of the Ni sorption samples and takovite (Figure 3.5). The structural parameters, derived from XAFS analyses, for the various sorption samples and takovite and $Ni(OH)_2$ are shown in Table 3.1. In the first coordination shell Ni is surrounded by six O atoms, indicating that Ni(II) is in an octahedral environment. The Ni–O distances for the Ni sorption samples are 2.02 to 2.03 Å and similar to those in takovite (2.03 Å). The Ni–O distances in crystalline $Ni(OH)_2(s)$ are distinctly longer (2.06 Å in this study). For the second shell, best fits were obtained by including both Ni and Si or Al as second-neighbor backscatter atoms. Since Si and Al differ in atomic number by 1 (atomic number = 14 and 13, respectively), backscattering is similar. They cannot easily be distinguished from each other as second-neighbor backscatters. There are 2.8 (for montmorillonite) to 5.8 (gibbsite) Ni second-neighbor (N) atoms, indicative of Ni surface precipitates. The Ni–Ni distances for the sorption samples were 3.00 to 3.03 Å, which are similar to those for takovite (3.01 Å), the mixed Ni–Al LDH phase, but much shorter than those in crystalline $Ni(OH)_2$ (3.09 Å). There are also 1.8 to 2.7 Si/Al second-neighbor atoms at 3.02 to 3.07 Å. The bond distances are in good agreement with the Ni–Al distances observed in takovite (3.03 Å) (Sparks, 2002, 2005).

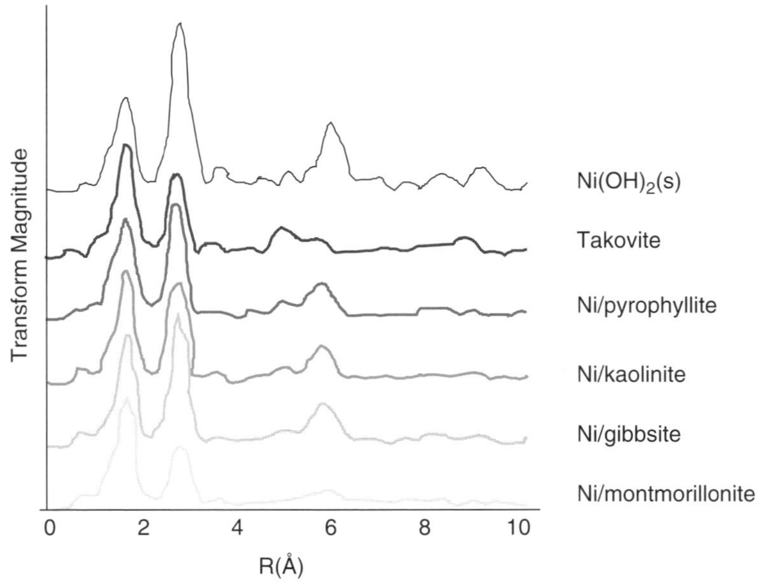

Figure 3.5. Radial structure functions (RSFs) produced by forward Fourier transforms of Ni sorbed on pyrophyllite, kaolinite, gibbsite, and montmorillonite, compared to the spectrum of crystalline $Ni(OH)_{2(S)}$ and takovite. The spectra are uncorrected for phase shift. (From Scheidegger et al., 1997.)

TABLE 3.1. Structural Information Derived from XAFS Analysis for Ni Sorption on Various Sorbonents and for Known Ni Hydoxides[a]

		Ni–O			Ni–Ni			Ni–Si/Al			N(Ni)/
	Γ(μmol/m^2)	R(Å)	N	$2\sigma^2$	R(Å)	N	$2\sigma^2$	R(Å)	N	$2\sigma^2$	N(Si/Al)
$Ni(OH)_2$	—	2.06	6.0	0.011	3.09	6.0	0.010				
Takovite	—	2.03	6.0	0.01	3.01	3.1	0.009	3.03	1.1	0.009	2.8
Pyrophyllite	3.1	2.02	6.1	0.01	3.00	4.8	0.009	3.02	2.7	0.009	1.8
Kaolinite	19.9	2.03	6.1	0.01	3.01	3.8	0.009	3.02	1.8	0.009	2.2
Gibbsite	5.0	2.03	6.5	0.01	3.02	5.0	0.009	3.05	1.8	0.09	2.7
Montmorillonite	0.35	2.03	6.3	0.01	3.03	2.8	0.011	3.07	2.0	0.015	1.4

Source: Scheidegger et al. (1997).

[a]Interatomic distances (*R*, Å), coordination numbers (*N*), and Debye–Waller factors ($2\sigma^2$, nm^2). The values reported are accurate to within R ± 0.02 Å, N_{Ni-O} ± 20%, N_{Ni-Ni} ± 40%, and $N_{Ni-Si/Al}$ ± 40%.

Mixed Co–Al and Zn–Al hydroxide surface precipitates can also form on aluminum-bearing metal oxides and phyllosilicates (Towle et al., 1997; Thompson et al., 1999a,b; Ford and Sparks, 2000). This is not surprising, as Co^{2+}, Zn^{2+}, and Ni^{2+} all have radii similar to Al^{3+}, enhancing substitution in the mineral structure and formation of a coprecipitate. However, surface precipitates have not been observed with Pb^{2+}, as Pb^{2+} is too large to substitute for Al^{3+} in mineral structures (Sparks, 2002, 2005).

Metal hydroxide precipitate phases can also form in the presence of non-Al-bearing minerals (Scheinost et al., 1999; Sparks, 2002, 2005). Using diffuse reflectance spectroscopy (DRS), which is quite sensitive for discriminating between Ni–O bond distances, it was shown that α-Ni $(OH)_2$ formed upon Ni^{2+} sorption to talc and silica (Figure 3.6).

The mechanism for the formation of metal hydroxide surface precipitates is not clearly understood. It is clear that the type of metal ion determines whether metal hydroxide surface precipitates form, and the type of surface precipitate formed (i.e., metal hydroxide or mixed metal hydroxide) is dependent on the sorbent type. The precipitation could be explained by the combination of several processes (Yamaguchi et al., 2001). First, the electric field of the mineral surface attracts metal ions (e.g., Ni) through adsorption, leading to a local supersaturation

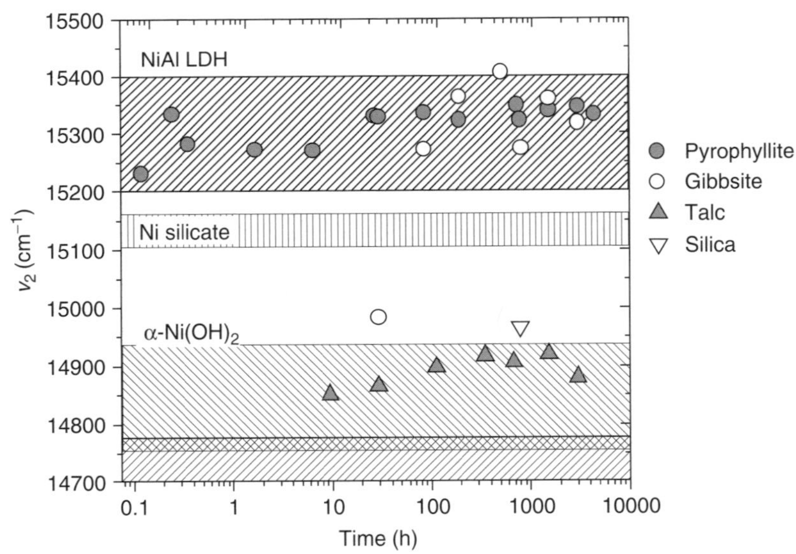

Figure 3.6. Fitted ν_2 band positions of the Ni-reacted minerals (dots and triangles) over time using diffuse reflectance spectroscopy (DRS). The ν_2 band is attributed to the crystal field splitting induced within the incompletely filled 3d electronic shell of Ni^{2+} through interaction with the negative charge of nearest-neighbor oxygen atoms. For talc and silica, the ν_2 band was at ~14,900 cm^{-1}, indicating the formation of an α-$Ni(OH)_2$ phase, whereas for Al-containing pyrophyllite and gibbsite, it appeared at ~15,300 cm^{-1}, indicating an Ni–Al LDH phase. (From Scheinost et al., 1999.)

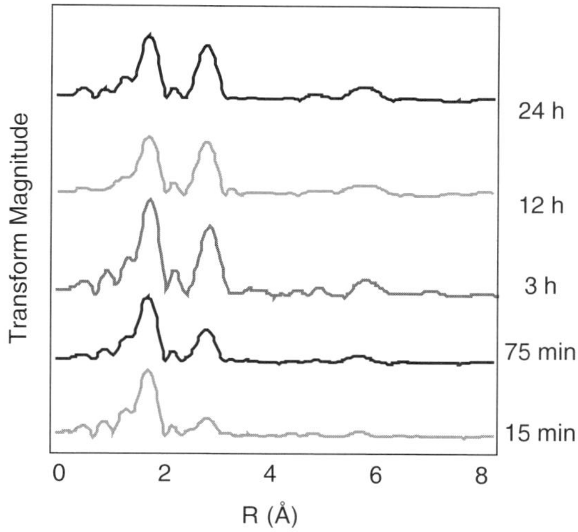

Figure 3.7. Radial structure functions (achieved from XAFS analyses) for Ni sorption on pyrophyllite for reaction times up to 24 hours, demonstrating the appearance and growth of the second shell (peak at ≈ 2.8 Å) contributions due to surface precipitation and growth of a mixed Ni–Al phase. (From Scheidegger et al., 1998.)

at the mineral–water interface. Second, the solid phase may act as a nucleation center for polyhydroxy species and catalyze the precipitation process (McBride, 1994). Third, the physical properties of water molecules adsorbed at the mineral surface are different from those of free water (Sposito, 1989), causing a lower solubility of metal hydroxides at the mineral–water interface. With time, Al, which is released by weathering of the mineral surface, slowly diffuses into the octahedral layer of the mineral and partially replaces the metal (e.g., Ni) in the octahedral sites. A Ni–Al LDH is formed, which is favored thermodynamically over α-Ni hydroxide (Sparks, 2002, 2005).

The formation of metal hydroxide surface precipitates and subsequent residence time effects on natural sorbents can greatly affect metal release and hysteresis. It has generally been thought that the kinetics of formation of surface precipitates was slow. However, recent studies have shown that metal hydroxide precipitates can form on time scales of minutes. In Figure 3.7 one can see that mixed Ni–Al hydroxide precipitates formed on pyrophyllite within 15 minutes, and they grew in intensity as time increased. Similar results have been observed with other soil components and with soils (Scheidegger et al., 1998; Roberts et al., 1999; Sparks, 2002, 2005).

The formation and subsequent "aging" of the metal hydroxide surface precipitate can have a significant effect on metal release. In Figure 3.8 one sees that as residence time (aging) increased from 1 hour to 2 years, Ni release from pyrophyllite, as a percentage of total Ni sorption, decreased from 23 to ~0% when HNO_3 (at pH 6.0) was employed as a dissolution agent for 14 days. This

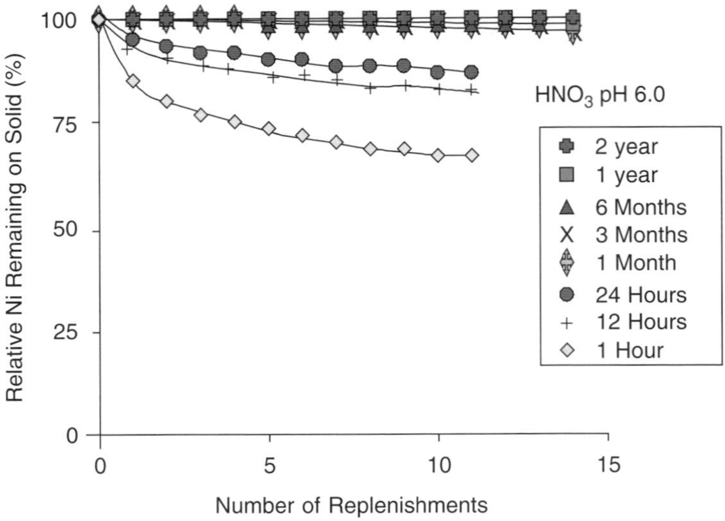

Figure 3.8. Dissolution of Ni from surface precipitates formed on pyrophyllite at residence times of 1 hour to two years. The figure shows the relative amount of Ni^{2+} remaining on the pyrophyllite surface following extraction for 24-hour periods (each replenishment represents a 24-hour extraction) with HNO_3 at pH 6.0. (From Scheckel and Sparks, 2001.)

enhanced stability is due to the transformation of the metal–Al hydroxide precipitates to a metal–Al phyllosilicate precursor phase as residence time increases. This transformation occurs via a number of steps (Figure 3.9). There is diffusion of Si originating from weathering of the sorbent into the interlayer space of the LDH, replacing anions such as NO_3^-. Polymerization and condensation of the interlayer Si slowly transforms the LDH into a precursor metal–Al phyllosilicate. The type of anion in the interlayer clearly affects the stability of the LDH phase. Peltier et al. (2005) measured enthalpies of formation for Ni LDH phases with nitrate and sulfate interlayers using acid solution calorimetry and compared them to published data on carbonate interlayer LDHs (Allada et al., 2005). Differences in the identity of the anion interlayer resulted in substantial changes in the enthalpies of formation of the LDH phases, in the order of increasing solubility nitrate < sulfate < carbonate. Substitution of silica for carbonate resulted in an even more exothermic enthalpy of formation, confirming that silica substitution increases the stability of the LDH precipitates (Peltier et al., 2005). The metal stabilization that occurs in surface precipitates on Al-free sorbents (e.g., talc) may be due to Ostwald ripening, resulting in increased crystallization (Scheckel and Sparks, 2001; Sparks, 2002, 2005).

Thus, with time, one can see that metal sorption on soil minerals can often result in a continuum of processes from adsorption to precipitation to solid-phase transformation (Figure 3.10), particularly in the case of metals such as Co, Ni, and Zn. The formation of metal surface precipitates could be an important mechanism

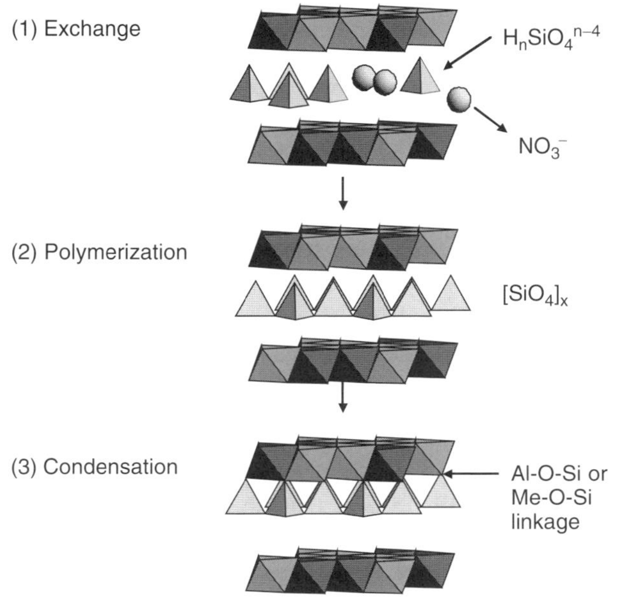

Figure 3.9. Hypothetical reaction process illustrating transformation of an initially precipitated Ni–Al LDH into a phyllosilicate-like phase during aging. The initial step involves the exchange of dissolved silica for nitrate within the LDH interlayer followed by polymerization and condensation of silica onto the octahedral Ni–Al layer. The resulting solid possesses structural features common to 1 : 1 and 2 : 1 phyllosilicates. (From Ford et al., 2001.)

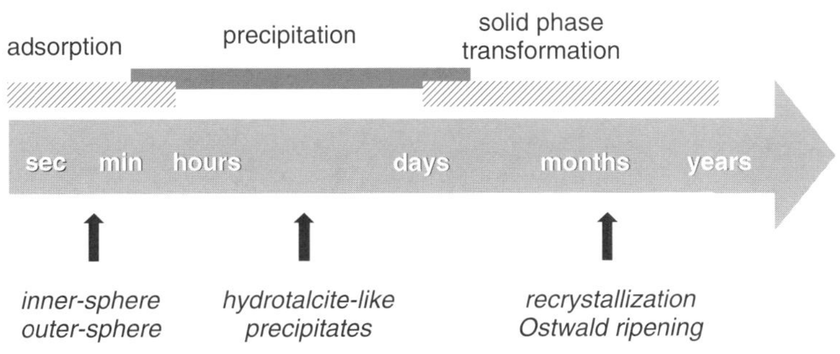

Figure 3.10. Changes in sorption processes with time showing a continuum from adsorption to precipitation to solid-phase transformations.

for sequestering metals in soils such that they are less mobile and bioavailable. Such products must be considered when modeling the fate and mobility of metals such as Co^{2+}, Mn^{2+}, Ni^{2+}, and Zn^{2+} in soil and water environments (Sparks, 2002, 2005).

3.3. MACROSCOPIC SORPTION–DESORPTION KINETIC METHODS

The majority of sorption kinetic studies have utilized either batch or flow-through methods coupled with aqueous measurements for determination of the concentrations of species of interest. More recent work has focused on molecular-scale approaches, including spectroscopic and microscopic techniques that allow for observations at increased spatial and temporal resolution to be made, often in situ and in real time. Complementary to both macroscopic and molecular-scale observations has been the utilization of theoretical techniques, such as molecular mechanics and quantum mechanics, to model surface complexes computationally. It has been through the integration of macroscopic, molecular-scale, and theoretical approaches that some of the most profound observations of sorption–desorption phenomena over the past decades have been made.

Macroscopic techniques for assessing kinetically controlled reactions have typically been burdened by relatively slow data collection rates. Typically, these data collection rates range from tens of seconds to minutes. The two main types of macroscopic experiments use either batch or flow-through methods (Amacher, 1991; Sparks et al., 1996).

Considerable sorption occurs before the first measurement can be made, particularly if batch and flow techniques are employed where the fastest that a measurement can be made is about 15 seconds. For such rapid reactions, chemical relaxation techniques, and preferably real-time molecular-scale techniques, can be used. The latter are discussed later in the chapter. One might ask why it is important to measure such reactions if they are so rapid. Since the reactions are occurring so far from equilibrium, back reactions are insignificant and one can determine chemical reaction rates, devoid of mass transfer processes. Therefore, chemical kinetic measurements are being made, and details about molecular processes and mechanisms can be ascertained.

Batch methods can be separated further into two general types, experiments that sacrifice the entire volume of an individual reactor and experiments that remove sample aliquots from a single larger reactor. Both types of experiment require that the material of interest be placed in a vessel and stirred continuously to ensure that the effect of transport processes is minimized. In the *sacrificial method*, at certain time during the course of an experiment, a small-volume reactor is sacrificed and used for analysis. This method eliminates the concentrating effect of removing sample aliquots; however, it requires a matrix of experimental vessels to define the system. The *aliquot method* does restrict the number of samples able to be withdrawn from the system; however, it is less labor intensive in terms of the experimental matrix and allows for easier alteration of system conditions.

Several disadvantages exist with either of the foregoing batch methods. First, slow data collection rates make measuring rapid reactions difficult. Typically, reactions are either over, or in a significantly different phase, prior to the collection of enough data points to define the system. A second drawback of batch methods is the potential effect of accumulating products during the reaction of interest. This is a particularly important problem when studying desorption phenomena, as increasing solution concentrations of the desorbing species can inhibit further reaction. Finally, to prevent transport effects from being significant, batch methods must be stirred continuously, often causing sample and/or reactor material to be abraded and even shed into solution in the case of reactor material. The presence of new surfaces, and/or new species, can significantly alter system conditions (Sparks, 1989, 1999, 2002). Even with serious disadvantages, numerous excellent studies have been performed using batch methods and have produced the seminal work from which advanced studies have been able to generate increasingly more resolved observations.

To overcome the disadvantages of batch methods, several types of flow methods were developed to collect kinetic information on aqueous systems (Sparks et al., 1996; Sparks, 1999, 2002). There are many advantages to flow methods, including the ability to more accurately simulate environmental soil-solution ratios, removal of desorbed species or reaction products, and increased exposure of the adsorbent and solution species. The classical flow method, or *continuous-flow technique*, involves the flow of experimental solution across a membrane filter impregnated with adsorbent material. After in-line reaction and filtration, the reacted experimental solution can be fraction collected and analyzed. The data collection rate for this technique is on the order of 30 to 60 seconds per sample. Although this technique is an improvement over batch methods, it has a significant drawback related to the flow regime established at the pressed colloid–filter interface. Flow conditions can exist such that transport effects become significant due to the thickness of the colloidal film that the experimental solution must pass through. A simple way to overcome this problem is the use of stirred-flow methods (Bartal et al., 1990; Eick et al., 1990; Sparks et al., 1996; Sparks, 1999, 2002; Vasudevan et al., 2002; Martin-Garin et al., 2003).

The *stirred-flow method* is an integration of the best characteristics of continuous-flow methods with the ability of batch methods to overcome diffusion effects. The system consists of a reaction chamber that can be stirred magnetically and is ported such that experimental solution can be flowed thorough the reaction volume and filtered upon exiting the cell. Therefore, the system does not lose any particulate material and the experimental solution can be fraction collected for analysis. The only drawback with this method is the restricted data collection time (Sparks et al., 1996; Sparks, 1999, 2002).

3.4. MOLECULAR-SCALE APPROACHES TO ASSESS SORPTION–DESORPTION

There are hundreds of analytical techniques available to study environmentally relevant systems. However, there are only a few techniques capable of probing relevant systems while in the hydrated state (in situ) and with a time resolution adequate to study processes in real time. Of significant interest are those techniques that are capable of obtaining data rapidly in order to capture *quick kinetics*.

It is this combination that scientists are currently searching for, and utilizing, to advance studies of sorption–desorption processes in natural soil systems, where rapid kinetics have often gone overlooked. As stated above, the acquisition of time-resolved data is critical for elucidating the pathways that govern kinetically controlled processes. Next, we review several of the most novel techniques currently being used to study sorption–desorption phenomena either in situ or in real time, or both. Some of these techniques are so new that they have not yet found widespread or even minimal use in the Earth and environmental sciences. Where appropriate, we illustrate examples from the environmental sciences; however, because some of these techniques represent the future of this type of investigation, examples have been taken from more disparate literature.

A useful example of using a high-time resolution technique to perform rapid kinetic studies was performed by Fendorf et al. (1993) using electron paramagnetic resonance (EPR) spectroscopy to analyze the sorption and desorption of Mn^{2+} onto the surface of birnessite (δ-MnO_2). The EPR stop-flow (EPR-SF) technique has a time resolution of approximately 20 ms, which was sufficient to analyze the sorption reaction, which was complete within ≈ 1 s of reactant mixing (Figure 3.11). Fendorf et al. (1993) determined a first-order rate dependence on Mn^{2+} and determined the forward rate constant for the sorption reaction ($k_f = 3.74 \times 10^{-3}$ s^{-1}) and the reverse reaction rate constant for the desorption reaction ($k_r = 3.08 \times 10^{-4}$ s^{-1}). Although the EPR-SF technique was incredibly powerful for this system, it has the inherent drawback that the reactants, products, or both must be paramagnetic. This study does, however, show that sorption–desorption phenomena can be measured and reaction pathways can be inferred by using appropriate techniques with high time resolution.

One of the more commonly used techniques in the Earth and environmental sciences to assess both sorbent and sorbate phases is generally described by the overarching term *x-ray adsorption spectroscopy* (XAS). More specifically, the use of extended x-ray absorption fine structure spectroscopy (EXAFS) has allowed scientists to study the bonding environments of adsorbed and structural species to ascertain the geometry of complexes at mineral surfaces as well as the structure of three-dimensional phases such as precipitates. EXAFS has been particularly important for studying systems where high-temporal resolution was not necessary, because the typical EXAFS experiment requires collection times on the order of 15 minutes to hours. With the focus of this chapter being on the development and use of techniques capable of measuring rapid kinetics, we focus briefly on the development, and limited use to date, of quick-EXAFS (QEXAFS) and energy-dispersive EXAFS (EDE) (Dent et al., 1999; Pascarelli et al., 1999; Newton et al., 2002; Ressler, 2003). These techniques have seen limited use in the Earth and environmental sciences. However, we believe that this technique will be one of the critical developments in the ability to study the rapid kinetics that describe many chemical systems in the environment.

The general advantage of QEXAFS is the ability to collect an EXAFS spectrum (i.e., a co-addition of scans with a high signal-to-noise ratio) in seconds rather than tens of minutes or even hours. The technology that allows this increase in

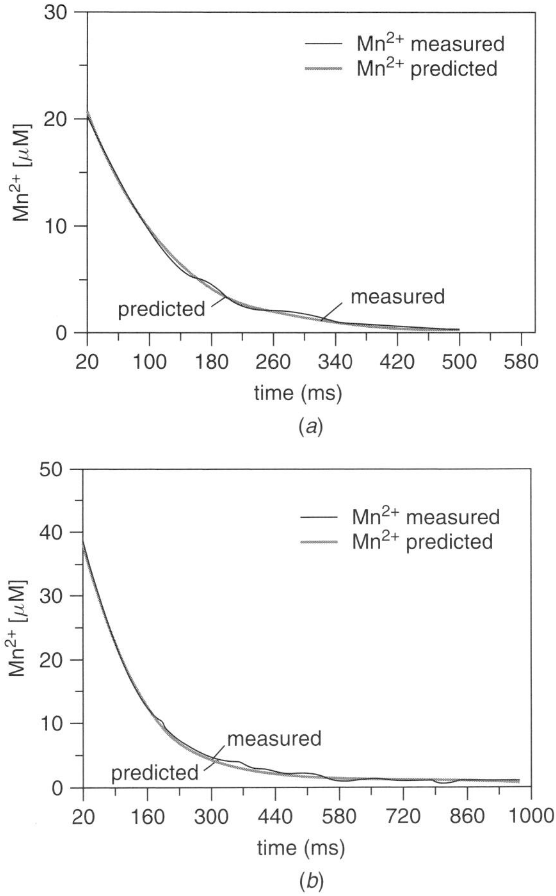

Figure 3.11. Time dependence predicted for Mn^{2+} sorption on δ-MnO_2, using the rate constants determined, compared with the sorption rate curve measured for initial Mn^{2+} concentrations of (*a*) 25 μM and (*b*) 40 μM. (From Fendorf et al., 1993.)

temporal resolution rests in the ability of the QEXAFS setup to scan through the energy range of interest rapidly. A typical EXAFS setup requires a double crystal monochromator which is fixed at a specific position to select for the appropriate x-ray energy. This technology requires that the monochromator be moved physically, set at the appropriate x–y position, and checked through computer feedback, all prior to the collection of a data point for a single energy but one that will make up the overall spectrum. This process in then repeated for the number of steps necessary to describe the energy range of interest and can take several minutes for a single spectrum to be collected. The QEXAFS system takes advantage of moving the monochromator at a constant velocity and therefore eliminates the time loss associated with step scanning (Dent, 2002). Full scan collections of less than 1 minute are common, and it is possible to achieve less than 1-second resolution in some specific systems. Although we do not discuss the EDE technique at length in this chapter, it is important to illustrate the capabilities that this technique may possess in the future. Both EXAFS and QEXAFS are limited by the slow collection times inherent in serial data collection. EDE eliminates this drawback by utilizing a bent polychromator and a position-sensitive detector that allow for parallel data collection on the millisecond to microsecond time scale (Dent, 2002).

Because the same energy can be used to collect QEXAFS spectra and x-ray diffraction (QXRD), the two have become an incredibly powerful tool for doing high-time resolution studies. Although to the best of the authors' knowledge, there are no studies directly related to Earth and environmental sciences that utilize this technique, it is appropriate to show an example from the catalysis literature illustrating the technique's potential.

Using coupled QEXAFS and QXRD, Sankar et al. (1993) studied the production of cordierite ($Mg_2Al_4Si_5O_{18}$) from a Mg-exchanged zeolite using a ZnO catalyst (Figure 3.12). Sankar et al. (1993) were able to track the changes during synthesis at high temporal resolution over a wide temperature range showing

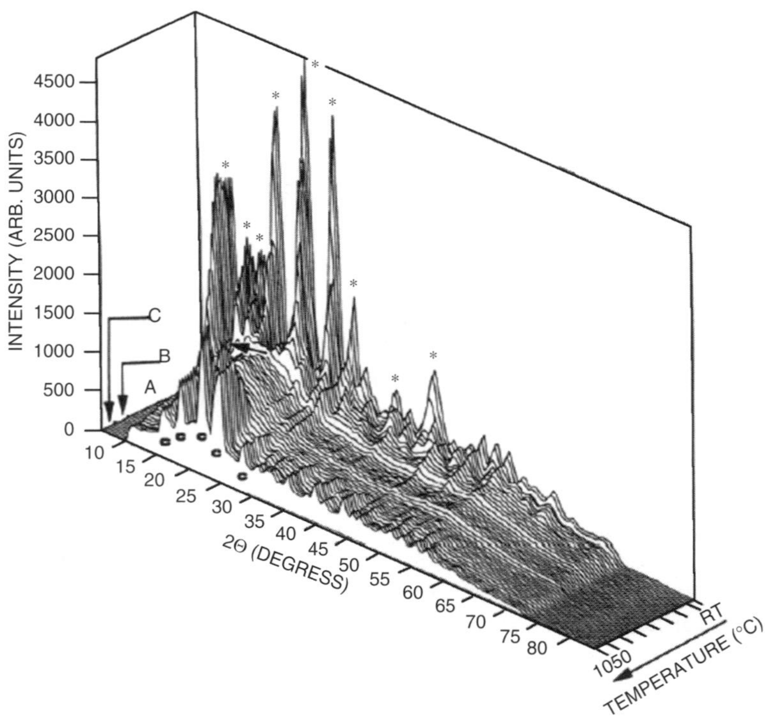

Figure 3.12. Combined QEXAFS/XRD following the formation of cordierite from Mg-exchanged zeolite. The major peak of stuffed quartz (region B) is indicated by an arrow. Characteristic peaks from the hexagonal form of cordierite (region C) are labeled "c." (From Sankar et al., 1993.)

the progression from the starting phase (Mg-exchanged zeolite) though transition phases (dehydrated zeolite and amorphous phases) to a final product (cordierite) and ultimately showing how the presence of ZnO affects the overall process (formation of cordierite 100°C lower than without a catalyst). With increased use we believe that this technique will find a place in the environmental sciences and allow scientists to study the reaction kinetics of sorption processes that have not yet been investigated adequately.

The use of infrared spectroscopy in the Earth and environmental sciences has been widespread for decades; however, until development of the attenuated total reflectance (ATR) technique, the primary use was ex situ material characterization (Chen and Gardella, 1998; Tejedor-Tejedor et al., 1998; Degenhardt and McQuillan, 1999; Peak et al., 1999; Wijnja and Schulthess, 1999; Arai and Sparks, 2001; Kirwan et al., 2003). For the study of environmental systems, the strength of the ATR–Fourier transform infrared (FTIR) technique lies in its intrinsic surface sensitivity. Spectra are collected only from absorptions of an evanescent wave with a maximum penetration depth of several micrometers from the internal reflection element into the solution phase (Harrick, 1967). This short optical path length allows one to overcome any absorption due to an aqueous phase associated with the sample while maintaining a high sensitivity to species at the mineral–water interface (McQuillan, 2001). Therefore, ATR–FTIR represents a technique capable of performing in situ spectroscopic studies in real time.

Recent developments in the use of rapid mirror motion as well as more efficient data computation have allowed FTIR spectra collection to achieve 10-ms resolution without the use of step-scan spectroscopy. This means that a full mid-IR spectrum (4000 to 400 cm^{-1}) can be collected in $\approx$ 10 ms. A relatively high signal-to-noise ratio can be achieved by co-adding 32 scans, thus producing a high-quality spectrum in 320 ms. The technological strategy that makes this possible is the rapid movement of the Michelson interferometer's moving mirror, which is made possible by detectors with rapid response time. Furthermore, SERIES software (Thermo Electron) has been developed which allows for the collection of interferograms without computational processing during data collection. All of the fast Fourier transforms are computed after the experiment has been run, saving significant time during data collection.

To the best of the authors' knowledge, there have not been any reports in the peer-reviewed literature using rapid-scan technology for studies in the Earth and environmental sciences. We therefore draw on two examples, one from the polymer literature and one from studies currently being performed in the Environmental Soil Chemistry Research Group at the University of Delaware.

To investigate the kinetics that control the rate of network connection of a highly cross-linked photopolymer system, Lovell et al. (2001) utilized rapid scan near-infrared (NIR) spectroscopy to study the polymerization of a dimethylacrylate dental resin. The research exploited the Thermo Electron rapid-scan capabilities to analyze the system with a time resolution of $\approx$ 30 ms. This was sufficiently faster than traditional techniques, which required data collection at the 2-second time scale and would thus miss the reaction of interest that reacts to completion within 10 seconds. The study concludes that by using rapid-scan NIR it is possible to study dental resin polymerization kinetics accurately, including complex network formation in mixed polymer systems.

At the University of Delaware, research is currently being performed to understand the fate and toxicity of arsenic (As) in natural soil systems, particularly those affected by poultry litter fertilization. The fundamental chemistry of interest pertains to the natural attenuation of As in soils by oxidation of As(III) to the less mobile and toxic As(V) form. This redox chemistry is thought to take place primarily at the surface of reactive manganese (Mn) oxide minerals [e.g., birnessite (δ-MnO_2)]. Because the prevalent As(V) species is the oxyanion arsenate ($H_2AsO_4^-$/$HAsO_4^{2-}$, depending on the pH value of the system), it is particularly useful to probe the process of the oxidation reaction using ATR–FTIR. By using the emergence and growth of the arsenate infrared absorbance peak as the progress variable for the oxidation reaction and plotting its intensity versus time, it is possible to obtain reaction-rate information (Figure 3.13). Studies are currently being performed using rapid-scan technology in an attempt to probe this reaction with high time resolution to try to establish the pathways involved in the oxidation of As(III) to As(V), the influence of different Mn oxide mineral types, the role of different Mn valence states at the mineral surface, as well as the effect of competing aqueous species which are present in natural soil systems [e.g., species such as sulfate (SO_4^{2-}), phosphate ($H_2PO_4^-$/HPO_4^{2-}, depending on the pH value of the system), and low-molecular-mass organic acids].

Because developments in studying the sorption–desorption process will surely advance from both techniques, which allow for the collection of rapid kinetic data and also the ability to perform experiments in situ with high sensitivity to the solid–solution interface, it is appropriate to review some of the developments over the past decade in the use of optical second-harmonic-generation (SHG) spectroscopy. Briefly, this technique utilizes a second-order nonlinear optical process that probes interfaces directly because SHG is symmetry forbidden (i.e., does not occur when there is symmetry in a system) in a medium with inversion symmetry (symmetrical behavior based on a mirror plane). The property rests on the inability of centrosymmetric meterials (i.e., materials having bulk inversion symmetry) to produce SHG, while noncentrosymmetric interfaces (i.e., interfaces between liquid and solid, solid and gas, gas and liquid), where bulk inversion symmetry is necessarily lost, are capable of generating the second harmonic effect (Shen, 1989). Therefore, this technique is a surface-sensitive probe capable of analyzing interfaces even in systems that have an overwhelming bulk contribution (Kim et al., 2003). Some of the most obvious advantages for using SHG include high surface sensitivity (detection of submonolayer coverage at interfaces), applicability to any system accessible by light, nondestructive nature, and high spatial and temporal (picosecond to subpicosecond) resolution (Shen, 1989).

Only a few studies have used SHG to study environmental systems, with the work by Mifflin et al. (2003) being one of the best examples. Mifflin et al. (2003) investigated the kinetics involved in the adsorption and desorption of chromate to

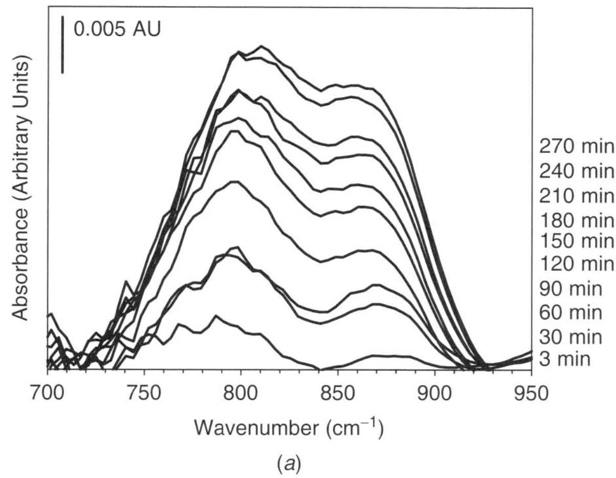

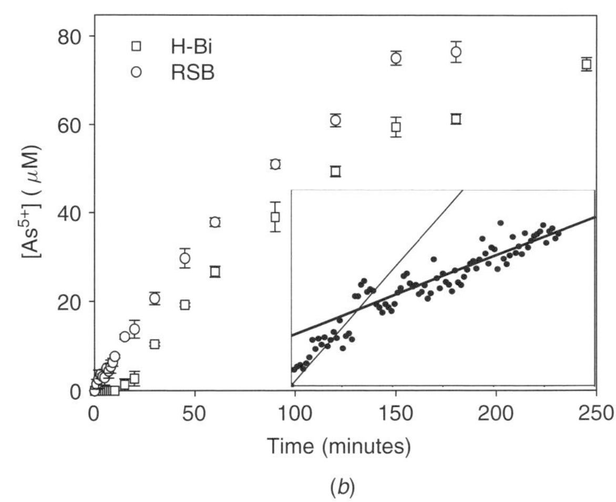

Figure 3.13. (*a*) Time-resolved attenuated total reflectance (ATR) Fourier transform infrared (FTIR) spectra collected during arsenic (As) oxidation on random stacked birnessite (RSB). Peaks represent the oxidation product, arsenate, adsorbed at the RSB surface. (*b*) As oxidation kinetic data collected on RSB (○) on hexagonal birnessite (H-Bi) (□) during a batch experiment. Inset shows the peak height versus time plot for the spectra seen in the top panel, illustrating the higher time resolution achievable with rapid-scan ATR-FTIR spectroscopy. (From Borda and Sparks, unpublished data, 2006.)

fused quartz under aqueous conditions (Figure 3.14). Fused silica was chosen as a proxy to model the mobility of chromate in soil systems that are rich in silica phases. The study concludes that SHG is an acceptable technique for studying this system and provides an adsorption rate constant [(3)(1) $\times$ 10^3 s^{-1} M^{-1}], a desorption rate constant [(0.9)(7) $\times$ 10^{-3} s^{-1}], standard free energy of chromate adsorption (32 kJ mol^{-1}) and a corresponding chromate binding constant (3.3 $\times$

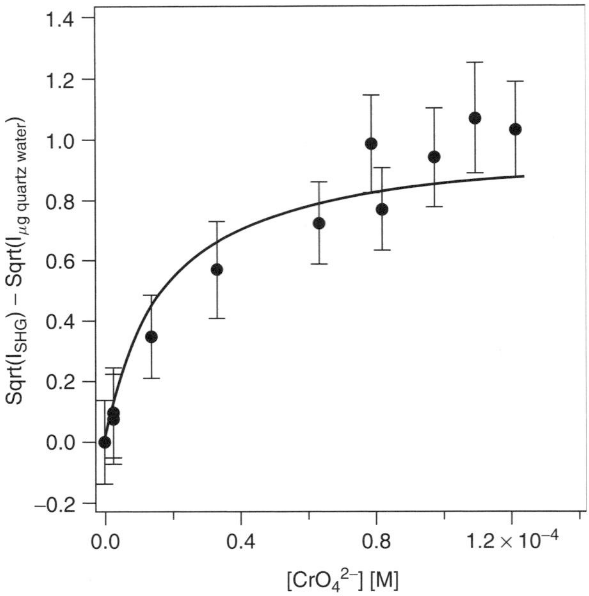

Figure 3.14. Adsorption isotherm of CrO_4^{2-} at the quartz–water interface at pH 4 bulk solution. $T = 300$ K, $\lambda_{SHG} = 290$ nm. (From Mifflin et al., 2003; copyright © 2003 American Chemical Society.)

10^{-5} M^{-1}) (Mifflin et al., 2003). The authors suggest that working on a more natural material will yield more appropriate results for natural systems; however, these initial findings strongly support previous work that showed chromate to be highly mobile in most soil environments.

3.5. CONCLUSIONS AND FUTURE DIRECTIONS

As the Earth and environmental science communities develop a new view of sorption processes centered on the concept of a continuum of sorption reactions, it will be necessary that the experimental techniques that are used assess these processes accurately. The authors of this chapter believe strongly that the evolution of techniques with high temporal and spatial resolution represents the future of experimental work in the Earth and environmental sciences and, more important, represent the cutting edge of scientific discovery.

Because it is widely recognized that a significant amount of reactivity takes places rapidly during sorption processes, it has become necessary to employ new experimental techniques that can adequately measure these processes with high temporal sensitivity. Techniques such as rapid-scan infrared spectroscopy and quick-EXAFS have shown great promise in the early stages of their development and application in our field. It will be over the next decade(s) that these techniques, as well as others, will attain their full potential and that significant findings will be made regarding the rapid kinetics that control adsorption at reactive mineral surfaces. This will open the door for increased understanding of reaction mechanisms and the development of more specific and productive models to predict the fate of important species in natural systems.

Because it is not only the ability to investigate rapid processes that will allow for a greater understanding of sorption processes, we must also look toward techniques that achieve high spatial resolution. Over the past number of years, it has been the development and application of techniques with high spatial resolution that have shown us the incredible heterogeneity that exists on small scales within natural materials. This is particularly the case with natural soil materials, where the composition, both mineralogically and microbiologically, varies on the order of nanometers and perhaps even angstroms. The recent push to investigate reactivity in the "critical zone" again illustrates the need for scientists working in the Earth and environmental sciences to adopt techniques that allow them to gain insight about reactivity, and changes in reactivity, on very small scales. The critical zone includes a number of important reactive interfaces, including, but not limited to, the plant–soil, mineral–water, mineral–microbe, and air–mineral interfaces. The richness of chemistry that is occurring in this zone, which in any one system may be on the order of millimeters, can only be investigated by techniques that allow scientists to study these important interfaces directly (in situ) and with spatial and temporal precision.

Finally, when one talks about the Earth and environmental sciences, particularly if one talks about the study of the critical zone, it is crucial to discuss the importance of collaborative research. Because natural systems involve the foci of many different disciplines and the important chemistry that drives these systems exists at the interfaces between them, it is imperative that scientists from a number of backgrounds work together to understand these complex systems. Scientists from differing fields are commonly equipped with differing tools to investigate systems of interest. To effectively understand something as complicated and interconnected as a natural system will require all of the tools and expertise that are available. There is a very bright future for Earth and environmental research, and the authors of this chapter believe it is only the beginning of a push to understand the complexity and importance of the chemistry that controls the fate and mobility of contaminants and nutrients in natural systems.

REFERENCES

Ainsworth, C. C., Pilou, J. L., Gassman, P. L., and Van Der Sluys, W. G. (1994). Cobalt, cadmium, and lead sorption to hydrous iron oxide: residence time effect. *Soil Sci. Soc. Am. J.* **58**, 1615–1623.

Alexander, M. (2000). Aging, bioavailability, and overestimation of risk from environmental pollutants. *Environ. Sci. Technol.* **34**, 4259–4265.

Allada, R. K., Peltier, E., Navrotsky, A., Casey, W. H., Johnson, A., Thompson-Berbeco, H., and Sparks, D. L. (2005). Calorimetric determination of the enthalpies of formation of hydrotalcite-like solids and their use in the geochemical modeling of metals in natural waters. *Clays Clay Miner.* **54**, 409–417.

Allmann, R. (1970). Double layer structures with layer ions $[Me(II)_{1-x}\ Me(III)_x\ (OH)_2]^{x+}$ of brucite type. *Chimia* **24**, 99–108.

Amacher, M. C. (1991). Methods of obtaining and analyzing kinetic data. In *Rates of Soil Chemical Processes*, ed. Sparks, D. L., and Suarez, D. L., Soil Science Society of America, Madison, WI, 19–59.

Arai, Y., and Sparks, D. L. (2001). ATR-FTIR spectroscopic investigation on phosphate adsorption mechanisms at the ferrihydrite–water interface. *J. Colloid Interface Sci.* **241**, 317–326.

Bargar, J. R., Brown, G. E., and Parks, G. A. (1995). XAFS study of Pb(II) sorption at the α-Al_2O_3–water interface. *Physica B* **209**, 455–456.

Barrow, N. J. (1986). Testing a mechanistic model: II. The effects of time and temperature on the reaction of zinc with a soil. *J. Soil Sci.* **37**, 277–286.

Bartal, A., Sparks, D. L., Pesek, J. D., and Feigenbaum, S. (1990). Analyses of adsorption-kinetics using a stirred-flow chamber, 1: Theory and critical tests. *Soil Sci. Soc. Am. J.* **54**, 1273–1278.

Bleam, W. F., and McBride, M. B. (1985). Cluster formation versus isolated-site adsorption: a study of Mn(Ii) and Mg(Ii) adsorption on Boehmite and Goethite. *J. Colloid Interface Sci.* **103**, 124–132.

Bleam, W. F., and McBride, M. B. (1986). The chemistry of adsorbed Cu(Ii) and Mn(Ii) in queous titanium-dioxide suspensions. *J. Colloid Interface Sci.* **110**, 335–346.

Bruemmer, G. W., Gerth, J., and Tiller, K. G. (1988). Reaction kinetics of the adsorption and desorption of nickel, zinc and cadmium by goethite, I: Adsorption and diffusion of metals. *J. Soil Sci.* **39**, 37–52.

Bunzl, K., Schmidt, W., and Sansoni, B. (1976). Kinetics of ion exchange in soil organic matter, IV: Adsorption and desorption of Pb^{2+}, Cu^{2+}, Zn^{2+}, and Ca^{2+} by peat. *J. Soil Sci.* **27**, 32–41.

Charlet, L., and Manceau, A. (1992). X-ray absorption spectroscopic study of the sorption of Cr(III) at the oxide–water interface, II: Adsorption, coprecipitation and surface precipitation on ferric hydrous oxides. *J. Colloid Interface Sci.* **148**, 443–458.

Chen, J. X., and Gardella, J. A. (1998). Quantitative ATR FT-IR analysis of surface segregation of polymer blends of polystyrene/poly(dimethylsiloxane)-co-polystyrene. *Appl. Spectrosc.* **52**, 361–366.

Chisholm-Brause, C. J., O'Day, P. A., Brown, G. E., Jr., and Parks, G. A. (1990a). Evidence for multinuclear metal-ion complexes at solid/water interfaces from x-ray absorption spectroscopy. *Nature* **348**, 528–530.

Chisholm-Brause, C. J., Roe, A. L., Hayes, K. F., Brown, G. E., Jr., Parks, G. A., and Leckie, J. O. (1990b). Spectroscopic investigation of Pb(II) complexes at the γ-Al_2O_3/water interface. *Geochim. Cosmochim. Acta* **54**, 1897–1909.

Degenhardt, J., and McQuillan, A. J. (1999). In situ ATR-FTIR spectroscopic study of adsorption of perchlorate, sulfate, and thiosulfate ions onto chromium(III) oxide hydroxide thin. *Langmuir* **15**, 4595–4602.

Dent, A. J. (2002). Development of time-resolved XAFS instrumentation for quick EXAFS and energy-dispersive EXAFS measurements on catalyst systems. *Top. Catal.* **18**, 27–35.

Dent, A., Evans, J., Newton, M., Corker, J., Russell, A., Abdul Rahman, M. B., Fiddy, S., Mathew, R., Farrow, R., Salvini, G., and Atkinson, P. (1999). High-quality energy-dispersive XAFS on the 1 s timescale applied to electrochemical and catalyst systems. *J. Sync. Rad.* **6**, 381–383.

Eick, M. J., Bartal, A., Sparks, D. L., and Feigenbaum, S. (1990). Analyses of adsorption-kinetics using a stirred-flow chamber, 2: Potassium calcium exchange on clay-minerals. *Soil Sci. Soc. Am. J.* **54**, 1278–1282.

Elzinga, E. J., and Sparks, D. L. (1999). Nickel sorption mechanisms in a pyrophyllite-montmorillonite mixture. *J. Colloid Interface Sci.* **213**, 506–512.

Farley, K. J., Dzombak, D. A., and Morel, F. M. M. (1985). A surface precipitation model for the sorption of cations on metal oxides. *J. Colloid Interface Sci.* **106**, 226–242.

Fendorf, S. E., and Sparks, D. L. (1994). Mechanisms of chromium(III) sorption on silica, 2: Effect of reaction conditions. *Environ. Sci. Technol.* **28**, 290–297.

Fendorf, S. E., Sparks, D. L., Franz, J. A., and Camaioni, D. M. (1993). Electron-paramagnetic resonance stopped-flow kinetic-study of manganese(II) sorption desorption on Birnessite. *Soil Sci. Soc. Am. J.* **57**, 57–62.

Fendorf, S. E., Lamble, G. M., Stapleton, M. G., Kelley, M. J., and Sparks, D. L. (1994). Mechanisms of chromium(III) sorption on silica, 1: Cr(III) surface structure derived by extended x-ray absorption fine structure spectroscopy. *Environ. Sci. Technol.* **28**, 284–289.

Ford, R. G., and Sparks, D. L. (2000). The nature of Zn precipitates formed in the presence of pyrophyllite. *Environ. Sci. Technol.* **34**, 2479–2483.

Ford, R. G., Scheinost, A. C., and Sparks, D. L. (2001). Frontiers in metal sorption/precipitation mechanisms on soil mineral surfaces. In *Advances in Agronomy*, ed. Sparks, D. L., Academic Press, San Diego, CA, 41–62.

Fuller, C. C., Davis, J. A., and Waychunas, G. A. (1993). Surface chemistry of ferrihydride, 2: Kinetics of arsenate adsorption and coprecipitation. *Geochim. Cosmochim. Acta*, **57**, 2271–2282.

Grossl, P. R., Eick, M. J., Sparks, D. L., Goldberg, S., and Ainsworth, C. C. (1997). Arsenate and chromate retention mechanisms on goethite, 2: Kinetic evaluation using a pressure-jump relaxation technique. *Environ. Sci. Technol.* **31**, 321–326.

Harrick, N. J. (1967). *Internal Reflectance Spectroscopy*, Wiley, New York.

Juillot, F., Morin, G., Ildefonse, P., Trainor, T. P., Benedetti, M., Galoisy, L., Calas, G., and Brown, G. E., Jr. (2003). Occurrence of Zn/Al hydrotalcite in smelter-impacted soils from northern France: evidence from EXAFS spectroscopy and chemical extractions. *Am. Mineral.* **88**, 509–526.

Kim, J., Chou, K. C., and Somorjai, G. A. (2003). Structure and dynamics of acetonitrile at the air/liquid interface of binary solutions studied by infrared–visible sum frequency generation. *J. Phys. Chem. B* **107**, 1592–1596.

Kirwan, L. J., Fawell, P. D., and van Bronswijk, W. (2003). In situ FTIR-ATR examination of poly(acrylic acid) adsorbed onto hematite at low pH. *Langmuir* **19**, 5802–5807.

Lovell, L. G., Berchtold, K. A., Elliott, J. E., Lu, H., and Bowman, C. N. (2001). Understanding the kinetics and network formation of dimethacrylate dental resins. *Polym. Adv. Technol.* **12**, 335–345.

Lövgren, L., Sjöberg, S., and Schindler, P. W. (1990). Acid/base reactions and Al(III) complexation at the surface of goethite. *Geochim. Cosmochim. Acta* **54**, 1301–1306.

Martin-Garin, A., Van Cappellen, P., and Charlet, L. (2003). Aqueous cadmium uptake by calcite: A stirred flow-through reactor study. *Geochim. Cosmochim. Acta* **67**, 2763–2774.

McBride, M. B. (1991). Processes of heavy and transition metal sorption by soil minerals. In *Interactions at the Soil Colloid–Soil Solution Interface*, ed. Bolt, G. H., Boodt, M. F. D., Hayes, M. H. B., and McBride, M. B., Kluwer Academic, Dordrecht, The Netherlands, 149–176.

McBride, M. B. (1994). *Environmental Chemistry of Soils*, Oxford University Press, New York.

McQuillan, A. J. (2001). Probing solid–solution interfacial chemistry with ATR-IR spectroscopy of particle films. *Adv. Mater.* **13**, 1034–1038.

Mifflin, A. L., Gerth, K. A., and Geiger, F. M. (2003). Kinetics of chromate adsorption and desorption at fused quartz/water interfaces studied by second harmonic generation. *J. Phys. Chem. A* **107**, 9620–9627.

Nachtegaal, M., Marcus, M. A., Sonke, J. E., Vangronsveld, J., Livi, K., Van Der Lelie, D., and Sparks, D. L. (2005). Effects of in situ remediation on the speciation, and bioavailability, of zinc in a smelter contaminated soil. *Geochim. Cosmochim. Acta* **69**, 4649–4664.

Newton, M. A., Dent, A. J., and Evans, J. (2002). Bringing time resolution to EXAFS: recent developments and application to chemical systems. *Chem. Soc. Rev.* **31**, 83–95.

O'Day, P. A., Brown, G. E., and Parks, G. A. (1994). X-ray absorption spectroscopy of cobalt(II) multinuclear surface complexes and surface precipitates on kaolinite. *J. Colloid Interface Sci.* **165**, 269–289.

Papelis, C., Brown, G. E., Jr., Parks, G. A., and Leckie, J. O. (1996). X-ray absorption studies of cadmium and selenate absorption on aluminum oxide. *Langmuir* **11**, 2041–2048.

Pascarelli, S., Neisius, T., and De Panfilis, S. (1999). Turbo-XAS: dispersive XAS using sequential acquisition. *J. Sync. Rad.* **6**, 1044–1050.

Peak, D., Ford, R. G., and Sparks, D. L. (1999). An in situ ATR-FTIR investigation of sulfate bonding mechanisms on goethite. *J. Colloid Interface Sci.* **218**, 289–299.

Peltier, E., Allada, R., Navrotsky, A., and Sparks, D. L. (2005). Nickel solubility and precipitation in soils: a thermodynamic study. *Clays Clay Miner.* **54**, 153–163.

Pignatello, J. J. (2000). The measurement and interpretation of sorption and desorption rates for organic compounds in soil media. In *Advances in Agronomy*, ed. Sparks, D. L., Academic Press, San Diego, CA, 1–73.

Ressler, T. (2003). Application of time-resolved in-situ x-ray absorption spectroscopy in solid-state chemistry. *Anal. Bioanal. Chem.* **376**, 584–593.

Roberts, D. R., Scheidegger, A. M., and Sparks, D. L. (1999). Kinetics of mixed Ni–Al precipitate formation on a soil clay fraction. *Environ. Sci. Technol.* **33**, 3749–3754.

Roe, A. L., Hayes, K. F., Chisholm-Brause, C. J., Brown, G. E., Jr., Parks, G. A., Hodgson, K. O., and Leckie, J. O. (1991). In situ x-ray absorption study of lead ion surface complexes at the goethite/water interface. *Langmuir* **7**, 367–373.

Sankar, G., Wright, P. A., Natarajan, S., Thomas, J. M., Greaves, G. N., Dent, A. J., Dobson, B. R., Ramsdale, C. A., and Jones, R. H. (1993). Combined Quexafs-Xrd: a new technique in high-temperature materials chemistry—an illustrative in-situ study of the zinc oxide–enhanced solid-state production of Cordierite from a precursor Zeolite. *J. Phys. Chem.* **97**, 9550–9554.

Scheckel, K. G., and Sparks, D. L. (2001). Dissolution kinetics of nickel surface precipitates on clay mineral and oxide surfaces. *Soil Sci. Soc. Am. J.* **65**, 685–694.

Scheidegger, A. M., and Sparks, D. L. (1996). Kinetics of the formation and the dissolution of nickel surface precipitates on pyrophyllite. *Chem. Geol.* **132**, 157–164.

Scheidegger, A. M., Lamble, G. M., and Sparks, D. L. (1996). Investigation of Ni sorption on pyrophyllite: an XAFS study. *Environ. Sci. Technol.* **30**, 548–554.

Scheidegger, A. M., Lamble, G. M., and Sparks, D. L. (1997a). The kinetics of nickel sorption on pyrophyllite as monitored by x-ray absorption fine structure (XAFS) spectroscopy. *J. Phys. IV* **4**, C2-773 to C2-775.

Scheidegger, A. M., Lamble, G. M., and Sparks, D. L. (1997b). Spectroscopic evidence for the formation of mixed-cation, hydroxide phases upon metal sorption on clays and aluminum oxides. *J. Colloid Interface Sci.* **186**, 118–128.

Scheidegger, A. M., Strawn, D. G., Lamble, G. M., and Sparks, D. L. (1998). The kinetics of mixed Ni–Al hydroxide formation on clay and aluminum oxide minerals: a time-resolved XAFS study. *Geochim. Cosmochim. Acta* **62**, 2233–2245.

Scheinost, A. C., Ford, R. G., and Sparks, D. L. (1999). The role of Al in the formation of secondary Ni precipitates on pyrophyllite, gibbsite, talc, and amorphous silica: a DRS study. *Geochim. Cosmochim. Acta* **63**, 3193–3203.

Shen, Y. R. (1989). Optical 2nd harmonic-generation at interfaces. *Annu. Rev. Phys. Chem.* **40**, 327–350.

Sparks, D. L. (1989). *Kinetics of Soil Chemical Processes*, Academic Press, San Diego, CA.

Sparks, D. L. (1999). Kinetics of soil chemical phenomena: future directions. In *Future Prospects for Soil Chemistry*, ed. Huang, P. M., Sparks, D. L., and Boyd, S. A., Soil Science Society of America, Madison, WI, 81–102.

Sparks, D. L. (2002). *Environmental Soil Chemistry*, Vol. 2, Academic Press, San Diego, CA, 352.

Sparks, D. L. (2005). Metal and oxyanion sorption on naturally occurring oxide and clay mineral surfaces. In *Environmental Catalysis*, ed. Grassian, V. H., Taylor & Francis, Boca Raton, FL, 3–36.

Sparks, D. L., and Zhang, P. C. (1991). Relaxation methods for studying kinetics of soil chemical phenomena. In *Rates of Soil Chemical Processes*, ed. Sparks, D. L., and Suarez, D. L., Soil Science Society of America, Madison, WI, 61–94.

Sparks, D. L., Fendorf, M., Toner, C. V., IV, and Carski, T. H. (1996). Kinetic methods and measurements. In *Methods of Soil Analysis: Chemical Methods*, ed. Sparks, D. L., Soil Science Society of America, Madison, WI, 1275–1307.

Sposito, G. (1984). *The Surface Chemistry of Soils*, Oxford University Press, New York.

Sposito, G. (1986). Distinguishing adsorption from surface precipitation. In *Geochemical Processes at Mineral Surfaces*, ed. Davis, J. A., and Hayes, K. F., American Chemical Society, Washington, DC, 217–229.

Sposito, G. (1989). *The Chemistry of Soils*, Oxford University Press, New York.

Strawn, D. G., and Sparks, D. L. (1999). The use of XAFS to distinguish between inner- and outer-sphere lead adsorption complexes on montmorillonite. *J. Colloid Interface Sci.* **216**, 257–269.

Strawn, D. G., and Sparks, D. L. (2000). Effects of soil organic matter on the kinetics and mechanisms of Pb(II) sorption and desorption in soil. *Soil Sci. Soc. Am. J.* **64**, 144–156.

Stumm, W., and Morgan, J. J. (1981). *Aquatic Chemistry*, Wiley-Interscience, New York, 780.

Taylor, R. M. (1984). The rapid formation of crystalline double hydroxy salts and other compounds by controlled hydrolysis. *Clay Miner.* **19**, 591–603.

Tejedor-Tejedor, M. I., Paredes, L., and Anderson, M. A. (1998). Evaluation of ATR-FTIR spectroscopy as an "in situ" tool for following the hydrolysis and condensation of alkoxysilanes under rich H_2O conditions. *Chem. Mater.* **10**, 3410–3421.

Thompson, H. A., Parks, G. A., and Brown, G. E., Jr., (1999a). Ambient-temperature synthesis, evolution, and characterization of cobalt–aluminum hydrotalcite-like solids. *Clays Clay Miner.* **47**, 425–438.

Thompson, H. A., Parks, G. A., and Brown, G. E., Jr., (1999b). Dynamic interactions of dissolution, surface adsorption, and precipitation in an aging cobalt(II)–clay–water system. *Geochim. Cosmochim. Acta* **63**, 1767–1779.

Towle, S. N., Bargar, J., Brown, G. E., Jr., and Parks, G. A. (1997). Surface precipitation of Co(II) (aq) on Al_2O_3. *J. Colloid Interface Sci.* **187**, 62–82.

Trivedi, P., Axe, L., and Tyson, T. A. (2002). Understanding diffusion mechanisms in microporous iron oxides with x-ray absorption spectroscopy. *Abstr. Pap. Am. Chem. Soc.* **223**, U604–U604.

Vasudevan, D., Cooper, E. M., and Van Exem, O. L. (2002). Sorption-desorption of lonogenic compounds at the mineral–water interface: study of metal oxide–rich soils and pure-phase minerals. *Environ. Sci. Technol.* **36**, 501–511.

Wijnja, H., and Schulthess, C. P. (1999). ATR-FTIR and DRIFT spectroscopy of carbonate species at the aged gamma-Al_2O_3/water interface. *Spectrochim. Acta A*, **55**, 861–872.

Yamaguchi, N. U., Scheinost, A. C., and Sparks, D. L. (2001). Surface-induced nickel hydroxide precipitation in the presence of citrate and salicylate. *Soil Sci. Soc. Am. J.* **65**, 729–736.

4

SPECTROSCOPIC TECHNIQUES FOR STUDYING METAL–HUMIC COMPLEXES IN SOIL

N. SENESI AND E. LOFFREDO

Università di Bari, Bari, Italy

Biophysico-Chemical Processes of Heavy Metals and Metalloids in Soil Environments,
Edited by Antonio Violante, Pan Ming Huang, and Geoffrey Michael Gadd

4.1. INTRODUCTION

The distribution and behavior of metal ions in soils is extremely complex and is governed by a variety of reactions, including complexation with organic and inorganic compounds, ion exchange, adsorption and desorption processes, precipitation and dissolution of solids, and acid–base equilibria. In particular, humic substances (HSs) [i.e., humic acids (HAs), fulvic acids (FAs), and humins], which represent the major (between 70 and 80%) natural organic components in soils, are able to interact with metal ions to form water-soluble, colloidal, and water-insoluble complexes of varying properties and widely differing chemical and biological stabilities (Schnitzer, 1978; Stevenson, 1994; Senesi and Loffredo, 2005). Accumulated evidence suggests that most processes in which metals are involved in soils, including mobility and transport, fixation and accumulation, chemical and biochemical reactivity, bioavailability, and toxicity, are affected by their interaction with HSs. The topic is of considerable practical interest also because of the continuous and increasing release of various heavy metals to soil by numerous modern agricultural practices.

Metal–HS complexation in soil is of practical and theoretical significance for several reasons. For example, chemical weathering of rocks and minerals, and related soil genesis and evolution, are known to involve metal complexation reactions with HSs. The bioavailability of several metal ions, especially trace elements, is strongly influenced by complexation with the soluble and insoluble fractions of HSs. For instance, at pH values commonly found in soils, nutrient metal ions that would ordinarily be converted to insoluble forms may be maintained in solution by complexation with HSs, thus increasing their bioavailability. On the contrary, the concentration of a toxic metal ion may be reduced to a nontoxic level through complexation to insoluble fractions of soil HSs, whereas soluble HSs complexes can function as metal carriers in the transport to ground- and surface-water bodies, thereby rendering the water unfit for several uses. Further, complexation of metal ions by HSs may represent a challenging problem for their analytical determination in soil samples.

Humic substances contain a large number of complexing sites per molecule, and thus behave as a natural "multiligand" complexing system that is distinguished from "simple" ligands such as inorganic anions (Buffle, 1988). The most common stoichiometries assessed are HS : metal=1 : 1 and 2 : 1, often involving the formation of metal chelates. Aromatic carboxyls and phenolic groups play a prominent role in the 1 : 1 binding of metal ions by soil HAs and FAs, by forming chelates that involve either two COOH groups in a phtalate-type site or one phenolic group and one COOH group in a salicylate-type site. Nonaromatic carboxyl and hydroxyl sites, including pyruvate- and glycolate-type sites and carbohydrate moieties, may also be involved in metal-ion binding by soil HAs and FAs. Other possible combinations involve two phenolic OH, quinone, NH_2, and sulfydryl groups, and conjugated ketonic structures (Stevenson, 1976a; Chen and Stevenson, 1986; Senesi and Loffredo, 2005). Metal ions may also coordinate with ligands belonging to two (or more) HS molecules, forming 2 : 1 complexes and/or chelates, and eventually producing chain structures that may result in the aggregation and precipitation as the chain grows at high metal-to-HS ratios (Stevenson, 1976a).

Two main types of complexes may be formed between metal ions and HSs: (1) inner-sphere complexes, resulting in the formation of bonds with some covalent character between the ligand atom(s) and the metal ion, both completely or partially dehydrated; and (2) outer-sphere complexes, resulting in electrostatic attraction between the ligand(s) and the metal ion that remains completely hydrated.

The electronic and steric environment of the ligand site, such as its chemical network, geometry, and conformation, and the physical and chemical characteristics of the surrounding medium, including pH, ionic strength, and metal concentration, can exert a marked influence on the overall complexation process (Buffle, 1988). In a given HS macromolecule, identical coordinating groups can be bound to different types of aliphatic chains and aromatic rings of various structures, which can exert differing electronic effects. The steric microenvironment of the binding site, particularly its size, will depend on the geometry, steric conformation, and flexibility of the entire complexant molecule. This is influenced by the formation disruption of hydrogen bonds and metal bridges, which can vary with pH, ionic strength, and concentration of the metal to be complexed. The hydration of hydrophilic sites and electrostatic effects (i.e., the electric field determined by the extent of ionization of major acidic complexing groups) can also influence the formation process and stability of complexes (Buffle, 1988). The relative importance of these different effects varies with the degree of site occupation by metal cations and represents the fundamental difference between HS complexants and simple ligands.

4.2. COMPLEXATION CAPACITY

In general, the complexation capacity (CC) of HSs is defined as the capacity of HSs for binding metal ions (Stevenson, 1994). To a good approximation, the CC of HSs may be considered as the weighted average of the complexation capacities of the individual ligands $(CC)_i$ in the system, according to (Perdue, 1989)

$$CC = \frac{\sum (CC)_i [\text{weight}]_i}{\sum [\text{weight}]_i} \tag{4.1}$$

where $[\text{weight}]_i$ is a weighting factor related to the relative abundance of the ligand in the multiligand system.

The CC of HSs is a function of various factors, including pH, ionic strength, HS concentration and properties, and nature of the metal ion (e.g., Perdue, 1989; MacCarthy and Perdue, 1991; Stevenson, 1994). In general, the CC of HSs is reported to increase with increasing pH from acidity up to about neutrality, to decrease at higher ionic strength, to increase at higher HS concentrations, and to vary with the nature of the metal ion (Perdue, 1989). For any given pH and ionic strength, trivalent cations are bound in greater amounts than are divalent cations.

The maximum CC of HSs should, however, be regarded as a constant quantity that is characteristic of the HS sample rather than a variable parameter (MacCarthy and Perdue, 1991). According to these authors, the apparent variations of the CC of HSs would, in fact, result from the inability to saturate all the ligand sites under some experimental conditions, and this effect would arise from the variability in the conditional concentration quotient (see Section 4.3) and the influence of dilution on complex formation.

4.3. STABILITY CONSTANTS

The stability constant is probably the most important quantitative parameter for the characterization of a metal–ligand complex in that it provides a numerical index of the affinity of the metal cation for the ligand and allows the development of quantitative models able to predict the speciation of metal ions in the system studied. Several different theoretical and experimental approaches have been attempted for the determination of stability constants of metal–HS complexes and modeling metal–HS complexation reactions. Data analysis and interpretation is, however, still controversial, due to the intrinsically complex and ill-defined nature of HSs. The multiligand, polyelectrolitic nature of HS macromolecules results in the inability to describe quantitatively the types, concentrations, and strengths of the several nonidentical binding sites in HSs and in the impossibility to ascertain and measure the stoichiometry of metal–HS complexation (MacCarthy and Perdue, 1991).

Because of the extreme complexity and wide controversy existing on the subject, only a general overview of the theoretical and experimental approaches is provided in this chapter. For a more detailed information, the reader is referred to the several reviews and book chapters available on the topics of stability constants and conceptual and mathematical models of metal–HS complexation (e.g., Perdue and Lytle, 1983; Sposito, 1986; Stevenson and Fitch, 1986; Perdue, 1989; MacCarthy and Perdue, 1991; Tipping, 1998; Perdue, 2001).

The overall reaction of a single metal ion (M) and a single ligand (L) to form the complex M_mL_n,

$$mM + nL \rightleftharpoons M_mL_n \tag{4.2}$$

can be described by (MacCarthy and Perdue, 1991)

$$K = \frac{\{M_mL_n\}}{\{M\}^m\{L\}^n} = \frac{[M_mL_n]}{[M]^m[L]^n}\frac{\gamma_{M_mL_n}}{(\gamma_M)^m(\gamma_L)^n} = K_c\Gamma \tag{4.3}$$

where K is the true *overall thermodynamic stability (equilibrium) constant*, braces $\{\cdot\}$ and brackets $[\cdot]$ denote activities and concentrations, respectively, γ-values are activity coefficients, K_c is the *concentration quotient*, and Γ is the activity coefficient ratio, which depends on the ionic concentration (ionic strength) of the solution.

Competition from side reactions, especially the acid–base chemistry of both the metal ion and the ligand (i.e., the hydrolysis of the metal ion to produce hydroxy complexes and the protonation of the ligand) is another factor that affects the extent of complexation of M by L and must be given due consideration. In general, protonation of ligands occurs at low pH and hydrolysis of metals occurs at high pH. Thus, the most favorable condition for the complexation of M by L is at intermediate pH values.

Metal–ligand complexation reactions are usually investigated at constant pH and ionic strength using experimental techniques that are able to distinguish the metal–ligand complex from the uncomplexed metal (i.e., from all forms of the ligand that are not bound to the metal ion. Thus, a *conditional stability constant*, or more precisely, a *conditional concentration quotient*, K_c^*, can be written (MacCarthy and Perdue, 1991)

$$\begin{aligned} K_c^* &= \frac{[M_mL_n]}{([M]^m/(\alpha_M)^m)([L]^n/(\alpha_L)^n)} \\ &= (\alpha_M)^m(\alpha_L)^n\frac{[M_mL_n]}{[M]^m[L]^n} = (\alpha_M)^m(\alpha_L)^nK_c \end{aligned} \tag{4.4}$$

where α_M and α_L are the side reaction coefficients of the metal ion and the ligand, respectively. Equation (4) thus describes the formation and stability of the complex relative to all other forms of the metal and ligand that may exist under actual experimental conditions. Further, Eq. (4.4) shows that the conditional concentration quotient, K_c^*, includes a pH-invariant but ionic strength–dependent term, K_c, and pH-dependent terms, α_M and α_L.

Assuming for simplicity a 1 : 1 stoichiometry for all ML_i complexes, the complexation reaction of a single metal ion M with a multiligand system such as HS containing a number of ligands L_i can be described by an *average stability constant*, or, more appropriately, an *average concentration quotient* (MacCarthy and Perdue, 1991):

$$\overline{K}_c = \frac{\sum[ML_i]}{[M]\sum[L_i]} \tag{4.5}$$

where $\sum[ML_i]$ and $\sum[L_i]$ are, respectively, the sum of the concentrations of all 1 : 1 complexes and all uncomplexed ligands in the mixture.

An *average conditional concentration quotient*, also called a *stability function*, $\overline{K}_c^*$, can be defined for the overall binding of a single metal ion to multiligand systems such as HSs (MacCarthy and Perdue, 1991) and calculated directly from experimental data as (Perdue, 1989)

$$\overline{K}_c^* = \frac{C_M - [M]}{[M](C_L - C_M + [M])} \tag{4.6}$$

where C_M and C_L are the total stoichiometric concentrations of the metal and ligand in the system examined and [M] is the concentration of the free metal ion. The value of $(C_M - [M])$ corresponds to the sum of the concentrations of all complexes formed between M and the multiligand system and is calculated by neglecting the presence of inorganic metal complexes. The quantity $(C_L - C_M + [M])$ represents the sum of the concentrations of all binding sites that are not associated with M and is calculated assuming an average 1 : 1 metal-to-ligand stoichiometry for the total of binding sites. Stability functions can also be defined for complexes of stoichiometries other than 1 : 1 but this aspect is not addressed here.

Average $\overline{K}_c^*$ values are ultimately functions of ionic strength, pH, and the degree of saturation of the multiligand system with metal ion. At a given pH and ionic strength, $\overline{K}_c^*$ will decrease steadily as the total metal-to-ligand ratio (C_M/C_L) increases. At low metal-to-ligand ratios, the occurrence of preferential reactions of stronger ligands will determine the functional nature of $\overline{K}_c^*$ (Perdue, 1989).

4.4. MODELING AND MODELS

The large body of field and laboratory data accumulated over the last four decades on the binding of metal ions and protons by HSs has prompted efforts for the development of chemical speciation models that could encapsulate the ample knowledge and information acquired and allow applications to real-world situations. The prevalent effort of any modeling approach is the quantitative description of the relative concentrations and strengths of the many nonidentical binding sites contained in HSs. The various models applied to describe metal–HS complexation can be classified as discrete ligand models and continuous multiligand models.

In the *discrete ligand approach*, only a few ligands or binding sites are required to fit the experimental data. Among several models of this type proposed, a simplified one denoted as Model V has been developed (Tipping and Hurley, 1992), which includes (1) site heterogeneity (i.e., discrete sites with a range of affinities), together with the formation of bidentate sites; (2) electrostatic effects;

and (3) competition among protons and metal ions. Model V, and its predecessor, Model IV, have been used with some success to describe published data sets on a range of metals. More recently, the inadequacies of Model V have been demonstrated and discussed, and an improved model (Model VI), also based on the discrete-site/electrostatic formulation but moved toward a more distributional approach in describing binding sites for metals, has been presented and applied to those data sets analyzed previously with Model V and to additional data sets (Tipping, 1998).

Continuous multiligand distribution models are based on the assumption that a large number of heterogeneous ligands with a range of binding affinities are involved in metal binding with formation of complexes characterized by a continuous (e.g., Gaussian) distribution of the stability constants (Gamble et al., 1980; Perdue and Lytle, 1983; Altman and Buffle, 1988). The initial, noncompetitive Gaussian distribution models that did not account for proton–metal competition were then extended in the competitive Gaussian distribution model (Dobbs et al., 1989). Later, this model was chosen to describe cation binding by HSs in MINTEQA2, a computer program for performing chemical equilibrium calculations (e.g., Allison and Perdue, 1995).

A series of models have been developed that characterize HS binding site heterogeneity by using a continuous distribution of equilibrium constants, together with electrostatic submodels (e.g., Benedetti et al., 1995; Milne et al., 1995; Kinniburgh et al., 1996). The most recent of these models is the consistent NICA (nonideal competitive adsorption)–Donnan model (Kinniburgh et al., 1999), which introduces an additional degree of scaling to the previous NICA–Donnan model, which allows for variable stoichiometry of binding.

Different types of models have been compared in several reviews (e.g., Perdue and Lytle, 1983; Dzombak et al., 1986; Buffle, 1988; Nifant'eva et al., 1999). Sposito (1986) provided a detailed review of a variety of models and illustrated the advantages of conceptual simplicity and calculational convenience of the *quasiparticle model*. This model includes previous concepts and models and describes mathematically an aqueous system containing HS by replacing it with a set of hypothetical, average, noninteracting molecules whose behavior mimics that of the actual HS system. In a recent review (Perdue, 2001), three of the most modern and successful models for the description of competitive proton and metal binding by HSs have been discussed comparatively, focusing on similarities and differences in their conceptual approaches. These are the competitive Gaussian distribution model (Dobbs et al., 1989), Model V (Tipping and Hurley, 1992), and the NICA–Donnan model (Koopal et al., 1994; Kinniburgh et al., 1996).

In conclusion, although the continuous models have some distinct advantages, the literature survey indicates that the most popular are still the discrete models, which offer various types of approaches that enable the explanation of complexation of metal ions by HSs.

4.5. ULTRAVIOLET–VISIBLE SPECTROSCOPY

In the ultraviolet (UV, 200 to 400 nm)–visible (Vis, 400 to 800 nm) region of the electromagnetic spectrum, absorption bands arise from electronic transitions from bound states (outer valence orbitals) to excited states (McCoustra, 1990). In organic molecules such as HSs, these exceptionally low energy transitions are associated with the presence of chromophores (i.e., conjugated double bonds and aromatic and related molecules with delocalized electronic orbitals). For metal–organic complexes, absorption should occur in the UV–Vis region from d $\rightarrow$ d orbital transitions associated with transition metals. However, these transitions are generally obscured by HS absorbance and cannot be evaluated.

The UV–Vis spectra of HSs show little structure, appearing largely as broad, coarsely structured absorption bands. This is because of heterogeneous substitution that results in chromophores with overlapping bands and spectral shifts due to slight differences in the macromolecular structures. However, UV–Vis spectroscopy has been used successfully to evaluate the stoichiometry of metal–HS complex formation based on Job's method. This method consists of measuring the variation of optical densities in the visible range (400 to 800 nm) of aqueous solutions containing different ratios of metal ion to complexing agent while maintaining a constant total concentration of both reactants. For example, the ions Cu^{2+}, Fe^{3+}, and Al^{3+} are shown to form 1 : 1 complexes with a soil FA at pH 3, whereas at pH 5, Cu^{2+} and Fe^{3+} form 2 : 1 molar complexes with the FA while the Al^{3+}–FA complex composition remains at 1 : 1 (Schnitzer and Skinner, 1963).

Comprehensive reviews on the application of UV–Vis spectroscopy to the study of HS and metal–HS complexes include those of Bloom and Leenheer (1989) and Stevenson (1994).

4.6. FLUORESCENCE SPECTROSCOPY

The absorption of visible and ultraviolet radiation raises a molecule from the ground electronic and vibrational state to excited states. Fluorescence is the most important radiative photoprocess by which excited electronic states decay, and consists in the emission of less energetic (lower wavelength) photons than the photons absorbed to produce the excited state. Two types of conventional fluorescence spectra, emission and excitation spectra, are useful for studies of metal–organic complexation. The emission spectrum is recorded by measuring the intensity of radiation emitted as a function of wavelength for a fixed excitation wavelength. The excitation spectrum is obtained by measuring the emission intensity at a fixed wavelength while varying the excitation wavelength.

HSs of any source and nature fluoresce due to the presence of conjugated double bonds and aromatic rings bearing various functional groups (Senesi, 1990a). However, only a small fraction of HS molecules that absorb radiation actually undergo fluorescence. Metal ions, especially paramagnetic ions, are generally able to quench the fluorescence of organic ligands by enhancing the rate of some nonradiative processes that compete with fluorescence. A review of the application of fluorescence to HS and metal–HS studies was provided by Senesi (1990a).

Paramagnetic transition metal ions such as Cu^{2+}, Fe^{3+}, Fe^{2+}, Co^{2+}, Ni^{2+}, Mn^{2+}, Cr^{3+}, and VO^{2+} may effectively quench the fluorescence of HS ligands, whereas the quenching effect for diamagnetic metal ions such as Pb^{2+} and Al^{3+} is much less pronounced, and Cd^{2+}, which forms much weaker complexes, and cations such as K^+, Na^+, Ca^{2+}, and Ba^{2+} show no quenching effect (Saar and Weber, 1980; Weber, 1988; Senesi, 1990a). The greater quenching ability of Cu^{2+} is attributed to its capacity to form strong inner-sphere complexes that are involved in binding several weak sites, causing conformational changes that make available additional internal binding sites of FAs (Ryan and Weber, 1982). In contrast, Mn^{2+} (and probably Co^{2+}) would form outer-sphere complexes with FAs not involving weakly acidic phenolic sites, where the ion is farther from the fluorophore, and thus exhibits low quenching ability (Ryan et al., 1983). Fluorescence quenching by Mn^{2+} is also shown to decrease as the molecular weight of soil FA increases as a function of the nature of FAs (Shestakov et al., 1987).

Two binding mechanisms are proposed to explain adsorption of Cu^{2+} by a soil FA at pH 3.5, 5, and 7 and various Cu/FA ratios (Figure 4.1) (Bartoli et al., 1987): (1) a proteolytic reaction (strictly complexation), occurring up to a Cu/FA ratio of 0.6, and (2) a charge neutralization, from a Cu/FA ratio of 0.6 to 1.1. Above a 1 : 1 ratio, the almost constant fluorescence suggests that Cu^{2+} is no longer adsorbed on FAs. Additional data indicate that the most strongly acidic carboxyl groups react first with Cu^{2+}, as would occur in bidentate chelation sites

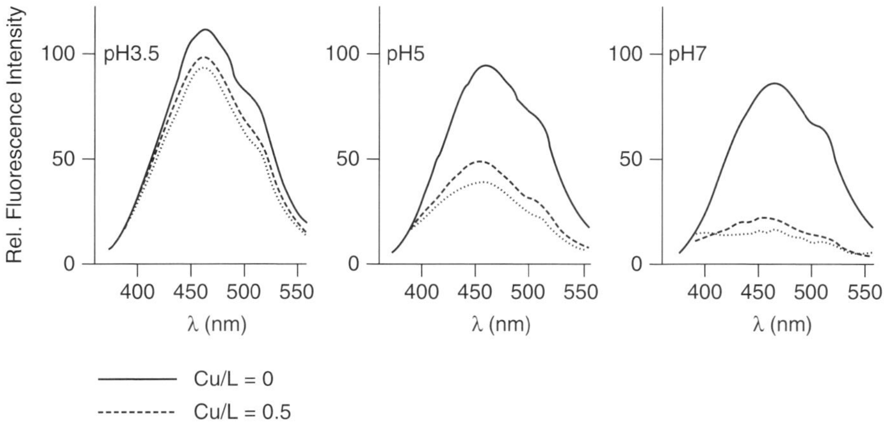

Figure 4.1. Fluorescence emission spectra ($\lambda_{exc} = 355$ nm) of a soil fulvic acid (FA) and its complexes with Cu^{2+} obtained at three pHs and two different Cu^{2+}: FA ligand (L) ratios. (From Bartoli et al., 1987.)

with salicylic- or phthalic-type structures (Underdown et al., 1981). However, a comparative ISE, ASV, and fluorescence quenching study showed that no single ligand could adequately describe Cu^{2+}–FA complexing over the entire pH range (Gregor et al., 1989). Thus, a mixed mode of coordination is proposed, with the dominant binding sites varying with pH and metal-to-ligand ratio. Comparison of FA–metal binding curves with those for model ligands indicates that potentially fluorescing groups such as salicylates or phthalates seem unlikely to be involved directly in Cu^{2+} complexing at low pH. Aromatic amino acid groups such as tyrosine and phenylamine moieties and citrate and malonate moieties in FA are suggested to provide important binding sites for Cu^{2+} at pH 3 to 7. At high pH, either polydentate fluorescing moieties are involved directly in complexing, or weakly complexing fluorescent centers become involved because of their close proximity to a strong donor site (i.e., within one or two carbon atoms of the complexing moieties) (Gregor et al., 1989).

Wavelength shifts of fluorescence emission maximum and/or excitation peaks are also often observed upon interaction of HSs with some paramagnetic metal ions (Senesi, 1990a). For example, the position and intensity of the peak at higher wavelength (465 nm) of a soil FA remain unchanged as more of either Cu^{2+} or Fe^{3+} is complexed, whereas the intensity of the lower wavelength peak (360 nm) decreases and shifts to a longer wavelength (390 nm) at both pH 4 and 6 (Figure 4.2) (Ghosh and Schnitzer, 1981).

Titration curves of HS fluorescence quenching versus concentration of added metal quencher have been used to obtain the CC values of HS ligands and the stability constants of HS–metal complexes (Saar and Weber, 1980, 1982; Underdown et al., 1981; Ryan et al., 1983; Weber, 1983; Dobbs et al., 1989; Grimm et al., 1991; Hernandez et al., 2006; Plaza et al., 2005, 2006). Two fluorescence techniques, lanthanide ion probe spectroscopy (LIPS) and fluorescence quenching of HSs by Cu^{2+}, have been used in conjunction with a continuous distribution model to study metal–HS complexation (Susetyo et al., 1991). In the LIPS technique, the HS samples are titrated by Eu^{3+} ions, and the titration plot of the ratio of the intensities of two emission lines of Eu^{3+} is used to estimate the amount of bound and free species of the probe ion. In the other technique, titration curves of fluorescence intensity quenched by Cu^{2+} versus the logarithm of total added Cu^{2+} are used.

A limiting aspect of fluorescence spectroscopy is that quantitative results obtained by different researchers using different procedures are generally not comparable (i.e., complexing capacities of HS appear to be dependent on the method of measurement). Also, the source of HS and the procedure used for its isolation, in addition to many experimental factors, including concentration of HS, ionic strength of solution, pH, temperature, and the method of data manipulation for the computation of stability constants, can influence the results (Saar and Weber, 1982).

Another limitation of fluorescence spectroscopy is that the method is very effective only with strongly binding, paramagnetic metal ions such as Cu^{2+}. This limitation can, however, be overcome by the use of fluorescent probes, such

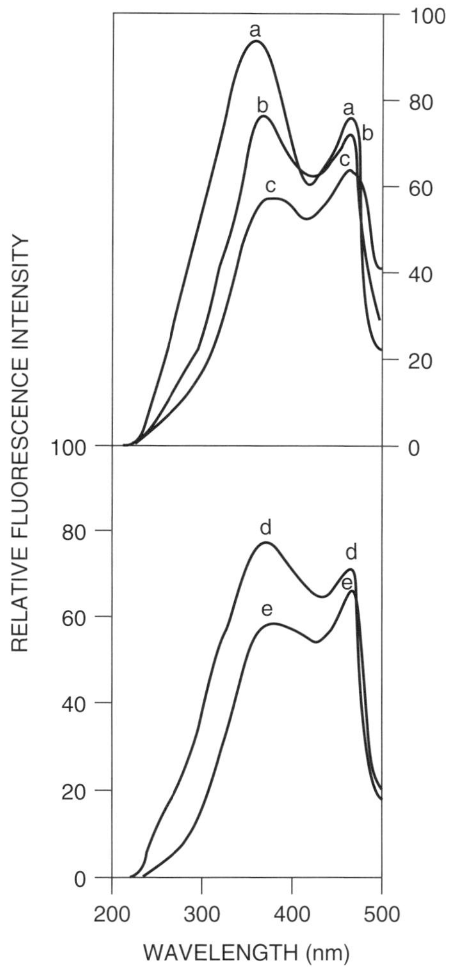

Figure 4.2. Fluorescence excitation spectra at pH 6 of a soil fulvic acid (*a*) and its complexes with Cu^{2+} [% Cu: 3.51 (*b*) and 7.22 (*c*)] and Fe^{3+} [% Fe: 3.47 (*d*) and 6.27 (*e*)]. (Adapted from Ghosh and Schnitzer, 1981.)

as the lanthanide ion probe that shows particularly promising for studies of metal binding by HS in environmental conditions, and at natural concentrations of both metal and HS.

Fluorescence spectroscopy has, however, several advantages over most other methods for studying metal–HS complexation in aqueous media. The method is relatively rapid since no separation is required between bound and free metal ion; thus, errors associated with the separation step in most speciation methods are avoided. Unlike most other methods, it allows direct measurement of the CC of the ligand through a determination of the concentration of free ligands, thus differentiating free and bound ligands. Neither supporting electrolyte nor buffer nor adsorbing material is required to be added to the samples. The method is even more sensitive than ASV and ISE potentiometry and is sensitive enough for application to unmodified natural organic ligands without preconcentration. In conclusion, fluorescence spectroscopy represents an excellent complement to other, indirect CC measurement techniques based on the determination of free metal ion concentration with respect to the bound metal fraction.

4.7. INFRARED SPECTROSCOPY

The most interesting portion of the infrared (IR) spectrum for the structural and analytical study of molecules is the medium IR region, between 4000 and 400 cm^{-1} (2.5 and 25 nm). Energy absorbed by an organic molecule in this region is converted into energy of molecular vibration, and the IR spectrum of the organic molecule consists of vibrational bands.

Despite the molecular complexity of HSs, the most striking features of IR spectra of HSs are their overall simplicity and similarity (Schnitzer, 1978; Stevenson, 1994), but these are more apparent than real. Simplicity (i.e., band broadness) generally results from the extended overlapping of very similar absorptions arising from individual functional groups of the same type but with different chemical environments.

The formation of metal–organic complexes may be studied by the shifts observed for vibrational frequencies of ligand functional groups, such as CH, NH, OH, C=O, C=C, and aromatic rings. These can be used to identify the absorbing functional groups involved in the complexation and, possibly, to provide information on the type of interaction occurring between the metal and the ligand in the complex formed (Nakamoto, 1986).

A wide number of books and reviews are available on the application of IR techniques to HS and metal–HS studies (MacCarthy and Rice, 1985; Bloom and Leenheer, 1989; Senesi, 1992; Stevenson, 1994; Senesi and Loffredo, 2005).

4.7.1. Carboxylate Groups

Ample evidence has been provided of the prominent role played by COOH groups in metal ion complexation by HAs and FAs (reviewed in Senesi, 1992; Stevenson, 1994). The C=O stretching absorption band at about 1710 cm^{-1} and the C–O stretching and O–H deformation absorption at about 1200 cm^{-1} are strongly reduced in intensity or disappear upon ionization of the COOH groups following reaction of HAs and FAs with several divalent and trivalent metal ions, including Cu^{2+}, Mn^{2+}, Zn^{2+}, Pb^{2+}, Co^{2+}, Ni^{2+}, Ca^{2+}, Mg^{2+}, Sr^{2+}, Fe^{3+}, and Al^{3+} (Figure 4.3). Simultaneously, bands near 1600 and 1380 cm^{-1} arising, respectively, from the asymmetric and symmetric stretching vibrations of the COO^{-} groups are reinforced or appear (Figure 4.3). The COO^{-}-to-COOH absorption ratio in the IR spectra of a number of metal–HA complexes prepared at the

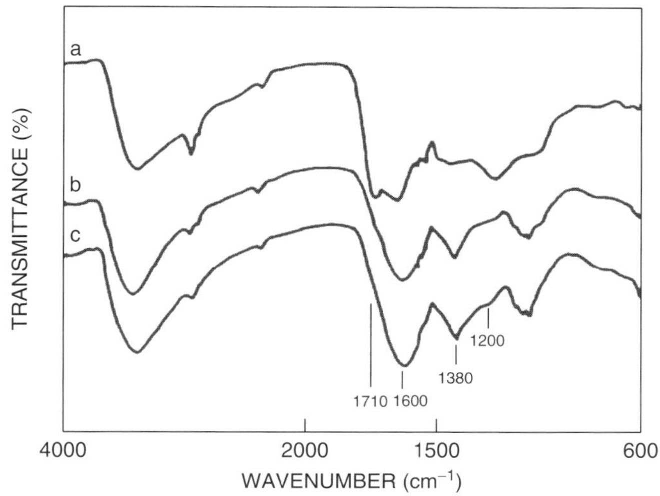

Figure 4.3. Infrared (IR) spectra of a soil humic acid (HA) (*a*) and its complexes with Cu^{2+} (Cu^{2+}/HA = 0.04) (*b*), and Cu^{2+} and Fe^{3+} [($Cu^{2+} + Fe^{3+}$)/HA = 0.08] (*c*). (Adapted from Senesi et al., 1986.)

same HA/metal ratio is shown to depend on the nature of the metal ion complexed and to vary in the decreasing order $Fe^{3+} > Cu^{2+} > Al^{3+} > Ca^{2+} > Mg^{2+}$. These results indicate a decreasing conversion of COOH groups to COO^- groups involved in metal complexation.

The frequency of the asymmetric and symmetric streching vibrations of COO^- may provide information on the ionic versus covalent character of carboxylate–metal bonding. With increasing covalency, the COO^- asymmetrical stretching band shifts to a higher frequency and the symmetrical stretching band to a lower frequency (Nakamoto, 1986). For example, bonds with high covalent character for Cu^{2+} in HAs and FAs are formed preferentially at low levels of metal ion, whereas bonding becomes increasingly ionic as the system is saturated with the metal (Piccolo and Stevenson, 1982). Frequency shifts in the 1600-cm^{-1} region are, however, variable and slight, and interpretation is complicated by interference from other groups.

The nature of the carboxylate binding site in HSs can be evaluated by measuring the separation between the two frequencies of the antisymmetric and symmetric stretching vibrations for the metal-complexed COO^- group (near 1600 and 1380 cm^{-1}, respectively) with respect to the uncomplexed carboxylate ion. This separation is larger in unidentate complexes, smaller in bidentate (chelate) complexes, and comparable in bridging complexes (Nakamoto, 1986). A large separation is measured for Cu^{2+}, Fe^{3+}, Co^{2+}, and Zn^{2+} complexes of HAs and FAs, thus suggesting the formation of unidentate metal complexes (Boyd et al., 1981a; Prasad et al., 1987). These results are also consistent with metal chelation sites involving either two adjacent COOH groups (phtalate-type sites) or a

COOH and an adjacent phenolic OH group (salicylate-type sites) (Boyd et al., 1981a; Prasad et al., 1987).

4.7.2. Other Functional Groups

The involvement of phenolic and/or alcoholic hydroxyl groups in metal-ion binding by HSs is suggested by the net shift of the OH band, from 3500 to 3400 cm^{-1} in HAs and FAs, to 3300 to 3200 cm^{-1} in HA and FA complexes with Zn^{2+}, Fe^{3+}, Cu^{2+}, Pb^{2+}, and Mn^{2+} (Banerjee and Mukherjee, 1972; Tan, 1978; Piccolo and Stevenson, 1982). The extent of the OH stretching shift toward lower frequencies depends on the type of metal and follows the order Mn $<$ Co $<$ Cu $<$ Fe (Banerjee and Mukherjee, 1972). Further, modifications of the typical band at 1070 cm^{-1} ascribed to polysaccharide components of FAs suggest that OH groups of these structures could also be involved in metal complexation (Prasad and Sinha, 1983).

Although frequency shifts in the region 1660 to 1600 cm^{-1} are hardly identifiable, due to the broad absorption, the shift to a lower frequency measured for the band at 1610 cm^{-1} in Cu^{2+}–FA complexes is attributed to the C=O group vibration in conjugated ketones weakened by resonance between C–O–Cu and C=O–Cu in the complexes (Piccolo and Stevenson, 1982). IR evidence is also provided for metal-binding sites involving amide N (and possibly amide C=O) and sulfonic groups (SO_3H) in N- and S-rich HS (Sposito et al., 1976; Prasad et al., 1987). The very sharp IR bands observed in the region 1130 to 1080 cm^{-1} and at between 890 and 697 cm^{-1} for HA and FA complexes with Cu^{2+}, Zn^{2+}, Al^{3+}, or Fe^{3+} are attributed to the metal–oxygen vibration of bound hydroxylated and/or hydrated metal ions (Juste and Delas, 1967, 1970; Tan, 1978; Piccolo and Stevenson, 1982).

4.7.3. Concluding Comments

Although the complexity and heterogeneity of HSs is still a cause of ambiguity and uncertainty in the interpretation of IR spectra of HS–metal complexes, IR spectrometry can provide useful information about the nature and reactivity of HS structural components that can bind metal ions and the molecular arrangement of the binding sites in HS–metal complexes. The utilization of IR spectroscopy in conjunction with selective group blocking techniques and, especially, together with chemical derivatization methods has the potential for considerably enhancing the quality of IR spectra and facilitating their interpretation, thus providing more informative IR data on metal–HS complexation. IR spectroscopy applied to HS–metal complexes in a deuterated water solution and Fourier transform (FT)-IR technique in common aqueous solution is providing unique information on the interacting species observed in the equilibrium state in a mildly perturbed system and in close-to-environmental conditions. Further, the use of diffuse reflectance Fourier transform infrared (DRIFT) spectroscopy applied to metal–HS complex powders is enhancing studies of metal-ion binding by HSs.

4.8. ELECTRON SPIN RESONANCE OR ELECTRON PARAMAGNETIC RESONANCE SPECTROSCOPY

The electron spin resonance (ESR) [or electron paramegnetic resonance (EPR)] spectrum consists of the first derivative of the energy absorption that occurs when a sample containing unpaired electrons distributed into two energy levels in thermodynamic equilibrium in a static magnetic field of value H_0 is impinged on by an incident electromagnetic radiation of frequency ν_0 perpendicular to the magnetic field, provided that the energy supplied equals the difference in energy between the two electron states. This is known as the *resonance condition*.

The ESR signal is highly dependent on the nature of the local environment about the absorbing electron; that is, the position of the ESR signal and the overall ESR spectral pattern depend on the environment conditions in the vicinity of the electron. The most important types of interactions in the spin system that affect the position and pattern of the ESR spectrum are the electron Zeeman, nuclear hyperfine, and ligand superhyperfine interactions (Wertz and Bolton, 1972).

The *electron Zeeman effect* arises from the interaction of unpaired electrons with the external magnetic field and determines the position at which resonance occurs [i.e., the deviation of the g-factor from the free electron value ($g = 2.00232$)]. Species with axial symmetry, such as Cu^{2+} and V^{4+} (i.e., with one principal axis of symmetry, conventionally the z-axis, and equivalent x› and y-axes), exhibit two g-values, usually labeled $g_{\|}$ ($= g_{zz}$, i.e., the g-value along the z or symmetry axis) and $g_{\perp}$ ($= g_{xx} = g_{yy}$, i.e., the g-value perpendicular to the z-axis in the $x-y$ plane). Anisotropy of the g-tensor is often averaged for paramagnetic species in solution by rapid rotation of the metal ion and a single isotropic g-value is exhibited, g_{1so} or $g_0 = \frac{1}{3}\,(g_{xx} + g_{yy} + g_{zz})$.

A *nuclear hyperfine interaction* arises from the magnetic moments of the unpaired electron and its nucleus, if this is magnetic [i.e., it has a nonzero spin ($I \neq 0$), such as the nuclei of Cu ($I = \frac{3}{2}$), Mn ($I = \frac{5}{2}$) , and V ($I = \frac{7}{2}$)]. Nuclear spin causes a splitting of the ESR signal into $2I + 1$ components (i.e., four lines for Cu, six for Mn, and eight for V). The splitting of the hyperfine components is, in general, approximated by $A/g\beta$, where A is the magnitude of the nuclear hyperfine interaction, the *hyperfine coupling constant*, which, like g, exhibits an orientation dependence.

A *ligand superhyperfine interaction* can occur if the ligand atoms have nuclear spin, such as ^{14}N ($I = 1$). A number of components may thus result for each ligand nucleus, which leads to very complex ESR patterns, particularly if the ligands are not identical.

Three types of spectral parameters can thus be obtained from analysis of paramagnetic metal spectra: (1) the g-value(s) of the metal(s) present in the sample; (2) the hyperfine coupling constant, A, if the metal nucleus has a nonzero spin and a nuclear hyperfine structure is apparent; and (3) ligand superhyperfine splitting if the unpaired electron of the metal ion is delocalized by partial covalence onto magnetic nuclei of surrounding ligands and if the related ligand nuclear superhyperfine structure is observed. Rarely are resonance lines due to the "forbidden transitions" observed (e.g., for Mn^{2+}).

To determine these parameters accurately and rigorously, the experimental ESR spectrum should be compared to a computer-simulated spectrum calculated using trial parameters, and a convenient mathematical representation and description of the ESR spectrum should be provided by use of the operator *spin Hamiltonian* (Wertz and Bolton, 1972). In practice, the g-values and hyperfine and superhyperfine constants, A, can be obtained relatively simply, although not rigorously, by direct computation from data derived accurately from the experimental ESR spectrum and from spectrometer setting values used in the measurement, according to conventional equations (Senesi, 1992).

The ESR technique has been used extensively to elucidate the chemical and geometrical properties of naturally occurring and laboratory-prepared complexes formed by HAs and FAs of various origin and nature with paramagnetic transition metal ions of major chemical and biological importance to agriculture and environment, including Fe, Cu, Mn, V, and Mo (reviewed by Senesi, 1990b, 1992, 1996; Senesi and Loffredo, 2005). ESR analysis of metal–HS associations can provide useful and in some cases unique information about oxidation states of metals bound, symmetry and type of coordination sites in HSs, binding mechanisms of metals to HSs, and identity of ligand atoms and groups involved in metal complexing. The paramagnetic metal "probe" addition method has also been applied widely to the study of the "residual" binding capacity of natural HAs and FAs to ascertain additional molecular and quantitative aspects of metal complexation by HAs and FAs. These include the nature of binding sites involved in various experimental conditions, the degree of mobility of bound metals, and the stability of complexes formed toward competitive physical and chemical treatments, including proton and metal ion exchange (Senesi, 1990b, 1992, 1996; Senesi and Loffredo, 2005).

4.8.1. Ferric Iron Complexes

The ESR spectra of soil HAs and FAs generally exhibit an asymmetrical, isotropic resonance line with a value of about $g = 4.2$, consistent with high-spin (five unpaired d-electrons) Fe^{3+} ions held in tetrahedral or octahedral sites of a low-symmetry (rhombic) ligand field (Figure 4.4) (Senesi et al., 1977, 1986, 1989a, 1991; McBride, 1978; Schnitzer and Ghosh, 1982; Goodman and Cheshire, 1987). The ESR signal of these Fe^{3+} complexes does not specify the chemical nature of the ligands, although g-values near 4.2 are consistent with Fe^{3+} complexes with O functional groups, possibly carboxylic acids and/or polyphenols. This form of Fe exhibits considerable resistance to proton and metal exchange, and to chemical reduction, thus suggesting that Fe^{3+} is strongly bound and protected in inner-sphere complexes in HSs (Senesi et al., 1977).

A very broad signal with a value of about $g = 2$ is often exhibited by soil HSs (Figure 4.4*a*) (Senesi et al., 1977; McBride, 1978; Goodman and Cheshire, 1987). Very probably, this signal consists of an envelope of several resonances arising from extended spin–spin coupling of various neighboring paramagnetic metal ions, probably high-spin Fe^{3+} ions in octahedral sites with no or only

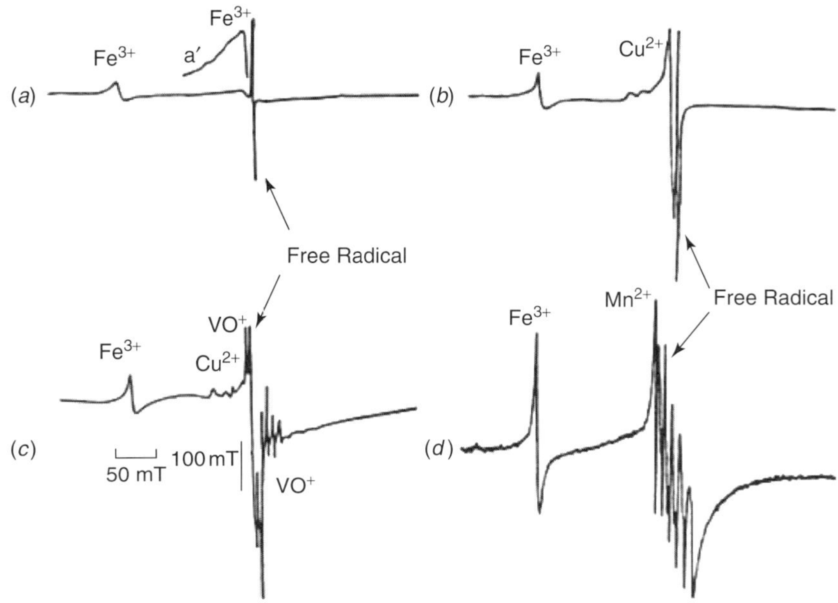

Figure 4.4. Representative wide scan range (800 mT) ESR spectra at 77 K of a Mollisol humic acid (HA) from the IHSS reference and standard collection (*a*; *a'*, higher gain), a Paleosol HA (*b*), a loam soil HA (*c*), and a decomposing leaf litter aqueous extract from a forest soil (*d*). (From Senesi, 1996.)

small axial distorsion from cubic symmetry. Iron in such sites is easily reduced by chemical agents and easily extracted by complexing agents, thus suggesting that it is weakly bound on external surfaces of HSs (Senesi et al., 1977).

Two weak resonances at low-field positions (g-values of about 9 and 6) have sometimes been observed in ESR spectra of HSs (Senesi et al., 1989a,1991). The former resonance probably arises from Fe^{3+} in sites with nearly orthorhombic symmetry, and the latter from high-spin Fe^{3+} in largely distorted, axially symmetric crystal fields. No direct ESR evidence of Fe^{2+} species has been obtained in soil HSs.

HSs also possess a high residual binding capacity toward Fe^{3+} ion that can form complexes stable against various physical and chemical treatments. The intensity of the resonance at $g = 2$ relative to that at $g = 4.2$ increases with increasing Fe^{3+} addition to a soil FA, indicating that most of the added Fe is bound to surface octahedral sites (Senesi et al., 1977).

4.8.2. Divalent Copper Complexes

HSs often exhibit an anisotropic, rigid-limit spectrum of the axial type in the region corresponding to about $g = 2$, with partially resolved $g_{||}$ and $g_{\perp}$ components of the g-value (Figures 4*b*,*c*, and 5*b*,*c*), which is ascribed to the presence of

Cu^{2+} ions (Goodman and Cheshire, 1976; Cheshire et al., 1977; Lakatos et al., 1977a; McBride, 1978; Boyd et al., 1981b; Schnitzer and Ghosh, 1982; Senesi and Sposito, 1984; Senesi et al., 1985, 1986, 1989b). Since the nuclear spin of both Cu isotopes, ^{63}Cu (69.2%) and ^{65}Cu (30.8%), has a value of $I = \frac{3}{2}$, the ESR spectrum should be split into four (i.e., $2I + 1$) features at both $g_{||}$ and $g_{\perp}$. However, only the component at $g_{||}$ is generally resolved partially into a quadruplet; the splitting of the $g_{\perp}$ component is generally too small to be resolved.

The ESR parameters of Cu^{2+}–HS complexes, either naturally occurring or obtained synthetically by Cu^{2+} ion doping of HS samples, are in any case consistent with a $d_{x^2-y^2}$ ground state for Cu^{2+} held in inner-sphere complexes in the HS matrix, with ligands arranged in a square-planar (distorted octahedral) coordination site (tetragonal symmetry). The experimental values of these parameters generally indicate binding sites for Cu^{2+} in HSs involving either only O functional groups (carboxyls, phenolic hydroxyls, carbonyls, and often, water molecules), or both O and N ligands, or even only N atoms (i.e., a tetraporphyrin site) (McBride, 1978; Boyd et al., 1981b; Schnitzer and Ghosh, 1982; Senesi and Sposito, 1984; Senesi et al., 1985, 1986, 1989b, 1991). The participation of N in the binding of Cu^{2+} by HSs is also supported by the resolved pattern observed in some cases at $g_{\perp}$, derived from superhyperfine coupling of the Cu unpaired electro to N ligand nuclei ($I = 1$) (Goodman and Cheshire, 1976).

Measured ESR spectral parameters also provide evidence of a high-covalent-bond contribution (i.e., delocalization of the unpaired electron toward the ligands) for Cu^{2+} in HSs. The combined effects of the crystal field splittings and covalency lead to the following empirical order of decreasing $g_{||}$ and $g_{\perp}$ factors for different donor atoms: O > N > S. Similarly, the values of the hyperfine coupling constant $A_{||}$ and $A_{\perp}$, which are related directly to the degree of covalency and g-factors, increase, and the absolute values of $A_{||}$ and $A_{\perp}$ obviously decrease with increasing electronegativity of the ligand atoms (O < N < S) (Jameson, 1981). However, small deviations from square-planar coordination will shift $A_{||}$ to smaller values, while larger deviations will shift $g_{||}$ to smaller values (closer to $g = 2$) (Peisach and Blumberg, 1974).

A good correlation exists between the $g_{||}$ and $A_{||}$ values and either the Cu^{2+} loading or the coordination environment of model and natural organic Cu complexes (McBride, 1989). There is evidence that covalent bonding, as indicated by low g-values and large hyperfine constants, is favored by low Cu^{2+} loading in HSs, where complexation to amine-type N groups is preferred to O-containing ligands. In contrast, ESR parameters measured for Cu^{2+} at high loadings in HSs indicate not very covalent but rigid binding of Cu^{2+}, largely to O-containing ligands, probably as inner-sphere complexes with a higher degree of mobility of Cu^{2+}.

An accurate analysis of the Cu^{2+} ESR pattern at $g_{||}$, possibly with the aid of a computer-simulated spectrum, reveals that in some cases it consists of two (or more) superimposed quadruplets, often very difficult to be differentiated, each component arising from a different type of local environment for Cu^{2+} (i.e., from different classes of binding sites for Cu^{2+} in HSs) (Senesi, 1990b, 1992, 1996).

A large amount of ESR evidence has been obtained regarding the capacity of Cu^{2+} ions to replace Mn^{2+}, VO^{2+}, and Fe^{3+} from FAs and HAs when these materials are treated with excess Cu^{2+} ions, and to form complexes of various stability toward protons or competing cations (Senesi et al., 1977, 1985, 1986, 1991).

4.8.3. Vanadyl Ion Complexes

More complicated but relatively well resolved spectra exhibiting a richly structured pattern with a value of about $g = 2$ are observed for some soil HAs or FAs (Figure 4.4*c*) and/or their fractions. Analysis made on an enlarged spectrum of the region (Figure 4.5*c*) indicates the presence of two distinct, overlapping rigid-limit spectra of the axial type. One comprises the typical anisotropic pattern of complexed Cu^{2+} discussed previously, the other consists of the partial superimposition of two hyperfine octuplets corresponding to the parallel and perpendicular components of V^{4+} (nuclear spin, $I = \frac{7}{2}$) (Goodman and Cheshire, 1975; Cheshire et al., 1977; Lakatos et al., 1977a; McBride, 1978; Senesi et al., 1989b).

The complexity of the spectrum may require computer simulation for accurate determination of the ESR parameters, which are consistent with a VO^{2+} ion rigidly bound in HSs as an inner-sphere complex in a square-planar coordination site. Vanadyl complexes occurring naturally in soil HSs are characterized by relatively strong ligand fields and high covalency, being consistent with oxygen ligands, mostly phenolate or possibly water molecules and more rarely, N ligands. This renders VO^{2+} ions difficult to remove from these sites, even with acid leaching. In contrast, complexes obtained by VO^{2+} doping of HSs involve weaker ligand fields and lower covalency, indicating that VO^{2+} is bound primarily to surface carboxylate groups while it remains partially hydrated, resulting in relatively labile and exchangeable forms (McBride, 1978; Templeton and Chasteen, 1980; Senesi et al., 1986, 1989b, 1991).

ESR spectroscopy has been used to study the dynamics of motion, molecular conformation, aggregation properties, stability constants, and stoichiometries in aqueous solution of VO^{2+} complexes with two gel-filtrated FA fractions of different molecular weight (MW) (FA-I and FA-II) (Templeton and Chasteen, 1980). ESR results indicate that both FA fractions involve similar binding sites of low symmetry, consisting of four O ligands bound in the first coordination sphere (inner sphere) of VO^{2+} ions. The higher-MW fraction forms a complex approximated as $(VO)_2$ $(FA\text{-}I)_6$, whereas the lower-MW fraction forms a simpler complex, VO–FA-II. Comparative evaluation of ESR parameters suggests that the ligand fields existing about the metal can be modeled by the complexes bis(phtalato)(salicylato)oxovanadium (IV) and mono(salicylato)oxova nadium (IV), respectively (Templeton and Chasteen, 1980).

4.8.4. Divalent Manganese Complexes

The ESR spectra of some soil HAs (Figures 4.4*d* and 4.5*d*) and fractionated soil FAs feature a well-resolved isotropic pattern with a *g*-value around 2, consisting

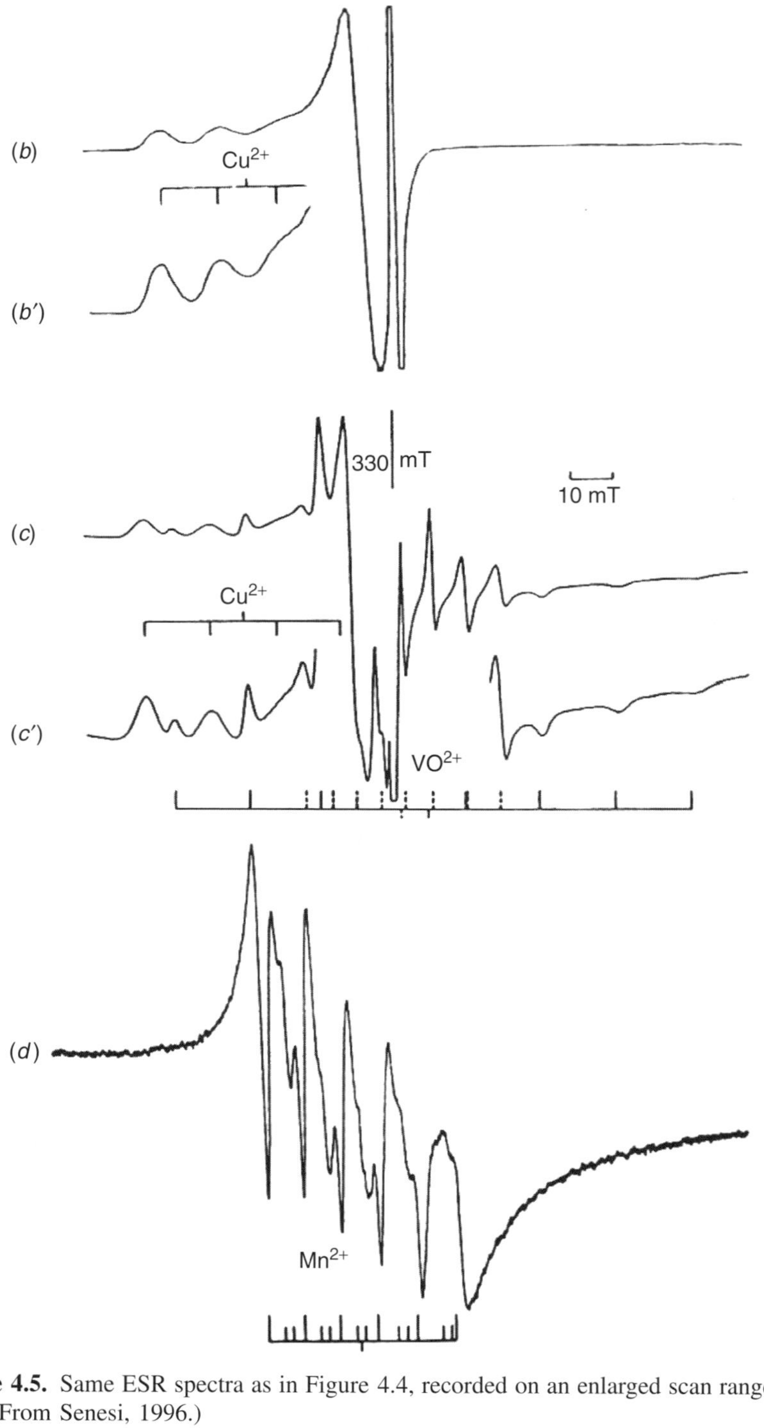

Figure 4.5. Same ESR spectra as in Figure 4.4, recorded on an enlarged scan range (200 mT). (From Senesi, 1996.)

of six almost equally spaced principal lines and possibly 10 secondary lines (corresponding to forbidden transitions) of lesser intensity. The ESR parameters of such spectra are consistent with high-spin hexahydrated Mn^{2+} ($I = \frac{5}{2}$) bound in outer-sphere complexes by electrostatic forces to six O atoms of negatively charged carboxylate and phenolate groups in a distorted octahedral environment (Cheshire et al., 1977; Gamble et al., 1977; McBride, 1978; Senesi et al., 1991).

At low pH most Mn^{2+} is adsorbed by a soil FA in the form of outer-sphere complexes, but at pH > 8, or $T > 50^\circ$C, Mn^{2+} can enter inner-sphere multiligand complexation sites (McBride, 1982). These results indicate that the type and stability of Mn^{2+}–HS complexes and, in turn, their ease of exchangeability and bioavailability in natural systems are strongly dependent on pH and temperature. Further, HSs isolated from various sources exhibit a high residual complexing capacity for added Mn^{2+} that can be bound in water-stable forms, but unlike Fe^{3+} and Cu^{2+}, it may be displaced completely by protons or strongly complexed metal ions (Senesi et al., 1991).

Weighted-average equilibrium constants of water-soluble Mn^{2+}–FA complexes have been determined by ESR spectroscopy on the basis of the linear functionality existing between the height of hyperfine peaks of Mn^{2+} and the concentration of free Mn^{2+} (Gamble et al., 1977). The increasing relaxation line broadening observed in the ESR spectrum of Mn^{2+} with increasing addition of complexing ligand could be related directly to increasing complexation of Mn^{2+} by FAs. The K_c values measured by ESR are in excellent agreement with those determined by an ion-exchange method, but the ESR method is faster, more sensitive, and more convenient than the ion-exchange procedure.

4.8.5. Molybdenum Complexes

ESR evidence has been obtained suggesting that HA can reduce molybdate to Mo(V) and complex the Mo(V) species in strongly acidic media (Lakatos et al., 1977a; Goodman and Cheshire, 1982). The ESR spectrum of a peat soil HA complex with Mo(V) enriched in ^{95}Mo (nuclear spin, $I = \frac{5}{2}$) features two distinct components, each split into two six-line hyperfine patterns at $g_{\parallel}$ and $g_{\perp}$, consistent with two different axially symmetric Mo(V)–HA complexes.

Treatment with 0.1 M HCl of ^{95}Mo(V)-enriched HA produces a low-intensity six-line ESR spectrum probably arising from a Mo(III) species (Goodman and Cheshire, 1982). This result suggests that Mo(III) species can be formed and remain stable in the solid state even in aerobic conditions when protected in HS complexes.

4.8.6. Concluding Comments

Major advantages of ESR spectroscopy are its high sensitivity and the ability to measure spectra directly with minimal or no sample preatreatment. ESR provides evidence that small quantities of metal ions can bond selectively in inner-sphere complexes at the most preferred sites for the metal, whereas in the presence of high amounts of metal added, the high degree of site occupation generally results in a loss of relative selectivity. ESR analysis also show that high pH values that generate a greater availability of negatively charged O ligands favor inner-sphere complexation for metals that are retained as hydrated ions at lower pH. Inner-sphere coordination is also preferred when competing water ligands are removed by dehydration, thus forcing the metal to enter into direct bonding with HS ligands. ESR data confirm that the more electronegative the metal ion, the stronger the metal is bound to HS and the higher the degree of bond covalency. The ESR approach also allows, in principle, the determination of the free ion concentration and therefore of the degree of complexation of the metal ion.

The intrinsic limitation of the ESR technique is its applicability only to paramagnetic metal ions that give a detectable ESR signal, such as Cu, Fe, Mn, V, and Mo. The major limitation of the ESR experiment is its inability to resolve signal component lines that may overlap to such an extent to result in merging of individual resonant lines or spin packets into a single overall broad line or envelope with a loss of information. Line broadening is determined by either a homogeneous or an inhomogeneous mechanism. The first effect arises from microwave-power saturation, which produces broad spectra for metal ions such as Fe^{3+}; thus, the choice of power is critical to avoid saturation. The second effect is caused by interactions with neighboring paramagnetic species of the same or different type, or neighboring nuclei. Temperature is also a critical parameter in an ESR experiment. Since sensitivity for paramagnetic species increases with lowering sample temperature, according to Curie's law, ESR measurements are often made at either liquid N (77 K) or liquid He (4.2 K) temperature, which may reduce some type of broadening.

ESR-related spectroscopies that hold the potential to overcome some resolution limitations and yield more information than the classical ESR approach about the chemical environment of paramagnetic metal ions are the electron–nuclear double resonance (ENDOR) (Kevan and Kispert, 1976) and electron-spin echo envelope modulation (ESEEM) (Kevan and Schwartz, 1979) spectroscopies. Either ENDOR or ESEEM represents by principle a useful tool in extending resolution of the ESR experiment. However, the sensitivity of ENDOR and ESEEM is much lower than that of ESR, and interpretation of ENDOR and ESEEM spectra is not a simple matter, especially if ligands are not well characterized, as is the case for HSs. Both ENDOR and ESEEM techniques have not yet been applied to strictly metal–HS complexes, but the sensitivity and ease of carrying out experiments are improving rapidly, so major scientific activity may be anticipated to occur in this area of ESR spectroscopy.

4.9. MÖSSBAUER SPECTROSCOPY

Mössbauer spectroscopy measures the resonant absorption of nuclear gamma rays involved with transitions between the ground and excited states of atomic nuclei with nonzero angular momenta. The precise energy of such transitions is influenced by the chemical environment of the nuclei and any external magnetic field and electric field gradient. The Mössbauer effect is highly isotopic specific, and only approximately 30 isotopes are Mössbauer-active, including iron, nickel, zinc, and mercury, but the most easily studied metal is iron (Gutlich et al., 1978).

The principal energy-dependent parameters that can be obtained from a Mössbauer spectrum are the isomer shift, the quadrupole coupling constants, the magnetic field, and the peak width. The magnitude of the *isomer shift* is proportional to the difference in electron density at the metal ion nucleus in the sample and in reference, usually the metallic form. Factors influencing the isomer shift include also the total population of 3d orbitals. The *quadrupole splitting*, or magnetic hyperfine interaction, is determined by the interaction between the quadrupole moment of the excited state and the electric field gradient at the metal nucleus. The latter derives primarily from a combination of charges originating from the electronic environment (valence electrons) of the metal ion and from surrounding atoms. In the case of iron, this effect gives rise to a doublet in the Mössbauer spectrum. The *magnetic hyperfine field* is proportional to any magnetic field experienced by the nucleus. In the case of iron, the interaction between the nuclear magnetic dipole moment and any magnetic field at the nucleus removes completely the degeneracy of nuclear energy levels and produces six peaks in the spectrum with intensity ratios 3 : 2 : 1 : 1 : 2 : 3 if the magnetic domains are randomly oriented. Finally, the value of the *peak width* can provide useful information on the possible presence of unresolved components in the Mössbauer spectrum. The experimental values of Mössbauer parameters may provide information on the type of coordination, symmetry, and chemical nature of the groups bound to the high-spin iron ions.

All Mössbauer investigations so far reported on metal–HS complexes have used the isotope ^{57}Fe (reviewed by Senesi, 1992). The Mössbauer spectrum obtained on a natural soil HAs containing 0.57% Fe could be computer-fitted to three doublets exhibiting parameters, suggesting the presence of three sites for Fe^{3+} ions in the HAs, two with octahedral and one with tetrahedral coordination (Figure 4.6*a*) (Senesi et al., 1977). Chemical reduction with hydrazine of the HA sample results in a Mössbauer spectrum in which one of the two doublets assigned to Fe^{3+} ions in octahedral sites disappears [(3a in Figure 4.6*a*), and a new doublet appears featuring much higher values of the isomer shift and quadrupole splitting, typical of Fe^{2+} ions in the same type of site [(3b) in Figure 4.6*b*] (Senesi et al., 1977).

Results of an extended Mössbauer study of several Fe^{2+}–HA complexes suggest the presence of partially hydrated high-spin Fe^{2+} ions held in elongated octahedral arrangements in ionic inner-sphere chelate-type complexes with carboxylate and phenolic and alcoholic hydroxyl groups and N-containing groups (Lakatos et al., 1977b). However, Mössbauer evidence of partial oxidation of Fe^{2+} to Fe^{3+} is obtained on exposure to air of some Fe^{2+}–HA complexes. On the contrary, Mössbauer evidence has shown that about one-half of the iron in a $^{57}Fe^{3+}$–HA slurry at pH 1 is present in two components with quadrupole splitting values consistent with Fe^{2+} ions partly in completely hydrated outer-sphere

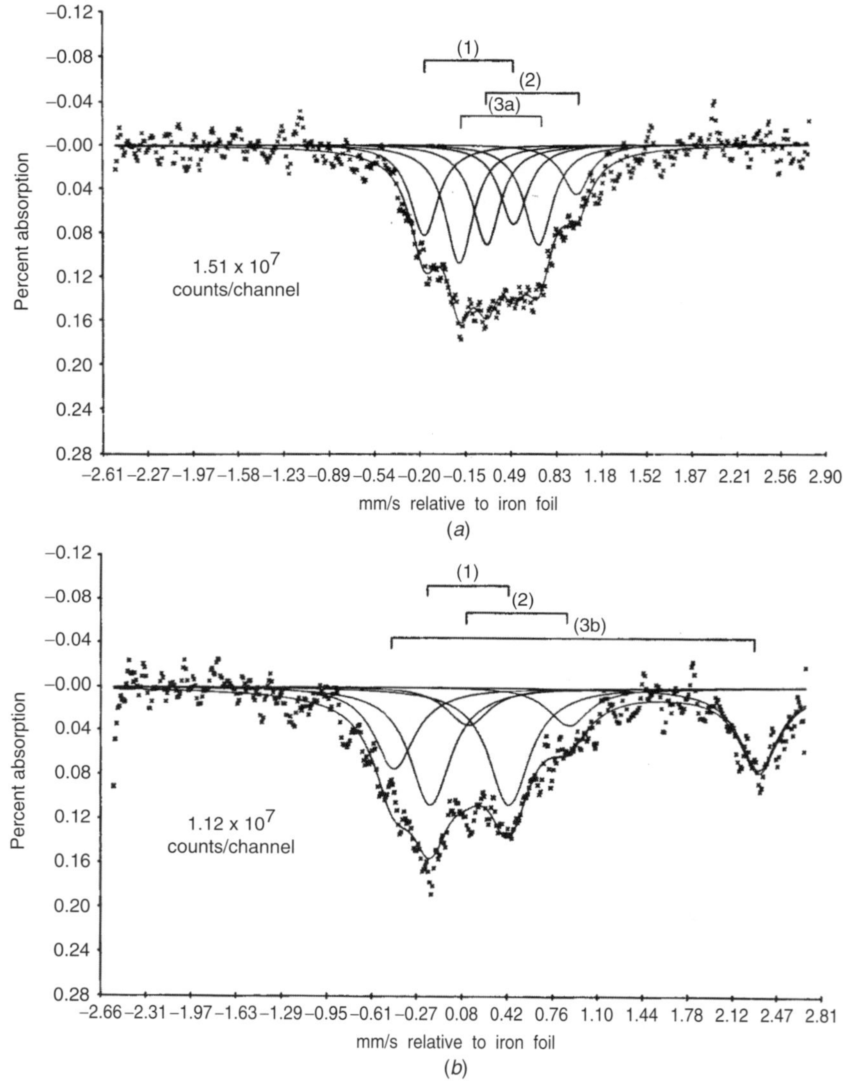

Figure 4.6. Mössbauer spectra of (*a*) a natural soil humic acid (HA) and (*b*) the same HA after reduction with hydrazine. Doublets (1), (2), and (3) refer to three different sites for iron in HA. Doublets (3a) and (3b) refer, respectively, to Fe^{3+} and Fe^{2+} ions in similar octahedral sites. (From Senesi et al., 1977.)

complexes with HAs, and partly associated directly with O ligands of HAs in inner-sphere complexes (Goodman and Cheshire, 1979).

In other Mössbauer studies conducted on various naturally occurring and laboratory-prepared Fe^{3+}–HA and Fe^{3+}–FA complexes at various pH values and recording temperatures, one or two Fe^{3+} doublets, a weak Fe^{2+} doublet, and one or two magnetically split sextets are obtained (reviewed in Senesi, 1992). These results suggest the occurrence of iron in these materials, partly as ferric and ferrous HA and FA complexes, and partly as inorganic species.

In conclusion, the main advantage of Mössbauer spectroscopy is to distinguish unambiguously between high-spin Fe^{3+} and Fe^{2+} ions on the basis of major differences observed in both the isomer shift and quadrupole splitting of these ions. However, the intrinsic limitation of the technique is its high specificity for the ^{57}Fe isotope. Despite the low natural abundance of this iron isotope (2.19%), it is also possible to obtain reasonable Mössbauer spectra with unenriched natural Fe–HS complexes. Caution must be exercised, however, in interpretation of Mössbauer data, and model compounds should be investigated for comparison, as several binding sites exist that may give rise to similar spectra.

4.10. NUCLEAR MAGNETIC RESONANCE

Nuclear magnetic resonance (NMR) spectroscopy has been applied to elucidate metal-binding mechanisms to organic ligands mainly by two approaches: by measuring the effects of metal complexation on either the relaxation times of ^{1}H of water molecules solvating the metal cation; or on the chemical shifts of NMR-active metal ions (e.g., Cd, Al, and Pb) (e.g., Connors, 1987; Wilson, 1989; Macomber, 1998). Relatively few and sparse studies have been performed by NMR on metal–HS complexes. A comprehensive and updated review has been provided by Kingery et al. (2001) on the various applications of NMR spectroscopy to the study of metal–HS interactions.

4.10.1. Solvation Water (^{1}H) Relaxation

In NMR experiments, relaxation is the process by which nuclear spin states are subjected to an external magnetic field establish equilibrium. The relaxation times of ^{1}H in water molecules coordinated to a paramagnetic ion are altered relative to those of water molecules in the bulk solution (Deczky and Langford, 1978). Coordination of the metal ion to a nonsolvent ligand may thus lead to either a decrease or an increase of observed relaxation rates. A decrease may result from the replacement of coordinated water by the ligand and/or an increase in electronic relaxation times, whereas an increase may be due to a reduction of relaxation times of protons in the solvation shell of the paramagnetic ion. Since the width at half-height of the NMR peak of water is proportional to the value of the relaxation time observed, addition of an organic ligand to a water solution of a paramagnetic metal ion may produce either line narrowing or line broadening, which correspond, respectively, to an increase or a decrease of the relaxation time observed.

Only two studies are available in the literature on the effects of HS complexation on relaxation times of ^{1}H in the solution shells of paramagnetic ions. A slight but definite decrease in line widths of the water NMR signal is observed upon addition of a soil FA to a Mn^{2+} water solution (Gamble et al., 1976). Comparison made with NMR data of Mn^{2+} complexes with model simple ligands is in favor of outer-sphere binding of hexahydrated Mn^{2+} ion by FA in a slightly distorted octahedral symmetry. Different behavior is suggested by NMR data for Fe^{3+} complexation by the same soil FA (Gamble et al., 1976). The great reduction of the line width of water NMR signal measured upon addition of FA to Fe^{3+} solution would imply either inner-sphere coordination of Fe^{3+} or a reduction of symmetry of the high-spin d^5 center.

In a later, more detailed study, the outer-sphere complexation of hexahydrated Mn^{2+} by FA is confirmed by ascribing the ^1H relaxation mechanism mainly to distortion of the octahedral hydrated-metal symmetry occurring by collisions with water molecules outside the complex (Deczky and Langford, 1978). In the presence of FA, the marked increase observed for the longitudinal relaxation rate dependence on Cu^{2+} ion concentration compared to Cu^{2+}-aquo ion and Cu^{2+}-bipyridine complex is attributed to an inner-sphere complex of Cu^{2+} with FA (Deczky and Langford, 1978).

4.10.2. NMR-Active Metal Ions

Chemical shifts of NMR-active metal ions, such as Cd, Al, and Pb, are affected by their local environments and can thus provide useful information on the metal coordination number and type and number and geometric arrangement of ligands complexing the metal, also as a function of metal/ligand molar ratios, pH, and other system properties.

Cadmium-113 NMR offers a useful means for characterizing Cd^{2+}–HS complexes. Although most recent studies on ^{113}Cd–NMR of Cd^{2+}–HS complexes have been performed on aquatic HSs, evidence by ^{113}Cd-NMR is provided of two kinds of Cd^{2+}–soil HA complexes involving either 1 or 10 Cd ions per HA molecule (Pommery et al., 1988). The involvement of N-containing groups as well as of O-containing groups is suggested by ^{113}Cd-NMR in Cd complexation by a soil FA in acid conditions (Chung et al., 1996).

4.10.3. Stability Constants and Structure-Binding Relationship

The ability of NMR to discriminate between free metal ions and their complexes with organic ligands renders this technique a potentially powerful analytical tool for measuring stability constants of HS–metal complexes. In the NMR experiment, a nucleus that can partition between two magnetically nonequivalent sites is said to undergo chemical exchange between the sites. The features of the observable resonance signal thus depend on the rate of the exchange process (Connors, 1987).

For example, in the simple case of 1 : 1 complex formation, several possibilities of chemical exchange rate may exist (Connors, 1987). Slow exchange possibly involves strongly bound complexes, and fast exchange generally refers to weak binding. For a system in slow-exchange conditions, the ratio of NMR spectral peak areas of nuclei in a free ligand site and in a metal ligand site is calculated, and the stability constant of the complex can be obtained knowing the concentration of M (Connors, 1987).

Although no studies are available in the literature on the use of NMR to obtain the stability constants of soil HS–metal complexes, some NMR experiments have been conducted to relate the structure of soil HS to stability constants. A positive correlation is reported between aromaticity of several HAs determined by solution-state ^{13}C-NMR and stability constants of their complexes with Cd^{2+} and Zn^{2+} ions (Ashley, 1996). Even though the carboxylic contents are similar, HAs with higher aromaticities exhibit higher stability constants than those with lower aromaticity, suggesting the involvement of phenolic groups in the complexation of these metals.

Analysis of distortion enhancement by polarization transfer (DEPT) ^{13}C-NMR and quaternary-C only (QUAT) subspectra of a soil HA and a soil FA suggest that carboxylate groups, of which the FA is richer than the HA, play a prominent role in the binding of Eu^{3+} to HS (Shin et al., 1996). This result is consistent with the higher binding constant measured for Eu^{3+} with FA than with HA, which is richer than FA in aromatic carbon and aliphatic methine and quaternary carbons.

Dipolar dephasing CPMAS ^{13}C-NMR has been applied to relate the structure of a soil HA and a sediment HA to the stability constant of Cd^{2+}–HA complexes (Sohn and Rajski, 1990). The stability constant of the Cd^{2+}–sediment HA is greater than that of Cd^{2+}–soil HA, probably due to the higher content and stronger binding of amino acid N groups, in addition to carboxylic groups, of the former HA with respect to the latter HA. Further, greater metal stability constants of Zn^{2+}–HA and Cd^{2+}–HA complexes are found to be associated with a higher degree of substituted aromatic C groups, possibly due to chelation effects (Cameron and Sohn, 1992).

4.10.4. Concluding Comments

To date, few, if any, of the NMR studies on metal–HS complexation have attempted an examination of metal–N spin couplings, due primarily to the low sensitivity and low natural abundance of ^{15}N nuclei. However, with high-resolution instrumentation, natural abundance ^{15}N-NMR may provide an additional useful technique for examining metal–HS complexation.

Two-dimensional NMR methods that yield a two-dimensional frequency spectrum have not yet been attempted successfully to study metal interactions with HS, although instances of successful applications of these approaches can be found in the study of metalloproteins (Kingery et al., 2001). Mononuclear (^{1}H) two-dimensional NMR experiments, such as total correlation spectroscopy (TOCSY), can show ^{1}H–^{1}H coupling throughout the complete spin system, and exchange protons can provide information on sites to which metal attach. For example, N-containing units in HS that bind the metal can be identified since the amido protons from these structures will exchange and disappear from the spectrum.

The through-space nuclear overhauser effect (NOE) can provide information on sites where metals interact even if the metals do not form stable bonds through which spin coupling can be transferred. For instance, metal–HS interactions can be studied by NOE spectroscopy (NOESY) by measuring interactions between the protons within the HS molecules before and after the addition of metals to understand the conformational changes that occur within the molecules (Kingery et al., 2001). An alternative approach is to measure the heteronuclear overhauser effect (HOE) directly between the metal ion and the HS proton in close proximity by HOE spectroscopy (HOES), as has been demonstrated for organo-Li complexes (Bauer, 1995).

The inverse detection heteronuclear multiple quantum coherence (HMQC) experiment is another approach to two-dimensional NMR techniques, which consists of a transfer of chemical shift and coupling information from relatively insensitive nuclei such as ^{13}C and some metals, to more sensitive nuclei such as ^{1}H. The advantage of this method is a substantial increase in the sensitivity obtained, due to the greater natural abundance of ^{1}H (Kingery et al., 2001).

In conclusion, the powerful two-dimensional NMR techniques are expected to evolve as very promising tools for the study of both HS structures and their interactions with metals.

4.11. SYNCHROTON-BASED X-RAY ABSORPTION SPECTROSCOPY

The intense x-ray beams produced by synchroton sources have enabled development of nondestructive x-ray absorption spectroscopy (XAS) techniques that can provide information on oxidation states and local chemical environments of elements in a variety of environmental materials (Fendorf and Sparks, 1996; Fendorf, 1999; Senesi and Loffredo, 2005). XAS theory is discussed in detail in several books (e.g., Teo, 1986; Köningsberger and Prins, 1988; Lytle, 1988).

Most elements are XAS-active and can absorb monochromatized x-rays specifically. Thus, XAS consists of recording absorption by a sample of x-rays as a function of the wavelength. The spectral scan is performed in the vicinity of an x-ray absorption edge (K, L, or M) of the target element chosen. Electron transitions allowed for XAS are at the K-edge, 1s $\rightarrow$ np; at the L-edge, 2p $\rightarrow$ nd; and at the M-edge, 3d $\rightarrow$ nf. Thus, spectral features of different elements possibly present in the sample do not overlap since K edges are separated by several hundred electron volts, and the actual position of the absorption edge depends, first, on the element absorbing the x-rays, and second, on its oxidation state. The XAS technique can therefore be used to investigate complex materials by successive tuning of the absorption edge of each spectroscopically active element present in the sample.

By convention the XAS technique is distinguished by two methods: x-ray absorption near-edge structure (XANES) spectroscopy and extended x-ray absorption fine structure (EXAFS) spectroscopy. The XANES spectra are obtained in the range from about 10 eV below to about 60 eV above the absorption edge, and can provide qualitative information about the oxidation state of an excited atom, its coordination geometry, and its bonding environment through comparison with model compounds (e.g., Bianconi, 1988; Fendorf, 1999).

The EXAFS analyzes the measured oscillatory structure that appears at higher energies, from about 50 to about 1000 eV above the absorption edge, and can reveal the local atomic environment surrounding the excited atom (Fendorf, 1999; Bloom et al., 2001). In particular, the structural information provided by EXAFS includes average interatomic distances and number and chemical identities of the atoms within a 5-Å radius of the atom that absorbs the x-ray photon. A number of steps are generally required to process and analyze an EXAFS spectrum. Briefly, these consist of (1) merging of scans; (2) isolation of the fine-structure scattering curve, $\chi(E)$, from the absorption edge by background removal; (3) conversion from a photon energy (eV) scale, $\chi(E)$, to a photoelectron kinetic energy or wavevector ($Å^{-1}$) scale, k; (4) isolation of the scattering curve, $\chi(k)$, which requires a number of operations that can be accomplished by several computer software packages; (5) Fourier transformation of the weighted scattering curve, $k^3\chi(k)$, to obtain the experimental radial structure function (RSF); and finally, (6) comparison through a curve-fitting process by an adequate EXAFS analysis program of the RSF and Fourier-filtered scattering curves to computed spectra derived from model structures. A common and highly regarded computer software package used for computing theoretical scattering curves, $\chi(k)$, and for RSF calculation for the model structure is FEFF (Rehr, 1993; Rehr et al., 1994; Zabinsky et al., 1994).

Only recently have XANES and EXAFS techniques been used to obtain more detailed and additional information on the chemical structure of the binding sites in soil HS–metal complexes, and a comprehensive review on this subject has been provided (Bloom et al., 2001).

4.11.1. Copper Complexes

Binding of Cu^{2+} by soil HSs at various pHs and by soil and peat HAs has been studied in detail by XANES and EXAFS spectroscopies by several authors (Davies et al., 1997; Hersterberg et al., 1997; Xia et al., 1997a,b). The similar pattern obtained for K-edge XANES spectra of Cu^{2+}–HS complexes, and their first derivatives, at pH 4, 5, and 6 (Figure 4.7), suggest that Cu^{2+} binding sites in soil HS are similar at these pH values (Xia et al., 1997a). The small preedge bump for the 1s → 3d transition and the splitting in α and β peaks for the 1s → 4p transition, which are apparent especially in the first derivative XANES spectra of Cu^{2+} in HS at all three pH values (Figure 4.7), suggest a reduced symmetry (i.e., a tetragonal distortion of the octahedral binding site of Cu^{2+} in HS), which is consistent with previous ESR results (Senesi, 1990b, 1992, 1996).

The RSF plots derived from EXAFS spectra and the corresponding FEFF simulations for Cu^{2+} bound to HS at pH 4, 5, and 6 show no change with pH (Figure 4.8) (Xia et al., 1997a). All RSF plots show a major peak centered at 1.5 Å, which arise from atoms in the first coordination shell, and a minor peak

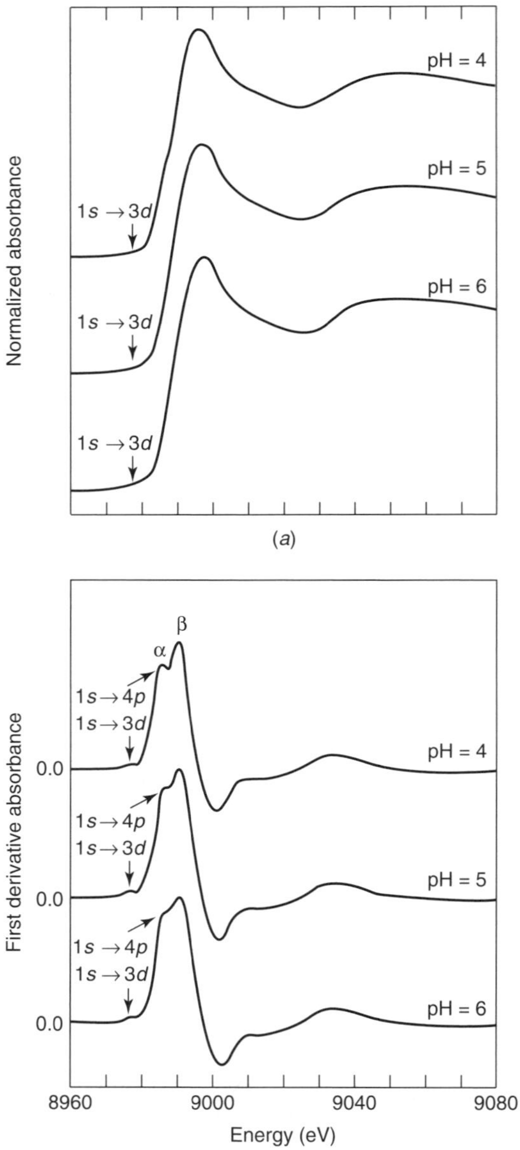

Figure 4.7. XANES spectra (*a*) and their first derivatives (*b*) of Cu^{2+}–humic substance complexes. (From Xia et al., 1997a.)

centered at about 2.2 Å, which represents scattering from atoms in the second coordination shell.

Given the similarities of the scattering of photons and photelectrons by O and N atoms, it is not possible to distinguish bonding to O from bonding to N.

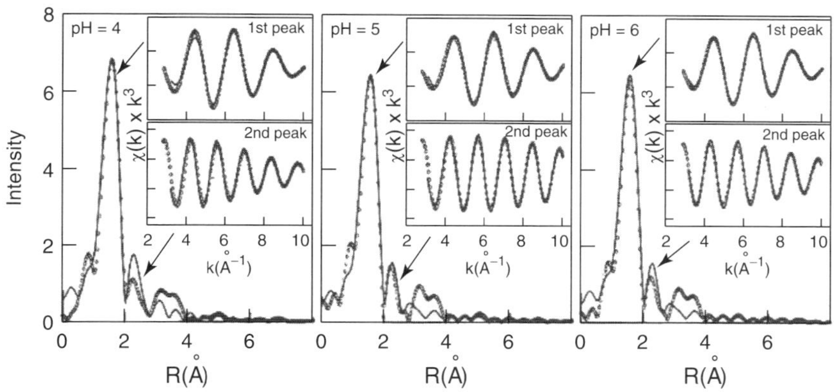

Figure 4.8. Experimental radial structure function (RSF) for Cu^{2+}–humic substance complexes at pH 4, 5, and 6 (dots) and FEFF simulations (solid line) for an adjusted model of the coordination site derived from bond network analysis. The inset shows plots of experimental (dots) and fitted (solid line) inverse Fourier-transformed scattering curves for the first atomic shell (Cu–O) and second atomic shell (Cu–C). (From Xia et al., 1997a.)

However, as the amount of N in soil HS is too small, O atoms from either water molecules or acidic HS functional ligands are considered to be the primary first-shell atoms. The position of the minor peak in the RSF plots suggests that C atoms rather than O atoms are more likely to appear in the second coordination shell of Cu^{2+} bound to HS. The presence of C atoms in the second shell is proof of inner-sphere complexation, while the number of C atoms in this shell can be an indication of the average number of acidic HS functional groups coordinating the Cu^{2+} ion. However, caution should be used in evaluation of the second-shell coordination numbers since experimental data usually have a large margin of error (>20%). This is especially true for the C atom, which has a weak backscattering amplitude, particularly if the bond lengths are variable and/or heterogeneous, as with HS ligands. Further, the average second-shell coordination number calculated by EXAFS can mean either that it is valid for all metal atoms bound or that it is an average of a wide range of coordination numbers.

Analysis of experimental results and model calculations leads to the conclusion that at pH 4 to 6 the average environment of Cu^{2+} bound to soil HS is a tetragonally distorted octahedral binding site involving four equatorial O atoms at an average distance of 1.94 Å and two axial O atoms at an average distance of 2.02 Å in the first coordination shell, and possibly four C atoms at an average distance of 3.13 Å in the second coordination shell (Xia et al., 1997a). Average first-shell Cu–O bond lengths of 1.92 to 1.93 are measured for Cu^{2+}–soil HA suspensions at various Cu/HA ratios and at pH 5.6 and 7.3 (Hersterberg et al., 1997).

Almost similar results are obtained in a XANES and EXAFS study by Davies et al. (1997, 2001) on Cu^{2+} complexes with three soil and peat HAs. No different metal-binding sites are revealed at different metal-loading levels, and no evidence of metal clustering is obtained even in heavily metal-loaded HAs, suggesting that metal ions are widely separated. None of the Fourier-transformed EXAFS spectra of Cu^{2+}–HS complexes show features expected for second-shell atoms, even at 77 K. The authors conclude that Cu^{2+} in HAs has four nearest-neighbor O or N atoms at 1.97 ± 0.02 Å, in good agreement with the slightly shortest distance estimates of Xia et al. (1997a).

Similar to ESR results, no evidence for Cu^{2+} bonding to reduced organic S in a reduced S-rich HA was obtained by EXAFS (Alcacio et al., 2001). Based on XANES data, these authors found that Cu^{2+} was mostly bound to O atoms of HA and not to N atoms. Further, some differences found in Cu^{2+}–O (axial) bond lengths suggested that the local molecular configuration of Cu^{2+}–HS complexes may differ depending on the source and type of HS.

4.11.2. Lead Complexes

In an EXAFS study of a soil contaminated by alkyltetravalent Pb compounds, Pb is found in the divalent state and complexed to salicylate- and catechol-type functional groups of HS (Manceau et al., 1996). The best spectral resemblance to reference compounds is obtained for a mixture of 60% salicylate and 40% catechol functional groups. Varying the proportion of salicylate and catechol forms within 10 to 20% has minor effects on the general shape of the combined EXAFS spectrum (Manceau et al., 1996).

Binding of Pb^{2+} by a soil HS at pH 4, 5, and 6 has been studied in detail in the freeze-dried state by XANES and EXAFS spectroscopies (Xia et al., 1997a). The similarity of L_{III}-edge XANES spectra of Pb^{2+}–HS complexes, and of their first derivatives, suggest that Pb^{2+}-binding sites in soil HS are similar at these pH values. The experimental RSF plots and FEFF simulations for Pb^{2+} bound to HS at pH 4, 5, and 6 show small changes with varying pH. As in the case of Cu^{2+}–HS complexes, all RSF plots show a major peak arising from atoms in the first shell and centered at 1.9 Å (at low pH) and 1.7 Å (at higher pH), and a minor peak centered at 2.7 Å arising from atoms in the second shell.

The best simulations for Pb^{2+}–Hs complexes are obtained for Pb^{2+} bound to four O atoms in the first shell at average distances decreasing from 2.46 Å to 2.32 Å with increasing pH from 4 to 6, and two C atoms at an average distance of 3.26 Å in the second shell. These result is an indication of inner-sphere bonding of Pb^{2+} to two O-containing functional groups of HS (Xia et al., 1997a). The greater width of the first-shell O peaks for Pb^{2+}–HS complexes, with respect to Cu^{2+}–HS complexes, indicates a higher degree of either chemical heterogeneity or distortion in the Pb-binding sites than in Cu-binding sites.

4.11.3. Zinc, Nickel, and Cobalt Complexes

Similar to Cu^{2+}, the XANES spectra and their first derivatives of Ni^{2+} and Co^{2+} complexed by a soil HS at pH 4 feature an absorption in the pre-edge region normally due to 1s → 3d transitions, but no splitting of peaks α and β

(Xia et al., 1997b). For Zn^{2+} complexes with the same HS sample the pre-edge features cannot be distinguished because of the inadequate number of data points collected in this region (Xia et al., 1997b). In any case, however, the experimental results indicate an average octahedral binding environment for these metal ions in soil HS.

The experimental RFS plots for Zn, Ni, and Co all have a major peak centered at 1.6 to 1.8 Å arising from atoms in the first coordination shell, and a much smaller peak centered at about 2.3 to 2.7 Å representing scattering from atoms in the second coordination shell (Xia et al., 1997b). Model fitting with FEFF indicates inner-sphere octahedral coordination with six O atoms in the first shell for Ni–and Co–HS complexes, whereas for Zn^{2+} the best fit is a coordination with 4 O and 2 S atoms. The bond distances for Ni–O and Co–O are 2.10 and 2.04 Å, respectively. The distances for Zn–O and Zn–S bonds are 2.13 and 2.33 Å, respectively.

Based on the position of the second peak of the RFS, Xia et al. (1997b) conclude that similar to Cu^{2+} complexes, C atoms rather than O atoms are more likely to be present in the second coordination shell of Zn^{2+}, Ni^{2+}, and Co^{2+} bound to soil HS, indicating inner-sphere complexation. The number of C atoms in the second shell, which may represent the average number of acidic HS functional ligands coordinating the metal, is (1) one for Co, at a bond distance of 2.87 Å; and (2) two for Ni and Zn, at a bond distance of 2.94 and 3.29 Å, respectively.

EXAFS spectra at the Zn-K edge and RSFs of Zn^{2+}–soil HA complexes obtained at various Zn^{2+} concentrations show that at low concentrations Zn^{2+}forms inner-sphere complexes of sixfold coordination to O ligands of HA (Sarret et al., 1997). As Zn^{2+} concentration is raised, the number of different ligand sites increases, the coordination being exclusively octahedral. At very high metal concentrations, however, the major part of Zn (80 to 90%) is bound as an outer-sphere complex (Sarret et al., 1997).

4.11.4. Ferric Iron and Manganese(II) Complexes

Complexes of Fe^{3+} and Mn^{2+} ions with three soil and peat HAs have also been studied by XANES and EXAFS spectroscopies (Davies et al., 1997, 2001). No second-shell peaks are observed in EXAFS spectra of Fe^{3+}- and Mn^{2+}–HA complexes. Further, no evidence in EXAFS spectra is observed for oxidation number changes in these redox-active metals as a function of varying metal loading.

Both Fe^{3+} and Mn^{2+} ions show six nearest-neighbor O or N atoms at a distance of 2.01 ± 0.05 and 2.20 ± 0.04 Å, respectively (Davies et al., 1997, 2001). The authors suggest that the highest uncertainty of interatomic distance for Fe^{3+} would indicate either that Fe^{3+} occupies more than one binding site with different nearest-neighbor atom distances or that Fe^{3+} occupies a single distorted-octahedral site.

4.11.5. Mercury(II) Complexes

The Hg–L_{III} edge XAS spectrum obtained by Wang et al. (1997) after addition of a small quantity of Hg^{2+} (5 mmol kg^{-1}) to a HA isolated from a soil contaminated with Hg yields a RSF plot showing a broad peak that is FEFF-fitted to an Hg–ligand bond length of 2.05 Å interpreted as an Hg–O bond. However, the EXAFS scan is not continued to higher energies and the spectral noise is high.

In another study (Skyllberg et al., 1997) of Hg^{2+} added to a soil HA at a ratio of 1 mol of Hg^{2+} for each mole of total organic S, the Hg–L_{III} edge spectrum yields an RSF plot with two peaks. The most prominent peak, at 2.1 Å, is consistent with the length for Hg –O or Hg–N bonds, whereas the minor peak at 2.5 Å exactly matches an Hg–S bond length. Using variable ratios of Hg to reduced S, only the Hg–O peak is visible when Hg^{2+} greatly exceeds reduced S, whereas both Hg–O and Hg–S peaks appear at ratios of less than 1 (Bloom et al., 2001).

A detailed EXAFS study of Hg^{2+} complexation by the same soil HA at a ratio of about 3 mol of Hg per mole of reduced S in the HA confirms the participation of reduced S in the binding of Hg^{2+} (Xia et al., 1999). The experimental RSF plot shows the same two peaks observed by Skyllberg et al. (1997) but with greater relative intensity for the Hg–S peak and better peak separation. A two-coordinate binding environment with one O atom and one S atom at distances of 2.02 and 2.38 A, respectively, in the first coordination shell is suggested for Hg^{2+} complexed by HA. Model calculations show that a second shell could contain one C atom and a second S atom at 2.78 and 2.93 A, respectively. Thus, in addition to thiol S, the possible contribution of disulfide/disulfane S is suggested for the bonding of Hg to soil HS (Xia et al., 1999). The appearance of C atom in the second shell suggests that one O-containing ligand such as COOH or phenol OH, rather than H_2O, is bound to Hg^{2+}.

Hersterberg et al. (2001) have shown that the shift from Hg–O/N bonding to Hg–S bonding in HA was a continuous function of either the total-S/Hg ratio or the reduced-S/Hg ratio.

4.11.6. Concluding Comments

XAS in the XANES and EXAFS regimes is a powerful, nondestructive, and noninvasive tool for studying specifically the chemistry of several elements in a complex matrix such as HS. In particular, these techniques allow the study of trace metal ion complexation by HS without any limitation due to the type of metal species, which needs not to be paramagnetic as in ESR spectroscopy. Further, additional details can be obtained on the dentality of the central ions and bond lengths and distances, which cannot be revealed by ESR that lacks a scattering component.

Further, the strong specific bonding sites involved in binding of some metals at low metal/C ratios in soil HS can be studied at more realistic, naturally occurring metal loading. However, synchroton light sources at large synchroton facilities are necessary to produce the intense x-ray beams needed for contemporary XAS.

4.12. EXPERIMENTAL METHODS OTHER THAN SPECTROSCOPIC

A variety of separation and nonseparation methods have been used to speciate metal ions in the presence of HS, assess complexing capacity of HS, and calculate conditional stability constants and quotients of metal–HS complexes. Because of the extreme amplitude of the topic, only a brief overview of the methods most commonly applied to soil HS is provided here. A number of reviews and book chapters have been published on the topic, to which the reader can refer for details (Saar and Weber, 1982; Stevenson and Fitch, 1986; Weber, 1988; Dabek-Zlotorzynska et al., 1998; Nifant'eva et al., 1999).

Most commonly used separation approaches include (1) proton-release titration (Van Dijk, 1971; Stevenson, 1976b, 1977; Takamatsu and Yoshida, 1978), (2) cation exchange with synthetic resins (e.g., Zunino et al., 1972; Crosser and Allen, 1977), (3) gel filtration or permeation chromatography (GPC) (e.g., Mantoura and Riley, 1975; Mantoura et al., 1978; Krajnc et al., 1995); (4) high-performance cation-exchange chromatography (HP-CEC) (Sutheimer and Cabaniss, 1997), (5) conventional dialysis (e.g., Zunino and Martin, 1977; Weber, 1983) and equilibrium dialysis–ligand exchange (EDLE) (Van Loon et al., 1992; Glaus et al., 2000), (6) ultrafiltration (UF) (Nifant'eva et al., 1999), and (7) capillary electrophoresis (CE) (Dabek-Zlotorzynska et al., 1998).

Most important nonseparation, other than spectroscopic methods, include (1) ion-selective electrode (ISE) potentiometry (e.g., Bresnahan et al., 1978; Saar and Weber, 1979, 1980; Buffle et al., 1980; Bhat et al., 1981; Langford et al., 1983); and (2) voltammetric techniques such as anodic stripping voltammetry (ASV) (e.g., Buffle and Greter, 1979; Bhat et al., 1981; Weber, 1983; Turner et al., 1987; Van den Hoop et al., 1995).

Two major disadvantages of dialysis and chromatographic methods, as well as ISE and ASV, are adsorption of interacting species and disturbing of complexation equilibria. Further, ISE and ASV are applicable adequately only to a limited typology of metal ions (e.g., Cu^{2+}, Cd^{2+}, Pb^{2+}, and Zn^{2+}). An important advantage of UF methods is that they do not take as much experimental time as does dialysis, and do not disturb the complexation equilibria.

In conclusion, the main problem is that the comparability of results obtained by different groups of researchers is hindered by the different natures of the methodological approaches and instrumental techniques employed and parameters measured, and by the variability of system conditions used to obtain data.

REFERENCES

Alcacio, T. E., Hesterberg, D., Chou, J. W., Martin, J. D., Beauchemin, S., and Sayers, D. E. (2001). Molecular scale characteristics of Cu(II) bonding in goethite–humate complexes. *Geochim. Cosmochim. Acta* **65**, 1355–1366.

Allison, J. D., and Perdue, E. M. (1995). Modeling metal–humic interactions with MINTEQA2. In *Humic Substances in the Global Environment and Implications on*

Human Health, ed. Senesi, N., and Miano, T. M., Elsevier, Amsterdam, The Netherlands, 927–942.

Altmann, R. S., and Buffle, J. (1988). The use of different equilibrium functions for interpretation of metal binding in complex ligand systems: its relation to site occupancy and site affinity distribution. *Geochim. Cosmochim. Acta* **52**, 1505–1519.

Ashley, J. T. F. (1996). Adsorption of Cu(II) and Zn(II) by estuarine, riverine, and terrestrial humic acids. *Chemosphere* **33**, 2175–2187.

Banerjee, S. K., and Mukherjee, S. K. (1972). Studies of the infrared spectra of some divalent transitional metal humates. *J. Indian Soc. Soil Sci.* **20**, 91–94.

Bartoli, F., Hatira, A., Andre, J. C., and Portal, J. M. (1987). Propriétés fluorescentes et colloidales d'une solution organique de podzol au cours du processus de complexation par le cuivre. *Soil Biol. Biochem.* **19**, 355–362.

Bauer, W. (1995). NMR of organolithium compounds: general aspects and application of two-dimensional heteronuclear Overhauser effect spectroscopy (HOESY). In *Lithium Chemistry*, ed. Sapse, A.-M., and Schleyer, P. V. R., Wiley-Interscience, New York, 125–172.

Benedetti, M. F., Milne, C. J., Kinniburgh, D. G., van Riemsdijk, W. H., and Kopal, L. K. (1995). Metal-ion binding to humic substances: application of the nonideal competitive adsorption model. *Environ. Sci. Technol.* **29**, 446 –457.

Bhat, G. A., Saar, R. A., Smart, R. B., and Weber, J. H. (1981). Titration of soil-derived fulvic acid by copper(II) and measurement of free copper(II) by anodic stripping voltammetry and copper(II) selective electrode. *Anal. Chem.* **53**, 2275 –2280.

Bianconi, A. (1988). XANES spectroscopy. In *X-ray Absorption: Principles, Applications, Techniques of EXAFS, SEXAFS and XANES*, ed. Köningsberger, D. C., and Prins, R., Wiley-Interscience, New York, 573–662.

Bloom, P. R., and Lenheer, J. A. (1989). Vibrational electronic and high-energy spectroscopic methods for characterizing humic substances. In *Humic Substances II: In Search of Structure*, ed. Hayes, M. H. B., MacCarthy, P., Malcolm, R. L., and Swift, R., Wiley, Chichester, West Sussex, England, 409–446.

Bloom, P. R., Bleam, W. F., and Xia, K. (2001). X-ray spectroscopy applications for the study of humic substances. In *Humic Substances and Chemical Contaminants*, ed. Clapp, C. E., Hayes, M. H. B., Senesi, N., Bloom, P. R., and Jardine, P. M., Soil Science Society of America, Madison, WI, 317–350.

Boyd, S. A., Sommers, L. E., and Nelson, D. W. (1981a). Copper(II) and iron(III) complexation by the carboxylate group of humic acid. *Soil Sci. Soc. Am. J.* **45**, 1241–1242.

Boyd, S. A., Sommers, L. E., Nelson, D. W., and West, D. X. (1981b). The mechanism of copper(II) binding by humic acid. An electron spin resonance study of a copper(II)–humic acid complex and some adducts with nitrogen donors. *Soil. Sci. Soc. Am. J.* **45**, 745–749.

Bresnahan, W. T., Grant, C. L., and Weber, J. H. (1978). Stability constants for the complexation of copper(II) ions with water and soil fulvic acid measured by an ion selective electrode. *Anal. Chem.* **50**, 1675–1679.

Buffle, J. (1988). *Complexation Reactions in Aquatic Systems: An Analytical Approach*, Ellis Horwood, Chichester, West Sussex, England.

Buffle, J., and Greter, F. L. (1979). Voltammetric study of humic and fulvic substances, II: Mechanism of reaction of the Pb–fulvic complexes on the mercury electrode. *J. Electroanal. Chem.* **101**, 231–251.

Buffle, J., Deladoey, P., Greter, F. L., and Haerdi, W. (1980). Study of the complex formation of copper(II) by humic and fulvic substances. *Anal. Chim. Acta* **116**, 255–274.

Cameron, D. F., and Sohn, N. L. (1992). Functional group content of soil and sedimentary humic acids determined by CP/MAS ^{13}C NMR related to conditional $Zn^{2+}$and Cd^{2+} formation constants. *Sci. Total Environ.* **113**, 121–132.

Chen, Y., and Stevenson, F. J. (1986). Soil organic matter interactions with trace elements. In *The Role of Organic Matter in Modern Agriculture*, ed. Chen, Y., and Avnimelech, Y., Nijhoff, Dordrecht, The Netherlands, 73–116.

Cheshire, M. V., Berrow, M. L., Goodman, B. A., and Mundie, C. M. (1977). Metal distribution and nature of some Cu, Mn and V complexes in humic and fulvic acid fractions of soil organic matter. *Geochim. Cosmochim. Acta* **41**, 1131–1138.

Chung, K. H., Rhee, S. W., Shin, H. S., and Moon, C. H. (1996). Probe of cadmium(II) binding on soil fulvic acid investigated by ^{113}Cd NMR spectroscopy. *Can. J. Chem.* **74**, 1360–1365.

Connors, K. A. (1987). *Binding Constants: The Measurement of Molecular Complex Stability,* Wiley-Interscience, New York.

Crosser, M. L., and Allen, H. E. (1977). Determination of complexation capacity of soluble ligands by ion exchange equilibrium. *Soil Sci.* **123**, 176–181.

Dabek-Zlotorzynska, E., Lai, E. P. C., and. Timerbaev, A. R (1998). Capillary electrophoresis: the state-of-art in metal speciation studies. *Anal. Chim. Acta* **359**, 1–26.

Davies, G., Fataftah, A., Cherkasskiy, A., Ghabbour, E. A., Radwan, A., Jansen, S. A., Kolla, S., Paciolla, M. D., Sein, L. T., Jr., Buermann, W., Balasubramanian, M., Budnick, J., and Xing, B. (1997). Tight metal binding by humic acids and its role in biomineralisation. *J. Chem. Soc. Dalton Trans.* **1997**, 4047–4060.

Davies, G., Ghabbour, E. A., Cherkasskiy, A., and Fataftah, A. (2001). Tight metal binding by solid phase peat and soil humic acids. In *Humic Substances and Chemical Contaminants*, ed. Clapp, C. E., Hayes, M. H. B., Senesi, N., Bloom, P. R., and Jardine, P. M., Soil Science Society of America, Madison, WI, 371–396.

Deczky, K., and Langford, H. (1978). Application of water nuclear magnetic resonance relaxation times to study of metal complexes of the soluble soil organic matter fraction fulvic acid. *Can. J. Chem.* **56**, 1947–1951.

Dobbs, J. C., Susetyo, W., Carreira, L. A., and Azarraga, L. V. (1989). Competitive binding of protons and metal ions in humic substances by lanthanide ion probe spectroscopy. *Anal. Chem.* **61**, 1519–1524.

Dzombak, D. A., Fish, W., and Morel, M. M. (1986). Metal–humate interaction, 1: Discrete ligands and continuous distribution models. *Environ. Sci. Technol.* **20**, 669–675.

Fendorf, S. E. (1999). Fundamental aspects and applications of x-ray absorption spectroscopy in clay and soil science. In *Synchrotron X-ray Methods in Clay Science*, Vol. 9, ed. Schulze, D. G., et al., CMS Workshop Lectures, Clay Minerals Society, Boulder, CO, 20–67.

Fendorf, S. E., and Sparks, D. L. (1996). X-ray absorption fine structure spectroscopy. In *Methods of Soil Analysis*, 3: *Chemical Methods*, SSSA Book Series 5, ed. Sparks, D. L., et al., Soil Science Society of America, Madison, WI, 377–416.

Gamble, D. S., Langford, C. H., and Tong, J. P. K. (1976). The structure and equilibria of a manganese(II) complex of fulvic acid studied by ion exchange and nuclear magnetic resonance. *Can. J. Chem.* **54**, 1239–1245.

Gamble, D. S., Schnitzer, M., and Skinner, D. S. (1977). Mn(II)–fulvic acid complexing equilibrium measurements by electron spin resonance spectrometry. *Can. J. Soil Sci.* **57**, 47–53.

Gamble, D. S., Underdown, A. W., and Langford, C. H. (1980). Copper(II) titration of fulvic acid ligand sites with theoretical, potentiometric, and spectrophotometric analysis. *Anal. Chem.* **52**, 1901–1908.

Ghosh, K., and Schnitzer, M. (1981). Fluorescence excitation spectra and viscosity behavior of a fulvic acid and its copper and iron complexes. *Soil Sci. Soc. Am. J.* **45**, 25–29.

Glaus, M. A., Hummel, W., and Van Loon, L. R. (2000). Trace metal–humate interactions., I. Experimental determination of conditional stability constants. *Appl. Geochem.* **15**, 953–973.

Goodman, B. A., and Cheshire, M. V. (1975). The bonding of vanadium in complexes with humic acid: an electron paramagnetic resonance study. *Geochim. Cosmochim. Acta* **39,** 1711–1713.

Goodman, B. A., and Cheshire, M. V. (1976). The occurrence of copper–porphyrin complexes in soil humic acids. *J. Soil. Sci.* **27**, 337–347.

Goodman, B. A., and Cheshire, M. V. (1979). A Mössbauer spectroscopic study of the effect of pH on the reaction between iron and humic acid in aqueous media. *J. Soil Sci.* **30**, 85–91.

Goodman, B. A., and Cheshire, M. V. (1982). Reduction of molybdate by soil organic matter: EPR evidence for formation of both Mo(V) and Mo(III). *Nature (London)* **299**, 618–620.

Goodman, B. A., and Cheshire, M. V. (1987). Characterization of iron-fulvic acid complexes using Mossbauer and EPR spectroscopy. *Sci. Total Environ.* **62**, 229–240.

Gregor, J. E., Powell, H. K. J., and Town, R. M. (1989). Evidence for aliphatic mixed mode coordination in copper(II)–fulvic acid complexes. *J. Soil Sci.* **40**, 661–673.

Grimm, D. M., Azarraga, L. V., Carreira, L. A., and Susetyo, W. (1991). Continuous multiligand distribution model used to predict the stability constant of Cu(II) metal complexation with humic material from fluorescence quenching data. *Environ. Sci. Technol.* **25**, 1427–1431.

Gutlich, P., Link, R., and Trautwein, A. (1978). *Mössbauer Spectroscopy and Transition Metal Chemistry,* Springer-Verlag, New York.

Hernandez, D., Plaza, C., Senesi, N., and Polo, A. (2006). Copper(II) and zinc(II) binding to humic acids from pig slurry and amended soils by fluorescence spectroscopy. *Environ. Pollut.* **143**, 212–220.

Hersterberg, D., Sayers, D. E., Zhou, W., Robarge, W. P., and Plummer, G. M. (1997). XAFS characterization of copper in model aqueous systems of humic acid and illite. *J. Phys. IV Fr.* **7**(C2), 833–834.

Hersterberg, D., Chou, J. W., Hutchinson, K. J., and Sayers, D. E. (2001). Bonding of Hg(II) to reduced organic sulfur in humic acid as affected by S/Hg ratio. *Environ. Sci. Technol.* **35**, 2741–2745.

Jameson, R. F. (1981). Coordination chemistry of copper with regard to biological systems. In *Metal Ions in Biological Systems*, Vol. 12, ed. Siegel, H., Marcel Dekker, New York, 1–30.

Juste, C., and Delas, J. (1967). Influence de l'addition d'aluminium, de fer, de calcium, de magnésium ou de cuivre sur la mobilité électrophorétique, le spectre d'absorption infrarouge et la solubilité d'un composé humique. *Ann. Agron.* **18**, 403–427.

Juste, C., and Delas, J. (1970). Étude de quelques propriètés des complexes formés par les acides humiques et les cations. *Bull. Assoc. Fr. Etude Sol.* **4**, 39–49.

Kevan, L., and Kispert, L. (1976). *Electron Spin Double Resonance Spectroscopy*, Wiley-Interscience, New York.

Kevan, L., and Schwartz, R. N. (1979). *Time Domain Electron Spin Resonance*, Wiley-Interscience, New York.

Kingery, W. L., Simpson, A. J., Han, F., and Xing, B. (2001). Nuclear magnetic resonance studies of metal interactions with humic substances. In *Humic Substances and Chemical Contaminants,* ed. Clapp, C. E., Hayes, M. H. B., Senesi, N., Bloom, P. R., and Jardine P. M., Soil Science Society America, Madison, WI, 397–426.

Kinniburgh, D. G., Milne, C. J., Benedetti, M. F., Pinheiro, J. P., Filius, J., and Koopal, L. (1996). Metal ion binding by humic acid: application of the NICA–Donnan model. *Environ. Sci. Technol.* **30**, 1687–1698.

Kinniburgh, D. G., van Riemsdijk, W. H., Koopal, L. K., Borkovec, M., Benedetti, M. F., and Avena, M. J. (1999). Ion binding to natural organic matter: competition, heterogeneity, stoichiometry and thermodynamic consistency. *Colloids Surf. A* **151**, 147–166.

Köningsberger, D., and Prins, R. (1988). *Principles, Applications and Techniques of EXAFS, SEXAFS and XANES Spectroscopy,* Wiley-Interscience, New York.

Koopal, L. K., Van Riemsdijk, W. H., de Wit, J. C. M., and Benedetti, M. F. (1994). Analytical isotherm equations for multicomponent adsorption to heterogeous surfaces. *J. Colloid Interface Sci.* **166**, 51–60.

Krajnc, M., Štupar, J., and Milićev, S. (1995). Characterization of chromium and copper complexes with fulvic acids isolated from soils in Slovenia. *Sci. Total Environ.* **159**, 23–31.

Lakatos, B., Tibai, T., and Meisel, J. (1977a). ESR spectra of humic acids and their metal complexes. *Geoderma* **19**, 319–338.

Lakatos, B., Korecz, L., and Meisel, J. (1977b). Comparative studies on the Mössbauer parameters of iron humates and polyuronates. *Geoderma* **19**, 149–157.

Langford, C. H., Gamblex, D. S., Underdown, A. W., and Lee, S. (1983). Interaction of metal ions with a well characterized fulvic acid. In *Aquatic and Terrestrial Humic Materials*, ed. Christman, R. F., and Gjessing, E. T., Ann Arbor Science, Ann Arbor, MI, 219–237.

Lytle, F. W. (1988). Experimental x-ray absorption spectroscopy. In *Applications of Synchrotron Radiation*, ed. Winick, H., et al., Gordon and Breach Science, New York, 135–223.

MacCarthy, P., and Perdue, E. M. (1991). Complexation of metal ions by humic substances: fundamental considerations. In *Interactions at the Soil Colloid–Soil Solution Interface*, NATO ASI Series, Vol. 190, ed. Bolt, G. H., De Boodt, M. F., Hayes, M. H. B., and McBride, M. B., Kluwer, Dordrecht, The Netherlands, 469–492.

MacCarthy, P., and Rice, J. A. (1985). Spectroscopic methods (other than NMR) for determining functionalities in humic substances. In *Humic Substances in Soil, Sediment and Water: Geochemistry, Isolation, and Characterization*, ed. Aiken, G. R., McKnight, D. M., Wershaw, R. L., and MacCarthy, P., Wiley-Interscience, New York, 527–559.

Macomber, R. S. (1998). *A Complete Introduction to Modern NMR Spectroscopy*, Wiley-Interscience, New York.

Manceau, A., Boisset, M. C., Sarret, G., Hazemann, J. L., Mench, M., Cambier, P., and Prost, R. (1996). Direct determination of lead speciation in contaminated soils by EXAFS spectroscopy. *Environ. Sci. Technol.* **30**, 1540–1552.

Mantoura, R. F. C., and Riley, J. P. (1975). The use of gel filtration in the study of metal binding by humic acids and related compounds. *Anal. Chim. Acta* **78**, 193–200.

Mantoura, R. F. C., Dickson, A., and Riley, J. P. (1978). The complexation of metals with humic materials in natural waters. *Estuar. Coast. Mar. Sci.* **6**, 387–408.

McBride, M. B. (1978). Transitxion metal binding in humic acids: an ESR study. *Soil Sci.* **126**, 200–209.

McBride, M. B. (1982). Electron spin resonance investigation of Mn^{2+} complexation in natural and synthetic organics. *Soil Sci. Soc. Am. J.* **46**, 1137–1143.

McBride, M. B. (1989). Reactions controlling heavy metal solubility in soils. In *Advances in Soil Science,* Vol. 14, ed. Stewart, B. A., Springer-Verlag, New York, 1–56.

McCoustra, M. R. S. (1990). Electronic absorption spectroscopy: theory and practice. In *Perspectives in Modern Chemical Spectroscopy*, ed. Andrews, D. L., Springer-Verlag, Berlin, 88–101.

Milne, C. J., Kinniburgh, D. G., De Wit, J. C. M., van Riemsdijk, W. H., and Koopal, L. K. (1995). Analysis of metal ion binding by a peat humic acid using a simple electrostatic model. *Geochim. Cosmochim. Acta* **59**, 1101–1112.

Nakamoto, K. (1986). *Infrared and Raman Spectra of Inorganic and Coordination Compounds,* Wiley-Interscience, New York.

Nifant'eva, T. I., Shkinev, V. M., Spivakov, B. Ya. and Burba, P. (1999). Membrane filtration studies of aquatic humic substances and their metal species: a concise overview, 2: Evaluation of conditional stability constants by using ultrafiltration. *Talanta* **48**, 257–267.

Peisach, J., and Blumberg, W. E. (1974). Structural implications derived from the analysis of the electron paramagnetic resonance spectra of natural and artificial copper proteins. *Arch. Biochem. Biophys.* **165**, 691–708.

Perdue, E. M. (1989). Effects of humic substances on metal speciation. In *Aquatic Humic Substances Influence on Fate and Treatment of Pollutants*, Advances in Chemistry Series, vol. 219, ed. Suffet, I. H., and MacCarthy, P., American Chemical Society, Washington, DC, 282–295.

Perdue, E. M. (2001). Modeling concepts in metal–humic complexation. In *Humic Substances and Chemical Contaminants*, ed. Clapp, C. E., Hayes, M. H. B., Senesi, N., Bloom, P. R., and Jardine, P. M., Soil Science Society of America, Madison, WI, 305–316.

Perdue, E. M., and Lytle, C. R. (1983). A critical examination of metal–ligand complexation models: application to defined multiligand mixtures. In *Aquatic and Terrestrial Humic Materials*, ed. Christman, R. F., and Gjessing E. T., Ann Arbor Science, Ann Arbor, MI, 295–313.

Piccolo, A., and Stevenson, F. J. (1982). Infrared spectra of Cu^{2+}, Pb^{2+}, and Ca^{2+} complexes of soil humic substances. *Geoderma* **27**, 195–208.

Plaza, C., D'Orazio, V., and Senesi, N. (2005). Copper(II) complexation of humic acids from the first generation of EUROSOILS by total luminescence spectroscopy. *Geoderma* ***125***, 177–186.

Plaza, C., Brunetti, G., Senesi, N., and Polo, A. (2006). Molecular and quantitative analysis of metal ion binding to humic acids from sewage sludge and sludge-amended soils by fluorescence spectroscopy. *Environ. Sci. Technol.* **40**, 917–923.

Pommery, J., Ebenga, J. P., Imbenotte, M., Palavitt, G., and Erb, F. (1988). Determination of the complexing ability of a standard humic acid with cadmium ions. *Water Res.* **22**, 185–189.

Prasad, B., and Sinha, M. K. (1983). Physical and chemical characterization of molecularly homogeneous fulvic acid fractions and their metal complexes. *J. Indian Soc. Soil Sci.* **31**, 187–191.

Prasad, B., Dkhar, G. D. and Singh, A. P. (1987). Cobalt(II), iron(III) and zinc(II) complexation by fulvic acids isolated from north-eastern Himalayan forest and cultivated soils. *J. Indian Soc. Soil Sci.* **35**, 194–197.

Rehr, J. J. (1993). Recent development in multiple-scattering calculations of XAFS and XANES. *Jpn. J. Appl. Phys. Suppl.* **32(2)**, 8–12.

Rehr, J. J., Booth, C. H., Bridges, F., and Zabinsky, S. I. (1994). X-ray absorption fine-structure in embedded atoms. *Phys. Rev. B.* **49**, 12347–12350.

Ryan, D. K., and Weber, J. H. (1982). Fluorescence quenching titration for determination of complexing capacities and stability constants of fulvic acid. *Anal. Chem.* **54**, 986–990.

Ryan, D. K., Thompson, C. P., and Weber, J. H. (1983). Comparison of Mn^{2+}, Co^{2+} and Cu^{2+} binding to fulvic acid as measured by fluorescence quenching. *Can. J. Chem.* **61**, 1505–1509.

Saar, R. A., and Weber, J. H. (1979). Complexation of cadmium(II) with water- and soil-derived fulvic acids: effect of pH and fulvic acid concentration. *Can. J. Chem.* **57**, 1263–1268.

Saar, R. A., and Weber, J. H. (1980). Comparison of spectrofluorometry and ion-selective electrode potentiometry for determination of complexes between fulvic acid and heavy-metal ions. *Anal. Chem.* **52**, 2093–2100.

Saar, R. A., and Weber, J. H. (1982). Fulvic acid: modifier of metal-ion chemistry. *Environ. Sci. Technol.* **16**, 510A–517A.

Sarret, G., Manceau, A., Hazemann, J. L., Gomez, A., and Mench, M. (1997). EXAFS study of the nature of zinc complexation sites in humic substances as a function of Zn concentration. *J. Phys. IV Fr.*, **7** (C2), 799–802.

Schnitzer, M. (1978). Humic substances: chemistry and reactions. In *Soil Organic Matter*, ed. Schnitzer, M., and Khan, S. U., Elsevier, Amsterdam, The Netherlands 1–64.

Schnitzer, M., and Ghosh, K. (1982). Characteristics of water soluble fulvic acid–copper and fulvic acid–iron complexes. *Soil Sci.* **134**, 354–363.

Schnitzer, M., and Skinner, S. I. M. (1963). Organo-metallic interactions in soils, I: Reactions between a number of metal ions and the organic matter of a Podzol Bh horizon. *Soil Sci.* **96**, 86–93.

Senesi, N. (1990a). Molecular and quantitative aspects of the chemistry of fulvic acid and its interaction with metal ions and organic chemicals, II: The fluorescence spectroscopy approach. *Anal. Chim. Acta* **232**, 77–106.

Senesi, N. (1990b). Application of electron spin resonance (ESR) spectroscopy in soil chemistry. In *Advances in Soil Science*, Vol. 14, ed. Stewart, B. A., Springer-Verlag, New York, 77–130.

Senesi, N. (1992). Metal–humic substance complexes in the environment: molecular and mechanistic aspects by multiple spectroscopic approach. In *Biogeochemistry of Trace Metals*, ed. Adriano, D. C., Lewis publishers, Boca Raton, FL, 429–496.

Senesi, N. (1996). Electron spin (or paramagnetic) resonance spectroscopy. In *Methods of Soil Analysis,* Part 3, *Chemical Methods*, SSSA Book Series no.5, Soil Science Society of America, Agronomy Society of America, Madison, WI.

Senesi, N., and Loffredo, E. (2005). Metal ion complexation by soil humic substances. In *Chemical Processes in Soil,* ed. Sparks, D. L., and Tabatabai, M. A., Soil Science Society of America, Madison, WI, 563–618.

Senesi, N., and Sposito, G. (1984). Residual copper(II) complexes in purified soil and sewage sludge fulvic acids: an electron spin resonance study. *Soil Sci. Soc. Am. J.* **48,** 1247–1253.

Senesi, N., Griffith, S. M., Schnitzer, M., and Townsend, M. G. (1977). Binding of Fe^{3+} by humic materials. *Geochim. Cosmochim. Acta* **41**, 969–976.

Senesi, N., Bocian, D. F., and Sposito, G. (1985). Electron spin resonance investigation of copper(II) complexation by soil fulvic acid. *Soil Sci. Soc. Am. J.* **49**, 114–119.

Senesi, N., Sposito, G., and Martin, J. P. (1986). Copper(II) and iron(III) complexation by soil humic acids: and IR and ESR study. *Sci. Total Environ.* **55**, 351–362.

Senesi, N., Miano, T. M., Provenzano, M. R., and Brunetti, G. (1989a). Spectroscopic and compositional comparative characterization of some IHSS reference and standard fulvic and humic acids of various origin. *Sci. Total Environ.* **81–82**, 143–156.

Senesi, N., Sposito, G., Holtzclaw, K. M., and Bradford, G. R. (1989b). Chemical properties of metal–humic acid fractions of a sewage-sludge amended aridisol. *J. Environ. Qual.* **18**, 186–194.

Senesi, N., Sposito, G., Bradford, G. R., and Holtzclaw, K. M. (1991). Residual metal reactivity of humic acids extracted from soil amended with sewage sludge. *Water Air Soil Pollut.* **55**, 409–425.

Shestakov, E. I., Karpukhin, A. I., Fadeev, V. V., and Chubarov, V. V. (1987). Fluorescence intensity of organic compounds containing manganese in podzolic soils. *Izv. Timiryazevsk.* **2**, 82–85.

Shin, H. S., Rhee, S. W., Lee, B. H., and Moon, C. H. (1996). Metal binding sites and partial structures of soil fulvic and humic acids compared: aided by Eu(III) luminescence spectroscopy and DEPT/QUAT ^{13} C NMR pulse techniques. *Org. Geochem.* **24**, 523–529.

Skyllberg, U. L., Bloom, P. R., Nater, E. A., Xia, K., and Bleam, W. F. (1997). Binding of mercury(II) by reduced sulfur in soil organic matter. In *Extended Abstracts, 4th International Conference on the Biogeochemistry of Trace Elements*, June 23–26 1997, Berkeley, CA, U.S. Army Cold Regions Research and Engineering Laboratory, Hanover, NH, 285–286.

Sohn, M., and Rajski, S. (1990). The adsorption of Cd(II) from seawater by humic acids of various sources of origin. *Org. Geochem.* **15**, 439–447.

Sposito, G. (1986). Sorption of trace metals by humic materials in soils and natural waters. *CRC Crit. Rev. Environ. Control* **16**, 193–229.

Sposito, G., Holtzclaw, K. M., and Baham, J. (1976). Analytical properties of the soluble metal-complexing frantions in sludge–soil mixtures, II: Comparative structural chemistry of fulvic acid. *Soil Sci. Soc. Am. J.* **40**, 691–698.

Stevenson, F. J. (1976a). Binding of metal ions by humic acids. In *Environmental Biogeochemistry,* Vol. 2, *Metals Transfer and Ecological Mass Balances*, ed. Nriagu, J. O., Ann Arbor Science, Ann Arbor, MI, 519–540.

Stevenson, F. J. (1976b). Stability constants of Cu^{2+}, Pb^{2+}, and Cd^{2+} complexes with humic acids. *Soil Sci. Soc. Am. J.* **40**, 665–672.

Stevenson, F. J. (1977). Nature of divalent transition metal complexes of humic acid as revealed by a modified potentiometric titration method. *Soil Sci.* **123**, 10–17.

Stevenson, F. J. (1994). *Humus Chemistry: Genesis, Composition, Reactions*, 2nd ed., Wiley-Interscience, New York.

Stevenson, F. J., and Fitch, A. (1986). Chemistry of complexation of metal ions with soil solution organics. In *Interactions of Soil Minerals with Natural Organic Microbes*, ed. Huang, P. M., and Schnitzer, M., Soil Science Society of America, Madison, WI, 2–58.

Susetyo, W., Carreira, L. A., Azarraga, L. V., and Grimm, D. M. (1991). Fluorescence techniques for metal–humic interactions. *Fresenius' J. Anal. Chem.* **339**, 624–635.

Sutheimer, S. H., and Cabaniss, S. E. (1997). Aluminum binding to humic substances determined by high performance cation exchange chromatography. *Geochim. Cosmochim. Acta* **61**, 1–9.

Takamatsu, T., and Yoshida, T. (1978). Determination of stability constants of metal–humic acid complexes by potentiometric titration and ion-selective electrodes. *Soil Sci.* **125**, 377–386.

Tan, K. H. (1978). Formation of metal–humic acid complexes by titration and their characterization by differential thermal analysis and infrared spectroscopy. *Soil Biol. Biochem.* **10**, 123–129.

Teo, B. K. (1986). *EXAFS: Basic Principles and Data Analysis*, Springer-Verlag, Berlin.

Templeton, G. D., and Chasteen, N. D. (1980). Vanadium–fulvic acid chemistry: conformational and binding studies by electron spin probe techniques. *Geochim. Cosmochim. Acta* **44**, 741–752.

Tipping, E. (1998). Humic ion-binding model VI: an improved description of the interactions of protons and metal ions with humic substances. *Aquat. Geochem.* **4**, 3–48.

Tipping, E., and Hurley, M. A. (1992). A unifying model of cation binding by humic substances. *Geochim. Cosmochim. Acta* **56**, 3627–3641.

Turner, D. R., Varney, M. S., Whitfield, M., Mantoura, R. F. C., and Riley, J. P. (1987). Electrochemical studies of copper and lead complexation by fulvic acid, II: A critical comparison of potentiometric and polarographic measurements. *Sci. Total Environ.* **60**, 17–34.

Underdown, A. W., Langford, C. H., and Gamble, D. S. (1981). The fluorescence and visible absorbance of Cu(II) and Mn(II) complexes of fulvic acid: the effect of metal ion loading. *Can. J. Soil Sci.* **61**, 469–474.

Van den Hoop, M. A. G. T., van Leeuwen, H. P., Pinheiro, J. P., Mota, A. M., and Simoes Goncalves, M. de L. (1995). Voltammetric analysis of the competition between calcium and heavy metals for complexation by humic material. *Colloids Surf. A* **95**, 305–313.

Van Dijk, H. (1971). Cation binding of humic acids. *Geoderma* **5**, 53–67.

Van Loon, L. R., Granacher, S., and Harduf, H. (1992). Equilibrium dialysis-ligand exchange: a novel method for determining conditional stability constants of radionuclide–humic acid complexes. *Anal. Chim. Acta* **268**, 235–246.

Wang, Z., Hesterberg, D., Zhou, W., Sayers, D. E., and Robarge, W. P. (1997). Extended x-ray absorption fine structure study of Hg speciation in a flood plain soil. In *Proc. 3rd International Conference on the Biogeochemistry of Trace Elements*, Paris, May 15–19, 1995. Les Colloques 85. CD-ROM, INRA, Paris.

Weber, J. H. (1983). Metal ion speciation studies in the presence of humic materials. In *Aquatic and Terrestrial Humic Materials*, ed. Christman, R. F., and Gjessing, E. T., Ann Arbor Science, Ann Arbor, MI, 315–331.

Weber, J. H. (1988). Binding and transport of metals by humic materials. In *Humic Substances and Their Role in the Environment*, ed. Frimmel, F. H., and Christman, R. F., Wiley-Interscience, Chichester, West Sussex, England, 165–178.

Wertz, J. E., and Bolton, J. R. (1972). *Electron Spin Resonance: Elementary, Theory and Practical Applications*, McGraw-Hill, New York.

Wilson, M. A. (1989). Solid-state nuclear magnetic resonance spectroscopy of humic substances: basic concepts and techniques. In *Humic Substances II: In Search of Structure*, ed. Hayes, M. H. B., MacCarthy, P., Malcolm, R. L., and Swift, R., Wiley, Chichester, West Sussex, England, 309–338.

Xia, K., Bleam, W. F., and Helmke, P. A. (1997a). Studies of the nature of Cu^{2+} and Pb^{2+} binding sites of soil humic substances using x-ray absorption spectroscopy. *Geochim. Cosmochim. Acta* **61**, 2211–2221.

Xia, K., Bleam, W. F., and Helmke, P. A. (1997b). Studies of the nature of binding sites of first row transition elements bound to aquatic and soil humic substances using x-ray absorption spectroscopy. *Geochim. Cosmochim. Acta* **61**, 2223–2235.

Xia, K., Skyllberg, U. L., Bleam, W. F., Bloom, P. R., Nater, E. A., and Helmke, P. A. (1999). X-ray absorption spectroscopic evidence for the complexation of Hg(II) by reduced sulfur in soil humic substances. *Environ. Sci. Technol.* **33**, 257–261.

Zabinsky, S. I., Rehr, J. J., Ankudinov, A., Albers, R. C., and Eller, M. J. (1994). Multiple-scattering calculations of x-ray absorption spectra. *Phys. Rev. B* **52**, 2995–3009.

Zunino, H., and Martin, J. P. (1977). Metal-binding organic macromolecules in soil, 2. *Soil Sci.* **123**, 188–202.

Zunino, H., Galindo, G., Peirano, P., and Aguilera, M. (1972). Use of the resin exchange method for the determination of stability constants of the metal–soil organic matter complexes. *Soil Sci.* **114**, 229–233.

5

FACTORS AFFECTING THE SORPTION–DESORPTION OF TRACE ELEMENTS IN SOIL ENVIRONMENTS

A. Violante
Università di Napoli Federico II, Portici (Napoli), Italy
G. S. R. Krishnamurti
North Vancouver, British Columbia, Canada
M. Pigna
Università di Napoli Federico II, Portici (Napoli), Italy

Biophysico-Chemical Processes of Heavy Metals and Metalloids in Soil Environments,
Edited by Antonio Violante, Pan Ming Huang, and Geoffrey Michael Gadd

5.1. INTRODUCTION

Trace elements may be present in solution with positive or negative charges and in different redox states. They occur predominantly in cationic form [Pb, Cu, Zn, Ni, Cd, Hg, Cr(III), and Co], but some trace elements are present in anionic form [As, Se, Cr(VI), Mo, and B]. Redox reactions, both biotic and abiotic, are of paramount importance in controlling the oxidation state, and thus mobility, phytoavailability, and toxicity of many trace elements, including Cr, Se, Co, Pb, As, Ni, and Cu (Huang and Germida, 2002; Sparks, 2003).

The term *trace elements* includes essential (e.g., Cu, Mn, Se, Zn, Co) and nonessential (e.g., Hg, Cd, Pb) elements. The elements essential for plants and animals are required in low concentrations and are termed *micronutrients*, but at high concentrations they may be toxic for plants, animals, and humans (Bolan and Duraisamy, 2003). Mobility of trace elements in soil environments is affected by sorption–desorption reactions, which are the predominant factors that control the bioavailability of metals. In aquatic environments, processes such as sorption to and desorption from solid phases as well as chemical complexation with inorganic and organic ligands control the bioavailability. As the fate of many trace elements is determined by surface properties of the particulate matter with which they are associated, their chemistry, bioavailability, and transport depend on the degree of partitioning between solid and solution phases. Thus, an understanding of the mechanisms involved in sorption–desorption and mobility is a prerequisite in determining the bioavailability of metal ions in both terrestrial and aquatic environments.

To properly understand the fate of trace elements in soils, and particularly to comprehend their mobility with time, kinetic investigations are necessary (Sparks, 1995). Their sorption by soils is often observed to be a multistep process involving an initial fast sorption followed by slow sorption, probably by diffusion into pores of inner soil surfaces (Kinniburgh and Jackson, 1981), due to the presence of surface sites of different reactivity and site preferences (Ainsworth et al., 1994).

Although first-order models have been used widely to describe the kinetics of chemical reactions on natural materials, a number of other simple kinetic models, such as zero-order, second-order, Elovich, power function, and parabolic diffusion models have also been employed. The final forms of these equations are given in Table 5.1. Complete details and applications of these models can be obtained in work by Sparks (1990, 1999, 2003).

The aim of this chapter is to provide the current state of knowledge on the factors that affect the mobility of trace elements in soil environments. Special attention is given to the influence of inorganic and organic ligands, including nutrients and root exudates, on the sorption–desorption processes of trace elements in cationic and anionic forms on/from soil components and soils.

TABLE 5.1. Linear Forms of Kinetic Equations Used[a]

Zero-order:	$C_t = a_1 - k_1 t$
First-order:	$\ln C_t = a_2 - k_2 t$
Second-order:	$1/C_t = a_3 + k_3 t$
Parabolic diffusion:	$q_t = a_4 + D t^{1/2}$
Power function:	$\ln q_t = \ln k + v \ln t$
Elovich:	$q_t = (1/\beta) \ln \alpha\beta + (1/\beta) \ln t$

[a] C_t is the amount of the element remaining in the soil (for desorption) or in solution (for sorption) at the end of the reaction time t; q_t is the amount of the element released (for desorption) or sorbed (for sorption) in time t; a_1, a_2 and a_3 are constants related to the amounts of the element in solution (for sorption) or in the soil (for desorption) at time 0; a_4 is the amount of element sorbed (for sorption) or released (for desorption) at time 0; k, k_1, k_2, k_3 and k_4 are constants; and D is an "apparent" diffusion coefficient.

5.2. SORPTION OF TRACE ELEMENTS ONTO SOIL COMPONENTS

The soil components responsible for trace element sorption include soil humic substances, phyllosilicates, carbonates, and variable-charge minerals (hydroxides and oxyhydroxides of Fe, Al, Mn, and Ti, short-range-ordered aluminosilicates such as allophanes and imogolite, and phyllosilicates coated by OH–Al or OH–Fe species), but they differ greatly in their sorption capacities, in their cation-and anion-exchange capacities, and in the binding energies of their sorption sites (Jackson, 1998; Sparks, 2003; Violante et al., 2005a). In fact, even a single mineral (e.g., a noncrystalline Al hydroxide) has different types of sorption sites, spanning a range of binding energies. The importance of the variable-charge minerals of soils in sorbing nutrients and pollutants has been widely recognized (Dixon and Weed, 1989; Cornell and Schwertmann, 1996; Sposito, 1996; Jackson, 1998; Kampf et al., 2000; Violante et al., 2002a; Sparks, 2003). The term *variable charge* is used to describe constituents whose charge varies with the pH of the soil solution. Because of their large surface area, disordered network, and high charge density, poorly crystalline metal oxides and aluminosilicates react readily with anions, cations, and organic molecules.

Sorption of trace elements onto soil components is greatly affected by pH, ionic factors, nature of the sorbents, redox reactions, and so on, but the sorption of elements in cationic form differs greatly from that of elements in anionic form. The presence of organic and inorganic ligands (including nutrients) in soil environments has a very important role in the sorption–desorption processes of trace elements.

Depending on the circumstances, organic and inorganic ligands can enhance, inhibit, or have no effect on the sorption of trace elements. Organic ligands are particularly abundant at the soil–plant interface. The rhizosphere is enriched with biomolecules of plant and microbial origins that include organic acids, sugars, amino acids, lipids, flavanoids, proteins, carbohydrates, coumarins, and other substances. Among them, the organic acids are the most significant as substrates for microbial metabolism. The most commonly found organic acids in the rhizosphere are acetic, butyric, citric, fumaric, malic, malonic, oxalic, propionic, and succinic acids (Huang and Schnitzer, 1986; Marschner, 1998; Chang et al., 2002; Violante et al., 2002a,b).

The behavior of foreign ligands on the sorption of elements in cationic form is quite different from that toward elements in anionic form. In fact, the solubility and mobility of cations is affected by complexation to inorganic and organic ligands, depending on whether the metal complexes that they form remain in solution or are themselves sorbed by the sorbent, and on whether sorbed complexes are bound more or less strongly than the free cations (Jackson, 1998). Complexation reactions of trace elements in cationic form with organic and inorganic ligands are particularly important in determining sorption–desorption onto and from soil components as well as their toxicity and phytoavailability in soil environments. An important characteristic of a cation is its strong affinity for OH^- ions. At low pH values the dominant species in aqueous solution is the "free" aquo cation [e.g., $Zn(OH_2)_6{}^{3+}$], but with rising pH it is hydrolyzed to an hydroxylated cation and then to a simple uncharged complex [e.g., $ZnOH^+$ or $Zn(OH)_2$] and under alkaline conditions, to anionic hydroxy species [e.g., $Zn(OH)_3{}^-$ or $Zn(OH)_4{}^{2-}$]. For some elements the free cation predominates throughout the acidic pH range (Zn, Cd, Co), but for others (Hg, Pb) the free cation exists only under extremely acidic conditions (Jackson, 1998). Within the pH range of most natural environments the principal hydrated and hydroxylated forms of most trace elements are cationic, anionic, or uncharged.

According to Lund (1990), the toxicity or bioavailability of a trace element is related to the activity of the free aquo ion. The presence of organic acids in the rhizosphere has a profound influence in affecting trace element availability (Chang et al., 2002). However, this hypothesis may be not valid in all situations; many data are reported in the literature on the uptake of Me–organic chelates by plants (Huang and Germida, 2002, and references therein). Krishnamurti et al. (2004) have shown the bioavailability of Cd–organic complexes to soil algae, which contradicts the long-held notion that Cd–organic complexes are not bioavailable to soil biota although they may increase the mobility of Cd.

Organic and inorganic ligands may compete for common sites on soil components (mainly metal oxides and short-range-ordered aluminosilicates) and affect the sorption of other anions, both nutrients and pollutants (Nagarajah et al., 1970; Lopez-Hernandez et al., 1986; Naidu and Harter, 1998; Neubauer et al., 2002; Violante et al., 2002c). It is well known that variable-charge minerals react readily with inorganic and organic ligands and organic molecules with both low-molecular-mass organic ligands (LMMOLs) and biopolymers (e.g., enzymes, polysaccharides, toxins, DNA, RNA) (Huang and Schnitzer, 1986; Boyd and Mortland, 1990; Violante and Gianfreda, 2000; Huang and Germida, 2002; Violante et al., 2002c). The competition among anions for sorption sites depends on the affinity of the anions for the surfaces of the sorbents as well as the nature and surface properties of the soil components and soils.

5.2.1. Trace Elements in Cationic Form

Except for some noncrystalline minerals that have very high specific surface charge density with highly reactive sites, humic substances appear to have the greatest capacity for sorption of trace elements in cationic form. A body of evidence has demonstrated that humic matter and metal oxides are much more effective scavengers of trace elements in cationic form than even the most efficient sorbent among phyllosilicates, indicating that specific sorption and other complexation processes are the dominant binding mechanisms (Jackson, 1998; Huang and Germida, 2002; Sparks, 2003).

Trace elements in cationic form are probably not dominantly sorbed on 001 faces of phyllosilicates because they are always vastly outnumbered by other cations with which they compete (Jackson, 1998). They may be strongly sorbed only on the edges of the phyllosilicates. However, clay minerals also have an important role as carriers of associated oxides and humic substances forming organomineral complexes, which present peculiar sorption capacities different from those of each single soil constituent (Jackson, 1998; Violante and Gianfreda, 2000; Violante et al., 2002c).

a. Sorption onto Organic Matter Soil organic matter has many different functional groups that act as Lewis bases in binding metals. These include (1) hard bases (ligands preferred by hard cations): carboxylic, phenolic, ester, alcoholic, ether, phosphate, and sulfate; (2) transition bases (ligands preferred by borderline cations): amines and amides; and (3) soft bases (ligands preferred by soft cations): sulfydryl, sulfide, disulfide, and thioether. The high degree of selectivity of soil organic matter for most trace elements in cationic form indicates that they form inner-sphere complexes with the functional groups, forming an internal five- or six-membered ring on structures (Senesi, 1992; Senesi and Loffredo, 1998; Huang and Germida, 2002; Sparks, 2003). The total binding capacity of humic acids for metal inorganic ions is about 200 to 600 mmol kg^{-1}. About 33% of this total is due to retention on cation complexing sites.

Complexation reactions have the following effects: (1) metal ions are prevented from being precipitated; (2) complexing agents can act as carriers for trace elements in soil solution; and (3) their toxicity is often reduced by complexation. The stability constant (K) of trace element complexes with humic acids increases with increasing pH and decreasing ionic strength. Trace elements in cationic form may be grouped into four categories. To the first group belong Fe, Al, and Pb, which are almost completely complexed to substances which themselves bind strongly to mineral surfaces. Chromium and Cu, which are complexed to a large degree as well but to substances that are loosely bound to mineral surfaces, belong to the second group. The ions that belong to the third group (Ni, Co, Cd) form weak complexes, whereas there is no evidence of complexation with natural organic substances for ions in the fourth group. The following stability series summarizes the results: Cr > Fe > Al > Pb ≫ Cu > Ni > Co ≫ Cd > Zn ≫ Mn = Ca = Mg.

Organic materials, either soluble or insoluble, have proven to be effective in reducing the oxidation state of some trace elements (Fendorf, 1995; Huang and Germida, 2002; Bolan et al., 2003; Sparks, 2003). Details of the interactions between soil organic matter and trace elements are reported in Chapter 4.

b. Sorption onto Variable Charge Minerals Variable-charge minerals selectively sorb polyvalent cations even when their surfaces are positively charged [solution pH values lower than the point of zero charge (PZC) of the sorbent]. Most transition cations (Pb, Cu, Cr, Ni, Co, Zn, Al, Fe, Mn) are often sorbed as inner-sphere complexes more strongly than are alkaline earth cations. Spectroscopic techniques such as electron spin resonance (ESR) and Extended x-ray absorption fine structure spectroscopy (EXAFS) have been used for the identification of metal complexes at the surfaces of Al, Fe or Mn oxides, silicate clays, and soil organic matter, as reported extensively in Chapters 3 and 4. The mechanism of metal ion association with hydrous oxide surfaces involves an ion-exchange process in which the sorbed cations replace bound protons. Usually, specifically sorbed cations raise the PZC values of variable-charge minerals. pH affects the sorption of trace elements either by changing the number of sites available for sorption or by changing the concentration of cation species [Me^{2+}, $MeOH^{+}$, $Me(OH)_2$].

A plot of cation sorption versus pH is sigmoidal (Figure 5.1). Sorption, which increases from 0 to 100% of the amount added over a narrow region of 1 to 2 pH units, is termed the *sorption edge*. The pH at which 50% of the total sorption has occurred is called pH_{50}. The lower the pH_{50} value of a trace element for a sorbent, the stronger is the element–surface complex. The pH_{50} position of the sorption edge for a given trace element is related to the hydrolysis or acid–base characteristics. In addition to pH, sorption of metals is dependent on sorptive concentration, surface coverage, and the types of sorbents (Sparks, 2003).

Experiments with various synthetic Fe, Al, and Mn oxides showed that the affinity of trace elements for Mn oxide was usually much greater than that for Fe or Al oxides. Pickering (1979) recorded the following affinity series for freshly precipitated Fe, Al, and Mn oxides.

Fe-oxide:	Pb > Cu > Zn > Ni > Cd > Co > Sr > Mg
Al-oxide:	Cu > Pb > Zn > Ni > Co > Cd > Mg > Sr
Mn-oxide:	Cu > Pb > Mn > Zn > Ni

However, the nature, crystallinity (Kinniburg and Jackson, 1976, 1981; McKenzie, 1980), crystal size, and surface charge of metal oxides and mixed metal oxides (e.g., Fe–Al oxides; Violante et al., 2003) also play an important role in the sorption selectivity of trace elements in cationic form. McBride (1982) compared the sorption behavior of different Al precipitation products of different crystallinity. The Cu sorption capacity followed the order noncrystalline Al-hydroxide > poorly crystalline boehmite > gibbsite. Iron and Mn oxides are

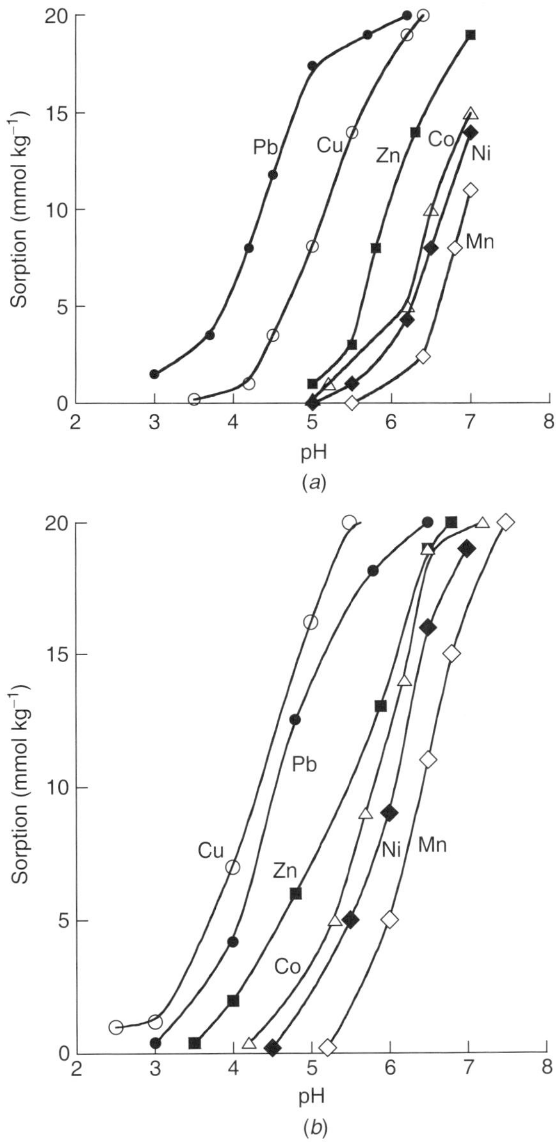

Figure 5.1. Effect of pH on the sorption of metal cations onto (*a*) hematite and (*b*) goethite when they were added at a rate of 20 mmol kg^{-1} of sorbate. (Redrawn from McKenzie, 1980.)

able to catalyze the oxidation of trace elements such as Co, Pb, Ni, Cr, and Cu (Huang, 2000).

Sorbed trace element species may range from simple cations to complex polynuclear hydroxides formed by hydrolysis and condensation in solution or by nucleation on the mineral surfaces (McBride, 1991). Indeed, for most trace elements, except Al, Fe, and Mn, direct precipitation from solution is unlikely because of their very low concentrations, even in polluted soils. In the last decade extensive research has been carried out on the heterogeneous precipitation of trace elements on the surfaces of minerals using modern spectroscopy techniques such as synchrotron-based x-ray absorption spectroscopy (Robarge, 1998; Sparks, 2003, and references therein). Reported observations include zinc precipitates on calcite, cobalt precipitates on Al_2O_3, polynuclear chromium(III) hydroxide structures on silica (Fendorf et al., 1994) and goethite, and the formation of mixed nickel–aluminum hydroxides on pyrophyllite (Scheidegger et al., 1996). Some studies have demonstrated the reduction of transition metals on the surfaces of Fe(II)-bearing minerals (Huang and Germida, 2002). Details of many of these studies are dealt with in Chapter 3.

c. Sorption onto Microorganisms Evidence on the sorption of trace elements on microorganisms have been reported. Yee and Fein (2001) demonstrated that Cd sorption onto various gram-positive and gram-negative bacterial species was pH dependent and that the sorption edge behavior was similar to that of trace elements onto oxides. Free-living bacteria and their extracellular macromolecular products (e.g., fibrils) can accumulate trace elements and may have mineral coatings with bound metals on their surfaces (Beveridge, 1989a,b; Jackson and Leppard, 2002, and references therein). As reported by Jackson and Leppard (2002), trace element accumulation may occur by (1) passive and active uptake by bacterial cells, with immobilization in the cytoplasm; (2) direct sorption, or surface complexation, by the cell wall and extracellular macromolecular products such as acid polysaccharides in colloidal fibrils; (3) precipitation of slightly soluble trace element compounds, such as NiS, on the cell surface; and (4) sorption and coprecipitation by mineral coatings, such as iron and manganese oxyhydroxides, ferrous sulfide, and clay minerals formed in situ by passive or biologically mediated precipitation or sorbed by the cell wall. The microbial accumulation of trace elements by bacterial cells is probably widespread and frequent in nature. Details on the interaction between trace elements and microorganisms are reported in Chapter 2.

d. Kinetics of Sorption Trace element sorption kinetics on oxides, hydroxides, and humic substances depend on the type of surface and trace element being studied, but generally are rapid. Half-times for bivalent cation (e.g., Pb, Zn, Cu, and Cd) sorption on peat ranged from 5 to 15 (Bunzl et al., 1976).

Cadmium sorption in soils is known to be a fast process, with 95% of the Cd sorption taking place in the first 10 minutes and reaching equilibrium in 1 hour (Christensen, 1984). Although the initial sorption of trace elements is rapid, further sorption is usually quite slow, which was ascribed to inter or intraparticle diffusion in pores, sites of low reactivity, and surface precipitation (Waychunas et al., 1993; Sparks, 1999). An important factor affecting the degree of slow sorption of trace elements is the resident time of the sorbate with the sorbent.

The kinetic reactions occurring in the sorption of Ni, Cd, and Zn on goethite during a period of 2 hours to 42 days at pH 6 were hypothesized to occur via a three-step mechanism using a Fickian diffusion model: (1) sorption of trace elements on external surfaces; (2) solid-state diffusion of trace elements from external to internal sites; and (3) trace element binding and fixation at positions inside the goethite particle (Bruemmer et al., 1988).

Slow sorption has also been ascribed to conversion of the sorbate from a high-energy state to a low-energy state. For example, sorption–desorption of Cu from soil was observed to be biphasic, which was attributed to high- and low-energy bonding sites (Lehman and Harter, 1984). Incubations for up to 4 days showed continued uptake of Cu and a decrease in the fraction released within the first 3 minutes, which was referred to as the *low-energy sorbed fraction*.

Ainsworth et al. (1994) observed that oxide aging did not cause hysteresis of trace element cation sorption–desorption. Aging the hydrous ferric oxide with trace elements cations resulted in hysteresis with Cd and Cu, but little hysteresis was observed with Pb. The extent of reversibility with aging for Co, Cd, and Pb was inversely proportional to the ionic radius of the ions (i.e., Co $<$ Cd $<$ Pb). The authors attibuted the hysteresis to Co and Cd incorporation into a recrystallizing solid (probably goethite) via isomorphic substitution, not to micropore diffusion.

Use of pressure-jump relaxation and other relaxation techniques have been shown to offer much in the study of sorption measurements on soil components (Sparks and Zhang, 1991; Sparks, 1995). An especially attractive approach for ascertaining sorption mechanisms on soils would be to combine relaxation approaches with in situ surface spectroscopic techniques. However, there are a few examples in the literature of studies where sorption reactions on soil components have been hypothesized via kinetic experiments and verified in separate spectroscopic investigations (Fuller et al., 1993; Waychunas et al., 1993; Fendorf et al., 1997; Grossl et al., 1997; Scheidegger et al., 1997).

5.2.1.1. Simultaneous Sorption of Trace Elements Few studies have been carried out on the competition in sorption of trace elements to soil components and soils. Competition in sorption between two or more trace elements is of paramount importance for understanding their relative affinity for a given sorbent.

Benjamin and Leckie (1981) found a nearly complete lack of competition of Cd, Cu, Zn, and Pb for sorption sites of a noncrystalline Fe precipitate. The lack of competition of these cations may be related to low surface coverage and/or precipitation effects and only partially to the presence of different binding sites having high selectivity for specific trace element ions (Kretzschmar and Voegelin, 2001).

Cowan et al. (1991) studied Cd sorption on iron oxides in the presence of alkaline-earth elements and demonstrated that Cd sorption was decreased with increasing Ca concentration. Competition between Cd and Zn for sorption sites on MnO_2, and Cu, Pb, and Zn on goethite has been demonstrated mainly at high surface loading of the sorbents with sorbed cations (Kretzschmar and Voegelin, 2001).

Sarkar et al. (1999) showed that the presence of small concentrations of Pb and Ni decreased the sorption of Hg at pH values of maximum sorption: of 38 and 31% on quartz and 14 and 11% on gibbsite, respectively. Recently, Violante et al. (2003) carried out experiments on the competitive sorption of Cu and Zn on a ferrihydrite. They demonstrated that Cu has a greater affinity for the surfaces of ferrihydrite and thus inhibits the sorption of Zn on common sites and is also able to remove Zn previously sorbed onto them.

Few studies have been carried out on the competition of three or more trace elements onto soil components or soils. Elliott et al. (1986) studied the competitive sorption of Cd, Cu, Pb, and Zn onto four soils with different chemical properties. For two mineral soils, sorption under acidic conditions (pH 5.0) followed the sequence Pb > Cu > Zn > Cd, which corresponds to the order of increasing pK for the first hydrolysis product. For two soils with high organic matter content, the order was Pb > Cu > Cd > Zn. Saha et al. (2002) examined the simultaneous sorption of Cd, Zn, and Pb on montmorillonite (Mt) and on hydroxyaluminum (HyAl)–and hydroxyaluminosilicate (HyAlSi)–montmorillonite complexes. The presence of HyAl and HyAlSi polymers on montmorillonite greatly increased the sorption of all three trace elements. The overall sorption behavior of the trace elements showed that the strength of sorption followed the order Pb $\gg$ Zn > Cd among the trace elements and HyAl–Mt $\cong$ HyAlSi–Mt $\gg$ Mt among the sorbents.

Competitive sorption of trace elements to organic soil components has also been studied. Kinniburgh et al. (1996) demonstrated that Cd sorption on a humic acid was reduced by Ca, but in contrast, Cu sorption was poorly reduced. Mandel et al. (2000) showed clear competitive effects of Ca and Mg on Ni sorption to a soil fulvic acid. Many studies have showed evidence that there may be differences in competition between selected trace elements depending on the functional group composition of the humic substances (Kretzschmar and Voegelin, 2001).

Capasso et al. (2004) have studied the sorption of Cu, Zn, and Cr(III) added alone or as a mixture of two or three cations onto polymerin, a humic-acid-like fraction of olive oil mill wastewater. The presence of both Zn and Cu did not affect the capacity of Cr(III) to be sorbed onto the humiclike sample (Figure 5.2*a*). In contrast, the presence of Zn and/or Cr(III) reduced Cu sorption. At an equilibrium concentration of 4 meq L^{-1}, the presence of Zn reduced Cu sorption by 15%, and the concomitant presence of Zn and Cr(III) decreased Cu sorption by about 50% (Figure 5.2*b*), showing that Cr(III) was much more effective than Zn in preventing Cu sorption. At the same equilibrium concentration of 4 meq L^{-1} Zn, the presence of Cu or Cu and Cr(III) decreased the sorption of Zn by about 70 and 79%, respectively, indicating that both Cu and Cr(III) strongly inhibited Zn sorption on the organic matter (Figure 5.2*c*).

Recently, Agbenin and Olojo (2004) studied the competitive sorption of Cu and Zn by a Bt horizon of a Savanna Alfisol as affected by pH and selective removal of Fe oxides and organic matter. These authors showed that the distribution coefficient K_d was five times greater for Cu than Zn. The removal of organic matter from the soil reduced $K_{d,\mathrm{Cu}}$ 40-fold compared to the natural soil, whereas

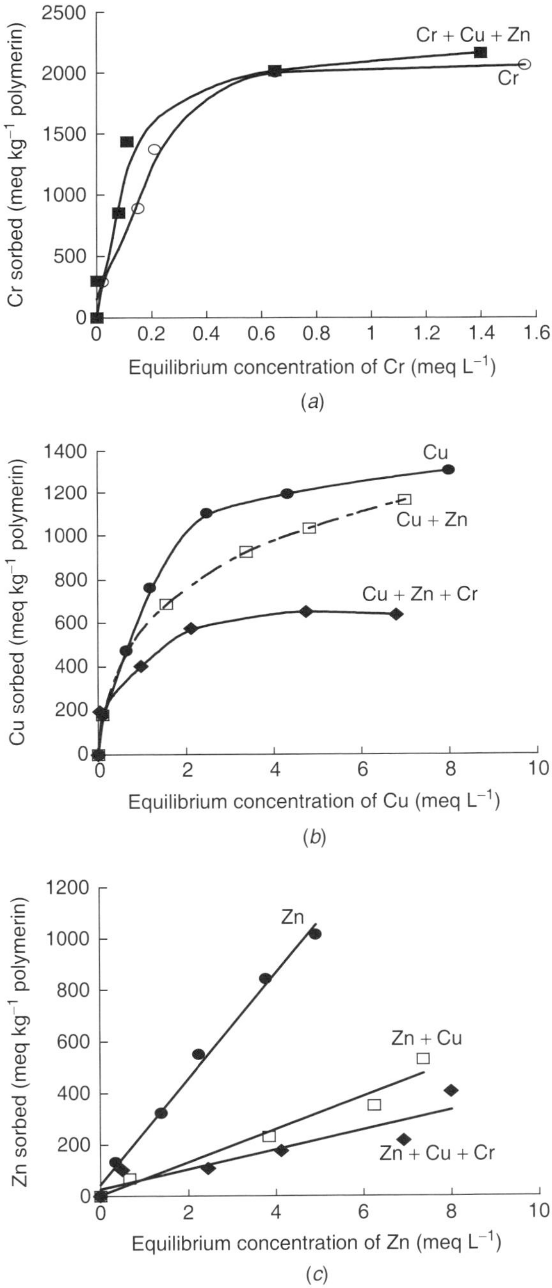

Figure 5.2. Sorption onto a humiclike material (polymerin) of (*a*) Cr(III) alone and as affected by the presence of equimolar concentrations of Zn + Cu, (*b*) Cu alone and as affected by the presence of equimolar concentrations of Zn and Zn + Cr(III), and (*c*) Zn alone and as affected by the presence of equimolar concentrations of Cu and Cu + Cr(III). (From Capasso et al., 2004.)

$K_{d,\mathrm{Zn}}$ was reduced by half. Removing amorphous hydrous oxides reduced $K_{d,\mathrm{Cu}}$ 100-fold and $K_{d,\mathrm{Zn}}$ by 20-fold as compared to natural soil. The metal sorption sites in the amorphous hydrous oxides and organic matter were more selective for Cu than for Zn.

5.2.1.2. Effect of Inorganic and Organic Ligands Many factors, such as pH, surface properties of the sorbents, number of sites available for sorption, and nature and charge of Me–L species in solution influence trace element sorption onto soil inorganic components (phyllosilicates and variable-charge minerals) in the presence of inorganic and biological ligands (Kinniburgh and Jackson, 1976, 1981; McBride, 1989; Goldberg et al., 1996a; McBride et al., 1997; Jackson, 1998; Violante et al., 2003).

a. Sorption onto Phyllosilicates Inorganic and organic ligands (e.g., organic acids) which form strong complexes with trace element cations usually prevent or reverse their association with negatively charged sorbents, as clay minerals, by forming stable dissolved or dispersed negatively charged complexes with the cations. Chloride ions form complexes in soil solution with some trace elements (e.g., Hg, Cd). They tend to displace OH^- ions from Hg(II), forming various dissolved complexes such as $HgCl_2$, $HgCl_3^-$, and $HgCl^+$, unless the pH is high enough for OH^- ions to prevail (Jackson, 1998). It has been ascertained that increasing concentrations of Cl^- strongly inhibited Hg sorption onto quartz (Leckie and James, 1974; Sarkar et al., 1999). Sulfate and phosphate reduced Hg(II) sorption on quartz due to the formation of $Hg(OH)_2SO_4^{2-}$, $Hg(OH)_2H_2PO_4^-$, and $Hg(OH)_2HPO_4^{2-}$ aqueous species (Sarkar et al., 1999). McBride (1991) showed that various LMMOLs (phthalate, fulvate, and citrate) strongly inhibited (in the order cited) the sorption of Cu (Figure 5.3) and Cd by montmorillonite. Zhou et al. (1999) found that Cu sorption on kaolinite at pH 4.0 was increased very slightly by increasing the oxalate/Cu molar ratio from 0 to 0.1 and was then drastically reduced at greater ratios. Farrah and Pickering (1976a,b), Pickering (1979), and Sakurai and Huang (1995) also found that oxalate strongly reduced Cu and Cd sorption on kaolinite and montmorillonite, whereas Wu et al. (2003) found that Pb sorption onto montmorillonite decreased with increasing concentration of EDTA and citric acid.

On the contrary, the presence of certain foreign ligands, occurring naturally in the rhizosphere, such as siderophores produced by microorganisms and phytosiderophores exuded by plants, may promote the sorption of trace elements onto phyllosilicates. They may also help to modify the mobility and then the phytoavailability of trace elements at the soil–root interface. Concentrations of microbial siderophores ranging up to 240 μg per kilogram of soil have been measured in bioassay. Microbial siderophores are outstanding in their specificity for Fe(III) (Kraemer et al., 2002; Reichman and Parker, 2005). For example, the stability constant of the 1 : 1 Fe(III) complex of hydroxamate siderophore desferrioxamine-B (DFOB) is about 16 orders of magnitude larger than that for its 1 : 1 Cu(II) complex (Kraemer et al., 2002).

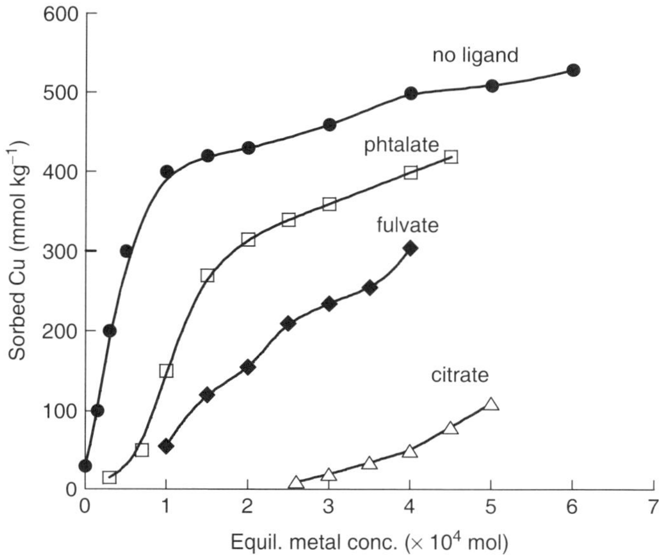

Figure 5.3. Effects of various organic complexing ligands on the sorption of Cu by montmorillonite. (Modified from McBride, 1991.)

Neubauer et al. (2000, 2002) studied the influence of DFOB on the sorption of Cu, Zn, and Cd onto montmorillonite (Figure 5.4) and kaolinite. DFOB showed a strong affinity for montmorillonite at pH < 7.5 (Figure 5.4*a*). At low pH and in the absence of DFOB only small fractions of these trace elements were sorbed on montmorillonite (Figure 5.4*b* to *d*). With increasing pH, sorption of trace elements increased, first for Cu and then for Zn and Cd. In the alkaline pH range, these elements (pH > 7 for Cu and pH > 8 for Zn and Cd) were easily removed from solution. The presence of DFOB (a DFOB/Me molar ratio of 3 to 4) promoted the sorption of Cu in the entire pH range (Figure 5.4*b*), whereas for Zn and Cd, sorption edges were shifted to lower pH values and maximum sorption was reached at about pH 7 for Zn (Figure 5.4*c*) and pH 8 for Cd (Figure 5.4*d*). Electrostatic interactions are the reasons for enhanced trace element sorption onto montmorillonite. In the case of Zn and Cd, positive-charged metal–DFOB complexes ($MeLH_2^+$) were the dominant species at about pH 7 and 8, respectively, and promoted sorption at pH < 7 or 8. The formation of neutral metal–DFOB complexes at pH > 7.5 (Zn) and 8 (Cd) was responsible for the decrease of sorption of these trace elements in alkaline systems.

b. Sorption onto Variable-Charge Minerals The processes, which affect the sorption of trace element cations onto variable-charge minerals in the presence of complexing agents, are particularly complex. In fact, complexing ligands may prevent sorption of trace element cations by forming soluble complexes or by being sorbed onto the surfaces of these minerals, blocking some sorption sites,

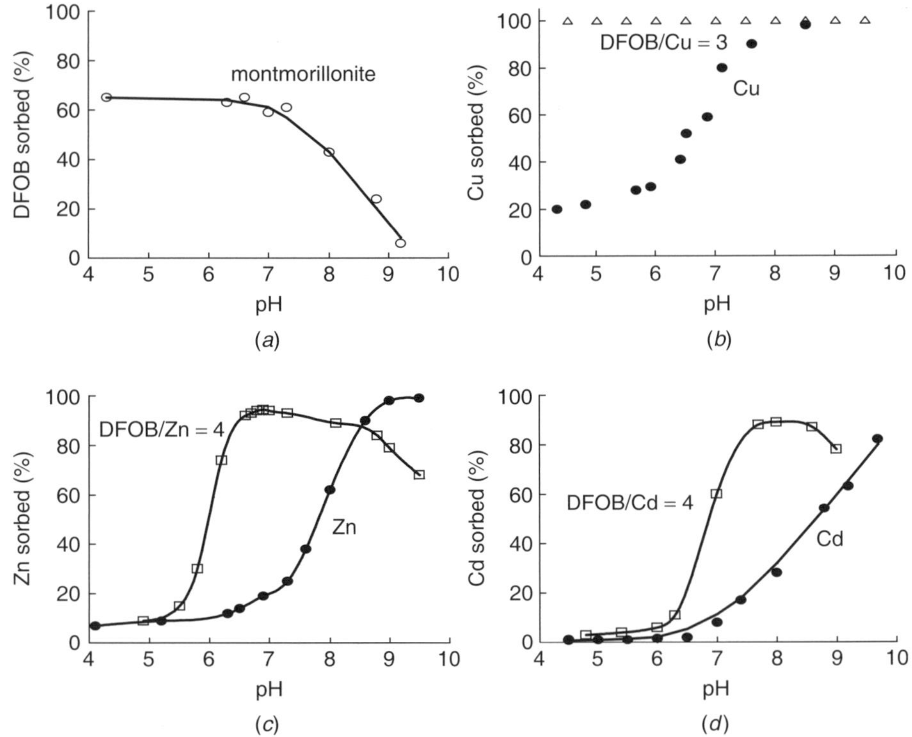

Figure 5.4. Sorption of desferrioxamine B (DFOB) on Na-montmorillonite in the absence of trace elements (*a*). Sorption of Cu (*b*), Zn (*c*), and Cd (*d*) as a function of pH in the absence and presence of DFOB. The initial concentration of the trace elements was 87.5 μmol L^{-1}. (Modified from Neubauer et al., 2002.)

or may facilitate their sorption, under certain conditions, by decreasing the positive charge of the sorbents. Enhanced sorption has been observed for Cd, Cu, and Zn on Al and Fe oxides in the presence of sulfate or of phosphate (Bolland et al., 1977; Kuo, 1986) due to the increased negative charge brought to the surfaces by these inorganic ligands, whereas high levels of phosphate sorption on a noncrystalline metal oxide inhibited Cu sorption, probably due to blocking of surface sites (McBride, 1985). According to many authors (McBride, 1989; Jackson, 1998, and references therein), some ligands enhance trace element sorption on oxides of Al, Fe, and Si by forming stable surface–metal–ligand complexes (*ternary complexes*). Chairidchai and Ritchie (1992) found that the effectiveness of an organic ligand in influencing trace element sorption on soil minerals or soils is affected by the PZC of a sorbent, the pH of the soil solution, and the quantity of complex formed. When the pH is above the PZC of a sorbent the organic ligand decreases the trace element sorption, but the opposite occurs when the pH is below the PZC of a sorbent. Murphy and Zachara (1995) proposed that anionic organic ligands enhance trace element sorption at pH values below the intersection of the ligand sorption envelope with the free-trace elements sorption edge, whereas they will diminish sorption above this pH value. Certainly, the ligand/metal ratio, the nature of organic and inorganic ligands and trace elements, and the surface properties of the sorbent seem to be critical in determining whether trace element sorption at surfaces is enhanced or inhibited.

Zhou et al. (1999) and Violante et al. (2003) studied the influence of increasing concentrations of oxalate and tartrate on the sorption of Pb or Cu at pH 4.0 to 4.5 on different crystalline (goethite and bayerite) and short-range-ordered oxides (noncrystalline Al precipitation products, ferrihydrite, mixed Fe–Al gels), at organic ligand/Pb or Cu molar ratios (r_L) ranging from 0.1 to 15.0. Figure 5.5 shows that the quantities of Pb sorbed onto ferrihydrite increased with increasing r_L up to 10, whereas Cu sorption on goethite strongly increased about three fold by increasing r_L from 0 to 2.5 and then rapidly decreased at greater r_L values. The increase of sorption of Pb on ferrihydrite by increasing oxalate/Pb molar ratio up to 10 must be attributed to the large surface area of this oxide, which allows the sorption of large amounts of Pb–Lx complexes; in contrast, on goethite the decrease in the sorption of Cu at an organic ligand/Cu molar ratio > 2.5 is due to the lower surface area of this crystalline metal oxide.

Yamaguchi et al. (2002) studied the influence of surface area of gibbsite on Ni sorption in the ternary Ni–gibbsite–citrate system over time. At low surface areas and corresponding high surface loadings, a Ni–Al layered double hydroxide (LDH) precipitate formed. At high surface area and lower surface loading, formation of an inner-sphere surface complex prevailed and a small amount of Ni–Al LDH formed only after an extended aging period. Citrate reduced the amount of Ni sorbed, but the effect was more pronounced for the gibbsite with a low surface area than for that with a large surface area. Furthermore, citrate prevented the formation of a LDH phase.

Neubauer et al. (2002) found that the sorption of Cu and Zn to ferrihydrite and goethite was strongly affected by certain siderophores (e.g., DFOB), but their

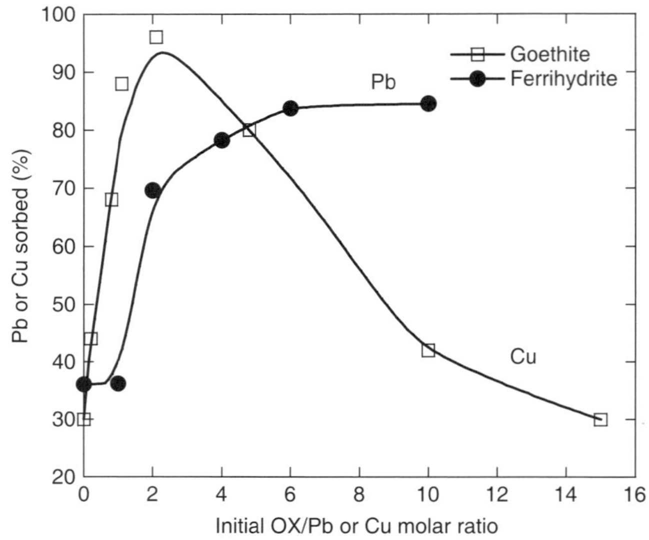

Figure 5.5. Effect of increasing concentrations of oxalate (OX) on the sorption of Cu onto goethite and Pb onto ferrihydrite. The initial concentration of Cu and Pb was 40 and 117 mmol kg^{-1}, respectively. (Modified from Zhou et al., 1999 and Violante et al., 2003.)

behavior was different from that of LMMOLs (e.g., oxalate) described previously (Figure 5.5). Between pH 4 and 8, DFOB completely prevented Cu sorption on ferrihydrite (Figure 5.6*a*). A strong mobilizing effect was also observed for Zn (Figure 5.6*b*). According to these authors, positively charged DFOB complexes were dominant up to pH 9.0 for Cu and up to pH 8.0 for Zn. Since the complexes are not attracted by positively charged ferrihydrite, DFOB decreased sorption of Cu and Zn, in contrast to montmorillonite suspensions (as discussed before; Figure 5.4).

Kraemer et al. (2002) demonstrated that DFOB had a strongly depleting effect on Eu(III) sorption by goethite and boehmite above pH 5. The dramatic reduction in sorbed Eu(III) was attributed to the formation of a strong cationic complex that was repelled from the positively charged surface of either goethite and boehmite.

5.2.1.3. Implication in Trace Element Bioavailability Many studies have been conducted on the sorption of trace elements in cationic form onto natural soil samples, showing in these cases that ligand ions can inhibit, promote, or have no effect on their sorption. The influence of inorganic and organic ligands on the mobility of trace elements is affected by the chemical, physicochemical, and mineralogical properties of soils (Mench and Martin, 1991).

Naidu and Harter (1998) and Bolan et al. (1999), using soils having varied amounts of variable-charge components, showed that there was a significant increase in sorption of Cd by soils (Oxisols, Xeralf) in the presence of sulfate

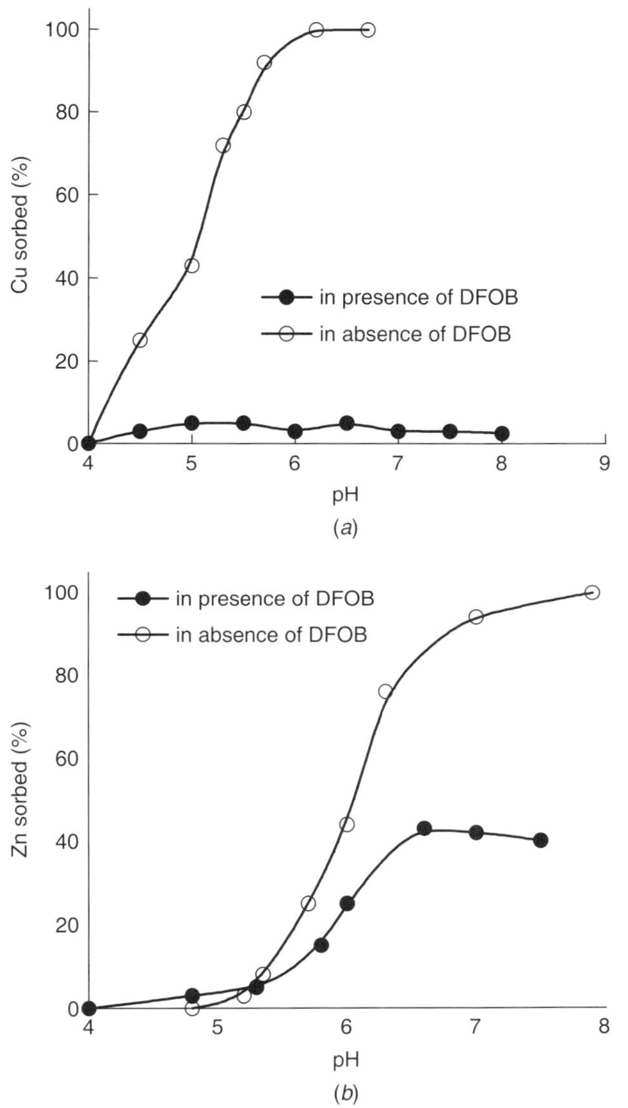

Figure 5.6. Percent of (*a*) Cu and (*b*) Zn sorbed onto ferrihydrite without (○) and with (●) 300 mM DFOB. Total metal concentrations were approximately 10 μmol L^{-1}. The ionic strength was 0.1 ($NaNO_3$). (Redrawn from Neubauer et al., 2002.)

or phosphate, but only a small effect of increasing phosphate sorption on Cd sorption by soils dominated by permanent-charge silicate clay minerals.

Recently, Bolan and Duraisamy (2003) have shown some case studies on the role of inorganic and organic soil amendments on immobilization and phytoavailability of trace elements. They showed evidence that lime is effective in reducing the phytoavailability of Cd and Cr(III), phosphate compounds are effective for Cd, and organic amendments are effective for Cu and Cr(III). The mechanisms proposed were: enhanced trace element sorption through increased surface charge (e.g., phosphate-induced trace element sorption), increased formation of complexes (e.g., Cd–phosphate and Cu–organic matter complexes), precipitation (e.g., chromic hydroxide), and finally, reduction of trace elements from higher-valency mobile form to lower-valency immobile form [e.g., Cr(VI) to Cr(III)].

The presence of phosphate was shown to increase Cd concentration in solution, with Cd present as Cd–phosphate complexes. This was cited as an explanation for a decrease in Cd sorption in the presence of phosphate (Krishnamurti et al., 1999b; Table 5.2), as the free Cd activity, rather than total Cd in solution, is a controlling factor in Cd sorption (O'Connor et al., 1984). On the other hand, phosphate was reported to enhance the sorption of Cu by forming bridges between the mineral surface and the metal (Pickering, 1979).

Peat and phosphate amendments are known to inhibit Cd uptake by plants. The mobility of Cd in phosphorus-amended soil is reduced significantly with a decrease in exchangeable fraction and a corresponding increase in carbonate and oxide fractions (Hettiarichchi et al., 1997).

There is controversy within the literature as to the importance and precise nature of Cd–organic associations in soil solutions. Cadmium usually seems to be present to a large extent in free ionic form or in inorganic complexes with chloride, sulfate, and phosphate. Neal and Sposito (1986) found that the sorption of Cd at a soil solution concentration between 0.001 and 0.1 μmol Cd kg^{-1} was inhibited by the formation of Cd–organic associations in aqueous solution. Later, Naidu and Harter (1998) studied the effect of pH and different organic ligands on sorption and extractability of Cd by soils varying widely in chemical characteristics. For all soils, the amount of Cd extracted decreased with increasing pH, irrespective of the nature of the organic ligand. The ability of ligand ions to desorb Cd followed the sequence maleate > citrate > fumarate > succinate = tartrate > malonate > oxalate > salicylate > acetate. Using

TABLE 5.2. Percent Cd Sorbed by Luseland Soil as a Function of Phosphate Concentration[a]

Phosphate Concentration (mol L^{-1})	% Cd Sorbed		
	15 min	30 min	60 min
0	97	97	98
0.10	83	85	86
0.50	58	60	64
1.00	38	45	51

Source: Krishnamurti et al. (1996b).

[a] Cd concentration: 8.9×10^{-5} mol L^{-1}; phosphate added as monoammonium phosphate.

experimental studies designed to separate the pH effects from ligand ion effects, it was found that at high pH values, Cd–ligand ion complexation was essential for the solubilization of Cd. They also concluded that the sorption curve slope depends on the relative energies of the metal–ligand and metal–surface bonds and on the ligand concentrations. They also found that in soils where ligand ions enhanced sorption of Cd, zeta potential studies provided evidence of soil surface charge reversal though ligand ion binding to soils. On the other hand, no significant effect on barley grain Cd concentration was observed when Cd was applied to a soil in various forms (Singh and Myhr, 1998). Also, an increased Cd availability to crops on a sewage-sludge-amended soil does not indicate the immobilization of Cd under the influence of increased organic matter content (Tichy et al., 1997).

Lorenz et al. (1997) have stated that the free ionic concentration in soil solution did not predict the concentration of Cd and Zn in plants better than does their total concentration in soil solution, which suggests that analysis of Cd and Zn speciation in soil solution is of little practical importance when their bioavailability is assessed.

Chelation and complexing are the key reactions governing Cu behavior in soils. Sorption of Cu by soils is related to the formation of organic complexes and is highly dependent on pH (Kabata-Pendias, 2001). Due to the great affinity of Cu for organic complexing, soluble Cu–organic forms appear to comprise most of the Cu solution over a wide range of pH in soils (Sauvé et al., 1997; Krishnamurti and Naidu, 2002). Organic complexing of Cu has a prominent practical implication in governing the bioavailability and the mobility of Cu in soils. The bioavailability of soluble forms depends most probably on both the molecular weight and amount of Cu complexes. Compounds of low molecular mass liberated during the decay of plant and animal residues as well as those applied with sewage sludges may greatly increase the Cu bioavailability (Kabata-Pendias, 2001).

5.2.2. Trace Elements in Anionic Form

Trace elements that exist in anionic form are sorbed primarily by chemisorption at reactive sites of metal oxides and allophanes and at the edges of phyllosilicates (Cornell and Schwertmann, 1996; Kampf et al., 2000; Violante et al., 2002c). Usually, they are not sorbed on soil organic matter, but certain elements (e.g., borate, arsenate, arsenite) are found to be bound to soil organic matter (Thanabalasingan and Pickering, 1986; McBride, 2000). Indeed, some organic anions may bond indirectly to organic groups through a bridging hydrolytic species of Al and Fe. Carbonates are also important sorbing surfaces (Goldberg and Glaubig, 1988a,b).

Sorption of anions onto variable-charge minerals and soils varies with pH. With increasing pH, within a certain range, sorption decreases (due to a decrease of positive charge of minerals), or increases to a maximum close to the pK_a for anions of monoprotic conjugate acids and then decreases. Slope breaks have been observed at pK_a values for anions of polyprotic conjugate acids (Hingston, 1981). Arsenite and selenite may be sorbed more easily at high pH values because they form weak acids at low pH values and consequently, may be dissociated only in alkaline environments (Sparks, 2003). Anions may be sorbed specifically or nonspecifically. Ligands which are specifically sorbed replace OH^- or OH_2 groups from the surfaces of variable-charge minerals. These reactions are promoted at low pH, which causes OH^- groups to accept protons, OH_2 group being an easier ligand to displace than OH^-. Specific sorption is also termed *inner-sphere sorption* because it involves direct coordination to the surface metal atom. Nonspecific sorption is also termed *outer-sphere sorption* and is influenced by the ionic strength of the system. Specifically sorbed anions usually lower the PZC of metal oxides; thus, the PZC of a particular oxide may give rise to different values depending on the type and extent of foreign ion sorption.

Trace elements, which form inner-sphere complexes, are molybdate, arsenate, arsenite, and selenite. Chemical behavior of arsenate is similar to that of phosphate and may form different surface complexes on inorganic soil components: monodentate, bidentate–binuclear, and bidentate–mononuclear complex in different proportions depending on pH and surface coverage (Hsia et al., 1994; Sun and Doner, 1996; Fendorf et al., 1997; Manning et al., 1998; O'Reilly et al., 2001). According to Fuller et al. (1993) and Waychunas et al. (1993), arsenate was sorbed predominantly as inner-sphere bidentate complexes, regardless of whether arsenate was sorbed postmineralization of the ferrihydrite or was present during precipitation. Several studies have suggested that arsenate is sorbed more than arsenite in a wide range of pH. However, literature studies have found that arsenite is sorbed more than arsenate at high solution pH, particularly onto Fe oxides. Arsenate and arsenite sorption on amorphous Fe oxide and Al oxide showed very little ionic strength dependence in the range 0.02 to 0.1 mol L^{-1} as a function of solution pH, but arsenite sorption showed decreasing sorption with increasing ionic strength in the range 0.1 to 1.0 mol L^{-1}, indicative of an outer-sphere sorption mechanism (Goldberg, 2002). The oxidation of arsenite to arsenate can be catalyzed on the surfaces of Mn oxides (Oscarson et al., 1981).

Soil selenium content is significantly correlated with total carbonate, free Fe oxide, extractable Al and Fe, and clay content. Selenite sorption onto iron oxides and a calcareous montmorillonitic soil was much greater than selenate (Goldberg and Glaubig, 1988a). Using EXAFS measurements, Hayes et al. (1987) ascertained that selenate formed a weakly bonded outer-sphere complex and selenite formed a strongly bonded inner-sphere complex when sorbed on goethite. However, Manceau and Charlet (1994), also using EXAFS spectroscopy, found that selenate may form inner-sphere complexes onto goethite in a manner analogous to sulfate.

The sorption mechanism of chromate is unclear. Zachara et al. (1989) suggested that chromate forms an outer-sphere complex on the surfaces of Fe and Al oxides. However, spectroscopic studies have shown that chromate forms inner-sphere complexes (both bidentate and monodentate) on goethite (Fendorf et al., 1997). This anion has a smaller shared charge than do arsenite and arsenate, creating a weaker bond on sorption (McBride, 1994) and consequently, exhibits a steeper reduced sorption at near-neutral pH values than that of arsenate (Grossl et al., 1997). In the Grossl et al. (1997) study double relaxation was reported for both chromate and arsenate sorption–desorption over the pH range 5.5 to 7.5, which indicates a mixture of monodentate and bidentate surface complexes from both kinetic and spectroscopic experiments.

Detailed information on the affinity sequence of various anions as either nutrients or toxins on soil components is still scanty. Competition in sorption among anions may give useful information.

5.2.2.1. Competitive Sorption Competition in sorption between trace elements in anionic form has also received attention. However, a systematic investigation of the relative competition for sorption onto variable-charge minerals and soils among various anions with different binding affinities is rather limited. Roy et al. (1986) found that the sorption of arsenate by Catlin soil (containing mainly illite and 2 : 1 layer minerals, plus kaolinite and chlorite) appeared to be independent of molybdate, while the presence of arsenate lowered molybdate sorption. However, these authors also found that the sorption of arsenate by two soils containing kaolinite (mainly), chlorite, and metal oxides was reduced in the presence of molybdate, whereas arsenate did not compete strongly with molybdate sorption. Manning and Goldberg (1996a) studied the effects of pH and competing molybdate and arsenate ions onto goethite and gibbsite. Molybdate at 50% surface coverage decreased the sorption of arsenate only at pH < 6.0, whereas arsenate reduced molybdate sorption within a wider pH range (2 to 9 for goethite and 2 to 8 for gibbsite). Their data suggested that arsenate occupies a fraction of the pH-dependent molybdate sorption sites on both goethite and gibbsite and that another distinct fraction of sites has a higher affinity for molybdate than arsenate at low pH. These authors (Manning and Goldberg, 1996b) also explored the presence of molybdate at 10-fold greater concentrations than arsenate and found that there were only slight decreases in arsenate sorption.

Goldberg (2002) found no evidence of any competition in sorption of arsenate and arsenite on Al or Fe oxides and montmorillonite, but only a small and apparent competitive effect of equimolar arsenate on arsenite sorption on kaolinite and illite. The minor competitive effect in this study was due to the small concentrations of As far from site saturation. In fact, Jain et al. (1999) showed evidence that arsenate prevents arsenite sorption on ferrihydrite.

The competitive effect of selenite on arsenate sorption on a calcareous montmorillonite soil was studied by Goldberg and Glaubig (1988a) using equilibrating solutions containing both anions in approximately equimolar amounts. Arsenate sorption was unaffected by the presence of selenite over a wide range of pH (1 to 13). These results indicate that these anions sorb on separate sites or, more probably, sorption sites were not limiting at the low concentrations used. In another work the same authors (Goldberg and Glaubig, 1988b) found no reduction in selenite sorption on the same calcareous soil in the presence of an equimolar amount of arsenate up to pH 9, but above pH 9 selenite sorption was much reduced by arsenate.

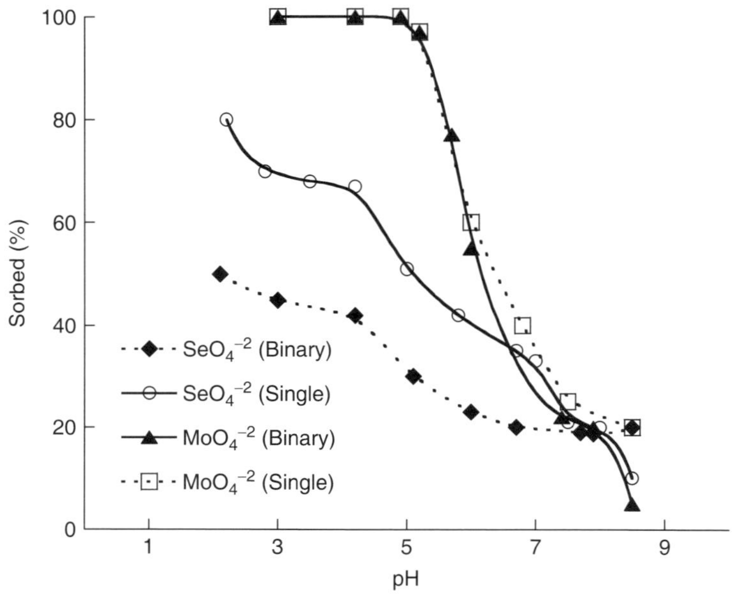

Figure 5.7. Competitive adsorption of molybdate and selenate on γ-Al_2O_3 as a function of pH (γ-Al_2O_3 = 30 g L^{-1}, $[MoO_4^{2-}] = [SeO_4^{2-}] = 5 \times 10^{-3}$ M). (Redrawn from Wu et al., 2001.)

Wu et al. (2001) found that molybdate sorption was affected insignifically in the presence of equimolar amounts of selenate; however, selenate sorption was significantly reduced in the presence of molybdate at pH < 7.0, where a 30% decrease in sorption was noticed (Figure 5.7).

5.2.2.2. Effect of Organic and Inorganic Ligands The presence of organic and inorganic ligands that interact with soil components, mainly variable-charge minerals, also affects the sorption of trace elements in anionic form by competing for available sorption sites and/or reducing the surface charge of the sorbents (Barrow, 1992). The competition depends on the affinity of the anions for the surfaces of the sorbents as well as the nature and surface properties of the minerals and soils. Certainly, competition may easily be observed at sufficient high surface coverage (discussed below). Goldberg et al. (1996b) found negligible competitive effects of sulfate, molybdate, and phosphate (sulfate < molybdate < phosphate) on boron sorption onto kaolinite, montmorillonite, and two arid soils, because they did not achieve site saturation in their experiments. These authors concluded that B-sorbing sites are specific to B and act independent of competing anions.

Mesuere and Fish (1992) found that oxalate diminished the sorption of chromate onto goethite most effectively at low pH and when sorbate concentrations were near surface-saturation levels. Chromate significantly inhibited oxalate sorption over a wide range of pH (from 4 to 10). Balistrieri and Chao (1987) suggested that for a given anion concentration ratio, the competition sequence with selenite on goethite is phosphate > silicate > citrate > molybdate > bicarbonate/carbonate > oxalate > fluoride > sulfate. Later, Dynes and Huang (1997) showed that the ability of 12 LMMOLs to inhibit selenite sorption on poorly crystalline Al hydroxides was oxalate > malate > citrate > succinate > glycolate > aspartate > salycilate > *p*-hydroxybenzoate > glycine = formiate = acetate. Generally, the larger the stability constant of the Al–organic solution complexes ($K_{\text{Al-L}}$), the more effective the organic acid was in competing with selenite for the sorption sites of the Al hydroxides. However, some of the organic acids competed less successfully than expected based on their $K_{\text{Al-L}}$ values. This was attributed to the stereochemical and electrostatic effects originating from both the surface of the Al hydroxides and the organic acids, which lowered the ability of some organic acids to compete with selenite for the sorption sites of the Al hydroxides.

Few studies have been carried out on the effect of silicic acid on the sorption of trace elements onto soil components, despite the fact that silicic acid is a ligand, which is ubiquitous in natural environments. Its concentrations in soils and natural waters range from 0.04 to 0.380 mM, with some as high as 0.814 mM. The kinetics of sorption of chromate, arsenite, and arsenate in the presence of sorbed silicic acid have been examined (Swedlund and Webster, 1999; Waltham and Eick, 2002; Garman et al., 2004). Garman et al. (2004) demonstrated that the rate and total quantity of chromate sorption onto goethite in the presence of silicic acid decreased mainly at pH > 4 (Table 5.3). The inhibition of chromate sorption ranged from 0 to 83.3% and 3.1 to 63.9% for 0.05 and 0.1 mM chromate, respectively. The molar ratio of chromate sorbed to silicic acid desorbed was less than 1, demonstrating the presence of excess surface sites for oxyanions sorption.

The sorption of silicic acid (added 60 hours before arsenic) also decreased the rate and the total amount of arsenic sorbed (Waltham and Eick, 2002; Table 5.4). The amount of As(III) sorbed decreased as the surface concentration of silicic acid increased. Furthermore, the inhibition of arsenite sorbed ranged from about 4% at a pH of 6 and 0.1 mM silicic acid up to 40% at a pH of 8 and 1 mol L^{-1} silicic acid. In all experiments except at a pH value of 8 and 1 mM silicic acid, the molar ratio of As(III) sorbed to silicic acid desorbed was greater than 1 (Table 5.4), indicating a greater quantity of arsenite sorbed compared with silicic acid desorbed. In contrast, silicic acid reduced the rate of As(V) sorption, which decreased by increasing pH and silicic acid concentration, but the total quantity of As(V) sorbed remained nearly constant, indicating that arsenate was able to replace silicate. Swendlund and Webster (1999) observed a reduction in As(V) sorption onto ferrihydrite at pH > 6 and attributed it to the polymerization of silicic acid.

The effect of dissolved organic carbon [humic (HA), fulvic (HF), or citric (CA) acid] on the sorption of arsenate and arsenite onto goethite and ferrihydrite was studied by Grafe et al. (2001, 2002). Arsenate sorption on goethite decreased in the presence of HA > FA > CA, while arsenite sorption was decreased in the presence of CA > FA ≈ HA. Onto ferrihydrite arsenate sorption was decreased only in the presence of CA, while arsenite sorption was reduced in the presence

TABLE 5.3. Rate and Quantity of Chromate Sorption in the Presence of Silicic Acid as a Function of pH

Si Concentration (mmol L^{-1})	pH	Si Sorbed (μmol m^{-2})	Cr(VI) Sorbed (μmol m^{-2})	Cr(VI) Sorbed in the Presence of Si (μmol m^{-2})	Si Desorbed (μmol m^{-2})	Cr(VI)/Si[a]	Cr(VI) Inhibition (%)
				0.05 mmol L^{-1} Cr(VI)			
0.10	4	0.81	0.65	0.65	0.147	4.42	0.0
0.10	6	0.90	0.65	0.56	0.195	3.56	13.9
0.10	8	1.00	0.36	0.09	0.012	7.5	75.0
1.0	4	2.33	0.65	0.60	0.31	1.94	7.7
1.0	6	2.81	0.65	0.33	0.14	2.35	49.2
1.0	8	3.31	0.36	0.06	0.03	2.00	83.3
				0.10 mmol L^{-1} Cr(VI)			
0.10	4	0.81	1.29	1.25	0.474	2.63	3.1
0.10	6	0.90	1.21	0.964	0.243	3.96	20.3
0.10	8	1.00	0.54	0.198	0.03	6.60	63.3
1.0	4	2.33	1.29	1.10	0.29	3.79	14.7
1.0	6	2.81	1.21	0.50	0.25	2.00	58.7
1.0	8	3.31	0.54	0.195	0.08	2.43	63.9

Source: Garman et al. (2004).

[a]Molar ratio of Cr(VI) adsorbed to Si desorbed.

TABLE 5.4. Quantity of As(III) and Silicic Acid Sorbed and Desorbed as a Function of pH

pH	Si Concentration mmol L^{-1}	Si Sorbed (μmol m^{-2})	As(III) Sorbed (μmol m^{-2})	As (III) Sorbed in the Presence of Si (μmol m^{-2})	Si Desorbed (μmol m^{-2})	As(III)/Si[a]	% As(III) Inhibition
4	0.10	0.81	1.08	0.97	0.50	1.95	10.4
6	0.10	0.90	1.09	1.05	0.30	3.56	3.7
8	0.10	1.00	1.11	1.04	0.32	1.60	6.3
4	1.0	2.33	1.08	0.70	0.44	1.60	34.8
6	1.0	2.81	1.09	0.71	0.42	1.84	34.8
8	1.0	3.31	1.11	0.67	0.69	0.97	40.0

Source: Waltham and Eick (2002).
[a] As(III) sorbed/Si desorbed molar ratios.

of FA and CA but not in the presence of HA. The exact reason for these results is unclear. Probably, differences in the surface coverage of the organic ligands and As(III) or As(V) on the sorbents (discussed below) and differences in the surface charge of the Fe oxides after sorption of the ligands greatly influenced the sorption of both organic ligands and arsenic ions.

The effect of phosphate on the sorption–desorption of arsenic in soil environments has received great attention, being phosphate commonly used as crop fertilizer (Smith et al., 1998; Frankenberger, 2002, and references therein; Violante and Pigna, 2002). The literature on arsenic–phosphate interactions in soils is very confusing. In fact, phosphate could displace sorbed and fixed arsenic from sorbing complexes, thereby increasing the arsenic in soil solution. Peryea (1998) reported increased arsenic solubility and thus the phytoavailability on P-fertilizer application to soils. On the contrary, application of phosphate was reported to decrease bioavailability of arsenic in soils (Hanada et al., 1975).

Competition in sorption between phosphate and arsenate may vary greatly on different soil minerals and on soils characterized by different mineralogical and chemical properties (Roy et al., 1986; Smith et al., 1998; Frankenberger, 2002; Violante and Pigna, 2002), although it has been established that the chemical behavior of arsenate is similar to that of phosphate. Violante and Pigna (2002) have carried out extensive work on the competitive sorption of phosphate and arsenate on selected phyllosilicates, metal oxides, and soil samples. They found that Mn, Fe, and Ti oxides and phyllosilicates particularly rich in Fe (nontronite, ferruginous smectites) were more effective than phosphate in sorbing arsenate after 24 hours of reaction, but more phosphate than arsenate was sorbed on noncrystalline Al precipitation products, gibbsite, boehmite, allophane, and kaolinite. Competitiveness between the anions also changed at different pH values. The arsenate sorbed/phosphate sorbed molar ratio usually decreased by increasing the pH of the systems, indicating that phosphate inhibits arsenate sorption more in neutral and alkaline systems than in acidic systems.

Sulfate was poor at preventing arsenate and molybdate sorption onto metal oxides and soils (Wu et al., 2001; Violante et al., 2005b), but reduced the sorption of selenate significantly (Wu et al., 2001). The alleviating effect of sulfate application on arsenic toxicity has also been reported (review by Kitagishi and Yamane, 1981).

5.2.3. Effect of Time and Surface Coverage on Competitive Sorption Between Trace Elements and Organic or Inorganic Ligands

Time of reaction and surface coverage have a great influence on the competitive sorption between trace elements and organic and inorganic ligands. Figure 5.8*a* shows the effect of time on the sorption of arsenate and phosphate onto an Andisol when anions were added alone or as a mixture at a surface coverage of about 100% (Violante and Pigna, 2002). The amounts of the anions sorbed increased with time, but the arsenate sorbed/phosphate sorbed molar ratio (rf) increased continously with time from 0.25 after 0.02 hour to 0.42 after 3 hours

(Figure 5.8*b*) to 0.51 after 24 hours (not shown). More recently, Pigna et al. (2004) found that when phosphate and arsenate were added simultaneously on synthetic hematites, the rf values initially increased (from 0 to 8 hours) and were greater than 1 and then slowly decreased up to 1. The extent of competition between the oxyanions must be related to sorption kinetics. Certainly, an initial faster sorption of an oxyanion onto the surface of a given sorbent affects the sorption of the other. For example, phosphate sorption onto the Andisol was initially faster than arsenate sorption. However, with time arsenate sorption increased, and consequently, its competitiveness with phosphate also increased with time (Figure 5.8). The opposite occurred when using hematite as sorbent. Probably, an initial reduction in surface charge because of the initial sorption of phosphate or arsenate may reduce the rate of anion sorption differently, which may be responsible for the residence time effect observed.

Competition for sorption sites appears evident by increasing the surface coverage of the sorbents. Table 5.5 shows the inhibition of arsenate sorption onto an Andisol at pH 5 in the presence of increasing concentrations of phosphate when the amounts of arsenate added to the soil sample were respectively at about 35, 70, and 100% surface coverage. At about 35% of surface coverage, arsenate sorption was very poorly inhibited, clearly because many sites were available. However, by increasing the surface coverage to about 70 to 100% the inhibition of phosphate on arsenate sorption increased because of the greater competition of the ligands for the sorption sites available. Certainly, both competition for sorption sites and change in the surface charge of the sorbents concur to explain the competition in sorption between ions.

5.2.4. Effects of Component Addition Sequence on Trace Element Sorption

Most competitive sorption studies have been carried out adding the ions contemporaneously. In natural environments, however, it is more likely that the ions will come in contact with a sorbent sequentially (i.e., the solid is exposed to one ion first, with the second ion coming in contact with a solid at a later time). The sorption of trace elements in cationic or anionic form is strongly influenced by the order of addition of organic and inorganic ligands and trace elements on the sorbents.

Bryce et al. (1994) demonstrated that in the ternary Ni–EDTA–ferrihydrite system (initial Ni/EDTA molar ratio of 1) the fraction of nickel sorbed was dependent on the component addition sequence (Figure 5.9*a*), but the fraction of sorbed EDTA was not. EDTA sorption could always be described in terms of the binary EDTA–ferrihydrite system, with the fraction sorbed decreasing sharply above pH 6 (Figure 5.9*b*). When Ni and EDTA were preequilibrated at pH 7, which favored the formation of 1 : 1 Ni–EDTA^{2-} complex, the fraction of Ni and EDTA sorbed onto ferrihydrite after 2, 24, and 48 hours was near 100% at pH < 6 but decreased sharply to approximately 0% at pH > 7 (Figure 5.9*c*). In this system, equilibrium was achieved within 2 hours and the fraction of Ni sorbed at a given pH was equivalent to the fraction of EDTA sorbed, suggesting

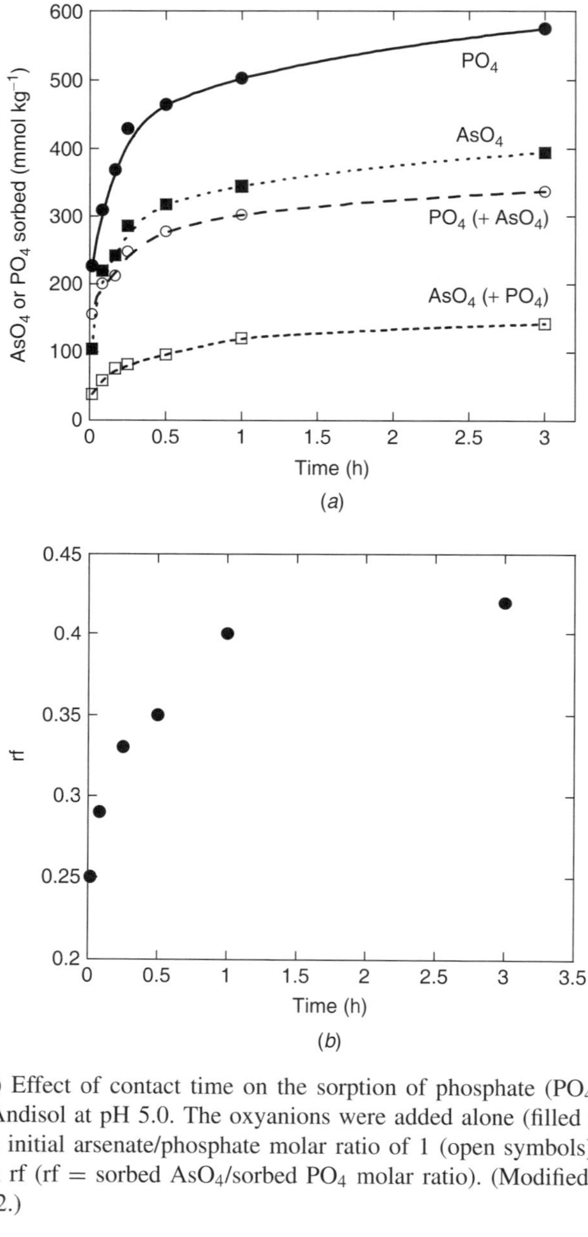

Figure 5.8. (*a*) Effect of contact time on the sorption of phosphate (PO_4) and arsenate (AsO_4) on an Andisol at pH 5.0. The oxyanions were added alone (filled symbols) or as a mixture at an initial arsenate/phosphate molar ratio of 1 (open symbols). (*b*) Effect of contact time on rf (rf = sorbed AsO_4/sorbed PO_4 molar ratio). (Modified from Violante and Pigna, 2002.)

that EDTA controls sorption. Nickel sorption in the metal-first addition sequence was significantly different from the preequilibrated system (Figure 5.9*d*). At increasing times after the addition of EDTA to the Ni/ferrihydrite system, little change was observed in Ni sorption at pH < 7, but at pH > 7 the fraction of Ni sorbed decreased slowly as a solution Ni–$EDTA^{2-}$ complex formed. After

TABLE 5.5. Effect of Increasing Amounts of Phosphate (PO_4) in Preventing Arsenate (AsO_4) Sorption on an Andisol at pH 5.0[a]

	≈ 35% Surface Coverage AsO_4 Added (233 mmol kg^{-1})		≈ 70% Surface Coverage AsO_4 Added (466 mmol kg^{-1})		≈ 100% Surface Coverage AsO_4 Added (666 mmol kg^{-1})	
Initial AsO_4/ PO_4 Molar Ratio	Inhibition[b] (%)	rf[c]	Inhibition (%)	rf	Inhibition (%)	rf
0	0	0	0	0	0	0
5.00	0	5.00	1	5.00	1	5.00
2.00	0	2.00	8	2.00	15	1.67
1.00	0	1.00	19	0.83	36	0.62
0.67	2	0.67	38	0.44	58	0.31
0.50	13	0.43	47	0.28	67	0.21

Source: Violante and Pigna (2002).

[a] Arsenate was added in order to have about 35, 70, and 100% of surface coverage of the sorbent.

[b] AsO_4 inhibition (%) = (AsO_4 sorbed alone−AsO_4 sorbed in the presence of PO_4/AsO_4 sorbed alone) × 100.

[c] rf stands for sorbed AsO_4/sorbed PO_4 molar ratio.

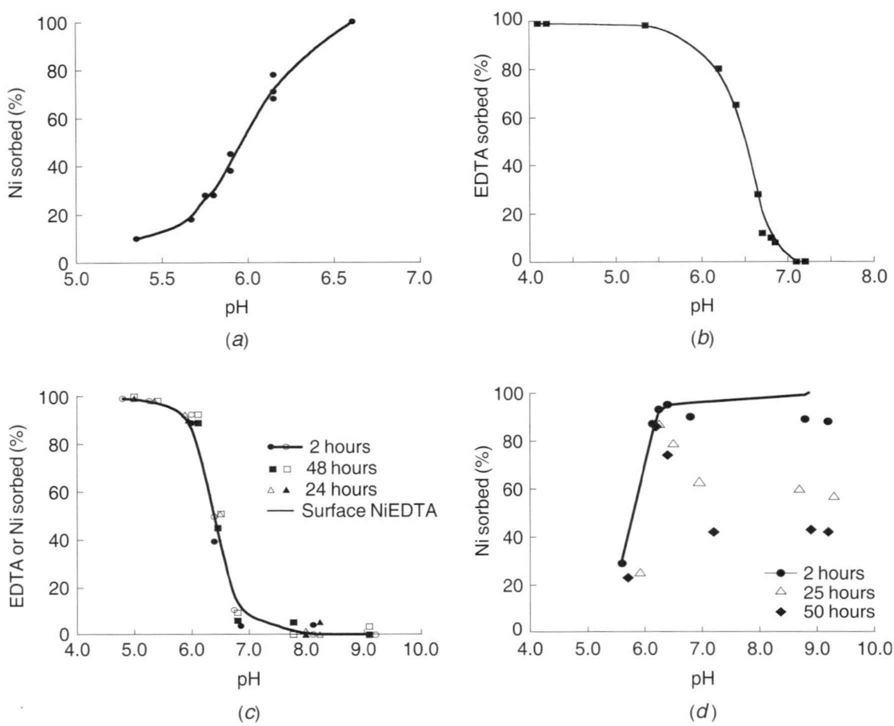

Figure 5.9. (*a*) Percent nickel sorbed onto ferrihydrite as a function of pH (experimental conditions: nickel 10^{-5} mol L^{-1}, Fe^{3+} 0.009 mol L^{-1}, $NaNO_3$ 0.1 mol L^{-1}); (*b*) EDTA sorbed (experimental conditions: EDTA 10^{-5} mol L^{-1}, Fe^{3+} 0.009 mol L^{-1}, $NaNO_3$ 0.1 mol L^{-1}); (*c*) percent nickel and EDTA sorbed, premixed addition sequence (experimental conditions: nickel 10^{-5} mol L^{-1}, EDTA 10^{-5} mol L^{-1}, Fe^{3+} 0.009 mol L^{-1}, $NaNO_3$, 0.1 mol L^{-1}, times are from contact with ferrihydrite; (*d*) percent nickel sorbed: metal first addition sequence (experimental conditions: 10^{-5} mol L^{-1} total nickel equilibrated with Fe^{3+} 0.009 mol L^{-1}, in $NaNO_3$ 0.1 × mol L^{-1} at pH 7.0 prior to the addition of EDTA 10^{-5} mol L^{-1}. Measurements at 2, 26, and 50 hours indicate time elapsed since EDTA addition. (Modified from Bryce et al., 1994.)

2 hours, approximately 15% of Ni was found in solution, increasing up to 50% desorbed after 50 hours. After 50 hours, Ni sorbed was much greater than the amount fixed in the preequilibrated Ni–EDTA system at pH $\geq$ 7 (Figure 5.9*c*), showing that the processes controlling desorption are kinetically slow. Also when EDTA was equilibrated individually with ferrihydrite prior to the addition of Ni, the system did not attain equilibrium. After 2 hours at pH $\geq$ 7, Ni sorption was initially > 80%, but it desorbed slowly with time.

The influence of LMMOLs on the sorption of Pb onto different metal oxides as affected by the sequence of addition of Pb and LMMOLs [oxalate (OX) or tartrate (TR)] on the sorbents was studied by Violante et al. (2003). These authors determined the amounts of Pb sorbed at pH 4.0 when added alone (Pb) or in the presence of the LMMOLs (OX or TR/Pb molar ratio of 4) when Pb was added as a mixture with LMMOLs [Pb + OX (or TR) system], 30 minutes. before LMMOLs [Pb before OX (TR) system] or 30 minutes after LMMOLs [OX (TR) before Pb system]. They found that larger amounts of Pb were sorbed when LMMOLs were added before Pb and usually according to the following sequence: OX (or TR) before Pb > Pb before OX (or TR) > Pb + OX (or TR) > Pb.

Studies on the sequence of addition of inorganic and organic ligands on the sorption of trace elements in anionic form have also been carried out (Hongshao and Stanforth, 2001; Violante and Pigna, 2002; Violante et al., 2005a). Hongshao and Stanforth (2001) demonstrated that when phosphate and arsenate were added simultaneously onto goethite, the two ions were sorbed about equally, with the total surface coverage being slightly greater than for either ion alone. When added sequentially the extent of exchange for the first ion depended on the equilibration time before the second ion was introduced: The longer the equilibration time, the greater the exchange. These authors found that when added sequentially, the extent of exchange for the first ion depended on the reaction time before the second ion was introduced: The longer the reaction time, the greater the exchange. They gave evidence that there is a fraction of the surface sites that are occupied very rapidly but which are not released once occupied.

5.3. DESORPTION OF TRACE ELEMENTS

In contrast to sorption studies, relatively little information is available on the desorption of trace elements from soils or soil components as affected by organic and inorganic ligands. Desorption studies have showed biphasic reaction processes for sorption and desorption of trace elements (Sparks, 1990): a fast reaction followed by a slow reaction.

The presence of inorganic ligands such as phosphate has a significant impact on the desorption of trace elements. Krishnamurti et al. (1999a,b) reported release characteristics which showed an increase in Cd desorption in the presence of phosphate (Table 5.6). Further, the irregularity in the fit of the kinetic data by the parabolic diffusion and Elovich models was indicative of desorption of Cd by mol L^{-1} NH_4Cl occurring from sorption sites with a wide range of variation in activation energies, as suggested earlier (Hingston, 1981).

Because of the importance of organic acids, resulting from organic matter decomposition and from the root exudates on the solubility of trace elements in the rhizosphere (Mench and Martin, 1991), it was demonstrated that LMMOLs have the ability to desorb Cd from soils, with malate, fumarate, and succinate being the most effective (Krishnamurti et al., 1997; Naidu and Harter, 1998) (Table 5.7).

5.3.1. Desorption Kinetics and Bioavailability

Leaching and extraction tests are used widely to assess trace element mobility and phytoavailability, as reviewed by Krishnamurti and Naidu in Chapter 11.

TABLE 5.6. Kinetics of Cd Desorption from Soils by mol L^{-1} NH_4Cl[a] and Phosphate

	Cd Desorbed in the Initial 30 min (μmol kg^{-1})		Overall Diffusion Coefficient k_d (μmol kg^{-1} $h^{-0.5}$)			Grain Cd[c] (μg kg^{-1})	
Soil	Absence of Phosphate	Presence of Phosphate	Absence of Phosphate	Presence of Phosphate	CAI[b] (μg kg^{-1})	Kyle	Arcola
Jedburgh	5.5×10^{-3}	57.8×10^{-3}	3.2×10^{-3}	7.9×10^{-3}	10	42	28
Luseland	188.0×10^{-3}	1150.0×10^{-3}	46.7×10^{-3}	80.3×10^{-3}	169	384	159

Source: Krishnamurti et al. (1999a,b).

[a]Desorption kinetics, in the presence and absence of 0.1 mol L^{-1} monoammonium phosphate during Cd adsorption by the soils, described by the parabolic diffusion model: $q = a + k_d t^{1/2}$, where q is the amount of Cd desorbed in time t (hours), a is a constant, and k_d is overall diffusion coefficient.

[b]Cadmium availability index (M NH_4Cl-extractable Cd: Krishnamurti et al., 1995).

[c]Cadmium content in the grain of durum wheat cultivars.

TABLE 5.7. Kinetics of Cd Desorption from Soils by Low-Molecular-Mass Organic Ligands (LMMOLs)

	Cd Released by Renewal[a] of LMMOL (μmol kg^{-1})						Overall Diffusion Coefficient k_d[b] (μmol $kg^{-1}h^{-0.5}$)						Grain Cd[d] (μg kg^{-1})	
Soil	Acetic	Citric	Fumaric	Oxalic	Succinic	Total	Acetic	Citric	Fumaric	Oxalic	Succinic	CAI[c] (μg kg^{-1})	Kyle	Arcola
Jedburgh	0.031	0.064	0.109	0.060	0.013	0.277	0.036	0.196	0.041	0.026	0.009	12	42	28
Waitville	0.048	0.166	0.153	0.152	0.016	0.535	0.046	0.049	0.050	0.036	0.019	22	68	40
Luseland	0.093	0.312	0.360	0.215	0.125	1.105	0.112	0.200	0.199	0.079	0.090	143	384	159

Source: Krishnamurti et al. (1997).

[a]Renewal of each of the 10^{-2} mol L^{-1} LMMOL three times after every 2-hour reaction period.

[b]Desorption kinetics of Cd by LMMOL from the soils described by the parabolic diffusion model: $q = a + k_d t^{1/2}$, where q is the amount of Cd desorbed in time t(hours), a is a constant, and k_d is the overall diffusion coefficient.

[c]Cadmium availability index (mol L^{-1} NH_4Cl-extractable Cd: Krishnamurti et al., 1995).

[d]Cadmium content in the grain of durum wheat cultivars.

As measurements in these protocols are made in equilibrium conditions, only thermodynamic information is obtained. However, kinetic extraction–desorption studies are a more correct approximation to the distribution of species in natural media (Aulitiia and Pickering, 1988; Bermond et al., 1998; Ortiz-Viana et al., 1999; Fangueiro et al., 2002; Gismera et al., 2004). The desorption rate constants of the trace element in sediments and soils can be related to its mobility and toxicity.

The kinetic data on desorption of Cd by phosphate, as related to the amount of Cd released during the initial 30-minute reaction period and the overall diffusion coefficients obtained from the desorption kinetics of Cd by mol L^{-1} NH_4Cl from the soils, reflect well the phytoavailable Cd in the two soils, as shown by the Cd availability index and the grain Cd content of two durum wheat cultivars, Kyle and Arcola, grown on the two soils (Table 5.6).

The kinetics of Cd release, as influenced by the LMMOLs, play an important role in plant Cd uptake. The kinetic rate constant of Cd release, as obtained from desorption kinetics of Cd by LMMOLs and the amount of Cd released by renewal of LMMOLs from the soil, followed the same trend as the cadmium availability index and Cd grain content of durum wheat grown on the soils (Table 5.7). These reports highlight the significance of Cd desorption kinetics in understanding Cd dynamics and phytoavailability.

5.3.1.1. Effect of Residence Time on Desorption Some researchers found that trace elements [Ni, Pb, As(V)] reacted with metal oxides and pyrophyllite over longer times resulted in either irreversible or reversible sorption mechanisms. Violante et al. (2003) studied the effect of residence time on the sorption of Zn onto ferrihydrite in the presence of Cu. As Cu has a greater affinity than Zn for the surfaces of ferrihydrite, Cu was added from 1 to 336 hours after Zn at a Zn/Cu molar ratio of 2. Zinc sorption increased, particularly when Cu was added 6 to 336 hours after Zn. A possible explanation of these findings is that trace elements initially sorbed on the surfaces of variable-charge minerals slowly form precipitates with time. As discussed before, sorption is considered to be the predominant sorption mechanism responsible for trace element uptake on mineral surfaces within the first few hours, while surface precipitation is considered to be a much slower process, occurring on a time scale of hours to days (McBride, 1994; Scheidegger et al., 1997; Sparks, 1999; Borda and Sparks, Chapter 3, this volume). Clearly, Cu added many hours or days after Zn addition cannot replace Zn ions that have formed precipitates on the surfaces of the ferrihydrite.

O'Reilly et al. (2001) studied the effect of sorption residence time on arsenate desorption by phosphate (phosphate/arsenate molar ratio of 3) from goethite at different pH values. Initially, desorption was very fast (35% arsenate desorbed at pH 6.0 within 24 hours), and then it slowed down. Total desorption increased with time, reaching about 65% total desorption after five months. These authors found no measurable effect of aging on desorption of arsenate in the presence of phosphate. In fact, the amount of arsenate desorbed one month after a residence time of 0.7 hour was 50% compared with 48% after a sorption residence time

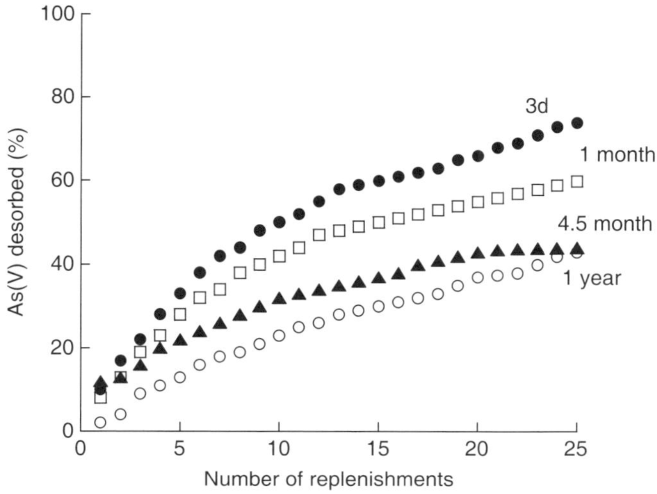

Figure 5.10. Residence-time effects on As(V)-desorption from aged As(V) reacted aluminum oxide at pH 4.5. (Redrawn from Arai and Sparks, 2002.)

of 7 months. Desorption results at pH 4.0 were similar to desorption behavior at pH 6.0. However, later, Arai and Sparks (2002) demonstrated that the longer the residence time (3 days to one year), the greater the decrease in arsenate desorption by phosphate from a bayerite. Figure 5.10 shows continuous arsenate release after 25 replenishment cycles from differently aged As(V) reacted by aluminum oxide at pH 4.5. The extent of arsenate release in 3-day and one-month aged samples after 25 days of desorption was much greater than for 4.5-month to one-year aged samples.

We found (unpublished data, 2007) similar results when studying the effect of residence time on arsenate desorption by phosphate (phosphate/arsenate molar ratio of 3) from an Andisol sample containing 42% allophanic materials (Vacca et al., 2002). The surface coverage of arsenate was 60%. The quantities of arsenate desorbed by phosphate after 170 hours decreased from 53% to 35% and 22%, when phosphate was added, respectively, 1, 5, or 15 days after arsenate addition. By increasing the time of reaction to 30 days the further removal of arsenate desorbed was negligible (<5%). The initial pH strongly affected the amount of arsenate desorbed from the Andisol. Usually, the higher the pH, the larger the amounts of arsenate desorbed by phosphate.

5.4. CONCLUSIONS AND FUTURE DIRECTIONS

The sorption–desorption processes of trace elements on or from soil components is affected by many factors, such as pH, nature of the sorbents, redox reactions, and presence and concentration of organic and inorganic ligands, including humic and fulvic acids, root exudates, and nutrients. The behavior of foreign ligands on the sorption of toxic elements in cationic form is quite different from that toward elements in anionic form. In fact, complexation reactions of trace elements in cationic form with organic and inorganic ligands have an important role to play in their sorption–desorption processes as well as in their toxicity and phytoavailability, whereas competition for available sites and/or reduction of the surface charge of the sorbents between foreign ligands and trace elements in anionic form affect primarily their mobility. Competition in sorption between two or more trace elements is of great importance in understanding their relative affinity for a given sorbent, but unfortunately, few studies have been carried out on this subject. The sorption of trace elements is strongly influenced by the sequence of addition of organic or inorganic ligands onto the sorbents.

The factors that affect the toxicity or bioavailability of a trace element in soil environments are still obscure. The hypothesis that the toxicity or bioavailability of a trace element is related to the activity of the free aquo ion is not valid in all situations, because the uptake of metal–organic chelates by plants and algae have been demonstrated. Detailed research is needed to understand the role of metal–organic species, which are accepted as dominant aqueous ionic species in soil solution in assessing bioavailability. Time of reaction and the surface coverage have a great influence on the competitive sorption between trace elements and organic and inorganic ligands. However, the residence time effect on sorption–desorption studies has not been given as much importance in understanding the mobility and bioavailability of trace elements.

Leaching and extraction tests are widely used for assessing trace element mobility and phytoavailability. As measurements in these protocols are made in equilibrium conditions, only thermodynamic information is obtained. However, kinetic extraction–desorption studies are a more correct approximation to the distribution of species in natural media. The desorption rate constants of the trace elements in sediments and soils can be related to their mobility and toxicity. Detailed studies are needed in this area.

Acknowledgments

This study was supported by the Italian Research Program of National Interest (PRIN), year 2004. DISSPA Number 106.

REFERENCES

Agbenin, J. O., and Olojo, L. A. (2004). Competitive adsorption of copper and zinc by a Bt horizon of a savanna Alfisol as affected by pH and selective removal of hydrous oxides and organic matter. *Geoderma* **119**, 85–95.

Ainsworth, C. C., Pilon, J. L., Gassman, P. L., and Van Der Sluys, W. G. (1994). Cobalt, cadmium, and lead adsorption to hydrous iron oxide: residence time effect. *Soil Sci. Soc. Am. J.* **58**, 1615–1623.

Arai, Y., and Sparks, D. L. (2002). Residence time effects on arsenate surface speciation at the aluminum oxide–water interface. *Soil Sci.* **167**, 303–314.

Auliitia, T. U., and Pickering, W. F. (1988). Sediment analysis: lability of selectively extracted fractions. *Talanta* **35**, 559–566.

Balistrieri, L. S., and Chao, T. T. (1987). Selenium adsorption by goethite. *Soil Sci. Soc. Am. J.* **51**, 1145–1151.

Barrow, N. J. (1992). The effect of time on the competition between anions for sorption. *J. Soil Sci.* **, 43**, 424–428.

Benjamin, M. M., and Leckie, J. O. (1981). Multiple-site adsorption of Cd, Cu, Zn and Pb on amorphous iron oxyhydroxide. *J. Colloid Surf. Sci.* **79**, 209–221.

Bermond, A., Yousfi, I., and Ghestem, J. P. (1998). Kinetic approach to the chemical speciation of trace metals in soils. *Analyst (London)* **123**, 785–789.

Beveridge, T. J. (1989a). Metal ions and bacteria. In *Metal Ions and Bacteria*, ed. Beveridge, T. J., and Doyle, R. J., Wiley, New York, 1–29.

Beveridge, T. J. (1989b). Role of cellular design in bacterial metal accumulation and mineralization. *Annu. Rev. Microbiol.* **43**, 147–171.

Bolan, N. S., and Duraisamy V. P. (2003). Role of inorganic and organic soil amendments on immobilisation and phytoavailability of heavy metals: a review involving specific case studies. *Aust. J. Soil Res.* **41**, 533–555.

Bolan, N. S., Naidu, R., Syers, J. K., and Tillman, R. W. (1999). Surface charge and solute interactions in soils. *Adv. Agron.*, **67**, 87–140.

Bolan, N. S., Adriano, D. C., Natesan, R., and Koo, B.-J. (2003). Effects of organic amendments on the reduction and phytoavailability of chromate in mineral soil. *J. Environ. Qual.* **32**, 120–128.

Bolland, M. D. A., Posner, A. M., and Quirk, J. P. (1977). Zinc adsorption by goethite in the absence and presence of phosphate. *Aust. J. Soil. Res.* **15**, 279–286.

Boyd, S. A., and Mortland, M. M. (1990). Enzyme interactions with clays and clay–organic matter complexes. In *Soil Biochemistry*, Vol. 6, ed. Bollag, J.-M., and Stotzky, G., Marcel Dekker, New York, 1–28.

Bruemmer, G. W., Gerth, J., and Tiller, K. G. (1988). Reaction kinetics of adsorption and desorption of nickel, zinc and cadmium by goethite, I: Adsorption and diffusion of metals. *J. Soil Sci.* **39**, 37–52.

Bryce, A. L., Kornicker, W. A., and Elzerman, A. W. (1994). Nickel adsorption to hydrous ferric oxide in the presence of EDTA: effects of component addition sequence. *Environ. Sci. Technol.* **28**, 2353–2359.

Bunzl, K., Schmidt, W., and Sansoni, B. (1976). Kinetics of ion exchange in soil organic matter, IV: Adsorption and desorption of Pb^{2+}, Cu^{2+}, Cd^{2+}, Zn^{2+} and Ca^{2+} by peat. *J. Soil Sci.* **27**, 32–41.

Capasso, R., Pigna, M., De Martino, A., Pucci, M., Sannino, F., and Violante, A. (2004). Potential remediation of waters contaminated with Cr(III), Cu and Zn by sorption on the organic polymeric fraction of olive mill wastewater (polymerin) and its derivatives. *Environ. Sci. Technol.* **38**, 5170–5176.

Chang, A. C., Page A. L., and Koo, B.-J. (2002). Biogeochemistry of phosphorus, iron, and trace elements as influenced by soil–plant–microbial interactions. In *Soil Mineral–Organic Matter–Microorganism Interactions and Ecosystem Health: Ecological Significance of the Interactions Among Clay Minerals, Organic Matter and Soil Biota*, ed. Violante, A., Huang, P. M., Bollag, J.-M., and Gianfreda, L., Developments in Soil Science, vol. 28B, Elsevier, New York, 43–57.

Chardichai, P., and Ritchie, G. S. P. (1992). The effect of pH on zinc adsorption by a lateritic soil in the presence of citrate and oxalate. *J. Soil Sci.* **43**, 713–728.

Christensen, T. H. (1984). Cadmium soil sorption at low concentrations, III: Prediction and observation of mobility. *Water Air Soil Pollut.* **26**, 255–264.

Cornell, R. M., and Schwertmann, U. (1996). *The Iron Oxides: Structure, Properties, Reactions and Uses*, VCH, New York.

Cowan, C. E., Zachara, J. M., and Resch, C. T. (1991). Cadmium adsorption on iron oxides in the presence of alkaline-earth elements. *Environ. Sci. Technol.* **25**, 437–446.

Dixon, J. B., and Weed, S. B. (1989). *Minerals in Soil Environments*, Soil Science Society of America, Madison, WI, 1244.

Dynes, J. J., and Huang, P. M. (1997). Influence of organic acids on selenite sorption by poorly ordered aluminum hydroxides. *Soil Sci. Soc. Am. J.* **61**, 772–783.

Elliott, H. A., Liberati, M. R., and Huang, C. P. (1986). Competitive adsorption of heavy metals by soils. *J. Environ. Qual.* **15**, 214–219.

Fangueiro, D., Bermond, A., Santos, E., Carapuca, H., and Duarte, A. (2002). Heavy metal mobility assessment in sediments based on a kinetic approach of the EDTA extraction: search for optimal experimental conditions. *Anal. Chim. Acta* **459**, 245–256.

Farrah, H., and Pickering, W. F. (1976a). The sorption of copper species by clays, I: Kaolinite. *Aust. J. Chem.* **29**, 1167–1176.

Farrah, H., and Pickering, W. F. (1976b). The sorption of copper species by clays, II: Illite and montmorillonite. *Aust. J. Chem.* **29**, 1177–1184.

Fendorf, S. E. (1995). Surface reactions of chromium in soils and waters. *Geoderma* **67**, 55–71.

Fendorf, S. E., Lamble, G. M., Stapleton, M. G., Kelley, M. J., and Sparks, D. L. (1994). Mechanisms of chromium(III) sorption on silica, I: Cr(III) surface structure derived by extended x-ray absorption fine structure spectroscopy. *Environ. Sci. Technol.* **28**, 284–289.

Fendorf, S. E., Eick, M. J., Grossl, P. R., and Sparks, D. L. (1997). Arsenate and chromate retention mechanisms on goethite, I: Surface structure. *Environ. Sci. Technol.* **31**, 315–320.

Frankenberger, W. T., Jr. (2002). *Environmental Chemistry of Arsenic*, Marcel Dekker, New York.

Fuller, C. C., Davis, J. A., and Weychunas, G. A. (1993). Surface chemistry of ferrihydrite, II: Kinetics of arsenate adsorption and coprecipitation. *Geochim. Cosmochim. Acta* **57**, 2271–2282.

Garman, S. M., Luxton, T. P., and Eick, M. J. (2004). Kinetics of chromate adsorption on goethite in the presence of sorbed silicic acid. *J. Environ. Qual.* **33**, 1703–1708.

Gismera, M. J., Lacal, J., da Silva, P., Garcia, R., Sevilla, M. T., and Procopio, J. R. (2004). Study of metal fractionation in river sediments: a comparison between kinetic and sequential extraction procedures. *Environ. Pollut.* **127**, 175–182.

Goldberg, S. (2002). Competitive adsorption of arsenate and arsenite on oxides and clay minerals. *Soil Sci. Soc. Am. J.* **66**, 413–421.

Goldberg, S., and Glaubig, R. A. (1988a). Anion sorption on a calcareous montmorillonitic soil: selenium. *Soil Sci. Soc. Am. J.* **52**, 954–958.

Goldberg, S., and Glaubig, R. A. (1988b). Anion sorption on a calcareous montmorillonitic soil: arsenic. *Soil Sci. Soc. Am. J.* **52**, 1297–1300.

Goldberg, S., Davis, J. A., and Hem, J. D. (1996a). The surface chemistry of aluminum oxides and hydroxides. In *The Environmental Chemistry of Aluminum*, 2nd ed., ed. Sposito, G., Lewis Publishers, Boca Raton, FL, 271–331.

Goldberg, S., Forster, H. S., Lesch, S. M., and Heich, E. L. (1996b). Influence of anion competition on boron adsorption by clay minerals and soils. *Soil Sci.* **161**, 99–103.

Grafe, M., Eick, M. J., and Grossl, P. R. (2001). Adsorption of arsenate(V) and arsenite(III) on goethite in the presence and absence of dissolved organic carbon. *Soil Sci. Soc. Am. J.* **65**, 1680–1687.

Grafe, M., Eick, M. J., and Grossl, P. R. (2002). Adsorption of arsenate and arsenite on ferrihydrite in the presence and absence of dissolved organic carbon. *J. Environ. Qual.* **31**, 1115–1123.

Grossl, P. R., Eick, M. J., Sparks, D. L., Goldberg, S., and Ainsworth, C. C. (1997). Arsenate and chromate retention on goethite, II: Kinetic evaluation using a pressure-jump relaxation technique. *Environ. Sci. Technol.* **31**, 321–326.

Hanada, S., Nakano, M., Saitoh, H., and Mochizuki, T. (1975). Studies on the pollution of apple orchard surface soils and its improvement in relation to inorganic spray residues, I. *Bull. Fac. Agric. Hirosaki Univ.* **25**, 13–17.

Hayes, K. F., Roe, A. L., Brown, G. E., Hodgson, K. O., Lee, J. O., and Parks, G. A. (1987). In situ x-ray absorption study of surface complexes: selenium oxyanions on α-FeOOH. *Science (Washington, DC,)* **238**, 783–786.

Hettiarichchi, G. M., Pierzynski, G. M., Zwonitzer, J., and Lambert, M. (1997). Phosphorus sources and rate effects on cadmium, lead and zinc bioavailability in a metal-contaminated soil. *Proc. 4th International Conference on the Biogeochemistry of Trace Elements*, Berkley, CA, June 23–26, 1997, 463–472.

Hingston, F. J. (1981). A review of anion adsorption. In *Adsorption of Inorganics at Solid–Liquid Interfaces*, ed. Anderson, M. A., and Rubin, A. J., Ann Arbor Science, Ann Arbor, MI, 51–84.

Hongshao, Z., and Stanforth, R. (2001). Competitive adsorption of phosphate and arsenate on goethite. *Environ. Sci. Technol.* **24**, 4753–4757.

Hsia, T. H., Lo, S. L., Lin, C. F., and Lee, D. Y. (1994). Characterization of arsenate adsorption on hydrous iron oxide using chemical and physical methods. *Colloids Surf. A Physicochem. Eng. Asp.*, **85**, 1–7.

Huang, P. M. (2000). Abiotic catalysis. In *Handbook of Soil Science*, ed. Sumner, M. E., CRC Press, Boca Raton, FL, B303–B332.

Huang, P. M., and Germida, J. J. (2002). Chemical and biochemical processes in the rhizosphere: metal pollutants. In *Interactions Between Soil Particles and Microorganisms: Impact on the Terrestrial Ecosystem*, ed. Huang, P. M., Bollag, J.-M., and Senesi, N., Wiley, New York, 381–438.

Huang, P. M., and Schnitzer M. (eds). (1986). *Interactions of Soil Minerals with Natural Organics and Microbes*, Spec. Publ. 17, Soil Science Society of America, Madison, WI.

Jackson, T. A. (1998). The biogeochemical and ecological significance of interactions between colloidal minerals and trace elements. In *Environmental Interactions of Clays*, ed. Parker, A., and Rae, J. E., Springer-Verlag, Berlin, 93–205.

Jackson, T. A., and Leppard, G. G. (2002). Energy dispersive x-ray microanalysis and its applications in biological research. In *Soil Mineral–Organic Matter–Microorganism*

Interactions and Ecosystem Health: Dynamics, Mobility and Transformations of Pollutants and Nutrients, Developments in Soil Science, Vol. 28, ed. Violante, A., Huang, P. M., Bollag, J.-M, and Gianfreda, L., Elsevier, New York, 219–260.

Jain, A., Raven, K. P., and Loeppert, R. H. (1999). Arsenite and arsenate adsorption on ferrihydrite: surface charge reduction and net OH^- release stoichiometry. *Environ. Sci. Technol.* **33**, 1179–1184.

Kabata-Pendias, A. (2001). *Trace Elements in Soils and Plants, 3rd ed.*, CRC Press, Boca Raton, FL.

Kampf, N., Scheinost, A. C., and Schultze, D. G. (2000). Oxide minerals. In *Handbook of Soil Science*, ed. Sumner, M. E., CRC Press, Boca Raton, FL, F125–F168.

Kinniburgh, D. G., and Jackson, M. L. (1976). Adsorption of alkaline earth, transition and heavy metal cations by hydrous oxides gels of iron and aluminum. *Soil Sci. Soc. Am. J.* **40**, 796–799.

Kinniburgh, D. G., and Jackson, M. L. (1981). Cation adsorption by hydrous metal oxides and clays. In *Adsorption of Inorganics at Solid-Liquid Interfaces*, ed. Anderson, M. A., and Rubin, A. S., Ann Arbor Science, Ann Arbor, MI, 91–160.

Kinniburgh, D. G., Milne, C. J., Benedetti, M. F., Pinheiro, J. P., Filius, J., Koopal, L. K., and van Riemsdijk, W. H. (1996). Metal ion binding by humic acid: application of the NICA-Donnan model. *Environ. Sci. Technol.* **30**, 1687–1698.

Kitagishi, K., and Yamane, I. (1981). *Heavy Metal Pollution in Soils of Japan*, Japan Science Society Press, Tokyo.

Kraemer, S. M., Xu, J., Raymond, K. N., and Sposito, G. (2002). Adsorption of Pb(II) and Eu(III) by oxide minerals in the presence of natural and synthetic hydroxamate siderophore. *Environ. Sci. Technol.* **36**, 1287–1291.

Kretzschmar, R., and Voegelin, A. (2001). Modeling competitive sorption and release of heavy metals in soils. In *Heavy Metals Release in Soils*, ed. Selim, H. M., and Sparks, D. L., Lewis Publishers, Boca Raton, FL, 55–88.

Krishnamurti, G. S. R., and Naidu, R. (2002). Solid-solution speciation and phytoavailability of copper and zinc in soils. *Environ. Sci. Technol.* **36**, 2645–2651.

Krishnamurti, G. S. R., Huang, P. M., Van Rees, K. C. J., Kozak, L. M., and Rostad, H. P. W. (1995). A new soil test method for the determination of plant-available cadmium in soils. *Commun. Soil Sci. Plant Anal.* **26**, 2857–2867.

Krishnamurti, G. S. R., Cieśliński, G., Huang, P. M., and Van Rees, K. C. J. (1997). Kinetics of cadmium release from soils as influenced by organic acids: implication in cadmium availability. *J. Environ. Qual.* **26**, 271–277.

Krishnamurti, G. S. R., Huang, P. M., and Kozak, L. M. (1999a). Desorption kinetics of cadmium from soils using M ammoinium nitrate and M ammonium chloride. *Commun. Soil Sci. Plant Anal.* **30**, 2785–2800.

Krishnamurti, G. S. R., Huang, P. M., and Kozak, L. M. (1999b). Sorption and desorption kinetics of cadmium from soils: influence of phosphate. *Soil Sci.* **164**, 888–898.

Krishnamurti, G. S. R., Megharaj, M., and Naidu, R. 2004. Bioavailability of cadmium–organic complexes to soil alga: an exception to the free ion model. *J. Agric. Food Chem.* **52**, 3894–3899.

Kuo, S. (1986). Concurrent sorption of phosphate and zinc, cadmium, or calcium by hydrous ferric oxide. *Soil Sci. Soc. Am. J.* **50**, 1040–1044.

Leckie, J. O., and James, R. O. (1974). Control mechanisms for trace metals in natural waters. In *Aqueous-Environmental Chemistry of Metals*, ed. Rubin, A. J., Ann Arbor Science, Ann Arbor, MI, 1–76.

Lehman, R. G., and Harter, R. D. (1984). Assessment of copper–soil bond strength by desorption kinetics. *Soil Sci. Soc. Am. J.* **48**, 769–772.

Lopez-Hernandez, D., Siegert, G., and Rodriguez, J. V. (1986). Competitive adsorption of phosphate with malate and oxalate by tropical soils. *Soil Sci. Soc. Am. J.* **50**, 1460–1462.

Lorenz, S. E., Hamon, R. E., Holm, P. E., Dominiques, H. C., Sequira, E. M., Christensen, T. H., and McGrath, S. P. (1997). Cadmium and zinc in plants and soil solutions from contaminated soils. *Plant Soil* **189**, 21–31.

Lund, W. (1990). Speciation analysis: Why and how? *Fresenius' J. Anal. Chem.* **337**, 557–564.

Manceau, A., and Charlet, L. (1994). The mechanism of selenate adsorption on goethite and hydrous ferric oxide. *J. Colloid Interface Sci.* **168**, 87–93.

Mandel, R., Salam, M. S. A., Murimboh, J., Hassan, N. M., Chakrabarti, C. L., Back, M. H., and Gregorie, D. C. (2000). Competition of Ca(II) and Mg(II) with Ni(II) for binding by a well-characterized fulvic acid in model solutions. *Environ. Sci. Technol.* **34**, 2201–2208.

Manning, B. A., and Goldberg, S. (1996a). Modeling competitive adsorption of arsenate with phosphate and molybdate on oxide minerals. *Soil Sci. Soc. Am. J.* **60**,121–131.

Manning, B. A., and Goldberg, S. (1996b). Modeling arsenate competitive adsorption on kaolinite, montmorillonite and illite. *Clays Clay Miner.* **44**, 609–623.

Manning, B. A., Fendorf, S. E., and Goldberg, S. (1998). Surface structures and stability of arsenic(III) on goethite: spectroscopic evidence for inner-sphere complexes. *Environ. Sci. Technol.* **32**, 2383–2388.

Marschner, H. (1998). *Mineral Nutrition of Higher Plants*, 2nd ed., Academic Press, London.

McBride, M. B. (1982). Cu^{2+}-adsorption characteristics of aluminum hydroxide and oxyhydroxides. *Clays Clay Miner.* **30**, 21–28.

McBride, M. B. (1985). Sorption of copper(II) on aluminum hydroxide as affected by phosphate. *Soil Sci. Soc. Am. J.* **49**, 843–846.

McBride, M. B. (1989). Reactions controlling heavy metal solubility in soils. *Adv. Soil Sci.* **10**, 1–56.

McBride, M. B. (1991). Processes of heavy and transition metal sorption by soil minerals. In *Interactions at the Soil Colloid–Soil Solution Interface*, ed. Bolt, G. H., De Boodt, M. F., Haynes, M. H. B., and McBride, M. B., Vol. 190, Kluver Academic, Dordrecht, The Netherlands, 149–176.

McBride, M. B. (1994). *Environmental Soil Chemistry*, Oxford University, Press, New York.

McBride, M. B. (2000). Chemisorption and precipitation reactions. In *Handbook of Soil Science*, ed. Sumner, M. E., CRC Press, Boca Raton, FL, B265–B302.

McBride, M. B., Sauve, S., and Hendershot, W. (1997). Solubility control of Cu, Zn, Cd and Pb in contaminated soils. *Eur. J. Soil Sci.* **48**, 337–346.

McKenzie, R. M. (1980). The adsorption of lead and other heavy metals on oxides of manganese and iron. *Aust. J. Soil Res.* **21**, 505–513.

Mench, M., and Martin, E. (1991). Mobilization of cadmium and other metals from two soils by root exudates of *Zea maize* L., *Nicotiana tabacum* L. and *Nicotiana rustica* L. *Plant Soil* **132**, 187–196.

Mesuere, K., and Fish, W. (1992). Chromate and oxalate adsorption on goethite, 2: Surface complexation modeling of competitive adsorption. *Environ. Sci. Technol.* **26**, 2365–2370.

Murphy, E. M., and Zachara, J. M. (1995). The role of sorbed humic substances on the distribution of organic and inorganic contaminants in groundwater. *Geoderma* **67**, 103–124.

Nagarajah, S., Posner, A. M., and Quirk, J. P. (1970). Competitive adsorption of phosphate with polygalacturonate and other organic anions on kaolinite and oxide surface. *Nature (London)* **228**, 83–84.

Naidu, R., and Harter, R. D. (1998). Effect of different organic ligands on cadmium sorption by and extractability from soils. *Soil Sci. Soc. Am. J.* **62**, 644–650.

Neal, R. H., and Sposito, G. (1986). Effects of soluble organic matter and sewage sludge amendments on cadmium adsorption by soils at low cadmium concentrations. *Soil Sci.* **142**, 164–172.

Neubauer, U., Nowak, B., Furrer, G., and Schulin, R. (2000). Heavy metal sorption on clay minerals affected by the siderophore desferrioxamine B. *Environ. Sci. Technol.* **34**, 2749–2755.

Neubauer, U., Furrer, G., and Schulin, R. (2002). Heavy metal sorption on soil minerals affected by the siderophore desferrioxamine B: the role of Fe(III) (hydr)oxides and dissolved Fe(III). *Eur. J. Soil Sci.* **53**, 45–55.

O'Connor, G. A., O'Connor, C., and Cline, G. R. (1984). Sorption of cadmium by calcareous soils: influence of solution composition. *Soil Sci. Soc. Am. J.* **48**, 1244–1247.

O'Reilly, S. E., Strawn, D. G., and Sparks, D. L. (2001). Residence time effects on arsenate adsorption/desorption mechanisms on goethite. *Soil Sci. Soc. Am. J.* **65**, 67–77.

Ortiz-Viana, M. M., da Salva, M. P., Agraz, R., Procopio, J. R., Sevilla, M. T., and Hernandez, L. (1999). Comparison of two kinetic approaches for copper speciation using ion-exchange modified carbon paste electrodes. *Anal. Chim. Acta* **382**, 179–188.

Oscarson, D. W., Huang P. M., Defose, C., and Herbillon, A. (1981). Oxidative power of Mn(IV) and Fe(III) oxides with respect to As(III) in terrestrial and aquatic environments. *Nature (London)* **291**, 50–51.

Peryea, F. J. (1998). Phosphate starter fertilizer temporarily enhances soil arsenic uptake by apple trees grown under field conditions. *Hort. Sci.* **33**, 826–829.

Pickering, W. F. (1979). Copper retention by sediment/soil components. In *Copper in the Environment Vol. I, Ecological Cycling*, ed. Nriagu, J. O., Wiley, New York, 217–253.

Pigna, M., Colombo, M., and Violante A. (2004). Competitive adsorption of arsenate and phosphate on synthetic hematites. *Proc. Ital. Congr. Agric. Chem.* (SICA), **21**, 70–76.

Reichman, S. M., and Parker, D. R. (2005). Metal complexation by phytosiderophores in the rhizosphere. In *Biogeochemistry of Trace Elements in the Rhizosphere*, ed. Huang, P. M., and Gobran, G. R., Elsevier, Amsterdam The Netherlands, 129–156.

Robarge, W. P. (1998). Precipitation/dissolution reactions in soils. In *Soil Physical Chemistry*, 2nd ed., ed. Sparks, D. L., CRC Press, Boca Raton, FL, 192–238.

Roy, W. R., Hassett, J. J., and Griffin, R. A. (1986). Competitive coefficient for the adsorption of arsenate, molybdate, and phosphate mixtures by soils. *Soil Sci Soc. Am. J.* **50**, 1176–1182.

Saha, U. K., Taniguchim, S., and Sakurai, K. (2002). Simultaneous adsorption of cadmium, zinc, and lead on hydroxyaluminum–and hydroxyaluminosilicate–montmorillonite complexes. *Soil Sci. Soc. Am. J.* **66**, 117–128.

Sakurai, K., and Huang, P. M. (1995). Cadmium adsorption on the hydroxyaluminium–montmorillonite complex as influenced by oxalate. In *Environmental Impact of Soil Componenet Interactions, Vol. II, Metals, Other Inorganics and Microbial Activities*, ed. Huang, P. M., Berthelin, J., Bollag, J.-M., McGill, W. B., and Page, A. L., Lewis Publishers, Boca Raton, FL, 39–46.

Sarkar, D., Essington, M. E., and Misra, K. C. (1999). Adsorption of mercury(II) by variable charge surfaces of quartz and gibbsite. *Soil Sci Soc. Am. J.* **63**, 1626–1636.

Sauvé, S., McBride, M. B., Norwell, W. A., and Hendershot, W. (1997). Copper solubility and speciation of in situ contaminated soils: effects of copper levels, pH and organic matter. *Water Air Soil Pollut.* **100**, 133–149.

Scheidegger, A. M., Lamble, G. M., and Sparks, D. L. (1996). Investigation of Ni sorption on pyrophyllite: an XAFS study. *Environ. Sci. Technol.* **30**, 548–554.

Scheidegger, A. M., Lamble, G. M., and Sparks, D. L. (1997). Spectroscopic evidence for the formation of mixed-cation hydroxide phases upon metal sorption on clays and aluminum oxides. *J. Colloid Interface Sci.* **62**, 2233–2245.

Senesi, N. (1992). Metal-humic substance complexes in the environment: molecular and mechanistic aspects by multiple spectroscopic approach. In *Biogeochemistry of Trace Elements*, ed. Adriano, D. C., Lewis Publishers, Boca Raton, FL, 429–451.

Senesi, N., and Loffredo, E. (1998). The chemistry of soil organic matter. In *Soil Physical Chemistry*, 2nd ed., ed. Sparks, D. L., CRC Press, Boca Raton, FL, 239–370.

Singh, B. R., and Myhr, K. (1998). Cadmium uptake by barley as affected by Cd sources and pH levels. *Geoderma* **84**, 185–194.

Smith, E., Naidu, R., and Alston, A. M. (1998). Arsenic in the soil environment: a review. *Adv. Agron.* **64**, 149–195.

Sparks, D. L. (1990). *Kinetics of Soil Chemical Processes*, Academic Press, San Diego, CA.

Sparks, D. L. (1995). Kinetics of metal sorption reactions. In *Metal Speciation and Contamination of Soil*, ed. Allen, H. E., Huang, C. P., Bailey, G. W., and Bowers, A. R., Lewis Publishers, Boca Raton, FL, 36–58.

Sparks, D. L. (1999). Kinetics and mechanisms of chemical reactions at the soil mineral/water interface. In *Soil Physical Chemistry*, 2nd ed, ed. Sparks, D. L., CRC Press, Boca Raton, FL, 135–191.

Sparks, D. L. (2003). *Environmental Soil Chemistry*, 2nd ed. Academic Press, San Diego, CA.

Sparks, D. L., and Zhang, P. C. (1991). Relaxation methods for studying kinetics of soil chemical phenomena. In *Rates of Soil Chemical Processes*, ed. Sparks, D. L., and Suarez, D. L., Spec. Publ. 27, Soil Science Society of America, Madison, WI, 61–94.

Sposito, G. (1996). *The Environmental Chemistry of Aluminum*, CRC Press, Boca Raton, FL.

Sun, X., and Doner, H. E. (1996). An investigation of arsenate and arsenite bonding structures on goethite by FTIR. *Soil Sci.* **161**, 865–872.

Swedlund, P. J., and Webster, J. G. (1999). Adsorption and polymerization of silicic acid on ferrihydrite, and its effect on arsenic adsorption. *Water Res.* **33**, 3413–3422.

Thanabalasingan, P., and Pickering, W. F. (1986). Arsenic sorption by humic acids. *Environ. Pollut.* **12**, 233–246.

Tichy, R., Nydl, V., Kuzel, S., and Kolar, L. (1997). Increased cadmium availability to crops on a sewage sludge amended soil. *Water Air Soil Pollut.* **94**, 361–372.

Vacca, A., Adamo, P., Pigna, M., and Violante, P. (2002). Properties and classification of selected soils from the Roccamonfina volcano, central-southern Italy. *Soil Sci. Soc. Am. J.* **67**, 198–207.

Violante, A., and Gianfreda, L. (2000). Role of biomolecules in the formation of variable-charge minerals and organo-mineral complexes and their reactivity with plant nutrients and organics in soil. In *Soil Biochemistry*, Vol. 10, ed. Bollag, J.-M., and Stotzky, G., Marcel Dekker, New York, 207–270.

Violante, A., and Pigna, M. (2002). Competitive sorption of arsenate and phosphate on different clay minerals and soils. *Soil Sci. Soc. Am. J.* **66**, 1788–1796.

Violante, A., Huang, P. M., Bollag, J.-M., and Gianfreda, L. (2002a). *Soil Mineral–Organic Matter–Microorganism Interactions and Ecosystem Health: Dynamics, Mobility and Transformations of Pollutants and Nutrients*, Developments in Soil Science, Vol. 28A, Elsevier, New York.

Violante, A., Huang, P. M., Bollag, J.-M., and Gianfreda, L. (2002b). *Soil Mineral–Organic Matter–Microorganism Interactions and Ecosystem Health: Ecological Significance of the Interactions Among Clay Minerals, Organic Matter and Soil Biota*, Developments in Soil Science, Vol. 28B, Elsevier, New York.

Violante, A., Krishnamurti, G. S. R., and Huang, P. M. (2002c). Impact of organic substances on the formation of metal oxides in soil environments. In *Interactions Between Soil Particles and Microorganism: Impact on the Terrestrial Ecosystem*, ed. Huang, P. M, Bollag, J.-M. and Senesi, N., Wiley, New York, 133–188.

Violante, A., Ricciardella, M., and Pigna, M. (2003). Adsorption of heavy metals on mixed Fe–Al oxides in the absence or presence of organic ligands. *Water Air Soil Pollut.* **145**, 289–306.

Violante, A., Ricciardella, M., Pigna, M., and Capasso, R. (2005a). Effects of organic ligands on the sorption of trace elements onto metal oxides and organo-mineral complexes. In *Biogeochemistry of Trace Elements in the Rhizosphere*, ed. Huang, P. M., and Gobran, G. R., Elsevier, Amsterdam, The Netherlands, 157–182.

Violante, A., Pigna, M., and Del Gaudio, S. (2005b). Adsorption–desorption processes of arsenate in soil environments. In *Soil Abiotic and Biotic Interactions and the Impact on the Ecosystem and Human Welfare*, ed. Huang, P. M., Bollag, J.-M., Violante, A., and Vityakon P., Science Publishers, Enfield, NH, 269–299.

Waltham, C. A., and Eick, M. J. (2002). Kinetics of arsenic adsorption on goethite in the presence of sorbed silicic acid. *Soil Sci. Soc. Am. J.* **66**, 818–825.

Waychunas, G. A., Rea, B. A., Fuller, C. C., and Davis, J. A. (1993). Surface chemistry of ferrihydrite, I: EXAFS studies of the geometry of coprecipitates and adsorbed arsenate. *Geochim. Cosmochim. Acta* **57**, 2251–2269.

Wu, C. H., Shang, L. L., Cheng, F. L., and Chao, Y. K. (2001). Modeling competitive adsorption of molybdate, sulfate and selenate on $\gamma -Al_2O_3$ by the triple-layer model. *J. Colloid Interface Sci.* **233**, 259–264.

Wu, Z., Gu, Z., Wang, X., Evans, L., and Guo, H. (2003). Effects of organic acids on adsorption of lead onto montmorillonite, goethite and humic acid. *Environ. Pollut.* **121**, 469–475.

Yamaguchi, N. U., Scheinost, A. S., and Sparks, D. L. (2002). Influence of gibbsite surface area and citrate on Ni sorption mechanisms at pH 7.5. *Clays Clay Miner.* **50**, 784–790.

Yee, N., and Fein, J. B. (2001). Cd adsorption onto bacterial surfaces: A universal adsorption edge? *Geochim. Cosmochim. Acta*, **65**, 2037–2042.

Zachara, J. M., Ainsworth, C. C., Cowan, C. E., and Resch, C. T. (1989). Adsorption of chromate by subsurface soil horizons. *Soil Sci. Soc. Am. J.* **53**, 418–428.

Zhou, D., De Cristofaro, A., He, J. Z., and Violante, A. (1999). Effect of oxalate on adsorption of copper on goethite, bayerite and kaolinite. In *Clays for Our Future*, ed. Kodama, H., Mermut, A. R., and Torrance, J. K., *Proc. 11th Clay Conference*, Ottawa, ON, Canada, 1998, 523–529.

6

MODELING ADSORPTION OF METALS AND METALLOIDS BY SOIL COMPONENTS

S. GOLDBERG

George E. Brown, Jr., Salinity Laboratory, USDA-ARS, Riverside, California

L. J. CRISCENTI

Sandia National Laboratories, Albuquerque, New Mexico

Biophysico-Chemical Processes of Heavy Metals and Metalloids in Soil Environments,
Edited by Antonio Violante, Pan Ming Huang, and Geoffrey Michael Gadd

6.1. INTRODUCTION

Adsorption is the accumulation of a chemical species at the interface between a solid phase and a fluid phase. In adsorption, a two-dimensional molecular surface structure is retained. Adsorption processes in soils have been described using a variety of approaches. Empirical adsorption models provide descriptions of experimental adsorption data without any theoretical basis. Chemical adsorption models, on the other hand, provide a molecular description of adsorption reactions.

The purpose of this chapter is to present several of the most common models used to describe metal and metalloid ion adsorption by soil components. Empirical models used in soil chemistry are described and their limitations discussed. Common chemical models used to describe metal adsorption on soil minerals are described and their advantages over empirical approaches discussed. Methods for obtaining model parameters are provided. Methods for establishing adsorption mechanisms and surface speciation are addressed. Limitations and approximations in the application of chemical models to natural systems are presented.

6.2. DESCRIPTION OF MODELS

6.2.1. Empirical Models—Historical Development

Adsorption reactions by soil minerals and soils have been described historically using empirical adsorption isotherm equations. As their name implies, it is understood that they are used for experiments at constant temperature, which unless indicated, otherwise is standard temperature, $T = 298$ K. An adsorption isotherm is a plot of the concentration adsorbed to a solid surface versus the concentration in aqueous solution for different total concentrations of a chemical species. Typically, adsorption isotherm equations are very good at describing experimental data, despite their lack of theoretical basis. Popularity of these equations stems in part from their simplicity and from the ease of estimation of their adjustable parameters.

6.2.1.1. Distribution Coefficient The simplest and most widely used adsorption isotherm equation is a linear function. This adsorption isotherm equation is

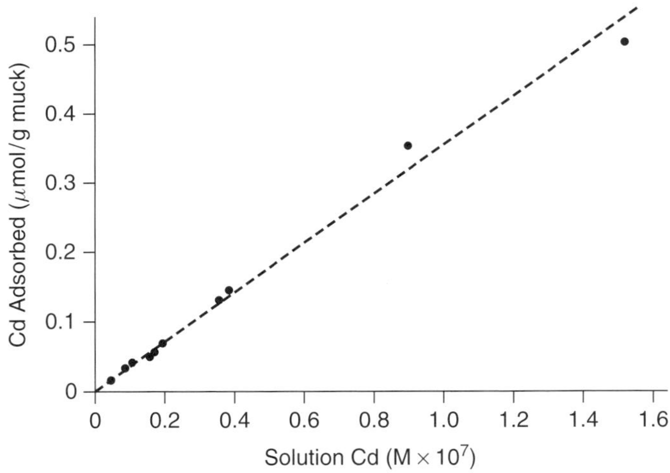

Figure 6.1. Linear adsorption of cadmium by the organic component of a muck soil. (From Turner et al., 1984.)

written in terms of the distribution coefficient, K_d:

$$x = K_d c \tag{6.1}$$

where x is the amount of chemical species adsorbed per unit mass of solid and c is the equilibrium solution concentration of the chemical species. Because of the linear assumption, the distribution coefficient usually describes adsorption data only over a very restricted solution concentration range. Figure 6.1 shows the ability of the distribution coefficient to describe linear adsorption of cadmium by the organic component of a muck soil (Turner et al., 1984). In this example, adsorption is linear over the solution metal concentration range investigated.

6.2.1.2. Freundlich Adsorption Isotherm Equation The Freundlich adsorption isotherm equation is the oldest of the nonlinear isotherms, and its use implies heterogeneity of adsorption sites. The Freundlich isotherm equation is

$$x = K c^{\beta} \tag{6.2}$$

where K is an affinity parameter and β is a dimensionless heterogeneity parameter; the smaller the value of β, the greater the heterogeneity (Kinniburgh, 1985). The Freundlich equation reduces to a linear adsorption isotherm when $\beta = 1$. Although the Freundlich adsorption isotherm is strictly valid only for metal adsorption at low aqueous metal concentration (Sposito, 1984), it has often been used to describe metal adsorption by soils over the entire metal concentration range investigated. The Freundlich equation does not obey Henry's law at low metal concentration, nor does it reach an adsorption maximum at high metal concentration (Kinniburgh, 1985). The ability of the Freundlich adsorption isotherm

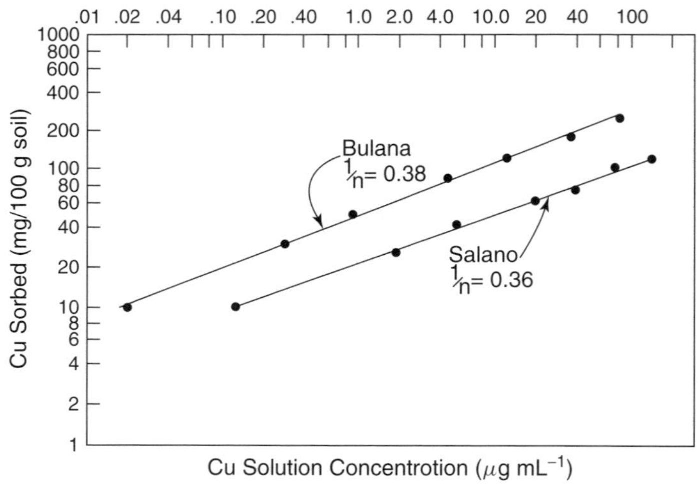

Figure 6.2. Fit of the Freundlich adsorption isotherm equation to copper adsorption by two soils. (From Kurdi and Doner, 1983.)

to describe metal adsorption data is indicated in Figure 6.2 for copper adsorption by two California soils (Kurdi and Doner, 1983).

6.2.1.3. Langmuir Adsorption Isotherm Equation The Langmuir adsorption isotherm equation was developed originally to describe gas adsorption onto clean surfaces and can be derived theoretically based on rates of evaporation and condensation (Adamson, 1976). The Langmuir adsorption isotherm equation is

$$x = \frac{x_m K c}{1 + K c} \tag{6.3}$$

where x_m is the maximum adsorption of chemical species per unit mass of solid and K is an affinity parameter related to the bonding energy of the species to the surface. The Langmuir isotherm is derived assuming a finite number of uniform adsorption sites and the absence of lateral interaction between adsorbed species. Despite the fact that these assumptions are violated in soils, the Langmuir equation has often been used to describe metal adsorption reactions on soil minerals and soils. In many studies, the Langmuir isotherm equation is only able to describe adsorption for low solution metal concentrations. A successful application of the Langmuir isotherm equation is presented in Figure 6.3 for adsorption of copper by a calcareous soil (Cavallaro and McBride, 1978).

6.2.1.4. Multisite Langmuir Adsorption Isotherm Equation The multisite Langmuir adsorption isotherm equation was formulated for the simultaneous adsorption of a gas by more than one type of surface site. The multisite Langmuir

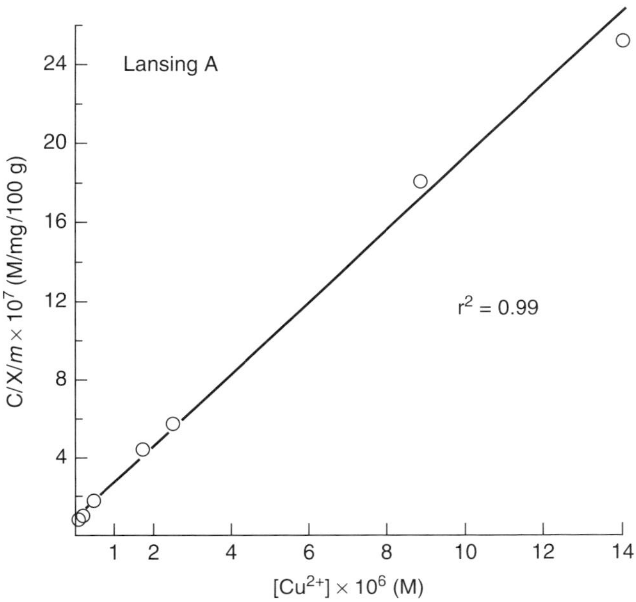

Figure 6.3. Fit of the Langmuir adsorption isotherm equation to copper adsorption by a soil. (From Cavallaro and McBride, 1978.)

isotherm equation is

$$x = \sum_{i=1}^{n} \frac{x_{m_i} K_i c}{1 + K_i c} \tag{6.4}$$

where n is the number of types of surface sites. Because of the increase in the number of adjustable parameters, the fit to metal adsorption data with the multisite Langmuir isotherm is often much improved over the single Langmuir isotherm. The high quality of fit obtained with the multisite Langmuir equation can be observed in Figure 6.4 for zinc adsorption by a Georgia soil (Shuman, 1975).

6.2.1.5. Limitations of the Empirical Approach Although theoretically impossible, the Langmuir adsorption isotherm equation can mathematically describe experimental data for a precipitation reaction (Veith and Sposito, 1977). Adherence of experimental data to an adsorption isotherm equation provides no information about the chemical reaction mechanism (Sposito, 1982). Any reaction process for which the distribution coefficient, K_d, is a finite, decreasing function of the amount adsorbed, x, and extrapolates to zero at a finite value of x, can be represented mathematically using a two-surface Langmuir adsorption isotherm equation (Sposito, 1982). Although adsorption isotherm equations are often excellent at describing metal adsorption, they must be considered simply as numerical relationships used to fit data. Independent experimental evidence

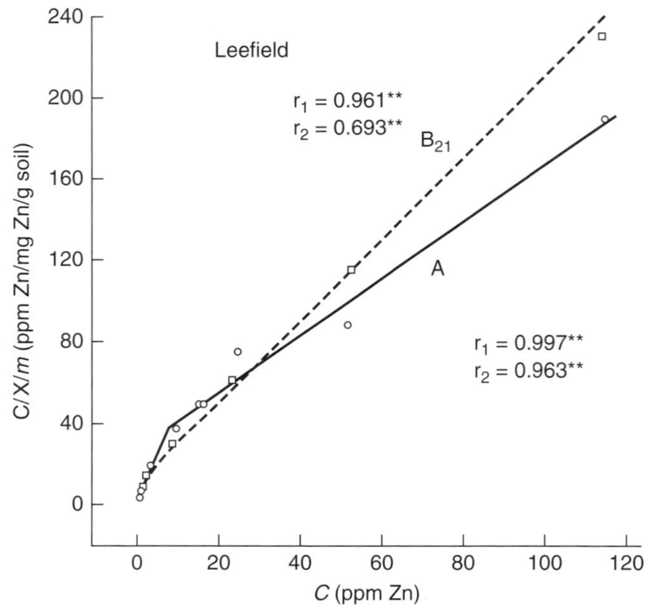

Figure 6.4. Fit of the two surface Langmuir adsorption isotherm equation to zinc adsorption by two soil horizons. (From Shuman, 1975.)

for an adsorption process must be available before any chemical significance can be attributed to isotherm equation parameters. Since use of these equations constitutes a curve-fitting procedure, isotherm equation parameters are valid only for the chemical conditions under which the experiment was conducted. Use of these equations for prediction of metal adsorption behavior under changing conditions of solution pH, solution ionic strength, and solution metal concentration is impossible.

6.2.2. Chemical Surface Complexation Models

Various chemical surface complexation models have been developed to describe potentiometric titration and metal adsorption data at the oxide–mineral solution interface. Surface complexation models provide molecular descriptions of metal adsorption using an equilibrium approach that defines surface species, chemical reactions, mass balances, and charge balances. Thermodynamic properties such as solid-phase activity coefficients and equilibrium constants are calculated mathematically. The major advancement of the chemical surface complexation models is consideration of charge on both the adsorbate metal ion and the adsorbent surface. In addition, these models can provide insight into the stoichiometry and reactivity of adsorbed species. Application of these models to reference oxide minerals has been extensive, but their use in describing ion adsorption by clay minerals, organic materials, and soils has been more limited.

Surface complexation models of the solid–solution interface share at least six common assumptions: (1) surfaces can be described as planes of constant electrical potential with a specific surface site density; (2) equations can be written to describe reactions between solution species and the surface sites; (3) the reactants and products in these equations are at local equilibrium and their relative concentrations can be described using mass law equations; (4) variable charge at the mineral surface is a direct result of chemical reactions at the surface; (5) the effect of surface charge on measured equilibrium constants can be calculated; and (6) the intrinsic (i.e., charge and potential independent) equilibrium constants can then be extracted from experimental measurements (Dzombak and Morel, 1990; Koretsky, 2000).

Four chemical surface complexation models that have been applied to metal and metalloid adsorption by reference minerals or soil systems are described. The first three models (i.e., the constant capacitance model, the diffuse double layer model, and the triple layer model) use the same approach in writing chemical reactions between the bulk solution and the solid surface. Each of these models assumes that all of the sites on a solid surface can be described by average surface site characteristics. If two or more surface site types are used to fit adsorption data with these models, the site types are still generic in nature, with no specific correlation to the solid surface structure. In contrast, the fourth model (i.e., the CD-MUSIC model) distinguishes between different surface site types based on crystal chemical considerations. The first three models differ in their approach and level of detail used to describe the charge and electric potential gradients at the solid–solution interface. The CD-MUSIC model has been implemented with two different descriptions of the electric potential gradient from the solid surface into bulk solution. In all four models, the surface sites are amphoteric surface hydroxyl groups that protonate and deprotonate as a function of pH.

6.2.2.1. Constant Capacitance Model The constant capacitance model of the oxide–solution interface (Schindler et al., 1976; Stumm et al., 1980) contains the following assumptions:

1. All ions, including protons and hydroxyls, adsorb in one surface plane, forming inner-sphere surface complexes.
2. No surface complexes are formed with ions from the background electrolyte.
3. The constant ionic medium reference state determines the activity coefficients of the aqueous species.
4. Surface complexes exist in a chargeless environment in the standard state.
5. The relationship between surface charge, σ, and surface potential, Ψ, is linear and given by

$$\sigma = \frac{CSa}{F}\Psi \tag{6.5}$$

where C is the capacitance (F m^{-2}), S is the surface area (m^2 g^{-1}), a is the particle concentration (g L^{-1}), F is the Faraday constant (C mol_c^{-1}), σ has units of mol_c L^{-1}, and Ψ has units of V.

The protonation and dissociation reactions of the surface functional group are postulated as

$$SOH + H^+ \rightleftharpoons SOH_2{}^+ \tag{6.6}$$

$$SOH \rightleftharpoons SO^- + H^+ \tag{6.7}$$

where SOH represents the surface functional group and S a metal ion in the oxide mineral that is bound to a reactive surface hydroxyl. In clay or soil systems, the surface functional group can also represent an aluminol or silanol group at the edge of a clay mineral particle. Surface complexation reactions used in the constant capacitance model for metal adsorption include

$$SOH + M^{m+} \rightleftharpoons SOM^{(m-1)} + H^+ \tag{6.8}$$

$$2SOH + M^{m+} \rightleftharpoons (SO)_2M^{(m-2)} + 2H^+ \tag{6.9}$$

where M represents a metal ion, m^+ the charge on the metal ion, Eq. (6.8) monodentate, and Eq. (6.9) bidentate binding of the metal ion with the surface oxygen ligands.

The intrinsic equilibrium constant expressions describing the above reactions are

$$K_+(\text{int}) = \frac{[SOH_2{}^+]}{[SOH]a_{H^+}} \exp\left(\frac{F\Psi}{RT}\right) \tag{6.10}$$

$$K_-(\text{int}) = \frac{[SO^-]a_{H^+}}{[SOH]} \exp\left(\frac{-F\Psi}{RT}\right) \tag{6.11}$$

$$K_M^1(\text{int}) = \frac{[SOM^{(m-1)}]a_{H^+}}{[SOH]a_{M^{m+}}} \exp\left[\frac{(m-1)F\Psi}{RT}\right] \tag{6.12}$$

$$K_M^2(\text{int}) = \frac{[(SO)_2M^{(m-2)}]a_{H^+}^2}{[SOH]^2 a_{M^{m+}}} \exp\left[\frac{(m-2)F\Psi}{RT}\right] \tag{6.13}$$

where R is the molar gas constant (J mol^{-1} K^{-1}), T is the absolute temperature (K), and a represents the activity of an aqueous species. Concentrations (mol L^{-1}), represented by square brackets, are used rather than activities for the surface species because activity coefficients for these species are assumed to be equal to 1. The electrostatic potential terms $\exp(-F\Psi_i/RT)$ have been interpreted as surface activity coefficients that correct for the effect of surface charge on surface complexation (e.g., Dzombak and Morel, 1990) or as work contributions to the overall Gibbs electrochemical potential of reaction (Sverjensky and Sahai, 1996).

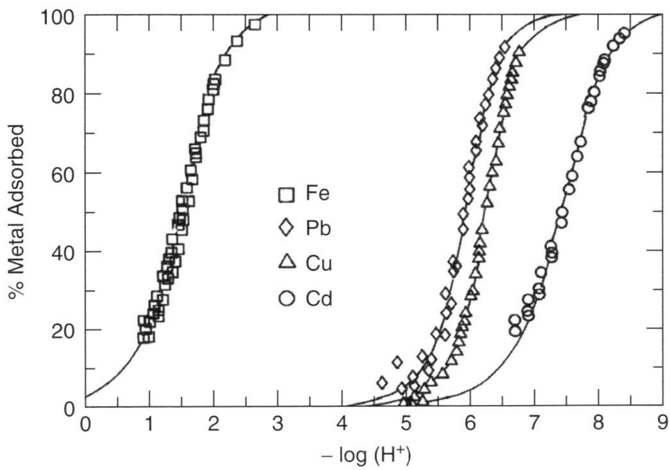

Figure 6.5. Fit of the constant capacitance model to metal adsorption on silica. (From Schindler et al., 1976.)

Recent work by Kulik (2002a,b) and Sverjensky (2003) to define useful standard states for surface species is discussed in Section 6.2.2.5a.

The mass balance for the surface functional group is

$$[SOH]_T = [SOH] + [SOH_2{}^+] + [SO^-] + [SOM^{(m-1)}] + 2[(SO)_2M^{(m-2)}] \quad (6.14)$$

and the charge balance is

$$\sigma = [SOH_2{}^+] - [SO^-] + (m-1)[SOM^{(m-1)}] + (m-2)[(SO)_2M^{(m-2)}] \quad (6.15)$$

This set of equations can be solved by hand or with a computer program using the mathematical approach of Westall (1980). The fit of the constant capacitance model to metal adsorption by silica is shown in Figure 6.5.

6.2.2.2. Diffuse Layer Model The diffuse layer model of the oxide–solution interface (Huang and Stumm, 1973; Dzombak and Morel, 1990) contains the following assumptions:

1. All ions, including protons and hydroxyls, adsorb in one surface plane, forming inner-sphere surface complexes.
2. No surface complexes are formed with ions from the background electrolyte.
3. The infinite dilution reference state is used for aqueous species; a zero surface charge reference state is used for surface species.
4. Two planes of charge represent the interfacial region: the surface plane and a diffuse layer plane that represents the closest distance of approach for the background electrolyte ions.

5. The relationship between surface charge and surface potential is

$$\sigma_d = \frac{Sa}{F} \operatorname{sgn}\Psi_d \left\{ 2\varepsilon_o DRT \sum_i c_i \left[\exp\left(\frac{-Z_i F \Psi_d}{RT}\right) - 1\right] \right\}^{1/2} \tag{6.16}$$

where ε_o is the permittivity of vacuum, D is the dielectric constant of bulk water, $\operatorname{sgn}\Psi_d = 1$ if $\Psi_d > 0$ and $\operatorname{sgn}\Psi_d = -1$ if $\Psi_d < 0$ (where d represents the diffuse plane), and c_i and z_i are the concentration and charge of solution species i, respectively.

In the diffuse layer model the surface reactions include Eqs. (6.6) and (6.7) for protonation and dissociation of the surface functional groups. In the two-site version of the model, surface complexation with metals is considered to occur on at most two types of sites: a small set of high-affinity "strong" sites, S^sOH, and a large set of low-affinity "weak" sites, S^wOH, analogous to Eq. (6.8) (Dzombak and Morel, 1990):

$$S^sOH + M^{m+} \rightleftharpoons S^sOM^{(m-1)} + H^+ \tag{6.17}$$

$$S^wOH + M^{m+} \rightleftharpoons S^wOM^{(m-1)} + H^+ \tag{6.18}$$

Although the diffuse layer model has the capability to consider bidentate metal complexes, such species are not usually considered.

The intrinsic equilibrium constants for the diffuse layer model are similar to those for the constant capacitance model where Ψ is replaced by Ψ_d. Equations (6.10) and (6.11) describe surface protonation and dissociation, respectively. Metal surface complexation is described by two constants similar to that defined in Eq. (6.12) for strong and weak sites:

$$K_M^S(\text{int}) = \frac{[S^sOM^{(m-1)}]a_{H^+}}{[S^sOH]a_{M^{m+}}} \exp\left[\frac{(m-1)F\Psi_d}{RT}\right] \tag{6.19}$$

$$K_M^W(\text{int}) = \frac{[S^wOM^{(m-1)}]a_{H^+}}{[S^wOH]a_{M^{m+}}} \exp\left[\frac{(m-1)F\Psi_d}{RT}\right] \tag{6.20}$$

The mass balances for the surface functional groups S^sOH, S^wOH, and S_T are

$$[S^sOH]_T = [S^sOH] + [S^sOH_2^+] + [S^sO^-] + [S^sOM^{(m-1)}] \tag{6.21}$$

$$[S^wOH]_T = [S^wOH] + [S^wOH_2{}^+] + [S^wO^-] + [S^wOM^{(m-1)}] \tag{6.22}$$

$$S_T = [S^sOH]_T + [S^wOH]_T \tag{6.23}$$

and the charge balance is

$$\sigma_d = [S^sOH_2{}^+] + [S^wOH_2{}^+] - [S^sO^-] - [S^wO^-] + (m-1)[S^sOM^{(m-1)}] + (m-1)[S^wOM^{(m-1)}] \tag{6.24}$$

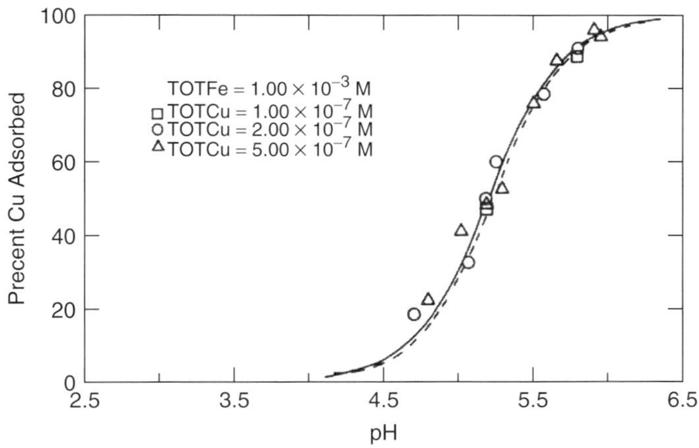

Figure 6.6. Fit of the diffuse layer model to copper adsorption by hydrous ferric oxide. The solid line represents the optimal fit for these data. The dashed line represents the fit corresponding to the best overall estimate of the Cu surface complexation constant obtained from 10 Cu adsorption edges. (From Dzombak and Morel, 1990.)

This set of equations can be approximated with hand calculations or solved using a computer program such as the one described by Dzombak and Morel (1990). An example fit of the diffuse layer model is indicated in Figure 6.6 for copper adsorption to hydrous ferric oxide.

6.2.2.3. Triple Layer Model The triple layer model of the oxide–solution interface (Yates et al., 1974; Davis and Leckie, 1978; Davis et al., 1978; Hayes and Leckie, 1987) contains the following assumptions:

1. Three planes of charge represent the interfacial region: the *o*-plane, the β-plane, and the *d*-plane.
2. Protons and hydroxyl ions adsorb in the *o*-plane.
3. Metals adsorb in either the *o*-plane or the β-plane.
4. Ions from the background electrolyte adsorb in the β-plane.
5. The diffuse layer plane, the *d*-plane, represents the closest distance of approach of dissociated charge.
6. Different reference and standard states have been applied to both aqueous and surface species (see below).
7. The relationships between the surface charges, σ_o and σ_d and the surface potentials, Ψ_o, Ψ_β, and Ψ_d are Eq. (6.16) and

$$\sigma_o = \frac{C_1 Sa}{F}(\Psi_o - \Psi_\beta) \qquad (6.25)$$

$$\sigma_d = \frac{C_2 Sa}{F}(\Psi_d - \Psi_\beta) \qquad (6.26)$$

where C_1 (F m^{-2}), the inner-layer capacitance, relates the charge at the innermost plane of adsorption to the potential drop between the *o*- and the β-planes, and C_2, the outer-layer capacitance, relates the charge at the β-plane to the potential drop between the β-and *d*-planes.

In the triple layer model the surface reactions for protonation and dissociation of the surface functional group are Eqs. (6.6) and (6.7) as written for the constant capacitance model, where Ψ is replaced by Ψ_o. The reactions for adsorption of the background electrolyte in the β-plane are

$$\mathrm{SOH} + \mathrm{C}^+ \rightleftharpoons \mathrm{SO}^- - \mathrm{C}^+ + \mathrm{H}^+ \tag{6.27}$$

$$\mathrm{SOH} + \mathrm{H}^+ + \mathrm{A}^- \rightleftharpoons \mathrm{SOH_2}^+ - \mathrm{A}^- \tag{6.28}$$

where C^+ is the cation and A^- is the anion of the background electrolyte and the dashes indicate that adsorption occurs in the β-plane. Metal ions can adsorb either in the *o*-plane forming surface complexes such as those described by Eqs. (6.8) and (6.9) or in the β-plane by the following surface complexation reactions:

$$\mathrm{SOH} + \mathrm{M}^{m+} \rightleftharpoons \mathrm{SO}^- - \mathrm{M}^{m+} + \mathrm{H}^+ \tag{6.29}$$

$$\mathrm{SOH} + \mathrm{M}^{m+} + \mathrm{H_2O} \rightleftharpoons \mathrm{SO}^- - \mathrm{MOH}^{(m-1)} + 2\mathrm{H}^+ \tag{6.30}$$

In the triple layer model one of the *o*-plane metal surface complexes is represented as bidentate, Eq. (6.9), while one of the β-plane metal surface complexes is represented as a hydroxy–metal surface species, Eq. (6.30). Davis and Leckie (1978) considered the hydroxy–metal complexation reaction to be more consistent with their experimental data. Often, an additional metal surface complex containing the background electrolyte anion is postulated to form in the β-plane:

$$\mathrm{SOH} + \mathrm{M}^{m+} + \mathrm{A}^- \rightleftharpoons \mathrm{SOHM}^{m+} - \mathrm{A}^- \tag{6.31}$$

The triple layer model has been used with different standard and reference states for both aqueous and surface species (e.g., Davis et al., 1978; Hayes and Leckie, 1987). These differences can result in different "best-fit" surface complexes for the same experimental adsorption data. For example, Hayes and Leckie (1987) expressed both the chemical potentials for aqueous and surface species by the expression

$$\mu_i = \mu_i^o + 2.303RT \log m_i + zF\phi_i \tag{6.32}$$

defining the standard state for both solution and surface species as 1 mol L^{-1} at zero surface charge and no ionic interaction. The reference state for all species was chosen to be infinite dilution relative to the aqueous phase and zero surface charge. The assumption that Eq. (6.32) applies to all aqueous ions is equivalent to abandoning the use of any form of aqueous ion activity coefficients. The outcome is particularly serious when fitting adsorption data at several ionic strengths. Using this formulation, Hayes and Leckie (1987) found that the surface species $SOCd^+$ and $SOPb^+$ best fit their adsorption data for Cd^{2+} and Pb^{2+} onto goethite from solutions ranging in ionic strength from 0.001 to 1.0 M $NaNO_3$. However, when the activity coefficients for aqueous ions were given by a version of an extended Debye–Hückel equation for 1 : 1 electrolytes (Helgeson et al., 1981), these same data were best fit with surface complexes involving the electrolyte anion $SOHCd^+ - NO_3^-$ and $SOHPb^+ - NO_3^-$ (Criscenti and Sverjensky, 1999). Hayes (1987) found that his pressure-jump kinetic data were also best described by this type of complex.

Some triple layer model intrinsic equilibrium constants for surface complexation in the o-plane are Eqs. (6.10) to (6.13) as in the constant capacitance model, where Ψ is replaced by Ψ_o. Possible intrinsic equilibrium constants for surface complexation in the β-plane are

$$K_{C+}(\text{int}) = \frac{[SO^- - C^+]a_{H^+}}{[SOH]a_{C^+}} \exp\left[\frac{F(\Psi_\beta - \Psi_o)}{RT}\right] \tag{6.33}$$

$$K_{A-}(\text{int}) = \frac{[SOH_2{}^+ - A^-]}{[SOH]a_{H^+}\, a_{A^-}} \exp\left[\frac{F(\Psi_o - \Psi_\beta)}{RT}\right] \tag{6.34}$$

$$K_M^1(\text{int}) = \frac{[SO^- - M^{m+}]a_{H^+}}{[SOH]a_{M^{m+}}} \exp\left[\frac{F(m\Psi_\beta - \Psi_o)}{RT}\right] \tag{6.35}$$

$$K_M^2(\text{int}) = \frac{[SO^- - MOH^{(m-1)}]a_{H^+}^2}{[SOH]a_{M^{m+}}} \exp\left[\frac{F((m-1)\Psi_\beta - \Psi_o)}{RT}\right] \tag{6.36}$$

$$K_M^3(\text{int}) = \frac{[SOHM^{m+} - A^-]}{[SOH]a_{M^{m+}}\, a_{A^-}} \exp\left[\frac{F(m\Psi_o - \Psi_\beta)}{RT}\right] \tag{6.37}$$

The mass balance for the surface functional group when these reactions are considered is

$$\begin{aligned}[SOH]_T = {} & [SOH] + [SOH_2{}^+] + [SO^-] + [SOM^{(m-1)}] \\ & + 2[(SO)_2M^{(m-2)}] + [SO^- - M^{m+}] + [SO^- - MOH^{(m-1)}] \\ & + [SOHM^{m+} - A^-] + [SO^- - C^+] + [SOH_2{}^+ - A^-]\end{aligned} \tag{6.38}$$

and the charge balances are

$$\sigma_o + \sigma_\beta + \sigma_d = 0 \tag{6.39}$$

$$\begin{aligned}\sigma_o = {} & [SOH_2{}^+] + (m-1)[SOM^{(m-1)}] + (m-2)[(SO)_2M^{(m-2)}] \\ & + m[SOHM^{m+} - A^-] + [SOH_2{}^+ - A^-] \\ & - [SO^-] - [SO^- - M^{m+}] - [SO^- - MOH^{(m-1)}] - [SO^- - C^+]\end{aligned} \tag{6.40}$$

$$\begin{aligned}\sigma_\beta = & m[SO^- - M^{m+}] + (m-1)[SO^- - MOH^{(m-1)}] \\ & + [SO^- - C^+] - [SOHM^{m+} - A^-] - [SOH_2{}^+ - A^-]\end{aligned} \quad (6.41)$$

The full set of equations can be solved with a computer program using the mathematical approach outlined by Westall (1980). Caution must be used to ensure that the computer code does not implement the standard and reference states proposed by Hayes and Leckie (1987). A fit of the triple layer model to silver adsorption on amorphous iron oxide is presented in Figure 6.7.

6.2.2.4. Charge Distribution Multisite Complexation (CD-MUSIC) Model The charge distribution multisite complexation model (CD-MUSIC) model of the oxide–solution interface (Hiemstra et al., 1989; Hiemstra and van Riemsdijk, 1996) contains the following assumptions:

1. The surface has multiple types of sites.
2. Protons and hydroxyl ions form inner-sphere surface complexes.
3. Two or three electrostatic planes represent the mineral–water interface.
4. Inner-sphere surface complexes have a spatial distribution of charge attributed to two different electrostatic planes, the surface *o*-plane and the intermediate 1-plane.
5. Outer-sphere ion pairs such as those formed with the background electrolyte are located in an outer 2-plane that corresponds to the start of the diffuse double layer.

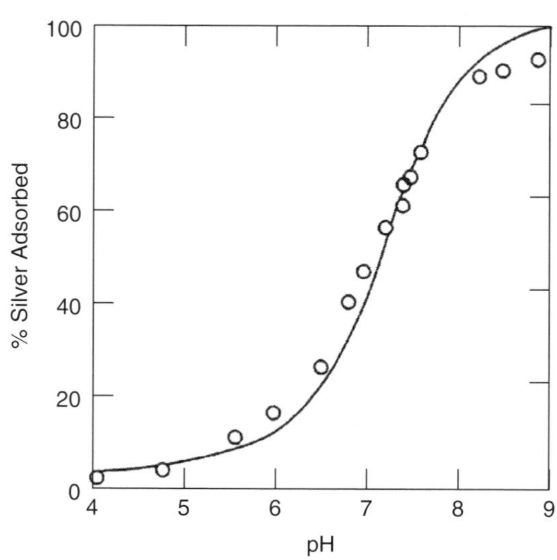

Figure 6.7. Fit of the triple layer model to silver adsorption by amorphous iron oxide. (From Davis and Leckie, 1978.)

6. The charge distribution of the central ion in the surface complex is estimated from the bond valence, ν:

$$\nu = \frac{z}{\mathrm{CN}} \tag{6.42}$$

where z is the charge of the cation and CN is its coordination number. The local formal charge in the surface complex is calculated as the bond valence contribution, s, for each metal and oxygen atom within the oxide structure (Brown and Altermatt, 1985; Hiemstra and van Riemsdijk, 2002; Hiemstra et al., 1996, 2004):

$$s = e^{(R_o - R)/B} \tag{6.43}$$

where R is the metal–oxygen bond length, R_o is an element specific length determined by analysis of the bond valence structure of many crystals, and B is a constant (37 pm). Both R_o and B are empirically determined parameters.

7. The relationships between the surface charges σ_o and σ_1 and the surface potentials Ψ_o, Ψ_1, and Ψ_2 are Eq. (6.16) and

$$\sigma_o = \frac{C_1 Sa}{F}(\Psi_o - \Psi_1) \tag{6.44}$$

$$\sigma_o + \sigma_1 = \frac{C_2 Sa}{F}(\Psi_1 - \Psi_2) \tag{6.45}$$

In the CD-MUSIC model the reactive surface functional group is defined as $S_nOH^{(n\cdot\nu-1)}$ where n is the number of metal ions in the solid phase coordinated with the reactive surface hydroxyl. The protonation reactions of the surface functional group are

$$\mathrm{S}_n\mathrm{O}^{(n\cdot\nu-2)} + \mathrm{H}^+ \rightleftharpoons \mathrm{S}_n\mathrm{OH}^{(n\cdot\nu-1)} \tag{6.46}$$

$$\mathrm{S}_n\mathrm{OH}^{(n\cdot\nu-1)} + \mathrm{H}^+ \rightleftharpoons \mathrm{S}_n\mathrm{OH}_2^{(n\cdot\nu)} \tag{6.47}$$

Surface complexation with metals produces inner-sphere surface complexes such as

$$\mathrm{S}_n\mathrm{O}^{(n\cdot\nu-2)} + \mathrm{M}^{m+} \rightleftharpoons \mathrm{S}_n\mathrm{OM}^{(n\cdot\nu+m^+-2)} \tag{6.48}$$

$$\mathrm{S}_n\mathrm{OH}^{(n\cdot\nu-1)} + \mathrm{M}^{m+} \rightleftharpoons \mathrm{S}_n\mathrm{OHM}^{(n\cdot\nu+m^+-1)} \tag{6.49}$$

Ion pairing reactions form outer-sphere surface complexes with the background electrolyte:

$$\mathrm{S}_n\mathrm{OH}^{(n\cdot\nu-1)} + \mathrm{C}^+ \rightleftharpoons \mathrm{S}_n\mathrm{OH}^{(n\cdot\nu-1)} - \mathrm{C}^+ \tag{6.50}$$

$$\mathrm{S}_n\mathrm{OH}_2^{(n\cdot\nu)} + \mathrm{A}^- \rightleftharpoons \mathrm{S}_n\mathrm{OH}_2^{(n\cdot\nu)} - \mathrm{A}^- \tag{6.51}$$

The mass balance for the surface functional group when these reactions are considered is:

$$\begin{aligned}[S_nOH^{(n\cdot\nu-1)}]_T = {} & [S_nOH^{(n\cdot\nu-1)}] + [S_nO^{(n\cdot\nu-2)}] + [S_nOH_2^{(n\cdot\nu)}] \\ & + [S_nOM^{(n\cdot\nu+m^+-2)}] + [S_nOHM^{(n\cdot\nu+m^+-1)}] \\ & + [S_nOH^{(n\cdot\nu-1)} - C^+] + [S_nOH_2^{(n\cdot\nu)} - A^-]\end{aligned} \quad (6.52)$$

and the charge balance is (Venema et al., 1996a)

$$\sigma_o + \sigma_1 + \sigma_2 = -\sigma_d \quad (6.53)$$

The charge balance expressions for the surface plane, σ_o, and the intermediate plane, σ_1, are dependent on the charge distribution, as are the intrinsic surface complexation constants. The changes in charges in these planes are given by (Venema et al., 1997)

$$\Delta\sigma_o = \Delta n_H + mf \quad (6.54)$$

$$\Delta\sigma_1 = (1 - f)m \quad (6.55)$$

where Δn_H is the change in the number of protons in the o-plane upon metal adsorption and f is the fraction of the charge of the metal ion assigned to the o-plane. The above set of equations has been solved using the computer program ECOSAT (Keizer and van Riemsdijk, 1995), FITEQL (Tadanier and Eick, 2002), or ORCHESTRA (Meeussen, 2003). The fit of the CD-MUSIC model to cadmium adsorption on goethite is shown in Figure 6.8.

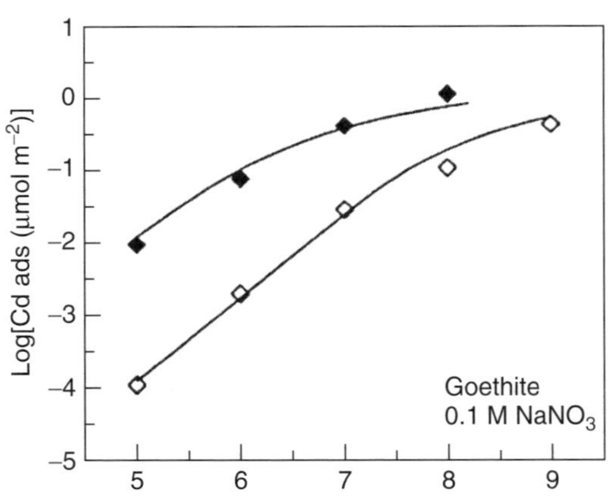

Figure 6.8. Fit of the CD-MUSIC model to cadmium adsorption by goethite. Open symbols represent total dissolved Cd at le-6 mol L^{-1}. Closed symbols represent total dissolved Cd at le-4 mol L^{-1}. (From Venema et al., 1996a.)

6.2.2.5. Obtaining Model Parameter Values

a. Standard States for the Activities of Surface Species A variety of standard states for the activities of surface species have been defined explicitly or implied by the way that equilibrium adsorption constants within a surface complexation model framework have been established (Sverjensky, 2003). The choice of standard state affects the equilibrium constant values resulting from fitting adsorption data with a surface complexation model. The standard states that have been used for surface species include the hypothetical 1.0 M standard state (implicitly assumed in Sections 6.2.2.1 to 6.2.2.4), the hypothetical 1.0 m standard state, and the hypothetical 1.0 mol per kilogram of solid standard state. Kulik (2002a,b) examined the standard states used for the Gibbs free energies of individual surface and aqueous species. He focused on the drawbacks to the hypothetical 1.0 M standard state for surface species, such as the built-in dependence on the site density of the solid (Kulik, 2002a). In fact, the molarities of both surface sites and species depend on site densities, surface areas, and solid concentrations (Sverjensky, 2003). These dependencies result in fitted equilibrium constants which cannot be compared directly without correcting for differences in the quantity or properties of the solids used in the experiments.

Sverjensky (2003) proposed new standard states, leading to equilibrium constants independent of the surface area, site density, and the amount of the solid sorbent. These new standard states are dependent only on surface site occupancy and can be used with any surface complexation model. Different standard states are defined for the activities of the sorbent sites and the sorbate species. The theoretical relationships that apply for all adsorption reactions are developed below using Eq. (6.6) as an example reaction.

For the sorbent sites, the standard state refers to unit activity of the sites on a surface where all sorbent sites are SOH, at any temperature and pressure.

$$\overline{\mu}_{\mathrm{SOH}} = \mu^{\#}_{\mathrm{SOH}} + RT \ln \lambda_{\mathrm{SOH}} X_{\mathrm{SOH}} \tag{6.56}$$

where $\overline{\mu}_{\mathrm{SOH}}$ is the electrochemical potential of SOH, $\mu^{\#}_{\mathrm{SOH}}$ is the standard chemical potential for the sorbent sites, λ_{SOH} is the rational activity coefficient for SOH, X_{SOH} is the mole fraction of SOH (the number of moles of SOH per mole of surface sites), and $\lambda_{\mathrm{SOH}} \rightarrow 1$ as $\mathrm{X}_{\mathrm{SOH}} \rightarrow 1$. The SOH sites are depicted to follow Raoult's law at high mole fractions. Therefore, the standard state molarity for sorbent sites is not unity. Instead, the standard state molarity depends on the site density, surface area, and amount of the actual sorbent solid:

$$M^{\#}_{\mathrm{SOH}} = \frac{N_s}{N_A} A_s C_s \tag{6.57}$$

where $M^{\#}_{\mathrm{SOH}}$ represents the standard state molarity of sorbent sites SOH, N_s is the surface site density on the sth solid sorbent (sites m^{-2}), A_s is the BET surface area on the sth solid sorbent (m^2 g^{-1}), C_s is the amount of the sth sorbent solid (g L^{-1}), and N_A is Avogadro's number (6.022×10^{23} sites mol^{-1}).

For the sorbate, the standard state refers to the unit activity of surface species on a completely saturated surface (i.e., all sorbent sites are occupied by sorbate) with zero potential at any temperature and pressure referenced to infinite dilution. The electrochemical potential for SOH_2^+ is

$$\overline{\mu}_{SOH_2^+} = \mu^{\ddagger}_{SOH_2^+} + RT \ln \lambda_{SOH_2^+} X_{SOH_2^+} + F\Psi_{SOH_2^+} \tag{6.58}$$

where $\mu^{\ddagger}_{SOH_2^+}$ is the standard chemical potential for the sorbate species, SOH_2^+ and $\lambda_{SOH_2^+} \rightarrow 1$ as $\Psi_{SOH_2^+} \rightarrow 0$ and $X_{SOH_2^+} \rightarrow 0$. Henry's law is followed at low mole fractions. In the standard state, the sorbate species will have an abundance determined by a *hypothetical* site density, surface area, and amount of solid sorbent. Sverjensky (2003) selected the following standard state values for all solids: $N^{\ddagger} = 10 \times 10^{18}$ sites m^{-2}, $A^{\ddagger} = 10\,\text{m}^2\,\text{g}^{-1}$, and $C^{\ddagger} = C_s$. Then

$$M^{\ddagger}_{SOH_2^+} = \frac{N^{\ddagger}}{N_A} A^{\ddagger} C^{\ddagger} \tag{6.59}$$

Using these new standard states for the sorbent sites and sorbate species, the equilibrium constant for the surface protonation reaction, Eq. (6.6), is given by

$$K_1^{\theta} = \frac{X_{>SOH_2^+}}{X_{>SOH} a_{H^+}} 10^{F\Psi_o/2.303RT} = \frac{M_{>SOH_2^+}}{M_{>SOH} a_{H^+}} \frac{N_s A_s}{N^{\ddagger} A^{\ddagger}} 10^{F\Psi_o/2.303RT} \tag{6.60}$$

These new equilibrium constants are independent of the properties of the solid and the amount of solid. Without this dependency, equilibrium constants derived from adsorption data on different samples of the same solid, and from adsorption data on different types of solids, can be compared directly.

b. Surface Site Density The total surface site density, N_s, is an important parameter in surface complexation models related to the total number of reactive functional groups. Surface site density values can be obtained using a wide variety of experimental methods, calculated from crystal dimensions, or optimized to fit experimental adsorption data (Davis and Kent, 1990). Experimental methods include potentiometric titration, tritium exchange, maximum ion adsorption, and infrared spectroscopy. Reviews of measured site densities are provided by James and Parks (1982), Davis and Kent (1990), and Koretsky et al. (1998). For one mineral, results from diverse methods can vary up to an order of magnitude with crystallographic calculations yielding the lowest and tritium exchange the highest values (Goldberg, 1991). While goodness-of-fit was found to be insensitive to changes in value of surface site density from 1 to 100 sites nm^{-2} (Hayes et al., 1991), the actual values of the surface complexation constants changed.

To allow the development of self-consistent parameter databases for surface complexation models, Davis and Kent (1990) recommended a surface site density value of 2.31 sites nm^{-2} for natural materials. This value closely approximates surface site densities for iron and manganese oxides and the edges of clay minerals. It was used successfully to describe boron (Goldberg et al., 2000) and molybdenum (Goldberg et al., 1998, 2002) adsorption on oxides, clay minerals, and soils using the constant capacitance model. An alternative reference site density of 12.05 sites nm^{-2} was proposed by Kulik (2002a) for all mineral–water surfaces because it corresponds roughly to the density of water molecules in a surface monolayer and represents a maximal density of monodentate surface complexes. The standard state proposed by Kulik (2002a) for a surface species is when 1 mol of adsorbate occupies all sites of reference total density on all of the surface of 1 mol of the adsorbent suspended in 1 kg of water–solvent at $P = 1$ bar and at a defined T, in the absence of external fields and at zero surface potential. In contrast, Sverjensky (2003) chose the standard state properties of a surface species to be $N^{\ddagger} = 10$ sites nm^{-2}, $A^{\ddagger} = 10\ m^2\ g^{-1}$, and $C^{\ddagger} = C_s$ (see Section 6.2.2.5a).

Crystal chemical considerations have been used to calculate the densities of different surface site types for specific minerals (e.g., Jones and Hockey, 1971; Yates, 1975; Hiemstra et al., 1987, 1989; Barron and Torrent, 1996). Koretsky et al. (1998) studied the predominant cleavage or growth faces for a suite of minerals, including goethite, hematite, corundum, kaolinite, albite, anorthite, and quartz. Using calculated bond strengths and charges, "ideal" slices through the crystal structures were defined to be charge-neutral or nearly charge-neutral slices produced with a minimum total strength of bonds severed. Setting the number of broken bonds at the surface equal to the number of reactive surface sites, or considering partial charges of coordinatively unsaturated atoms at the surface, gave the best agreement with available experimentally determined site densities from tritium exchange experiments. In addition, the types of surface hydroxyl groups predicted using this approach were in qualitative agreement with those observed from surface infrared spectroscopy (Koretsky et al., 1998; Koretsky, 2000).

c. Capacitances It has often been stated that capacitances cannot be determined experimentally, and therefore their values must be chosen to optimize model fit to data. This is not strictly correct. Capacitance values for the constant capacitance model can be obtained from the slopes of plots of the conditional protonation-dissociation constants versus surface charge as described by Goldberg (1992). The capacitance value, C_+, obtained from plotting conditional protonation constants is not usually equal to the value, C_-, obtained from plotting conditional dissociation constants. Since computer models for parameter optimization, such as FITEQL (Herbelin and Westall, 1996), require a single value of capacitance, it is usually chosen to optimize model fit to experimental data. This simplification is generally valid since surface complexation constant values are not very sensitive to changes in capacitance value (Goldberg and Sposito, 1984).

In the triple layer model, values of the capacitance, C_1, can be obtained experimentally from the slopes of plots used in linear (Davis et al., 1978), double (James et al., 1978), or electrokinetic (Sprycha, 1984) extrapolations as described by Goldberg (1992). As for the constant capacitance model, capacitance values, C_{1+}, obtained from extrapolations below the zero point of charge, are not usually equal to the values, C_{1-}, obtained from extrapolations above the zero point of charge. Electrokinetic extrapolation is the only method for determining the value of capacitance C_2 experimentally (Sprycha, 1984). This method assumes that the zeta potential, ζ, is equal to the diffuse layer potential, Ψ_d. Because of these experimental difficulties, capacitance values in the triple layer model have almost universally been treated as adjustable parameters. The capacitance, C_1, is adjusted to optimize fit to experimental data and the value of the capacitance, C_2, is fixed at 0.2 F m^{-2}. Use of this value for C_2 in the triple layer model has been criticized because of its uncertain physical interpretation (Hiemstra and van Riemsdijk, 1991).

Sverjensky (2005) suggested that while the triple layer model fits to surface charge data are not dependent on the value for C_2, the value for C_2 is important for predicting zeta potentials. The small value of 0.2 F m^{-2} implies a rather large distance between the β-and the *d*-planes of the model. Sverjensky (2005) proposed that the separation of these two planes is influenced by the size of the electrolyte cation in the β-plane and assumed that $C_1 = C_2$. Using this assumption, good agreement between predicted and experimental zeta potentials was found for rutile in LiCl and CsCl solutions (Kallay et al., 1994) and for hematite in $NaNO_3$ solutions (Schudel et al., 1997).

Sverjensky (2001) showed that the values for capacitance C_1 obtained for a wide variety of oxides and electrolyte types within the framework of the triple layer model (Sahai and Sverjensky, 1997a) fell into two groups. For rutile, anatase, and magnetite, values of C_1 increased with decreasing crystallographic radius of the electrolyte cation from Cs^+ to Li^+. For quartz, amorphous silica, goethite, hematite, and alumina, values of C_1 increased with decreasing hydrated electrolyte cation radius from Li^+ to Cs^+.

The distance between the *o*- and β-planes was influenced by both the size and state of hydration of the adsorbing electrolyte ions at the β-plane and the presence of water molecules between the planes. These planes do not represent physical distances away from the mineral surface, but rather, planes of constant electric potential. For minerals with high dielectric constants, such as rutile, anatase, and magnetite, the work required for the removal of waters of solvation from cations near the mineral surfaces is negligible (James and Healy, 1972). For these minerals, $1/C_1$ was found to be correlated to the crystallographic radius of the electrolyte cation. On the other hand, for minerals of lower dielectric constant, such as hematite, goethite, alumina, quartz, and amorphous silica, for which the work required to remove waters of solvation near the surface is larger (James and Healy, 1972), C_1 increased in the order of decreasing hydrated radius (Sverjensky, 2001). It should be noted that in this study, regardless of the mineral involved, the electrolyte cations bind into the β-plane of the triple layer model. In this approach, the assignment to the β-plane, a plane of constant electrical potential, does not imply outer-sphere (fully solvated) complexation. Instead, the physical distances to the *o*-plane and the β-plane are substrate dependent.

With his model, Sverjensky (2001) predicted different distances for the adsorption of different electrolyte cations (i.e., $Rb^+ = 3.3$ Å, $Sr^{2+} = 2.9$ Å) at the rutile–water interface that compared well to the distances reported from x-ray standing-wave experiments (Fenter et al., 2000). The model also suggested that trace amounts of metals (e.g., Sr^{2+}, Ca^{2+}) other than the electrolyte cations should form inner-sphere complexes if adsorbed to the β-plane of rutile and similar solids, and form outer-sphere complexes if adsorbed to the β-plane of quartz, goethite, and similar solids. These predictions were consistent with the results of x-ray standing-wave and EXAFS studies (Axe et al., 1998; Fenter et al., 2000; O'Day et al., 2000; Sahai et al., 2000).

In the CD-MUSIC model, C_1 is obtained from titration data and C_2 is chosen to provide a good fit to the salt dependency of specifically adsorbing ions (Hiemstra and van Riemsdijk, 1996). It should be noted that the diffuse layer model does not contain any capacitance parameters.

d. Protonation–Dissociation Constants Values for the intrinsic protonation and dissociation constants can be obtained from the same extrapolation techniques as those used to obtain capacitance values. In the constant capacitance model, the surface complexation constant values are the intercepts obtained by extrapolating alkimetric or acidimetric titration curves to zero net surface charge. Values of intrinsic protonation–dissociation constants obtained for the constant capacitance model using linear extrapolation are compiled in Goldberg (1992).

In the triple layer model, values for the intrinsic protonation and dissociation constants, as well as values for the intrinsic surface complexation constants for the background electrolyte, can be obtained from linear, double, or electrokinetic extrapolations to zero surface charge and zero and infinite electrolyte concentration. Values of intrinsic protonation–dissociation constants and intrinsic surface complexation constants for background electrolytes obtained for the triple layer model using the various extrapolations are compiled in Goldberg (1992). Use of graphical extrapolation methods has been criticized because the triple layer parameter values obtained are not unique (Koopal et al., 1987).

Values for protonation–dissociation constants and surface complexation constants for background electrolytes can also be determined by optimization of potentiometric titration data using a computer program. Computer optimization produces bias-free parameters and provides quality-of-fit criteria and parameter standard deviations. Nonlinear least-squares optimization using the FITEQL program was used to obtain surface complexation constants for the diffuse layer model (Dzombak and Morel, 1990). These authors optimized individual potentiometric titration data sets at each ionic strength and then weighted the individual optimum values to obtain overall best estimates using the following equation:

$$\overline{\log K(\mathrm{int})} = \sum \frac{(1/\sigma_{\log K(\mathrm{int})})_i}{\sum (1/\sigma_{\log K(\mathrm{int})})_i} [\log K(\mathrm{int})]_i \qquad (6.61)$$

where $(\sigma_{\log K(\mathrm{int})})_i$ is the standard deviation calculated for log K(int) of the ith data set. Individual protonation-dissociation constants and overall best estimates for hydrous ferric oxide are presented in Dzombak and Morel (1990). In the CD-MUSIC model, the intrinsic constant for the surface functional group was optimized by trial and error using the ECOSAT computer program (Venema et al., 1996b). In this approach, site densities of the two crystal planes were derived from crystal structure and shape.

Values for protonation–dissociation constants of generic surface sites can also be predicted using solvation and crystal chemical theory (Sverjensky and Sahai, 1996). Using the standard states defined above (Sverjensky, 2003), the values of $\log K_1^\theta$ and $\log K_2^\theta$ can be calculated from $\log K_{\mathrm{ZPC}}^\theta$ and $\log K_n^\theta$, which are defined next. At the zero point of charge ($\mathrm{pH_{ZPC}}$), the surface of a metal oxide has a net zero charge and the surface equilibrium can be expressed by

$$\mathrm{SO^- + 2H^+ \rightleftharpoons SOH_2{}^+} \tag{6.62}$$

and

$$\mathrm{pH_{ZPC}} = 0.5 \log K_{\mathrm{ZPC}}^\theta = 0.5\left(|\log K_1^\theta| + |\log K_2^\theta| \right) \tag{6.63}$$

The overall equation for equilibrium surface protonation is

$$\mathrm{SO^- + SOH_2{}^+ \rightleftharpoons 2SOH} \tag{6.64}$$

and the equilibrium constant expression is

$$\log K_n^\theta = \log \frac{a_{\mathrm{SOH}}^2}{a_{\mathrm{SO^-}}\, a_{\mathrm{SOH_2{}^+}}} = \log K_2^\theta - \log K_1^\theta \tag{6.65}$$

Sverjensky and Sahai (1996) proposed that the standard Gibbs free energy of the νth surface protonation reaction ($\Delta G_{r,\nu}^0$) can be broken into three terms according to

$$\Delta G_{r,\nu}^0 = \Delta G_{s,\nu}^0 + \Delta G_{pi,\nu}^0 + \Delta G_{ii,\nu}^0 \tag{6.66}$$

where $\Delta G_{s,\nu}^0$ represents a Born solvation contribution, $\Delta G_{pi,\nu}^0$ an electrostatic interaction term, and $\Delta G_{ii,\nu}^0$ a term intrinsic to the aqueous proton. The Born solvation term is treated by building on earlier studies of metal adsorption (James and Healy, 1972). The proton interaction term is built by summing an attractive interaction between the proton and the surface oxygen with a repulsive interaction between the proton and the underlying metal of the solid sorbent (Yoon et al., 1979).

Predictive equations for the equilibrium surface protonation constants derived using the approach of Sverjensky and Sahai (1996) are

$$\log K_{\mathrm{ZPC}}^\theta = \frac{-\Delta\Omega_{r,\mathrm{ZPC}}}{2.303RT}\frac{1}{\varepsilon_s} - B_{\mathrm{ZPC}}\frac{s}{r_{\mathrm{H^+}}} + \log K_{ii,\mathrm{ZPC}}'' \tag{6.67}$$

and

$$\log K_n^{\theta} = -B_n \frac{s}{r_{H^+}} + \log K''_{ii,n} \tag{6.68}$$

where ε_s and s/r_{H^+} represent the dielectric constant and the Pauling bond strength per angstrom, respectively, for the *s*th solid. The term involving the dielectric constant of the solid arises from the solvation term in Eq. (6.66), and the terms involving the Pauling bond strength arise from the electrostatic interaction term. The dielectric constant parameter also builds implicitly in a dependence on ionicity of the metal–oxide bond (Sahai, 2002) via the relationships with dielectric constant, polarizability, and ionicity. The $\Delta\Omega_{r,\mathrm{ZPC}}$, B_{ZPC}, B_n, $\log K''_{ii,\mathrm{ZPC}}$, and $\log K''_{ii,n}$ terms are coefficients obtained by calibrating the equations with experimentally derived equilibrium constants. The symbol $\Delta\Omega_{r,\mathrm{ZPC}}$ in Eq. (6.67), represents a Born solvation coefficient for the reaction

$$\Delta\Omega_{r,\mathrm{ZPC}} = \Omega_{>\mathrm{SOH_2}^+} - \Omega_{>\mathrm{SO}^-} = \Omega^{\mathrm{abs}}_{>\mathrm{SOH_2}^+} - \Omega^{\mathrm{abs}}_{>\mathrm{SO}^-} \tag{6.69}$$

where the symbols represent the conventional ($\Omega_{>\mathrm{SOH_2}^+}$, $\Omega_{>\mathrm{SO}^-}$) and absolute Born coefficients ($\Omega^{\mathrm{abs}}_{>\mathrm{SOH_2}^+}$, $\Omega^{\mathrm{abs}}_{>\mathrm{SO}^-}$). The relationship between the conventional and absolute Born coefficients for the *j*th surface species is given by

$$\Omega_j^{\mathrm{abs}} = \Omega_j + \Omega^{\mathrm{abs}}_{>\mathrm{SOH}} \tag{6.70}$$

based on the convention that $\Omega^{\mathrm{abs}}_{>\mathrm{SOH}} = 0.0$. The absolute solvation coefficient of the *j*th surface species is calculated using

$$\Omega_j^{\mathrm{abs}} = \frac{\eta Z^2 j}{4 R_{e,j}} \tag{6.71}$$

where $R_{e,j}$ represents the effective electrostatic radius for the surface species, Z_j represents the charge on the *j*th species, and $\eta = 166.027$ kcal Å mol^{-1} (Sverjensky, 1993). The effective electrostatic radius of surface species is defined as

$$R_{e,j} = r_{x,j} + \gamma_Z \tag{6.72}$$

where $r_{x,j}$ represents a crystallographic radius and γ_Z represents a constant for cations or anions for a given charge. Sverjensky and Sahai (1996) showed that values for the coefficients $\Delta\Omega_{r,\mathrm{ZPC}}$, B_{ZPC}, B_n, $\log K''_{ii,\mathrm{ZPC}}$, and $\log K''_{ii,n}$ in Eqs. (6.67) and (6.68) can be found within the framework of the constant capacitance, diffuse layer, or triple layer model by regressing experimentally determined values for $\log K^{\theta}_{\mathrm{ZPC}}$ and $\log K^{\theta}_n$. These equations can then be used to determine surface protonation–dissociation constants for other metal oxides given the dielectric constants and the Pauling bond strength per angstrom for the metal–oxygen bonds within these solid phases.

According to Bickmore et al. (2003, 2004), bond relaxation may actually be a key to understanding surface acidity constants. With ab initio methods, Bickmore et al. (2003, 2004) calculated the molecular structures of (hydr)oxide solution monomers and the relaxation of mineral surfaces. Bond-valence methods for the prediction of solution and surface functional group acidity constants were reevaluated based on these calculated structures. It was found that the acidity constants for oxyacids and hexaaquo cations are correlated to the Lewis base strength, S_b, of a conjugate base and metal–oxygen bond ionicity, I, and cannot be treated with the same model. The acidity constants for both $-OH$ and $-OH_2$ surface functional groups, were found to be correlated to S_b and I through the same regression equation as the acidity constants for solution oxyacids:

$$\mathrm{p}K_a^{\mathrm{int}}(\pm 0.83) = 61.5 S_b + 22.5 I - 19.2 \tag{6.73}$$

e. Metal Surface Complexation Constants Metal surface complexation constants can be evaluated graphically, although this necessitates the simplifying assumption that $\Psi = 0$. This limitation can be overcome by computer optimization of the constants. Metal surface complexation constants for all models have been obtained using various computer programs, including MICROQL (Westall, 1979), FITEQL (Herbelin and Westall, 1996), HYDRAQL (Papelis et al., 1988), and ECOSAT (Keizer and van Riemsdijk, 1995). All of the models contain different basic assumptions for the solid–solution interface. Therefore, surface complexation constants obtained with one model must never be used in any other model. Model parameter values are interdependent, so that values obtained for the surface complexation constants are dependent on the values chosen for surface site density and capacitances. Scientists must be aware of all input parameter values used when extracting surface complexation constant values from the literature.

In the absence of experimental data, the equilibrium constants for electrolyte cation and anion adsorption for the triple layer model can also be estimated based on free energy correlation type equations using the internally consistent parameter sets developed by Sverjensky and Sahai (1996), Sahai and Sverjensky (1997a,b), and Koretsky et al. (1998) and expanded in more recent papers (Criscenti and Sverjensky, 1999, 2002; Sverjensky, 2005). These parameters can be calculated using the computer program, GEOSURF (Sahai and Sverjensky, 1998). The standard Gibbs free energy for adsorption of the νth electrolyte cation or anion is broken down into three terms, representing a Born solvation contribution, an electrostatic interaction between the adsorbate and the surface, and an energy intrinsic to the aqueous adsorbate. This expression is analogous to Eq. (6.66) for surface protonation. The equilibrium constants for the electrolyte cation and anion can be predicted from expressions similar to Eq. (6.67) for $\log K_{\mathrm{ZPC}}^{\theta}$:

$$\log K_{\mathrm{M^+}}^{\theta} = \frac{-\Delta\Omega_{r,\mathrm{M^+}}}{2.303RT}\left(\frac{1}{\varepsilon_s}\right) - B_{\mathrm{M^+}}\left(\frac{s}{r_{M^+}}\right) + \log K_{ii,\mathrm{M^+}}'' \tag{6.74}$$

$$\log K_{\mathrm{L^+}}^{\theta} = \frac{-\Delta\Omega_{r,\mathrm{L^-}}}{2.303RT}\left(\frac{1}{\varepsilon_s}\right) - B_{\mathrm{L^-}}\left(\frac{s}{r_{L^-}}\right) + \log K_{ii,\mathrm{L^-}}'' \tag{6.75}$$

where s represents the Pauling bond strength (Sverjensky and Sahai, 1996) and r_{M^+} and r_{L^-} represent the distances by which the adsorbing ions are repulsed by the underlying cation of the solid and the surface oxygen, respectively (Sverjensky, 2005). Values of r_{M^+} are estimated by prediction from crystal structure analysis and a theoretical analysis of capacitances (Sverjensky, 2001), while values for r_{L^-} are approximated by adding a characteristic distance for each surface to an effective crystallographic radius for L^-. These values are used in regression calculations to obtain the repulsion coefficients B_{M^+} and B_{L^-} and the ion-specific $\log K''_{ii,M^+}$ and $\log K''_{ii,L^-}$. Figure 6.9 illustrates the results of several linear regressions using $\log K^{\theta}_{M^+}$ and $\log K^{\theta}_{L^-}$ values determined by fitting experimental surface charge data using the triple layer model. Figure 6.9*c* shows the inverse correlation between $\Delta\Omega_{r,M^+}$ and the effective electrostatic radius of the ion, $R_{e,j}$. Figure 6.9*d* illustrates that $\log K''_{ii,M^+}$ can be correlated to the aqueous-phase equilibrium metal hydrolysis constant, $\log K_{M(OH)}$. Analogous correlations can be made for the monovalent anions (Sverjensky, 2005).

In the diffuse layer model, all intrinsic metal surface complexation constants were optimized with the FITEQL program for both the strong and weak sites using the best estimates of the protonation constant, $\overline{\log K_+(\text{int})} = 7.29$, and the dissociation constant $\overline{\log K_-(\text{int})} = -8.93$ obtained with Eq. (6.61) (Dzombak and Morel, 1990). Thus, individual values of $\log K^{s}_{M_i}(\text{int})$ and $\log K^{w}_{M_i}(\text{int})$ and best estimates of $\overline{\log K^{s}_{M_i}(\text{int})}$ and $\overline{\log K^{w}_{M_i}(\text{int})}$ are unique in that they represent a self-consistent thermodynamic database for metal adsorption on hydrous ferric oxide.

Another standardized database for the diffuse layer model was developed for montmorillonite by Bradbury and Baeyens (2005). Surface complexation constants for strong and weak sites and cation exchange were fit to adsorption data for various metals using constant site densities and protonation–dissociation constants in a nonelectrostatic modeling approach. Linear free energy relationships were developed to predict surface complexation constants for additional metals from their aqueous hydrolysis constants.

Recognizing the utility of a standardized database of surface complexation constants, Kulik (2002a) recommended the normalization of intrinsic constants to a reference site density of 12.05 sites nm^{-2}:

$$\log K_{\text{norm}} = \log K(\text{int}) + \log \frac{N_s}{12.05} \tag{6.76}$$

This calculation allows a comparison of surface complexation constants for a particular mineral obtained with a specific surface complexation model. A set of constants normalized in this manner can be used to provide a self-consistent surface speciation.

The Rossendorf expert system for surface and sorption thermodynamics, RES^3T (Brendler et al., 2003, 2004), is a mineral-specific digitized thermodynamic database that contains mineral properties, specific surface areas, ion sorption data, surface complexation reactions, and bibliographic information based

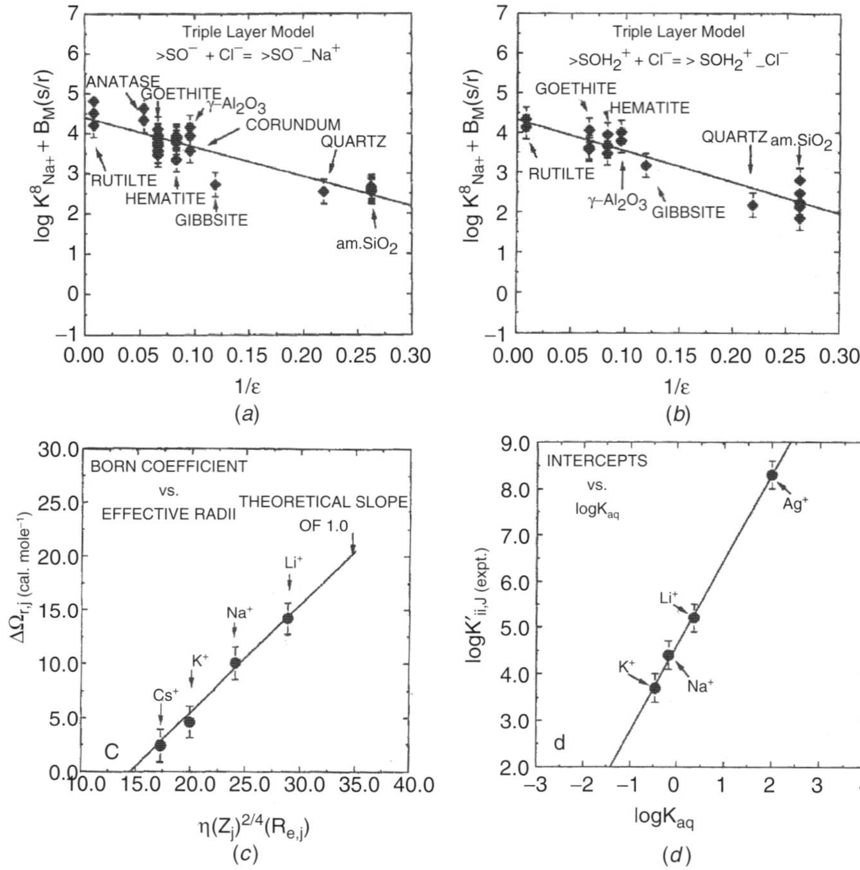

Figure 6.9. (*a*) Values of log K^{θ}_{M+} as a function of the dielectric constant ($1/\varepsilon_s$) and values of the bond strength per Å (s/r_{M^+}); *(b)* values of log K^{θ}_{L-} as a function of the dielectric constant ($1/\varepsilon_s$) and the repulsive term in Å ($2/r_{L^+}$); (*c*) values of $\Delta\Omega_{r,M^+}$ as a function of the effective electrostatic radius ($1/R_{e,j}$) and values of $r_{x,j}$; (*d*) linear free energy correlations for values of logK''_{ii,M^+} with the aqueous-phase equilibrium association constants $K_{M(OH)}{}^{o}$. (From Sverjensky, 2005.)

presently on 1460 literature references. Surface complexation approaches considered include the constant capacitance, diffuse layer, triple layer, and CD-MUSIC models. Data and parameter sets can be exported to other modeling applications. The normalization calculation of Eq. (6.76) is contained as an option in the RES^{3}T database.

The Common Thermodynamic Database Project, CTDP (van der Lee and Lomenech, 2004), is another thermodynamic database that includes sorption equilibrium constants for the constant capacitance model, the diffuse layer model, and the triple layer model. At present the database includes 330 surface species for nine solids and is formatted in XML, a code independent electronic format.

Criscenti and Sverjensky (1999, 2002) continued to build the internally consistent set of triple layer model equilibrium constants developed by Sverjensky and Sahai (1996) and Sahai and Sverjensky (1997a,b) by reexamining sets of adsorption edge and isotherm data for divalent metal cation adsorption onto oxide surfaces. In contrast to previous investigations, they found that the adsorption of transition and heavy metals on solids such as goethite, γ-Al_2O_3, corundum, and anatase, which have dielectric constants between 10 and 22, was best described by surface complexes of the metal with the electrolyte anion. Metal (M^{2+}) adsorption from $NaNO_3$ solutions is described by

$$SOH + M^{2+} + NO_3^- \rightleftharpoons SOHM^{2+} - NO_3^- \tag{6.77}$$

from $NaClO_4$ solutions by

$$SOH + M^{2+} + ClO_4^- \rightleftharpoons SOHMClO_4^+ \tag{6.78}$$

from NaCl solutions by

$$SOH + M^{2+} + Cl^- \rightleftharpoons SOHMCl^+ \tag{6.79}$$

Adsorption of these same metals onto solids such as quartz and silica, with low dielectric constants between 4 and 5, may be accompanied by the electrolyte anion in $NaClO_4$ solutions, but in $NaNO_3$ and NaCl solutions, metal adsorption occurred as SOM^+ or SOMOH. The large Born solvation free energies on low dielectric constant solids opposed the coadsorption of the electrolyte anion. Using the triple layer model with metal–anion surface complexes, isotherms of metal adsorption over a range of surface coverages could be described with a single-site model, presenting an alternative to the concept that changes in isotherm slope reflect the filling of different types of surface sites (Criscenti and Sverjensky, 2002). Because the equilibrium constants reported by Criscenti and Sverjensky (1999, 2002) preceded the work of Sverjensky (2003) on standard and reference states for mineral–water interfaces, they should be modified according to Eq. (6.60). The choice of surface complexes that best fit the adsorption edge and isotherm curves should not be affected by more recent studies (Sverjensky, 2001, 2003, 2005).

6.2.2.6. Establishing Ion Adsorption Mechanisms and Surface Speciation In using surface complexation models, the user must specify the adsorption mechanisms and types of surface complexes for all adsorbing metal and metalloid ions. To preserve the chemical significance of surface complexation models, adsorption mechanisms should be established from independent experiments. Spectroscopic techniques can be used to provide direct experimental evidence of metal and metalloid adsorption mechanisms. Indirect experimental procedures of establishing adsorption mechanisms include point of zero charge shifts and ionic strength dependence.

a. Spectroscopic Techniques

VIBRATIONAL SPECTROSCOPY Infrared and Raman spectroscopies have proven to be useful techniques for studying the interactions of ions with surfaces. Direct evidence for inner-sphere surface complex formation of metal and metalloid anions has come from vibrational spectroscopic characterization. Both Raman and Fourier transform infrared (FTIR) spectroscopies are capable of examining ion adsorption in wet systems. Chromate (Hsia et al., 1993) and arsenate (Hsia et al., 1994) were found to adsorb specifically on hydrous iron oxide using FTIR spectroscopy. Raman and FTIR spectroscopic studies of arsenic adsorption indicated inner-sphere surface complexes for arsenate and arsenite on amorphous iron oxide, inner-sphere and outer-sphere surface complexes for arsenite on amorphous iron oxide, and outer-sphere surface complexes for arsenite on amorphous aluminum oxide (Goldberg and Johnston, 2001). These surface configurations were used to constrain the surface complexes in application of the constant capacitance and triple layer models (Goldberg and Johnston, 2001).

NUCLEAR MAGNETIC RESONANCE SPECTROSCOPY Nuclear magnetic resonance (NMR) spectroscopy can be applied to aqueous samples and can distinguish between inner- and outer-sphere ion surface complexes. The adsorption behavior of the cations Cs^+ and Na^+ was studied on the surfaces of silica, boehmite, kaolinite, and illite (Kim and Kirkpatrick, 1997). Cesium was adsorbed both as inner-and outer-sphere surface complexes and in the diffuse layer, while Na was adsorbed only as outer-sphere surface complexes and in the diffuse layer. The adsorbed Na ions were fully hydrated, while the Cs ions had direct contact with the surface oxygen atoms.

ELECTRON SPIN RESONANCE SPECTROSCOPY Electron spin resonance (ESR) is a technique that can also be used on aqueous samples and has been used to study the adsorption of copper, manganese, and chromium on aluminum oxides and hydroxides. Copper(II) was found to adsorb specifically on amorphous alumina and microcrystalline gibbsite forming at least one Cu–O–Al bond (McBride, 1982; McBride et al., 1984). Manganese(II) adsorbed on amorphous aluminum hydroxide was present as a hydrated outer-sphere surface complex (Micera et al., 1986). Electron spin resonance combined with electron spin-echo experiments revealed that chromium(III) was adsorbed as an outer-sphere surface complex on hydrous alumina that gradually converted to an inner-sphere surface complex over 14 days of reaction time (Karthein et al., 1991).

X-RAY ABSORPTION SPECTROSCOPY X-ray absorption spectroscopy (XAS) includes x-ray absorption near-edge (XANES) and extended x-ray absorption fine structure (EXAFS) spectroscopy. An advantage of XAS spectroscopy is that adsorption experiments can be carried out in aqueous systems (Fendorf et al., 1994). X-ray absorption spectroscopy has been used to examine the sorption of both cations and anions to oxide and silicate minerals found in soils, with an emphasis on ions that are potential contaminants in the environment.

Cationic contaminants include numerous heavy metals and transition metals and several alkaline earth and alkali metals. The adsorption reactions of Pb^{2+}, Cd^{2+}, Co^{2+}, Hg^{2+}, Cu^{2+}, Zn^{2+}, Ni^{2+}, ${UO_2}^{2+}$, Sr^{2+}, Cs^{+}, and NpO_2^{+} onto different oxide, hydroxide, and aluminosilicate minerals have all been investigated using XAS. The nature of the surface complexes formed has been found to be a function of crystal structure, sorbing cation, ligands present in solution, and surface coverage.

X-ray absorption spectroscopy has provided evidence for inner-sphere bidentate surface complexes of Pb^{2+} on Al_2O_3 (Chisholm-Brause et al., 1990a; Bargar et al., 1996, 1997a; Strawn et al., 1998), goethite (Roe et al., 1991; Bargar et al., 1997b), hematite (Bargar et al., 1997b), and amorphous iron oxide (Trivedi et al., 2003). Using grazing-incidence XAFS on single crystals of α-Al_2O_3, Bargar et al. (1996) reported that inner-sphere Pb^{2+} complexes formed preferentially on the (0001) surface, while outer-sphere Pb^{2+} complexes formed on the $(11\bar{0}2)$ surface. On montmorillonite, the type of Pb^{2+} surface complex that formed was a function of ionic strength (Strawn and Sparks, 1999). At low ionic strengths, Pb^{2+} adsorption to montmorillonite was pH independent and XAS suggested that an outer-sphere complex formed. At higher ionic strength, Pb^{2+} adsorption became pH dependent and XAS suggested that inner-sphere complexes formed. Dyer et al. (2003) were able to describe Pb^{2+} adsorption on amorphous iron oxide with the triple layer model using the bidentate mononuclear and monodentate mononuclear surface species observed in the XAS study of Trivedi et al. (2003).

Randall et al. (1999) studied the structure and composition of Cd^{2+} complexes sorbed on several iron oxyhydroxide minerals: goethite, lepidocrocite, akagenite, and schwertmannite using EXAFS. In all cases, adsorbed Cd^{2+} formed inner-sphere complexes over a wide range of solution pH and Cd^{2+} concentration. However, the bonding mechanism differed between minerals and depended on the availability of different types of adsorption sites at the mineral surface. For example, Cd^{2+} sorbed to goethite by the formation of bidentate surface complexes at corner-sharing sites, while Cd^{2+} sorbed to lepidocrocite by the formation of surface complexes at bi-and/or tridentate edge-sharing sites. Manceau et al. (2000) also found Cd^{2+} to form bidentate surface complexes at corner-sharing sites on goethite but mononuclear, fully hydrated surface complexes on lepidocrocite. Venema et al. (1996b) described Cd^{2+} adsorption on goethite with the CD-MUSIC model using the surface species observed in the EXAFS study of Spadini et al. (1994).

Cobalt adsorption onto γ-Al_2O_3 (Chisholm-Brause et al., 1990b), corundum (Towle et al., 1999), and kaolinite (O'Day et al., 1994) has been studied using XAS. For Co^{2+} adsorption onto kaolinite, O'Day et al. (1994) found that at low surface coverages, Co^{2+} sorbed as an inner-sphere bidentate complex. Using grazing-incidence XAFS on single crystals of corundum, Towle et al. (1999) found that Co^{2+} adsorbed in an inner-sphere fashion, forming a tridentate complex on the (0001) surface and a tetradentate complex on the $(11\bar{0}2)$ surface. Katz and Hayes (1995) describe Co^{2+} adsorption by α-Al_2O_3 with the triple layer model, using the surface species observed by Chisholm-Brause et al. (1990b) using EXAFS spectroscopy.

X-ray absorption spectroscopy has been used to investigate Zn^{2+} adsorption onto ferrihydrite, goethite, a mixture of α-Al_2O_3 and γ-Al_2O_3, and hydrous manganese oxide. Trivedi et al. (2001a,b) concluded that Zn^{2+} adsorbed to both hydrous ferric oxide and hydrous manganese oxide as an outer-sphere complex but formed an inner-sphere complex on goethite. In contrast, Waychunas et al. (2002) found that Zn^{2+} formed inner-sphere tetrahedrally coordinated bidentate surface complexes on ferrihydrite at low surface coverages. Trainor et al. (2000) found that on alumina at low surface coverages, Zn^{2+} also formed inner-sphere, bidentate, tetrahedrally coordinated surface complexes.

Collins et al. (1999a) found that Hg^{2+} sorbed to goethite as an inner-sphere bidentate complex. Cheah et al. (1998) found that Cu^{2+} sorbed to amorphous silica and γ-Al_2O_3 as monomeric and monodentate inner-sphere surface complexes. However, bidentate complexes may also form on γ-Al_2O_3. Using polarized EXAFS, Dähn et al. (2003) determined that Ni^{2+} sorbed to montmorillonite edge sites as an inner-sphere mononuclear surface complex. Inner-sphere surface complexes were observed with XAS for Cr^{3+} adsorption on manganese (Manceau and Charlet, 1992) and iron oxides (Charlet and Manceau, 1992).

Strontium adsorption onto soil minerals is an important retardation mechanism for $^{90}Sr^{2+}$. Chen et al. (1998) investigated the adsorption of Sr^{2+} onto kaolinite, illite, hectorite, and montmorillonite over a range of ionic strengths and from two different electrolyte solutions, $NaNO_3$ and $CaCl_2$. In all cases, the EXAFS spectra suggested Sr^{2+} adsorbed to clay minerals as an outer-sphere mononuclear complex. Sahai et al. (2000) also found that on amorphous silica, goethite, and kaolinite substrates, Sr^{2+} adsorbed as a hydrated surface complex above pH 8.6. On the other hand, Collins et al. (1998) concluded from EXAFS spectra that Sr^{2+} adsorbed as an inner-sphere complex on goethite.

Bostick et al. (2002) studied Cs^+ adsorption onto vermiculite and montmorillonite with EXAFS and found that Cs^+ formed both inner-and outer-sphere complexes on both aluminosilicates. The inner-sphere complexes bound to the siloxane groups in the clay structure. Combes et al. (1992) found that NpO_2^+ adsorbed onto goethite as a mononuclear surface complex. Waite et al. (1994) were successful in describing uranyl adsorption to ferrihydrite with the diffuse layer model using the inner-sphere, mononuclear, bidentate surface complex observed with EXAFS.

X-ray absorption fine structure studies on metal and metalloid anions include arsenate and chromate (Fendorf et al., 1997) on goethite, arsenate (Waychunas et al., 1993) on goethite and hydrous iron oxide, and arsenate on gibbsite (Ladeira et al., 2001). Mixtures of inner-and outer-sphere surface complexes were observed for arsenite on $\gamma-Al_2O_3$ (Arai et al., 2001). Manning and Goldberg (1996) postulated a mixture of bidentate and monodentate surface complexes in modeling arsenate adsorption on goethite consistent with the results of Waychunas et al. (1993). Using EXAFS, Manning et al. (1998) observed a bidentate binuclear bridging complex for arsenite adsorbed on goethite. This complex was incorporated into the constant capacitance model, and an excellent fit to arsenite adsorption data was obtained. Grossl et al. (1997) used the EXAFS results of Fendorf et al. (1997) to describe arsenate and chromate adsorption on goethite using the constant capacitance model.

X-RAY REFLECTIVITY X-ray reflectivity measurements can provide important information about mineral-water interfaces in situ by accurately determining the position of an adsorbed monolayer relative to the substrate surface. By measuring x-ray reflectivity of calcite, with and without lead, Sturchio et al. (1997) established that the lead ions were located in the surface atomic layer. X-ray reflectivity measurements found rubidium to be specifically adsorbed to the rutile surface at the tetradentate site (Zhang et al., 2004). These authors were able to include this information in the CD-MUSIC model to obtain an accurate description of rubidium adsorption.

b. Point of Zero Charge Shifts The point of zero charge, PZC, of a solid can be obtained directly from electrokinetic measurements and colloidal stability experiments or indirectly from potentiometric titration data. Electrophoretic mobility is a measure of the movement of charged particles in an electric field with zero electrophoretic mobility indicating the condition of zero surface charge. Specific inner-sphere ion adsorption produces shifts in PZC and reversals of electrophoretic mobility with increasing ion concentration (Hunter, 1981). Shifts in PZC have been observed following arsenate (Anderson et al., 1976; Harrison and Berkheiser, 1982; Suarez et al., 1998), arsenite (Pierce and Moore, 1980; Suarez et al., 1998), chromate (Lumsdon et al., 1984), and molybdate (McKenzie, 1983; Goldberg et al., 1996) adsorption on oxide minerals. Consistent with electrophoretic mobility results, an inner-sphere surface complex was postulated to describe molybdate adsorption on oxides and clay minerals with the constant capacitance model (Goldberg et al., 1996). While all adsorption that produces a shift in the PZC is inner-sphere, adsorption that does not shift the PZC may be either inner-sphere or outer-sphere.

c. Ionic Strength Effects Use of ionic strength dependence of adsorption to distinguish between inner-and outer-sphere surface complexes has been advocated for metal ions (Hayes and Leckie, 1987). Ions showing little ionic strength dependence in their adsorption behavior, such as lead, cadmium (Hayes and Leckie, 1987), and arsenate (Hsia et al., 1994; Goldberg and Johnston, 2001), were considered specifically adsorbed as strong inner-sphere surface complexes. Ions showing ionic strength dependence in their adsorption behavior, such as barium (Hayes, 1987) and arsenite (Goldberg and Johnston, 2001), were considered to be weakly bound as outer-sphere surface complexes. McBride (1997) refined this concept, indicating that ions that show decreasing adsorption with increasing ionic strength are adsorbed outer-sphere, whereas ions that show little ionic strength dependence or show increasing adsorption with increasing ionic strength are adsorbed inner-sphere. Greater adsorption with increasing ionic strength results from the higher activity of counter ions available in solution to compensate the surface charge generated by specific ion adsorption (McBride, 1997). Criscenti and Sverjensky (1999) showed that transition and heavy metals that were thought not to exhibit ionic strength dependence in adsorption behavior, in fact exhibited a dependence that was a function of the solution electrolyte. In $NaNO_3$ solutions, transition and heavy metal ions exhibited no ionic strength dependence in their adsorption behavior; in NaCl solutions, the metals consistently exhibited decreasing adsorption with increasing ionic strength; and in $NaClO_4$ solutions, they exhibited increasing adsorption with increasing ionic strength. In general, for divalent cations, transition and heavy metal adsorption can be described with inner-sphere complexes, and alkaline earth metal adsorption can be described with outer-sphere surface complexes. However, this generalization may not hold up as more experiments are conducted in electrolyte solutions other than $NaNO_3$.

d. Ab initio and Molecular Modeling Ab initio and molecular modeling can be used to investigate the stoichiometries and relative adsorption energies of viable surface complexes and to set bounds on the uncertainties associated with different surface complexation models (Criscenti, 2004). Ab initio and molecular mechanics modeling provides significant information regarding: (1) the stoichiometry of adsorbing species in terms of both the number of bonds an adsorbing species will form with a mineral surface and metal–ligand pairing at the surface, (2) the structure of water at the mineral–water interface and the corresponding structure in the presence of an adsorbing species, and (3) the relative adsorption energies of different possible surface species. Ab initio and molecular modeling simulations can also be valuable tools to examine how various factors, such as: (1) short- and long-range solvation of the ion both in solution and at the mineral surface, (2) the periodic structure of the mineral surface, and (3) the structure and dielectric constant of interfacial water, influence the adsorption process as a whole (Criscenti, 2004).

Quantum mechanics calculations on small clusters of atoms can be used in combination with x-ray absorption fine-structure (EXAFS) data to determine the mechanism of metal binding on oxide surfaces. For example, Collins et al. (1999a) found several geometries for Hg^{2+} adsorption to goethite that were consistent with EXAFS data, but only one that gave calculated Hg–Fe distances consistent with those observed. In this geometry, Hg^{2+} adsorption occurred via two oxygen atoms bound to edge-sharing Fe sites on the (110) surface. This same group of researchers (Collins et al., 1998, 1999b; Randall et al., 1999) combined quantum mechanics calculations and EXAFS data to constrain the geometry of Cd^{2+} and Sr^{2+} surface complexes on goethite and found that these metals were bound to the same type of site as mercury. Peacock and Sherman (2004a) used ab initio calculations in support of their EXAFS data to determine the stoichiometry of adsorbed Cu^{2+} species on several iron oxides. These surface complex stoichiometries were then used to fit experimental Cu^{2+} adsorption data within a diffuse layer or triple layer model framework. The same type of investigation was conducted for vanadium adsorption to goethite (Peacock and Sherman, 2004b). In this case, fits to sorption edge data using the surface complexes determined from ab initio cluster calculations and EXAFS data were more nebulous.

Classical molecular mechanics approaches have also been used to investigate the adsorption of metal ions to different mineral surfaces from aqueous solution. For example, using a molecular dynamics approach with interatomic force fields, Cygan et al. (1998) examined Cs^+ adsorption onto fully hydroxylated kaolinite surfaces from chloride solutions. Using a different force field approach, Steele et al. (2000) investigated the adsorption of Cu^{2+}, Zn^{2+}, and Cd^{2+} to the (001) muscovite surface. The calculations showed that these metals do not form strong bonds with the smooth basal plane of muscovite. However, upon the introduction of edge-like defects, both Cu^{2+} and Zn^{2+} cations were bound strongly to the defect sites with bond lengths and coordination numbers in agreement with experiment.

A large research effort has been under way to develop a comprehensive picture of the interface between aqueous solution and the (110) surface of rutile (α-TiO_2). This effort has combined molecular-scale and macroscopic approaches, including experimental measurements, quantum mechanics calculations, molecular simulations, and Gouy–Chapman–Stern models (Zhang et al., 2004). Ab initio calculations and molecular dynamics simulations, validated through direct comparison with x-ray standing-wave measurements, were used to predict ion distributions not measured experimentally. Surface oxygen proton affinities computed using the CD-MUSIC model have been improved by the incorporation of ab initio bond lengths and partial charges. All cations considered (Na^+, Rb^+, Ca^{2+}, Sr^{2+}, Zn^{2+}, Y^{3+}, Nd^{3+}) were found to adsorb as inner-sphere species directly to surface oxygen atoms, while the specific binding geometries and reaction stoichiometries were dependent on ionic radius. This investigation illustrated the success of using different types of modeling to investigate adsorption reactions.

6.3. ADVANTAGES OF SURFACE COMPLEXATION MODELS

The major advantage of the surface complexation models over the empirical approaches is that they have the potential to be predictive in nature and applicable to more than one field site. However, this potential has yet to be realized in a general way. Surface complexation models account for the surface charge arising from the charging of the reactive surface functional group and from metal surface complexation reactions. The models define specific surface species, chemical reactions, mass balances, and charge balances. Goodness-of-fit to experimental adsorption data cannot be used as evidence for the presence of any postulated surface complex. Chemical significance is optimized if as many parameters as possible are obtained experimentally. If there is no independent experimental evidence allowing the determination of the exact structure of adsorbed surface complexes, the use of models having chemical simplicity and a small number of adjustable parameters is preferable.

6.4. APPLICATIONS TO NATURAL SYSTEMS AND LIMITATIONS OF SURFACE COMPLEXATION MODELS

6.4.1. Single- or Two-Site Assumption

In most surface complexation models, metal adsorption edge and isotherm data can be described by assuming that the metal adsorbs to one or at most two average sets of reactive surface sites. This is clearly a gross simplification since natural materials are complex multisite mixtures having a variety of surface functional groups. Experimental evidence indicates that even synthetic oxide surfaces contain multiple types of surface sites that may undergo metal adsorption reactions (Rochester and Topham, 1979a,b; Benjamin and Leckie, 1980, 1981). Thus, even surface complexation constants determined in pure mineral systems are most likely average composite values. However, even in the CD-MUSIC model, which is able to account for the many types of surface sites available on individual minerals, metal adsorption is often described with only two dominant site types (Venema et al., 1996b). If, in the presence of two or more competing metal ions, the site heterogeneity of oxide surfaces manifests itself as selective metal adsorption by some fraction of the total adsorption sites, then the resulting surface complexation constants for each metal will exhibit some surface composition dependence being average values based on only a fraction of the surface functional groups. The presence of organic matter also complicates description of soil systems since humic and fulvic materials can act both as adsorbates on oxide surfaces and as adsorbents for metal ions.

6.4.2. Reactive Surface Area

In the application of surface complexation models to clay minerals or to soils dominant in clays, the assumption is often made that metal ion adsorption occurs primarily on the aluminol and silanol groups of clay edges. The effect of the permanent charge sites on the adsorption process may not be considered. This simplification may be inappropriate, particularly for metal and metalloid anions, since repulsive electrostatic forces emanating from clay faces may spill over and affect the adsorption process on clay edges (Secor and Radke, 1985).

Surface area is a relative measurement whose value is dependent on scale and measurement technique. The surface area of various amorphous aluminum oxides was affected by various factors, including aging, drying, heating, reaction in aqueous solution, and concentration of the starting reagents during synthesis (Goldberg et al., 2001). Amorphous aluminum oxides of widely differing initial surface areas measured in the dry state had surface areas of comparable magnitude upon reaction in aqueous solution and subsequent redrying; initially, high surface areas decreased and initially low surface areas increased. These results cast doubt on the use of initial BET surface areas as chemical characterization parameters indicative of mineral reactivity and as input parameters into surface complexation models. It is likely that other amorphous oxides may exhibit similar behavior and that evaluation of these systems, especially ferrihydrite, is necessary.

For goethites, an inverse relationship between surface area and adsorption capacity was attributed to variability in surface site density (Hiemstra and van Riemsdijk, 1996; Villalobos et al., 2003). The higher reactive site density of a lower surface area goethite formed by rapid reagent addition was attributed to a higher surface area percentage of (001) faces as observed with atomic force microscopy (Gaboriaud and Ehrhardt, 2003). The higher surface area goethite formed by slow reagent addition had a significantly smaller percentage of (001) faces.

6.4.3. Component Additivity

One approach for modeling adsorption on complex mineral assemblages such as soils and sediments is *component additivity* (CA). This approach attempts to predict adsorption on a complex mineral assemblage using the results of a surface characterization of the assemblage and experimental data and model parameters obtained for adsorption by pure reference minerals (Davis et al., 1998). The simplest mineral assemblage is a binary mixture. Constant capacitance modeling of adsorption of various cations and anions using the additivity concept was unsuccessful for binary mixtures of $Al(OH)_3/Fe(OH)_3$ (Anderson and Benjamin, 1990a) and $SiO_2/Fe(OH)_3$ (Anderson and Benjamin, 1990b). It was found that Al and Si dissolved and reprecipitated on $Fe(OH)_3$ blocking many of its adsorption sites.

A major advantage of a successful CA approach is that the surface complexation model parameters are transferable from one field site to another. Unfortunately, at present, CA approaches suffer from significant problems (Davis et al., 2004). In the estimation of surface site types and surface area abundances, it is assumed that the relative mineral surface areas can be estimated from bulk weight abundances and that the surface area of minerals in the assemblage is the same as that of reference minerals used in the laboratory. The lack of fundamental data on the effects of competitive adsorption of solutes common in soil solution and groundwater could lead to overestimation of adsorption due to competitive ion effects on adsorption in the natural system. The lack of fundamental data on the effects of solutes common in soil solution and groundwater on surface charge and surface potentials introduces the potentially erroneous assumption that these ions are not potential determining. Some computer optimization programs (eg., FITEQL) require use of the same surface complexation model, capacitance(s), and surface potentials for all component minerals (Davis et al., 2004).

A CA approach described zinc adsorption on aquifer sand material from Cape Cod by assuming that the aluminum and iron phases present in the quartz grain coatings have a surface area and site density similar to those of poorly crystalline materials (Davis et al., 1998). A similar approach provided only a semiquantitative prediction of uranium adsorption on an alluvial aquifer sediment from Naturita, Colorado, depending on the assumptions made about the relative amounts of surface area of quartz, ferrihydrite, and goethite (Davis et al., 2004). In both of these studies, the surface complexation model considered

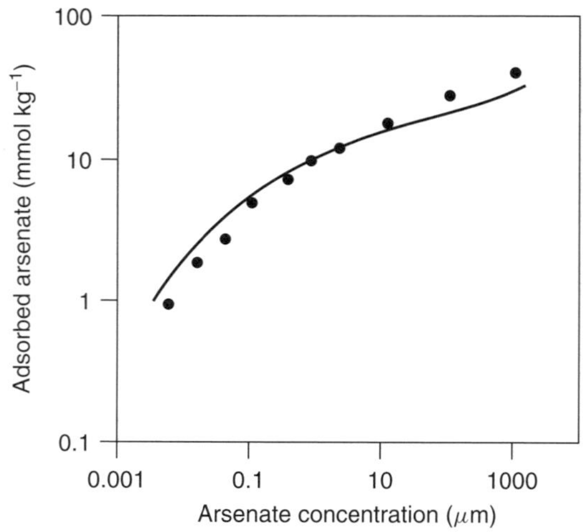

Figure 6.10. Ability of the component additivity approach to describe arsenate adsorption by a soil. (From Gustaffson, 2001.)

was a nonelectrostatic model that did not include electrostatic terms or protonation–dissociation constants. Because of the exclusion of the electrostatic terms, the mass action equations are not expected to provide accurate representations of reaction stoichiometry at the molecular scale (Davis et al., 2004).

Gustaffson (2001) used an electrostatic CA approach with the CD-MUSIC model to describe arsenate adsorption on a spodic B horizon containing allophane and ferrihydrite. He used the arsenate surface complexation constants obtained on gibbsite to represent allophane. The fit of the model was generally very good, although the fit was not quantitative over the entire range of arsenic surface coverages (see Figure 6.10).

An alternative type of component additivity approach is to use different adsorption models for different adsorbents (Weng et al., 2001, 2002). These authors described Cd^{2+}, Zn^{2+}, and Ni^{2+} binding in soils by combining the diffuse layer model for hydrous ferric oxide, the CD-MUSIC model for goethite, the Donnan model for illite, and the nonideal competitive adsorption (NICA) Donnan model (Kinniburgh et al., 1999) for humic acid.

6.4.4. Generalized Composite

In the generalized composite (GC) approach it is assumed that the adsorption behavior of a complex mineral assemblage can be described by surface complexation reactions written with generic surface functional groups that represent average properties of the assemblage as a whole rather than as specific mineral phases (Davis et al., 1998). The GC approach uses the overall surface area value and fits surface complexation constants to the experimental adsorption data. The number of site types and surface complexes are chosen to provide good data simulation (Davis et al., 1998). Generalized composite approaches successfully described zinc adsorption on Cape Cod (Davis et al., 1998) and uranium adsorption on Naturita (Davis et al., 2004) aquifer sediments. A GC approach developed from laboratory studies using Naturita sediment could predict in situ K_D values for uranium measured in field experiments of the same sample in contact with groundwater of variable chemical conditions (Curtis et al., 2004). The quality of the prediction is shown in Figure 6.11. In the GC approaches described above, the surface complexation model was nonelectrostatic.

Generalized composite approaches have also been used in application of the constant capacitance model to describe molybdenum (Goldberg et al., 1998) and arsenate adsorption by soil (Goldberg and Glaubig, 1988) and sediments (Gao et al., 2006) and the triple layer model to describe calcium and magnesium adsorption by soil (Charlet and Sposito, 1989). In these applications the electrostatic terms and protonation–dissociation reactions were retained.

The predictive capability of the constant capacitance model to describe metal and metalloid adsorption has been tested (Goldberg et al., 2002; 2005). In this approach a general regression model was used to predict surface complexation constants from easily measured soil chemical properties such as surface area, cation exchange capacity, organic carbon content, inorganic carbon content, aluminum oxide content, and iron oxide content. This approach provided a completely independent model evaluation and was able to predict molybdenum (Goldberg et al., 2002) and arsenate (Goldberg et al., 2005) adsorption on numerous diverse soils having a wide range of chemical characteristics. The predictive capability of this approach is shown in Figure 6.12 for both monodentate and bidentate molybdenum surface configurations and in Figure 6.13 for arsenate adsorption by three soil samples.

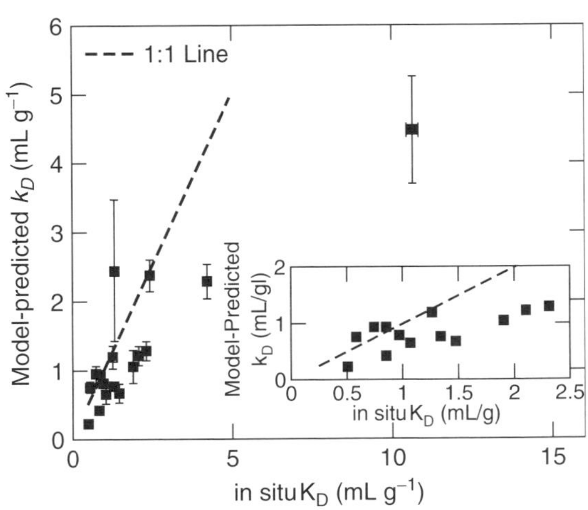

Figure 6.11. Ability of the generalized composite approach to predict K_D values for uranium adsorption on aquifer sediment. (From Curtis et al., 2004.)

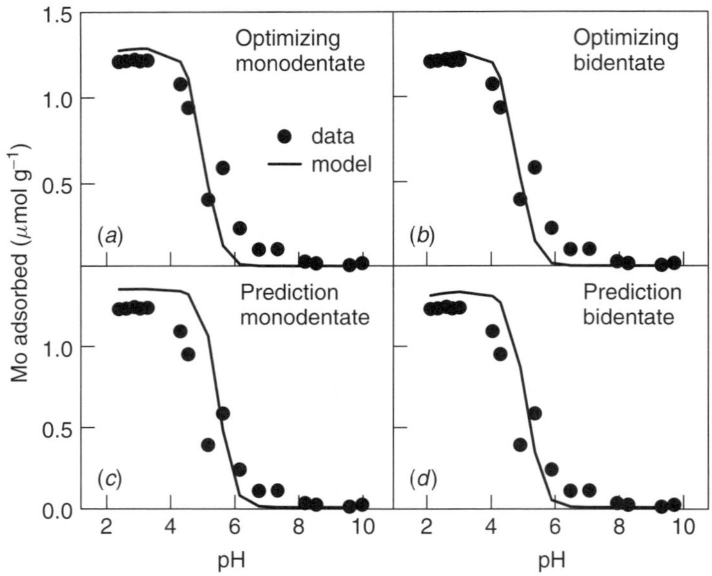

Figure 6.12. Constant capacitance model fits to and predictions of molybdenum adsorption by a soil. (From Goldberg et al., 2002.)

An intercomparison of CA and GC diffuse layer modeling approaches was carried out by 12 teams from eight countries using uranyl adsorption data on the weathered schists from the Koongarra uranium deposit in Australia (Payne et al., 2004). Exclusively predictive CA simulations were unable to provide satisfactory description of the data, and reoptimization of some parameter values was necessary. The GC models provided better simulation of the data but optimized a greater number of model parameters. The ability of either model to describe uranium adsorption on other rock samples from the same field site was unsatisfactory.

6.5. SUMMARY

Various empirical and chemical models of metal adsorption were presented and discussed. Empirical model parameters are only valid for the experimental conditions under which they were determined. Surface complexation models are chemical models that provide a molecular description of metal and metalloid adsorption reactions using an equilibrium approach. Four such models, the constant capacitance model, the diffuse layer model, the triple layer model, and the CD-MUSIC model, were described. Characteristics common to all the models are equilibrium constant expressions, mass and charge balances, and surface activity coefficient electrostatic potential terms. Various conventions for defining the standard state activity coefficients for the surface species have been

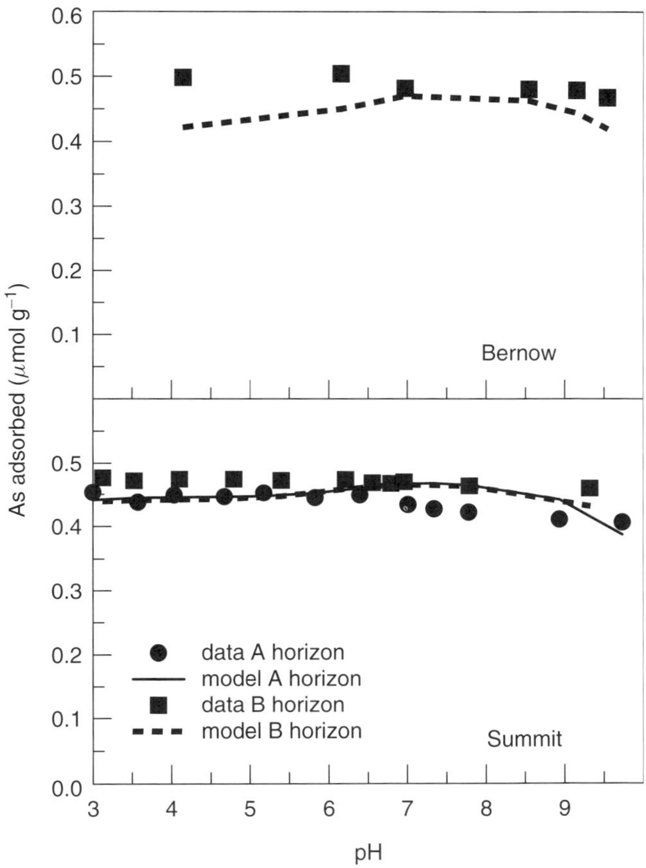

Figure 6.13. Constant capacitance model predictions of arsenate adsorption by three soil samples. (Adapted from Goldberg et al., 2005.)

described. Methods for determining parameter values for surface site density, capacitances, protonation–dissociation constants, and metal surface complexation constants were presented. Experimental methods of establishing metal surface configuration include: vibrational spectroscopy, nuclear magnetic resonance spectroscopy, electron spin resonance spectroscopy, x-ray absorption spectroscopy, and x-ray reflectivity. Metal surface speciation can also be inferred indirectly from point of zero charge shift and ionic strength dependence experiments and ab initio molecular modeling calculations. Applications of the surface complexation models to heterogeneous systems using the component additivity and the generalized composite approaches were described. Continuing research is needed to develop consistent and realistic protocols for describing metal adsorption reactions. The availability of standardized model parameter databases is critical.

Acknowledgments

L. J. Criscenti acknowledges support from the U.S. Department of Energy, Office of Basic Energy Sciences, Division of Chemical Sciences, Geosciences, and Biosciences and the Nuclear Regulatory Commission, Office of Nuclear Regulatory Research. Sandia National Laboratories is a multi-program laboratory operated by the Sandia Corporation, a Lockheed Martin Company for the U.S. Department of Energy's National Nuclear Security Administration under Contract DE-AC04-94AL85000.

REFERENCES

Adamson, A. W. (1976). *Physical Chemistry of Surfaces*, 3rd ed., Wiley, New York.

Anderson, P. R., and Benjamin, M. M. (1990a). Modeling adsorption in aluminum-iron oxide suspensions. *Environ. Sci. Technol.* **24**, 1586–1592.

Anderson, P. R., and Benjamin, M. M. (1990b). Constant capacitance surface complexation model. Adsorption in silica–iron binary oxide suspensions. *Am. Chem. Soc. Symp. Ser.* **416**, 272–281.

Anderson, M. A., Ferguson, J. F., and Gavis, J. (1976). Arsenate adsorption on amorphous aluminum hydroxide. *J. Colloid Interface Sci.* **54**, 391–399.

Arai, Y., Elzinga, E. J., and Sparks, D. L. (2001). X-ray absorption spectroscopic investigation of arsenite and arsenate at the aluminum oxide–water interface. *J. Colloid Interface Sci.* **235**, 80-88.

Axe, L., Bunker, G. B., Anderson, P. R., and Tyson, T. A. (1998). An XAFS analysis of strontium at the hydrous ferric oxide surface. *J. Colloid Interface Sci.* **199**, 44–52.

Bargar, J. R., Towle, S. N., Brown, G. E., and Parks, G. A. (1996). Outer-sphere Pb(II) adsorbed at specific surface sites on single crystal α-alumina. *Geochim. Cosmochim. Acta* **60**, 3541–3547.

Bargar, J. R., Brown, G. E., and Parks, G. A. (1997a). Surface complexation of Pb(II) at oxide–water interfaces. 1. XAFS and bond-valence determination of mononuclear and polynuclear Pb(II) sorption products on aluminum oxides. *Geochim. Cosmochim. Acta* **61**, 2617–2637.

Bargar, J. R., Brown, G. E., and Parks, G. A. (1997b). Surface complexation of Pb(II) at oxide–water interfaces. 2. XAFS and bond-valence determination of mononuclear Pb(II) sorption products and surface functional groups on iron oxides. *Geochim. Cosmochim. Acta* **61**, 2639–2652.

Barron, V., and Torrent, J. (1996). Surface hydroxyl configurations of various crystal faces of hematite and goethite. *J. Colloid Interface Sci.* **177**, 407–410.

Benjamin, M. M., and Leckie, J. O. (1980). Adsorption of metals at oxide interfaces: effects of the concentrations of adsorbate and competing metals. In *Contaminants and Sediments*, ed. Baker, R. A., Vol. 2, Ann Arbor Science, Ann Arbor, MI, 305–322.

Benjamin, M. M., and Leckie, J. O. (1981). Multiple-site adsorption of Cd, Cu, Zn, and Pb on amorphous iron oxyhydroxide. *J. Colloid Interface Sci.* **79**, 209–221.

Bickmore, B. R., Rosso, K. M., Nagy, K. L., Cygan, R. T., and Tadanier, C. J. (2003). Ab initio determination of edge surface structures for dioctahedral 2 : 1 phyllosilicates: implications for acid–base reactivity. *Clays Clay Miner.* **51**, 359–371.

Bickmore, B. R., Tadanier, C. J., Rosso, K. M., Monn, W. D., and Eggett, D. L. (2004). Bond-valence methods for pK_a prediction: critical reanalysis and a new approach. *Geochim. Cosmochim. Acta* **68**, 2025–2042.

Bostick, B. C., Vairavamurthy, M. A., Karthikeyan, K. G., and Chorover, J. (2002). Cesium adsorption on clay minerals: an EXAFS spectroscopic investigation. *Environ. Sci. Technol.* **36**, 2670–2676.

Bradbury, M. H., and Baeyens, B. (2005). Modelling the sorption of Mn(II), Co(II), Ni(II), Zn(II), Cd(II), Eu(III), Am(III), Sn(IV), Th(IV), Np(V), and U(VI) on montmorillonite: linear free energy relationships and estimates of surface binding constants for some selected heavy metals and actinides. *Geochim. Cosmochim. Acta* **69**, 875–892.

Brendler, V., Vahle, A., Arnold, T., Bernhard, G., and Fanghänel, T. (2003). RES^3T–Rossendorf expert system for surface and sorption thermodynamics. *J. Contam. Hydrol.* **61**, 281–291.

Brendler, V., Richter, A., Nebelung, C., and Vahle, A. (2004). *Development of a Mineral-Specific Sorption Database for Surface Complexation Modeling.* Part I, *Final Report.* Part II, *Manual*, Project PtWt+E 02E9471, FZ Rossendorf, Dresden, Germany.

Brown, I. D., and Altermatt, D. (1985). Bond-valence parameters obtained from a systematic analysis of the inorganic crystal structure database. *Acta Crystallogr. B* **41**, 244–247.

Cavallaro, N., and McBride, M. B. (1978). Copper and cadmium adsorption characteristics of selected acid and calcareous soils. *Soil Sci. Soc. Am. J.* **42**, 550–556.

Charlet, L., and Manceau, A. A. (1992). X-ray absorption spectroscopic study of the sorption of Cr(III) at the oxide-water interface. II. Adsorption, coprecipitation, and surface precipitation of hydrous ferric oxide. *J. Colloid Interface Sci.* **148**, 443–458.

Charlet, L., and Sposito, G. (1989). Bivalent ion adsorption by an oxisol. *Soil Sci. Soc. Am. J.* **53**, 691–695.

Cheah, S. F., Brown, G. E., and Parks, G. A. (1998). XAFS spectroscopy study of Cu(II) sorption on amorphous SiO_2 and γ-Al_2O_3: effect of substrate and time on sorption complexes. *J. Colloid Interface Sci.* **208**, 110–128.

Chen, C.-C., Papelis, C., and Hayes, K. F. (1998). Extended x-ray absorption fine structure (EXAFS) analysis of aqueous Sr^{II} ion sorption at clay–water interfaces. In *Adsorption of Metals by Geomedia: Variables, Mechanisms, and Model Applications, Proc. American Chemical Society Symposium* ed. Jenne, E. A., Academic Press, San Diego, CA, 333–348.

Chisholm-Brause, C. J., Hayes, K. F., Roe, A. L., Brown, G. E., Parks, G. A., and Leckie, J. O. (1990a). Spectroscopic investigation of Pb(II) complexes at the γ-Al_2O_3/water interface. *Geochim. Cosmochim. Acta* **54**, 1897–1909.

Chisholm-Brause, C. J., O'Day, P. A., Brown, G. E., and Parks, G. A. (1990b). Evidence for multinuclear metal-ion complexes at solid/water interfaces from x-ray absorption spectroscopy. *Nature* **348**, 528–531.

Collins, C. R., Sherman, D. M., and Ragnarsdóttir, K. V. (1998). The adsorption mechanism of Sr^{2+} on the surface of goethite. *Radiochim. Acta* **81**, 201–206.

Collins, C. R., Sherman, D. M., and Ragnarsdóttir, K. V. (1999a). Surface complexation of Hg^{2+} on goethite: mechanism from EXAFS spectroscopy and density functional calculations. *J. Colloid Interface Sci.* **219**, 345–350.

Collins, C. R., Ragnarsdóttir, K. V., and Sherman, D. M. (1999b). Effect of inorganic and organic ligands on the mechanism of cadmium sorption to goethite. *Geochim. Cosmochim. Acta* **63**, 2989–3002.

Combes, J. M., Chisholm-Brause, C. J., Brown, G. E., Parks, G. A., Conradson, S. D., Eller, I. R., Triay, P. G., Hobart, D. E., and Meijer, A. (1992). EXAFS spectroscopic study of neptunium(V) sorption at the α-FeOOH/water interface. *Environ. Sci. Technol.* **26**, 376–382.

Criscenti, L. J. (2004). Adsorption processes: At what spatial scale do we need to understand them? *Proc. 11th International Symposium on Water–Rock Interaction*, WRI-11, Saratoga Springs, NY, June 27–July 2, 2004, ed. Wanty, R. B., and Seal, R. R., A.A. Balkema, New York, 909–916.

Criscenti, L. J., and Sverjensky, D. A. (1999). The role of electrolyte anions (ClO_4^-, NO_3^-, Cl^-) in divalent metal (M^{2+}) adsorption on oxide and hydroxide surfaces in salt solutions. *Am. J. Sci.* **299**, 828–899.

Criscenti, L. J., and Sverjensky, D. A. (2002). A single-site model for divalent transition and heavy metal adsorption over a range of metal concentrations. *J. Colloid Interface Sci.* **253**, 329–352.

Curtis G. P., Fox, P., Kohler, M., and Davis, J. A. (2004). Comparison of in situ uranium K_D values with a laboratory determined surface complexation model. *Appl. Geochem.* **19**, 1643–1653.

Cygan, R. T., Nagy, K. L., and Brady, P. V. (1998). Molecular models of cesium sorption on kaolinite. In *Adsorption of Metals by Geomedia: Variables, Mechanisms, and Model Applications*, ed. Jenne, E. A., Academic Press, San Diego, CA, 383–399.

Dähn, R., Scheidegger, A. M., Manceau, A., Schlegel, M. L., Baeyens, B., Bradbury, M. H., and Chateigner, D. (2003). Structural evidence for the sorption of Ni(II) atoms on the edges of montmorillonite clay minerals: a polarized x-ray absorption fine structure study. *Geochim. Cosmochim. Acta* **67**, 1–15.

Davis, J. A., and Kent, D. B. (1990). Surface complexation modeling in aqueous geochemistry. *Rev. Miner.* **23**, 177–260.

Davis, J. A., and Leckie, J. O. (1978). Surface ionization and complexation at the oxide/water interface. II. Surface properties of amorphous iron oxyhydroxide and adsorption of metal ions. *J. Colloid Interface Sci.* **67**, 90–107.

Davis, J. A., James, R. O., and Leckie, J. O. (1978). Surface ionization and complexation at the oxide/water interface. I. Computation of electrical double layer properties in simple electrolytes. *J. Colloid Interface Sci.* **63**, 480–499.

Davis, J. A., Coston, J. A., Kent, D. B., and Fuller, C. C. (1998). Application of the surface complexation concept to complex mineral assemblages. *Environ. Sci. Technol.* **32**, 2820–2828.

Davis, J. A., Meece, D. E., Kohler, M., and Curtis, G. P. (2004). Approaches to surface complexation modeling of uranium(VI) adsorption on aquifer sediments. *Geochim. Cosmochim. Acta* **68**, 3621–3641.

Dyer, J. A., Trivedi, P., Scrivner, N. C., and Sparks, D. L. (2003). Lead sorption onto ferrihydrite. 2. Surface complexation modeling. *Environ. Sci. Technol.* **37**, 915–922.

Dzombak, D. A., and Morel, F. M. M. (1990). *Surface Complexation Modeling. Hydrous Ferric Oxide,* Wiley, New York.

Fendorf, S. E., Sparks, D. L., Lamble, G. M., and Kelley, M. J. (1994). Applications of x-ray absorption fine-structure spectroscopy to soils. *Soil Sci. Soc. Am. J.* **58**, 1583–1595.

Fendorf, S. E., Eick, M. J., Grossl, P., and Sparks, D. L. (1997). Arsenate and chromate retention mechanisms on goethite. I. Surface structure. *Environ. Sci. Technol.* **31**, 315–320.

Fenter, P., Cheng, L., Rihs, S., Machesky, M. L., Bedzyk, M. J., and Sturchio, N. C. (2000). Electrical double-layer structure at the rutile–water interface as observed in situ with small-period x-ray standing waves. *J. Colloid Interface Sci.* **225**, 154–165.

Gaboriaud, F., and Ehrhardt, J.-J. (2003). Effects of different crystal faces on the surface charge of colloidal goethite (α-FeOOH) particles: an experimental and modeling study. *Geochim. Cosmochim. Acta* **67**, 967–983.

Gao, S., Goldberg, S., Herbel, M. J., Chalmers, A. T., Fujii, R., and Tanji, K. K. (2006). Sorption processes affecting arsenic solubility in oxidized surface sediments from Tulare Lake. *Chem. Geol.* **228**, 33–43.

Goldberg, S. (1991). Sensitivity of surface complexation modeling to the site density parameter. *J. Colloid Interface Sci.* **45**, 1–9.

Goldberg, S. (1992). Use of surface complexation models in soil chemical systems. *Adv. Agron.* **47**, 233–329.

Goldberg, S., and Glaubig, R. A. (1988). Anion sorption on a calcareous, montmorillonitic soil–arsenic. *Soil Sci. Soc. Am. J.* **52**, 1297–1300.

Goldberg, S., and Johnston, C. T. (2001). Mechanisms of arsenic adsorption on amorphous oxides evaluated using macroscopic measurements, vibrational spectroscopy, and surface complexation modeling. *J. Colloid Interface Sci.* **234**, 204–216.

Goldberg, S., and Sposito, G. (1984). A chemical model of phosphate adsorption by soils. I. Reference oxide minerals. *Soil Sci. Soc. Am. J.* **48**, 772–778.

Goldberg, S., Forster, H. S., and Godfrey, C. L. (1996). Molybdenum adsorption on oxides, clay minerals, and soils. *Soil Sci. Soc. Am. J.* **60**, 425–432.

Goldberg, S., Su, C., and Forster, H. S. (1998). Sorption of molybdenum on oxides, clay minerals, and soils: mechanisms and models. In *Adsorption of Metals by Geomedia: Variables, Mechanisms, and Model Applications,* ed. Jenne, E. A., *Proc. American Chemical Society Symposium*, Academic Press, San Diego, CA, 401–426.

Goldberg, S., Lesch, S. M., and Suarez, D. L. (2000). Predicting boron adsorption by soils using soil chemical parameters in the constant capacitance model. *Soil Sci. Soc. Am. J.* **64**, 1356–1363.

Goldberg, S., Lebron, I., Suarez, D. L., and Hinedi, Z. R. (2001). Surface characterization of amorphous aluminum oxides. *Soil Sci. Soc. Am. J.* **65**, 78–86.

Goldberg, S., Lesch, S. M., and Suarez, D. L. (2002). Predicting molybdenum adsorption by soils using soil chemical parameters in the constant capacitance model. *Soil Sci. Soc. Am. J.* **66**, 1836–1842.

Goldberg, S., Lesch, S. M., Suarez, D. L., and Basta, N. T. (2005). Predicting arsenate adsorption by soils using soil chemical parameters in the constant capacitance model. *Soil Sci. Soc. Am. J.* **69**, 1389–1398.

Grossl, P., Eick, M. J., Sparks, D. L., Goldberg, S., and Ainsworth, C. C. (1997). Arsenate and chromate retention mechanisms on goethite. II. Kinetic evaluation using a pressure-jump relaxation technique. *Environ. Sci. Technol.* **31**, 321–326.

Gustafsson, J. P. (2001). Modelling competitive anion adsorption on oxide minerals and an allophane-containing soil. *Eur. J. Soil Sci.* **52**, 639–653.

Harrison, J. B., and Berkheiser, V. E. (1982). Anion interactions with freshly prepared hydrous iron oxides. *Clays Clay Miner.* **30**, 97–102.

Hayes, K. F. (1987). Equilibrium, spectroscopic, and kinetic studies of ion adsorption at the oxide/aqueous interface. Ph.D. dissertation, Stanford University, Stanford, CA.

Hayes, K. F., and Leckie, J. O. (1987). Modeling ionic strength effects on cation adsorption at hydrous oxide/solution interfaces. *J. Colloid Interface Sci.* **115**, 564–572.

Hayes, K. F., Redden, G., Ela, W., and Leckie, J. O. (1991). Surface complexation models: an evaluation of model parameter estimation using FITEQL and oxide mineral titration data. *J. Colloid Interface Sci.* **142**, 448–469.

Helgeson, H. C., Kirkham, D. H., and Flowers, G. C. (1981). Theoretical prediction of the thermodynamic behavior of aqueous electrolytes at high pressures and temperatures. IV. Calculation of activity coefficients, osmotic coefficients, and apparent molal and standard and relative partial molal properties to 600°C and 5 KB. *Am. J. Sci.* **281**, 1249–1516.

Herbelin, A. L., and Westall, J. C. (1996). *FITEQL: A Computer Program for Determination of Chemical Equilibrium Constants from Experimental Data,* Rep. 96-01, Version 3.2, Department of Chemistry, Oregon State University, Corvallis, OR.

Hiemstra, T., and van Riemsdijk, W. H. (1991). Physical chemical interpretation of primary charging behavior of metal (hydr)oxides. *Colloids Surf.* **59**, 7–25.

Hiemstra, T., and van Riemsdijk, W. H. (1996). A surface structural approach to ion adsorption: the charge distribution (CD) model. *J. Colloid Interface Sci.* **179**, 488–508.

Hiemstra, T., and van Riemsdijk, W. H. (2002). On the relationship between surface structure and ion complexation of oxide–solution interfaces. In *Encyclopedia of Surface and Colloid Science*, ed. Hubbard, A. T., Marcel Dekker, New York, 3773–3799.

Hiemstra, T., van Riemsdijk, W. H., and Bruggenwert, M. G. M. (1987). Proton adsorption mechanism at the gibbsite and aluminum oxide solid/solution interface. *Neth. J. Agric. Sci.* **35**, 281–293.

Hiemstra, T., van Riemsdijk, W. H., and Bolt, G. H. (1989). Multisite proton adsorption modeling at the solid/solution interface of (hydr)oxides: a new approach. I. Model description and evaluation of intrinsic reaction constants. *J. Colloid Interface Sci.* **133**, 91–104.

Hiemstra, T., Rahnemaie, R., and van Riemsdijk, W. H. (2004). Surface complexation of carbonate on goethite: IR spectroscopy, structure and charge distribution. *J. Colloid Interface Sci.* **278**, 282–290.

Hsia, T. H., Lo, S. L., Lin, C. F., and Lee, D. Y. (1993). Chemical and spectroscopic evidence for specific adsorption of chromate on hydrous iron oxide. *Chemosphere* **26**, 1897–1904.

Hsia, T. H., Lo, S. L., Lin, C. F., and Lee, D. Y. (1994). Characterization of arsenate adsorption on hydrous iron oxide using chemical and physical methods. *Colloids Surf. A: Physicochem. Eng. Asp.* **85**, 1–7.

Huang, C. P., and Stumm, W. (1973). Specific adsorption of cations in hydrous γ-Al_2O_3. *J. Colloid Interface Sci.* **43**, 409–420.

Hunter, R. J. (1981). *Zeta Potential in Colloid Science. Principles and Applications,* Academic Press, London.

James, R. O., and Healy, T. W. (1972). Adsorption of hydrolyzable metal ions at the oxide–water interface. III. A thermodynamic model of adsorption. *J. Colloid Interface Sci.* **40**, 65–81.

James R. O., and Parks, G. A. (1982). Characterization of aqueous colloids by their electrical double-layer and intrinsic surface chemical properties. *Surf. Colloid Sci.* **12**, 119–216.

James, R. O., Davis, J. A., and Leckie, J. O. (1978). Computer simulation of the conductometric and potentiometric titrations of the surface groups on ionizable latexes. *J. Colloid Interface Sci.* **65**, 331–343.

Jones, P., and Hockey, J. A. (1971). Infra-red studies of rutile surfaces. 2. Hydroxylation, hydration and structure of rutile surfaces. *Trans. Faraday Soc.* **67**, 2679–2685.

Kallay, N., Colic, M., Fuerstenau, D. W., Jang, H. M., and Matijevic, E. (1994). Lyotropic effect in surface charge, electrokinetics, and coagulation of a rutile dispersion. *Colloid Polym. Sci.* **272**, 554–561.

Karthein, R., Motschi, H., Schweiger, A., Ibric, S., Sulzberger, B., and Stumm, W. (1991). Interactions of chromium(III) complexes with hydrous δ-Al_2O_3: rearrangements in the coordination sphere studied by electron spin resonance and electron spin-echo spectroscopies. *Inorg. Chem.* **30**, 1606–1611.

Katz, L. E., and Hayes, K. F. (1995). Surface complexation modeling. I. Strategy for modeling monomer complex formation at moderate surface coverage. *J. Colloid Interface Sci.* **170**, 477–490.

Keizer, M. G., and van Riemsdijk, W. H. (1995). *ECOSAT: A Computer Program for the Calculation of Speciation and Transport in Soil–Water Systems,* Tech. Rep., Department of Soil Science and Plant Nutrition, Wageningen Agricultural University, Wageningen, The Netherlands.

Kim, Y., and Kirkpatrick, R. J. (1997). ^{23}Na and ^{133}Cs NMR study of cation adsorption on mineral surfaces: local environments, dynamics, and effects of mixed cations. *Geochim. Cosmochim. Acta* **61**, 5199–5208.

Kinniburgh, D. G. (1985). *ISOTHERM. A Computer Program for Analyzing Adsorption Data*, Rep. WD/ST/85/02, Version 2.2, British Geological Survey, Wallingford, Berkshire, England.

Kinniburgh, D. G., van Riemsdijk, W. H., Koopal, L. K., Borkovec, M., Benedetti, M. F., and Avena, M. J. (1999). Ion binding to natural organic matter: competition, heterogeneity, stoichiometry and thermodynamic consistency. *Colloids Surf. A: Physicochem. Eng. Asp.* **151**, 147–166.

Koopal, L. K., van Riemsdijk, W. H., and Roffey, M. G. (1987). Surface ionization and complexation models: a comparison of methods for determining model parameters. *J. Colloid Interface Sci.* **118**, 117–136.

Koretsky, C. (2000). The significance of surface complexation reactions in hydrologic systems: a geochemist's perspective. *J. Hydrol.* **230**, 127–171.

Koretsky, C. M., Sverjensky, D. A., and Sahai, N. (1998). Surface site types on oxide and silicate minerals based on crystal chemistry: implications for site densities, multi-site adsorption, surface infrared spectroscopy, and dissolution kinetics. *Am. J. Sci.* **298**, 349–438.

Kulik, D. A. (2002a). Sorption modelling by Gibbs energy minimisation: towards a uniform thermodynamic database for surface complexes of radionuclides. *Radiochim. Acta* **90**, 815–832.

Kulik, D. A. (2002b). A Gibbs energy minimization approach to model sorption equilibria at the mineral–water interface: thermodynamic relations for multi-site surface-complexation. *Am. J. Sci.* **302**, 227–279.

Kurdi, F., and Doner, H. E. (1983). Zinc and copper sorption and interaction in soils. *Soil Sci. Soc. Am. J.* **47**, 873–876.

Ladeira, A. C. Q., Ciminelli, V. S. T., Duarte, H. A., Alves, M. C. M., and Ramos, A. Y. (2001). Mechanism of anion retention from EXAFS and density functional calculations: arsenic(V) adsorbed on gibbsite. *Geochim. Cosmochim. Acta* **65**, 1211–1217.

Lumsdon, D. G., Fraser, A. R., Russell, J. D., and Livesey, N. T. (1984). New infrared band assignments for the arsenate ion adsorbed on synthetic goethite (α-FeOOH). *J. Soil Sci.* **35**, 381–386.

Manceau, A., and Charlet, L. (1992). X-ray absorption spectroscopic study of the sorption of Cr(III) at the oxide–water interface. I. Molecular mechanism of Cr(III) oxidation on Mn oxides. *J. Colloid Interface Sci.* **148**, 425–442.

Manceau, A., Nagy, K. L., Spadini, L., and Ragnarsdottir, K. V. (2000). Influence of anionic layer structure of Fe-oxyhydroxides on the structure of Cd surface complexes. *J. Colloid Interface Sci.* **228**, 306–316.

Manning, B. A., and Goldberg, S. (1996). Modeling competitive adsorption of arsenate with phosphate and molybdate on oxide minerals. *Soil Sci. Soc. Am. J.* **60**, 121–131.

Manning, B. A., Fendorf, S. E., and Goldberg, S. (1998). Surface structures and stability of arsenic(III) on goethite. Spectroscopic evidence for inner-sphere complexes. *Environ. Sci. Technol.* **32**, 2383–2388.

McBride, M. B. (1997). A critique of diffuse double layer models applied to colloid and surface chemistry. *Clays Clay Miner.* **45**, 598–608.

McBride, M. B. (1982). Cu^{2+}-adsorption characteristics of aluminum hydroxide and oxyhydroxides. *Clays Clay Miner.* **30**, 21–28.

McBride, M. B., Fraser, A. R., and McHardy, W. J. (1984). Cu^{2+} interaction with microcrystalline gibbsite. Evidence for oriented chemisorbed copper ions. *Clays Clay Miner.* **32**, 12–18.

McKenzie, R. M. (1983). The adsorption of molybdenum on oxide surfaces. *Aust. J. Soil Res.* **21**, 505–513.

Meeussen, J. C. L. (2003). ORCHESTRA: an object-oriented framework for implementing chemical equilibrium models. *Environ. Sci. Technol.* **37**, 1175–1182.

Micera, G., Dallacchio, R., Deiana, S., Gessa, C., Melis, P., and Premoli, A. (1986). Manganese(II)–aluminum hydroxide interaction: an ESR study. *Colloids Surf.* **17**, 395–400.

O'Day, P. A., Parks, G. A., and Brown, G. E. (1994). Molecular structure and binding sites of cobalt(II) surface complexes on kaolinite from x-ray absorption spectroscopy. *Clays Clay Miner.* **42**, 337–355.

O'Day, P. A., Newville, M., Neuhoff, P. S., Sahai, N., and Carroll, S. A. (2000). X-ray absorption spectroscopy of strontium(II) coordination. I. Static and thermal disorder in crystalline, hydrated, and precipitated solids and in aqueous solution. *J. Colloid Interface Sci.* **222**, 184–197.

Papelis, C., Hayes, K. F., and Leckie, J. O. (1988). *HYDRAQL: A Program for the Computation of Chemical Equilibrium Composition of Aqueous Batch Systems Including Surface-Complexation Modeling of Ion Adsorption at the Oxide/Solution Interface,* Tech. Rep. 306, Department of Civil Engineering, Stanford University, Stanford, CA.

Payne, T. E., Davis, J. A., Ochs, M., Olin, M., and Tweed, C. J. (2004). Uranium adsorption on weathered schist—intercomparison of modelling approaches. *Radiochim. Acta* **92**, 651–661.

Peacock, C. L., and Sherman, D. M. (2004a). Copper(II) sorption onto goethite, hematite, and lepidocrocite: a surface complexation model based on ab initio molecular geometries and EXAFS spectroscopy. *Geochim. Cosmochim. Acta* **68**, 2623–2637.

Peacock, C. L., and Sherman, D. M. (2004b). Vanadium(V) adsorption onto goethite (α-FeOOH) at pH 1.5 to 12: a surface complexation model based on ab initio molecular geometries and EXAFS spectroscopy. *Geochim. Cosmochim. Acta* **68**, 1723–1734.

Pierce, M. L., and Moore, C. B. (1980). Adsorption of arsenite on amorphous iron hydroxide from dilute aqueous solution. *Environ. Sci. Technol.* **14**, 214–216.

Randall, S. R., Sherman, D. M., Ragnarsdòttir, K. V., and Collins C. R. (1999). The mechanism of cadmium surface complexation on iron oxyhydroxide minerals. *Geochim. Cosmochim. Acta* **63**, 2971–2987.

Rochester, C. H., and Topham, S. A. (1979a). Infrared study of surface hydroxyl groups on goethite. *J. Chem. Soc. Faraday Trans. I* **75**, 591–602.

Rochester, C. H., and Topham, S. A. (1979b). Infrared study of surface hydroxyl groups on haematite. *J. Chem. Soc. Faraday Trans. I* **75**, 1073–1088.

Roe, A. L., Hayes, K. F., Chisholm-Brause, C. J., Brown, G. E., Parks, G. A., Hodgson, K. O., and Leckie, J. O. (1991). In situ x-ray absorption study of lead ion surface complexes at the goethite–water interface. *Langmuir* **7**, 367–373.

Sahai, N. (2002). Is silica really an anomalous oxide? Surface acidity and aqueous hydrolysis revisited. *Environ. Sci. Technol.* **36**, 445–452.

Sahai, N., and Sverjensky, D. A. (1997a). Evaluation of internally consistent parameters for the triple-layer model by the systematic analysis of oxide surface titration data. *Geochim. Cosmochim. Acta* **61**, 2801–2826.

Sahai, N., and Sverjensky, D. A. (1997b). Solvation and electrostatic model for specific electrolyte adsorption. *Geochim. Cosmochim. Acta* **61**, 2827–2848.

Sahai, N., and Sverjensky, D. A. (1998). GEOSURF: a computer program for modeling adsorption on mineral surfaces from aqueous solution. *Comput. Geosci.* **24**, 853–873.

Sahai, N., Carroll, S. A., Roberts, S., and O'Day, P. A. (2000). X-ray absorption spectroscopy of strontium(II) coordination. II. Sorption and precipitation at kaolinite, amorphous silica, and goethite surfaces. *J. Colloid Interface Sci.* **222**, 198–212.

Schindler, P. W., Fürst, B., Dick, R., and Wolf, P. U. (1976). Ligand properties of surface silanol groups. I. Surface complex formation with Fe^{3+}, Cu^{2+}, Cd^{2+}, and Pb^{2+}. *J. Colloid Interface Sci.* **55**, 469–475.

Schudel, M., Behrens, S. H., Holthoff, H., Kretzschmar, R., and Borkovec, M. (1997). Absolute aggregation rate constants of hematite particles in aqueous suspensions: a comparison of two different surface morphologies. *J. Colloid Interface Sci.* **196**, 241–253.

Secor, R. B., and Radke, C. J. (1985). Spillover of the diffuse double layer on montmorillonite particles. *J. Colloid Interface Sci.* **103**, 237–244.

Shuman, L. M. (1975). The effect of soil properties on zinc adsorption by soils. *Soil Sci. Soc. Am. Proc.* **39**, 454–458.

Spadini, L., Manceau, A., Schindler, P. W., and Charlet, L. (1994). Structure and stability of Cd^{2+} surface complexes on ferric oxides. I. Results from EXAFS spectroscopy. *J. Colloid Interface Sci.* **168**, 73–86.

Sposito, G. (1982). On the use of the Langmuir equation in the interpretation of "adsorption" phenomena. II. The "two-surface" Langmuir equation. *Soil Sci. Soc. Am. J.* **46**, 1147–1152.

Sposito, G. (1984). *The Surface Chemistry of Soils*, Oxford University Press, Oxford.

Sprycha, R. (1984). Surface charge and adsorption of background electrolyte ions at anatase/electrolyte interface. *J. Colloid Interface Sci.* **102**, 173–185.

Steele, H. M., Wright, K., Nygren, M. A., and Hillier, I. H. (2000). Interactions of the (001) surface of muscovite with Cu(II), Zn(II), and Cd(II): a computer simulation study. *Geochim. Cosmochim. Acta* **64**, 257–262.

Strawn, D. G., and Sparks, D. L. (1999). The use of XAFS to distinguish between inner- and outer-sphere lead adsorption complexes on montmorillonite. *J. Colloid Interface Sci.* **216**, 257–269.

Strawn, D. G., Scheidegger, A. M., and Sparks, D. L. (1998). Kinetics and mechanisms of Pb(II) sorption and desorption at the aluminum oxide water interface. *Environ. Sci. Technol.* **32**, 2596–2601.

Stumm, W., Kummert, R., and Sigg, L. (1980). A ligand exchange model for the adsorption of inorganic and organic ligands at hydrous oxide interfaces. *Croat. Chem. Acta* **53**, 291–312.

Sturchio, N. C., Chiarello, R. P., Cheng, L., Lyman, P. F., Bedzyk, M. J., Qian, Y., You, H., Yee, D., Geissbuhler, P., Sorenson, L. B., Liang, Y., and Baer, D. R. (1997). Lead adsorption at the calcite–water interface: synchrotron x-ray standing wave and x-ray reflectivity studies. *Geochim. Cosmochim. Acta* **61**, 251–263.

Suarez, D. L., Goldberg, S., and Su, C. (1998). Evaluation of oxyanion adsorption mechanisms on oxides using FTIR spectroscopy and electrophoretic mobility. *Am. Chem. Soc. Symp. Ser.* **715**, 136–178.

Sverjensky, D. A. (1993). Physical surface-complexation models for sorption at the mineral–water interface. *Nature* **364**, 776–780.

Sverjensky, D. A. (2001). Interpretation and prediction of triple-layer model capacitances and the structure of the oxide–electrolyte–water interface. *Geochim. Cosmochim. Acta* **65**, 3643–3655.

Sverjensky, D. A. (2003). Standard states for the activities of mineral surface sites and species. *Geochim. Cosmochim. Acta* **67**, 17–28.

Sverjensky, D. A. (2005). Prediction of surface charge on oxides in salt solutions: revisions for 1 : 1 (M^+L^-) electrolytes. *Geochim. Cosmochim. Acta* **69**, 225–257.

Sverjensky, D. A., and Sahai, N. (1996). Theoretical prediction of single-site surface-protonation equilibrium constants for oxides and silicates in water. *Geochim. Cosmochim. Acta* **60**, 3773–3797.

Tadanier, C., and Eick, M. J. (2002). Formulating the charge distribution multisite surface complexation model using FITEQL. *Soil Sci. Soc. Am. J.* **66**, 1505–1517.

Towle, S. N., Bargar, J. R., Brown, G. E., and Parks, G. A. (1999). Sorption of Cu(II) on metal oxide surfaces. II. Identification of $Co(II)_{(aq)}$ adsorption sites on the (0001) and (1(1) $\bar{0}$2) surfaces of α-Al_2O_3 by grazing incidence XAFS spectroscopy. *J. Colloid Interface Sci.* **217**, 312–321.

Trainor, T. P., Brown, G. E., and Parks, G. A. (2000). Adsorption and precipitation of aqueous Zn(II) on alumina powders. *J. Colloid Interface Sci.* **231**, 359–372.

Trivedi, P., Axe, L., and Tyson, T. A. (2001a). XAS studies of Ni and Zn sorbed to hydrous manganese oxide. *Environ. Sci. Technol.* **35**, 4515–4521.

Trivedi, P., Axe, L., and Tyson, T. A. (2001b). An analysis of zinc sorption to amorphous versus crystalline iron oxides using XAS. *J. Colloid Interface Sci.* **244**, 230–238.

Trivedi, P., Dyer, J. A., and Sparks, D. L. (2003). Lead sorption onto ferrihydrite. 1. A macroscopic and spectroscopic assessment. *Environ. Sci. Technol.* **37**, 908–914.

Turner, M. A., Hendrickson, L. L., and Corey, R. B. (1984). Use of chelating resins in metal adsorption studies. *Soil Sci. Soc. Am. J.* **48**, 763–769.

van der Lee, J., and Lomenech, C. (2004). Towards a common thermodynamic database for speciation models. *Radiochim. Acta* **92**, 811–818.

Veith, J. A., and Sposito, G. (1977). On the use of the Langmuir equation in the interpretation of "adsorption" phenomena. *Soil Sci. Soc. Am. J.* **41**, 697–702.

Venema, P., Hiemstra, T., and van Riemsdijk, W. H. (1996a). Comparison of different site binding models for cation sorption: description of pH dependency, salt dependency, and cation–proton exchange. *J. Colloid Interface Sci.* **181**, 45–59.

Venema, P., Hiemstra, T., and van Riemsdijk, W. H. (1996b). Multisite adsorption of cadmium on goethite. *J. Colloid Interface Sci.* **183**, 515–527.

Venema, P., Hiemstra, T., and van Riemsdijk, W. H. (1997). Interaction of cadmium with phosphate on goethite. *J. Colloid Interface Sci.* **192**, 94–103.

Villalobos, M., Trotz, M. A., and Leckie, J. O. (2003). Variability in goethite surface site density: evidence from proton and carbonate sorption. *J. Colloid Interface Sci.* **268**, 273–287.

Waite, T. D., Davis, J. A., Payne, T. E., Waychunas, G. A., and Xu, N. (1994). Uranium(VI) adsorption to ferrihydrite: application of a surface complexation model. *Geochim. Cosmochim. Acta* **58**, 5465–5478.

Waychunas, G. A., Rea, B. A., Fuller, C. C., and Davis, J. A. (1993). Surface chemistry of ferrihydrite. 1. EXAFS studies of the geometry of coprecipitated and adsorbed arsenate. *Geochim. Cosmochim. Acta* **57**, 2251–2269.

Waychunas, G. A., Fuller, C. C., and Davis, J. A. (2002). Surface complexation and precipitate geometry for aqueous Zn(II) sorption on ferrihydrite. I. X-ray absorption extended fine structure spectroscopy analysis. *Geochim. Cosmochim. Acta* **66**, 1119–1137.

Weng, L., Temminghoff, E. J. M., and van Riemsdijk, W. H. (2001). Contribution of individual sorbents to the control of heavy metal activity in sandy soil. *Environ. Sci. Technol.* **35**, 4436–4443.

Weng, L., Temminghoff, E. J. M., Lofts, S., Tipping, E., and van Riemsdijk, W. H. (2002). Complexation of dissolved organic matter and solubility control of heavy metals in a sandy soil. *Environ. Sci. Technol.* 36, 4804–4810.

Westall, J. C. (1979). *MICROQL. I. A Chemical Equilibrium Program in BASIC. II. Computation of Adsorption Equilibria in BASIC*, Tech. Rep., Swiss Federal Institute of Technology, EAWAG, Dübendorf, Switzerland.

Westall, J. (1980). Chemical equilibrium including adsorption on charged surfaces. *Am. Chem. Soc. Adv. Chem. Ser.* **189**, 33–44.

Yates, D. E. (1975). The Structure of the oxide/aqueous electrolyte interface. Ph.D. dissertation, University of Melbourne, Melbourne, VIC, Australia.

Yates, D. E., Levine, S., and Healy, T. W. (1974). Site-binding model of the electrical double layer at the oxide/water interface. *J. Chem. Soc. Faraday Trans. I* **70**, 1807–1818.

Yoon, R. H., Salman, T., and Donnay, G. (1979). Predicting points of zero charge of oxides and hydroxides. *J. Colloid Interface Sci.* **70**, 483–493.

Zhang Z., Fenter, P., Cheng, L., Sturchio, N. C., Bedzyk, M. J., Predota, M., Bandura, A., Kubicki, J. D., Lvov, S. N., Cummings, P. T., Chioalvo, A. A., Ridley, M. K., Bénézeth, P., Anovitz, L., Palmer, D. A., Machesky, M. L., and Wesolowski, D. J. (2004). Ion adsorption at the rutile–water interface: linking molecular and macroscopic properties. *Langmuir* **20**, 4954–4969.

PART II

TRANSFORMATIONS AND DYNAMICS OF METALS AND METALLOIDS AS INFLUENCED BY SOIL–ROOT–MICROBE INTERACTIONS

7

BIOGEOCHEMISTRY OF METALS AND METALLOIDS AT THE SOIL–ROOT INTERFACE

P. Hinsinger

INRA–SupAgro–UMR Biogéochimie du Sol et de la Rhizosphère, Montpellier, France

F. Courchesne

Université de Montréal, Montréal, Quebec, Canada

Biophysico-Chemical Processes of Heavy Metals and Metalloids in Soil Environments,
Edited by Antonio Violante, Pan Ming Huang, and Geoffrey Michael Gadd

7.1. INTRODUCTION

In his review of the recent achievements and future frontiers in soil chemistry, Sparks (2001) pointed out "the understanding of rhizosphere chemistry as a promising area of research." The concept of the *rhizosphere*, which is now more than a century old, was actually first put forward by Hiltner in 1904 to describe "the soil compartment influenced by the root." However, interest in unraveling the biogeochemical processes occurring at this unique soil–root interface is rather recent, the first reports on the specificity of rhizosphere chemistry dating to the late 1960s. Even more recent is the interest of the scientific community in studying the fate of trace elements such as heavy metals and metalloids in the rhizosphere. Most of the published literature on this matter is less than 10 years old, and excellent pieces of research are currently being conducted in an increasing number of laboratories worldwide. This is the reason why Pierzinski (2003), in his keynote lecture, which drew up the conclusions of the 7th ICOBTE (International Conference on the Biogeochemistry of Trace Elements), stressed the rhizosphere as being central for future areas of research on the biogeochemistry of trace elements. Although the aim of the present chapter is to give a comprehensive view of state-of-the-art knowledge relevant to the role of the rhizosphere in the biogeochemistry of trace elements, we believe that we are still at the eve of understanding the role of rhizosphere processes and properties. In short, many data still need to be acquired to gain a more complete picture of the mobility and bioavailability of heavy metals and metalloids at the soil root–interface.

Major technical limitations of the past are currently being overcome by a range of cutting-edge methods, each representing a new avenue to further our understanding of the biogeochemistry of trace elements in the rhizosphere. This comprises the development of innovative microanalytical techniques for evidencing or monitoring minute changes at environmentally relevant spatial and temporal scales. New biological approaches are also provided by molecular biology, such as the use of genetically engineered plants and microorganisms that can be used for understanding and quantifying the specific traits of rhizosphere biogeochemistry (e.g., via the use of biosensors). The fast development of sophisticated numerical models in soil biogeochemistry is another promising avenue that will help foster our understanding of the complex interactions that occur in the rhizosphere and ultimately govern the transfer of trace elements from the soil into the plant.

In this chapter we define the spatial and temporal characteristics of the rhizosphere and relate these to major physical processes that govern the transport of water and trace elements at the soil–root interface. In the following sections we cover the various types of biogeochemical gradients that occur in the rhizosphere: elemental concentrations, pH, pCO_2, pO_2, redox potential, and organic ligand concentrations, which alter the speciation and ultimately determine the mobility and bioavailability of heavy metals and metalloids at the soil–root interface. We focus primarily on the direct role of plant roots; the role of rhizosphere microorganisms, including symbionts such as mycorrhizal fungi, is addressed in separate chapters (see Chapters 2 and 9). Nevertheless, a few references to the contributions of both roots and rhizosphere microorganisms are given here to stress that the estimation of their relative importance is a quite difficult task.

7.2. DEFINITIONS AND CONCEPTS

Some of the key terms and concepts used in this chapter are defined and presented below in an attempt to establish their meaning as clearly as possible and to acknowledge the range of meanings available in the scientific literature.

7.2.1. Rhizosphere

Active plant roots influence and transform the properties of the soil and the soil solution in their vicinity. The *rhizosphere* has been variously conceptualized as the soil–root interface (Wenzel et al., 2001), the environment of plant roots (Lorenz et al., 1997), the portion of the soil in contact with or adhering to roots (Pierret et al., 1999), the zone where root activity significantly influences soil biological properties (Marschner, 1991), the soil in which the microbial activity is influenced by plant roots (Russell, 1982), and the soil volume surrounding roots and containing high densities of microbionts (Bakker et al., 2004). In this context, the rhizosphere is probably best defined as the volume of soil surrounding living plant roots and that is influenced by root activity (Hinsinger, 1998; Hinsinger et al., 2005). A more elaborate presentation of the spatial extension and of the temporal dynamics of the soil–root interface is proposed in the following section.

7.2.2. Bioavailability

Bioavailability is a somewhat fuzzy concept that is commonly defined as the portion of the total element content that can interact and be taken up readily by a given organism. The elements available are not only those elements that are in chemical forms accessible to the biota, but also the compounds likely to be convertible to such forms (Soil Science Society of America, 1997) during the time scale relevant to the organism of interest. Clearly, the latter proposition emphasizes the dynamic nature of the concept of bioavailability. Although from an operational point of view, bioavailability is most often estimated via soil tests, which at best provide an indication of chemical availability (see below), it is clear from a conceptual point of view that bioavailability is implicitly dependent on the organism that is targeted. Indeed, Harmsen et al. (2005) stressed that "regarding the organisms, a 'bio-influenced' zone can be defined, which is that part of the surrounding soil and pore water with which the organism interacts ... thus, there could be numerous bioavailabilities depending on the type of target organisms and timescale." In the case of plants, the rhizosphere is the bio-influenced zone of soil which is thus fully relevant to the processes and factors that determine the bioavailabilities of a trace element to various plant species.

The total trace element content of a soil has often been used as a measure of bioavailability. However, this approach has been seriously challenged because trace elements are present in various forms both in the soil solution and in soils, with each form having a distinct potential for uptake by the biota (Krishnamurti and Naidu, 2002). In agriculture, one-step extraction procedures (DTPA, EDTA, $NaNO_3$) are used to predict *phytoavailability*, the size of the pool of a trace element susceptible to depletion by plants during growth (McLaughlin et al., 2000). In ecotoxicology, the activities of free metal ions in the soil solution are often used (Sauvé et al., 1997; McLaughlin et al., 2000; Renella et al., 2004; Tye et al., 2004), while other approaches, such as the DGT (diffusive gradient in thin films), have been suggested as a better surrogate of bioavailable metals in soils for metals such as copper (Zhang et al., 2001). The recent development of bacterial biosensors for metals such as Cu (Tom-Petersen et al., 2001) and Hg (Bontidean et al., 2004) is another tool that should expand in the future, as it is based on the response of an organism. It is thus more relevant to the estimation of metal bioavailability than any chemical test. However, as shown clearly by Bontidean et al. (2004) for Hg, the bioavailability of a metal as indicated by such biosensors can differ substantially from that estimated by plant uptake. The use of biosensors for evaluating the bioavailability of trace elements for higher plants deserves further critical assessment.

A commonly used approach to estimating bioavailability consists of simulating, using greenhouse or growth-chamber experiments, the field conditions under which plants grow and to establish the relationships existing between the speciation of trace elements in soils and their uptake by plants (Chaignon and Hinsinger, 2003; Wang et al., 2003). For such bioassays, bioavailability is equated to the actual amount of an element entering an organism over a given duration of contact (Harmsen et al., 2005). Bulk soil is generally used as the reference material for the estimation of the bioavailability of trace elements in soils. However, because differences in the concentrations and forms of trace elements are known to exist between the bulk soil and the rhizosphere, the relevance of measurements from the bulk soil for estimating the bioavailability of trace elements at the soil–root interface is considered to be limited (Rengel and Marschner, 2005).

7.2.3. Speciation

The *speciation* of an element refers to its quantitative distribution among the various chemical species that this element can form in solution. This concept was developed for aqueous chemistry and has been extended to solid-state speciation to account for the distribution of elements among the various compounds to which they can be bound. Speciation is often depicted operationally as a process leading to identification of the different chemical and physical forms of an element existing in a given sample or system, including organic and inorganic compounds, oxidation states, or isotopic composition (Kot and Namiesnik, 2000).

The various elemental forms generally recognized in soil solutions comprise free uncomplexed ions, ion pairs, ions complexed with organic anions, and ions complexed with organic macromolecules and inorganic colloids. In the solid phase the elements are found on the exchange complex, complexed by soil organic matter, sorbed onto or occluded within oxides and clay minerals, co-precipitated with secondary pedogenic minerals (e.g., Al, Fe, Mn oxides, carbonates and phosphates, sulfides) or as part of the crystal lattices of primary minerals. If the total dissolved trace element concentrations can be determined by a variety of analytical techniques, few species of trace elements can be measured directly, such as the free metal ion activities determined by ion-selective electrodes (e.g., for Cu^{2+}). The detailed speciation of trace elements in waters cannot be determined and thus needs to be calculated using speciation programs such as GEOCHEM, MINEQL, and WHAM. In the case of the soil solid phase, speciation is generally defined operationally using sequential or independent extraction procedures. For a range of reasons associated with soil properties, plant characteristics, and trace element behavior, no method has yet been accepted universally for the estimation of speciation of trace elements and for the assessment of their bioavailability to plants and soil organisms (Nolan et al., 2003). Recent advances in analytical chemistry, such as the development of x-ray absorption spectroscopy (i.e., XANES, EXAFS), have enabled researchers to gain a better knowledge of the types of solid-phase complexes involving trace elements (e.g., Panfili et al., 2005; Skinner et al., 2005).

7.3. SPATIAL AND TEMPORAL EXTENSION OF THE SOIL–ROOT INTERFACE

The soil–root interface is a soil microenvironment that is extremely dynamic in both space and time. Here we review some of the main aspects of this variability.

7.3.1. Spatial Extension

The lateral extension of the soil–root interface from the outer surface of the root toward the encasing bulk soil is obviously not a fixed value and is difficult to quantify with certainty. The rhizocylinder volume extends several micrometers up to a few centimeters away from the root, as a function of plant species and of the physical, chemical, and mineralogical properties of the soil (Hinsinger et al., 2005). The relative competence of roots to modify the surrounding soil will indeed differ with plant species because of the distinct nature of their exudates, nutrient acquisition strategies, and the architecture of their root systems. For example, extension of the elemental depletion zone in soils is much larger for plants with root hairs (Figure 7.1) than for those without, a root component that effectively expands the rhizocylinder and exploration of the soil (Marschner, 1995). For plants with root hairs, those with the longer total root hair length are favored because of the increased surface area of contact between the root and the surrounding soil. The intensity of the rhizospheric activity and the nature of rhizosphere properties (e.g., the spatial distribution of exudates along roots) is also known to differ as a function of position along roots (tips vs. mature zones) and of root age classes (Römheld, 1991; Marschner et al., 2004). The extension of the reach of roots within the soil can be augmented through mycorrhizal symbiosis to create an environment termed the *mycorrhizosphere* (Smith and Read, 1997; Pennisi, 2004). More than 80% of plant species are mycorrhized and the contribution of mycorrhizal hyphae to the spatial extension of the nutrient depletion zone has been illustrated clearly for P by Li et al. (1991), for instance. Whereas root hairs extend up to about 1 to 2 mm at most, mycorrhizal hyphae can extend up to several centimeters from the root surface and thus help the plant to access to water, nutrients, and possibly pollutants that would otherwise not be accessible to the roots.

Some soil materials are more prone than others to transformation or to registering the transformations imposed by roots, as evidenced clearly for the role of proton buffering capacity in rhizosphere pH changes (Nye, 1986; Hinsinger et al., 2003). In general, and for a given plant type, fine-textured soils rich in weatherable minerals such as chlorite, trioctahedral micas, or amphibole, will be transformed more readily by root activity, and potentially over a longer distance away from the root, than will a coarse sandy quartzitic soil. The capacity of the soil solids to buffer the chemical changes imposed by roots will also contribute to define the extension of the rhizosphere. Moreover, the interpretation of the lateral extent of the rhizosphere depends on our capacity to measure the effects of root activity on soil materials and on the soil property of interest. For some properties, such as macronutrient depletion or root-related secondary precipitates, the rhizo-

Figure 7.1. Root of *Thlaspi caerulescens* with associated root hairs in close contact with soil particles. (From Whiting et al., 2003.)

sphere–bulk soil boundary could be quite sharp, whereas it remains progressive or even diffuse for others, such as microbial counts or water content.

The spatial extension of the rhizosphere also depends largely on the architecture of the root system and on root growth and development. Schenk and Jackson (2002) analyzed a database of vertical root profiles containing 475 profiles from 209 geographic locations. They found that over 90% of the soil profiles had at least 50% of their root biomass within the first 30 cm of the profile, including the organic horizons, and that 95% of all roots were in the upper 2 m. However, many tree species have large taproots extending down several meters in depth, notably in areas where the water supply is limited (Canadell et al., 1996). From a geographic perspective, these authors showed that the depth distribution of roots in soil is strongly controlled by environmental constraints such as climatic conditions, yet the architecture of the root systems and its variability among plant species are determined genetically (Lichtenegger and Kutschera-Mitter, 1991; Zobel, 1991). Some plants will, however, cope with unfavorable nutritional conditions by investing much of their root development in localized patches of higher nutrient availability in the soil (Robinson, 1994; Schöttelndreier and Falkengren-Grerup, 1999; Hodge, 2004), thus responding to environmental controls. The formation of dense clusters of lateral roots (*proteoid roots*) subjecting the soil to an intense chemical alteration also reflects the existence of nutrient (P, N, or Fe) deficiencies. For hyperaccumulator plants, it was demonstrated that the roots of *Thlaspi caerulescens* responded positively to elemental concentration in soils and that the plant allocated up to 70% of its total root biomass in Zn-enriched soil (Whiting et al., 2000).

7.3.2. Temporal Extension

The longevity of fine roots for trees varies from a few days or weeks to more than five years, depending on the environmental conditions prevailing at the microsite

(Vogt and Bloomfield, 1991; Fahey and Hughes, 1994). In the case of annual crop plants, the duration of roots is limited to weeks or months, whereas it can reach several years for perennials. Clearly, for most plants, fine roots represent relatively short-lived tissues, yet their impact on the properties of the surrounding soil is measurable and undeniable. This suggests that the growth and activity of fine roots can transform soil solids in a matter of months, or even weeks, in the field. In this respect, changes such as the transformation of micas into vermiculite were found to occur within a few days close to an artificially densified root network, showing the potential of roots to affect soils irreversibly in the short term (Hinsinger and Jaillard, 1993). The persistence of changes imposed by roots to the soil, once the root is dead and decomposed, will vary as a function of the nature of the change, with the impact on chemical properties being somewhat less permanent than that of mineralogical transformations. In some specific instances, soil features inherited from past root activity can last for very long periods. In an extreme case, the inheritance of calcareous rhizoconcretions and of drab-haloed root traces was documented in Antarctica in a Devonian paleosol that supported a well-drained forest (Retallack, 1997).

Under field conditions, repeated colonization by roots of the same areas in soil profiles is often observed, and this contributes to creating large centimeter- to decimeter-scale tonguing features. The recolonization of root-induced macropores of the previous crop by the current crop under corn is an example of such dynamics (Rasse and Smucker, 1998). These features are long-lived and can eventually become self-entrained sources of spatial heterogeneity within soils. Once formed, these low-density and organic matter–rich tongues can indeed evolve as channels favoring the preferential macroporous flow of water in soils, as illustrated by Price (1984). Because of their elevated organic matter content, the soil sheath at the periphery of these channels also becomes an attractive site for subsequent root generation. For example, a large proportion of the roots of wheat (up to 80%) were found to be preferentially located in the soil within 1 mm of macropores where microbial populations are at a maximum (Pierret et al., 1999; Stewart et al., 1999).

7.4. WATER UPTAKE AND TRANSPORT PROCESSES

7.4.1. Radial Fluxes Toward the Root Surface

Water uptake is the primary function of roots, and plants have evolved an entire diversity of root architectures and related strategies of carbon allocation to access this essential resource. Whenever a plant is transpiring (i.e., essentially during the day period), the uptake by roots is responsible for a gradient of water potential which generates fluxes of water toward the root (Hamza and Aylmore, 1992; Doussan et al., 2003). Such gradients may extend several centimeters from the root surface and are subjected to considerable spatial and temporal variation: Water influx is usually greater near the young apical part of the roots, which thus results in the steepest gradient along the root axis, as modeled by Doussan et al.

(2003). Temporal variations over short time scales have also been accounted for in this model. They are a result of the diurnal pattern of water uptake which peaks at midday and is zero at night. At night, however, the direction of water fluxes can revert, this *hydraulic lift* being responsible for water efflux from roots at the top of the root system, especially from those roots growing in the topsoil when it dries out (Caldwell et al., 1998).

The flux of water into roots supports radial transport of all the solutes contained in the soil solution in the rhizosphere. This transport process, called *mass flow*, contributes significant fluxes of solutes when their concentration in the soil solution is large and/or if their actual uptake rate by the plant is small (Barber, 1995; Hinsinger, 2004). This is typically the case for Ca and Mg in many agricultural soils, for which mass flow can even exceed the actual amount taken up by the root, thereby leading to a buildup of their concentration in the rhizosphere. Precipitates of Ca carbonate or sulfate have been observed around roots that are the direct consequence of such a process (Malzer and Barber, 1975; Hinsinger, 1998). In saline soils, the buildup of Na and Cl concentrations is similarly expected to occur (Sinha and Singh, 1974; Hamza and Aylmore, 1992). Via such processes, the radial flow of water in the rhizosphere may thus affect the fate of trace elements, although it has been little demonstrated except for the coprecipitation of trace metals with Ca carbonates (Assadian and Fenn, 2001). Changes in the speciation of Cd, which are largely dependent on Cl concentration, might also be expected under these circumstances. Beyond such indirect effects and because trace elements most often occur at low concentrations in the soil solution, their transport by mass flow is expected to be rather small. In the rhizosphere of radish, Lorenz et al. (1994) reported that mass-flow contributed about 40% of the uptake of Cd and Zn in a contaminated soil. There are few other documented reports, and it is thus difficult to draw general conclusions. This example nonetheless shows that mass flow should not be neglected when considering the transport of trace elements in the rhizosphere.

Whenever the mass flow is less than the actual uptake of an element, a decrease in the concentration of the element in the rhizosphere is expected. Such a depletion gradient is then the driving force for diffusion toward the root surface, which is the second important transport process likely to occur in the rhizosphere. From the previous example one can deduce that the diffusion of Cd and Zn contributed close to 60% of the metal uptake by radishes growing in a contaminated soil. Water dynamics still plays an important role in the diffusional transport of trace elements, which makes sense, as this process occurs in the soil solution contained in soil pores. Indeed, the diffusion flux is a function of the concentration gradient, the diffusion coefficient, and an impedance factor which itself is a function of the water content of the soil (Barber, 1995; Tinker and Nye, 2000; Jungk, 2002). This impedance factor, which is also termed the *tortuosity*, describes the actual path followed by ions moving through soil pores. It increasingly diverges from a straight line (the main diffusion path in a homogeneous liquid phase) with decreasing water content.

7.4.2. Modeling Fluxes and Transport Processes in the Rhizosphere

The two above-mentioned processes of solute transport and their links with water uptake and water content in the rhizosphere are fairly well known and accounted for in published models of plant nutrition (reviewed by, e.g., Barber, 1995; Tinker and Nye, 2000; Jungk, 2002). However, such models have been shown to diverge from experimental data, in particular for cases of low nutrient or water availability (Jaillard et al., 2000; Liebersbach et al., 2004). This suggests that they fail in accounting for all the processes involved in the acquisition of nutrients by whole-plant root systems. To name a few of their potential drawbacks, they account poorly for the actual root architecture (Dunbabin et al., 2006) and do not account for mycorrhizas, despite their well-known contribution to the acquisition of poorly mobile nutrients such as P and Zn (Smith and Read, 1997). Moreover, current models do not account for the heterogeneity of root functioning in space (e.g., for the higher uptake of water at apices) (Doussan et al., 2003). They do not account either for the temporal heterogeneity of root functioning, such as the diurnal variations in water uptake and the potential for outward fluxes due to hydraulic lift. Last but not least, they hardly account for any other rhizosphere processes apart from those related to water and nutrient uptake. Indeed, they do not account for other root-mediated processes, such as pH changes or the exudation of organic ligands, despite their well-documented effects on nutrient acquisition (e.g., Hinsinger, 1998). Such models have seldom been applied to trace elements, even those that are micronutrients. In a recent study by Sterckeman et al. (2004), it was found that the Barber–Cushman model described the experimental data obtained for Cd uptake by maize fairly adequately, except at the highest levels of soil contamination in the potentially toxic range (Figure 7.2). They also found fairly good agreement between predicted and experimental data for *Thlaspi caerulescens*, a Cd hyperaccumulator plant species. Further work is thus clearly needed to establish the role of water uptake and water content on the fate of trace elements in the rhizosphere.

7.4.3. Longitudinal Fluxes: Preferential Flow Paths

The longitudinal fluxes that may arise from the preferential flow paths of water and solutes along large root channels are even less well documented than the radial fluxes of trace elements in the rhizosphere. It is well known that roots tend to colonize former biopores, as indicated in Section 4.2. As many of those macropores, especially vertical/subvertical biopores created by earthworms or root activity, are prone to preferential flow during re-saturation events, they may complicate our understanding of (1) the directions of trace elements and water fluxes, and ultimately, (2) the origin of the solutes circulating in the root environment. This process has received virtually no attention so far, however, possibly because of the practical difficulties associated with its quantitative assessment.

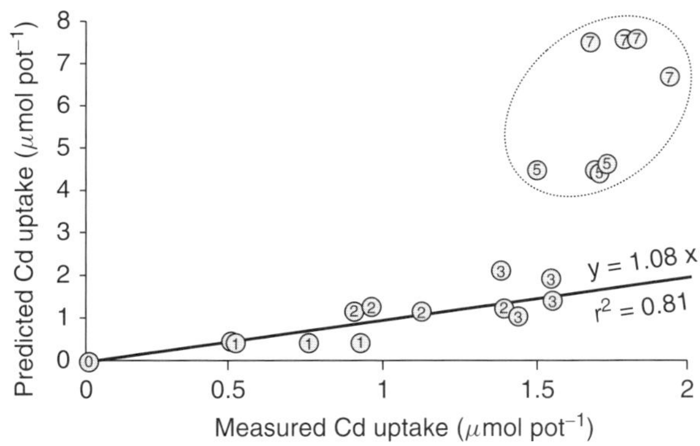

Figure 7.2. Predicted as a function of measured Cd uptake by maize after 24 days of cultivation time, according to the Barber–Cushman model. The numbers enclosed in circles stand for the level of Cd added to the soil (expressed as mg kg^{-1}). The model adequately predicts the uptake for Cd contamination levels up to 3 mg kg^{-1} (level of Cd added to the soil) but clearly overestimates the actual uptake of Cd for greater levels of soil contamination, which were thus discarded for the regression analysis. (Adapted from Sterckeman et al., 2004.)

7.5. METAL AND METALLOID CONCENTRATION GRADIENTS AT THE SOIL–ROOT INTERFACE

Chemical gradients exist at the soil–root interface and tend to organize along two main axes: radially from the root surface toward the bulk soil and longitudinally along the length of individual fine roots. These gradients occur in both the liquid and solid phases of the soil and are controlled by a series of physical, chemical, and biological processes that are associated with the presence of living plant roots. However, there is currently no consensus on whether rhizospheric processes augment or impoverish the metal pool in the vicinity of roots, with the result being strongly dependent on plant species and soil types, on the element of interest, and on nutritional conditions. In the following section we present field or laboratory examples of chemical gradients in heavy metal or metalloid concentrations observed for forest trees and agricultural plants.

7.5.1. Gradients in Solid-Phase Metals and Metalloids

Documenting concentration gradients in metals and metalloids at the soil–root interface is crucial to establishing their availability to roots. This assessment, in turn, helps to estimate metal uptake and their potential adverse effects on plants and for organisms, including humans, feeding on them. Although the bioavailability of metals is often best predicted by their dissolved concentration, the metals in the solid phase are key components because some plants take up most of their nutrients from the nonmobile fractions (less available than the water-soluble and exchangeable metals) of the solids (Knight et al., 1997) and because the replenishment of metals in the soil solution is ultimately controlled by the solid-phase pools (Zhang et al., 2001).

Many authors have shown that the bioavailable trace metal fraction is usually higher in the rhizosphere than in the bulk soil (Marschner, 1995; Marschner and Römheld, 1996; Hinsinger, 1998). For instance, Séguin et al. (2004) obtained an almost systematic increase in water-soluble Al, Cd, Cu, Fe, Mn, Ni, Pb, and Zn in the rhizosphere of trembling aspen (*Populus tremuloïdes* Minchx) in Cu-contaminated soils (Figure 7.3). Similar trends were observed under sugar maple (*Acer saccharum* Marsh.), balsam fir (*Abies balsamea* L.), and white birch (*Betula papyrifera*) in an uncontaminated area. Courchesne et al. (2001) measured a higher content of $BaCl_2$-exchangeable Al, Fe, Mn, Cu, and Zn in the rhizosphere than in the bulk soil of uncontaminated balsam fir and black spruce (*Picea mariana* Mill.) stands. These results suggest the probable existence of a larger pool of bioavailable heavy metals close to tree roots in both metal-contaminated and uncontaminated soils. In general, the total metal content is constant throughout; however, Smith and Pooley (1989) found that total Al followed a strongly decreasing gradient from the bulk soil to the root surface in an uncontaminated red spruce (*Picea rubens* Sarg.) forest. Total Zn exhibited an inverse trend, whereas Fe was invariant.

Data obtained using a different approach and pertaining to agricultural plants also reveal the existence of increasing gradients in available heavy metals. Indeed, a greenhouse experiment by Wang et al. (2002) indicated that the water-soluble and exchangeable Cd, Cr, Cu, Ni, Pb, and Zn fractions of an uncontaminated soil were at a maximum in the rhizosphere of wheat (*Triticum aestivum* L.). Loose-

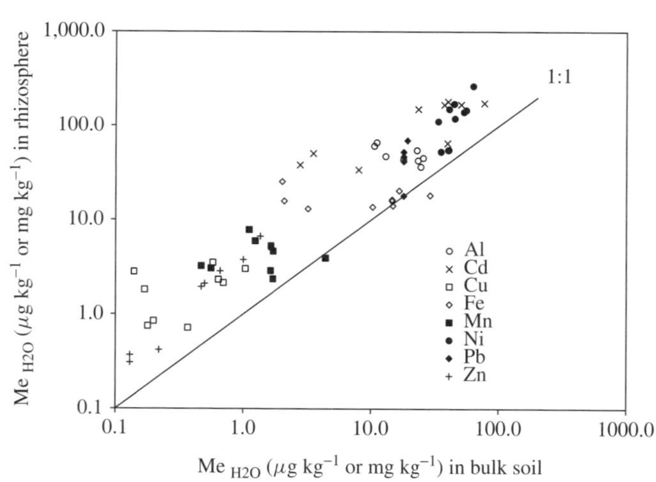

Figure 7.3. Trace metal concentration in water extracts from the bulk soil and the inner rhizosphere of trembling aspen (*Populus tremuloïdes* Minchx) at Rouyn-Noranda. (Modified from Séguin et al., 2004.)

more et al. (2004) also showed for tobacco (*Nicotiana tabacum* L. cv. SR1) grown in in a contaminated soil that the exchangeable Zn fraction increased in the rhizosphere. The amount of NH_4OAc-extractable Pb was found to be much higher in the rhizosphere of rice (*Oryza sativa* L.) than in the bulk materials of a contaminated soil (Lin et al., 2004). Contrary findings were reported, however, where the rhizosphere was depicted as an environment that had lost a significant fraction of its bioavailable metal pools. For example, Cherrey et al. (1999) found the rhizosphere of ryegrass (*Lolium perenne* cv. Aubisque) to be impoverished in EDTA, DTPA, or $CaCl_2$-extractable Cu for Cu-contaminated vineyard soils. Similarly, Chaignon et al. (2002a) observed lower Na–acetate- and $CaCl_2$-extractable Cu contents in the rhizosphere of tomato (*Lycopersicon esculentum* cv. St-Pierre) and oilseed rape (*Brassica napus* cv. Goeland) in an acidic Cu-contaminated vineyard soil. Hammer and Keller (2002) noted the decrease of several Cd pools (EDTA-, DTPA-, $NaNO_3$-extractable) in the rhizosphere following the growth of *Thlaspi caerulescens* in contaminated soils. The rhizospheres of kale (*Brassica oleracea acephala* L. cv. Winterbor) and candytuft (*Iberis intermedia* Guers.) grown in contaminated soil were also markedly depleted in Tl, with up to 40 to 50% of the uptake originating from non-plant-available Tl (Al-Najar et al., 2003).

Studies on the soil–root interface of trees using microscale chemical analyses support the results of macroscopic observations on metal gradients, here increased metal concentrations close to roots. For example, Naftel et al. (2002) used x-ray fluorescence (XRF) for micrometric-scale mapping of the spatial distribution of metals (Cu, Fe, Mn) within roots and in the rhizosphere of trembling aspen (*Populus tremuloïdes* Minchx) growing in a Cu-contaminated soil. They showed that the highest metal intensities occurred in a band about 30 μm across, Probably arising from soil particles on the outside of the root. In subsequent studies on the same cross sections of aspen roots with time-of-flight secondary-ion mass spectrometry (ToF-SIMS), Martin et al. (2004) identified a distinct concentration of Fe and Al atoms at the root surface in the form of soil encrustations (Figure 7.4). The most elevated Mn levels were found in the roots in both epidermis and cortex. Work by Skinner et al. (2005) on Cu and Mn using x-ray absorption near-edge structure (XANES) spectroscopy together with ToF-SIMS indicated that there is a reducing region which extends no more than 20 μm from the root surface. The distribution and oxidation state of Mn between the root plaque and the root interior were useful in interpreting metal transport mechanisms through the cell membrane.

Metals can be depleted or accumulate in the rhizosphere along a concentration gradient that increases or decreases from the root toward the bulk soil, depending on whether their rate of uptake is, respectively, greater or lower than the rate of supply by mass flow to the soil–root interface (Lorenz et al., 1994; Marschner, 1995). Beyond this effect, however, the concentration gradients measured in various solid-phase metal fractions at the soil–root interface can also be attributed to the effects of a variety of processes, such as rhizosphere acidification or alkalinization, adsorption or desorption reactions, and precipitation or dissolution phenomena, which are themselves associated with plant uptake and a range of

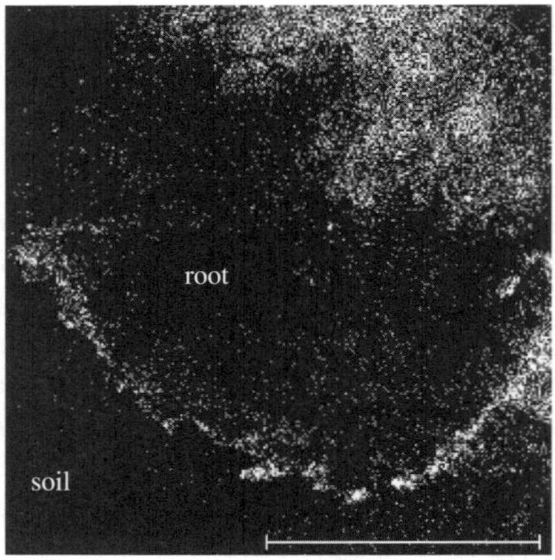

Figure 7.4. Time-of-flight secondary-ion mass spectroscopy (ToF-SIMS) image of the distribution of Fe on a cross section of a trembling aspen (*Populus tremuloïdes* Minchx) root embedded in clayey soil material. The Fe values are normalized to the total ion yield of the surface. The bar is 100 μm. (From Martin et al., 2004.)

root activities (Hinsinger, 2001). Microorganisms are also involved, but the discussion of their impact is beyond the scope of this chapter (see chapter 2). The existence of gradients in pH values and in dissolved or solid-phase organic matter at the soil–root interface is well documented and is discussed in more detail in other sections of this chapter. In terms of mineral weathering, the rhizosphere thus ends up being an environment more corrosive to mineral structures than that of the adjacent bulk soil. This action has been demonstrated in various studies (Leyval and Berthelin, 1991; Hinsinger et al., 1992; Arocena and Glowa, 2000; Bakker et al., 2004) for weatherable minerals such as micas, apatite, and amphiboles. Hence, the dissolution of soil solids triggered by acidification of the rhizosphere increases the mobilization of metals and their uptake by plants (Marschner, 1995), as demonstrated by Loosemore et al. (2004) for tobacco (*Nicotiana tabacum* L. cv. SR1) grown in a Zn-contaminated soil.

Another consequence of increased weathering is the accumulation of weathering products, (i.e., Fe and Al oxides) close to the root surface (Courchesne and Gobran, 1997; Martin et al., 2004). These secondary minerals, alone or in combination with organic matter, increase the potential cation- (CEC) or anion-exchange (AEC) capacity of the rhizosphere materials (Gobran and Clegg, 1996) and can thus augment the retention of heavy metals and metalloids, in either cationic or anionic form, depending on chemical conditions. The formation of oxides adds to the already increased CEC that is conferred to the rhizosphere by the accumulation of organic matter. The presence of metal oxides and organic matter in mildly acidic to alkaline systems is known to favor the retention of metals (Harter and Naidu, 2001). Gradients showing an increased metal concentration toward the root surface therefore develop. The Fe and Al oxides can also adsorb dissolved organic substances, a process that will further increase the organic matter level in the rhizosphere (Jones, 1998), together with the associated metal retention capacity. However, the acidification of the rhizosphere of an already acidic soil by root activity has the potential to reverse the adsorption reaction and to cause the desorption of metals, a common observation under acidifying conditions (Harter and Naidu, 2001). An increase in available metal close to roots can be measured under such conditions (McGrath et al., 1997).

7.5.2. Gradients in dissolved metals and metalloids

The literature shows that many fewer data are available on dissolved than on solid-phase heavy metals at the soil–root interface. To date, several approaches have been used to acquire soil solution samples from the rhizospheric environment, including water displacement, water extraction, centrifugation, and micro-suction cups. However, because of technical constraints, no single method appears to sample unaltered rhizosphere solution.

One of the earliest reports on gradients in dissolved metals in the rhizosphere was that of Lorenz et al. (1994), who used a water displacement technique. Their results indicated that total dissolved Cd and Zn were depleted from the rhizosphere by plant uptake following the growth of radish (*Raphanus sativus* cv. 'Crystal Ball') in a contaminated soil. Subsequent work with radish confirmed the depletion of Cd and Zn (Figure 7.5). It showed further that most of the total Cd and Zn was present as Cd^{2+} or Zn^{2+} in the bulk soil solutions and that the proportion of free ions was lower in the rhizosphere because of their complexation by dissolved organic substances (Lorenz et al., 1997). The depletion trend is in contrast with work of Griffiths et al. (1994), who observed that the concentrations of Al, Fe, Cu, Mn, and Zn in an uncontaminated soil were higher in hyphal mats than in nonmat soil solutions under Douglas fir [*Pseudotsuga menziesii* (Mirb.) Franco]. They attributed this increased metal availability to the accelerated mineral weathering induced by the organic acids released in the rhizosphere of Douglas fir.

A distinct and innovative experimental approach based on the use of micro-suction cups provided in situ evidence for an increase in free Al^{3+} in the rhizosphere of oak (*Quercus robur* L.) seedlings (Göttlein et al., 1999). The exudation of H^{+} was viewed as the main mechanism causing the release of Al in the vicinity of roots, a process of major relevance to the estimation of rhizotoxicity. In contrast, Dieffenbach and Matzner (2000) found that total dissolved Al and free Al^{3+} concentrations decreased in the rhizosphere of mature Norway spruce (*Picea abies* L. Karst.) using a field rhizotron setup equipped with micro-suction cups. They estimated that the Al gradient extended 1 to 5 mm away from long roots and that it could develop or disappear rapidly, in a matter of days or weeks. Using a similar rhizotron approach under growth chamber conditions, Arocena et al. (2004) showed that the rhizosphere depletion gradient observed

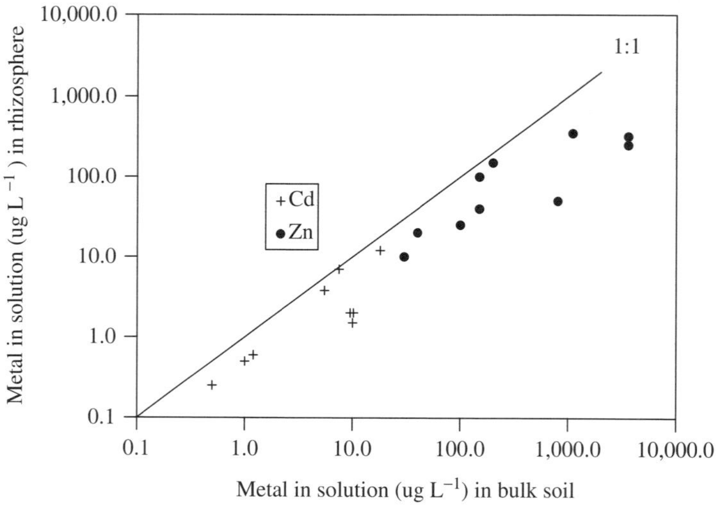

Figure 7.5. Cadmium and zinc concentration in soil solution from the bulk soil and the rhizosphere of radish (*Raphanus sativus*). (Modified from Lorenz et al., 1997.)

for free Al^{3+}, Fe^{3+}, and Mn^{2+} was induced in the A horizon of an uncontaminated forest soil by the symbiosis of Norway spruce with the ectomycorrhizal fungus *Piloderma croceum*. The colonization by *Piloderma* was considered to stimulate nutrient uptake by the roots of the spruce seedlings. Alternatively, the complexation of Al, Fe, and Mn by exudates might explain the reduced free ion concentrations close to roots.

Work by Cornu et al. (2007) using Rhizon microsamplers to collect the soil solution in the rhizosphere of tomato (*Lycopersicon esculentum* cv. St-Pierre) grown in small rhizoboxes filled with acidic Cu-contaminated soils showed a decrease in total dissolved Cu, which was accounted for largely by the pH increase recorded simultaneously. In addition, Legrand et al. (2005) pointed out the role of organic substances in complexing metals in the rhizosphere of Cu-contaminated soils. The Cu^{2+} activity measured using an ion-selective electrode was indeed shown to be similar in water extracts from the rhizosphere of white birch (*Betula papyrifera* Marsh.) and in the adjacent bulk soil, although total dissolved Cu was higher close to roots (Figure 7.6). The fraction of the total water-soluble Cu present as Cu^{2+} was, however, higher in the bulk soil, reflecting the complexation of Cu by organic substances at the soil–root interface. Under a range of forest tree species growing on contaminated and uncontaminated soils, the concentrations of dissolved Zn and of DPASV-labile Zn (Zn_L) in water extracts increased from the bulk soil toward the roots (Courchesne et al., 2006). Calculations with the WHAM speciation code indicated that free Zn^{2+} ion activities were higher in the rhizosphere than in the bulk soil but that the fraction of total dissolved Zn in water extracts that was present as free Zn^{2+} did not differ significantly.

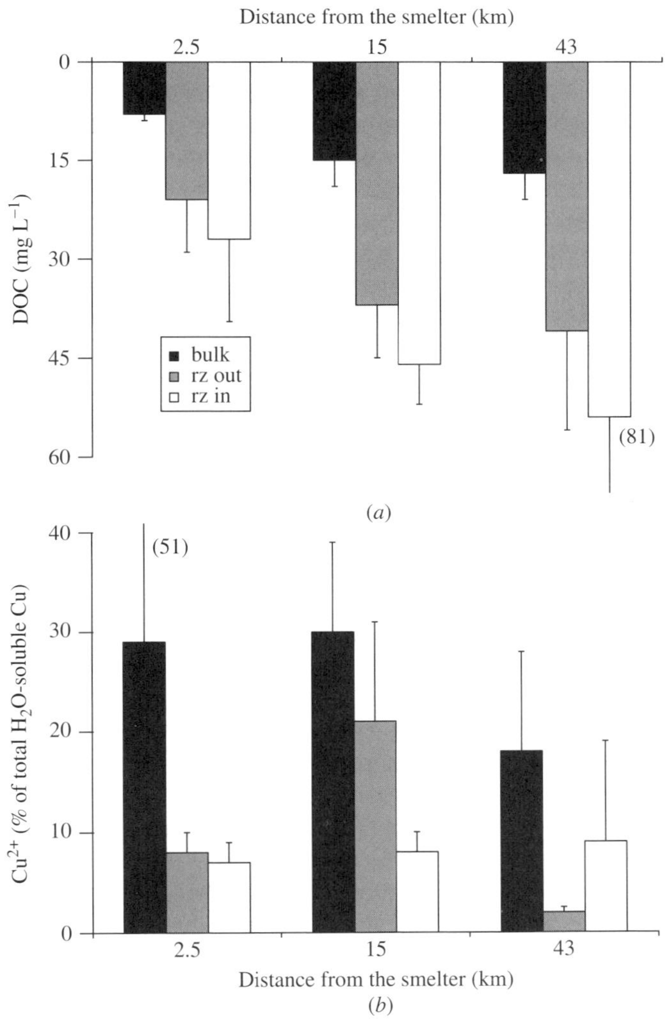

Figure 7.6. (*a*) Dissolved organic carbon (DOC) and (*b*) Cu^{2+} activity expressed as a percentage of total H_2O-soluble Cu in the bulk soil, the outer rhizosphere, and the inner rhizosphere along a soil contamination gradient at Sudbury. Mean values and standard deviations are for three field replicates (Modified from Legrand et al., 2005.)

Dissolved metals can thus increase or decrease close to roots depending on the experimental or site-specific conditions. This is similar to the observations made previously on solid-phase metals in the rhizosphere. However, it is obvious that more knowledge on the chemical composition of the rhizosphere solution is needed to better document the fine-scale chemical environment to which the roots are exposed.

7.5.3. Kinetics of metals dynamics in the rhizosphere

One of the major advances in the field of soil chemistry over the last few decades is the development and refinement of kinetic studies (Sparks, 2001). The quantification of reaction rates has helped elucidate the mechanisms of ion or organic chemical sorption and desorption, surface precipitation and dissolution, metal complexation by organic substances, and redox reactions in the bulk soil. Similar kinetic approaches have yet to be applied to the rhizosphere. A major scientific gap thus exists with respect to the availability of kinetic data pertaining to the chemical and biochemical reactions occurring at the soil–root interface. Because the rhizosphere is such a dynamic environment where the flow of energy and matter exceeds those measured in the bulk soil, it is reasonable to submit the hypothesis that reaction kinetics at the soil–root interface differ significantly from those measured in the adjacent soil matrix. Unfortunately, the data that would allow unequivocal testing of this hypothesis are presently unavailable. Management of the rhizospheric environment designed to, for example, enhance nutrient uptake by plants or optimize the phytoremediation of contaminated substrates would also greatly benefit from the availability of kinetic data. Similar benefits are expected for the successful numerical modeling of the biogeochemistry of the soil–root interface.

The literature does contain some references to studies on reaction kinetics in the rhizosphere, notably with respect to interactions between metals and dissolved organic substances. For example, the presence of low-molecular-mass organic acids (LMMOAs) secreted by plant roots and rhizosphere microbes (acetic, citric, fumaric, malic, oxalic, and succinic acids) was shown to enhance the rate of Cd release from contaminated tropical soils (Onyatta and Huang, 2003). The process was considered to be of importance for the fate of Cd in the rhizosphere because of the sustained release of LMMOAs by plant roots. Shenker et al. (1999) studied the behavior of rhizoferrin, a siderophore produced by the fungus *Rhizopus arrhizus* known to be an outstanding Fe carrier to plants. They concluded that a kinetic rather than an equilibrium approach should be used to determine the chelating efficiency of rhizoferrin for Fe^{3+} at the time scale of biological uptake processes. Similarly, Neubauer et al. (2002) stressed the importance of reaction kinetics to describe the effect of the siderophore desferrioxamine B (DFOB) on the sorption of Cd, Cu, and Zn to clay minerals. Matocha et al. (2001) measured the reductive dissolution kinetics of birnessite by catechol, an organic ligand that has been identified among siderophore molecules. They observed that the dissolution reaction was rapid, being essentially complete within seconds under conditions of excess catechol, and suggested that this could play a role in the availability of Mn to plants. Compared to microbial siderophores, even fewer kinetic data are available for plantborne organic ligands. However, Reichard et al. (2005) studied the kinetics of dissolution of a ubiquitous Fe oxyhydroxide, goethite, in the presence of phytosiderophores and oxalate. In doing so, they were able to elucidate the role of these ligands, especially phytosiderophores that are unique to grasses, in the mobilization of Fe from goethite surfaces. Finally, another important kinetic concern is the rapid biodegradation of these ligands in the rhizosphere. To that effect, Ström et al. (2001) found that the rate of biodegradation of malate, a common root exudate, was enhanced significantly in the rhizosphere of maize (*Zea mays* L. cv. Pioneer 3377) compared to the bulk soil. In short, a more systematic evaluation of the contrasts in reaction kinetics between the rhizosphere and its associated bulk soil should be a challenging future research endeavor.

7.6. pH AND pCO_2 GRADIENTS AT THE SOIL–ROOT INTERFACE

Gradients of pH extending up to two units above or below the bulk soil pH are among the best documented features of rhizosphere biogeochemistry. In comparison to pH gradients, pCO_2 gradients resulting from both root and rhizosphere microbial respiration have been little studied despite their potential effects on soil pH. Changes in rhizosphere pH values have various origins and can severely alter the speciation of most trace elements, both in the soil solution and in the solid phase. They ultimately have a considerable impact on the bioavailability of these trace elements for both rhizosphere microorganisms and plant roots.

7.6.1. Processes Responsible for pH and pCO_2 Gradients in the Rhizosphere

The major root-induced process responsible for significant pH changes in the rhizosphere is the production and consumption of protons by plant roots to compensate for unbalanced cation–anion uptake (Nye, 1986; Marschner, 1995; Hinsinger et al., 2003). Whenever roots take up more cations than anions, they compensate for this excess of positive charges entering the root by an equivalent number of positive charges being released as protons, thereby leading to rhizosphere acidification. Conversely, alkalinization occurs whenever roots take up more anions than cations. This process is responsible for considerable fluxes of protons, which largely explain the observed, upward, or downward pH gradients reported in the rhizosphere of many plant species (Hinsinger et al., 2003).

In comparison, the contribution of organic acids from plants is most often considered negligible. It should be stressed that the organic acids are expected to be released as organic anions given that they are dissociated in the neutral pH of the cytosol of root cells (Hinsinger et al., 2003). As described above for the uptake of nutrient ions, this means that the exudation of organic anions needs to be compensated for by an equivalent release of positive charges as protons, in most instances, or sometimes as cations, such as K^+ (Ryan et al., 2001; Hinsinger et al., 2003). What needs to be stressed, however, is that the effluxes of organic anions are fairly small for most plant species (Jones, 1998) compared to the fluxes of protons that are driven by the unbalanced uptake of cations and anions (Hinsinger et al., 2003). Among crops, large exudation rates of citrate (and protons) have been reported only for a few species, such as white lupin (*Lupinus albus* L.) and chickpea (*Cicer arietinum* L.). Soil microorganisms can also be involved in the production of massive amounts of organic acids and even inorganic acids (e.g., nitric acid in nitrifiers) (Robert and Berthelin, 1986). The contribution of microorganisms to pH gradients in the rhizosphere has, however, seldom been quantified. One well-documented case in this respect is that of ectomycorrhizal fungi. It has been shown that some of them can be responsible for considerable levels of oxalic acid (Casarin et al., 2004). In their work, Casarin et al. (2003) showed clearly that the rhizosphere pH of pine (*Pinus pinaster* Soland. in Ait.) seedlings infected by *Rhizopogon roseolus* was more than 1.5 units below that of nonmycorrhizal pine seedlings, but the ectomycorrhizal fungi *Hebeloma cylindrosporum* had hardly any effect on rhizosphere pH. They also showed a linear negative relationship between pH and oxalate concentration in the rhizosphere of pine roots infected with *R. roseolus* as well as close to the hyphae of *R. roseolus* grown in pure culture: pH = 6.90 –0.0236 (oxalate) expressed as μmol g soil^{-1} ($n = 21, r^2 = 0.922$).

Besides the fluxes of protons that occur to counterbalance charge imbalances, redox-coupled pH changes can also take place in the rhizosphere, as is the case for lowland rice. Kirk and Le Van Du (1997) showed that the precipitation of iron oxide occurring as a consequence of root-induced oxidation of the rhizosphere of rice was responsible for a significant proportion of the concurrent acidification.

A ubiquitous process that can contribute to significant pH changes is the respiration of roots and rhizosphere microorganisms and the subsequent buildup of pCO_2 (Hinsinger et al., 2003). This poorly documented process is expected to have a negligible effect in most acidic soils, given the pK_1 value of carbonic acid (6.36) and, conversely, a pronounced effect in neutral to calcareous soils. The rather flat gradient of pCO_2 measured by Gollany et al. (1993) in the rhizosphere of a grass grown in a calcareous soil reached values of about 100 mmol mol^{-1} (i.e., 300-fold that of atmospheric pCO_2). This would be expected to yield a pH decrease of about 1.6 units. Bidel et al. (2000) showed that respiration was especially large at the root tip, with an important contribution from microbial respiration as a consequence of microbial stimulation due to rhizodeposition.

When considering pH changes in the rhizosphere, it should be remembered that the corresponding fluxes of protons can vary considerably according to the buffering capacity of the soil (Hinsinger et al., 2003). A given efflux of protons can lead to virtually no change of pH in a strongly buffered soil such as a calcareous soil, whereas it can lead to a very significant decrease in pH in a poorly buffered soil. In both cases the protons will react with the soil solution and soil solid phase and might thereby alter the fate of trace elements. Thus, the actual fluxes of protons occurring at the root–soil interface are functionally more relevant than the corresponding pH gradients that can be measured in the rhizosphere.

7.6.2. Impact of pH and pCO_2 Gradients on the Speciation of Trace Metals

The pH is known to have a dramatic influence on the speciation of trace elements such as transition metals and metalloids in the soil solution. Thus, any change in rhizosphere pH should have a considerable impact. However, there is only limited direct evidence to support this because of the technical difficulties in sampling the soil solution at a spatial resolution that is appropriate to assess the changes in speciation.

Changes in the solubility of trace elements in the rhizosphere that arise at least partly from rhizosphere pH gradients are better documented. For example, it has been shown by Chaignon et al. (2002a) in an acidic vineyard soil contaminated by Cu-based fungicide sprays that $CaCl_2$-extractable Cu decreased with increasing rhizosphere pH for tomato (*Lycopersicon esculentum* cv. St-Pierre) and oilseed rape (*Brassica napus* cv. Goeland) fed with two sources of N (Figure 7.7). This fraction of soil Cu is often used as an estimate of soil solution Cu (Sauvé et al., 1997) and corresponds to the most soluble fractions of total soil Cu. Total Cu reached 165 mg kg^{-1} (dry weight basis) in soil studied by Chaignon et al. (2002a). Such a decrease in Cu solubility might be the result of Cu uptake by plant roots, which amounted to 6 and above 9 mg kg^{-1} for tomato and oilseed rape, respectively, without any significant difference between the two forms of N supplied. Even more convincing was a study by Chaignon et al. (2007) which showed clearly that rhizosphere alkalinization resulted in a decrease in $CaCl_2$-extractable Cu, according to $CaCl_2$–Cu $= 9.47–8.94(1 + \exp[-1.947(\text{rhizosphere pH} -3.837)])^{-1}$ ($n = 50, r^2 = 0.941$).

These results were obtained for oilseed rape grown in the same acidic soil as above but which had been subjected to various rates of lime application, in order to investigate a broader range of pH values. A substantial decrease in Cu solubility as a consequence of rhizosphere alkalinization was also found consistently for tomato grown in two acidic Cu-contaminated vineyard soils, including the one described above, in another study where soil solution Cu was sampled by Rhizon microsamplers (Cornu et al., 2007). Similarly, a decrease in

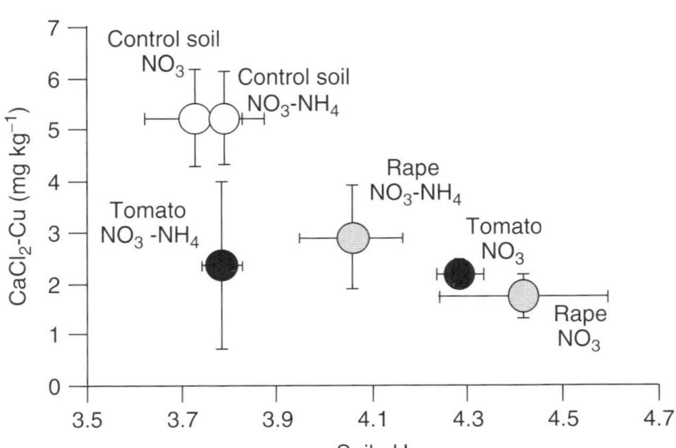

Figure 7.7. Decrease in $CaCl_2$-extractable Cu as a function of soil or rhizosphere pH in an acidic, Cu-contaminated vineyard soil. (Adapted from Chaignon et al., 2002a.)

soil solution Cd concentration following root-induced alkalinization occurred in the rhizosphere of two lupin species grown in an acidic contaminated soil (Römer et al., 2000). The acidification reported in the rhizosphere of maple trees (*Acer saccharum* Marsh.) sampled in situ may also have contributed some fraction of the increase in water-soluble Cu and Zn for Cu-contaminated soils sampled near a smelter in Canada (Séguin et al., 2004). However, these authors also pointed to a significant increase in water-extractable organic C in the rhizosphere, a process that might have contributed a fraction of the increased solubility of the metals.

Rhizosphere pH-controlled changes in the solid-phase speciation of trace elements have been further documented, although in only a few studies. For example, Loosemore et al. (2004) found that rhizosphere acidification was responsible for the increase in the exchangeable fraction of soil Zn that was observed in the rhizosphere of tobacco (*Nicotiana tabacum* L. cv. 'SR1') grown in a mildly acidic Zn-contaminated soil sampled 1.3 km from a Zn smelter (Figure 7.8). In another study on the rhizosphere of lowland rice (*Oryza sativa*), which is known to decrease its rhizosphere pH considerably as a consequence of both ammonium uptake and root-induced oxidation, it was reported that the fractionation of Zn was strongly affected (Kirk and Bajita, 1995). In this experiment with a noncontaminated soil, Zn extracted with ammonium oxalate, Cu acetate, and acid-soluble Zn increased in the rhizosphere, with a distinct maximum 1 to 2 mm from the root surface. However, the changes in redox potential might have contributed some proportion of the changes observed, and it should also be stressed that this was not a contaminated soil. Changes in the fractionation of various trace elements have been reported to occur in the rhizosphere in a few other studies (see Section 5), but their direct links with pH gradients have seldom been demonstrated.

7.6.3. Impact of pH and pCO_2 Gradients on the Bioavailability of Trace Elements

Changes in rhizosphere pH need to be accounted for when predicting the bioavailability of trace elements, as shown clearly by Loosemore et al. (2004) for Zn. In this study, the level of exchangeable Zn in the bulk contaminated soil was close to nil once it had been limed to the highest rate. However, despite such a liming effect on exchangeable Zn, the soil exhibited a substantial bioavailability of Zn as measured by the amount of Zn taken up by tobacco. This greater than expected uptake resulted from root-induced acidification, which, in turn, resulted in a large increase in exchangeable Zn in the rhizosphere of tobacco. The opposite effect was also found and was attributed to root-induced alkalinization of the rhizosphere of oilseed rape grown in an acidic Cu-contaminated soil. In this case, Cu bioavailability was lower than expected based on the bulk soil pH value (Chaignon et al., 2007). This study showed further that liming did not significantly reduce the bioavailability of Cu in this acidic soil, as the root-induced changes of rhizosphere pH counterbalanced the effect of liming to some extent. More important, these and a few other studies clearly show that the relationship between metal bioavailability and pH cannot be fully predicted from pH–metal

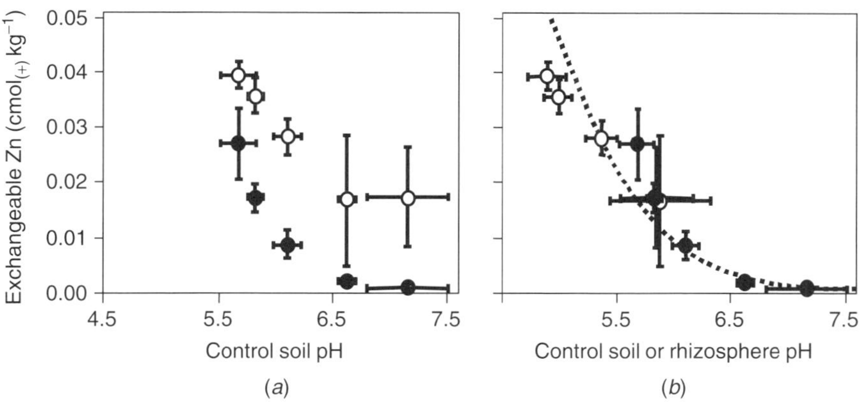

Figure 7.8. Exchangeable Zn in the control (unplanted) soil (closed symbols) and in the rhizosphere (open symbols) of tobacco (*Nicotiana tabacum* L. cv. SR1) as a function of (*a*) the control soil pH or (*b*) the control soil and rhizosphere pH, respectively. The mildly acidic Zn-contaminated soil had been sampled 1.3 km from a Zn smelter and limed to obtain a range of control soil pH. The relationship obtained for both rhizosphere and control soils could be described by the same model: [exchangeable Zn] $= 10^{-pH}/(10^{-pK} + 10^{-pH})$ [hydrolyzable Zn], with $pK = 4.90$ and [hydrolysable Zn] $= 0.13$ $cmol_+$ kg^{-1}. (Adapted from Loosemore et al., 2004.)

solubility relationships established for the bulk soil (Chaignon et al., 2002a; 2007; Loosemore et al., 2004; Cornu et al., 2007). The actual bioavailability of trace elements to plants might therefore be either higher or lower than expected, depending on the intensity and direction of the pH gradient in the rhizosphere.

One important feature of pH–trace metal relationships that is rarely accounted for when considering soil to plant transfer of metals is the effect of the root apoplasm. This compartment, corresponding to the three-dimensional network of root cell walls, is made of organic materials that bear a significant cation exchange capacity and tend to bind (via adsorption and possibly complexation) divalent and trivalent metal cations, as documented extensively for Al^{3+}. This effect is also pronounced for divalent metal cations such as Cu^{2+} because of the high affinity of root cell wall constituents for this metal. In addition, binding is pH dependent, as root cell walls have pH-dependent charges. The increased sorption of Cu in root apoplasm with increasing pH may also explain why the bioavailable fraction of Cu, and possibly other metals, has sometimes been shown to increase with increasing pH when assessed by measuring the total or apoplasmic metal content in plant roots (Chaignon et al., 2002a, 2007). Further understanding of this unique portion of the rhizosphere is clearly needed to predict the bioavailability of metals to plants, especially those metals that can result in severe rhizotoxicities such as Cu, particularly as it has been studied in relation to Al rhizotoxicity (Parker et al., 1998; Shomer et al., 2003).

The rhizosphere pH is also of crucial importance for determining the bioavailability of trace elements other than metals. The fate of trace elements that occur as oxyanions such as Mo or As is known to be fairly dependent on pH, via its effect on the surface charge of solids and ultimately on the adsorption–desorption of these oxyanions. For instance, rhizosphere acidification that occurred as a consequence of P-deficiency in canola (*Brassica napus* L.) grown on As-contaminated kaolinite resulted in enhanced bioavailability of both phosphate and arsenate, due to enhanced desorption of these oxyanions (Quaghebeur and Rengel, 2004).

7.7. REDOX POTENTIAL AND pO_2 GRADIENTS AT THE SOIL–ROOT INTERFACE

Gradients of redox potential (pe or Eh) in the rhizosphere are poorly documented compared to those of pH. These gradients are coupled with one another to some extent. Indeed, major processes responsible for changes in pO_2 and hence in pe are root and rhizosphere microbial respiration, which also affect pCO_2 and thus possibly the rhizosphere pH, as stressed in Section 7.6. Another reason for such coupling is that beside electrons, protons are involved in many redox reactions affecting, for instance, iron and manganese oxide minerals in soils (Hinsinger et al., 2003). The lack of quantitative data on pe and pO_2 changes in the rhizosphere is due largely to the technical difficulties in measuring these variables at an appropriate spatial scale. Nevertheless, pO_2 changes are expected to alter the speciation of those trace elements that can occur in various redox states, and also the fate of most trace elements via their impact on the dissolution–precipitation equilibria of iron and manganese oxides, which are know to be important trace element–bearing phases. Gradients of redox potential may ultimately have a major impact on the bioavailability of trace elements for both rhizosphere microorganisms and plant roots.

7.7.1. Processes responsible for decreasing redox potential and pO_2 gradients in the rhizosphere

There are only a few reports on the decrease in pe or pO_2 in the rhizosphere, probably because of the lack of suitable methodologies. To monitor temporal pe changes in the rhizosphere, Fischer et al. (1989) used a matrix of redox microelectrodes inserted in a plastic window panel on the side of a pot so that the growing plant roots could be observed and the root-induced changes in pe (Eh) measured once roots grew close to the electrodes. They showed that a steep decrease of Eh occurred soon after the root apex contacted the microelectrode (Figure 7.9). They also showed that Eh returned to its initial value after less than 24 hours of contact, suggesting that root-induced reduction occurred mostly in the region of the root tip. A ubiquitous root-induced process which is responsible for a decrease in pO_2 in the rhizosphere is the respiration of roots and rhizosphere microorganisms and the corresponding consumption of O_2. The microbial component of this process will be largely related to rhizodeposition, as the respiratory catabolism will result in a degradation of rhizodeposits which will evolve

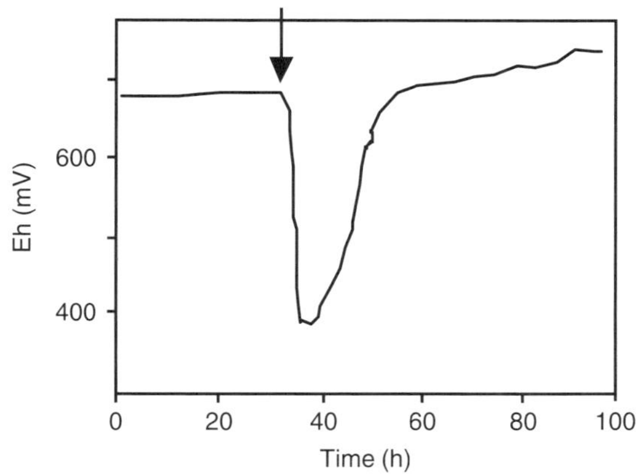

Figure 7.9. Temporal evolution of redox potential (Eh) in the vicinity of a growing root as measured by a microelectrode. The arrow indicates when the apex of the growing root first contacted the microelectrode. (Adapted from Fischer et al., 1989.)

CO_2 and consume O_2. Bidel et al. (2000) showed that respiration was larger at the root tip than along more basal portions of the roots, as a consequence of both greater respiration in the meristematic region of the root and of the stimulation of microbial respiration in the apical region of the root due to rhizodeposition.

In addition, some rhizodeposits and metabolites secreted by rhizosphere microorganisms are reductant molecules (e.g., phenolic compounds such as caffeic acid or carboxylic compounds like malic acid) (Brown and Ambler, 1973; Römheld and Marschner, 1983). These have been shown to be released by roots of Fe-deficient plants. However, it is now recognised that the Fe-reducing capacity of roots of most plant species is related to plasma membrane–bound enzymes (Fe reductases). Although there is ample evidence of Fe reduction occurring in plant roots as a consequence of such an enzymatic process, including that obtained by using redox-sensitive dye indicators (Dinkelaker et al., 1993), it is still unclear whether this process can lead to the establishment of substantial pe gradients in the rhizosphere.

7.7.2. Processes Responsible for Increasing Redox Potential and pO_2 Gradients in the Rhizosphere

Some plants can induce a release of O_2 at the surface of roots and thus an increase in pO_2 in the rhizosphere. This process is known to occur in an adaptation of plants to submerged soil conditions, as in wetland plants, and is well documented for lowland rice (*Oryza sativa*). To cope with anoxic or hypoxic conditions occurring in the soil or sediments, such plants have evolved a specialized structure, the *aerenchyma*, which conducts O_2 to root tissues from the atmosphere and the shoots. The portion of O_2 that is not consumed in the roots for respiration leaks through the root apoplasm (cell walls) and ultimately into the rhizosphere. Two pieces of evidence support this phenomenon. First, an increase in redox potential was shown to occur by using redox microelectrodes inserted at various distances from rice roots (Flessa and Fischer, 1992). However, this work showed that the increase in pe was significant only at the surface of rice roots. Second, indirect evidence for root-induced oxidation of the rhizosphere is based on the observation of iron plaque formation in root apoplasm and at the surface of roots of rice and wild wetland plants. This occurs as a consequence of the oxidation of Fe^{2+}, which is dominant in the bulk of the soil solution, to Fe^{3+} (Begg et al., 1994; Kirk and Bajita, 1995) and to the subsequent precipitation of iron oxides such as goethite or lepidocrocite, due to the low solubility of these Fe^{3+}-bearing minerals (Bacha and Hossner, 1977; Chen et al., 1980; Loeppert et al., 2005). Using redox-sensitive dye indicators, Trolldenier (1988) further reported that part of the oxidation in the rhizosphere of lowland rice was due to Fe^{2+}-oxidizing bacteria.

7.7.3. Impact of Redox Potential and pO_2 Gradients on the Speciation of Trace Elements

To our knowledge, there are no reports in the literature of redox-driven changes in trace element speciation in the rhizosphere soil solution compared to the bulk soil. However, there is some evidence for changes in solid-phase speciation of trace elements that involve redox reactions in the rhizosphere. Besides the formation of iron plaque in lowland rice and wetland plants, precipitates of manganese oxides at the surface of roots are frequently reported (e.g., Bacha and Hossner, 1977). These minerals are known to play an essential role in the fate of trace elements in soils (both heavy metals and metalloids that form oxyanions) as a consequence of adsorption or coprecipitation processes. This has been studied, extensively as As contamination of lowland rice has become an increasing concern for human health in Southeast Asia. Several reports showing the role of the iron plaque in the sequestration of As clearly demonstrate that root-induced changes of redox conditions can have a dramatic effect on the solid-phase speciation of As in the vicinity of roots (Liu et al., 2004a,b; Chen et al., 2005; Loeppert et al., 2005). Although more difficult to demonstrate, it is likely that a preliminary step in this process is root-induced oxidation of arsenite, which is theoretically the dominant dissolved species of As in the soil solution under reduced conditions. The resulting formation of arsenate favors strong adsorption onto the positively charged iron oxides that form the iron plaque in roots of lowland rice, as in the conceptual model devised by Meharg (2004). The use of synchrotron-based spectroscopic methods has enabled Hansel et al. (2002) and Blute et al. (2004) to (1) map As and Fe in the rhizosphere of aquatic plants such as cattail (*Typha latipholia*) and reed canary grass (*Phalaris arundinacea*), (2) establish the speciation of As associated with the Fe plaque [80% As(V) and 20% As(III) as determined by XANES in cattail roots] and (3) indicate that the mechanism of adsorption was likely to be dominant, although coprecipitation could also occur and explain the heterogeneous distribution of As observed in the Fe plaque.

The fractionation of soil Zn was also reported to be much affected in the rhizosphere of lowland rice grown in a noncontaminated soil. Kirk and Bajita (1995) showed that various fractions of Zn increased in the rhizosphere, with a distinct maximum 1 to 2 mm from the root surface. Although part of such changes in the solid-phase speciation of Zn could be due to concurrent root-induced acidification (see Section 7.6), root-induced changes in redox potential might also have contributed to the changes observed. It should, however, be stressed that this was not a contaminated soil. Another study on solid-phase speciation of Zn showed dramatic changes occurring as a consequence of plant growth. Using various synchrotron x-ray spectroscopic approaches, Panfili et al. (2005) showed that two candidate species for phytoremediation (phytostabilization) of Zn-contaminated dredge sediments resulted in a massive change in Zn fractionation. Indeed, while Zn sulfide was the dominant Zn-bearing solid phase in the control unplanted sediment, several months of growth with two grasses (*Agrostis tenuis* and *Festuca rubra*) resulted in the almost complete disappearance of Zn sulfide and in the formation of a range of new Zn solid phases (dominated by a Zn-bearing phosphate and kerolite–Zn-bearing hydrotalcite) which were indicative of a massive reoxidation of the dredge sediment. However, it is not clear whether this is a rhizosphere effect or the result of a bulk change in the structure of the sediment possibly combined with plant-induced temporal changes in the sediment's water content which ultimately favored its oxidation.

7.7.4. Impact of Redox Potential and pO_2 Gradients on the Bioavailability of Trace Elements

The impact of iron plaque formation as a result of the leakage of O_2 from lowland rice roots on the bioavailability of As to rice plants has been studied extensively given its potential consequence on human health in As-contaminated regions of Southeast Asia (Liu et al., 2004a,b; Meharg, 2004; Chen et al., 2005). Evaluating the uptake of As (both arsenate and arsenite) by rice roots with and without iron plaque, Chen et al. (2005) showed that the plaque resulted in a decreased uptake of arsenate (Figure 7.10) and, more surprisingly, in an increased uptake of arsenite. Overall, it is clear that the root-induced oxidation of the root apoplasm and of the root–soil interface results in a decreased bioavailability of As as well as of Fe and Mn. Indeed, it should be remembered that plants living in submerged soil in wetlands first have to cope with potential risks of Fe or Mn phytotoxicity. The precipitation of Fe and Mn oxides at the surface of roots as a result of O_2 leakage thus plays an essential role in decreasing the bioavailability of these metals.

In contrast to the above, for those plants living under oxic soil conditions, the availability of Fe is a major constraint given the poor solubility of Fe oxides. In this respect it is recognised that the most common strategy developed by higher plants for the acquisition of Fe is based on a combination of root-induced acidification and enzymatic Fe reduction (Marschner, 1995). The plasma membrane–bound reductase system is involved in the reduction of Fe^{3+} to Fe^{2+}, which is supposed to be the only bioavailable Fe species in all higher plants except grasses (which developed an alternative strategy based on the secretion of phytosiderophores). The enzymatic reduction of Fe in the root apoplasm is thus

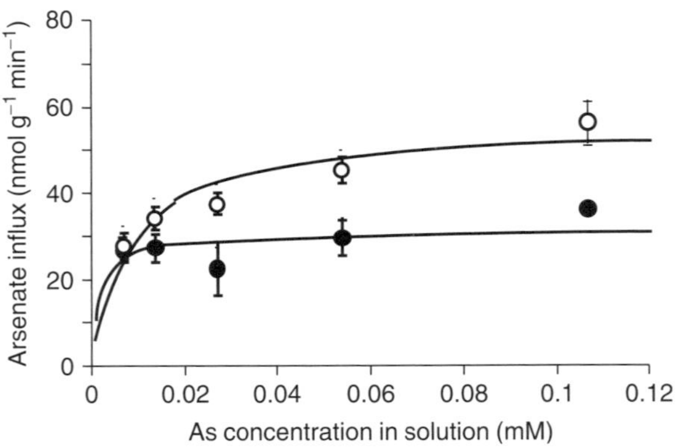

Figure 7.10. Concentration-dependent kinetics of arsenate uptake in rice (*Oryza sativa*) roots with (closed symbols) and without iron plaque (open symbols). The curves are fitted to Michaelis–Menten kinetics, with a poor fit with iron plaque. (Adapted from Chen et al., 2005.)

an essential process that governs the bioavailability of Fe. The same is likely to be true for Mn. Interestingly, the release of protons and the Fe reductase activity have been shown to be enhanced as a response to Fe deficiency, suggesting an efficient adaptative strategy. However, Cornu et al. (2007) showed that Fe deficiency in tomato (*Lycopersicon esculentum*) significantly increased the bioavailability of other metals, such as Cu in Cu-contaminated vineyard soils, although this study did not reveal more reducing or acidic conditions in the rhizosphere of Fe-deficient tomato plants than in Fe-sufficient plants. It is thus not possible to identify which process was responsible for the increase in Cu bioavailability observed in the rhizosphere of Fe-deficient tomato plantlets. Additional studies are clearly needed to verify to what extent root-induced redox processes are involved in changes of trace element bioavailability in the rhizosphere of higher plants.

7.8. ORGANIC SUBSTANCE GRADIENTS AT THE SOIL–ROOT INTERFACE

Organic compounds are released in the soil from decomposing organic tissues and by all the organisms contributing to the rhizosphere: plant roots, mycorrhizas, free-living fungi, bacteria, algae, and microfauna. As such, these substances are ubiquitous at the soil–root interface, although their abundance and production vary considerably in time and space. Because of the high reactivity of organic substances, they will contribute to almost every biogeochemical reaction operating in the rhizosphere. In this context, our objectives are to document gradients of organic substances at the soil–root interface, the nature and origin of the substances involved, and their influence on the speciation and bioavailability of heavy metals and metalloids.

7.8.1. Gradients of Organic Substances

The release by plants of organic C in the rhizosphere, rhizodeposition, represents a significant proportion of the net photosynthetic C. Being strongly dependent on plant species, this C flux generally varies from 20 to 30% of the net photosynthetic C of herbaceous plant species (Nguyen, 2003) and can reach up to 70% for tree species such as Douglas fir (Marschner, 1995; Warembourg et al., 2003). Microorganisms not only consume the C compounds produced by other organisms, they also produce and release organic substances, notably low-molecular-mass organic acids (LMMOAs), and thus add to the total flux of organic C in the rhizosphere. The main components of rhizodeposition are LMMOAs, sugars and polysaccharides (mucilages), proteins, enzymes, and sloughed-off cells (Table 7.1). The LMMOAs are a group of compounds that include carboxylic acids (citric, malic, etc.), amino acids (cysteine, glutamine, etc.), and phenolic acids (caffeic, etc.). They bear functional groups that once dissociated, allow LMMOAs to complex cations and to compete with anions on positively charged soil mineral surfaces. Mucilages are high-molecular-mass and gelatinous organic compounds secreted by rhizosphere microbes and root cap or epidermal cells that protect the root apex from desiccation, play an important role in maintaining soil–root contact, and in soil aggregation, protect the root against toxicity of metals such as Al and intervene in ion uptake (e.g., phosphate). Mixtures of soil materials and microbes with root mucilage and/or microbial exopolysaccharides are termed *mucigel*. The sloughed-off cells and decomposing root tissues do not represent root exudates but nonetheless, contribute significantly to rhizodeposition and therefore to the total input of organic substances in the rhizosphere.

The amount of organic C rhizodeposited in soils varies in space with vegetation types and plant species and in time as a function of plant growth stage and age (Hegde and Fletcher, 1996). Exudation is known further to follow a diurnal pattern characterized by larger depositional fluxes during the day and by changes in the composition of the deposits between day and night (Melnitchouck et al., 2005). The amount and composition of rhizodeposits vary further along roots, from basal zones to the apex where mucilages dominate, with exudates being released primarily in the zone of elongation behind the root tip (Marschner et al., 2004). In grasses, *phytosiderophores*, which are amino acids involved in the complexation of Fe and other heavy metals, are released in discrete time periods and root zones (Marschner, 1995; Hinsinger, 1998): the secretion of phytosiderophore occurs behind the root tips and follows a diurnal pattern with a distinct peak between 2 and 6 hours after the onset of light. Specialized roots such as cluster roots of white lupin or proteoid roots of *Proteaceae* species are sites of elevated rates of exudation of carboxylates and phenolics and show a temporal pattern of exudation. They exhibit such high rates only for a few days at maturity, while younger or senescent cluster–proteoid roots show lower rates (Neumann and Martinoia, 2002). Rhizodeposition is also strongly influenced by exogenous conditions and stresses (Bertin et al., 2003; Veneklaas et al., 2003;

TABLE 7.1. Some Organic Compounds Present in Root Exudates of Various Plant Species

Class of Compounds	Organic Substances
Aliphatic acids	Acetic, acetonic, butyric, citric, formic, fumaric, gluconic, glycolic, lactic malic, malonic, oxalic, phtalic, propionic, pyruvic, shikimic, succinic, tartaric, and valeric acid
Amino acids	Aminobutyric acid, asparagine, cystathionine, cystein, cystine, histidine, homoserine, glycine, leucine, lysine, methionine, mugineic acid, phytosiderophores, proteinogenic amino acids, proline, serine, valine
Aromatic acids	Caffeic, *p*-coumeric, ferulic, gallic, gentisic, *p*-hydroxybenzoic, protocatechuic, salicylic, sinapic, and syringic acid
Carbohydrates	Arabinose, desoxyribose, fructose, galactose, glucose, maltose, raffinose, rhamnose, ribose, sucrose and xylose
Enzymes	Amylase, invertase, phosphatase, protease
Fatty acids	Linoleic, linolenic, oleic, palmitic, and stearic acid
Phenols	Anthocyanins, flavanol, flavones, flavanones, and isoflavonoids
Sterols	Campestrol, cholesterol, sitosterol, and stigmasterol

Source: Modified from Bertin et al. (2003) and Dakora and Phillips (2002).

Gherardi and Rengel, 2004). Nutrient deficiency in the soil, mechanical constraints, and limited water supply all stimulate the release of exudates by plants (Nguyen, 2003).

Decreasing concentration gradients in organic substances from the root surface toward the bulk soil have been found for both trees and agricultural plants. Shen et al. (1996) measured higher levels of LMMOAs in the rhizosphere of beech (*Fagus sylvatica* L.) than in bulk soil. Lactic, acetic, and formic acids dominated the solution, with citric and oxalic acids being less abundant. The amounts of malonic, fumaric, and succinic acids were found to be higher in the rhizosphere of dwarf bamboo (*Yushania niitakeyamensis*) than in the bulk soil (Chen et al., 2001). A total of 12 and 20 organic components were identified in the rhizosphere of Norway spruce (*Picea abies* L. Karst.) and silver birch (*Betula pendula* Roth), respectively (Sandnes et al., 2005). In all cases, concentrations of organic acids (butyric, propionic, acetic) were generally low and the mean values were higher in the rhizosphere of birch than of spruce. Yet, as noted by Van Hees et al. (2003) for oxalate, the flux of LMMOAs through the soil is probably much faster than is suggested by the measurement of instantaneous concentration values. For crop plants such as durum wheat (*Triticum turgidum* var. *durum*), acetic and succinic acids accounted for 84 to 93% of LMMOAs in water extracts of rhizosphere soil (Cieślinski, et al., 1998). By contrast, no water-extractable LMMOAs were measured in the bulk soil extracts. The rhizosphere of rice was enriched in amino acids (e.g., histidine, valine, glycine) compared to control cultures (Bacilio-Jimenez et al., 2003). Finally, gradients of ectoenzymes such as acid phosphatase were reported by Tarafdar and Jungk (1987) for wheat and rape.

Jones and Brassington (1998) calculated that over 80% of the organic anions (citrate, malate, oxalate) entering the solution was sorbed to the soil within 10 minutes, They suggested that sorption greatly reduced the capacity of LMMOAs to mobilize nutrients from the rhizosphere. Not only are these substances sorbed, but they are also strongly degraded, with the rate of biodegradation increasing in the rhizosphere, where microbial counts are higher. Consequently, LMMOAs are typically present at low concentrations in the soil solution (<50 μM) and have hourly residence times due to continuous removal by microorganisms (Van Hees et al., 2005). The combination of these processes (release, sorption, and biodegradation) is expected to generate and maintain steep concentration gradients of metal-chelating organic substances, although these gradients have seldom been observed directly at the soil–root interface.

7.8.2. Functions of Organic Substances Involved in the Fate of Heavy Metals and Metalloids in the Rhizosphere

Root exudates and decomposing root tissues constitute a fundamental C substrate for microbes living in the rhizosphere, and, through stimulating microbial activity tend to increase the turnover rate of organic C at the soil–root interface (Helal and Sauerbeck, 1984). Rhizodeposited substances also contain macronutrients such as N, P, and S that become available for plant uptake on decomposition of the organic structures. As for trace elements and heavy metals, the sustained input of organic substances at the soil–root interface strongly affects their solubility, speciation, and mobility and thus, their bioavailability. The principal processes involved in the rhizosphere are the increased dissolution of solids, with subsequent release of heavy metals into solution, the complexation of dissolved metallic ions by organic ligands, and the desorption by ligand exchange of anionic species of metals and metalloids bound to soil exchange sites by dissolved organic anions (Jones, 1998).

7.8.2.1. Impact of Organic Substances on the Speciation of Trace Elements in the Rhizosphere There is ample evidence in the literature demonstrating that root or microbial exudates can enhance the weathering of solid phases and the dissolution of associated macronutrients such as P (Ström et al., 2002), as demonstrated for ectomycorrhizal roots or fungi (Casarin et al., 2004; Hofland et al., 2004). In comparison, this has been much less documented for metal cations, especially heavy metals (Rengel, 2002). A substantial proportion of recent research in this field has been dedicated to elucidating the implications of microbial and plant siderophores in the dissolution of Fe oxides (Inoue et al., 1993; Kraemer et al., 1999; Kraemer, 2004; Reichard et al., 2005). These have shown that such Fe-complexing compounds can efficiently dissolve Fe from poorly soluble Fe oxides, including goethite, which is a ubiquitous crystalline oxyhydroxide in soils. Reichard et al. (2005) have shown that the mechanism involved a ligand-promoted dissolution in the case of mugineic acid (phytosiderophore) and even reported a synergistic effect of oxalate and this phytosiderophore. There are fewer reports dealing with the potential impact of siderophores on the fate of metals other than Fe itself, although Fe oxides are known as major metal-bearing phases in soils. For instance, Kraemer et al. (1999) showed that above pH 6.5, a major proportion of Pb was complexed by the microbial siderophore desferrioxamine, which also resulted in decreased adsorption of Pb onto goethite.

For phytosiderophores, published stability constants of complex formation with a range of metals suggest that they might have a major impact on the speciation of metals such as Zn, Ni, and Cu. In the case of Cu^{2+}, the stability constant is even greater than that for Fe^{3+}, and it has been shown that the concentration of free Cu^{2+} ion in solution is much decreased in the presence of phytosiderophores (Figure 7.11) (Thomas et al., 2005). Malinowski et al. (2004) also reported that Cu was bound to extracellular root exudates, as shown by lower dissolved Cu^{2+} activity in the rhizosphere of tall fescue (*Festuca arundinacea* Schreb.) grown in a nutrient solution. The binding of Cu^{2+} by exudates was increased by P deficiency-induced released of exudates in the soil solution. This suggests that exudates other than phytosiderophores were involved in the binding of Cu^{2+}: These could be carboxylates, as they are known to be exuded in greater amounts under P-deficient conditions. The complexation of Mn^{2+} by malate was also reported to increase its mobility and bioavailability in the rhizosphere by preventing Mn oxidation in an uncontaminated soil (Jauregui and Reisenauer, 1982). Further evidence for the complexation of free heavy metal ions by dissolved organic substances in the rhizosphere have been provided by Hamon et al. (1995) (Figure 7.12), Lorenz et al. (1997), and Legrand et al. (2005).

Beyond their impact on the dissolution of metal-bearing mineral phases and on the complexation of metals, organic substances released by roots and microbes can also have an impact on the sorption or desorption of metals at the surface of mineral particles. On the one hand, organic substances released as exudates can compete with sorbed anionic species of metals (e.g., chromate) and metalloids (arsenate) and displace them from the exchange complex of soils to the rhizosphere solution. On the other hand, soluble exudates can enhance metal sorption through the formation of organometallic complexes, notably in the lower pH range (Neubauer et al., 2002). This interaction could, however, reduce the sorption of metals in the mildly acidic to alkaline pH range (as mentioned above in the work of Kraemer et al., 1999) through the development of repulsive forces between negatively charged soil particles and dissolved organometallic complexes.

7.8.2.2. Impact of Organic Substances on the Bioavailability of Trace Elements in the Rhizosphere The potential of plants to create chemical conditions assuring the accelerated dissolution of micronutrient metals is best exemplified by plants such as *Banksia* trees and white lupin (*Lupinus albus* L.) that can produce proteoid/cluster roots exuding large amounts of organic, notably carboxylic, acids (Dinkelaker et al., 1995). Indicators of micronutrient availability such as DTPA (diethylenetriaminepentaacetic acid)-extractable Fe, Mn, and Zn were found to

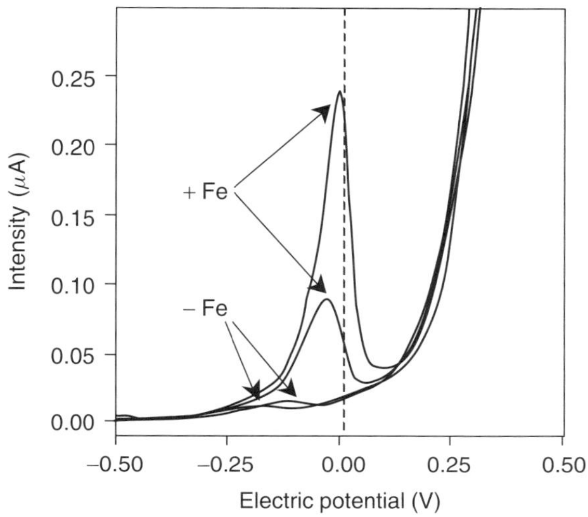

Figure 7.11. DPASV (differential pulse anodic stripping voltametry) polarograms obtained for the exudate solutions obtained from either Fe-deficient (−Fe) or Fe-sufficient (+ Fe) ryegrass (*Lolium perenne*), at two levels of addition of Cu (0.14 and 0.71 μM). The dashed line indicates the electric potential corresponding to the labile Cu species (presumably dominated by Cu^{2+}) that was obtained in the Cu solution without exudates (0.011 V). The shift of the peak toward more negative electric potential values and lower intensities is indicative of increasing complexation of Cu with decreasing Cu concentrations and increasing phytosiderophore secretion as a result of Fe deficiency. (Adapted from Thomas et al., 2005.)

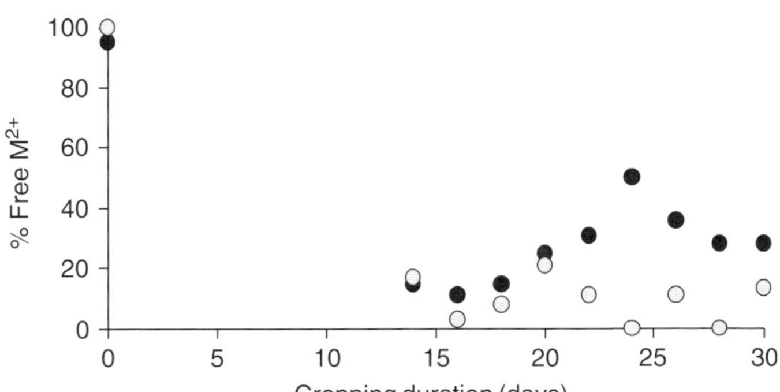

Figure 7.12. Temporal evolution of the speciation of Cd (closed symbols) and Zn (open symbols) in the soil solution collected in pots grown with radish (*Raphanus sativus*). The proportion of free metal ion (free M^{2+}) which is close to 100% before plant growth is considerably decreased as a consequence of the complexation of Cd and Zn by root exudates. (Adapted from Hamon et al., 1995.)

be five- to eightfold higher in the rhizosphere zone of cluster roots of white lupin than in the bulk soil (Dinkelaker et al., 1989). Treeby et al., (1989) conducted an experiment that clearly showed mobilization of the micronutrients Cu, Fe, Mn, and Zn by the root exudates of graminaceous species grown in a noncontaminated soil. For these plants, Fe deficiency increased the release of phytosiderophores and ultimately resulted in the increased bioavailability of Cu, Mn, and Zn. This was studied further by Chaignon et al. (2002b), who observed in a calcareous soil contaminated by heavy fungicide applications that Fe deficiency of wheat resulted in a roughly threefold increase in the phytosiderophore release rate and in the bioavailability of Cu and Zn (Figure 7.13). The dependence of metal release on the nutritional status of the plant was also revealed by the response of wheat to Zn deficiency (Rengel et al., 1998). Deficiency in Mn also appears to trigger the same effect, although to a lesser extent (Jauregui and Reisenauer, 1982). Whether the dissolved metals are actually taken up by the plants that provoked their release is still a matter of debate and depends largely on the plant species and soil chemical and physical properties. The case of phytosiderophores in graminaceous plant species is nonetheless a good example of where it has been established that the roots possess a transport system involved in the uptake of the Fe^{3+}_{-} and Zn^{2+}_{-} phytosiderophore complexes, a fact that points to the weakness of the free ion activity model (Thomas et al., 2005). In contrast, organic compounds released by roots and involved in the complexation of phytotoxic metals have also been reported to reduce the contamination of the plants. The preferential binding of Pb by exudates was indeed seen as a factor contributing to restrict its uptake by roots (Marschner, 1995). The complexation of metals by exudates may ultimately reduce the toxicity of metals to roots as proposed for Al binding by LMMOAs (Jones and Darrah, 1994). This point has been extensively studied in the case of Al detoxification by malate or citrate (Ma et al., 2001; Ryan et al., 2001). The complexation of metals by organic substances in the rhizosphere can thus have an entire range of effects and either increase or decrease their bioavailability and phytotoxicity.

7.9. CONCLUSIONS AND FUTURE DIRECTIONS

Considerable advances have been made in recent years in an understanding of the mobility of heavy metals and metalloids in the soil, while many of the processes and pathways involved in the acquisition and translocation of such elements in higher plants have been elucidated, down to a molecular level. In contrast, we should recognize that we are still at the eve of fully understanding the processes controlling the mobility and bioavailability of heavy metals and metalloids at the soil–root interface. Although this area of research has developed considerably over recent years, as revealed by the growing number of publications on the topic, unraveling the biogeochemistry of trace elements in the rhizosphere still constitutes a challenge for environmental scientists.

This chapter has shown that this unique interface is of prime interest for understanding the fate of trace elements in the soil–plant continuum and in

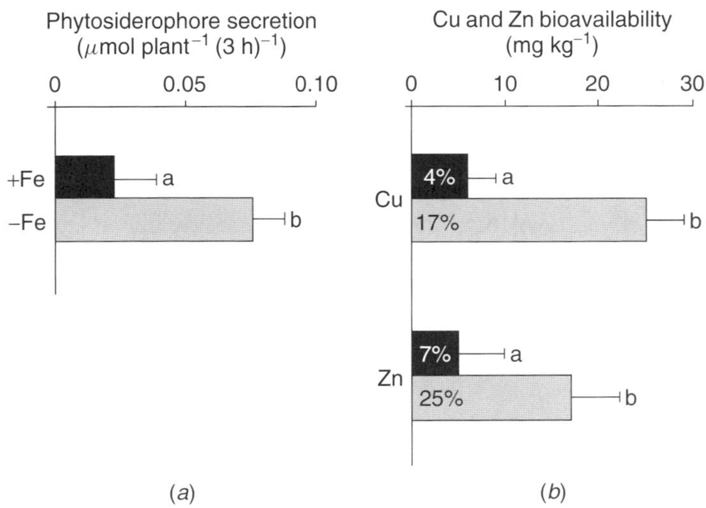

Figure 7.13. (*a*) Rate of phytosiderophore secretion from roots of Fe-deficient (−Fe, gray bars) or Fe-sufficient (+ Fe, black bars) wheat (*Triticum aestivum* L.), (*b*) bioavailability of Cu and Zn as estimated by the amounts of Cu and Zn taken up by 3-week-old wheat plants after 8 days of growth in a calcareous vineyard soil contaminated by fungicide applications (total soil Cu and Zn amounting to 150 and 70 mg kg^{-1}, respectively). The percentages refer to the fraction of total soil Cu and Zn taken up by wheat from the rhizosphere, which appears to be significantly increased, in a manner similar to phytosiderophore secretion, as a response to Fe deficiency. (Adapted from Chaignon et al., 2002b.)

terrestrial ecosystems. The dynamics of the rhizosphere is indeed conducive to the development of a range of chemical gradients that have complex temporal dynamics and spatial patterns. These comprise gradients in trace element concentrations, pH, pCO_2, pO_2, redox potential, and organic ligand concentrations which govern the speciation and ultimately the mobility and bioavailability of heavy metals and metalloids at the soil–root interface. Although much progress is currently being made by obtaining data at an ever-increasing spatial resolution due to the advent of a range of cutting-edge methods in analytical chemistry and molecular biology, a major challenge will be to achieve the adequate upscaling of observations, processes, and interactions to environmentally relevant scales for the management of contaminated sites. In this respect, numerical models coupling soil biogeochemical reactions, transport functions, plantborne processes, and root development will be invaluable tools to further our understanding of the complex interactions that occur in the rhizosphere, which ultimately govern the transfer of trace elements from the soil to the plant.

Although the present chapter focused on the direct role of plant roots, it has to be borne in mind that rhizosphere microorganisms are key players in the biogeochemistry of trace elements in the rhizosphere. These organisms are addressed in Chapters 2 and 9. However, another of our challenges for the future will be to develop a more holistic approach to studying rhizospheric interactions by integrating advances in microbial and plant ecology and physiology as well as in soil biogeochemistry in order to better predict and eventually manage the mobility and bioavailability of heavy metals and metalloids at the soil–root interface.

Acknowledgments

We wish sincerely to thank our graduate students and research associates for their pivotal contribution to this work: V. Chaignon, M. Clairotte, B. Cloutier-Hurteau, J. Y. Cornu, P. Legrand, A. Michaud, V. Séguin, and M.-C. Turmel.

REFERENCES

Al-Najar, H., Schulz, R., and Römheld, V. (2003). Plant availability of thallium in the rhizosphere of hyperaccumulator plants: a key factor for assessment of phytoextraction. *Plant Soil* **249**, 97–105.

Arocena, J. M., and Glowa, K. R. (2000). Mineral weathering in ectomycorrhizosphere of subalpine fir (*Abies lasiocarpa* (Hook.) Nutt.) as revealed by soil solution composition. *For. Ecol. Manag.* **132**, 61–70.

Arocena, J. M., Göttlein, A., and Raidl, S. (2004). Spatial changes of soil solution and mineral composition in the rhizosphere of Norway-spruce seedlings colonized by *Piloderma croceum*. *J. Plant Nutr. Soil Sci.* **167**, 479–486.

Assadian, N., and Fenn, L. B. (2001). Rhizosphere chemical changes enhance heavy metal absorption by plants growing in calcareous soils. In *Trace Elements in the Rhizosphere*, ed. Gobran, G. R., Wenzel, W. W., and Lombi, E., CRC Press, Boca Raton, FL 43–60.

Bacha, R. E. and Hossner, L. R. (1977). Characteristics of coatings formed on rice roots as affected by iron and manganese additions. *Soil Sci. Soc. Am. J.* **41**, 931–935.

Bacilio-Jimenez, M., Aguilar-Flores, S., Ventura-Zapata, E., Peres-Campos, E., Bouquelet, S., and Zeteno, E. (2003). Chemical characterization of root exudates from rice (*Oryza sativa*) and their effects on the chemotactic response of endophytic bacteria. *Plant Soil* **249**, 271–277.

Bakker, M. R., George, E., Turpault, M.-P., Zhang, J. L., and Zeller, B. (2004). Impact of Douglas-fir and Scots pine seedlings on plagioclase weathering under acidic conditions. *Plant Soil* **266**, 247–259.

Barber, S. A. (1995). *Soil Nutrient Bioavailability: A Mechanistic Approach*, 2nd ed., Wiley, New York.

Begg, C. B. M., Kirk, G. J. D., MacKenzie, A. F., and Neue, H.-U. (1994). Root-induced iron oxidation and pH changes in the lowland rice rhizosphere. *New Phytol.* **128**, 469–477.

Bertin, C., Yang, X., and Weston, L. A. (2003). The role of root exudates and allelochemicals in the rhizosphere. *Plant Soil* **256**, 67–83.

Bidel, L. P. R., Renault, P., Pagès, L., and Rivière, L. M. (2000). Mapping meristem respiration of *Prunus persica* (L.,) Batsch seedlings: potential respiration of the meristems, O_2 diffusional constraints and combined effects on root growth. *J. Exp. Bot.* **51**, 755–768.

Blute, N. K., Brabander, D. J., Hemond, H. F., Sutton, S., Newville, M. G., and Rivers, M. L. (2004). Arsenic sequestration by ferric iron plaque on cattail roots. *Environ. Sci. Technol.* **38**, 6074–6077.

Bontidean, I., Mortaria, A., Letha, S., Brown, N. L., Karlson, U., Larsenc, M. M., Vangronsveld, J., Corbisier, P., and Csöregi, E. (2004). Biosensors for detection of mercury in contaminated soils. *Environ. Poll.* **131**, 255–262.

Brown, J. C., and Ambler, J. E. (1973). "Reductants" released by roots of Fe-deficient soybeans. *Agron. J.* **65**, 311–314.

Caldwell, M. M., Dawson, T. E., and Richards, J. H. (1998). Hydraulic lift: consequences of water efflux from the roots of plants. *Oecologia* **113**, 151–161.

Canadell, J., Jackson, R. B., Ehleringer, J. R., Mooney, H. A., Sala, O. E., and Schulze, E. D. (1996). Maximum rooting depth of vegetation types at the global scale. *Oecologia* **108**, 583–595.

Casarin, V., Plassard, C., Souche, G., and Arvieu, J. C. (2003). Quantification of oxalate ions and protons released by ectomycorrhizal fungi in rhizosphere soil. *Agronomie* **23**, 461–469.

Casarin, V., Plassard, C., Hinsinger, P., and Arvieu, J. C. (2004). Quantification of ectomycorrhizal fungal effects on the bioavailability and mobilisation of soil P in the rhizosphere of *Pinus pinaster. New Phytol.* **163**, 177–185.

Chaignon, V., and Hinsinger, P. (2003). A biotest for evaluating copper bioavailability to plants in a contaminated soil. *J. Environ. Qual.* **32**, 824–833.

Chaignon, V., Bedin, F., and Hinsinger, P. (2002a). Copper bioavailability and rhizosphere pH changes as affected by nitrogen supply for tomato and oilseed rape cropped on an acidic and a calcareous soil. *Plant Soil* **243**, 219–228.

Chaignon, V., Di Malta, D., and Hinsinger, P. (2002b). Fe-deficiency increases Cu acquisition by wheat cropped in a Cu-contaminated, vineyard soil. *New Phytol.* **154**, 121–130.

Chaignon, V., Quesnoit, M., and Hinsinger, P. (2007). Rhizosphere pH, bioavailability and extractability of Cu in a Cu-contaminated acidic soil as affected by liming. *Environ. Poll.* (submitted).

Chen, C. C., Dixon, J. B., and Turner, F. T. (1980). Iron coatings on rice roots: mineralogy and quantity influencing factors. *Soil Sci. Soc. Am. J.* **44**, 635–639.

Chen, M.-C., Wang, M.-K., Chiu, C. Y., Huang, P. M., and King, H. B. (2001). Determination of low-molecular weight dicarboxylic acids and organic functional groups in the rhizosphere and bulk soil of *Tsuga* and *Yushania* in a temperate rain forest. *Plant Soil* **231**, 37–44.

Chen, Z., Zhu, Y. G., Liu, W. J., and Meharg, A. A. (2005). Direct evidence showing the effect of root surface iron plaque on arsenite and arsenate uptake into rice (*Oryza sativa*) roots. *New Phytol.* **165**, 91–97.

Cherrey, A., Chaignon, V., and Hinsinger, P. (1999). Bioavailability of copper in the rhizosphere of rape and ryegrass cropped in vineyard soils. In *Proc. 5th International Conference on the Biogeochemistry of Trace Elements*, ed. Wenzel, W. W., Adriano, D. C., Alloway, B., Doner, H. E., Keller, C., Lepp, N. W., Mench, M., Naidu, R., and Pierzynski, G. M., International Society for Trace Element Research, Vienna, Austria, 196–197.

Cieśliński, G., van Rees, K. C. J., Szmigielska, A. M., Krishnamurti, G. S. R., and Huang, P. M. (1998). Low-molecular-weight organic acids in the rhizosphere soils of durum wheat and their effect on cadmium bioaccumulation. *Plant Soil* **203**, 109–117.

Cornu, J. Y., Staunton, S., and Hinsinger P. (2007). Copper concentration in plants and in the rhizosphere as influenced by iron status of tomato (*Lycopersicon esculentum* L.). *Plant Soil* **292**, 63–77.

Courchesne, F., and Gobran, G. R. (1997). Mineralogical variations of bulk and rhizosphere soils from a Norway spruce stand, southwestern Sweden. *Soil Sci. Soc. Am. J.* **61**, 1245–1249.

Courchesne, F., Séguin, V., and Dufresne, A. (2001). Solid phase fractionation of metals in the rhizosphere of forest soils. In *Trace Elements in the Rhizosphere*, ed. Gobran, G. R., Wenzel, W. W., and Lombi, E., CRC Press, Boca Raton, FL, 189–206.

Courchesne, F., Kruyts, N., and Legrand, P. (2006). Labile zinc concentration and free copper ion activity in the rhizosphere of forest soils. *Environ. Toxicol. Chem.* **25**, 635–642.

Dakora, F. D., and Phillips, D. A. (2002). Root exudates as mediators of mineral acquisition in low-nutrient environments. *Plant Soil* **245**, 35–47.

Dieffenbach, A., and Matzner, E. (2000). In situ soil solution chemistry in the rhizosphere of mature Norway spruce (*Picea abies* (L.) Karst.) trees. *Plant Soil* **222**, 149–161.

Dinkelaker, B., Römheld, V., and Marschner, H. (1989). Citric-acid excretion and precipitation of calcium citrate in the rhizosphere of white lupin (*Lupinus albus* L.). *Plant Cell Environ.* **12**, 285–292.

Dinkelaker, B., Hahn, G., and Marschner, H. (1993). Non-destructive methods for demonstrating chemical changes in the rhizosphere, II: Application of methods. *Plant Soil* **155–156**, 73–76.

Dinkelaker, B., Hengeler, C., and Marschner, H. (1995). Distribution and function of proteoid roots and other root clusters. *Bot. Acta* **108**, 183–200.

Doussan, C., Pagès, L., and Pierret, A. (2003). Soil exploration and resource acquisition by plant roots: an architectural and modelling point of view. *Agronomie* **23**, 419–431.

Dunbabin, V. M., McDermott, S., and Bengough, A. G. (2006). Upscaling from rhizosphere to whole root system: modeling the effects of phospholipid surfactants on water and nutrient uptake. *Plant Soil* **283**, 57–72.

Fahey, T. J., and Hughes, J. W. (1994). Fine root dynamics in a northern hardwood forest ecosystem, Hubbard Brook Experimental Forest, NH. *J. Ecol.* **82**, 533–548.

Fischer, W. R., Flessa, H., and Schaller, G. (1989). pH values and redox potentials in microsites of the rhizosphere. *Z. Pflanzenernachr. Bodenkd.* **152**, 191–195.

Flessa, H., and Fischer, W. R. (1992). Plant-induced changes in the redox potentials of rice rhizospheres. *Plant Soil* **143**, 55–60.

Gherardi, M. J., and Rengel, Z. (2004). The effect of manganese supply on carboxylates by roots of lucerne (*Medicago sativa*). *Plant Soil* **260**, 271–282.

Gobran, G. R., and Clegg, S. A. (1996). Conceptual model for nutrient availability in the mineral soil–root system. *Can. J. Soil Sci.* **76**, 125–132.

Gollany, H. T., Schumacher, T. E., Rue, R. R., and Liu, S.-Y. (1993). A carbon dioxide microelectrode for in situ pCO_2 measurement. *Microchem. J.* **48**, 42–49.

Göttlein, A., Heim, A., and Matzner, E. (1999). Mobilization of aluminum in the rhizosphere soil solution of growing tree roots in an acidic soil. *Plant Soil* **211**, 41–49.

Griffiths, R. P., Baham, J. E., and Caldwell, B. A. (1994). Soil solution chemistry of ectomycorrhizal mats in forest soil. *Soil Biol. Biochem.* **26**, 331–337.

Hammer, D., and Keller, C. (2002). Changes in the rhizosphere of metal-accumulating plants evidenced by chemical extractants. *J. Environ. Qual.* **31**, 1561–1569.

Hamon, R. E., Lorenz, S. E., Holm, P. E., Christensen, T. H., and McGrath, S. P. (1995). Changes in trace metal species and other components of the rhizosphere during growth of radish. *Plant Cell Environ.* **18**, 749–756.

Hamza, M. A., and Aylmore, L. A. G. (1992). Soil solute concentration and water uptake by single lupin and radish plant roots, I: Water extraction and solute accumulation. *Plant Soil* **145**, 187–196.

Hansel, C. M., La Force, M. J., Fendorf, S., and Sutton, S. (2002). Spatial and temporal association of As and Fe species on aquatic plant roots. *Environ. Sci. Technol.* **36**, 1988–1994.

Harmsen, J., Rulkens, W., and Eijsackers, H. (2005). Bioavailability: concept for understanding or tool for predicting? *Land Contam. Reclam.* **13**, 161–171.

Harter, R. D., and Naidu, R. (2001). An assessment of environmental and solution parameter impact on trace-metal sorption by soils. *Soil Sci. Soc. Am. J.* **65**, 597–612.

Hegde, R. S., and Fletcher, J. S. (1996). Influence of plant growth stage and season on the release of root phenolics by mulberry as related to the development of phytoremediation technology. *Chemosphere* **32**, 2471–2479.

Helal, H. M., and Sauerbeck, D. (1984). Effects of plant roots on C and P metabolism in soil. *Plant Soil* **76**, 175–182.

Hiltner, L. (1904). Über neuere Erfahrungen und Probleme auf dem Gebiete der Bodenbakteriologie unter besonderer Berücksichtigung der Gründüngung und Brache. *Arb. Dtsch. Landwirtsch. Ges.* **98**, 59–78.

Hinsinger, P. (1998). How do plant roots acquire mineral nutrients? Chemical processes involved in the rhizosphere. *Adv. Agron.* **64**, 225–265.

Hinsinger, P. (2001). Bioavailability of trace elements as related to root-induced chemical changes in the rhizosphere. In *Trace Elements in the Rhizosphere*, ed. Gobran, G. R., Wenzel, W. W., and Lombi, E., CRC Press, Boca Raton, FL, 25–41.

Hinsinger, P. (2004). Nutrient availability and transport in the rhizosphere. In *Encyclopedia of Plant and Crop Science*, ed. Goodman, R. M., Marcel Dekker, New York, 1094–1097.

Hinsinger, P., and Jaillard, B. (1993). Root-induced release of interlayer potassium and vermiculitization of phlogopite as related to potassium depletion in the rhizosphere of ryegrass. *J. Soil Sci.* **44**, 525–534.

Hinsinger, P., Jaillard, B., and Dufey, J. E. (1992). Rapid weathering of a trioctahedral micas by the roots of ryegrass. *Soil Sci. Soc. Am. J.* **56**, 977–982.

Hinsinger, P., Plassard, C., Tang, C., and Jaillard, B. (2003). Origins of root-induced pH changes in the rhizosphere and their responses to environmental constraints: a review. *Plant Soil* **248**, 43–59.

Hinsinger, P., Gobran, G. R., Gregory, P. J., and Wenzel, W. W. (2005). Rhizosphere geometry and heterogeneity arising from root-mediated physical and chemical processes. *New Phytol.* **168**, 293–303.

Hodge, A. (2004). The plastic plant: root responses to heterogeneous supplies of nutrients. *New Phytol.* **162**, 9–24.

Hoffland, E., Kuyper, T. W., Wallander, H., Plassard, C., Gorbushina, A., Haselwandter, K., Holmström, S., Landeweert, R., Lundström, U. S., Rosling, A., Sen, R., Smits, M.

M., van Hees, P. A. W., and van Breemen, N. (2004). The role of fungi in weathering. *Front. Ecol. Environ.* **2**, 258–264.

Inoue, K., Hiradate, S., and Takagi, S. (1993). Interaction of mugineic acid with synthetically produced iron-oxides. *Soil Sci. Soc. Am. J.* **57**, 1254–1260.

Jaillard, B., Schneider, A., Mollier, A., and Pellerin, S. (2000). Modélisation du prélèvement minéral par les plantes fondée sur le fonctionnement bio-physico-chimique de la rhizosphère. In *Fonctionnement des peuplements végétaux sous contrainte environnementale*, INRA, Paris, 253–287.

Jauregui, M. A., and Reisenauer, H. M. (1982). Dissolution of oxides of manganese and iron by root exudate components. *Soil Sci. Soc. Am. J.* **46**, 314–317.

Jones, D. L. (1998). Organic acids in the rhizosphere: a critical review. *Plant Soil* **205**, 25–44.

Jones, D. L., and Brassington, D. S. (1998). Sorption of organic acids in acid soils and its implication in the rhizosphere. *Eur. J. Soil Sci.* **49**, 447–455.

Jones, D. L., and Darrah, P. R. (1994). The role of root-derived organic acids in the mobilization of nutrients from the rhizosphere. *Plant Soil* **166**, 247–257.

Jungk, A. (2002). Dynamics of nutrient movement at the soil–root interface. In *Plant Roots: The Hidden Half*, 3rd ed., Marcel Dekker, New York, 587–616.

Kirk, G. J. D., and Bajita, J. B. (1995). Root-induced iron oxidation, pH changes and zinc solubilization in the rhizosphere of lowland rice. *New Phytol.* **131**, 129–137.

Kirk, G. J. D., and Le Van Du (1997). Changes in rice root architecture, porosity, and oxygen and proton release under phosphorus deficiency. *New Phytol.* **135**, 191–200.

Knight, B., Zhao, F. J., McGrath, S. P., and Shen, Z. G. (1997). Zinc and cadmium uptake by the hyperaccumulator *Thlaspi caerulescens* in contaminated soils and its effects on the concentrations and chemical speciation of metals in soil solution. *Plant Soil* **197**, 71–78.

Kot, A., and Namiesnik, J. (2000). The role of speciation in analytical chemistry. *Trends Anal. Chem.* **19**, 69–79.

Kraemer, S. M. (2004). Iron oxide dissolution and solubility in the presence of siderophores. *Aquat. Sci.* **66**, 3–18.

Kraemer, S. M., Cheah, S. F., Zapf, R., Raymond, K. N., and Sposito, G. (1999). Effect of hydroxamate siderophores on Fe release and Pb(II) adsorption by goethite. *Geochim. Cosmochim. Acta* **63**, 3003–3008.

Krishnamurti, G. S. R., and Naidu, R. (2002). Solid-solution speciation and phytoavailability of copper and zinc in soils. *Environ. Sci. Technol.* **36**, 2645–2651.

Legrand, P., Turmel, M.-C., Sauvé, S., and Courchesne, F. (2005). Speciation and bioavailability of trace metals (Cd, Cu, Ni, Pb, Zn) in the rhizosphere of contaminated soils. In *Biogeochemistry of Trace Elements in the Rhizosphere*, ed. Huang, P. M., and Gobran, G. R., Elsevier, New York, 261–299.

Leyval, C., and Berthelin, J. (1991). Weathering of micas by roots and rhizospheric microorganisms of Pine. *Soil Sci. Soc. Am. J.* **55**, 1009–1016.

Li, X.-L., George, E., and Marschner, H. (1991). Extension of the phosphorus depletion zone in VA-mycorrhizal white clover in a calcareous soil. *Plant Soil* **136**, 41–48.

Lichtenegger, E., and Kutschera-Mitter, L. (1991). Spatial root types. In *Plant Roots and Their Environment*, ed. McMichael, B. L., and Persson, H., Elsevier Science, Amsterdam, The Netherlands, 359–365.

Liebersbach, H., Steingrobe, B., and Claassen, N. (2004). Roots regulate ion transport in the rhizosphere to counteract reduced mobility in dry soil. *Plant Soil* **260**, 79–88.

Lin, Q., Chen, Y. X., He, Y. F., and Tian, G. M. (2004). Root-induced changes of lead availability in the rhizosphere of *Oryza sativa* L. *Agric. Ecosyst. Environ.* **104**, 605–613.

Liu, W. J., Zhu, Y. G., Smith, F. A., and Smith, S. E. (2004a). Do iron plaque and genotypes affect arsenate uptake and translocation by rice seedlings (*Oryza sativa* L.) grown in solution culture? *J. Exp. Bot.* **55**, 1707–1713.

Liu, W. J., Zhu, Y. G., Smith, F. A., and Smith, S. E. (2004b). Do phosphorus nutrition and iron plaque alter arsenate (As) uptake by rice seedlings in hydroponic culture? *New Phytol.* **162**, 481–488.

Loeppert, R. H., White, N., Biswas, B., and Drees, R. (2005). Soil mineralogical control on As dynamics in rice paddy soils. In *Proc. 8th International Conference on the Biogeochemistry of Trace Elements*, Adelaide, Australia, 650–651.

Loosemore, N., Straczek, A., Hinsinger, P., and Jaillard, B. (2004). Zinc mobilisation from a contaminated soil by three genotypes of tobacco as affected by soil and rhizosphere pH. *Plant Soil* **260**, 19–32.

Lorenz, S. E., Hamon, R. E., and McGrath, S. P. (1994). Differences between soil solutions obtained from rhizosphere and non-rhizosphere soils by water displacement and soil centrifugation. *Eur. J. Soil Sci.* **45**, 431–438.

Lorenz, S. E., Hamon, R. E., Holm, P. E., Domingues, H. C., Sequeira, E. M., Christensen, T. H., and McGrath, S. P. (1997). Cadmium and zinc in plants and soil solutions from contaminated soils. *Plant Soil* **189**, 21–31.

Ma, J. F., Ryan, P. R., and Delhaize, E. (2001). Aluminium tolerance in plants and the complexing role of organic acids. *Trends Plant Sci.* **6**, 273–278.

Malinowski, D. P., Zuo, H., Beleski, D. P., and Alloush, G. A. (2004). Evidence for copper binding by extracellular root exudates of tall fescue but not perennial ryegrass infected with *Neotyphodium* spp. Endophytes. *Plant Soil* **267**, 1–12.

Malzer, G. L., and Barber, S. A. (1975). Precipitation of calcium and strontium sulfates around plant roots and its evaluation. *Soil Sci. Soc. Am. Proc.* **39**, 492–495.

Marschner, H. (1991). Root-induced changes in the availability of micronutrients in the rhizosphere. In *Plant Roots: The Hidden Half*, ed. Waisel, Y., Eshel, A., and Kafkafi, U., Marcel Dekker, New York, 503–528.

Marschner, H. (1995). *Mineral Nutrition of Higher Plants*, 2nd ed., Academic Press, London.

Marschner, H., and Römheld, V. (1996). Root-induced changes in the availability of micronutrients in the rhizosphere. In *Plant Roots: The Hidden Half*, 2nd ed., ed. Waisel, Y., Eshel, A., and Kafkafi, U., Marcel Dekker, New York, 557–579.

Marschner, P., Crowley, D., and Yang, C. H. (2004). Development of specific rhizosphere bacterial communities in relation to plants species, nutrition and soil type. *Plant Soil* **261**, 199–208.

Martin, R. R., Naftel, S., Macfie, S., Skinner, W., Courchesne, F., and Séguin, V. (2004). Time of flight secondary ion mass spectrometry studies on the distribution of metals between the soil, rhizosphere and roots of *Populus tremuloides* Minchx growing in forest soil. *Chemosphere* **54**, 1121–1125.

Matocha, C. J., Sparks, D. L., Amonette, J. E., and Kukkadapu, R. K. (2001). Kinetics and mechanism of birnessite reduction by catechol. *Soil Sci. Soc. Am. J.* **65**, 58–66.

McGrath, S. P., Shen, Z. G., and Zhao, F. J. (1997). Heavy metal uptake and chemical changes in the rhizosphere of *Thlaspi caerulescens* and *Thlaspi ochroleucum* grown in contaminated soils. *Plant Soil* **188**, 153–159.

McLaughlin, M. J., Zarcinas, B. A., Stevens, D. P., and Cook, N. (2000). Soil testing for heavy metals. *Commun. Soil Sci. Plant Anal.* **31**, 1661–1700.

Meharg, A. A. (2004). Arsenic in rice: understanding a new disaster for South-East Asia. *Trends Plant Sci.* **9**, 415–417.

Melnitchouck, A., Leinweber, P., Eckhardt, K.-U., and Beese, R. (2005). Qualitative differences between day- and night-time rhizodeposition in maize (*Zea mays* L.) as investigated by pyrolysis-field ionization mass spectrometry. *Soil Biol. Biochem.* **37**, 155–162.

Naftel, S. J., Martin, R. R., Courchesne, F., Séguin, V., and Protz, R. (2002). Studies of the effects of soil biota on metal biovailability. *Can. J. Anal. Sci. Spectr.* **47**, 36–40.

Neubauer, U., Furrer, G., and Schulin, R. (2002). Heavy metal sorption on soil minerals affected by the siderophore desferrioxamine b: the role Fe(III) (hydr)oxides and dissolved Fe(III). *Eur. J. Soil Sci.* **53**, 45–55.

Neumann, G., and Martinoia, E. (2002). Cluster roots: an underground adaptation for survival in extreme environments. *Trends Plant Sci.* **7**, 162–167.

Nolan, A. L., Lombi, E., and McLaughlin, M. J. (2003). Metal bioaccumulation and toxicity in soils: Why bother with speciation? *Aust. J. Chem.* **56**, 77–91.

Nguyen, C. (2003). Rhizodeposition of organic C by plant: mechanisms and controls. *Agronomie* **23**, 375–396.

Nye, P. H. (1986). Acid-base changes in the rhizosphere. In *Advances in Plant Nutrition*, Vol. 2, ed. Tinker, B., and Läuchli, A., Praeger Scientific, New York, 129–153.

Onyatta, J. O., and Huang, P. M. (2003). Kinetics of cadmium release from selected tropical soils from Kenya by low-molecular-weight organic acids. *Soil Sci.* **168**, 234–252.

Panfili, F., Manceau, A., Sarret, G., Spadini, L., Bert, V., Kirpichtchikova, T., Matthew, M., and Ahamdach, N. (2005). Evolution of Zn speciation induced by phytostabilization in a contaminated sediment, using scanning electron microscopy, x-ray fluorescence, EXAFS spectroscopy, and principal component analysis. *Geochim. Cosmochim. Acta* **69**, 2265–2284.

Parker, D. R., Pedler, J. F., Thomason, D. N., and Li, H. (1998). Alleviation of copper rhizotoxicity by calcium and magnesium at defined free metal-ion. *Soil Sci. Soc. Am. J.* **62**, 965–972.

Pennisi, E. (2004). The secret life of fungi. *Science* **304**, 1620–1622.

Pierret, A., Moran, C. J., and Pankhurst, C. E. (1999). Differentiation of soil properties related to the spatial association of wheat roots and soil macropores. *Plant Soil* **211**, 51–58.

Pierzyinski, G. (2003). Trace element chemistry, contamination, and ecotoxicity. In *Proc. 7th International Conference on the Biogeochemistry of Trace Elements University of Uppsala Ed.*, Uppsala, Sweden, June 16–19, 2003.

Price, A. G. (1984). Measurement and variability of physical properties and soil water distribution in a forest podzol. *J. Hydrol.* **161**, 347–364.

Quaghebeur, M., and Rengel, Z. (2004). Phosphate and arsenate interactions in the rhizosphere of canola (*Brassica napus*). *Funct. Plant Biol.* **31**, 1085–1094.

Rasse, D. P., and Smucker, J. M. (1998). Root recolonization of previous root channels in corn and alfalfa rotations. *Plant Soil* **204**, 203–212.

Reichard, P. U., Kraemer, S. M., Frazier, S. W., and Kretzschmar, R. (2005). Goethite dissolution in the presence of phytosiderophores: rates, mechanisms, and the synergistic effect of oxalate. *Plant Soil* **276**, 115–132.

Renella, G., Landi, L., and Nannipieri, P. (2004). Degradation of low molecular weight organic acids complexed with heavy metals in soils. *Geoderma* **122**, 311–315.

Rengel, Z. (2002). Genetic control of root exudation. *Plant Soil* **245**, 59–70.

Rengel, Z., and Marschner, P. (2005). Nutrient availability and management in the rhizosphere: exploiting genotypic differences. *New Phytol.* **168**, 305–312.

Rengel, Z., Römheld, V., and Marschner, H. (1998). Uptake of zinc and iron by wheat genotypes differing in zinc efficiency. *J. Plant Physiol.* **152**, 433–438.

Retallack, G. J. (1997). Early forest soils and their role in Devonian global change. *Science* **276**, 583–585.

Robert, M., and Berthelin, J. (1986). Role of biological and biochemical factors in soil mineral weathering. In *Interactions of Soil Minerals with Natural Organics and Microbes*, ed. Huang, P. M., and Schnitzer, M., Spec. Publ. 17, Soil Science Society of America, Madison, WI, 453–495.

Robinson, D. (1994). The responses of plants to non-uniform supplies of nutrients. *New Phytol.* **127**, 635–674.

Römer, W., Kang Dong Kyu, Egle, K., Gerke, J. and Keller, H. (2000). The acquisition of cadmium by *Lupinus albus* L., *Lupinus angustifolius* L., and *Lolium multiflorum* Lam. *J. Plant Nutr. Soil Sci.* **163**, 623–628.

Römheld, V. (1991). The role of phytosiderepohores in acquisition of iron and other micronutrients in graminaceous species: an ecological approach. *Plant Soil* **130**, 127–134.

Römheld, V., and Marschner, H. (1983). Mechanism of iron uptake by peanut plants, I: Fe(III) reduction, chelate splitting, and release of phenolics. *Plant Physiol.* **71**, 949-954.

Russell, R. S. (1982). *Plant Root Systems*, McGraw-Hill, New York.

Ryan, P. R., Delhaize, E., and Jones, D. L. (2001). Function and mechanism of organic anion exudation from plant roots. *Annu. Rev. Plant Physiol. Plant Mol. Biol.* **52**, 527–560.

Sandnes, A., Eldhuset, T. D., and Wollebaek, G. (2005). Organic acids in root exudates and soil solution of Norway spruce and silver birch. *Soil Biol. Biochem.* **37**, 259–269.

Sauvé S., McBride M. B., Norvell, W. A., and Hendershot, W. H. (1997). Copper solubility and speciation of in situ contaminated soils: effects of copper level, pH and organic matter. *Water Air Soil Pollut.* **100**, 133–149.

Schenk, H. J., and Jackson, R. B. (2002). The global biogeography of roots. *Ecol. Monogr.* **72**, 311–328.

Schöttelndreier, M., and Falkengren-Grerup, U. (1999). Plant induced alteration in the rhizosphere and the utilisation of soil heterogeneity. *Plant Soil* **209**, 297–309.

Séguin, V., Gagnon, C., and Courchesne, F. (2004). Changes in water extractable metals, pH and organic carbon concentrations at the soil–root interface of forested soils. *Plant Soil* **260**, 1–17.

Shen, Y., Ström, L., Jönsson, J.-Å., and Tyler, G., (1996). Low-molecular organic acids in the rhizosphere soil solution of beech forest (*Fagus sylvatica* L.) Cambisols determines by ion chromatogaphy using supported liquid-membrane enrichment technique. *Soil Biol. Biochem.* **28**, 1163–1169.

Shenker, M., Hadar, Y., and Chen, Y. (1999). Kinetics of iron complexing and metal exchange in solutions by rhizoferrin, a fungal siderophore. *Soil Sci. Soc. Am. J.* **63**, 1681–1687.

Shomer, I., Novacky, A. J., Pike, S. M., Yermiyahu, U., and Kinraide, T. B. (2003). Electrical potentials of plant cell walls in response to the ionic environment. *Plant Physiol.* **133**, 411–422.

Sinha, B. K., and Singh, N. T. (1974). Effect of transpiration rate on salt accumulation around corn roots in a saline soil. *Agron. J.* **66**, 557–560.

Skinner, W. M., Martin, R. R., Naftel, S. J., Macfie, S., Courchesne, F., and Séguin, V. (2005). Multi-technique studies of the distribution of metals between the soil, rhizosphere and roots of *Populus tremuloides* growing in forest soil. In *Proc. 8th International Conference on the Biogeochemistry of Trace Elements*, ed. McLaughlin, M., Naidu, R., and Lepp, N., Adelaide, Australia, 482–483.

Smith, W. H., and Pooley, A. S. (1989). Red spruce rhizosphere dynamics: spatial distribution of aluminum and zinc in the near-root soil zone. *For. Sci.* **35**, 1114–1119.

Smith, S. E., and Read, D. J. (1997). *Mycorrhizal Symbiosis*, Academic Press, San Diego, CA.

Soil Science Society of America (1997). *Glossary of Soil Science Terms*, SSI, Madison, WI.

Sparks, D. L. (2001). Elucidating the fundamental chemistry of soils: past and recent achievements and future frontiers. *Geoderma* **100**, 303–319.

Sterckeman, T., Perriguey, J., Cael, M., Schwartz, C., and Morel, J. L. (2004). Applying a mechanistic model to cadmium uptake by *Zea mays* and *Thlaspi caerulescens*: consequences for the assessment of the soil quantity and capacity factors. *Plant Soil* **262**, 289–302.

Stewart, J. B., Moran, C. J., and Wood, J. T. (1999). Macropore sheath: quantification of plant root and soil macropore association. *Plant Soil* **211**, 59–67.

Ström, L., Owen, A. G., Godbold, D. L., and Jones, D. L. (2001). Organic acid behaviour in a calcareous soil: sorption reactions and biodegradation rates. *Soil Biol. Biochem.* **33**, 2125–2133.

Ström, L., Owen, A. G., Godbold, D. L., and Jones, D. L. (2002). Organic acid P mobilization in the rhizosphere and uptake by maize roots. *Soil Biol. Biochem.* **34**, 703–710.

Tarafdar, J. C., and Jungk, A. (1987). Phosphatase activity in the rhizosphere and its relation to the depletion of soil organic phosphorus. *Biol. Fertil. Soils* **3**, 199–204.

Thomas, S., Mahammedi, D., Clairotte, M., Benedetti, M. F., Castrec-Rouelle, M., Persin, F., Peu, P., Martinez, J., and Hinsinger, P. (2005). Bioavailability and extractability of copper and zinc in a soil amended with pig slurry: effect of iron deficiency in the rhizosphere of two grasses. In *Biogeochemistry of Trace Elements in the Rhizosphere*, ed. Huang, P. M., and Gobran, G. R., Elsevier, Amsterdam, The Netherlands, 337–363.

Tinker, B. T., and Nye, P. H. (2000). *Solute Movement in the Rhizosphere*, Oxford University Press, Oxford.

Tom-Petersen, A., Hosbond, C., and Nybroe, O. (2001). Identification of copper-induced genes in *Pseudomonas fluorescens* and use of a reporter strain to monitor bioavailable copper in soil. *FEMS Microbiol. Ecol.* **38**, 59–67.

Treeby, M., Marschner, H., and Römheld, V. (1989). Mobilization of iron and other micronutrient cations from a calcareous soil by plant-borne microbial and synthetic metal chelator. *Plant Soil* **114**, 217–226.

Trolldenier, G. (1988). Visualization of oxidizing power of rice roots and of possible participation of bacteria in iron deposition. *Z. Pflanzenernachr. Bodenkd.* **151**, 117–121.

Tye, A. M., Young, S., Crout, N. M. J., Zhang, H., Preston, S., Zhao, F. J., and McGrath, S. P. (2004). Speciation and solubility of Cu, Ni and Pb in contaminated soils. *Eur. J. Soil Sci.* **55**, 579–590.

Van Hees, P. A. W., Godbold, D. L., Jentschke, G., and Jones, D. L. (2003). Impact of ectomycorrhizas on the concentration and biodegradation of simple organic acids in a forest soil. *Eur. J. Soil Sci.* **54**, 697–706.

Van Hees, P. A. W., Jones, D. L., Finlay, R., Godbold, D. L., and Lundström, U. S. (2005). The carbon we do not see: the impact of low molecular weight compounds on carbon dynamics and respiration in forest soils — a review. *Soil Biol. Biochem.* **37**, 1–13.

Veneklaas, E. J., Stevens, J., Cawthray, G. R., Turner, S., Grigg, A. M., and Lambers, H. (2003). Chickpea and white lupin rhizosphere carboxylates vary with soil properties and enhance phosphorus uptake. *Plant Soil* **248**, 187–197.

Vogt, K. A., and Bloomfield, J. (1991). Tree roots turnover and senescence. In *Plant Roots: The Hidden Half*, ed. Waisel, Y., Eshel, A., and Kafkafi, U., Marcel Dekker, New York, 287–306.

Wang, Z., Shan, X., and Zhang, S. (2002). Comparison between fractionation and bioavailability of trace elements in rhizosphere and bulk soils. *Chemosphere* **46**, 1163–1171.

Wang, W.-S., Shan, X.-Q., Wen, B., and Zhang, S.-Z. (2003). Relationships between the extractable metals in soils and metals taken up by maize roots and shoots. *Chemosphere* **53**, 523–530.

Warembourg, F. R., Roumet, C., and Lafont, F. (2003). Differences in rhizosphere carbon-partitioning among plant species of different families. *Plant Soil* **256**, 347–357.

Wenzel, W. W., Wieshammer, G., Fitz, W. J., and Puschenreiter, M. (2001). Novel rhizobox design to assess rhizosphere characteristics at high spatial resolution. *Plant Soil* **237**, 37–45.

Whiting, S. N., Leake, J. R., McGrath, S. P., and Baker, A. J. M. (2000). Positive responses to Zn and Cd by roots of the Zn and Cd hyperaccumulator *Thlaspi caerulescens*. *New Phytol.* **145**, 199–210.

Whiting, S. N., Broadley, M. R., and White, P. J. (2003). Applying a solute transfer model to phytoextraction: zinc acquisition by *Thlaspi caerulescens*. *Plant Soil* **249**, 45–56

Zhang, H., Zhao, F.-J., Sun, B., Davison, W., and McGrath, S. P. (2001). A new method to measure effective soil solution concentration predicts copper availability to plants. *Environ. Sci. Technol.* **35**, 2602–2607.

Zobel, R. W. (1991). Genetic control of root systems. In *Plant Roots: The Hidden Half*, ed. Waisel, Y., Eshel, A., and Kafkafi, U., Marcel Dekker, New York, 27–38.

8

BIOGEOCHEMICAL PROCESSES CONTROLLING THE CYCLING OF ARSENIC IN SOILS AND SEDIMENTS

S. FENDORF

Stanford University, Stanford, California

M. J. HERBEL

BioScreen Testing Services, Torrance, California

K. J. TUFANO AND B. D. KOCAR

Stanford University, Stanford, California

Biophysico-Chemical Processes of Heavy Metals and Metalloids in Soil Environments,
Edited by Antonio Violante, Pan Ming Huang, and Geoffrey Michael Gadd

8.1. INTRODUCTION

Arsenic is a naturally occurring trace element that poses a human health risk when incorporated into food or water supplies. The greatest risk imposed by arsenic results from contamination of drinking water, for which the World Health Organization recommends a maximum limit of 10 $\mu g\ L^{-1}$. Continued ingestion of drinking water that has hazardous levels of arsenic can lead to arsenicosis and cancers of the bladder, skin, lungs, and kidneys (Mandal and Suzuki, 2002). Unfortunately, arsenic-tainted drinking waters are a global threat (Mandal and Suzuki, 2002; Smedley and Kinniburgh, 2002) and presently are having a devastating impact on human health in Asia. In Bangladesh alone, in 2003, nearly 60 million people were consuming drinking water with arsenic concentrations exceeding the World Health Organization's recommended limit (Yu et al., 2003).

Arsenic contamination of the environment often results from human activities such as mining or pesticide application, but recently, natural sources of arsenic have demonstrated a devastating impact on water quality. Arsenic becomes problematic from a health perspective principally when it partitions into the aqueous rather than the solid phase. Processes favoring the partitioning of As into the aqueous phase, leading potentially to hazardous concentrations, vary extensively but can broadly be grouped into four categories: (1) ion displacement, (2) desorption (or limited sorption) at pH values > 8.5, (3) reduction of arsenate to arsenite, and (4) reductive dissolution of Fe and Mn (hydr)oxides (Cullen and Reimer, 1989; Smith et al., 1998; Smedley and Kinniburgh, 2002). Although various processes may liberate arsenic from solids, a transition from aerobic to anaerobic conditions, and commensurate arsenic and iron–manganese reduction, appears to be a dominant global means by which high concentrations of dissolved arsenic are generated.

Dissolved concentrations, and the resulting mobility, of arsenic within soils and sediments are the combined result of biogeochemical processes linked to hydrologic factors. Within the subsequent sections of this chapter, we explore and describe the biological and chemical processes that control the partitioning of arsenic between the solid and aqueous phases, with a particular emphasis on the impacts of redox transformations of arsenic and iron in terrestrial and aquatic environments.

8.2. AQUEOUS CHEMISTRY OF ARSENIC

Two oxidation states of arsenic, As(V) and As(III), predominate in surface and near-surface environments. In solution, arsenic exists primarily as oxyanions; arsenate [As(V) as H_3AsO_4] has pK_a values of 2.2, 6.9, and 11.5, while arsenite [As(III) as H_3AsO_3] has pK_a values of 9.2, 12.1, and 13.4 (Smith et al., 1998; Goldberg and Johnston, 2001). Thus, at circumneutral pH, $H_2AsO_4^-$, $HAsO_4{}^{2-}$, and $H_3AsO_3^0$ species dominate. Plant and microbial activity may methylate As(V) or As(III), forming, for example, dimethylarsenic acid (DMAA) and monomethylarsonous acid (MMAA) (Cullen and Reimer, 1989). However, methylated species are usually not abundant in aqueous solutions compared to inorganic forms of arsenic (Smith et al., 1998; Smedley and Kinniburgh, 2002) Thio- (Helz et al., 1995; Rochette et al., 2000; Wilkin et al., 2003) and carbonato- (Kim et al., 2000; Lee and Nriagu, 2003) complexes of arsenic also exist within natural waters; thiolated forms of arsenic may, in fact, represent an important reactive component within sulfidic environments (Wilkin et al., 2003).

8.3. ARSENIC RETENTION IN SOILS AND SEDIMENTS

8.3.1. Adsorption Processes

Partitioning of arsenic onto soil solids is dependent, foremost, on its oxidation state. In general, As(V) binds extensively and strongly to most mineral constituents of soils and sediments, while As(III) retention is more convoluted and dependent on specific soil chemical conditions (Table 8.1). As a consequence of arsenate's strong retention, arsenic tends to have a limited impact on environmental quality in aerobic soils except under extreme pH conditions ($pH > 8.5$). Surface complexes of arsenate on iron and aluminum oxides, examined using both infrared (Lumsdon et al., 1984; Sun and Doner, 1996) and extended x-ray absorption fine structure (EXAFS) spectroscopy (Waychunas et al., 1993; Manceau, 1995; Fendorf et al., 1997; Arai et al., 2001; Sherman and Randall, 2003), are dominated by bidentate binuclear (double-corner sharing) moieties. The dominant surface complex of arsenate is consistent with infrared studies of phosphate on iron (hydr)oxides (Parfitt et al., 1975; Arai and Sparks, 2001), a factor supporting the analogous strong retention of phosphate.

Aluminum hydroxides and aluminosilicate clay minerals may also retain appreciable concentrations of arsenate, and they exhibit a strong preference for arsenate relative to arsenite (Tables 8.1 and 8.2) (Xu et al., 1988; Manning and Goldberg, 1997a,b; Smith et al., 1998). Similarly, Mn oxides may impart a strong influence on arsenic binding. Reaction of arsenite solutions with Mn oxides such as birnessite results in extensive and rapid uptake (Oscarson et al., 1981). However, arsenic is retained as arsenate surface complexes (Manning et al., 2002), owing to arsenic oxidation by Mn(III/IV) (Oscarson et al., 1981; Driehaus et al., 1995; Manning et al., 2002).

Arsenic may also bind to organic matter in soils and sediments, (Thanabalsingam and Pickering, 1986a; Grafe et al., 2001, 2002; Redman et al., 2002; Ko et al., 2004) with As(V) and As(III) having maximum adsorption on humic acids at pH 5.5 and 8.0, respectively (Thanabalsingam and Pickering, 1986a). Arsenic(V) adsorbs onto solid-phase humic acids more extensively than As(III), with amine (NH_2) groups suspected as the primary functional group responsible for arsenic retention (Thanabalsingam and Pickering, 1986a). Arsenic adsorption by humic substances is also enhanced by cation addition, particularly Fe, Al, and Mn, whereby the cations act as bridging complexes for arsenate on humic acids (Lin et al., 2004). Nevertheless, compared to Fe, Al, or P, organic matter tends to be poorly correlated with total As (Chen et al., 2002), suggesting that its contribution to arsenic retention in soils and sediments is limited.

TABLE 8.1. Retention Maximum for Arsenic on Various Solids Common to Soils and Sediments Derived from Adsorption Isotherms at Fixed pH

Adsorbent	As(V) (mmol Kg^{-1})	pH	As(III) (mmol Kg^{-1})	pH	Reference
Al oxides					
Gibbsite	35	4.0	—	—	Approximate (Hingston et al., 1971)
	15	9.0	—	—	
Amorphorous Al hydroxide	1500	4.0	—	—	Hingston et al. (1976)
	600	9.0	—	—	
	1600	5.0	—	—	Anderson et al. (1976)
	1200	7.0			
	500	9.0			
Activated alumina	Two-fold higher for As(V) than for As(III)				Gosh and Yuan (1987)
	67	6–7	14	6.5–8.5	Gosh and Yuan (1987)
Bauxite	52	6–7	16	6.5–8.5	Gosh and Yuan (1987)
Aluminosilicates					
Montmorillonite	8	5.0	3	5.0	Using landfill leachate (Frost and Griffin, 1977)
Kaolinite	7	5.0	1	5.0	Frost and Griffin, (1977)

Fe (hydr)oxides					
Hydrous ferric oxide	3514	4.0	2675	8.0	Dixit and Hering (2003)
Goethite	173	4.0	173	8.0	Dixit and Hering (2003)
Magnetite			332	8.0	Dixit and Hering (2003)
2-line ferrihydrite	2000	4.6	$\geq$ 6000	4.6	Raven et al. (1998)
	1500	9.2	$\geq$ 6000	9.2	
2-line ferrihydrite on quartz sand	483	7.1	1206	7.1	This study
Others					
Birnessite (δ-MnO_2)	100	—	—	—	Lenoble et al. (2004)
Pyrolusite	10	6.5	—	—	Thanabalsingam and Pickering (1986b)
Cryptomelane	25	6.5	—	—	Oscarson et al. (1983)
Calcite ($CaCO_3$)	N.D.[a]	—	N.D.	—	Goldberg and Glaubig (1988)
Activated carbon	10	3–4	—	—	Gupta and Chen (1978)
Humic acids	90–110	5.5	—	5.5	Thanabalsingam and Pickering (1986a)

[a]N.D., not detected.

TABLE 8.2. Adsorption Envelopes for As(V) and As(III) on Various Soil Solids

Adsorbent	$As(V)_{aq}$[a] $As(V)_{ads}$[b]	pH[c] Maximum	$As(III)_{aq}$[a] $As(III)_{ads}$[b]	pH[c] Maximum	Reference
Al oxides					
Amorphous Al hydroxide	1600; 1700 (4.5)	4–7	—	—	Anderson et al. (1976)
	133; 900 (4.5)	4–7	—	—	
	20; 20 (2–10)	2–10	20; 16 (8.5)	7–9.5	Goldberg (2002)
Activated alumina	53.4; 26.5 (3–7)	3–7	26; 11 (8.2)	3–10	Gupta and Chen (1978)
Bauxite	53.4; 26.5 (3–7)	3–7	16; 9 (8.5)	3–10	Gupta and Chen (1978)
Aliminosilicates					
Montmorillonite	20; 0.41 (5, 12.5)	5–7, >11	—	—	Goldberg and Glaubig (1988)
	20; 0.35 (5)	5–7	20; 0.4 (3)	3–4	Goldberg (2002)
Kaolinite	20; 0.5 (5)	3–9	20; 0.25 (8–10)	7–11	Goldberg (2002)
Illite	20; 0.5 (4–6)	3–7	20; 0.22 (8–9)	7–10	Goldberg (2002)

Fe (hydr)oxides					
Hydrous ferric oxide	20; 40 (2–9)100; 2100 (4.0)	2–10 < 7	20; 40 (2–11)	2–11	Goldberg (2002); Dixit and Hering (2003)
Hydrous ferric oxide	—	—	100; 1500 (8–9)	5–9.5	Dixit and Hering (2003)
Goethite	100; 140 (4.0)	<8	100; 120 (8–9)	4–10	Dixit and Hering (2003)
Magnetite	—	—	100; 140 (9.0)	8–9.5	Dixit and Hering (2003)
2-line ferrihydrite					
Others					
Hydrous manganese oxide	10; 16 (3.0)	2.5–5	—	—	Thanabalsingam and Pickering (1986b)
Calcite ($CaCO_3$)	20; 2.0 (11)	9–13	—	—	Goldberg and Glaubig (1988)
Activated carbon	19.4; 5.3 (4)	3–5	—	—	Gupta and Chen (1978)
Humic Acids	5–100; 80–140 (6)	5–7	20; 110 (8)	7–9	Thanabalsingam and Pickering (1986a)

[a]Initial aqueous concentration (μM).
[b]Maximum adsorption (mmol kg^{-1} at pH indicated).
[c]pH range over which maximum adsorption occurred.

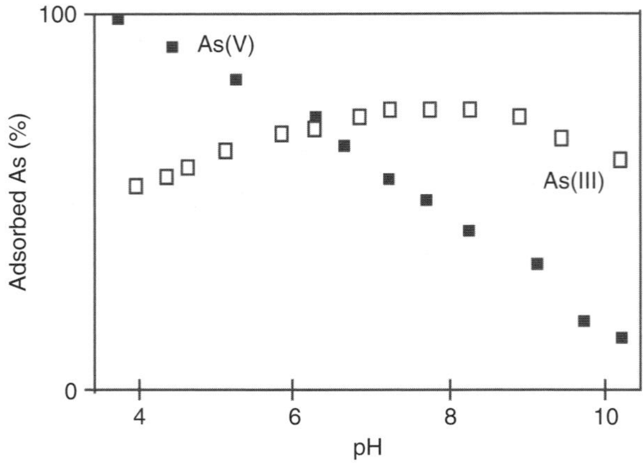

Figure 8.1. Magnitude and pH dependency of As(V) (arsenate) and As(III) (arsenite) adsorption on hydrous ferric oxide; reaction conditions were 50 μM arsenic reacted with 0.03 g L^{-1} hydrous ferric oxide. (Data from Dixit and Hering, 2003.)

In contrast to arsenate, arsenite exhibits a limited affinity for most soil minerals, with the exception of iron (hydr)oxides, for which it exhibits a high degree of retention (see Table 8.1). Ferric (hydr)oxides and magnetite, in fact, adsorb arsenite more extensively than arsenate at all but acidic conditions (Raven et al., 1998; Dixit and Hering, 2003), as illustrated for hydrous ferric hydroxide (Figure 8.1). Similar to arsenate, arsenite forms a bidentate binuclear complex, albeit with a slightly longer As–Fe distance, on goethite (α-FeOOH) (Manning et al., 1998; Ona-Nguema et al., 2005) and lepidocrocite (β-FeOOH) (Ona-Nguema et al., 2005); on ferrihydrite and hematite, arsenite again forms bidentate binuclear complexes but is also present in bidentate, mononuclear, and to a small degree, monodentate coordination on the mineral surface (Ona-Nguema et al., 2005). Despite the multitude of potential surface complexes on ferric (hydr)oxides, the observation of inner-sphere arsenite moieties appears to account for its extensive retention on such phases (Dixit and Hering, 2003). However, surface complexes of arsenite, although extensive, are far more labile than for its oxidized counterpart, arsenate (as described below, Section 8.4.2); adsorption and desorption processes of arsenic, in general, are described in Chapter 5.

8.3.2. Precipitation of Arsenic Phases

Both As(V) and As(III) may precipitate within soils and sediments, but the constituents inducing precipitation vary dramatically. Arsenate, similar to phosphate, tends to precipitate with hard multivalent cations such as aluminum and ferric iron under acidic conditions and calcium and magnesium under alkaline conditions; arsenate may also replace SO_4^{2-} or, in particular, PO_4^{3-} in minerals due to similar size and charge characteristics (Smedley and Kinniburgh, 2002). Various heavy metals (e.g., Mn, Cd, Pb) and alkali earth metals (Ca, Mg, Ba) also have the capacity to precipitate with arsenate (see eg., Sadiq, 1997), although these phases tend to be relatively soluble and thus have a limited impact on dissolved concentrations of arsenic, particularly in reducing environments (Rochette et al., 1998).

In contrast to arsenate, arsenite undergoes ligand displacement and incorporation into sulfidic solids. In fact, the solubility of As(III) is often controlled by sulfide precipitates, particularly in regions where sulfidogenesis occurs, limiting As(III) concentrations in extremely reducing environments (Moore et al., 1988). Under hydrothermal conditions with high Fe^{2+}, sulfides may coprecipitate with Fe and As as arsenopyrite (FeAsS) or arsenic-rich (arsenian) pyrite [$Fe(S,As)_2$]; at lower levels of ferrous iron, orpiment (As_2S_3) or realgar (AsS) may form. Dissimilatory SO_4^{2-} and As(V) reduction may similarly induce precipitation of orpiment and realgar under ambient surface conditions (Newman et al., 1997a; Ahmann et al., 1994). Furthermore, adsorption of As(III) onto metal sulfides is often followed by incorporation of As into mineral structures (Bostick and Fendorf, 2003; Bostick et al., 2003).

8.4. DESORPTION OF ARSENIC IN SOILS AND SEDIMENTS

In surface and subsurface environments, changes in water chemistry often result in release of As from solid phases through various desorption pathways. Introduction of waters containing high concentrations of competing ligands such as phosphate or waters that lead to dramatic shifts in pH can displace arsenic. The most destructive alteration from an environmental quality perspective results from incursion of dissolved labile organic carbon that stimulates microbial activity, leading to changes in arsenic speciation and/or enhanced dissolution of Fe(III) and Mn(III/IV) minerals hosting arsenic. Transition from aerobic to anaerobic conditions induced by organic carbon and limited oxygenation, in fact, appears to be the dominant condition leading to high levels of dissolved arsenic. The converse, exposure of reduced soils/sediments to aerated water, leads to oxidative dissolution of arsenic-bearing sulfides, such as arsenopyrite, realgar, and orpiment. However, oxygenation generally leads to the production of arsenate, having a high affinity for various minerals, inclusive of ferric (hydr)oxides. As a consequence, arsenic undergoes secondary partitioning back onto the solid phase (i.e., aeration leads to a shift in solids hosting arsenic but not to a partitioning of arsenic into the aqueous phase).

8.4.1. Ion Displacement

Competitive ion displacement can represent an important means by which arsenic is released to the aqueous phase, where it is subject to transport. Displacement and mobilization of As by phosphates is of particular concern (Manning and Goldberg, 1996; Reynolds et al., 1999; Violante and Pigna, 2002; Dixit and Hering, 2003), and regions where fertilizer or pesticide runoff and leaching occur are specifically at risk for this mobilization pathway (Peryea and Kammerack, 1997; Jain and Loeppert, 2000). Carbonate can also compete with arsenic for adsorption sites on mineral surfaces (Van Geen et al., 1994; Villalobos and Leckie, 2001; Appelo et al., 2002), but the extent of arsenate displacement may not be appreciable. Natural organic matter may compete as well with As and inhibit arsenic adsorption onto iron (hydr)oxides due to competitive adsorption reactions (Xu et al., 1991; Redman et al., 2002). Other anions, such as Cl^-, SO_4^{2-}, and NO_3^-, have minimal impact on As desorption, yet these ions can contribute to ionic strength and salinization effects on As retention in soils and sediments (Gupta and Chen, 1978; Smith et al., 1998) that are potentially important in the desorption of arsenite.

8.4.2. Desorption of Arsenite Compared to Desorption of Arsenate

The conception that arsenite is a more mobile species than arsenate has been displaced during the past decade on the basis of arsenic retention to various ferric (hydr)oxides (see Figure 8.1, for example). At circumneutral pH values and higher, arsenite generally adsorbs to a greater extent on ferric (hydr)oxides than arsenate (Manning et al., 1998; Raven et al., 1998; Dixit and Hering, 2003), and forms inner-sphere complexes comparable to those of arsenate (Manning et al., 1998; Ona-Nguema et al., 2005). However, unlike arsenate, which tends to have a high affinity for numerous solids, inclusive of metal oxides and aluminosilicates, appreciable binding of arsenite appears relatively specific to ferric (hydr)oxides (Manning and Goldberg, 1997a). Thus, the expectation arises that arsenic mobilization will be predicated on iron reduction within anaerobic environments, not on arsenic reduction.

Recognizing that arsenite forms inner-sphere complexes helps to rectify its extensive binding on iron (hydr)oxides. Caution should be used, however, in drawing conclusions regarding binding strength from the magnitude of retention when considering solids with multiple operational binding mechanisms. In addition to various inner-sphere complexes (which include monodentate, bidentate–binuclear, and bidentate–mononuclear complexes), outer-sphere complexes arising from electrostatic interactions and hydrogen bonding can result. The latter moieties, unfortunately, are not obvious within x-ray absorption spectra, the chief means by which the bonding environment of arsenite has been revealed, particularly if they are weak interactions resulting in long As–Fe distances. As a consequence, arsenate and arsenite desorption need to be examined explicitly to determine the propensity for arsenic to migrate in soils and sediments.

We have examined arsenic retention on goethite, used as a common and representative ferric (hydr)oxide, and ensuing desorption under hydrodynamic conditions. Arsenic, as arsenate or arsenite, was loaded onto goethite-coated quartz sand (nominally, 10 g Fe kg^{-1} sand) at various fractions of the maximum surface coverage, which was established by conducting adsorption isotherms at pH 7 and fitting the data using the Langmuir equation. The adsorption maximum for arsenite was 0.28 mol As kg^{-1} goethite–sand, while that for arsenate was 0.18 mol kg^{-1}, results consistent with the greater adsorption of arsenite than arsenate at circumneutral pH (e.g., Dixit and Hering, 2003). Arsenic-loaded goethite-coated sands were then packed into columns and resulting desorption trends monitored upon the injection (flow) of a simulated groundwater medium buffered at pH 7.

Despite the greater extent of arsenite retention on ferric (hydr)oxides relative to arsenate, greater quantities and resulting aqueous concentrations occur for arsenite desorption (Figure 8.2). In particular, higher surface coverage (expressed as proportion of the adsorption maximum) enhances the difference between desorption from arsenite-loaded and that from arsenate-loaded columns—with a much higher proportion of arsenic being desorbed as arsenite. At surface

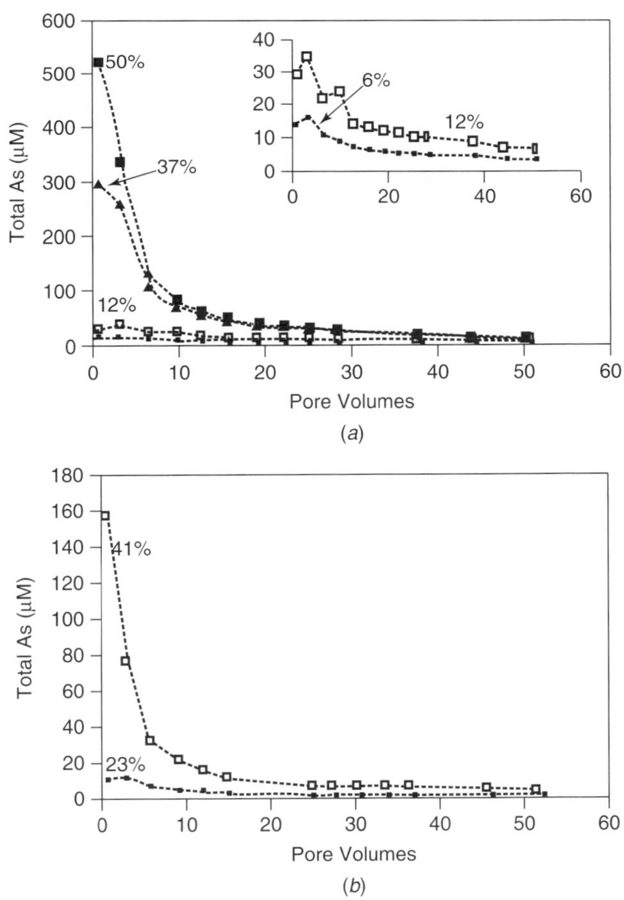

Figure 8.2. Desorption of (*a*) As(III) and (*b*) As(V) from goethite on the onset of dynamic flow conditions. Goethite-coated quartz sand was loaded with arsenic at various fractions of the adsorption maxima, as determined by batch incubations and Langmuir isotherm fits, and then subjected to continuous injection of minimal groundwater medium.

coverages near 50%, arsenite desorption results in initial aqueous concentrations exceeding 500 μM whereas for arsenate, concentrations are below 200 μM (Figure 8.2). For both arsenite and arsenate, desorption decreases exponentially with time (represented by pore volumes eluted through the column at a flow rate of ca. 3 pore volumes per day); by 20 pore volumes, both species of arsenic are below 50 μM aqueous concentration and are continuing to diminish. The rapid decline in desorption is indicative of a fixed surface population residing in a highly labile state which responds rapidly to disequilibrium. The difference in arsenite–arsenate desorption from ferrihydrite is also striking (Herbel and Fendorf, 2006). Rather than exhibiting a rapid decay in the concentration desorbing from ferrihydrite, aqueous concentrations remain greater than 70 μM even after 20 pore volumes (under conditions similar to those used for the goethite experiments described above). Furthermore, the cumulative loss of As(III) from ferrihydrite–sand columns (27.4% of the initial loading) was more than twice that of As(V) (10.8%) (Herbel and Fendorf, 2006).

Arsenate desorption from iron (hydr)oxides is measurable but limited; arsenite, by comparison, undergoes extensive release under hydrodynamic conditions. The extensive yet apparently weaker adsorption of arsenite can again be rectified simply by considering the multitude of potential surface complexes that result on mineral surfaces. Arsenite, in particular, is thought to bind on iron (hydr)oxides through multiple inner-sphere complexes (Ona-Nguema et al., 2005), having a range of binding strengths, in combination with outer-sphere and hydrogen-bonded moieties, giving rise to extensive but weak complexes. As a consequence, an appreciable fraction of arsenite appears to reside in weakly adsorbed complexes that desorb rapidly in response to decreases in aqueous concentrations. The observation that arsenite forms more labile complexes on ferric (hydr)oxides challenges the presumption that iron reduction is the primary factor liberating arsenic to the aqueous phase. Arsenic reduction, in fact, may have a pronounced role in destabilizing arsenic and allowing its transport within soils. Although iron reduction has been suggested as a means for arsenic desorption, and in fact would lead to the depletion of a prominent sorbent, arsenic reduction may be more influential. The expression of the *arrA* gene, responsible for respiratory arsenate reduction (Saltikov and Newman, 2003), may be useful as a proxy for arsenic migration (Malasarn et al., 2004), although the actual transport of arsenic will depend on a host of reactions far more extensive than just Fe(III) or As(V) reduction.

8.5. BIOGEOCHEMICAL PROCESSES

8.5.1. Arsenate Reduction: Detoxification

Because of arsenic's inherent toxicity, many organisms have evolved mechanisms to convert As to forms that are readily removed from the cell (Oremland and Stolz, 2003). Certain higher eukaryotic organisms (e.g., fungi) along with specific archaea and aerobic eubacteria can convert As(V) to As(III), which is then methylated to form less toxic species, such as MMA, DMA, or trimethylarsine (TMA) (Cullen and Reimer, 1989). The most common means of arsenic detoxification involves the ArsC enzyme system; As(V) is brought into the cytoplasm, reduction transpires, and then As(III) is pumped with a specific transporter across the cell membrane into the external milieu. Unlike respiratory reduction of As(V), the detoxification pathway requires energy. Even dissimilatory As(V) reducing microorganisms may have the ArsC enzyme system, as recently shown for *Shewanella* strain ANA-3 (Saltikov et al., 2003). Microbial As(V) detoxification mechanisms may affect the overall As speciation in contaminated environments; however, dissimilatory Fe(III) and As(V) reduction processes are more likely to mobilize arsenic within reducing environments (Jones et al., 2000; Langner and Inskeep, 2000).

Great strides in resolving the physiological and enzymatic mechanisms of As(V) uptake and reduction by As(V)-reducing organisms are presently being made. In arsenic-resistant bacteria, As(V) enters the cell via nonspecific outer membrane porins or through phosphate specific transporters (Oremland et al., 2002; Rosen, 2002; Oremland and Stolz, 2003). It can then undergo respiratory reduction in the periplasm or be transported into the cytoplasm for detoxification. The As(V)-detoxifying ArsC enzymes typically reside in the cytoplasm, and As(III) formed in the process is exported across the cell membrane using the ArsA and ArsB enzymes (Mukhopadhyay et al., 2002; Oremland and Stolz, 2003). In contrast, As(V) reductases (ArrA, ArrB, etc.) reside interior to the outer membrane, within the periplasm for gram-negative bacteria (Afkar et al., 2003; Oremland and Stolz, 2003; Saltikov and Newman, 2003).

8.5.2. Arsenate Reduction: Respiration

Over the last decade, at least 16 species of As(V)-respiring bacteria and two species of archaea have been isolated from a diversity of natural environments, including freshwater sediments, alkaline and saline lakes, and hot springs (Stolz and Oremland, 1999; Huber et al., 2000; Oremland et al., 2002; Oremland and Stolz, 2003). Dissimilatory As(V) reduction transpires in environments ranging from estuarine sediments to gastrointestinal tracts of animals, with the addition of arsenate and electron donor (Dowdle et al., 1996; Herbel et al., 2002), suggesting that As(V)-respiring microorganisms are widespread. Additionally, As-respiring bacteria are capable of reducing As(V) within solids such as scorodite (Newman et al., 1997b), with recent evidence suggesting that dissolution and subsequent uptake of As precedes reduction on the basis of the reductases residing within, rather than outside, the outer membrane (Saltikov and Newman, 2003). These prokaryotic organisms can link the reduction of soluble, adsorbed, or mineralized As(V) with the oxidation of a wide variety of electron donors, including lactate, acetate, pyruvate, glucose, and H_2, to gain energy for cell growth. To date, no obligate As(V)-respiring microbes have been identified; all of the As(V)-reducing organisms can respire on other electron acceptors inclusive of NO_3^-, Fe(III), and SO_4^{2-}. The ability to utilize a variety of electron acceptors and donors implies that the As(V)-respiring prokaryotes are opportunists and that they can thrive on other available electron acceptors in environments when As(V) concentrations become depleted.

8.5.3. Arsenic Desorption on Anaerobiosis

The greatest likelihood for As release in soils and sediments typically occurs upon transition from oxidizing to reducing conditions. Under saturated conditions, the rapid consumption of O_2 by aerobic microbes combined with the low solubility of O_2 induces anaerobic bacteria to utilize alternative electron acceptors. Arsenic may be displaced either through reduction of arsenate to arsenite or through mineralogical transformations (inclusive of dissolution) of the soil matrix.

8.5.3.1. Desorption on Metal Reduction Many bacteria and archaea can respire on Mn(III/IV) and Fe(III) oxides, leading to their dissolution, with the potential for concomitant displacement of arsenic into the aqueous phase (Cummings et al., 1999). In fact, within most soils and sediments, total As levels correlate with Fe content rather than Al or clay content (Smedley and Kinniburgh, 2002), and thus reductive dissolution–transformation of Fe(III) phases should have a major impact on arsenic. Respiratory reduction of Fe in sediments generally occurs in zones where O_2, NO_3^-, and Mn(IV) [all being oxidants of Fe(II) and alternative electron acceptors] are diminished (Lovley, 2000).

The most readily bioreducible Fe(III) (hydr)oxides are the high-surface-area least thermodynamically stable phases such as ferrihydrite (Schwertmann and Taylor, 1989; Lovley, 1991; Roden and Zachara, 1996). Reduction of Fe(III) within such phases, however, does not lead to simple congruent dissolution but rather initiates production of a convoluted assortment of secondary phases (Figure 8.3). Ferrous iron produced during Fe(III) reduction induces a transformation of ferrihydrite (or lepidocrocite) to more stable minerals, such as goethite (α-FeOOH) and magnetite (Fe_3O_4) (Zachara et al., 2002; Hansel et al., 2003). Hansel et al. (2003), for example, illustrate that dissimilatory respiration on Fe(III) within ferrihydrite results in a cascade of geochemical reactions that lead to a series of secondary solids induced by back-reaction with Fe(II) (Figure 8.3). Ferrihydrite (when supported by a quartz sand substrate) undergoes a mineralogical shift to goethite at Fe(II) loadings less than 1 mmol g^{-1}[resulting from 0.3 mM aqueous Fe(II) in their experiments] or to magnetite at higher Fe(II) loadings. Despite having a nearly equal affinity for As (Dixit and Hering, 2003), transformation of ferrihydrite to more crystalline phases (goethite, hematite, or magnetite) decreases the available surface area (Appelo et al., 2002) and thus diminishes their capacity to retain arsenic.

Congruent dissolution of Fe(III) (hydr)oxides would be expected to release arsenic until the surface site concentration is depleted below the level of adsorbed arsenic (in combination with other high-affinity anions, such as phosphate). Secondary mineralization of iron, however, convolutes this view, leading to the potential for arsenic to be incorporated on or in the newly forming solids.

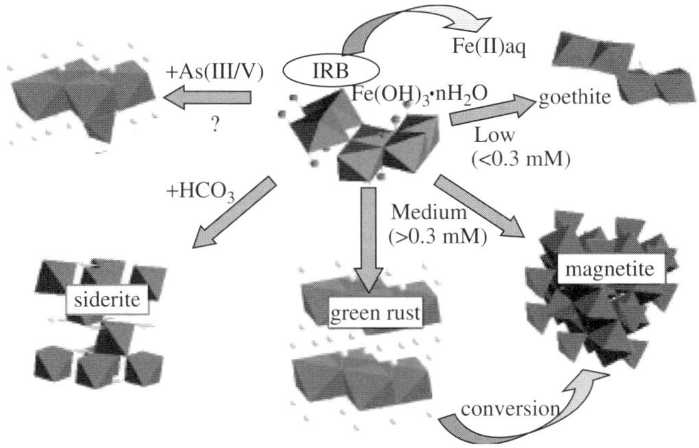

Figure 8.3. Ferrihydrite transformations upon reaction with Fe(II) produced biologically.

We recently observed elution trends of arsenic from ferrihydrite under iron-reducing conditions that are inconsistent with arsenic desorption upon reductively induced mineralogical transition (Herbel and Fendorf, 2006). Here we illustrate arsenic release from As(III)-loaded ferrihydrite-coated sand under hydrodynamic conditions (at a flow rate of one pore volume per day) in the absence (abiotic) and presence of *Shewanella putrefaciens*, strain CN32 (Figure 8.4). During the period of greatest As release, which occurs at the onset of flow, a rapid mineralogical transition occurs within the inoculated column as (principally) magnetite is generated at the expense of ferrihydrite. The initial pulse of As, however, also results in abiotic columns lacking biological activity or Fe(II). Furthermore, the extent of desorption is greatly suppressed upon the production of Fe(II) within biologically active columns, a factor supported by Fe(II) addition to abiotic systems (Herbel and Fendorf, 2006). It therefore appears that while reductive dissolution of ferric (hydr)oxides can lead to arsenic release, under conditions conducive to remineralization, arsenic desorption is suppressed rather than promoted. The recent observation of a ferrous–arsenite precipitate (Thoral et al., 2005) may account in part for arsenic retention under conditions of intense Fe(II) production. We have not observed such a solid within the bioreduced columns described here, however, but rather find arsenic incorporated in or adsorbed on the secondary solids.

8.5.3.2. Desorption on Iron and Arsenic Reduction The effects of biological (dissimilatory) Fe(III) and As(V) reduction on As transport under dynamic flow conditions in an Fe (hydr)oxide-dominated system illustrate the complexity of reactions influencing arsenic dynamics within soils. We examined the impact of a model freshwater anaerobic bacterium, *Sulfurosprillum barnesii* strain SES-3, that reduces both Fe(III) and As(V) through respiratory processes (Oremland et al., 1994; Zobrist et al., 2000), on the desorption of arsenic from ferrihydrite-coated sands (Herbel and Fendorf, 2005, 2006). As noted above for As(III)-loaded

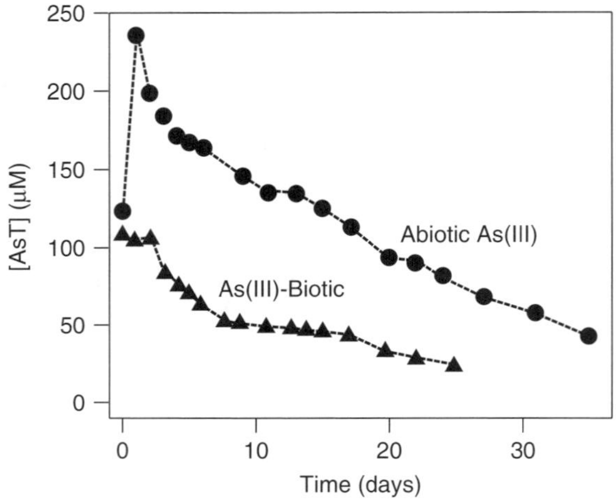

Figure 8.4. Arsenic elution from abiotic and biotic [*S. putrefaciens*, an Fe(III)-reducing bacterium] columns initially containing As(III)–ferrihydrite–quartz sand. The initial surface coverage (ca. 800 mg Kg^{-1} ferrihydrite-sand) is approximately 50% of the adsorption maximum. A flow rate of 1 pore volume per day was maintained with an artificial ground water medium. See Herbel and Fendorf (2006) for a detailed description of the medium.

ferrihydrite, arsenic release from As(V) loaded ferrihydrite is most pronounced during the initial influx of waters containing labile carbon, but decreases quickly once Fe(II) levels increase. Again more arsenic [as As(V) and As(III)] is released under advective flow in abiotic columns from ferrihydrite sands, and over longer periods of time, than in systems containing bacteria respiring on As(V) and Fe(III) (Figure 8.5). Owing to the generation of As(III), the enhanced retention of arsenic under Fe(III)-reducing conditions results from secondary mineralization of Fe into phases such as magnetite, as noted above for columns initially loaded with As(III).

Although initial trends of arsenic desorption are comparable between As(III)- and As(V) loaded ferrihydrite (Figures 8.4 and 8.5; Herbel and Fendorf, 2006), higher concentrations of dissolved As are sustained after 15 days of reaction from columns loaded initially with As(V) (sustained tailing concentrations of ~10 μM) than from an As(III) column (tailing concentrations $< 5\mu M$)—in both cases, the arsenic eluted was in the trivalent state. The increased tailing (longer-term desorption) of As in columns initially having As(V) relative to those having As(III) appears related to the pathway (desorption and transport across bacterial membranes) by which As(V) is enzymatically reduced.

Dissimilatory reduction of As(V) could, hypothetically, transpire either through the terminal reductase being located on the exterior of the bacterial membrane (the outer membrane in the case of gram-negative bacteria) or through arsenic desorption and transport within (or through) the bacterial membrane. Current findings on the location and behavior of As(V)-reducing enzymes in dissimilatory As(V)-reducing bacteria does not support the premise of an arsenate reductase located

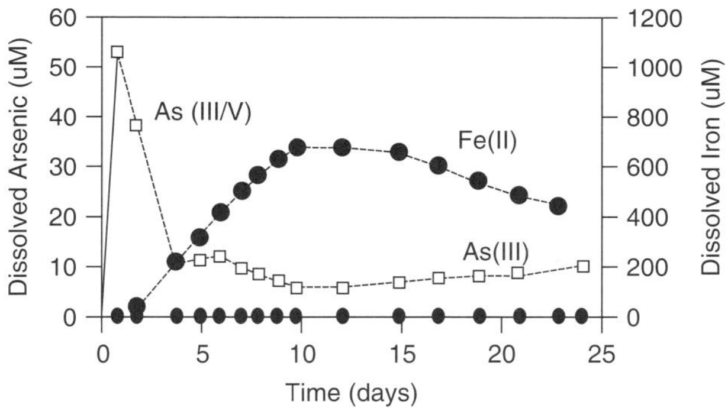

Figure 8.5. Total dissolved As and Fe(II) in effluent from columns having a ferrihydride-coated sand column loaded initially with As(V) (156 mg kg ferrihydrite-sand) and *Sulfurospirillum barnesii*, an As(V)- and Fe(III) reducing bacterium. (Data from Herbel and Fendorf, 2006.)

on the cell exterior (Krafft and Macy, 1998; Stolz and Oremland, 1999; Oremland and Stolz, 2003; Saltikov and Newman, 2003). An alternative and more likely mechanism thus requires As(V) to desorb from ferrihydrite, pass through (or into) a bacterial membrane, and undergo respiratory reduction. Although arsenate is (relatively) strongly retained on ferrihydrite, it is likely that bacteria maintain a gradient in As concentrations promoting desorption and diffusion of As(V) to the cell. Expulsion of As(III) into the external media then occurs via As(III)-specific transporter proteins. This scenario would facilitate As(III) transport away from the ferrihydrite surface and would enhance movement of As through the column. Furthermore, it would result in down-field migration and elution of arsenic, both of which are observed (Herbel and Fendorf, 2006), and therefore explain the greater concentration of arsenic tailing in columns where both As(V) and Fe(III) reduction transpires relative to those having only Fe(III) reduction (Figure 8.5).

8.6. CONCLUSIONS AND FUTURE DIRECTIONS

Arsenic retention on soil solids is critical for regulating the dissolved concentrations of this hazardous element, thus helping to limit its exposure to living organisms and migration within the environment. Strong partitioning of arsenic on soil solids is most prevalently disrupted by the onset of anaerobic conditions (anaerobiosis), leading to increased aqueous concentrations of arsenic. Variations in arsenic chemistry, compounded by biogeochemical transformations of the soil matrix, upon the onset of anaerobic conditions transpire to produce a convolution of reactions that have varying impacts on arsenic retention, a summary of which is given in Figure 8.6.

Arsenic desorption upon anaerobiosis has been ascribed to both the reduction of arsenic, from arsenate to arsenite, and iron(III), the latter leading to the

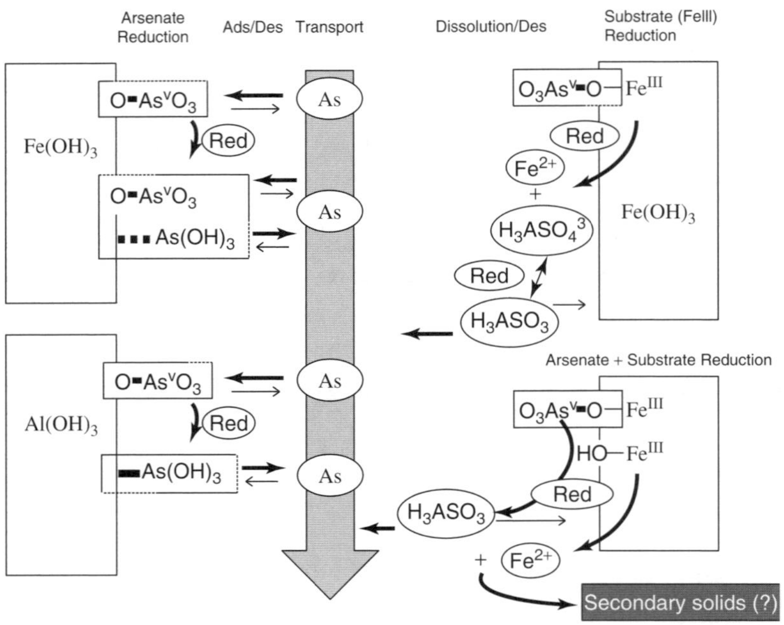

Figure 8.6. Postulated pathways and changes in arsenic partitioning (and transport) under As(V) (left side), Fe(III) (top right), or Fe(III) and As(V) reducing conditions. (Width of arrows denotes magnitude of reaction.)

reductive dissolution of ferric (hydr)oxides that act as principal sinks of arsenic (McGeehan and Naylor, 1992; Jones et al., 2000; McCreadie et al., 2000; Zobrist et al., 2000; Islam et al., 2004; Delemos et al., 2006). Recent advances in our knowledge of arsenic reduction mechanisms, binding affinities, and (bio)reductive transformation of Fe(III) (hydr)oxides, lead to a more complex view of arsenic dynamics under reducing conditions. We first consider the implications of As(V) reduction (left side of Figure 8.6). Reduction of As(V) to As(III) leads to a large alteration in the types of solids retaining arsenic, with a large specificity for Fe(III) (hydr)oxides relative to other soil solids (Manning and Goldberg, 1997a); in comparison to arsenate, the adsorption capacity of arsenite is small on Al (hydr)oxides and aluminosilicates (see Table 8.1). The extensive binding of arsenite on Fe (hydr)oxides is, in part, a result of labile complexes in combination with stronger inner-sphere complexes, contrasting the binding of arsenate, which is dominantly via strong inner-sphere complexation. As a consequence, As(III) desorption is rapid in response to a decrease in aqueous concentrations (i.e., a shift from equilibrium with respect to the surface complexation reactions).

Given recent advances in modes of bacterial arsenate reduction (Saltikov and Newman, 2003), the primary means by which arsenic is reduced in nonsulfidic systems, and probably in all systems that have pH values in the circumneutral range or higher (Rochette et al., 2000), we can further extend our view of arsenic retention upon reduction of arsenate to arsenite. The two modes of arsenic reduction, respiratory or detoxification, both require arsenic to be released from the solid phase. The primary reductases for respiratory reduction, ArrA and ArrB, both reside interior to the outer membrane of bacteria (Saltikov and Newman, 2003), and thus it is likely that As(V) must desorb from the surface, cross the outer membrane, and then undergo reduction. Arsenate reduction through detoxification is also restricted by membrane translocation, since this process takes place in the cytoplasm. In either case, it appears that for arsenate reduction to occur, it must desorb from mineral surfaces before reduction; only if the reductase(s) reside on the external portion of the outer membrane could reduction transpire in the absence of desorption, and this is not apparent for known reduction pathways. After As(V) is reduced, arsenite would be expelled from the cell and would undergo more extensive transport than its oxidized precursor.

Iron reduction will also take place under anaerobic conditions, and thus we need to consider its implications on arsenic partitioning and transport (right portion of Figure 8.6). Iron reduction will lead, in part, to dissolution of the mineral phase and thus diminish the number of adsorption sites. We therefore need to consider the ratio of arsenic to potential reactive surface sites on the Fe(III) (hydr)oxide surface. For a soil having 50 mg kg^{-1} As, 120 m^2 kg^{-1} Fe(III) (hydr)oxide is conservatively needed to adsorb all the As (and considering that all surface sites could be occupied by As); this translates to 5 g Fe kg^{-1} soil of goethite with a conservative specific surface area of 20 m^2 g^{-1}. For soils having a total of 10 g kg^{-1} Fe as goethite and 2 g kg^{-1} Fe as ferrihydrite, 50 mg kg^{-1} As would occupy less than 10% of the surface sites. Hence, depending on the specific solid-phase arsenic concentration and iron (hydr)oxide content/surface area, Fe(III) mineral transformations may affect dissolved concentrations of arsenic (conditions of high As/Fe) directly or have a minimal impact (low As/Fe conditions) until extensive (90% removal) dissolution–transformation has transpired.

We next need to consider the Fe transformation pathway and whether dissolution is the sole fate of the mineral or whether secondary minerals are generated that may sequester arsenic. Here, recent work illustrates that the intensity of iron reduction [indicated by the resulting dissolved Fe(II) concentration] is a key factor controlling iron mineralogical transformations (see, e.g., Hansel et al., 2003) and thus the fate of arsenic. For soils with labile iron oxides (typically ferrihdyrite, but also possibly lepidocrocite or proported nanogoethite), stimulation of iron reduction would lead to a sequence of iron transformations (as illustrated in Figure 8.3) and lead to arsenic incorporation and adsorption, as demonstrated here by the noted increased sequestration of arsenic under Fe(III)-reducing conditions of ferrihydrite (Figure 8.4). Additionally, generation of high concentrations of Fe(II) and As(III) can lead to the formation of a secondary Fe(II)–As(III)-bearing precipitate (Thoral et al., 2005).

While we have considered Fe(III) or As(V) reduction independently, each of which could have important implications on the fate and transport of arsenic, As(V) and Fe(III) reduction will typically transpire in concert (as projected in Figure 8.6, bottom right panel). Obligate As(V) reducing bacteria have not been identified to date; rather they have a capacity to respire on a diverse set of electron acceptors, and in particular, many are noted iron reducers (Oremland and Stolz, 2003). Furthermore, on the basis of As(III) nearly always being the dominant species of arsenic under iron-reducing conditions, we may speculate that As(V) reduction coincides with Fe(III) reduction. We therefore postulate that Fe–As-reducing bacteria promote simultaneous reduction of As(V) and Fe(III), resulting in parallel rather than sequential reduction.

Summarizing the fate of arsenic under anaerobic conditions, we should note two important points: (1) dissolved concentrations and migration of arsenic are typically promoted under reducing conditions, but (2) the intensity of iron reduction [measured by dissolved Fe(II) concentrations] and concomitant extent of iron bioreductive mineralization will be the primary variable influencing the extent of release. Arsenate reduction is the largest single factor promoting arsenic release but will be invariant under reducing conditions. The second point (the intensity of iron reduction controlling arsenic release) results from the potential for arsenic to be sequestered in secondary mineralization products or precipitates and that less intense reduction will promote arsenic transport, whereas more intense reduction will lead to As incorporation in or adsorption on secondary bioreduction products such as magnetite.

So what does this synopsis tell us about the expected fate of arsenic under anaerobic conditions? First, owing to the weaker surface complexes of As(III) versus As(V), arsenic migration will be greater under anaerobic than aerobic conditions (with the exception of environments having high dissolved concentrations of sulfide, which can lead to sequestration of As). Sedimentary basins of Southeast Asia experiencing high concentrations of As in groundwater, for example, have prolonged periods (months) of anaerobic conditions; arsenic within these environments would therefore be expected to be relatively mobile. However, the extent of arsenic release will be modified appreciably by the fate of Fe. Under conditions conducive to rapid and extensive Fe(III) reduction [anaerobic conditions, high concentrations of labile DOC, bioavailable Fe(III), and warm temperatures], secondary mineralization will help to sequester arsenic. Returning to the situation in Southeast Asia, surface soils (upper ca. 0.5 m) undergo rapid fluctuating redox conditions, leading to high-surface-area short-range-ordered Fe(III) (hydr)oxides such as ferrihydrite during the aerobic periods. Conditions conducive to intense iron reduction during monsoonal flooding would stimulate secondary Fe phases such as biogenic magnetite, providing a suitable host for arsenic even under anaerobic conditions. In this projection, arsenic would be cycled from one type of solid to another during aerobic–anaerobic transitions.

Two remaining factors need to be considered in addressing the ultimate fate of arsenic within the sedimentary basins of Southeast Asia. First, arsenic, while cycled in the surface soils, is partially leached into deeper portions of the profile (i.e., As is not conserved during the anaerobic cycle). Second, these basins are active and continue to receive sediment loads, leading to continual burial of arsenic deeper in the soil profile. Once deeper in the soil, the cycling of iron, and hence the fate of arsenic, is altered. Dampened fluctuation in anaerobic–aerobic transitions results in more crystalline Fe(III) phases during aerobic periods and, in combination with diminishing concentrations of labile DOC with depth, less intense Fe(III) reduction and secondary mineralization during anaerobic periods. In combination, subsurface processes will lead to enhanced As mobility, and thus transport to aquifers residing deeper in the sediment profile.

Acknowledgments

We are grateful for the valuable discussions and analytical contributions of Brandy Stewart, Guangchao Li, Yoko Masue, and Jodi Switzer Blum. We would also like to thank two anonymous reviewers for their thoughtful input, and Professor A. Violante for his constructive comments and handling of the manuscript. This research was supported by the US. National Science Foundation EMSI Program (grant number NSF-CHE-0431425).

REFERENCES

Afkar, E., Lisak, J., Saltikov, C. W., Basu, P., Oremland, R. S., and Stolz, J. F. (2003). The respiratory arsenate reductase from *Bacillus selenitireducens* strain MLS10. *FEMS Microbiol. Lett.* **226**, 107–112.

Ahmann, D., Roberts, L. R., Krumholz, L. R., and Morel, F. M. M. (1994). Microbe grows by reducing arsenic. *Nature* **371**, 750.

Anderson, M. A., Ferguson, J. F., and Gavis, J. (1976). Arsenate adsorption on amorphous aluminum hydroxide. *J. Colloid Interface Sci.* **54**, 391–399.

Appelo, C. A. J., Van der Weiden, M. J. J., Tournassat, C., and Charlet, L. (2002). Surface complexation of ferrous iron and carbonate on ferrihydrite and the mobilization of arsenic. *Environ. Sci. Technol.* **36**, 3096–3103.

Arai, Y., and Sparks, D. L. (2001). ATR-FTIR spectroscopic investigation of phosphate adsorption mechanisms at the ferrihydrite–water interface. *J. Colloid Interface Sci.* **241**, 317–326.

Arai, Y., Elzinga, E. J., and Sparks, D. L. (2001). X-ray absorption spectroscopic investigation of arsenite and arsenate adsorption at the aluminum oxide–water interface. *J. Colloid Interface Sci.* **235**, 80–88.

Bostick, B. C., and Fendorf, S. (2003). Arsenite sorption on troilite (FeS) and pyrite (FeS2). *Geochim. Cosmochim. Acta* **67**, 909–921.

Bostick, B. C., Fendorf, S., and Manning, B. A. (2003). Arsenite adsorption on galena (PbS) and sphalerite (ZnS). *Geochim. Cosmochim. Acta* **67**, 895–907.

Chen, M., Ma, L. Q., and Harris, W. G. (2002). Arsenic concentrations in Florida surface soils: influence of soil type and properties. *Soil Sci. Soc. Am. J.* **66**, 632–640.

Cullen, W. R., and Reimer, K. J. (1989). Arsenic speciation in the environment. *Chem. Rev.* **89**, 713–764.

Cummings, D. E., Caccavo, F., Jr., Fendorf, S., and Rosenzweig, R. F. (1999). Arsenic mobilization by the dissimilatory Fe(III)-reducing bacterium *Schewanella alga* BrY. *Environ. Sci. Technol.* **33**, 723–729.

Delemos, J. L., Bostick, B. C., Renshaw, C. E., Sturup, S., and Feng, X. H. (2006). Landfill-stimulated iron reduction and arsenic release at the Coakley Superfund site (NH). *Environ. Sci. Technol.* **40**, 67–73.

Dixit, S., and Hering, J. G. (2003). Comparison of arsenic(V) and arsenic(III) sorption onto iron oxide minerals: implications for arsenic mobility. *Environ. Sci. Technol.* **37**, 4182–4189.

Dowdle, P. R., Laverman, A. M., and Oremland, R. S. (1996). Bacterial dissimilatory reduction of arsenic(V) and arsenic(III) in anoxic sediments. *Appl. Environ. Microbiol.* **62**, 1664–1669.

Driehaus, W., Seith, R., and Jeckel, M. (1995). Oxidation of arsenic(III) with manganese oxides in water-treatment. *Water Res.* **29**, 297–305.

Fendorf, S. E., Eick, M. J., Grossl, P. R., and Sparks, D. L. (1997). Arsenate and chromate retention mechanisms on goethite, 1: Surface structure. *Environ. Sci. Technol.* **31**, 315–320.

Frost, R. R., and Griffin, R. A. (1977). Effect of pH on adsorption of arsenic and selenium from landfill leachate by clay minerals. *Soil Sci. Soc. Am. J.* **41**, 53–57.

Goldberg, S. (2002). Competitive adsorption of arsenate and arsenite on oxides and clay minerals. *Soil Sci. Soc. Am. J.* **66**, 413–421.

Goldberg, S., and Glaubig, R. A. (1988). Anion sorption on a calcareous, montmorillonitic soil: arsenic. *Soil Sci. Soc. Am. J.* **52**, 1297–1300.

Goldberg, S., and Johnston, C. T. (2001). Mechanisms of arsenic adsorption on amorphous oxides evaluated using macroscopic measurements, vibrational spectroscopy, and surface complexation modeling. *J. Colloid Interface Sci.* **234**, 204–216.

Gosh, M. M., and Yuan, J. R. (1987). Adsorption of inorganic arsenic and organoarsenicals on hydrous oxides. *Environ. Prog.* **6**, 150–157.

Grafe, M., Eick, M. J., and Grossl, P. R. (2001). Adsorption of arsenate(V) and arsenite(III) on goethite in the presence and absence of dissolved organic carbon. *Soil Sci. Soc. Am. J.* **65**, 1680–1687.

Grafe, M., Eick, M. J., Grossl, P. R., and Saunders, A. M. (2002). Adsorption of arsenate and arsenite on ferrihydrite in the presence and absence of dissolved organic carbon. *J. Environ. Qual.* **31**, 1115–1123.

Gupta, S. K., and Chen, K. Y. (1978). Arsenic removal by adsorption. *J. Water Pollut. Control Fed.* **50**, 493–506.

Hansel, C. M., Benner, S. G., Neiss, J., Dohnalkova, A., Kukkadapu, R. K., and Fendorf, S. (2003). Secondary mineralization pathways induced by dissimilatory iron reduction of ferrihydrite under advective flow. *Geochim. Cosmochim. Acta* **67**, 2977–2992.

Helz, G. R., Tossel, J. A., Charnock, J. M., Pattrick, A. D., Vaughan, D. J., and Garner, C. D. (1995). Oligomerization in As(III) sulfide solutions: theoretical constraints and spectroscopic evidence. *Geochim. Cosmochim. Acta* **59**, 5491–5604.

Herbel, M. J., and Fendorf, S. (2005). Transformation and transport of arsenic within ferric hydroxide coated sands upon dissimilatory reducing bacterial activity, (In *Advances in Arsenic Research: Integration of Experimental and Observational Studies and Implications for Mitigation*, ed. O'Day, P., Vlassopoulos, D., Meng, X., and Benning, L., American Chemical Society, Washington, DC, 77–90.

Herbel, M. J., and Fendorf, S. (2006). Biogeochemical processes controlling the speciation and transport of arsenic within iron coated sands. *Chem. Geol.* **228**, 16–32.

Herbel, M. J., Switzer Blum, J., Hoeft, S. E., Cohen, S. M., Arnold, L. L., Lisak, J., Stolz, J. F., and Oremland, R. S. (2002). Dissimilatory arsenate reductase activity and arsenate-respiring bacteria in bovine rumen fluid, hamster feces, and the termite hindgut. *FEMS Microbiol. Ecol.* **41**, 59–67.

Hingston, F. J., Posner, A. M., and Quirk, J. P. (1971). Competitive adsorption of negatively charged ligands on oxide surfaces. *Faraday Discuss. Chem. soc.* **52**, 334–342.

Huber, R. M., Sacher, M., Vollmann, A., Huber, H., and Rose, D. (2000). Respiration of arsenate and selenate by hyperthermophilic Archaea. *Syst. Appl. Microbiol.* **23**, 305–314.

Islam, F. S., Gault, A. G., Boothman, C., Polya, D. A., Charnock, J. M., Chatterjee, D., and Lloyd, J. R. (2004). Role of metal-reducing bacteria in arsenic release from Bengal delta sediments. *Nature* **430**, 68–71.

Jain, A., and Loeppert, R. H. (2000). Effect of competing anions on the adsorption of arsenate and arsenite by ferrihydrite. *J. Environ. Qual.* **29**, 1422–1430.

Jones, C., Langner, H., Anderson, K., McDermott, T., and Inskeep, W. (2000). Rates of microbially mediated arsenate reduction and solubilization. *Soil Sci. Soc. Am. J.* **64**, 600–608.

Kim, M.-J., Nriagu, J., and Haack, S. (2000). Carbonate ions and arsenic displacement by groundwater. *Environ. Sci. Technol.* **34**, 3094–3100.

Ko, I., Kim, J.-Y., and Kim, K.-W. (2004). Arsenic speciation and sorption kinetics in the As–hematite–humic acid system. *Colloid Surf.A* **234**, 43–50.

Krafft, T., and Macy, J. M. (1998). Purification and characterization of the respiratory arsenate reductate of *Crysiogenes arsenatis*. *Eur. J. Biochem.* **255**, 647–653.

Langner, H. W., and Inskeep, W. P. (2000). Microbial reduction of arsenate in the presence of ferrihydrite. *Environ. Sci. Technol.* **34**, 3131–3136.

Lee, J. S., and Nriagu, J. O. (2003). Arsenic carbonate complexes in aqueous systems. In *Biogeochemistry of Environmentally Important Trace Elements*. ed. Cai, Y., and Braids, O. C., American Chemical Society, Washington, DC, 33–41.

Lenoble, V., Laclautre, C., Serpaud, B., Deluchat, V., and Bollinger, J. C. (2004). As(V) retention and As(III) simultaneous oxidation and removal on a MnO_2–loaded polystyrene resin. *Sci. Total Environ.* **326**, 197–207.

Lin, H.-T., Wang, M. C., and Li, G.-C. (2004). Complexation of arsenate with humic substance in water extract of compost. *Chemosphere* **56**, 1105–1112.

Lovley, D. R. (1991). Dissimilatory Fe(III) and Mn(IV) reduction. *Microbiol. Rev.* **55**, 259–287.

Lovley, D. R. (2000). Fe(III) and Mn(IV) reduction. In *Environmental Microbe–Metal Interactions*. ed. Lovley, D. R., ASM Press, Washington, DC, 3–30.

Lumsdon, D. G., Fraser, A. R., Russel, J. D., and Livesey, N. T. (1984). New infrared band assignments for the arsenate ion adsorbed on synthetic goethite (α-FeOOH). *Soil Sci.* **35**, 381–386.

Malasarn, D., Saltikov, W., Campbell, K. M., Santini, J. M., Hering, J. G., and Newman, D. K. (2004). arrA is a reliable marker for As(V) respiration. *Science* **306**, 455.

Manceau, A. A. (1995). The mechanism of anion adsorption on iron oxides: evidence for the bonding of arsenate tetrahedra on free $Fe(O,OH)_6$ edges. *Geochim. Cosmochim. Acta* **59**, 3647–3653.

Mandal, B. K., and Suzuki, K. T. (2002). Arsenic round the world: a review. *Talanta* **58**, 201–235.

Manning, B. A., and Goldberg, S. (1996). Modeling competitive adsorption of arsenate with phosphate and molybdate on oxide minerals. *Soil Sci. Soc. Am. J.* **60**, 121–131.

Manning, B. A., and Goldberg, S. (1997a). Arsenic(III) and arsenic(V) absorption on three California soils. *Soil Sci.* **162**, 886–895.

Manning, B. A., and Goldberg, S. (1997b). Adsorption and stability of arsenic(III) at the clay mineral–water interface. *Environ. Sci. Technol.* **31**, 2005–2011.

Manning, B. A., Fendorf, S., and Goldberg, S. (1998). Surface structures and stability of arsenic(III) on goethite: spectroscopic evidence for inner-sphere complexes. *Environ. Sci. Technol.* **32**, 2382–2388.

Manning, B. A., Fendorf, S., Bostick, B. C., and Suarez, D. L. (2002). Arsenic(III) oxidation and arsenic(V) adsorption reactions on synthetic birnessite. *Environ. Sci. Technol.* **36**, 976–981.

McCreadie, H., Blowes, D. H., Ptacek, C. J., and Jambor, J. L. (2000). Influence of reduction reactions and solid-phase composition on porewater concentrations of arsenic. *Environ. Sci. Technol.* **34**, 3159–3166.

McGeehan, S. L., and Naylor, D. V. (1992). Simultaneous determination of arsenite, arsenate, selenite, and selenate by suppressed ion chromatography. *J. Environ. Qual.* **21**, 68–73.

Moore, J. N., Ficklin, W. H., and Johns, C. (1988). Partitioning of arsenic and metals in reducing sulfidic sediments. *Environ. Sci. Technol.* **22**, 432–437.

Mukhopadhyay, R., Rosen, B. P., Pung, L. T., and Silver, S. (2002). Microbial arsenic: from geocycles to genes and enzymes. *FEMS Microbiol. Rev.* **26**, 311–325.

Newman, D. K., Beveridge, T. J., and Morel, F. M. (1997a). Precipitation of arsenic trisulfide by *Desulfotomaculum auripigmentum*. *Appl. Environ. Microbiol.* **63**, 2022–2028.

Newman, D. K., Kennedy, E. K., Coates, J. D., Ahmann, D., Ellis, D. J., Lovley, D. R., and Morel, F. M. M. (1997b). Dissimilatory arsenate and sulfate reduction in *Desulfotomaculum auripigmentum* sp. nov. *Arch. Microbiol.* **168**, 380–388.

Ona-Nguema, G., Morin, G., Juillot, F., Calas, G., and Brown, G. E. J. (2005). EXAFS analysis of arsenite adsorption onto two-line ferrihydrite, hematite, goethite, and lepidocrocite. *Environ. Sci. Technol.* **39**, 9147–9155.

Oremland, R. S., and Stolz, J. F. (2003). The ecology of arsenic. *Science* **300**, 939–944.

Oremland, R. S., Switzer Blum, J., Culbertson, C. W., Visher, P. T., Miller, L. G., Dowdle, P. R., and Strohmeier, F. E. (1994). Isolation, growth, and metabolism of an obligately anaerobic, selenate-respiring bacterium, strain SES-3. *Appl. Environ. Microbiol.* **60**, 3011–3019.

Oremland, R. S., Newman, D. K., Kail, B. W., and Stolz, J. F. (2002). Bacterial respiration of arsenate and it significance in the environment In *Environmental Chemistry of Arsenic*, ed. Frankenberger, W. T., Jr., Marcel Dekker, New York, 273–295

Oscarson, D. W., Huang, P. M., and Liaw, W. K. (1981). Role of manganese in the oxidation of arsenite by freshwater lake sediments. *Clays Clay Miner.* **29**, 219–225.

Oscarson, D. W., Huang, P. M., Hammer, U. T., and Liaw, W. K. (1983). Oxidation and sorption of arsenite by manganese dioxide as influenced by surface coatings of iron and aluminum oxides and calcium carbonate. *Water Air Soil Pollut.* **20**, 233–244.

Parfitt, R. L., Atkinson, R. J., and Smart, R. S. C. (1975). Mechanism of phosphate fixation by iron oxides. *Soil Sci. Soc. Am. J.* **39**, 837–841.

Peryea, F. J., and Kammereck, R. (1997). Phosphate-enhanced movement of arsenic out of a lead arsenate-contaminated topsoil and through uncontaminated subsoil. *Water Air Soil Pollut.* **93**, 243–254.

Raven, K. P., Jain, A., and Loeppert, R. H. (1998). Arsenite and arsenate adsorption on ferrihydrite: kinetics, equilibrium, and adsorption envelopes. *Environ. Sci. Technol.* **32**, 344–349.

Redman, A. D., Macalady, D. L., and Ahmann, D. (2002). Natural organic matter affects arsenic speciation and adsorption on hematite. *Environ. Sci. Technol.* **36**, 2889–2896.

Reynolds, J. G., Naylor, D. V., and Fendorf, S. E. (1999). Arsenic sorption in phosphate-amended soils during flooding and subsequent aeration. *Soil Sci. Soc. Am. J.* **63**, 1149–1156.

Rochette, E. A., Li, G. C., and Fendorf, S. E. (1998). Stability of arsenate minerals in soil under biotically generated reducing conditions. *Soil Sci. Soc. Am. J.* **62**, 1530–1537.

Rochette, E. A., Bostick, B. C., Li, G., and Fendorf, S. (2000). Kinetics of arsenate reduction by dissolved sulfide. *Environ. Sci. Technol.* **34**, 4714–4720.

Roden, E., and Zachara, J. M. (1996). Microbial reduction of crystalline iron(III) oxides: influence of oxide surface area and potential for cell growth. *Environ. Sci. Technol.* **30**, 1618–1628.

Rosen, B. P. (2002). Biochemistry of arsenic detoxification. *FEBS Lett.* **529**, 86–92.

Sadiq, M. (1997). Arsenic chemistry in soils: an overview of thermodynamic predictions and field observations. *Water Air Soil Pollut.* **93**, 117–136.

Saltikov, C. W., and Newman, D. K. (2003). Genetic identification of a respiratory arsenate reductase. *Proc. Natl. Acad. Sci. USA* **100**, 10983–10988.

Saltikov, C. W., Cifuentes, A., Venkateswaran, K., and Newman, D. K. (2003). The ARS detoxification system is advantageous but not required for As(V) respiration by the genetically tractable *Schewanella* species strain ANA-3. *Appl. Environ. Microbiol.* **69**, 2800–2809.

Schwertmann, U., and Taylor, R. M. (1989). Iron oxides.*In Minerals in Soil Environments, 2nd ed. Book Ser. 1*, Soil Science Society of America, Madison, WI, 379–438.

Sherman, D. M., and Randall, S. R. (2003). Surface complexation of arsenic(V) to iron(III) hydroxides: structural mechanism from ab initio molecular geometries and EXAFS spectroscopy. *Geochim. Cosmochim. Acta* **67**, 4223–4230.

Smedley, P. L., and Kinniburgh, D. G. (2002). A review of the source, behaviour and distribution of arsenic in natural waters. *Appl. Geochem.* **17**, 517–568.

Smith, E., Naidu, R., and Alston, A. M. (1998). Arsenic in the soil environment: a review. *Adv. Agron.* **64**, 149–195.

Stolz, J. F., and Oremland, R. S. (1999). Bacterial respiration of arsenic and selenium. *FEMS Microbiol. Rev.* **23**, 615–627.

Sun, X., and Doner, H. E. (1996). An investigation of arsenate and arsenite bonding structure on goethite by FTIR. *Soil Sci.* **161**, 865–872.

Thanabalsingam, P., and Pickering, W. F. (1986a). Arsenic sorption by humic acids. *Environ. Pollut. Ser. B Chem. Phys.* **12**, 233–246.

Thanabalsingam, P., and Pickering, W. F. (1986b). Effect of pH on interaction between As(III) or As(V) and manganese oxide. *Water Air Soil Pollut.* **29**, 205–216.

Thoral, S., Rose, J., Garnier, J. M., van Geen, A., Refait, P., Traverse, A., Finda, E., Nahon, D., and Bottero, J. Y. (2005). XAS study of iron and arsenic speciation during Fe(II) oxidation in the presence of As(III). *Environ. Sci. Technol.* **39**, 9478–9485.

Van Geen, A., Robertson, A. P., and Leckie, J. O. (1994). Complexation of carbonate species at the geothite surface: implications for adsorption of metal ions in natural waters. *Geochim. Cosmochim. Acta* **58**, 2073–2086.

Villalobos, M., and Leckie, J. O. (2001). Surface complexation modeling and FTIR study of carbonate adsorption to goethite. *J. Colloid Interface. Sci.* **235**, 15–32.

Violante, A., and Pigna, M. (2002). Competitive sorption of arsenate and phosphate on different clay minerals and soils. *Soil Sci. Soc. Am. J.* **66**, 1788–1796.

Waychunas, G. A., Rea, B. A., Fuller, C. C., and Davis, J. A. (1993). Surface chemistry of ferrihydrite, 1: EXAFS studies of the geometry of coprecipiated and adsorbed arsenate. *Geochim. Cosmochim. Acta* **57**, 2251–2269.

Wilkin, R. T., Wallschlager, D., and Ford, R. G. (2003). Speciation of arsenic in sulfidic waters. *Geochem. Trans.* **4**, 1–7.

Xu, H., Allard, B., and Grimvall, A. (1988). Influence of pH and organic substance on the adsorption of As(V) on geologic materials. *Water Air Soil Pollut.* **40**, 293–305.

Xu, H., Allard, B., and Grimvall, A. (1991). Effects of acidification and natural organic materials on the mobility of arsenic in the environment. *Water Air Soil Pollut.* **57–58**, 269–278.

Yu, W. H., Harvey, C. M., and Harvey, C. W. (2003). Arsenic in Bangladesh: a geostatistical and epidemiological framework for evaluating health effects and potential remedies. *Water Resour. Res.* **39**, 1146.

Zachara, J. M., Kukkadapu, R. K., Fredrickson, J. K., Gorby, Y. A., and Smith, S. C. (2002). Biomineralization of poorly crystalline Fe(III) oxides by dissimilatory metal reducing bacteria (DMRB). *Geomicrobiol. J.* **19**, 179–207.

Zobrist, J., Dowdle, P. R., Davis, J. A., and Oremland, R. S. (2000). Mobilization of arsenite by dissimilatory reduction of adsorbed arsenate. *Environ. Sci. Technol.* **34**, 4747–4753.

9

MICROBIAL OXIDATION AND REDUCTION OF IRON IN THE ROOT ZONE AND INFLUENCES ON METAL MOBILITY

S. C. NEUBAUER

Baruch Marine Field Laboratory, University of South Carolina, Georgetown, South Carolina

D. EMERSON

American Type Culture Collection, Manassas, Virginia

J. P. MEGONIGAL

Smithsonian Environmental Research Center, Edgewater, Maryland

Biophysico-Chemical Processes of Heavy Metals and Metalloids in Soil Environments,
Edited by Antonio Violante, Pan Ming Huang, and Geoffrey Michael Gadd

9.1. INTRODUCTION

Iron (Fe) is the fourth most abundant element in the Earth's crust at 5% by mass (Weaver and Tarney, 1984). In soils, oxidized (ferric) iron is the most abundant element that can serve as a terminal electron acceptor for microbial metabolism, and there are microbes that can gain energy by the oxidation of ferrous iron, the product of iron reduction. This microbial cycling of iron in the environment influences the capacity of ecosystems to store soil carbon, to mineralize nutrients, and to produce greenhouse gases. In addition to these microbial processes, geochemical reactions are involved in the cycling of iron within the biosphere (Figure 9.1), and it is often difficult to distinguish between biotic and abiotic iron transformations.

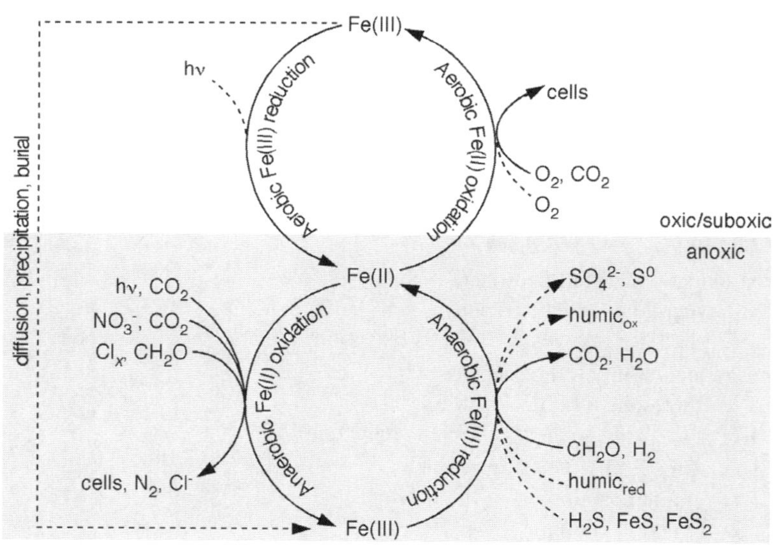

Figure 9.1. Iron cycle at circumneutral pH. Iron can be both oxidized and reduced under aerobic or anaerobic conditions, by microbial or abiotic processes. The significance of many of these reactions in soils is currently unknown. Solid lines represent microbial processes; dashed lines are chemical iron transformations.

The focus of this chapter is on the physiology, ecology, and biogeochemistry of the aerobic and anaerobic microbes involved in the oxidation and reduction of iron in the rhizosphere (root zone) of wetland plants. Because the wetland rhizosphere is an interface between oxic and anoxic environments, it is a site where microbial reactions can cycle iron between oxidized and reduced forms. Furthermore, other important microbial and chemical processes can interact with microbial iron transformations near plant roots. Iron plaque, a rusty-colored coating found on many wetland plant roots, often contains high concentrations of metals such as Cr, Pb, Cu, and Zn and the metalloids As and Si (Si is largely inert and will not be considered further). The sequestration of metals and metalloids can reduce the availability of these potentially toxic elements. However, iron plaque may not be a permanent sink for these elements. As plant roots senesce and iron reduction begins to dominate over iron oxidation, iron plaque can be dissolved and elements associated with the plaque released in a more bioavailable form. Microbial processes help iron cycle rapidly between oxidized and reduced forms, but the role of microbes in the exchange of elements between iron plaque and soil has not received sufficient study.

9.1.1. Iron Oxidation

The deposition of iron oxides by *Fe(II)-oxidizing bacteria* (FeOB) has been proposed as a mechanism for the deposition of banded iron formations early in Earth's history (Widdel et al., 1993; Emerson, 2000; Konhauser et al., 2002) and it may have been an important energy-generating pathway on Mars if that planet once supported life (Robbins and Iberall, 1991; McKay et al., 1996). Microbes that gain energy from the oxidation of Fe(II) to Fe(III) are a diverse group that have developed several distinct strategies that compete with the chemical oxidation of Fe(II). Rates of chemical Fe(II) oxidation are strongly influenced by pH and oxygen concentrations (Singer and Stumm, 1970):

$$\frac{-d[\mathrm{Fe(II)}]}{dt} = k \times [\mathrm{Fe(II)}] \times [\mathrm{OH^-}]^2 \times \mathrm{pO_2} \qquad (9.1)$$

At circumneutral pH, free Fe(II) is chemically unstable in the presence of O_2; the half-life for ferrous iron is on the order of 10 minutes in fully oxygenated solutions. Since the doubling time of known lithotrophic FeOB is on the scale of hours and the energy yield from Fe(II) oxidation is low (Ehrlich et al., 1991), the potential for a microbial contribution to iron oxide formation in circumneutral environments had been largely neglected. However, microbial Fe(II) oxidizers can be found at high abundances in circumneutral environments (Emerson and Weiss, 2004). Because abiotic Fe(II) oxidation rates are slower at lower O_2 concentrations (Liang et al., 1993), Fe(II) oxidizers growing at near-neutral pH are generally found in microaerobic zones at the interface between oxic and anoxic environments. A second strategy for accessing Fe(II) is to grow in an anoxic environment where microbial Fe(II) oxidation is coupled to anoxygenic photosynthesis (Widdel et al., 1993), nitrate reduction (Straub et al., 1996; 2004), or perchlorate reduction (Chaudhuri et al., 2001; Lack et al., 2002). In such environments, chemical oxidation is limited by the absence of O_2. The third strategy, which is used by organisms such as *Acidothiobacillus* (formerly, *Thiobacillus*) *ferrooxidans* and *Leptospirillum ferrooxidans*, is to grow at low pH (<3) where Fe(II) is chemically stable (Ehrlich et al., 1991; Blake and Johnson, 2000). These acidophilic organisms have been well studied due to their role at acid mine drainages. Although ombrotrophic bogs can have a pH below 3, they typically have low mineral contents and are not representative of wetlands where Fe cycling is likely to be important, most of which tend to be more circumneutral in pH. Thus, we do not consider Fe(II) oxidation at very low pH further here. Several recent reviews provide detailed information on various aspects of circumneutral FeOB (Emerson, 2000; Croal et al., 2004a; Edwards et al., 2004; Emerson and Weiss, 2004; Megonigal et al., 2004).

9.1.2. Iron Reduction

Dissimilatory Fe(III) reduction may have been one of the earliest forms of microbial respiration on the planet (Vargas et al., 1998). However, it was not until the last quarter of the twentieth century that it was confirmed that nonfermentative microbes could link Fe(III) reduction to the oxidation of H_2, acetate, and other organic compounds (Balashova and Zavarzin, 1980; Lovely and Phillips, 1988; Lovley et al., 1993). In this chapter, we use the term *Fe(III)-reducing bacteria* (FeRB) to refer to any microorganisms that gain energy through coupling dissimilatory reduction of ferric iron with respiration.

Collectively, *iron-reducing microbes* are a diverse group, including both Bacteria and Archaea, that can use a wide range of electron acceptors in addition to Fe(III). Individual species have been studied that can reduce O_2, NO_3^-, S^0, SO_4^{2-}, humic substances, and/or other metals including Mn(IV), U(VI), and Cr(VI) (Coleman et al., 1993; Lovley et al., 1993; 1996; Kashefi and Lovley, 2000; Lovley, 2000; Lovley and Coates, 2000; Thamdrup, 2000). The FeRB also show diversity in electron donor use. Members of the largest group of FeRB, the Geobacteraceae, are obligate anaerobes capable of coupling Fe(III) reduction to the complete oxidation of H_2, acetate, and simple organic acids (Lovley et al., 1987; Lovley and Phillips, 1988; Lovley, 2000). In contrast, other FeRB, including *Shewanella putrefaciens* and *S. alga*, can use H_2 or complex organic molecules (e.g., lactate, formate) as electron donors but are generally only capable of incomplete oxidation of these compounds to acetate (Lovley, 2000). For more information on the physiology of FeRB and their environmental importance, the reader is referred to recent reviews by Lovley (2000), Thamdrup (2000), Straub et al. (2001), Croal et al. (2004a), and Megonigal et al. (2004).

Like Fe(II) oxidation, Fe(III) reduction can proceed rapidly due to strictly chemical mechanisms (Figure 9.1). Iron(III) reduction by sulfides is the dominant abiotic reduction mechanism whereby sulfides accumulate following SO_4^{2-}

reduction (Jacobson, 1994):

$$2CH_2O + SO_4^{2-} \rightarrow H_2S + 2HCO_3^- \quad (9.2)$$

$$2FeOOH + 4H^+ + H_2S \rightarrow 2Fe^{2+} + S^0 + 4H_2O \quad (9.3)$$

$$6FeOOH + 10H^+ + S^0 \rightarrow 6Fe^{2+} + SO_4^{2-} + 8H_2O \quad (9.4)$$

Lovley (2000) suggested that H_2S reduction of Fe oxides is unimportant because H_2S and Fe(III) oxides usually do not occur simultaneously; in theory, biological Fe(III) reduction should deplete Fe(III) stocks before SO_4^{2-} reduction (i.e., H_2S production) begins. However, there are systems where nearly all Fe(III) reduction is chemically driven (Jacobson, 1994; Kostka and Luther, 1995). In such sulfidic environments, biological Fe(III) reduction may be limited to microsites where Fe(III) is abundant and H_2S is scarce. Biological Fe(III) reduction can be important in marine environments when H_2S–metal oxide interactions prevent H_2S accumulation even when SO_4^{2-} reduction is responsible for a considerable fraction of carbon metabolism. Iron(III) can also be reduced abiotically by organic compounds including phenols, quinones, and humic substances, even though Fe(III)-organic reactions are usually slow at circumneutral pH (LaKind and Stone, 1989; Lovley et al., 1996). However, the shuttling of electrons from humic acids to Fe(III) oxides can be significant at near-neutral pH and may play an important role in microbially mediated Fe(III) reduction (Lonergan et al., 1996; Lovley et al., 1996; Weiss et al., 2004). The aerobic photoreduction of Fe(III) may be locally important in some environments (Barbeau et al., 2001; Emmenegger et al., 2001) but is unlikely to play a significant role in subsurface root zone soils.

9.2. OVERVIEW OF THE RHIZOSPHERE

Plants influence soil microbial activity because they are both a strong source and sink for resources that microorganisms require. Root activity regulates the abundance of organic carbon, nutrients, and terminal electron acceptors, and it influences soil physical properties such as porosity. The influence of a root on the surrounding soil is greatest at the root surface and declines gradually with distance into the bulk soil (Cheng et al., 1996).

9.2.1. Definitions

The *rhizosphere* is the three-dimensional zone of influence that a root exerts on the surrounding soil. Its inner boundary is the *rhizoplane*, which is the two-dimensional root surface. The rhizosphere penetrates a variable distance into the soil as determined by the biological, chemical, and physical soil characteristics. The thickness or volume of the rhizosphere for a particular element or compound depends on the source–sink strength of the root, the source–sink strength of the soil, and the diffusion rate (van Bodegom et al., 2001a).

Quantifying rhizosphere biogeochemistry is difficult because the thickness of the rhizosphere is usually less than a few millimeters and varies widely in time and space (Pedersen et al., 1995). In situ measurements of the rhizosphere have been achieved with microelectrodes (Armstrong et al., 2000), precise spatial sampling (Gribsholt et al., 2003), and chemical probing (Cheng et al., 1994). However, in practice the rhizosphere is often defined operationally because quantification requires destructive sampling of roots, which disrupts chemical gradients in the soil. A common method for isolating upland rhizosphere soil is to gently shake the roots to remove loosely held soil; soil remaining on the roots is called *rhizosphere soil*. In this case, rhizosphere soil presumably represents particles bound by root and microbial exudates or fungal hyphae. This method is not effective in soils that are highly sandy or organic because they do not adhere well to roots. An operational definition for investigating rhizosphere iron cycling in wetlands is the root surface and its associated iron plaque (Weiss et al., 2003). This definition is conservative because it is likely that the influence of wetland plants on iron reduction and oxidation penetrates somewhat beyond the boundary of the Fe plaque. The advantage of this definition is that Fe plaque is easily delineated and separated from the bulk soil, providing an objective and reproducible means of sampling the rhizosphere iron pool.

9.2.2. Rhizosphere Processes

The most fundamental influence of wetland plants on rhizosphere microbiology is the introduction of organic carbon and molecular oxygen (O_2; Figure 9.2). Organic carbon is the dominant electron donor for soil microbial respiration globally, and plants are the major source of organic carbon in terrestrial systems. Plants supply carbon by exuding soluble compounds from live roots and the turnover of dead leaves, shoots, and roots. Microbial activity is generally carbon limited and therefore stimulated in the presence of plants (Megonigal et al., 2004).

Unvegetated, saturated sediments are anaerobic below a few millimeters of the surface because microbial O_2 demand exceeds the O_2 supply. Plants enhance the O_2 supply by delivering atmospheric O_2 (or photosynthetically produced O_2 in the case of isoetid plants) through stems to roots, where it can leak into the surrounding soil, a process known as *radial* O_2 *loss* (ROL; Armstrong, 1964). The rate of radial O_2 loss varies among plant species due to differences in the O_2 transport mechanism (e.g., passive diffusion vs. through-flow of gases) and root morphology (Colmer, 2003). Intraspecies variability occurs with diurnal and seasonal changes in plant activity (Flessa, 1994; Pedersen et al., 1995) and root age (Armstrong, 1971).

Molecular oxygen regulates both aerobic and anaerobic microbial metabolisms (Bodelier, 2003). Because of intense competition for O_2, aerobic rhizosphere bacteria may be adapted to low O_2 concentrations and periods of anoxia. Competition for O_2 among microbial species, and between biotic and abiotic processes, has received relatively little attention (Laanbroek, 1990) but is likely to influence rhizosphere oxidation rates. Aerobic processes that occur in the wetland rhizosphere are those found in other aerobic–anaerobic interface environments and

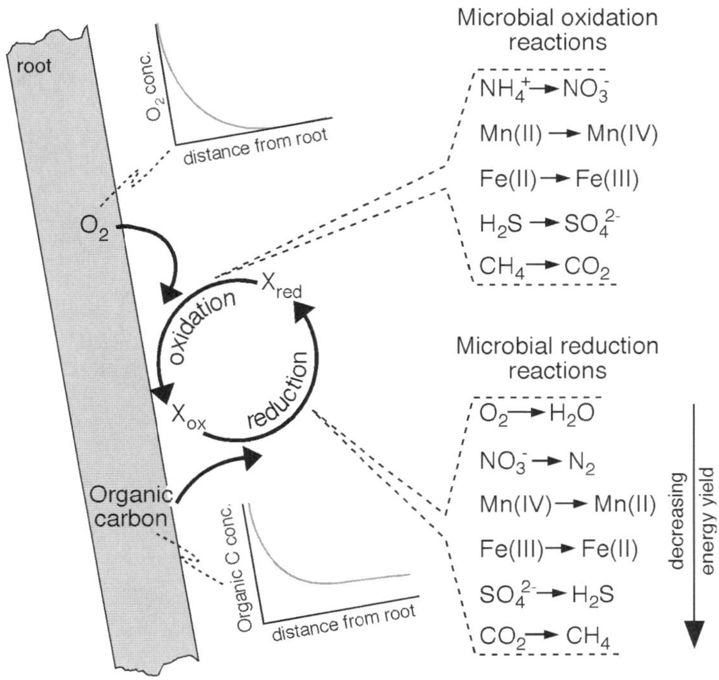

Figure 9.2. Plants influence microbial transformations of oxidized (X_{ox}) and reduced compounds (X_{red}) in the rhizosphere by supplying O_2 and organic carbon. Due to consumption within the soil, concentrations of both O_2 and organic C decrease with distance from the root.

include aerobic respiration and the oxidation of NH_4^+, Mn(II), Fe(II), H_2S, and CH_4 (Figure 9.2).

Aerobic respiration influnces anaerobic respiration by three mechanisms: interspecific competition for organic carbon and H_2, alternative respiration pathways, and regeneration of terminal electron acceptors. Because aerobic respiration yields more free energy than do other forms of respiration, aerobes outcompete anaerobes for organic carbon (Megonigal et al., 2004). A similar phenomenon occurs within species that have the capacity to use two or more alternative respiration pathways. For example, denitrifying bacteria switch from O_2 reduction to nitrate reduction in the absence of O_2 (Payne, 1981; Firestone, 1982). Finally, biotic and abiotic oxidation can regenerate terminal electron acceptor substrates such as NO_3^-, Fe(III), and SO_4^{2-}, thereby preventing electron acceptor limitation of respiration.

In the absence of O_2, competition among anaerobic microbes for electron donors sets up a series of alternative terminal electron-accepting processes in the order NO_3^- reduction, Mn(IV) reduction, Fe(III) reduction, SO_4^{2-} reduction, and methanogenesis (Figure 9.2; Ponnamperuma, 1972; Megonigal et al., 2004). To a first approximation, a single terminal electron-accepting process dominates microbial respiration until the supply of the required electron donor becomes limiting. However, heterogeneity in substrate abundance can cause competing pathways of microbial respiration to coexist in small areas such as the rhizosphere (Højberg et al., 1994).

The microbial community structure of the rhizosphere is distinct from the bulk soil (e.g., Mummey and Stahl, 2004), as expected from plant-driven effects on substrate availability. Wetland rhizospheres typically have higher densities of microorganisms, including methanogens and methane-oxidizing bacteria (Gilbert and Frenzel, 1995), iron-reducing and iron-oxidizing bacteria (FeRB and FeOB; Weiss et al., 2003), and sulfate-reducing bacteria (Rooney-Varga et al., 1998; Hines et al., 1999). There are also differences in microbial diversity (Bergholz et al., 2001; Derakshani et al., 2001; Eller and Frenzel, 2001; Burke et al., 2002). We do not yet understand how fine-scale differences in microbial community structure influence ecosystem-level biogeochemical cycling.

9.3. IRON OXIDATION IN THE RHIZOSPHERE

The presence of iron oxyhydroxide coatings (i.e., Fe plaque, often dominated by ferrihydrite) on the surface of wetland plant roots is visual evidence that subsurface iron oxidation is occurring in otherwise anoxic wetland soils and sediments. Oxygen delivered via radial O_2 loss may react with reduced iron in soil pore spaces to form oxidized iron that can be deposited on the plant roots as Fe plaque. Despite a long history of observing Fe plaque on wetland plant roots and understanding the basics of plaque formation [i.e., reaction of plant-transported O_2 with Fe(II) in soils and sediments], it was largely assumed that plaque formation is predominately an abiotic (i.e., chemical) process because the kinetics of chemical oxidation can be extremely rapid (Mendelssohn et al., 1995). However, recent evidence has demonstrated that populations of lithotrophic FeOB are associated with Fe plaque and may play a role in plaque deposition.

9.3.1. Microbially Mediated Rhizosphere Fe(II) Oxidation

The rhizosphere is home to a diverse microbial community, including aerobic heterotrophs (Gilbert and Frenzel, 1998), methane oxidizers (Bosse and Frenzel, 1997; Calhoun and King, 1997), and ammonium and nitrite oxidizers (Bodelier et al., 1996; Arth et al., 1998). Microscopy has also shown that microbial cells are associated with Fe plaque (Trolldenier, 1988; St-Cyr et al., 1993), but visual examinations alone cannot determine if cells are responsible for plaque formation. Trolldenier (1988) demonstrated that rusty-colored colonies formed when root plaque was inoculated into an iron-containing medium, but further characterization of these colonies was not attempted.

It was not until the late 1990s that the presence of lithotrophic FeOB in soils was demonstrated conclusively (Emerson et al., 1999). Since then, at least five strains of aerobic lithotrophic FeOB have been enriched from wetland soils. Four strains enriched by Weiss and colleagues were isolated from the rhizosphere of four different plant species from three different wetlands (Weiss, 2002). All were microaerobic obligately lithotrophic Fe oxidizers within the β-proteobacteria that were unable to grow on either organic or inorganic energy sources, except iron. Their physiology as obligate lithotrophic Fe-oxidizing bacteria is very similar to that of FeOB isolated from groundwater (Emerson and Moyer, 1997) and the Loihi hydrothermal vent system (Emerson and Moyer, 2002). In addition to the strains enriched by Emerson and colleagues, a wetland FeOB (strain TW2) was enriched from freshwater soils of an Alabama wetland (Sobolev and Roden, 2001). This organism grows in circumneutral microaerobic conditions and can oxidize Fe(II) for energy but can also grow with acetate as an electron donor. Strain TW2 is genotypically distinct from the FeOB enriched by Emerson, Weiss, and colleagues and lies within the β-proteobacteria (Sobolev and Roden, 2004). Edwards et al. (2003) isolated obligate chemolithoautotrophic FeOB from the deep sea that could grow under microaerobic conditions or anaerobically in the presence of nitrate, but it is unknown if these deep-sea FeOB have relatives that live in soil environments.

Aerobic lithotrophic FeOB appear to be a ubiquitous feature of wetland soils. Weiss et al. (2003) surveyed 13 wetland and aquatic habitats in the mid-Atlantic region of the United States and cultured FeOB successfully from every wetland sampled and from 23 of 25 plant species sampled. In a detailed analysis of rhizosphere FeOB on the roots of *Typha latifolia* (broadleaf cattail) and *T. angustifolia* (narrowleaf cattail) collected from multiple wetlands, Weiss et al. (2003) found that aerobic lithotrophic FeOB comprised approximately 1% of all microbial cells on the root surface (range <0.01 to 6.3% or 3.4×10^2 to 1.2×10^6 cells g^{-1}, $n = 6$). Furthermore, FeOB comprised a greater fraction of the microbial community on the root surface than in bulk soil, indicating that rhizosphere gradients of O_2 (via radial O_2 loss) and Fe(II) (diffusion from wetland porewater) are favorable for the growth of FeOB. These data suggest that FeOB influence Fe(II) oxidation rates.

There is little information on the presence of anaerobic FeOB in the rhizosphere. A preliminary survey of a Maryland tidal freshwater marsh soil indicated that nitrate-reducing anaerobic FeOB were present, albeit in low numbers ($\sim 10^3$ to 10^4 cells mL^{-1}; Emerson, unpublished data). In freshwater sediments, Straub and Buchholz-Cleven (1998) found greater numbers of anaerobic nitrate-reducing FeOB (1×10^5 to 5×10^8 cells g^{-1}) than anaerobic phototrophic FeOB (0.1 to 3.9×10^3 cells g^{-1}).

9.3.2. Microbial vs. Chemical Fe(II) Oxidation

As mentioned previously, it has long been assumed that the dominant route of Fe plaque formation is via chemical rather than biological pathways. In part, this is due to potentially rapid rates of chemical Fe oxidation under circumneutral conditions (see above). Although microbial cells were observed embedded within the Fe plaque (Trolldenier, 1988; St-Cyr et al., 1993), it was assumed that these cells catalyzed abiotic Fe(II) oxidation by providing Fe oxide nucleation sites rather than actively metabolizing Fe(II). For methodological reasons, it is not easy to distinguish between in situ rates of microbial and chemical Fe(II) oxidation. There are no known specific inhibitors of either process, and radioisotopes (e.g., ^{55}Fe) undergo rapid isotope exchange between Fe(II) and Fe(III) (Roden and Lovley, 1993). Recent work on the natural abundance of stable isotopic signatures of iron has shown that hydrous ferric oxides produced by an anaerobic phototrophic FeOB had $\delta^{56}Fe$ signatures that were 1.5‰ enriched relative to aqueous Fe(II) (Croal et al., 2004b). This fractionation was similar to that reported by Bullen et al. (2001) for abiotic Fe(II) oxidation under aerobic conditions, suggesting that it may be difficult to use stable isotopes to distinguish between biotic and abiotic oxidation except when Fe oxides are deposited in low-oxygen environments (Croal et al., 2004b). Finally, unlike historically well-studied FeOB such as *Gallionella ferruginea* and *Leptothrix ochracea*, which produce distinct tube- or sheath-shaped Fe oxides, the rhizosphere Fe(II) oxidizers identified earlier in this chapter do not produce visually distinct oxides, thus making it more difficult to identify the in situ activity of these organisms.

Because there have been no reliable methods for separating microbial and abiotic oxidation rates in situ, we have used pure cultures of rhizosphere-isolated Fe(II) oxidizers in laboratory settings to understand the importance of biological Fe(II) oxidation in the rhizosphere (Neubauer et al., 2002). In one set of experiments, a Fe(II)-oxidizing bacterium enriched from the rhizosphere (strain BrT) was grown in bioreactors that allowed precise control of key factors [e.g., pH, O_2, Fe(II) supply rate] that can influence Fe(II) oxidation rates (Neubauer et al., 2002). In these experiments, strain BrT accounted for 18 to 53% of the total Fe(II) oxidation. The FeOB also depressed rates of chemical Fe(II) oxidation, perhaps due to the temporary sequestration of Fe(II) on exopolymers produced by the FeOB. This mechanism could make Fe(II) temporarily unavailable for chemical Fe(II) oxidation and more available to FeOB. The net result of bacterial Fe(II) oxidation and the suppression of chemical Fe(II) oxidation was that the FeOB accelerated total Fe(II) oxidation rates by up to 20% versus cell-free treatments (Neubauer et al., 2002). We have also grown FeOB in microcosms where the only source of O_2 for Fe(II) oxidation was radial O_2 loss from the roots of *Juncus effusus* (soft rush). The results from those experiments were similar to those from bioreactor studies: the presence of the FeOB tended to accelerate total Fe(II) oxidation rates relative to a system without FeOB (i.e., chemical oxidation only). Other studies have demonstrated similarly that lithotrophic FeOB from wetland and nonwetland environments can mediate a significant fraction of total iron oxidation at circumneutral pH (Emerson and Revsbech, 1994; Sobolev and Roden, 2001; Hanert, 2002; James and Ferris, 2004).

In contrast to the results outlined above, van Bodegom et al. (2001b) determined that biological Fe oxidation was insignificant in rice paddy soils because MPN estimates of FeOB density were low [10^2 FeOB gdw^{-1} (grams dry weight) soil]. By comparison, Weiss et al. (2003) reported densities of 3.7×10^6 FeOB gdw^{-1} soil and 5.9×10^5 FeOB gdw^{-1} root in *Typha* spp. wetlands. It is possible that the soils studied by van Bodegom and colleagues simply did not have a large population of lithotrophic FeOB and that iron oxidation was dominated by chemical reactions. Alternatively, the use of air-dried soils that were rapidly shaken following rewetting may have been detrimental to the growth of microaerobic FeOB. Despite van Bodegom's findings, direct measurements (albeit in laboratory settings) and circumstantial evidence indicate that lithotrophic aerobic FeOB can contribute significantly to iron oxidation in the rhizosphere.

9.4. IRON REDUCTION IN THE RHIZOSPHERE

The rhizosphere is a favorable location for Fe(III)-reducing microbes (FeRB) because both electron acceptors [Fe(III) oxides] and electron donors (e.g., organic carbon) are supplied and/or regenerated in proximity to plant roots. The FeRB can reduce a variety of Fe(III) oxide types, including ferrihydrite [$Fe(OH)_3$], lepidocrocite (γ-FeOOH), goethite (α-FeOOH), hematite (α-Fe_2O_3), magnetite ($Fe^{II}Fe^{III}_2O_4$), and Fe(III) oxides bound in smectite clay (Kostka and Nealson, 1995; Roden and Zachara, 1996; Kostka et al., 2002a). The availability of surfaces for FeRB attachment is the most important feature in determining the reduction rate of a particular Fe(III) oxide type. Thus, poorly crystalline Fe oxides (e.g., ferrihydrite) that have higher surface areas are more reducible than crystalline minerals such as hematite. In a study comparing the mineralogy and reactivity of rhizosphere and nonrhizosphere soils, Weiss et al. (2004) showed that 66 ($\pm$ 7)% of the Fe oxides in the rhizosphere were poorly crystalline (defined as the 0.2 M acid ammonium oxalate extractable fraction) versus only 23 ($\pm$ 7)% in nonrhizosphere soils. Furthermore, the percentage of poorly crystalline Fe(III) oxides was correlated with the abundance of FeRB in the rhizosphere (Weiss et al., 2003). The importance of poorly crystalline Fe oxides in supporting Fe(III) reduction was demonstrated in both field and laboratory settings, with rhizosphere Fe(III) oxides (less crystalline) showing higher Fe(III) reduction rates than Fe oxides from bulk soil (more crystalline; Weiss et al., 2004; 2005). In addition, the higher reactivity of rhizosphere Fe oxides persisted following additions of organic carbon, humic substances, and the FeRB *Geobacter metallireducens* to incubation vessels, indicating that the differences in reactivity were related to iron oxide morphology. Over a number of years, plant roots can cause a progressive long-term increase in extractable Fe(III) pools relative to unvegetated sites (Kostka and Luther, 1995; Roden and Wetzel, 1996; Frenzel et al., 1999; Weiss et al., 2004) that may increase the importance of Fe(III) reduction (relative to other catabolic processes) in vegetated substrates.

The availability of organic carbon is a second variable that affects rates of Fe(III) reduction in wetland and aquatic sediments (Thamdrup, 2000; Roden and Wetzel, 2002; Megonigal et al., 2004). For example, Roden and Wetzel (2002) demonstrated that although Fe(III) reduction kinetics were affected primarily by Fe(III) availability, Fe(III) reduction rates also correlated directly with the availability of organic carbon. Plant roots can serve as sources of labile organic carbon (e.g., as root exudates and detrital inputs; Blaabjerg et al., 1998; Hines et al., 1994; Gribsholt and Kristensen, 2002) and therefore may accelerate Fe(III) reduction rates in the rhizosphere. Additionally, there may be a shift in the competitive balance between FeRB, SO_4^{2-} reducers, and methanogens as electron donors (i.e., organic carbon) are supplied by plant roots and the competitive pressure for electron donors is lessened.

9.4.1. Rhizosphere Fe(III) Reducers

A physiologically diverse group of anaerobic microbes are found in the rhizosphere, including SO_4^{2-} reducers (Rooney-Varga et al., 1997; Hines et al., 1999), methanogens (King, 1994; Lehmann-Richter et al., 1999; Chin et al., 2004), and Fe(III) reducers (King and Garey, 1999; Weiss et al., 2003). Although FeRB have been isolated and characterized from many soil and sedimentary environments, there has been very little work to describe the abundance and activities of these organisms in the rhizosphere. However, it is clear that FeRB can be a significant component of the total rhizosphere microbial community. On the roots of *Typha latifolia* and *T. angustifolia* collected from six wetlands in Maryland and Virginia, FeRB accounted for a median of 9.0% (range 3.5 to 32.1%; $n = 6$) of the rhizosphere microbial community (Weiss et al., 2003). By comparison, FeRB in bulk soil from the same sites were always less than 0.5% of the soil microbial community. Similarly, FeRB were orders of magnitude more abundant in a vegetated bioturbated salt marsh than in unvegetated undisturbed sediments elsewhere in the system (Kostka et al., 2002b). The FeRB quantified by Weiss et al. (2003) were not characterized but are probably members of the Geobacteracae, since the MPN counting procedure used only acetate as an organic carbon source. Other FeRB, including *Shewanella* spp, may have been present but not detected, due to their inability to utilize acetate.

In what may be the only other study to look specifically at FeRB associated with wetland plant roots, King and Garey (1999) reported that FeRB isolates from the roots of *Spartina alterniflora* (cordgrass) and *Typha latifolia* were morphologically similar to *Geobacter* spp. These isolates grew on multiple electron donors, including acetate, ethanol, and succinate, and reduced Fe(III) and the humic acid analog AQDS (King and Garey, 1999). Significant Fe(III) reduction activity was reported at ecologically relevant (for the Maine study sites) temperatures of 0 and 20°C, but the freshwater and marine strains were adapted to different salinity regimes.

9.4.2. Microbial vs. Chemical Fe(III) Reduction

Typically, Fe(III) reduction is measured as the accumulation of HCl- or ascorbate-extractable Fe(II) or the disappearance of Fe(III) in anaerobically incubated sediments (Lovley and Phillips, 1986; Roden and Wetzel, 1996; Kostka et al., 2002c), but this approach cannot distinguish between biological and chemical Fe(III) reduction. One common method for separating the contributions of biological and abiotic reduction mechanisms involves suppressing H_2S production [and therefore chemical Fe(III) reduction by sulfides] by adding molybdate (MoO_4^{2-}), a specific inhibitor of SO_4^{2-} reduction (Oremland and Capone, 1988). The Fe(III) reduction rate measured in the absence of molybdate represents total (biological + chemical) Fe(III) reduction, whereas rates in the presence of molybdate indicate only biological Fe(III) reduction. It is possible that molybdate interferes with processes other than SO_4^{2-} reduction, but results from the molybdate method are consistent with other geochemical data and are commonly used to separate enzymatic and nonenzymatic Fe(III) reduction (Megonigal et al., 2004). Rates of biological Fe(III) reduction have also been estimated using a mass balance approach where total metabolism (i.e., ΣCO_2 production) and SO_4^{2-} reduction rates are measured and Fe(III) reduction is calculated by difference (Canfield et al., 1993; Thamdrup and Canfield, 1996). This approach has been used most often in anaerobic incubations of subtidal marine sediments where methanogenesis can usually be ignored and denitrification is limited by low NO_3^- concentrations. The stable isotope signatures of ferric and ferrous iron may eventually be useful in distinguishing between biological and chemical Fe(III) reduction, but further research on the fractionation associated with biological and chemical processes is required. Pure culture work with *Shewanella alga* demonstrated a 1.3‰ fractionation between ferrihydrite and ferrous iron (Beard et al., 1999; 2003) and the amount of fractionation may depend on the rate of biotic Fe(III) reduction (Johnson et al., 2005). However, larger fractionations have been observed for abiotic Fe(III) reduction (Anbar et al., 2000). Interestingly, the isotopic fractionation reported for biological Fe(III) reduction is similar in direction and magnitude to the 1.5‰ fractionation reported by Croal et al. (2004b) for anaerobic phototrophic Fe(II) oxidation.

Thamdrup (2000) reviewed biological Fe(III) reduction in freshwater, estuarine, and marine sediments and found that enzymatic Fe(III) reduction accounted for 0 to ~50% of total carbon metabolism. Furthermore, the contribution of Fe(III) reduction was related to the concentration of poorly crystalline Fe(III) oxides; Fe(III) reduction increased in sediments with higher Fe oxide concentrations (Hines et al., 1997; Thamdrup, 2000; Roden and Wetzel, 2002). Recent work has found large populations of FeRB in vegetated salt marsh sediments (Lowe et al., 2000; Kostka et al., 2002b; Koretsky et al., 2003). In mineral salt marsh sediments, microbial processes accounted for over 90% of the Fe(III) reduction (Kostka et al., 2002b; Gribsholt et al., 2003), whereas microbial and chemical Fe(III) reduction were of equal importance in an organic-rich brackish soil (Neubauer et al., 2005). Koretsky et al. (2003) proposed that seasonal changes in rates of H_2S production can cause an oscillation between chemically dominated Fe(III) reduction in summer and biologically dominated Fe(III) reduction during other times of the year. These findings differ from the classical view that SO_4^{2-} reduction dominates anaerobic metabolism in saline systems and that most Fe(III) reduction is due to abiotic reduction by sulfides (Jacobson, 1994; Kostka and Luther, 1995). In freshwater wetlands, Fe(III) reduction is typically dominated by microbial processes due to low sulfide concentrations and can be the dominant anaerobic decomposition pathway (Roden and Wetzel, 1996; Neubauer et al., 2005).

As summarized by Jacobson (1994), biological Fe(III) reduction will be more important than chemical reduction when amorphous Fe(III) oxides are plentiful and continually regenerated, or H_2S production is low relative to the Fe(III) concentration. This first condition is likely to be met in the rhizosphere where radial O_2 loss drives Fe oxide formation. The second condition will be met in low-salinity wetlands, or in saline systems with mineral (i.e., iron-rich) sediments. However, even chemical reduction of Fe(III) is ultimately due to microbes since the H_2S that reduces the Fe is the result of a biological process, SO_4^{2-} reduction (Megonigal et al., 2004).

9.5. POTENTIAL FOR FeOB–FeRB INTERACTIONS

Aerobic–anaerobic interface zones are biogeochemical hot spots where oxidation and reduction reactions can occur in close proximity. This has been demonstrated for carbon (e.g., methane oxidation and methanogenesis; King, 1996), nitrogen (e.g., nitrification and denitrification; Reddy et al., 1989), and sulfur cycling processes (e.g., sulfide oxidation and sulfate reduction; Brune et al., 2000). As reviewed above, FeOB and FeRB are abundant in the rhizosphere, and Fe oxidation and reduction occur in the rhizosphere, but our inability to distinguish between in situ rates of biological versus chemical iron cycling has made it difficult to establish the unambiguous existence of a microbially driven rhizosphere iron cycle. In this section we describe some potential mechanisms for spatial and temporal coupling between microbial iron oxidation and reduction, with the understanding that chemical iron cycling can also be important under some conditions. All of the iron cycling methods described below are between microaerobic FeOB and anaerobic FeRB—in theory, coupling between anaerobic FeOB and FeRB need not occur near a root or other aerobic–anaerobic interface (Straub et al., 2004).

9.5.1. Spatial Coupling

Spatial separation between aerobic microbial Fe oxidation and anaerobic reduction would allow these groups of microbes to coexist in the rhizosphere. Radial O_2 loss from plant roots is the only source of molecular O_2 to otherwise anoxic subsurface wetland soils. Thus, aerobic FeOB should be located on or near the root surface, where both O_2 and Fe(II) are found at appropriate concentrations, whereas anaerobic FeRB should be found in the bulk soil away from the direct influence of root-derived O_2 (Figure 9.3*a* and *b*). This generality, however, may need to be modified, as recent findings indicate that the FeRB *Geobacter sulfurreducens,* which was formerly classified as an obligate anaerobe, can survive exposure to O_2 and even grow using O_2 as an electron acceptor (Lin et al., 2004). The principal products of microbial iron oxidation at circumneutral pH will be

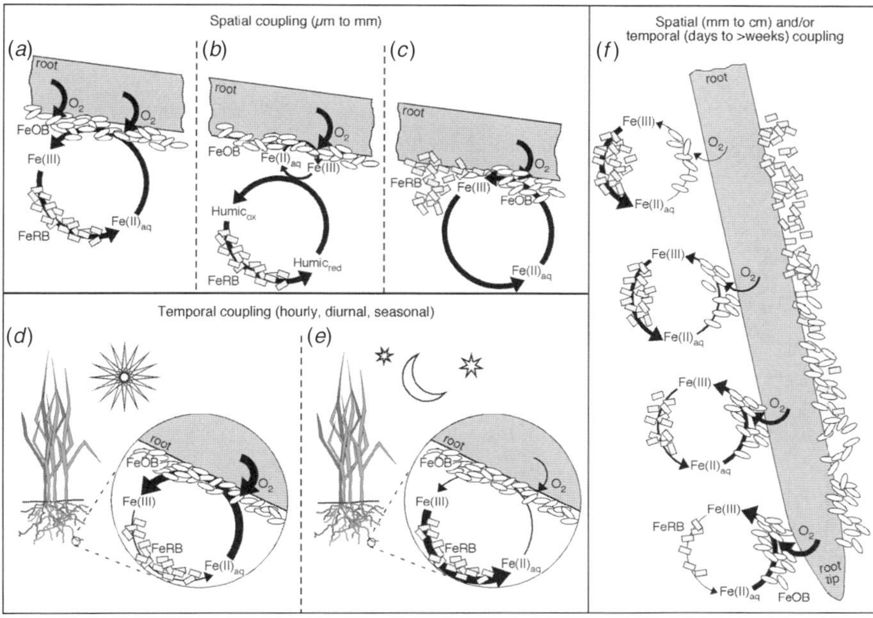

Figure 9.3. Mechanisms for spatial and temporal coupling between microbial Fe(II) oxidation and Fe(III) reduction. In each panel, arrow thickness indicates the relative importance of each process, not the gross rate. (*a–c*) Fine-scale spatial coupling on a scale of μm to mm. The Fe(II)-oxidizing bacteria (FeOB) colonize the root surface and utilize O_2 leaked from plant roots to oxidize pore-water Fe(II), while Fe(III)-reducing bacteria (FeRB) live a short distance from the root surface in anoxic soils/sediments or are found on the root surface but are spatially segregated from the FeOB. Movements of Fe(III) and/or humic acids link the activities of FeOB and FeRB. (*d,e*) Temporal coupling. Rates of radial O_2 loss can vary on hourly, diurnal, and seasonal time scales, with concomitant effects on the relative rates of Fe(II) oxidation and Fe(III) reduction. (*f*) Spatiotemporal coupling along the length of a root. Gradients in rates of Fe(II) oxidation and Fe(III) reduction can occur along the length of the root, or the same spot on a root can transition from net Fe(II) oxidation to net Fe(III) reduction as the root ages. The FeOB and FeRB can be segregated between the root surface and bulk soil–sediment (left side of root) or spatially segregated along the length of the root (right side of root). Throughout the figure, Fe(III) generically refers to both particulate $Fe(OH)_3$ oxides and $Fe(III)_{aq}$.

particulate Fe(III) hydroxides and oxides (Emerson and Revsbech, 1994; Emerson and Moyer, 1997; Edwards et al., 2003), but there is evidence that soluble Fe(III) may be produced in some instances (Sobolev and Roden, 2001). Particulate oxides will be localized around the root, but associations with microbially produced exopolymers may influence their availability for Fe reduction. Also, humic acids and other electron-shuttling compounds could link spatially separated FeOB and FeRB by transferring electrons from FeRB in anoxic soil to Fe oxides in the rhizosphere (Lovley et al., 1996; King, 2005; Figure 9.3*b*).

As revealed by direct in situ measurements (Christensen et al., 1994; Armstrong et al., 2000) and modeling approaches (Armstrong et al., 2000; van Bodegom et al., 2001a), O_2 released from wetland plant roots is typically consumed within 5 mm of the root surface. However, in sediments with a low reducing capacity, high rates of radial O_2 loss can result in significant oxygenation at distances greater than 1 cm from roots (e.g., Pedersen et al., 1995). At the limit of O_2 penetration, the reduction of microbially produced soluble and solid-phase Fe(III) can begin, resulting in the production of soluble Fe(II) that will diffuse back toward the root surface. There can also be zones of iron reduction within the larger oxidized rhizosphere due to localized (on a scale of microns) depletion of O_2 around organic-rich particles, as has been proposed for various laboratory and sedimentary environments (Jorgensen, 1977; Jahnke, 1985; Sobolev and Roden, 2004). In the oxidized rhizosphere, zones of iron oxidation and reduction could thus be separated by micrometers to millimeters, due to the differing tolerances of FeOB and FeRB to molecular O_2.

Analyses of individual 1-cm root segments have revealed that both FeOB and FeRB are present on the same root length (Weiss et al., 2003). One explanation for the simultaneous presence of FeOB and FeRB on the root surface is that a single 1-cm piece of root can have both oxidizing and reducing zones, resulting in the spatial segregation of FeOB and FeRB along the root surface (Figure 9.3*c*). The diameter of epidermal root cells is on the order of 10 to 20 μm (from micrographs in Smits et al., 1990; Armstrong et al., 1991) versus FeOB cell diameters of 0.4 to 0.9 μm (Weiss, 2002). Thus, we can speculate that clusters of FeOB thrive where O_2 is leaking out between epidermal cells, and FeRB thrive in microanoxic zones located on the root surface between these zones of O_2 leakage. To our knowledge, nobody has used microelectrodes to examine O_2 or redox variations across the root surface on such a fine spatial scale. The use of FISH (fluorescence in situ hybridization) to stain FeOB and FeRB differentially will be useful in resolving the spatial distribution of these microbes along the root surface.

9.5.2. Temporal Coupling

Temporal changes in iron cycling are driven by hourly, daily, and seasonal variations in rates of radial O_2 loss and organic matter inputs, which are influenced by plant development and rates of photosynthetic activity. Rates of radial O_2 loss vary diurnally as changes in temperature and humidity influence rates of pressurized ventilation and thermal diffusion within plant aerenchyma tissue (Colmer, 2003). Organic carbon inputs are linked to photosynthetic activity and change as a function of time of day and season; this connection has been well demonstrated in studies linking photosynthesis and CH_4 production (Whiting and Chanton, 1993; Megonigal et al., 1999; Lu et al., 2002). Diurnal changes in radial O_2 loss and organic carbon inputs should then influence rates of iron oxidation and reduction. The rhizosphere should support net iron oxidation when O_2 is released via radial O_2 loss (Figure 9.3*d*) and net iron reduction when O_2 is unavailable (Figure 9.3*e*). For example, exposing *Zostera marina* (eelgrass) plants to light resulted in decreases in pore-water concentrations of Fe^{2+} and H_2S due to rhizosphere oxidation; these changes were not observed in unvegetated sediments (Hebert and Morse, 2003). Rates of SO_4^{2-} reduction also increased during the day in seagrass communities because photosynthesis increased the delivery of organic exudates to the rhizosphere community (Blaabjerg et al., 1998). By extension, we might suspect that gross rates of both Fe(II) oxidation and Fe(III) reduction would increase during the day relative to night, although in situ rates of Fe oxidation and reduction in the rhizosphere have not been measured on such short scales.

Longer-term patterns of plant photosynthesis, radial O_2 loss, and root growth and death can create similar patterns of iron cycling on a seasonal scale. The *Spartina maritima* (small cordgrass)–dominated salt marsh soils of the Tagus Estuary in Portugal contain thick iron-rich rhizoconcretions ($\leq$ 0.4 cm) that form around plant roots and can persist in the sediment for several years (Vale et al., 1990; Sundby et al., 1998). Based on microelectrode measurements of Fe^{2+} and O_2, Sundby et al. (2003) proposed a mechanism for concretion formation in the Tagus Estuary marshes. Noting that rhizosphere iron oxidation can only occur when Fe^{2+} and O_2 are available, Sundby et al. (2003) hypothesized that rapid rates of biological and/or chemical Fe(II) oxidation occurred in spring as new roots invaded anoxic sediments with initially high Fe^{2+} concentrations (up to 1700 μmol L^{-1}) and undetectable pore-water [O_2]. During summer, when plant activity was greatest, Fe^{2+} was below the limit of detection (i.e., all available Fe^{2+} had been oxidized) and O_2 concentrations reached a maximum above 100 μmol L^{-1}. As plant activity decreases through the fall and winter, O_2 concentrations in the sediments decrease (also reported by Flessa, 1994). As radial O_2 loss decreases, anaerobic processes, including Fe(III) reduction, become more important.

Temporal changes in coupled iron oxidation–reduction over the course of a single growing season can have effects on other wetland processes. In a tidal freshwater marsh on the Patuxent River, Maryland, microbial Fe(III) reduction was the most important anaerobic decomposition pathway in early summer and accounted for over 90% of total anaerobic metabolism (Neubauer et al., 2005). By the end of the summer, methanogenesis was the dominant anaerobic pathway (77% of anaerobic metabolism). The importance of Fe(III) reduction early in the summer is coincident with peak aboveground plant biomass at the study site. This suggests that high rates of radial O_2 loss rapidly regenerated Fe(III) oxides, which were then immediately available for reduction. As rates of radial O_2 loss decreased with plant senescence, Fe(III) reduction became limited by the availability of poorly crystalline reducible Fe(III) oxides. Because the balance between Fe(III) reduction and methanogenesis is affected by competition between FeRB and methanogens for electron donors, decreasing rates of Fe(III) reduction relieved the competitive pressure on methanogens and allowed methanogenesis to increase in importance.

Neubauer et al. (2005) hypothesized that the highest rates of rhizosphere Fe oxidation and reduction occur simultaneously, whereas Sundby et al. (2003) proposed a seasonal separation between the processes. This difference may be related to rates of radial O_2 loss. In the Tagus Estuary, high summer rates of radial O_2 loss from *S. maritima* led to measurable root zone O_2 and the absence of porewater Fe(II). In contrast, the *Peltandra virginica*/*Pontederia cordata*–dominated tidal freshwater marsh studied by Neubauer and colleagues contained high porewater Fe(II) concentrations (up to 2700 μmol L^{-1}) throughout the summer, with little visible Fe plaque on the roots. Thus, there may be tight coupling between oxidation and reduction when rates of ROL are low, with a longer temporal disconnect between the processes at high ROL rates.

9.5.3. Spatiotemporal Coupling

Although radial O_2 loss is the ultimate source of O_2 to FeOB and other aerobes living in the rhizosphere, rates of radial O_2 loss are not uniform along the length of a root. As a root ages (i.e., with increasing distance from the root tip), suberization of the endodermis decreases root permeability and rates of radial O_2 loss in older roots (Armstrong, 1971; Smits et al., 1990; Armstrong et al., 1994; Christensen et al., 1994). Thus, gradients in rates of microbial iron oxidation and reduction may occur along the length of a root (Figure 9.3*f*), with relatively more iron oxidation occurring near the root tip (although little iron oxidation is likely at the gas-impermeable root tip). As a contrast, isoetids exhibit their greatest O_2 losses from the root base, due to high overall root permeability (Smits et al., 1990), and should therefore show opposite patterns of rhizosphere iron oxidation and reduction. There can also be a temporal disconnect between microbial iron oxidation and reduction where the same spot on a root shifts from net Fe oxidation to net Fe reduction (with a corresponding shift from active FeOB to active FeRB) as the root ages. Indeed, a temporal change in rates of ROL for a root segment is likely to lead to spatial changes in the distribution of iron oxidation and reduction along the length of a root.

9.6. INTERACTIONS OF IRON WITH OTHER METALS IN THE RHIZOSPHERE

Given the abundance of Fe in soils, the variety of mineral forms that it may take, and the complex interactions of those mineral forms with other metals, Fe minerals often play an essential role in the overall control of metal interactions in the soil. This is especially true in the rhizosphere, where as discussed above, microbially mediated cycling of Fe between oxidized and reduced forms can be rapid. Wetland plants can increase metal sequestration by driving the formation of Fe plaque, particularly in tidal wetland soils, where burial is rapid. The binding of metals to Fe plaque in the rhizosphere is one reason that wetland plants have been investigated for the phytoremediation of metal-contaminated sites (Weis and

Weis, 2004). If the rhizosphere plays an important role in the sequestration of metals in wetland soils, fluxes of potentially toxic metals out of the wetland to adjacent aquatic systems will, in theory, be reduced. However, plants may also increase metal mobilization through rhizosphere acidification and the oxidation of metal–sulfide complexes (Jacob and Otte, 2003). They may also represent only a temporary sink for metals if rhizosphere Fe plaque is reduced following plant senescence. Furthermore, metals can be exported from the ecosystem if contaminated plant parts are consumed by people or wildlife. The pathways and possible health effects of metal consumption have been especially well studied in Southeast Asia, where metal contamination (notably As) of rice crops is a serious public health issue (e.g., Meharg and Rahman, 2003; Meharg, 2004).

9.6.1. Metals in the Rhizosphere

For over 20 years it has been recognized that the rhizosphere can play an important role in the mobility of a number of metals and that the presence of Fe plaque can significantly affect the dynamics of these interactions (Weis and Weis, 2004). For example, the field observations of Taylor and Crowder (1983) showed that several heavy metals were manyfold more concentrated in the roots and rhizosphere of *Typha latifolia* than in the aboveground plant tissues. A study of *Phragmites australis* (common reed) in a landfill leachate system found that Fe formed a highly concentrated plaque on the roots that was enriched in other metals, such as Cu, Zn, Pb, Cd, Al, and Mn (Peverly et al., 1995). Although this study did not look expressly at iron–metal associations, the authors inferred that the Fe plaque might act as a filter to reduce translocation of these metals into the plant. An investigation by Sundby et al. (2005) of a Pb-contaminated salt marsh in Portugal found a striking seasonality to the presence of soluble Pb(II), with high concentrations in summer and levels below detection in the winter. Fe plaque was also present and appeared to be more abundant in summer. Although the authors did not measure plaque-bound Pb specifically, they suggested that Fe plaque was clearly not a major sink for soluble Pb(II), perhaps because the Pb was complexed with organic compounds. These and other field observations have led to a number of laboratory studies, most of which use individual species of plants grown hydroponically in the greenhouse to test the effects of Fe plaque on different metals that are added to the solution. Results from a number of these studies are summarized below.

There was a moderate adsorption of Al to Fe plaque, and this resulted in reduced Al uptake in plant tissue of *Phragmites* (Batty et al., 2002). Copper is concentrated in Fe plaque but will preferentially adsorb to Mn if Mn is present in the plaque (St-Cyr and Crowder, 1990; Ye et al., 2001). The sequestration of Cu in the root plaque appears to reduce uptake of Cu by the plant. However, as Cu concentrations build up on the Fe plaque, the amount of Cu transported into the plant tissue can increase (Ye et al., 2001). Zinc can also be concentrated in Fe plaque. For *Typha*, this effect is more pronounced in the field than in artificial laboratory microcosms (Ye et al., 1998). Several reports have shown that the presence of Fe plaque has a modest effect on ameliorating Zn toxicity in *T. latifolia* and *Oryza sativa* (rice), presumably by acting as a barrier to Zn uptake by the plant (Greipsson, 1995; Ye et al., 1998; Zhang et al., 1998). In the case of As, its fate is determined to a significant degree by oxidation state. Arsenite [As(III)], the more soluble form of As, does not appear to be affected by the presence of Fe plaque in rice. In fact, in soil microcosms its uptake was elevated in the presence of Fe plaque (Chen et al., 2005). The opposite appears to be true for arsenate [As(V)]; several studies using rice show that As(V) is concentrated in Fe plaque and its transport into plant tissues is reduced (Figure 9.4*a*; Liu et al., 2004a,b; Chen et al., 2005). Other metals that do not appear to be strongly affected by the presence of Fe plaque include Ni (Ye et al., 1997), Pb, and Cd (Ye et al., 1998), although these findings are limited both in scope and plant type.

The presence of phosphate can influence the interaction of metals with Fe plaque in the rhizosphere because phosphate has a strong affinity for Fe oxides and, by extension, Fe plaque (see also Chapter 5). For example, in a soil microcosm experiment with rice, phosphate inhibited arsenate uptake, but the inhibitory effect was lower in the presence of Fe plaque (Chen et al., 2005). Using rice grown in laboratory microcosms, Liu et al. (2004b) showed that phosphate inhibited As sorption to Fe plaque (Figure 9.4*b*), and caused a greater fraction of the As to be concentrated in the roots rather than on the root surface or in Fe plaque (see also Chapter 3).

Some recent investigations have applied advanced spectroscopic methods to nondestructively probe the associations between Fe oxides and other metals in Fe plaque with unprecedented spatial resolution. These techniques include x-ray fluorescence microscopy (XRF) and x-ray near-edge spectroscopy (XANES). Hansel et al. (2001) investigated the sequestration of Pb, Mn, and Zn to Fe plaque on the wetland plant *Phalaris arundinacea* (reed canarygrass) from a mine waste–contaminated site. They found that Pb was concentrated within the Fe plaque. However, extended x-ray absorption fine structure (EXAFS) data indicated that the Pb was not directly coordinated with the Fe oxides but was probably bound to organic matter in the form of a biofilm. Mn and Zn were also associated with Fe plaque, but chemically they appeared to exist as discrete carbonates. This work on Pb contrasts somewhat with that of Ye et al. (1998), who reported that Pb was not bound significantly to Fe plaque in *Typha*.

In another report, Hansel et al. (2002) studied Fe plaque associated with the rhizosphere of *Typha latifolia* and *Phalaris arundinacea* from a mine waste–contaminated site that was high in As. For both plants, the Fe and As concentrations associated with roots were approximately 10-fold concentrated on roots relative to their abundances in the bulk soil. XRF revealed that Fe plaque consisted primarily of ferrihydrite but also had appreciable levels of goethite and siderite; *T. latifolia* also had a significant amount of lepidocrocite. Their analysis suggested that As was sequestered fairly homogeneously within the Fe plaque. A XANES analysis indicated that approximately 80% of the As was As(V) and 20% As(III). Blute et al. (2004) reported similar ratios for As(V) and As(III)

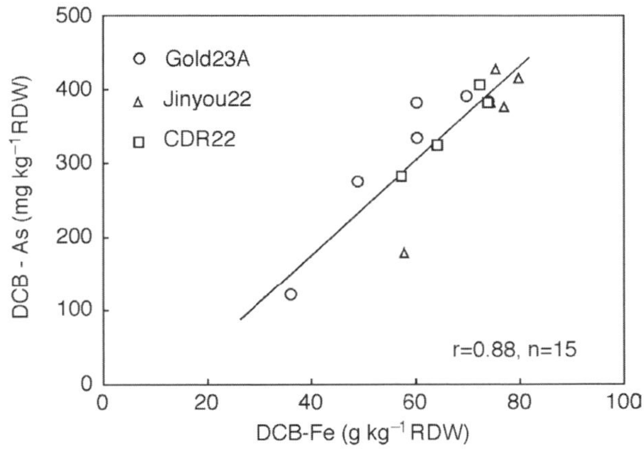

(*a*)

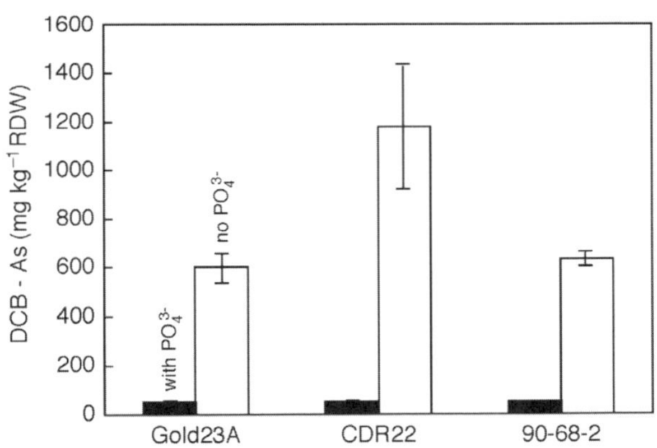

(*b*)

Figure 9.4. (*a*) The strong binding of arsenic to iron plaque is reflected in the significant correlation between dithionite–citrate–bicarbonate (DCB)-extractable iron and arsenic concentrations on rice roots. (*b*) Rice roots have significantly greater plaque As concentrations in the absence of PO_4^{3-} than when PO_4^{3-} is present, indicating that PO_4^{3-} inhibits As binding to iron plaque. Gold23A, Jinyou22, CDR22, and 90-68-2 are individual rice cultivars. RDW = root dry weight. [(*a*) From Liu et al., 2004a, with permission of the Society for Experimental Botany and Oxford University Press; (*b*) from Liu et al., 2004b, with permission of the New Phytologist Trust.]

from *Typha* roots taken from a different As-contaminated wetland. They used XRF and XANES to examine an Fe plaque coating approximately 40 μm thick and found that As was both concentrated in the plaque and distributed relatively uniformly throughout the plaque layer. There was little As in the plant tissue, although bulk measurements from the same site suggested that As was at least 50-fold more concentrated in the Fe plaque on the root surface.

9.6.2. Microbes and Iron–Metal Interactions

Interest in rehabilitating metal-contaminated sites or using natural systems as pollutant filters has lead to much interest in the physical, chemical, and biological mechanisms that control the distributions of metals and metalloids in the environment (e.g., Lovley and Coates, 1997; Gadd, 2000; Weis and Weis, 2004; Gadd, Chapter 2, this volume). Circumstantial evidence suggests that FeOB and FeRB may both play an important role in influencing iron–metal interactions in the rhizosphere, but this has not been demonstrated through direct experimentation. For example, most of the studies discussed above did not consider a role for microbes in affecting rhizosphere metal distributions. In many studies it is not clear if microbes were assumed to be present and part of the experimental system or if the authors assumed that all metal interactions were with the plant and abiotic soil parameters such as pH, clay content, and other metals. In any case, unless extraordinary efforts are taken to grow plants under aseptic conditions, microbes will be present. Studies conducted in the field will have "natural" microbial populations that evolved as part of the ecosystem and are adapted to the prevailing environmental conditions. In contrast, laboratory microcosms may be colonized by opportunistic microbial populations that although capable of surviving, are not adapted to living in metal-rich environments or utilizing iron as an energy source. Thus, careful consideration of the experimental design is necessary to determine the in situ role of microbes in influencing iron metal distributions. For example, Stolz and Greger (2002) studied the sequestration of As, Cd, Cu, Pb, and Zn in wetland plants and found significantly different patterns of metal binding between plants grown in hydroponic microcosms or in soils. The authors suggested that soil–plant interactions might well account for the differences in metal binding and translocation by these plants, but neither Fe plaque nor associated microbes were considered as possible factors.

Although the specific role of FeOB and FeRB in mobilizing metals and affecting metal distributions in the rhizosphere has not been studied, several groups have examined how microbes affect metal volatilization and accumulation in plants. For example, *Schoenoplectus* (*Scirpus*) *robustus* (salt marsh bulrush) showed higher accumulation of Hg in stems and roots when inoculated with a mixture of metal-tolerant bacteria isolated from the rhizosphere than when grown under sterile conditions (de Souza et al., 1999). Laboratory-based investigations suggest that FeOB can affect iron–metal interactions by generating Fe oxides, which are a strong sink for many metals, and by producing organic materials (e.g., cell surfaces, exopolymers) that can also bind metals. An investigation of bacteriogenic iron oxides from an Fe-oxidizing microbial mat community found that the Sr, Cs, Pb, and U were substantially partitioned in bacteriogenic oxides (Ferris et al., 2000). Furthermore, these authors suggested that the presence of bacterially produced organic polymers played an important role in metal sorption.

Iron(III)-reducing microbes can lead to increased mobility of many metals by reducing Fe oxides to which metals are bound. For example, Cummings et al. (1999) showed that Fe(III) reduction by the FeRB *Shewanella alga* mobilized arsenate [As(V)] that was bound to the iron oxides. Thus, seasonal cycles in metal availability may parallel seasonal changes in the balance between Fe(II) oxidation and Fe(III) reduction in the rhizosphere (Figure 9.3*d* and *e*). However, under some circumstances, arsenate can bind to Fe^{2+} produced by Fe(III) reduction to form low solubility ferrous arsenates, resulting in little net arsenic mobilization (Papassiopi et al., 2003). Direct reduction of Fe(III) by FeRB can also release metals that were substituted for iron in mineral complexes (e.g., Mn and Co substituted for Fe in goethite; Bousserrhine et al., 1999). Additionally, many "iron-reducing" microbes can directly reduce other elements, including As, Cr, and U, and therefore affect metal and metalloid solubility and mobility (Lovley, 2000; Lovley and Coates, 2000). Furthermore, humic compounds can shuttle electrons between FeRB and metals, including U(VI) and Cr(VI), leading to changes in metal solubility (Figure 9.3*b*; Lovley and Coates, 1997). This mechanism may be especially important in accessing metals that are protected in clay-encrusted microaggregates and are therefore unavailable for direct microbial reduction. Somewhat paradoxically, FeRB may also increase metal sorption to Fe oxides. A study looking at Sr binding to synthetically produced hydrous ferric oxides alone and in the presence of the FeRB *S. alga* found significantly more Sr binding to *S. alga* cultures and mixed *S. alga*–oxide complexes than to oxides only (Small et al., 1999). The interactions are complex, however, since the presence of calcium carbonates can significantly alter the binding of Sr to Fe oxides and Fe carbonate minerals (e.g., siderite) that form as a result of Fe reduction (Roden et al., 2002).

9.7. CONCLUSIONS AND FUTURE DIRECTIONS

Iron plaque in the rhizosphere can function as a sink for metals and metalloids, several of which can be toxic when in soluble form, so the rapid cycling of iron between oxidized and reduced forms may influence the bioavailability of other metals. As highlighted in this review, more work needs to be done in assessing the activity of natural rhizosphere communities of Fe-cycling microbes and the role that they play in iron–metal interactions. What populations are present? Is microbial community structure controlled by the plant or other soil ecosystem parameters such as percent organic matter or clay content? In fact, with the exception of rice (Liesack et al., 2000), we know little about the general structure of wetland rhizosphere microbial communities. In addition to Fe, other metals and metalloids, including As, Mn, Cr, U, and Se, are subject to microbially mediated redox chemistry. To what extent does microbial cycling influence the distributions of these and other metals in the rhizosphere? How important are microbial processes relative to chemical or physical factors (e.g., localized changes in soil pH)? Can we differentiate between metal interactions that result from microbially catalyzed redox reactions and those that result from passive binding to microbially produced exopolymers or other organic ligands? How does microbial activity influence phosphate dynamics, which in turn play a significant role in interactions with other metals? As illustrated in this chapter, there are many important yet unanswered (and in many cases, unasked!) questions about the role that microbes play in plant–microbe–soil systems to influence the transformations and distributions of metals and metalloids.

REFERENCES

Anbar, A. D., Roe, J., E., Barling, J. and Nealson, K. H. (2000). Nonbiological fractionation of iron isotopes. *Science* **288**, 126–128.

Armstrong, W. (1964). Oxygen diffusion from the roots of some British bog plants. *Nature* **204**, 801–802.

Armstrong, W. (1971). Radial oxygen losses from intact rice roots as affected by distance from the apex, respiration, and waterlogging. *Physiol. Plant.* **25**, 192–197.

Armstrong, W., Justin, S. H. F. W., Beckett, P. M., and Lythe, S. (1991). Root adaptation to soil waterlogging. *Aquat. Bot.* **39**, 53–73.

Armstrong, W., Strange, M. E., Cringle, S., and Beckett, P. M. (1994). Microelectrode and modelling study of oxygen distribution in roots. *Ann. Bot.* **74**, 287–299.

Armstrong, W., Cousins, D., Armstrong, J., Turner, D. W., and Beckett, P. M. (2000). Oxygen distribution in wetland plant roots and permeability barriers to gas-exchange with the rhizosphere: a microelectrode and modelling study with *Phragmites australis*. *Ann. Bot.* **86**, 687–703.

Arth, I., Frenzel, P., and Conrad, R. (1998). Denitrification coupled to nitrification in the rhizosphere of rice. *Soil Biol and Biochemi.* **30**, 509–515.

Balashova, V. V., and Zavarzin, G. A. (1980). Anaerobic reduction of ferric iron by hydrogen bacteria. *Microbiology* [English translation of *Mikrobiologiya*]. **48**, 635–639.

Barbeau, K., Rue, E. L., Bruland, K. W., and Butler, A. (2001). Photochemical cycling of iron in the surface ocean mediated by microbial iron(III)-binding ligands. *Nature* **413**, 409–413.

Batty, L. C., Baker, A. J. M., and Wheeler, B. D. (2002). Aluminium and phosphate uptake by *Phragmites australis*: the role of Fe, Mn and Al root plaques. *Ann. Bot.* **89**, 443–449.

Beard, B. L., Johnson, C. M., Cox, L., Sun, H., Nealson, K. H., and Aguilar, C. (1999). Iron isotope biosignatures. *Science* **285**, 1889–1892.

Beard, B. L., Johnson, C. M., Skulan, J. L., Nealson, K. H., Cox, L., and Sun, H. (2003). Iron isotope constraints on Fe cycling and mass balance in the oxygenated Earth. *Geology* **31**, 629–632.

Bergholz, P. W., Bagwell, C. E. and Lovell, C. R. (2001). Physiological diversity of rhizoplane diazotrophs of the saltmeadow cordgrass, *Spartina patens*: implications for host specific ecotypes. *Microb. Ecol.* **42**, 466–473.

Blaabjerg, V., Mouritsen, K. N., and Finster, K. (1998). Diel cycles of sulphate reduction rates in sediments of a *Zostera marina* bed (Denmark). *Aquat. Microb. Ecol.* **15**, 97–102.

Blake, R. C., II, and Johnson, D. B. (2000). Phylogenetic and biochemical diversity among acidophilic bacteria that respire on iron. In *Environmental Microbe–Metal Interactions*, ed. Lovley, D. R., ASM Press, Washington, DC, 53–78.

Blute, N. K., Brabander, D. J., Hemond, H. F., Sutton, S. R., Newville, M. G., and Rivers, M. L. (2004). Arsenic sequestration by ferric iron plaque on cattail roots. *Environ. Sci. and Technol.* **38**, 6074–6077.

Bodelier, P. L. E. (2003). Interactions between oxygen-releasing roots and microbial processes in flooded soils and sediments. In *Ecological Studies: Root Ecology*, ed. de Kroon, H., and Visser, E. J. W., Springer-Verlag, Berlin, 331–362.

Bodelier, P. L. E., Libochant, J. A., Blom, C. W. P. M., and Laanbroek, H. J. (1996). Dynamics of nitrification and denitrification in root-oxygenated sediments and adaptation of ammonia-oxidizing bacteria to low-oxygen or anoxic habitats. *Appl. and Environ. Microbiol.* **62**, 4100–4107.

Bosse, U., and Frenzel, P. (1997). Activity and distribution of methane-oxidizing bacteria in flooded rice soil microcosms and in rice plants (*Oryza sativa*). *Appl. Environ. Microbiol.* **63**, 1199–1207.

Bousserrhine, N., Gasser, U. G., Jeanroy, E., and Berthelin, J. (1999). Bacterial and chemical reductive dissolution of Mn-, Co-, Cr-, and Al-substituted goethite. *Geomicrobiol. J.* **16**, 245–258.

Brune, A., Frenzel, P. and Cypionka, H. (2000). Life at the oxic–anoxic interface: microbial activities and adaptations. *FEMS Microbiol. Rev.* **24**, 691–710.

Bullen, T. D., White, A. F., Childs, C. W., Vivit, D. V., and Schultz, M. S. (2001). Demonstration of a significant iron isotope fractionation in nature. *Geology* **29**, 699–702.

Burke, D. J., Hammerlynck, E. P., and Hahn, D. (2002). Interactions among plant species and microorganisms in salt marsh sediments. *Appl. Environ. Microbiol.* **68**, 1157–1164.

Calhoun, A., and King, G. M. (1997). Regulation of root-associated methanotrophy by oxygen availability in the rhizosphere of two aquatic macrophytes. *Appl. Environ. Microbiol.* **63**, 3051–3058.

Canfield, D. E., Thamdrup, B., and Hensen, J. W. (1993). The anaerobic degradation of organic matter in Danish coastal sediments: Fe reduction, Mn reduction, and sulfate reduction. *Geochim. Cosmochim. Acta*, **57**, 3867–3883.

Chaudhuri, S. K., Lack, J. G., and Coates, J. D. (2001). Biogenic magnetite formation through anaerobic biooxidation of Fe(II). *Appl. Environ. Microbiol.* **67**, 2844–2848.

Chen, Z., Zhu, Y.-G., Liu, W.-J., and Meharg, A. A. (2005). Direct evidence showing the effect of root surface iron plaque on arsenite and aresnate uptake in rice (*Oryza sativa*) roots. *New Phytol.* **165**, 91–97.

Cheng, W., Coleman, D. C., Carroll, C. R., and Hoffman, C. A. (1994). Investigating short-term carbon flows in the rhizospheres of different species, using isotopic trapping. *Agron. J.* **86**, 782–788.

Cheng, W., Zhang, Q. L., Coleman, D. C., Carroll, C. R., and Hoffman, C. A. (1996). Is available carbon limiting microbial respiration in the rhizosphere? *Soil Biol. Biochem.* **28**, 1283–1288.

Chin, K.-J., Lueders, T., Friedrich, M. W., Klose, M., and Conrad, R. (2004). Archaeal community structure and pathway of methane formation on rice roots. *Microb. Ecol.* **47**, 59–67.

Christensen, P. B., Revsbech, N. P., and Sand-Jensen, K. (1994). Microsensor analysis of oxygen in the rhizosphere of the aquatic macrophyte *Littorella uniflora* (L.) Ascherson. *Plant Physiol.* **105**, 847–852.

Coleman, M. L., Hedrick, D. B., Lovley, D. R., White, D. C., and Pye, K. (1993). Reduction of Fe(III) in sediments by sulphate-reducing bacteria. *Nature* **361**, 436–438.

Colmer, T. D. (2003). Long-distance transport of gases in plants: a perspective on internal aeration and radial oxygen loss from roots. *Plant Cell Environ.* **26**, 17–36.

Croal, L. R., Gralnick, J. A., Malasarn, D., and Newman, D. K. (2004a). The genetics of geochemistry. *Ann. Rev. Genet.* **38**, 175–202.

Croal, L. R., Johnson, C. M., Beard, B. L. and Newman, D. K. (2004b). Iron isotope fractionation by Fe(II)-oxidizing photoautotrophic bacteria. *Geochim. Cosmochim. Acta* **68**, 1227–1242.

Cummings, D. E., Caccavo F., Jr., Fendorf, S., and Rosenzweig, R. F. (1999). Arsenic mobilization by the dissimilatory Fe(III)-reducing bacterium *Shewanella alga* BrY. *Environ. Sci.Technol.* **33**, 723–729.

Derakshani, M., Lukow, T., and Liesack, W. (2001). Novel bacterial lineages at the (sub)division level as detected by signature nucleotide-targeted recovery of 16S rRNA genes from bulk soil and rice roots of flooded rice microcosms. *Appli. and Environ. Microbiol.* **67**, 62–631.

de Souza, M. P., Huang, C. P. A., Chee, N., and Terry, N. (1999). Rhizosphere bacteria enhance the accumulation of selenium and mercury in wetland plants. *Planta* **209**, 259–263.

Edwards, K. J., Rogers, D. R., Wirsen, C. O. and McCollom, T. M. (2003). Isolation and characterization of novel psychrophilic, neutrophilic, Fe-oxidizing, chemolithoautotrophic α- and γ-*Proteobacteria* from the deep sea. *Appl. Environ. Microbiol.* **69**, 2906–2913.

Edwards, K. J., Bach, W., McCollom, T. M., and Rogers, D. R. (2004). Neutrophilic iron-oxidizing bacteria in the ocean: their habitats, diversity, and roles in mineral deposition, rock alteration, and biomass production in the deep-sea. *Geomicrobiol. J.* **21**, 393–404.

Ehrlich, H. L., Ingledew, W. J., and Salerno, J. C. (1991). Iron- and manganese-oxidizing bacteria. In *Variations in Autotrophic Life*, ed. Shively, J. M., and Barton, L. L., Academic Press, San Diego, CA, 147–170.

Eller, G., and Frenzel, P. (2001). Changes in activity and community structure of methane-oxidizing bacteria over the growth period of rice. *Appl. Environ. Microbiol.* **67**, 2395–2403.

Emerson, D. (2000). Microbial oxidation of Fe(II) and Mn(II) at circumneutral pH. In *Environmental Microbe–Metal Interactions*, ed. Lovley, D. R., ASM Press, Washington, DC, 31–52.

Emerson, D., and Moyer, C. (1997). Isolation and characterization of novel iron-oxidizing bacteria that grow at circumneutral pH. *Appl. Environ. Microbiol.* **63**, 4784–4792.

Emerson, D., and Moyer, C. L. (2002). Neutrophilic Fe-oxidizing bacteria are abundant at Loihi Seamount hydrothermal vents and play a major role in Fe oxide deposition. *Appl. Environ. Microbiol.* **68**, 3085–3093.

Emerson, D., and Revsbech, N. P. (1994). Investigation of an iron-oxidizing microbial mat community located near Aarhus, Denmark: laboratory studies. *Appl. Environ. Microbiol.* **60**, 4032–4038.

Emerson, D., and Weiss, J. V. (2004). Bacterial iron oxidation in circumneutral freshwater habitats: findings from the field and the laboratory. *Geomicrobiol. J.* **21**, 405–414.

Emerson, D., Weiss, J. V., and Megonigal, J. P. (1999). Iron-oxidizing bacteria are associated with ferric hydroxide precipitates (Fe-plaque) on the roots of wetland plants. *Appl. Environ. Microbiol.* **65**, 2758–2761.

Emmenegger, L., Schönenberger, R., Sigg, L., and Sulzberger, B. (2001). Light-induced redox cycling of iron in circumneutral lakes. *Limnol. Oceanogr.* **46**, 49–61.

Ferris, F. G., Hallberg, R. O., Lyvén, B., and Pedersen, K. (2000). Retention of strontium, cesium, lead and uranium by bacterial iron oxides from a subterranean environment. *App. Geochem.* **15**, 1035–1042.

Firestone, M. K. (1982). Biological denitrification. In *Nitrogen in Agriculture*, ed. Stevenson, F. J., American Society of Agronomy, Madison, WI, 289–326.

Flessa, H. (1994). Plant-induced changes in the redox potential of the rhizospheres of the submerged vascular macrophytes *Myriophyllum verticillatum* L. and *Ranunculus circinatus* L. *Aquat. Bot.* **47**, 119–129.

Frenzel, P., Bosse, U., and Janssen, P. H. (1999). Rice roots and methanogenesis in a paddy soil: ferric iron as an alternative electron acceptor in the rooted soil. *Soil Biol. Biochem.* **31**, 421–430.

Gadd, G. M. (2000). Bioremedial potential of microbial mechanisms of metal mobilization and immobilization. *Curr Opin. Biotechnol.* **11**, 271–279.

Gilbert, B., and Frenzel, P. (1995). Methanotrophic bacteria in the rhizosphere of rice microcosms and their effect on porewater methane concentration and methane emission. *Biol. Fertil. Soils* **20**, 93–100.

Gilbert, B., and Frenzel, P. (1998). Rice roots and CH_4 oxidation: the activity of bacteria, their distribution and the microenvironment. *Soil Biol. Biochem.* **30**, 1903–1916.

Greipsson, S. (1995). Effect of iron plaque on roots of rice on growth of plants in excess zinc and accumulation of phosphorus in plants in excess copper or nickel. *J. Plant. Nutr.* **18**, 1659–1665.

Gribsholt, B., and Kristensen, E. (2002). Effects of bioturbation and plant roots on salt marsh biogeochemistry: a mesocosm study. *Mar. Ecol. Prog. Ser.* **241**, 71–87.

Gribsholt, B., Kostka, J. E., and Kristensen, E. (2003). Impact of fiddler crabs and plant roots on sediment biogeochemistry in a Georgia saltmarsh. *Mar. Ecol. Prog. Series.* **259**, 237–251.

Hanert, H. H. (2002). Bacterial and chemical iron oxide deposition in a shallow bay on Palaea Kameni, Santorini, Greece: microscopy, electron probe microanalysis, and photometry of in situ experiments. *Geomicrobiol. J.* **19**, 317–342.

Hansel, C. M., Fendorf, S., Sutton, S., and Newville, M. (2001). Characterization of Fe plaque and associated metals on the roots of mine-waste impacted aquatic plants. *Environ. Sci. Technol.* **35**, 3863–3868.

Hansel, C. M., La Force, M. J., Fendorf, S., and Sutton, S. (2002). Spatial and temporal association of As and Fe species on aquatic plant roots. *Environ. Sci. Technol.* 36, 1988–1994.

Hebert, A. B., and Morse, J. W. (2003). Microscale effects of light on H_2S and Fe^{2+} in vegetated (*Zostera marina*) sediments. *Mar. Chem.* **81**, 1–9.

Hines, M. E., Banta, G. T., Giblin, A. E., Hobbie, J. E., and Tugel, J. B. (1994). Acetate concentrations and oxidation in salt-marsh sediments. *Limnol. Oceanogr.* **39**, 140–148.

Hines, M. E., Faganeli, J., and Planinc, R. (1997). Sedimentary anaerobic microbial biogeochemistry in the Gulf of Trieste, northern Adriatic Sea: influences of bottom water oxygen depletion. *Biogeochemistry* **39**, 65–86.

Hines, M. E., Evans, R. S., Sharak Genthner, B. R., Willis, S. G., Friedman, S., Rooney-Varga, J. N., and Devereux, R. (1999). Molecular phylogenetic and biogeochemical studies of sulfate-reducing bacteria in the rhizosphere of *Spartina alterniflora*. *Appl. Environ. Microbiol.* **65**, 2209–2216.

Højberg, O., Revsbech, N. P., and Tiedje, J. M. (1994). Denitrification in soil aggregates analyzed with microsensors for nitrous oxide and oxygen. *Soil Sci. Soc. Am. J.* **58**, 1691–1698.

Jacob, D. L., and Otte, M. L. (2003). Conflicting processes in the wetland plant rhizosphere: Metal retention or mobilization? *Water Air Soil Pollut. Focus* **3**, 91–104.

Jacobson, M. E. (1994). Chemical and biological mobilization of Fe(III) in marsh sediments. *Biogeochemistry* **25**, 41–60.

Jahnke, R. (1985). A model of microenvironments in deep-sea sediments: formation and effects on porewater profiles. *Limnol. and Oceanogr.* **30**, 956–965.

James, R. E., and Ferris, F. G. (2004). Evidence for microbial-mediated iron oxidation at a neutrophilic groundwater spring. *Chem. Geol.* **212**, 301–311.

Johnson, C. M., Roden, E. E., Welch, S. A., and Beard, B. L. (2005). Experimental constraints on Fe isotope fractionation during magnetite and Fe carbonate formation coupled to dissimilatory hydrous ferric oxide reduction. *Geochim. Cosmoch. Acta.* **69**, 963–993.

Jorgensen, B. B. (1977). Bacterial sulfate reduction within reduced microniches of oxidized marine sediments. *Mar Biol.* **41**, 7–17.

Kashefi, K., and Lovley, D. R. (2000). Reduction of Fe(III), Mn(IV), and toxic metals at 100°C by *Pyrobaculum islandicum*. *Appl. Environ. Microbiol.* **66**, 1050–1056.

King, G. M. (1994). Associations of methanotrophs with the roots and rhizomes of aquatic vegetation. *Appl. Environ. Microbiol.* **60**, 3220–3227.

King, G. M. (1996). In situ analyses of methane oxidation associated with the roots and rhizomes of a bur reed, *Sparganium eurycarpum*, in a Maine wetland. *Appl. Environ. Microbiol.* **62**, 4548–4555.

King, G. M. (2005). Ecophysiology of microbial respiration. In *Respiration in Aquatic Ecosystems*, ed. del Giorgio, P. A., and Williams, P. J. L. B., Oxford University Press, New York, 18–35.

King, G. M., and Garey, M. A. (1999). Ferric iron reduction by bacteria associated with the roots of freshwater and marine macrophytes. *Appl. Environ. Microbiology.* **65**, 4393–4398.

Konhauser, K. O., Hamade, T., Raiswell, R., Morris, R. C., Ferris, F. G., Southam, G., and Canfield, D. E. (2002). Could bacteria have formed the Precambrian banded iron formations? *Geol* **30**, 1079–1082.

Koretsky, C. M., Moore, C. M., Lowe, K. L., Meile, C., Dichristina, T. J., and van Cappellen, P. (2003). Seasonal oscillation of microbial iron and sulfate reduction in saltmarsh sediments (Sapelo Island, GA, USA). *Biogeochemistry* **64**, 179–203.

Kostka, J. E., and Luther, G. W. III, (1995). Seasonal cycling of Fe in salt marsh sediments. *Biogeochemistry* **29**, 159–181.

Kostka, J. E., and Nealson, K. H. (1995). Dissolution and reduction of magnetite by bacteria. *Environ. Sci. Technol.* **10**, 2535–2539.

Kostka, J. E., Roychoudhury, A., and van Cappellen, P. (2002a). Rates and controls of anaerobic microbial respiration across spatial and temporal gradients in saltmarsh sediments. *Biogeochemistry* **60**, 49–76.

Kostka, J. E., Dalton, D. D., Skelton, H., Dollhopf, S., and Stucki, J. W. (2002b). Growth of iron(III)-reducing bacteria on clay materials as the sole electron acceptor and comparison of growth yields on a variety of oxidized iron forms. *Appl. Environ. Microbiol.* **68**, 6256–6262.

Kostka, J. E., Gribsholt, B., Petrie, E., Dalton, D., Skelton, H., and Kristensen, E. (2002c). The rates and pathways of carbon oxidation in bioturbated saltmarsh sediments. *Limnol. and Oceanogr.* **47**, 230–240.

Laanbroek, H. J. (1990). Bacterial cycling of minerals that affect plant growth in waterlogged soils: a review. *Aquat. Bot.* **38**, 109–125.

Lack, J. G., Chaudhuri, S. K., Chakraborty, R., Achenbach, L. A., and Coates, J. D. (2002). Anaerobic biooxodation of Fe(II) by *Dechlorosoma suillum*. *Microb. Ecol.* **43**, 424–431.

LaKind, J. S., and Stone, A. T. (1989). Reductive dissolution of goethite by phenolic reductants. *Geochi. Cosmochim. Acta* **53**, 961–971.

Lehmann-Richter, S., Grosßkopf, R., Liesack, W., Frenzel, P., and Conrad, R. (1999). Methanogenic archaea and CO_2-dependent methanogenesis on washed rice roots. *Environ. Microbiol.* **1**, 159–166.

Liang, L., McNabb, J. A., Paulk, J. M., Gu, B., and McCarthy, J. F. (1993). Kinetics of Fe(II) oxygenation at low partial pressure of oxygen in the presence of natural organic matter. *Environ. Sci. Technol.* **27**, 1864–1870.

Liesack, W., Schnell, S., and Revsbech, N. P. (2000). Microbiology of flooded rice paddies. *FEMS Microbiol. Ecol.* **24**, 625–645.

Lin, W. C., Coppi, M. V., and Lovley, D. R. (2004). *Geobacter sulfurreducens* can grow with oxygen as a terminal electron acceptor. *Appl. Environ. Microbiol.* **70**, 2525–2528.

Liu, W.-J., Zhu, Y.-G., Smith, F. A., and Smith, S. E. (2004a). Do phosphorus nutrition and iron plaque alter arsenate (As) uptake by rice seedlings in hydroponic culture? *New Phytol.* **162**, 481–488.

Liu, W.-J., Zhu, Y.-G., Smith, F. A., and Smith, S. E. (2004b). Do iron plaque and genotypes affect arsenate uptake and translocation by rice seedlings (*Oryza sativa* L.) grown in solution culture? *J. Exp. Bot.* **55**, 1707–1713.

Lonergan, D. J., Jenter, H. L., Coates, J. D., Phillips, E. J. P., Schmidt, T. M., and Lovley, D. R. (1996). Phylogenetic analysis of dissimilatory Fe(III) reducing bacteria. *J. Bacteriol.* **178**, 2402–2408.

Lovley, D. R. (2000). Fe(III) and Mn(IV) reduction. In *Environmental Microbe–Metal Interactions*, ed. Lovley, D.R., ASM Press, Washington, DC, 3–30.

Lovley, D. R., and Coates, J. D. (1997). Bioremediation of metal contamination. *Curr. Opin. Biotechnol.* **8**, 285–289.

Lovley, D. R., and Coates, J. D. (2000). Novel forms of anaerobic respiration of environmental relevance. *Curr. Opin. Microbiol.* **3**, 252–256.

Lovley, D. R., and Phillips, E. J. P. (1986). Organic matter mineralization with reduction of ferric iron in anaerobic sediment. *Appl. Environ. Microbiol.* **51**, 683–689.

Lovley, D. R., and Phillips, E. J. P. (1988). Novel mode of microbial energy metabolism: organic carbon oxidation coupled to dissimilatory reduction of iron or manganese. *Appl. Environ. Microbiol.* **54**, 1472–1480.

Lovley, D. R., Stolz, J. F., Nord, G. L., and Phillips, E. J. P. (1987). Anaerobic production of magnetite by a dissimilatory iron-reducing microorganism. *Nature* **330**, 252–254.

Lovley, D. R., Giovannoni, S. J., White, D. C., Champine, J. E., Phillips, E. J. P., Gorby, Y. A., and Goodwin, S. (1993). *Geobacter metallireducens* gen. nov. sp. nov., a microorganism capable of coupling the complete oxidation of organic compounds to the reduction of iron and other metals. *Arch. Microbiol.* **159**, 336–344.

Lovley, D. R., Coates, J. D., Blunt-Harris, E. L., Phillips, E. J. P., and Woodward, J. C. (1996). Humic substances as electron acceptors for microbial respiration. *Nature* **382**, 445–448.

Lowe, K., DiChristina, T. J., Roychoudhury, A., and van Cappellen, P. (2000). Microbiological and geochemical characterization of microbial Fe(III) reduction in salt marsh sediments. *Geomicrobiol. J.* **17**, 163–178.

Lu, Y., Watanabe, A., and Kimura, M. (2002). Contribution of plant-derived carbon to soil microbial biomass dynamics in a paddy rice microcosm. *Biol. Fertil. Soils* **36**, 136–142.

McKay, D. S., Gibson, E. K., Jr., Thomas-Keprta, K. L., Vali, H., Romanek, C. S., Clemett, S. J., Chillier, X. D. F., Maechling, C. R., and Zare, R. N. (1996). Search for past life on Mars: possible relic biogenic activity in Martian meteorite ALH84001. *Science* **273**, 924–930.

Megonigal, J. P., Whalen, S. C., Tissue, D. T., Bovard, B. D., Albert, D. B., and Allen, A. S. (1999). A plant–soil–atmosphere microcosm for tracing radiocarbon from photosynthesis through methanogenesis. *Soil Sci. Soc. Am. J.* **63**, 665–671.

Megonigal, J. P., Hines, M. E., and Visscher, P. T. (2004). Anaerobic metabolism: linkages to trace gases and aerobic metabolism. In *Biogeochemistry*, ed. Schlesinger, W. H., Elsevier-Pergamon Press, Oxford, 317–424.

Meharg, A. A. (2004). Arsenic in rice: understanding a new disaster for South-East Asia. *Trends Plant Sci.* **9**, 415–417.

Meharg, A. A., and Rahman, M. M. (2003). Arsenic contamination of Bangladesh paddy field soils: Implications for rice contribution to arsenic consumption. *Environ. Sci. Technol.* **37**, 229–234.

Mendelssohn, I. A., Kleiss, B. A., and Wakely, J. S. (1995). Factors controlling the formation of oxidized root channels: a review. *Wetlands* **15**, 37–46.

Mummey, D. L., and Stahl, P. D. (2004). Analysis of soil whole- and inner microaggregate bacterial communities. *Microb. Ecol.* **48**, 41–50.

Neubauer, S. C., Emerson, D., and Megonigal, J. P. (2002). Life at the energetic edge: kinetics of circumneutral iron oxidation by lithotrophic iron-oxidizing bacteria isolated from the wetland-plant rhizosphere. *Appl. Environ. Microbiol.* **68**, 3988–3995.

Neubauer, S. C., Givler, K., Valentine, S., and Megonigal, J. P. (2005). Seasonal patterns and plant-mediated controls of subsurface wetland biogeochemistry. *Ecology* **86**, 3334–3344.

Oremland, R. S., and Capone, D. G. (1988). Use of "specific" inhibitors in biogeochemistry and microbial ecology. *Adv. Microb. Ecol.* **10**, 285–383.

Papassiopi, N., Vaxevanidou, K., and Paspaliaris, I. (2003). Investigating the use of iron reducing bacteria for the removal of arsenic from contaminated soils. *Water Air Soil Pollut. Focus* **3**, 81–90.

Payne W. J. (1981). *Denitrification*, Wiley, New York.

Pedersen, O., Sand-Jensen, K., and Revsbech, N. P. (1995). Diel pulses of O_2 and CO_2 in sandy lake sediments inhabited by *Lobelia dortmanna*. *Ecology* **76**, 1536–1545.

Peverly, J. H., Surface, J. M., and Wang, T. (1995). Growth and trace metal absorption by *Phragmites australis* in wetlands constructed for landfill leachate treatment. *Ecol. Eng.* **5**, 21–35.

Ponnamperuma, F. N. (1972). The chemistry of submerged soils. In *Advances in Agronomy*, ed. Brady, N. C., Academic Press, New York, 29–96.

Reddy, K. R., Patrick W. H., Jr., and Lindau, C. W. (1989). Nitrification-denitrification at the plant root-sediment interface in wetlands. *Limnol. Oceanogr.* **34**, 1004–1013.

Robbins, E. I., and Iberall, A. S. (1991). Mineral remains of early life on Earth? On Mars? *Geomicrobiol. J.* **9**, 51–66.

Roden, E. E., and Lovley, D. R. (1993). Evaluation of ^{55}Fe as a tracer of Fe(III) reduction in aquatic sediments. *Geomicrobiol. J.* **11**, 49–56.

Roden, E. E., and Wetzel, R. G. (1996). Organic carbon oxidation and suppression of methane production by microbial Fe(III) oxide reduction in vegetated and unvegetated freshwater wetland sediments. *Limnol. Oceanogr.* **41**, 1733–1748.

Roden, E. E., and Wetzel, R. G. (2002). Kinetics of microbial Fe(III) oxide reduction in freshwater wetland sediments. *Limnol. Oceanogr.* **47**, 198–211.

Roden, E. E., and Zachara, J. M. (1996). Microbial reduction of crystalline iron(III) oxides: influence of oxide surface area and potential for cell growth. *Environ. Sci. Technol.* **30**, 1618–1628.

Roden, E. E., Leonardo, M. R., and Ferris, F. G. (2002). Immobilization of strontium during iron biomineralization coupled to dissimilatory hydrous ferric oxide reduction. *Geochim. Cosmochim. Acta* **66**, 2823–2839.

Rooney-Varga, J. N., Devereux, R., Evans, R. S., and Hines, M. E. (1997). Seasonal changes in the relative abundance of uncultivated sulfate-reducing bacteria in a salt marsh sediment and in the rhizosphere of *Spartina alterniflora*. *Appl Environ. Microbiol.* **63**, 3895–3901.

Rooney-Varga, J. N., Sharak Genthner, B. R., Devereux, R., Willis, S. G., Friedman, S. D., and Hines, M. E. (1998). Phylogenetic and physiological diversity of sulphate-reducing bacteria isolated from a salt marsh sediment. *Syst. Appl. Microbiol.* **21**, 557–568.

Singer, P. C., and Stumm, W. (1970). Acidic mine drainage: the rate-determining step. *Science* **167**, 1121–1123.

Small, T. D., Warren, L. A., Roden, E. E., and Ferris, F. G. (1999). Sorption of strontium by bacteria, Fe(III) oxide, and bacteria-Fe(III) oxide composites. *Environ. Sci. Technol.* **33**, 4465–4470.

Smits, A. J. M., Laan, P., Thier, R. H., and van der Velde, G. (1990). Root aerenchyma, oxygen leakage patterns and alcoholic fermentation ability of the roots of some nymphaeid and isoetid macrophytes in relation to the sediment type of their habitat. *Aquatic. Botany.* **38**, 3–17.

Sobolev, D., and Roden, E. E. (2001). Suboxic deposition of ferric iron by bacteria in opposing gradients of Fe(II) and oxygen at circumneutral pH. *Appl. Environ. Microbiol.* **67**, 1328–1334.

Sobolev, D., and Roden, E. E. (2004). Characterization of a neutrophilic, chemolithoautotrophic, Fe(II)-oxidizing ß-protobacterium from freshwater wetland sediments. *Geomicrobiol. J.* **21**, 1–10.

St-Cyr, L., and Crowder, A. A. (1990). Mn and Cu in the root plaque of *Phragmites australis* (Cav.) Trin. ex Steudel. *Soil Sci.* **149**, 191–198.

St-Cyr, L., Fortin, D., and Campbell, P. G. C. (1993). Microscopic observations of the iron plaque of a submerged aquatic plant (*Vallisneria americana* Michx). *Aquat. Bot.* **46**, 155–167.

Stolz, E., and Greger, M. (2002). Accumulation properties of As, Cd, Cu, Pb, and Zn by four wetland plant species growing on submerged mine tailings. *Environ. Exp. Bot.* **47**, 271–280.

Straub, K. L., Benz, M., Schink, B., and Widdel, F. (1996). Anaerobic, nitrate-dependent microbial oxidation of ferrous iron. *Appl. Environ. Microbiol.* **62**, 1458–1460.

Straub, K. L., and Buchholz-Cleven, B. E. E. (1998). Enumeration and detection of anaerobic ferrous iron-oxidizing, nitrate-reducing bacteria from diverse European sediments. *Appl. Environ. Microbiol.* **64**, 4846–4856.

Straub, K. L., Benz, M., and Schink, B. (2001). Iron metabolism in anoxic environments at near neutral pH. *FEMS Microbiol. Ecol.* **34**, 181–186.

Straub, K. L., Schonhuber, W. A., Buchholz-Cleven, B. E. E., and Schink, B. (2004). Diversity of ferrous iron-oxidizing, nitrate-reducing bacteria and their involvement in oxygen- independent iron cycling. *Geomicrobiol. J.* **21**, 371–378.

Sundby, B., Vale, C., Caçador, I., Catarino, F., Madureira, M. -J., and Caetano, M. (1998). Metal-rich concretions on the roots of salt marsh plants: mechanisms and rate of formation. *Limnol. Oceanogr.* **43**, 245–252.

Sundby, B., Vale, C., Caectano, M., and Luther G. W., III, (2003). Redox chemistry in the root zone of a salt marsh sediment in the Tagus Estuary, Portugal. *Aquat. Geochem.* **9**, 257–271.

Sundby, B., Caetano, M., Vale, C., Gobeil, C., Luther, G. W., III, and Nuzzio, D. B. (2005). Root-induced cycling of lead in salt marsh sediments. *Environ. Sci. Technol.* **39**, 2080–2086.

Taylor, G. J., and Crowder, A. A. (1983). Uptake and accumulation of heavy metals by *Typha latifolia* L. in wetlands of the Sudbury, Ontario region. *Can. J. Bot.* **61**, 63–73.

Thamdrup, B. (2000). Bacterial manganese and iron reduction in aquatic sediments. In *Advances in Microbial Ecology*, ed. Schink, B., Kluwer Academic/Plenum Publishers, New York, 41–84.

Thamdrup, B., and Canfield, D.E. (1996). Pathways of carbon oxidation in continental margin sediments off central Chile. *Limnol. Oceanogr.* **41**, 1629–1650.

Trolldenier, G. (1988). Visualization of oxidizing power of rice roots and of possible participation of bacteria in iron deposition. *Z. Pflanzerenaehr. Bodenkd.* **151**, 117 –121.

Vale C., Caterino, F., Cortesão, C., and Caçador, I. (1990). Presence of metal-rich rhizoconcretions on the roots of Spartina maritima from the salt marshes of the Tagus Estuary, Portugal. *Sci. Total Environ.* 97–98, 617–626.

van Bodegom, P., Goudriann, J., and Leffelaar, P. (2001a). A mechanistic model on methane oxidation in a rice rhizosphere. *Biogeochemistry*, **55**, 145–177.

van Bodegom, P., Stams, F., Mollema, L., Boeke, S., and Leffelaar, P. (2001b). Methane oxidation and the competition for oxygen in the rice rhizosphere. *Appl. Environ. Microbiol.* **67**, 3586–3597.

Vargas, M., Kashefi, K., Blunt-Harris, E. L., and Lovely, D. R. (1998). Microbiological evidence for Fe(III) reduction on early Earth. *Nature*, **395**, 65–67.

Weaver, B. L., and Tarney, J. (1984). Empirical approach to estimating the composition of the continental crust. Nature, **310**, 575–577.

Weis, J. S., and Weis, P. (2004). Metal uptake, transport and release by wetland plants: implications for phytoremediation and restoration. *Environ. Int.* **30**, 685–700.

Weiss, J. V. (2002). *Microbially-Mediated Iron Cycling in the Rhizosphere of Wetland Plants*, Ph.D. dissertation. George Mason University, Fairfax, VA.

Weiss, J. V., Emerson, D., Backer, S. M., and Megonigal, J. P. (2003). Enumeration of Fe(II)-oxidizing and Fe(III)-reducing bacteria in the root zone of wetland plants: implications for a rhizosphere iron cycle. *Biogeochemistry* **64**, 77–96.

Weiss, J. V., Emerson, D., and Megonigal, J. P. (2004). Geochemical control of microbial Fe(III) reduction potential in wetlands: comparison of the rhizosphere to non-rhizosphere soil. *FEMS Microbiol. Ecol.* **48**, 89–100.

Weiss, J. V., Emerson, D., and Megonigal, J. P. (2005). Rhizosphere iron(III) deposition and reduction in a *Juncus effusus* L.–dominated wetland. *Soil Sci. Soc. of Am. J.* **69**, 1861–1870.

Whiting, G. J., and Chanton, J. P. (1993). Primary production control of methane emission from wetlands. *Nature* **364**, 794–795.

Widdel, F., Schnell, S., Heising, S., Ehrenreich, A., Assmus, B., and Schink, B. (1993). Ferrous iron oxidation by anoxygenic phototrophic bacteria. *Nature* **362**, 834–835.

Ye, Z. H., Baker, A. J. M., Wong, M. H., and Willis, A. J. (1997). Copper and nickel uptake, accumulation and tolerance in *Typha latifolia* with and without iron on the root surface. *New Phytol.* **136**, 481–488.

Ye, Z., Baker, A. J. M., Wong, M.-H., and Willis, A. J. (1998). Zinc, lead and cadmium accumulation and tolerance in *Typha latifolia* as affected by iron plaque on the root surface. *Aquat. Bot.* **61**, 55–67.

Ye, Z. H., Cheung, K. C., and Wong, M. H. (2001). Copper uptake in *Typha latifolia* as affected by iron and manganese plaque on the root surface. *Can. J. Bot.* **79**, 314–320.

Zhang, X. K., Zhang, F. S., and Mao, D. R. (1998). Effect of iron plaque outside roots on nutrient uptake by rice (*Oryza sativa* L.): zinc uptake by Fe-deficient rice. *Plant. Soil* **202**, 33–39.

10

THE COMPLEXITY OF AQUEOUS COMPLEXATION: THE CASE OF ALUMINUM–AND IRON(III)–CITRATE

M. E. Essington

The University of Tennessee, Knoxville, Tennessee

Biophysico-Chemical Processes of Heavy Metals and Metalloids in Soil Environments,
Edited by Antonio Violante, Pan Ming Huang, and Geoffrey Michael Gadd

10.1. INTRODUCTION

Low-molecular-mass organic ligands (LMMOLs) comprise one of the more transitory but ever-present components of the soil solution pool of organic carbon. Their occurrence is localized primarily in the microenvironment surrounding soil microbes and in the rhizosphere (the volume of soil whose biogeochemical properties are affected directly by plant roots) because of the intense biological activity in this volume of soil. A wide variety of LMMOLs are known to exist in the soil solution. They range in character from formic acid, a single-carbon aliphatic acid, to complex multiring phenolic acids. Further, LMMOLs have been characterized as having a molecular-mass cutoff that ranges between 300 and 1000 daltons (van Hees et al., 2000, 2003; Strobel, 2001). In general, the soil solution is dominated by the more soluble aliphatic acid anions (e.g., formate, acetate, lactate, oxalate, malonate, malate, succinate, and citrate), which are found in the concentration range < 1 to 1100 μM (Jones, 1998; van Hees et al., 2000; Strobel, 2001).

Due to their innate ability to complex, and in some instances chelate metal cations, LMMOLs can influence the bioavailability and transport of metals. Aqueous complexation reactions involving LMMOLs and metal cations can enhance metal solubility by inhibiting adsorption and precipitation reactions (Xu and Ji, 2003; Essington et al., 2005) and the rate at which these processes occur (Dietzel and Böhme, 2005). Organic acids may also form specific complexes with mineral surface functional groups, affecting surface charge characteristics (Yao and Yeh, 1996), enhancing metal cation adsorption (Lackovic et al., 2004), competing with other specifically retained substances for surface sites (Kafkafi et al., 1988; Geelhoed et al., 1998; Wijnja and Schulthess, 2000; Gräfe et al., 2002), and inhibiting mineral crystallization and affecting mineral precipitation and dissolution directly (Huang and Violante, 1986; Jardine and Zelazny, 1995; Lebron and Suarez, 1999; Wada et al., 1999). Indeed, many LMMOLs in soils (principally, the di- and tricarboxylates) reside almost completely in the adsorbed phase (Jones, 1998; Strobel, 2001).

Both plant roots and soil microbes are known to exude greater amounts of LMMOLs when subjected to Fe, P, and K (and possibly Ca and Zn) deficiencies, and as a mechanism to mitigate Al toxicity effects (Inskeep and Silvertooth, 1988; Hocking, 2001; Jones et al., 2003). Of the LMMOLs known to occur in the rhizosphere, the citrate ligand (Figure 10.1) reportedly dominates in significance if not in concentration (Jones, 1998; van Hees et al., 2000; Haynes and Mokolobate, 2001; Hocking, 2001; Strobel, 2001). Numerous literature surveys document citrate concentrations in soil solutions and adsorbed phases as a function of horizon (and associated chemical and mineralogical characteristics) and vegetation type. In a review by Jones (1998), citrate was reported to range from a low of less than 0.001 to 4700 $\mu mol\ L^{-1}$ in soil solutions. Literature cited by Strobel (2001) found citrate to range from below detectable levels (undefined) to 370 $\mu mol\ L^{-1}$ in solution and from 7 to 92 $mmol\ kg^{-1}$ on soil solids, with the higher citrate concentrations found in surface organic horizons (O1 and O2) relative to subsurface mineral horizons (B and C horizons).

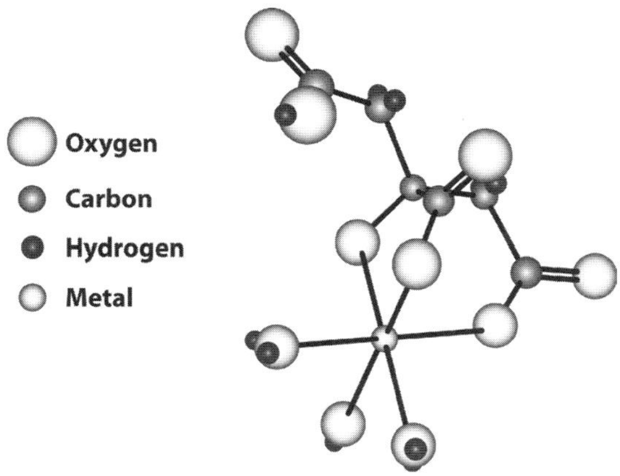

Figure 10.1. Structure of a metal (M^{3+})–citrate (cit^{3-}) complex, $Mcit^0(aq)$.

As indicated above, citrate influences both metal and ligand chemistry in the rhizosphere via aqueous complexation reactions and reactions at the solid–solution interface. Citrate is an important biomolecule in both plant and animal systems, particularly with respect to Fe(III) bioavailability and internal transport and in the amelioration of Al toxicity (Jones, 1998). Thus, this ligand has received considerable study. Yet despite this importance and relationship to citrate chelation capabilities, the aqueous Fe(III)–and Al–citrate complexation chemistry remains ambiguous. Consequently, in this chapter we focus on an examination of citrate aqueous metal complexation, with specific emphasis on the complexation of Fe(III) and Al.

10.2. CITRIC ACID DISSOCIATION AND THE HYDROLYSIS OF IRON(III) AND ALUMINUM

In the natural pH range of soil solutions, the citrate, Fe(III), and Al species present are dictated, in part, by acid–base equilibria (Figure 10.2). Thus, a quantifiable and critical understanding of citric acid dissociation and metal hydrolysis reactions is requisite to the assessment of Fe(III)–and Al–citrate speciation.

10.2.1. Citric Acid

Citric acid (2-hydroxypropane-1,2,3-tricarboxylic acid; Figure 10.1) is a multidentate ligand—a natural chelate—consisting of four potentially ionizable functional groups (three carboxyl groups and one hydroxyl). The nonionized citric acid may be denoted as H_3cit^0 (aq), in recognition of the three readily dissociable carboxyl groups. Some authors, however, prefer the $H_4cit^0(aq)$ notation to represent the nonionized acid, which recognizes the additional dissociation of the hydroxyl group in highly alkaline solutions (as well as to signify the role

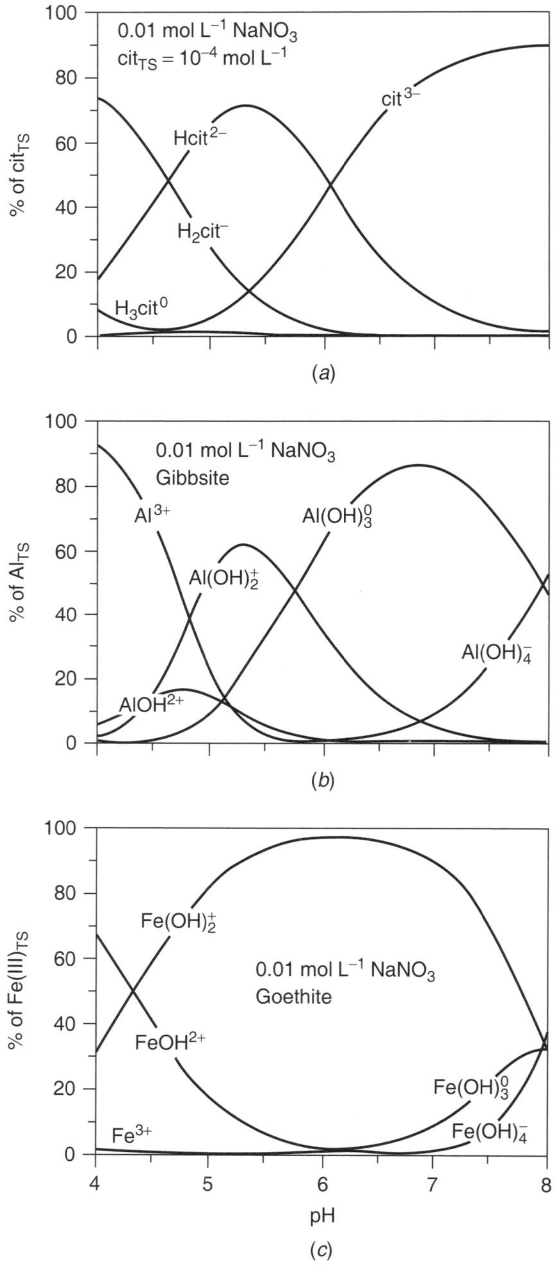

Figure 10.2. Acid-base speciation of (*a*) citrate, (*b*) aluminum, and (*c*) iron(III) as a function of solution pH at 25°C and in 0.01 mol L^{-1} $NaNO_3$ solutions. The distributions were computed using a total soluble citrate (cit_{TS}) concentration of 10^{-4} mol L^{-1} and total soluble aluminum (Al_{TS}) and iron(III) [$Fe(III)_{TS}$] concentrations controlled by gibbsite ($Al(OH)_3(s)$) and goethite [$FeOOH(s)$]. (Relevant thermodynamic data from May and Murray, 2000; Baes and Mesmer, 1986; Liu and Millero, 1999.)

of the hydroxyl moiety in metal chelation). If $H_3cit^0(aq)$ is considered the base species, an approach taken in this chapter, the completely ionized citrate ligand is represented by $H_{-1}cit^{4-}(aq)$.

The ionization of citric acid moieties may be described by the general dissociation reaction

$$H_{4-j}cit^{1-j}(aq) = H_{3-j}cit^{j-}(aq) + H^+(aq) \tag{10.1}$$

where j ranges between 1 and 4. The equilibria established in Eq. (10.1) are formally defined by the thermodynamic acid dissociation constants:

$$K_{01j} = \frac{(H_{3-j}cit^{j-})(H^+)}{(H_{4-j}cit^{1-j})} \tag{10.2}$$

where the parentheses denote activities. One may also define the conditional equilibrium constants as

$${}^{I}K_{01j} = \frac{[H_{3-j}cit^{j-}][H^+]}{[H_{4-j}cit^{1-j}]} \tag{10.3}$$

where the brackets denote molar concentrations and the ${}^{I}K_{01j}$ values are specific (conditional) to the ionic strength (superscript I in ${}^{I}K_{01j}$) of the medium in which the reactions occur. Although well studied, agreement among the reported ${}^{I}K_{01j}$ values ($j = 1$ to 3) appears to be restricted to the lower-ionic-strength determinations (Table 10.1). For example, mean log ${}^{0.1}K_{01j}$ values ($\pm\ \sigma$) are log ${}^{0.1}K_{011} = -2.94 \pm 0.05$, log ${}^{0.1}K_{012} = -4.36 \pm 0.02$, and log ${}^{0.1}K_{013} = -5.70 \pm 0.04$. However, log ${}^{1.0}K_{01j}$ values are log ${}^{1}K_{011} = -2.94 \pm 0.25$, log ${}^{1}K_{012} = -4.20 \pm 0.20$, and log ${}^{1}K_{013} = -5.36 \pm 0.29$. The greater variability in log ${}^{1.0}K_{01j}$ values, relative to ${}^{0.1}K_{01j}$ values, is evidenced in the standard deviations (σ values). Conversion of the citric acid conditional dissociation constants (for $I \leq 0.2$) to the zero ionic strength condition leads to the following mean log $K_{01j} \pm \sigma$ values: log $K_{011} = -3.16 \pm 0.04$, log $K_{012} = -4.78 \pm 0.04$, and log $K_{013} = -6.23 \pm 0.06$. Values for ${}^{I}K_{014}$ have not been determined critically, although log ${}^{I}K_{014}$ values have been reported to range between ~ -11 and ~ -13.8 (Rajan et al., 1978; Königsberger et al., 2000; Pierre and Gautier-Luneau, 2000). As indicated by Königsberger et al. (2000), the reasons for the relatively large discrepancy among the measured log ${}^{1.0}K_{01j}$ values is not clear, particularly in light of the significance of citric acid in biological systems, its ubiquity in nature, and the degree to which it has been studied. They point out that one potential source of error may be associated with the definition of the base species, $H_3cit^0(aq)$ versus $H_4cit^0(aq)$, and the propagation of errors during data analysis. However, it may be argued that incomplete description of the chemical environment, such as the omission of background electrolyte–citrate complexes [e.g., $Nacit^{2-}(aq)$, or $Kcit^{2-}(aq)$], or differences in the constants used to describe complex formation [K_{110} values for $Nacit^{2-}(aq)$ or $Kcit^{2-}(aq)$], may account for the variation in the ionization constants among the assorted literature sources.

TABLE 10.1. Citric Acid Ionization Constants from Selected Literature Sources[a]

Medium	$-\log^I K_{011}$	$-\log^I K_{012}$	$-\log^I K_{013}$	Reference
1 mol L^{-1} KNO_3	2.83(0.04)	4.13(0.03)	5.33(0.05)	Vanni and Gennaro (1974)
1 mol L^{-1} $NaClO_4$	2.811(0.007)	4.096(0.009)	5.171(0.008)	Ekström and Olin (1979)
	3.319(0.009)	4.487(0.009)	5.780(0.008)	Arena et al. (1980)
1 mol L^{-1} NaCl	2.794(0.003)	4.073(0.001)	5.159(0.001)	Königsberger et al. (2000)
0.6 mol L^{-1} NaCl	2.769(0.001)	4.081(0.001)	5.217(0.001)	Öhman and Sjöberg (1983)
0.2 mol L^{-1} KCl	2.87(0.01)	4.27(0.01)	5.57(0.01)	Lakatos et al. (2001)
0.15 mol L^{-1} KCl (37°C)	2.87(0.06)	4.37(0.04)	5.75(0.03)	Amico et al. (1979)
0.15 mol L^{-1} NaCl (37°C)	2.856(0.001)	4.248(0.001)	5.539(0.001)	Venturini and Berthon (1989)
0.15 mol L^{-1} NaCl	2.94(0.01)	4.34(0.02)	5.61(0.03)	Gorzsás et al. (2004)
0.12 mol L^{-1} NaCl	2.94(0.05)	4.32(0.01)	5.70(0.05)	Rajan et al. (1978)[b]
	2.92(0.06)	4.32(0.01)	5.70(0.06)	Rajan et al. (1981)
0.1 mol L^{-1} $NaClO_4$	2.93(0.01)	4.37(0.01)	5.64(0.01)	Ramamoorthy and Manning (1973)
0.1 mol L^{-1} KNO_3	2.81(0.08)	4.36(0.02)	5.70(0.02)	Field et al. (1974)
0.1 mol L^{-1} KCl	3.025(0.0015)	4.389(0.0015)	5.710(0.0015)	Motekaitis and Martell (1984)
	2.90(0.02)	4.35(0.01)	5.90(0.02)	Gregor and Powell (1986)
0.1 mol L^{-1} KNO_3	2.92	4.35	5.70	Harris et al. (2003)

[a] $-\log^I K_{01n}$ values are for 25°C (unless specified otherwise) and the reactions defined in Eq. (10.1). Values in parentheses are standard deviations.

[b] The authors also report a $-\log^I K_{014}$ value of 13.8(0.3).

TABLE 10.2. Equilibrium Solubility of Gibbsite [$Al(OH)_3(s)$] as Measured by the Total Soluble Concentration of Aluminum (Al_{TS}) in Nitrate and Citrate Background Electrolytes[a]

pH	0.05 mol L^{-1} $NaNO_3$ (mol L^{-1} Al_{TS})	0.05 mol L^{-1} Na_3cit (mol L^{-1} Al_{TS})
4.06	2.62×10^{-4}	8.31×10^{-3}
5.06	1.57×10^{-5}	2.41×10^{-3}
6.15	3.66×10^{-7}	1.16×10^{-3}

Source: Data from Essington et al. (2005).
[a] Gibbsite solubility determined after an 8-week equilibration at ambient temperatures (20 to 22°C).

The distribution of citric acid between the various dissociation products illustrates that the citrate ligand bears a formal charge throughout a broad range of "typical" soil pH values (Figure 10.2*a*). The singly and doubly deprotonated ligands predominate in the pH < 6 range, and the triply deprotonated species predominates when the solution pH is greater than approximately 6. One would then anticipate that the power of citrate to complex metals should increase with increasing pH. Evidence to this effect is provided by Essington et al. (2005) (Table 10.2). Their findings indicate that the equilibrium solubility of gibbsite [$Al(OH)_3(s)$] in the presence of citrate, as measured by the total concentration of soluble Al (Al_{TS}), is approximately 1.5 orders of magnitude greater than that in the presence of nitrate at pH 4. However, at pH 6 the solubility of gibbsite in citrate is approximately 3.5 orders of magnitude greater than that in nitrate. Similar results were obtained when monitoring $Fe(III)_{TS}$ concentrations controlled by goethite [$FeOOH(s)$] in the presence and absence of citrate and as a function of pH.

10.2.2. Aluminum

Aluminum and Fe(III) hydrolysis, specifically the formation of monomers, is represented by the reaction ($M^{3+} = Fe^{3+}$ or Al^{3+})

$$M^{3+}(aq) + jH_2O(l) = M(OH)_j^{3-j}(aq) + jH^+(aq) \tag{10.4}$$

where j ranges from 1 to 4. The thermodynamic and conditional hydrolysis constants are defined by (assuming the activity of water to be unity)

$$K_{10-j} = \frac{(M(OH)_j^{3-j})(H^+)^j}{(M^{3+})} \tag{10.5}$$

$$^{I}K_{10-j} = \frac{[M(OH)_j^{3-j}][H^+]^j}{[M^{3+}]} \tag{10.6}$$

Again, the parentheses denote activities and the brackets denote molar concentrations. Although several dimer and polymeric species have been identified or proposed [e.g., (Al or Fe)$_2$(OH)$_2^{4+}$(*aq*), (Al or Fe)$_3$(OH)$_4^{5+}$(*aq*), and $Al_{13}O_4(OH)_{24}^{7+}$(*aq*)], they generally have little significance in natural water systems, as the relatively high concentrations of Al_{TS} or Fe_{TS} required for formation ($> 10^{-4}$ to 10^{-3} mol L^{-1}) are incongruous with the very low concentrations supported by hydroxide and oxyhydroxide precipitates (Baes and Mesmer, 1986; Daniele et al., 1994; Bertsch and Parker, 1995).

Nordstrom and May (1995) surveyed the numerous studies that report measured hydrolysis constants for Al^{3+}(*aq*) at 25°C. They conclude that K_{10-1} is well established relative to accuracy and precision of measurement, with log K_{10-1} values from 22 studies ranging from −4.3 to −5.86 (with a mean log $K_{10-1} \pm \sigma$ of -4.985 ± 0.28). Less well characterized are values for K_{10-2} and K_{10-3}, as the $Al(OH)_2^+$(*aq*) and $Al(OH)_3^0$(*aq*) species occur in the pH range where Al exhibits very low solubility, as shown in Table 10.2. In a survey of 10 studies, log K_{10-2} values ranged from −8.56 to −10.55 (with a mean log $K_{10-2} \pm \sigma$ of -9.65 ± 0.79); and in a survey of five studies, log K_{10-3} values ranged from −15.0 to −17.25 (with a mean log $K_{10-3} \pm \sigma$ of -16.1 ± 1.0). Values for K_{10-4} are generally obtained from $Al(OH)_3$(*s*) or AlOOH(*s*) solubility studies in alkaline solutions (pH > 8). Thus, the reliability and uncertainty associated with K_{10-4} values are related directly to the reliability and uncertainty of solubility determinations, including the integrity and nature of the solids. In a survey of 23 studies, log K_{10-4} values ranged from −19.83 to −27.98, forming a bimodal distribution with modes of −22.2 and −23.3 (with an overall mean log K_{10-4} $\pm \sigma$ of -22.97 ± 1.94). The distribution predicted for Al_{TS} as a function of pH in a solution in equilibrium with gibbsite, and as computed using the hydrolysis constants of Baes and Mesmer (1986), is illustrated in Figure 10.2*b*. Note that the hydrolysis products [$Al(OH)_2^+$(*aq*) and $Al(OH)_3^0$(*aq*)] predominate throughout the pH range that is typical for soil solutions, and as discussed above, there is relatively little confidence in the K_{10-2} and K_{10-3} values used to predict the formation of these two species.

10.2.3. Iron(III)

The hydrolysis of Fe^{3+} has also been well studied, particularly in seawater and synthetic seawater solutions ($I \approx 0.7$), either through potentiometric and spectrophotometric studies [for the formation of $FeOH^{2+}$(*aq*) and $Fe(OH)_2^+$(*aq*)] or through studies involving the solubility of ferric hydroxides [for the formation of $Fe(OH)_2^+$(*aq*), $Fe(OH)_3^0$(*aq*), and $Fe(OH)_4^-$(*aq*)] (Table 10.3). In general, the K_{10-1} value for $FeOH^{2+}$(*aq*) formation is considered most reliable (Table 10.3). From the compilations of Liu and Millero (1999) and Byrne et al. (2000), log K_{10-1} values from 10 studies ranged from −2.1 to −2.21 (with a mean log $K_{10-1} \pm \sigma$ of -2.17 ± 0.03 at 25°C). However, considerably less reliability is attributed to values for K_{10-2}, K_{10-3}, and K_{10-4}. Byrne et al. (2000) surmised that log ${}^{0.7}K_{10-2}$ values are smaller than or approximately equal to −7, citing studies by Byrne and Kester (1976a,b, 1978, 1981) which generated

TABLE 10.3. Iron(III) Hydrolysis Constants from Selected Literature Sources[a]

Medium, I	$-\log^I K_{10-1}$	$-\log^I K_{10-2}$	$-\log^I K_{10-3}$	$-\log^I K_{10-4}$	Source
Pure water, 0	2.17	7.17	—	—	Kester et al. (1975)
Seawater, 0.68	2.71	7.07	13.62	24.2	Byrne and Kester (1976a,b)
Pure water, 0	—	—	12.92	24.4	Byrne and Kester (1976a)
$NaClO_4$, 0.68	2.77	7.28	—	—	Byrne and Kester (1976b)
NaCl, 0.68	2.76	7.00	—	—	Byrne and Kester (1976b)
Pure water, 0	2.19	5.67	>12	21.6	Baes and Mesmer (1986)
Seawater, 0.72	2.62	6.0	12.5	21.8	Millero et al. (1995)
Pure water, 0	2.20	5.54	11.80	21.6	Millero et al. (1995)
NaCl, 0.7	2.52	6.5	15	22.8	Liu and Millero (1999)
NaCl, 0.1	2.20	6.5	15	22.8	Liu and Millero (1999)
NaCl, 0.01	1.98	6.2	14	22.4	Liu and Millero (1999)
Pure water, 0	2.1(0.03)	6.3(0.1)	14.3(0.6)	22.3(0.2)	Liu and Millero (1999)
	2.18(0.01)	—	—	—	Byrne et al. (2000)
Seawater, 0.7	2.74(0.01)	6.86(0.04)	—	—	Byrne et al. (2000)
Seawater, 0.72	2.5(0.2)	6.5(0.3)	13.6(0.3)	—	Liu and Millero (2002)
Seawater, 0.7 (20°C)	—	6.4(0.15)	15.1(0.8)	22.7(0.08)	Sunda and Huntsman (2003)

[a] $-\log^I K_{10-j}$ values are for 25°C (unless specified otherwise) and the reactions defined in Eq. (10.4). Values in parentheses are standard deviations.

log $^{0.7}K_{10-2}$ values of −7.28, −8.6, and ≤ −7. More recent potentiometric and solubility studies (Table 10.3) resolved log $^{0.7}K_{10-2}$ values (in seawater or salt solutions) that ranged from −6.86 to −6.0. The determination of K_{10-3} values are most commonly performed through the interpretation of $Fe(OH)_3(s)$ solubility data. However, as discussed by Byrne et al. (2000), numerous experimental artifacts can confound the interpretation of solubility data, including the very low solubility of $Fe(OH)_3(s)$, the difficulty separating $Fe(OH)_3{}^0(aq)$ from $Fe(OH)_3(s)$, the dependence of $Fe(OH)_3(s)$ solubility on particle size (which is a function of pH), and uncertainties related to the integrity of the solid. Estimates of $^{0.7}K_{10-3}$ vary over several orders of magnitude, with log $^{0.7}K_{10-3}$ ranging from −15.1 to −12.5 (Table 10.3 and Kuma et al., 1992) while log K_{10-3} ranges from −14.3 to −11.8. Clear evidence of the formation of $Fe(OH)_4{}^0(aq)$ is illustrated in the ferric hydroxide solubility results of Liu and Millero (1999). Their log $^{0.7}K_{10-4}$ value of −22.8 (log $K_{10-4} = -22.3$) is consistent with those of Millero et al. (1995), Kuma et al. (1996), and Sunda and Huntsman (2003) (−21.8 to −22.7). The distribution of $Fe(III)_{TS}$ predicted as a function of pH in a solution in equilibrium with goethite and as computed using the hydrolysis constants of Liu and Millero (1999) is illustrated in Figure 10.2*c*. Again, as was noted for Al hydrolysis, there is relatively little confidence in the K_{10-2} and K_{10-3} values used to predict the hydrolytic Fe(III) species that predominate in the pH range of normal soil solutions.

10.3. CHEMICAL MODELS OF METAL–CITRATE AQUEOUS COMPLEXATION

A chemical model is a unique set of aqueous species, and their formation constants, that describe the solution chemistry of a metal–ligand pair throughout a range of chemical properties (e.g., pH, metal and ligand concentrations, and ionic strength). Numerous chemical models have been proposed to describe Fe(III)–and Al–citrate aqueous speciation. Most commonly, these models have been derived through potentiometric (acid–base) titrations, although NMR (^{1}H, ^{13}C, and ^{27}Al), ultraviolet, and electrospray mass spectroscopic techniques have also been employed. Unfortunately, there exists very little agreement among the models, with the exception of the $AlHcit^+(aq)$, $Alcit^0(aq)$, and $Fecit^0(aq)$ species, which are predicted to predominate in pH < 3 solutions.

The difficulty in assessing metal–citrate complexation may be attributed to a number of confounding factors. Both the metals [$Al^{3+}(aq)$ and $Fe^{3+}(aq)$] and the ligand [$cit^{3-}(aq)$] are polyprotic, having complex acid–base chemistry throughout a wide pH range (Figure 10.2), chemistry that requires the critical assessment and consideration of $cit^{3-}(aq)$ protonation, as well as $Al^{3+}(aq)$ and $Fe^{3+}(aq)$ hydrolysis reactions (this also implies the involvement of protons and hydroxides in the metal–citrate complexes). Both metals form sparingly soluble oxyhydroxy precipitates, which greatly restricts the experimentally accessible pH range and generally limits the applicability to acidic environments of the chemical speciation models developed. Further, there is relatively little confidence in the constants that describe $Al^{3+}(aq)$ and $Fe^{3+}(aq)$ hydrolysis in the pH range that encompasses typical soil solutions (pH 4 to 9). Indeed, most studies restrict their evaluations to strongly acidic solutions, the results of which have little practical application to natural water systems. Experimental conditions such as the molar ratio of total soluble (denote by the superscript TS) citrate to total soluble Al or Fe(III) [cit_{TS}/Al_{TS} or $cit_{TS}/Fe(III)_{TS}$], pH, and temperature affect the ultimate equilibrium state achieved under the experimental conditions defined. For example, high citrate/metal ratios tend to favor only the formation of metal–citrate monomers, while both dinuclear and trinuclear complexes have been identified when the ratios are close to unity. Direct spectroscopic and potentiometric titration studies have shown that the attainment of metal–citrate speciation equilibrium is restricted kinetically (i.e., not instantaneous, as is often assumed for aqueous complexation reactions). Mononuclear species predominate during short metal residence times (on the order of minutes), while di- and trinuclear species become significant after reaction times of hours to days.

The possible combinations and stoichiometries of Al^{3+} or Fe^{3+}, cit^{3-}, H^+, and OH^- that may occur in a metal–citrate species are numerous. In general, the formation of soluble metal–citrate complexes can be described by

$$m\text{M}^{3+}(aq) + n\text{cit}^{3-}(aq) + j\text{H}^+(aq) = \text{M}_m\text{H}_j\text{cit}_n{}^{3m+j-3n}(aq) \qquad (10.7)$$

where m, n, and j are stoichiometric coefficients, and $M^{3+}(aq)$ represents $Al^{3+}(aq)$ or $Fe^{3+}(aq)$. The equilibrium constants that describe the formation of the $M_mH_jcit_n{}^{3m+j-3n}(aq)$ species are defined as

$$K_{mnj} = \frac{(\text{M}_m\text{H}_j\text{cit}_n{}^{3m+j-3n})}{(\text{M}^{3+})^m(\text{H}^+)^j(\text{cit}^{3-})^n} \qquad (10.8)$$

where the parentheses denote activities. The conditional equilibrium constants (${}^{I}K_{mnj}$ values) are defined similarly, with brackets replacing parentheses to denote molar concentrations.

The stoichiometry of the proton in Eqs. (10.7) and (10.8) (denoted by j) is a collective term that represents three possible configurations for the $M_mH_j cit_n{}^{3m+j-3n}(aq)$ complex. The value of j is 1 when $Hcit^{2-}$ occurs in the complex, as in $MHcit^+(aq)$. In this instance, a single carboxyl and the hydroxyl are protonated on the citrate molecule. Such complexes are significant only when solution pH values are less than 4. More commonly, j is zero or negative, the latter representing either the occurrence of $H_{-1}cit^{4-}$ in the complex (all citrate moieties ionized) [as in $MH_{-1}cit^-(aq)$], or the occurrence of $H_{-1}cit^{4-}$ and a metal hydrolysis product in the complex [as in $MOH(H_{-1}cit)^{2-}(aq)$]. Chemical models derived from potentiometric acid–base titration studies cannot distinguish between the two potential proton sources (citrate hydroxyl or metal-bound water), as titrations are sensitive only to the gain or loss of protons (a macroscopic determination). Thus, without ancillary spectroscopic information, the metal–citrate complexes that are formed through deprotonation or hydrolysis are commonly represented as $M_m(OH)_j cit_n{}^{3m-j-3n}(aq)$ in the literature. For example, $MOHcit^-(aq)$ may represent $MH_{-1}cit^-(aq)$ or $(MOH)cit^-(aq)$; $M(OH)_2cit^{2-}(aq)$ may represent $(M(OH)_2)cit^{2-}(aq)$ or $MOH(H_{-1}cit)^{2-}(aq)$. Although the difference in notation, for example between $MOHcit^-(aq)$ and $MH_{-1}cit^-(aq)$, may appear to be semantic, the nomenclature supports the experimental observation that hydroxide is commonly consumed during the potentiometric titration of metal–citrate systems. It is also assumed that the macroscopic species $MOHcit^-(aq)$ is actually $MH_{-1}cit^-(aq)$, even though $MOH^{2+}(aq)$ formation occurs in acidic solutions, while $H_{-1}cit^{4-}(aq)$ is only predicted to form in strongly alkaline solutions. In essence, if protons are produced during the formation of metal–citrate species, the proton source is generally always identified as the citrate. Mechanistically, the central metal cation is bound by a terminal and the central carboxyl groups, and the hydroxyl moiety (Figure 10.1), forming three ring structures that include the metal and yielding a very stable complex.

10.3.1. Aluminum–Citrate Interactions

Early studies have established the formation of $Alcit^0(aq)$ in strongly acidic solutions. Equilibrium studies performed by Pattnaik and Pani (1961) were interpreted to confirm the formation of the neutral $Alcit^0(aq)$ species in pH < 3.1 solutions ($\log {}^{0.25}K_{110} = -4.71$ at 33°C in 0.25 mol L^{-1} $NaClO_4$). Rajan et al. (1981) reported that the $Alcit^0(aq)$ species predominated in the pH range 2.3 to 7.0 ($\log {}^{0.12}K_{110} = -5.0 \pm 0.4$ at 25°C in 0.12 mol L^{-1} NaCl). Using ^{27}Al NMR, Karlik et al. (1983) documented the existence of two Al species in a pH 2 solution: $Al(H_2O)_6{}^{3+}(aq)$ and presumably a 1 : 1 Al–citrate complex. They observed a gradual downfield shift and spreading of the single Al–citrate NMR peak with increasing pH. The authors attributed this to the fast exchange of Al between several, yet unidentified Al–citrate species. Lopez-Quintela et al. (1984), using stopped-flow and pressure-jump relaxation techniques with pH 1.4 to 2.7 solutions determined the thermodynamic formation constants for $Al(H_2cit)^{2+}(aq)$ (${}^{0.001}K_{112} = 8.2 \pm 1.5 \times 10^2$; $\log {}^{0.001}K_{112} = 2.9$), $Al(Hcit)^+(aq)$ (${}^{0.001}K_{111} = 3.6 \pm 1.6 \times 10^6$; $\log {}^{0.001}K_{111} = 6.6$), and $Alcit^0(aq)$ (${}^{0.001}K_{110} = 5.3 \pm 2.3 \times 10^{10}$; $\log {}^{0.001}K_{110} = 10.7$). They also observed several superimposed relaxation effects at pH values between 3 and 5, suggesting that additional Al–citrate species would be required to describe complexation.

Generally, at pH values greater than 3, interpretations afforded the many Al–citrate studies diverge. Pattnaik and Pani (1961) postulated that $Alcit^0(aq)$ was dibasic at pH values greater than 3.1 (losing two protons via hydrolysis or deprotonation), with titration curve inflection points at pH 3.49 and 6.79. The inflections were interpreted to indicate the formation of $Al(H_{-1}cit)^-(aq)$ and $AlOH(H_{-1}cit)^{2-}(aq)$. Wiese and Veith (1975) interpreted their titration results to support the existence of polynuclear complexes: $Al_3(OH)cit_3{}^-(aq)$

in the pH range 3 to 3.5, $Al_3(OH)_4cit_2^-(aq)$ in the pH range 3.8 to 4.8, and $Al_{10}(OH)_{15}cit_6^{3-}(aq)$ in the pH range 5.6 to 7.5. Rajan et al. (1981) used only the $Al(OH)_3cit^{3-}(aq)$ species to describe the titration results observed in the pH range 7.0 to 10.3. Unfortunately, the interpretations afforded these early titration results are highly suspect, particularly when solution pH is greater than approximately 3. The formation of mononuclear and polynuclear $Al^{3+}(aq)$ hydrolysis products were generally neglected, as was the likely precipitation of $Al(OH)_3(s)$.

10.3.1.1. Öhman Model Perhaps the first critical examinations of Al–citrate complexation were performed by Öhman and Sjöberg (1983). They conducted potentiometric titrations in the pH range 2.7 to 5 in 0.6 mol L^{-1} NaCl (25°C) solutions containing Al_{TS} and cit_{TS} concentrations that ranged between 0.25 and 8 mmol L^{-1} (cit_{TS}/Al_{TS} ratios of 1 : 1, 2 : 1, 4 : 1, 8 : 1, 16 : 1, and 32 : 1). In addition to predicting the log $^{0.6}K_{01j}$ values for citric acid (Table 10.1), they modeled their titration curves with a chemical model consisting of the mononuclear species $Al(Hcit)^+(aq)$, $Alcit^0(aq)$, and $Alcit_2^{3-}(aq)$, and the trinuclear species $Al_3(OH)_4cit_3^{4-}(aq)$ (Table 10.4). In their data analysis, they modeled $Al^{3+}(aq)$ hydrolysis by considering the $AlOH^{2+}(aq)$ and $Al(OH)_4^-(aq)$ monomers (log $^{0.6}K_{10-1} = -5.52$ and log $^{0.6}K_{10-4} = -23.46$) and the $Al_3(OH)_4^{5+}(aq)$ and $Al_{13}O_4(OH)_{24}^{7+}(aq)$ polynuclear species (log $^{0.6}K_{30-4} = -13.57$ and log $^{0.6}K_{13,0,-32} = -109.2$). They also surmised that the $Al_3(OH)_4cit_3^{4-}(aq)$ species was composed of both hydrolyzed $Al^{3+}(aq)$ and $H_{-1}cit^{4-}(aq)$ [e.g., $Al_3(OH)_{1+x}(H_{x-3})cit_3^{4-}(aq)$, $0 \leq x \leq 3$], arguing that although the pK_{014} value for $cit^{3-}(aq)$ dissociation is very high ($> \sim 13$) relative to the first hydrolysis constant for $Al^{3+}(aq)$ ($pK_{10-1} \sim 5$), deprotonation of the citrate hydroxyl group had previously been shown to start at pH values as low as 6.

The chemical model of Öhman and Sjöberg (1983) was later amended by Öhman (1988) to include the $Al(H_{-1}cit)^-(aq)$, $AlOH(H_{-1}cit)^{2-}(aq)$, and $Al_3(OH)_4(H_{-1}cit)_3^{7-}(aq)$ species. Öhman (1988) used the Al–citrate model from Öhman and Sjöberg (1983) as a starting point, but used a different model for $Al^{3+}(aq)$ hydrolysis (Baes and Mesmer, 1986). The Öhman model for Al–citrate complexation is described in Table 10.4. In addition to proposing the three new Al–citrate species, Öhman's experiments illustrated that equilibrium in the Al–citrate system was not instantaneous. At relatively short reaction times (upon mixing to generate equimolar cit_{TS}/Al_{TS} solutions), the formation of mononuclear species is favored; however, the trinuclear species $[Al_3(OH)(H_{-1}cit)_3^{4-}(aq)]$ predominates throughout the pH range 3 to 7 after reaction times of 5 min (Figure 10.3). Öhman (1988) also illustrated that Al–citrate complexation chemistry is dominated by Al–citrate monomers when cit_{TS}/Al_{TS} is high (> 100 : 1). For example, $Alcit^0(aq)$, $Al(H_{-1}cit)^-(aq)$, and $AlOH(H_{-1}cit)^{2-}(aq)$ are predicted to predominate throughout the pH range ≈ 3 to 8 when cit_{TS}/Al_{TS} is 100 : 1 (Figure 10.4a). Öhman (1988) also argued that $Al_3(OH)(H_{-1}cit)_3^{4-}(aq)$ correctly

TABLE 10.4. Chemical Models and the Conditional Formation Constants [$\log {}^{I}K_{mnj}$ ($\pm\ \sigma$) at 25°C] That Have Been Developed to Describe the Aqueous Speciation of Al–Citrate Systems[a]

Aqueous Complex (mnj)[b]	Complexation Model							
	Motekaitis and Martell (1984),[c] $I = 0.1$	Gregor and Powell (1986),[d] $I = 0.1$	Öhman and Ajöberg (1983) and Öhman (1988),[e] $I = 0.6$	Venturini and Berthon (1989),[f] $I = 0.15$	Lakatos et al. (2001),[g] $I = 0.2$	Harris et al. (2003),[h] $I = 0.1$	Essington et al. (2005),[i] $I = 0.15$	$I = 0.6$
$Alcit^{0}(aq)$	7.98	8.35	7.142	8.246	7.85	8.40	4.46	3.61
(110)	(0.001)	(0.01)	(0.009)	(0.004)	(0.02)	(0.04)	(0.03)	(0.03)
$AlHcit^{+}(aq)$	10.92	11.01	9.39	10.502	10.18	10.66	—	—
(111)	(0.001)	(0.01)	(0.02)	(0.008)	(0.03)	(0.09)		
$AlOHcit^{-}(aq)$	4.67	—	3.59	—	4.27	4.19	—	—
(11−1)	(0.001)		(0.13)		(0.09)	(—)		
$Alcit^{3}{}_{2}{}^{-}(aq)$	—	13.36	11.60	13.068	12.73	13.56	—	—
(120)		(0.02)	(0.12)	(0.040)	(0.10)	(0.15)		
$AlHcit^{2}{}_{2}{}^{-}(aq)$	—	17.32	—	—	—	—	—	—
(121)		(0.04)						
$Al(OH)_{2}cit^{2-}(aq)$	—	—	−2.64	—	−1.77	—	−4.40	−4.87
(11−2)			(0.14)		(0.02)		(0.02)	(0.02)
$Al(OH)_{3}cit^{3-}(aq)$	—	—	—	—	—	—	−10.66	−10.78
(11−3)							(0.02)	(0.02)

$AlOHcit^{4}{}_{2}{}^{-}(aq)$	—	7.29	—	6.777	7.81	7.65	—	—
(12–1)		(0.02)		(0.251)	(0.06)	(0.38)		
$Al(OH)_2cit^{5}{}_{2}{}^{-}(aq)$	—	0.11	—	−0.209	0.4	—	—	—
(12–2)		(0.03)		(0.049)	(—)			
$Al_2(OH)_2cit^{2}{}_{2}{}^{-}(aq)$	—	—	—	12.694	—	12.12	—	—
(22–2)				(0.011)		(0.12)		
$Al_3(OH)_4cit^{4}{}_{3}{}^{-}(aq)$	—	—	14.43	15.085	16.34	16.15	—	—
(33–4)			(0.02)	(0.064)	(0.03)	(0.42)		
$Al_3(OH)_7cit^{7}{}_{3}{}^{-}(aq)$	—	—	10.91	—	—	—	—	—
(33–7)			(0.09)					

[a] Stability constants are for the reaction defined by Eq. (10.6).
[b] The stoichiometric denotation associated with each complex (*mnj*), as defined by Eq. (10.7).
[c] cit_{TS}/Al_{TS} ratios of 1 : 1, 2 : 1, and 3 : 1, and titrations in 0.1 mol L^{-1} KNO_3.
[d] cit_{TS}/Al_{TS} ratio of approximately 5 : 1, and titrations in 0.1 mol L^{-1} KCl.
[e] cit_{TS}/Al_{TS} ratios of 1 : 1, 2 : 1, and 4 : 1, and titrations in 0.6 mol L^{-1} NaCl.
[f] cit_{TS}/Al_{TS} ratios ranging from 1 : 2 to 10 : 1, and titrations in 0.15 mol L^{-1} NaCl at 37°C.
[g] cit_{TS}/Al_{TS} ratios of 1 : 1 and 3 : 1, and titrations in 0.2 mol L^{-1} KCl.
[h] cit_{TS}/Al_{TS} ratios of 1 : 1 and 2 : 1, and titrations in 0.1 mol L^{-1} KNO_3 [model excludes $Al_2cit^{3}{}_{3}{}^{-}$ and includes $Al_2(OH)_2cit^{2}{}_{2}{}^{-}$].
[i] Al–citrate speciation in 0.05 mol L^{-1} Na_3cit and 0.45 mol L^{-1} $NaNO_3$+ 0.05 mol L^{-1} Na_3cit solutions in equilibrium with gibbsite (cit_{TS}/Al_{TS} ratios ranged from approximately 5 : 1 to 50 : 1).

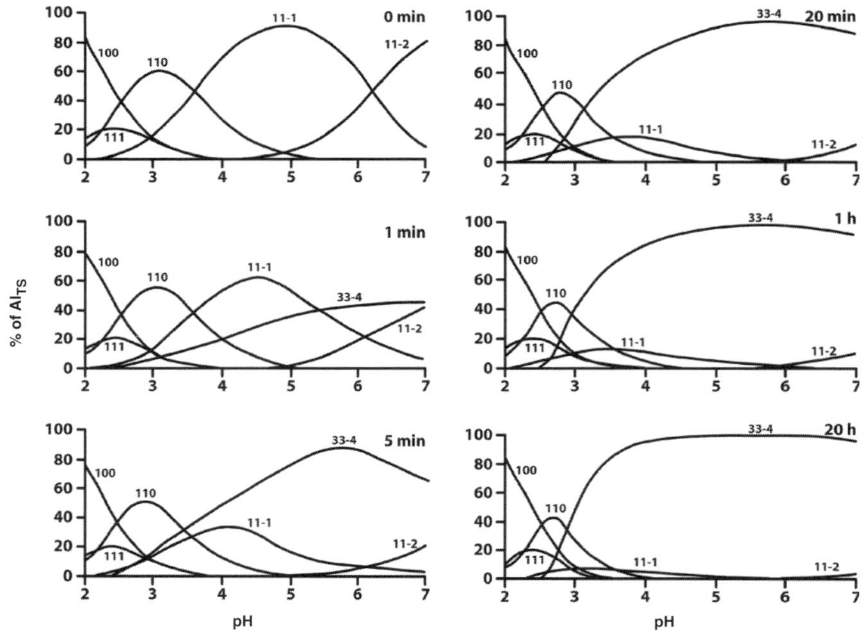

Figure 10.3. The predicted distribution of Al–containing species at 25°C according to the Al–citrate speciation model of Öhman (1988) (Table 10.4) as a function of pH and reaction time in 0.6 mol L^{-1} NaCl with $Al_{TS} = cit_{TS} = 0.01$ mol L^{-1} (modified from Öhman, 1988). The species are identified by stoichiometric coefficients according to Eq. (10.7).

indicated the structure of the trinuclear species [rather than $Al_3(OH)_4cit_3{}^{4-}(aq)$], owing to the strong dissociation tendency of the citrate hydroxyl.

10.3.1.2. Motekaitis and Martell Model Contrary to the Öhman model, Motekaitis and Martell (1984) found no need to invoke polynuclear Al–citrate species to describe their potentiometric titration data. Their minimalist model, consisting of $Al(Hcit)^+(aq)$, $Alcit^0(aq)$, and $Al(H_{-1}cit)^-(aq)$ (Table 10.4), was generated from titrations of Al–citrate solutions in 0.10 mol L^{-1} KCl, using cit_{TS}/Al_{TS} ratios of 1 : 1, 2 : 1, and 3 : 1 from pH 2 to the onset of Al precipitation (approximately pH 7.5). They employed 4-hour equilibration times and included in their computations the $Al^{3+}(aq)$ hydrolysis for all four monomers [$AlOH^{2+}(aq)$ to $Al(OH)_4{}^-(aq)$], as well as $Al_2(OH)_2{}^{4+}(aq)$ and $Al_3(OH)_4{}^{5+}(aq)$ [with log ${}^{0.1}K_{m,0,-j}$ values obtained from the compilations of Smith and Martell (1976)]. Motekaitis and Martell (1984) also proposed the mechanistic formation of $Al(H_{-1}cit)^-(aq)$ instead of $AlOHcit^-(aq)$, arguing that the formation of the former was complete at pH values below the first $Al^{3+}(aq)$ hydrolysis constant (log $K_{10-1} = -5.30$). In support of the Motekaitis and Martell model, Greenaway (1986) observed several distinct ^{27}Al NMR resonances in the pH range 2 to 10, which differed from the findings of Karlik et al. (1983). The findings of

Greenaway were attributed to slow exchange of Al between several species, most notably $Al(Hcit)^{+}(aq)$, $Alcit^{0}(aq)$, and $Al(H_{-1}cit)^{-}(aq)$ (although an additional resonance was observed, but remained unidentified). The predicted distribution of Al–citrate complexes, according to the Motekaitis and Martell model, shows that the $Al(H_{-1}cit)^{-}(aq)$ species predominates throughout a wide range of pH values (Figure 10.4*b*).

10.3.1.3. Gregor and Powell Model Gregor and Powell (1986) coupled potentiometric titration and ^{13}C NMR results to elucidate Al–citrate speciation. Their titration experiments were performed at 25°C in 0.1 mol L^{-1} KCl with cit_{TS}/Al_{TS} greater than 5 : 1 and in the pH range 2 to 7. The $Al^{3+}(aq)$ hydrolysis products $AlOH^{2+}(aq)$, $Al_3(OH)_4{}^{5+}(aq)$, and $Al_{13}(OH)_{32}{}^{7+}(aq)$ (log $^{0.1}K_{m0-j}$ values were not provided) were included in their data analysis, as was the $Kcit^{2-}(aq)$ species (log $^{0.1}K_{110} = 0.77$). The chemical model that described the titration data consisted of $AlHcit^{+}(aq)$, $AlHcit_2{}^{2-}(aq)$, $Alcit^{0}(aq)$, $Alcit_2{}^{3-}(aq)$, $Al(H_{-1}cit)cit^{4-}(aq)$, and $Al(H_{-1}cit)_2{}^{5-}(aq)$ (Table 10.4). They noted that the formation of the trinuclear species proposed by Öhman and Sjöberg (1983), $Al_3(OH)_4cit_3{}^{4-}(aq)$, was not supported by their data. However, as the authors indicated, the relatively high cit_{TS}/Al_{TS} (> 5 : 1) may not have been conducive to the formation of the trinuclear species.The equilibration time employed during the titrations is also not clear from Gregor and Powell (1986), only that the pH readings were stable. As noted by Öhman and Sjöberg (1983), if the equilibration times are short, the predicted formation of $Al_3(OH)_4cit_3{}^{4-}(aq)$ will be diminished relative to the mononuclear species.

The ^{13}C NMR studies of Gregor and Powell (1986) were performed on solutions having a cit_{TS}/Al_{TS} ratio of 4 : 1 with a solution pH that varied from 2 to 8. The NMR data were interpreted with reference to the titration-based chemical model, which predicts $Alcit^{0}(aq)$ to predominate in the pH range < 4.2; $Alcit_2{}^{3-}(aq)$ in the pH range 4.2 to 5.8; and $Al(H_{-1}cit)cit^{4-}(aq)$ in the pH range > 6 (Figure 10.4*c*). A downfield shift (relative to citric acid) in the NMR resonances of the terminal and central carboxyl moieties and the hydroxyl, and an upfield shift for CH_2, suggests the deprotonation and complexation of citric acid to form $Alcit^{0}(aq)$ in the pH range 2 to 4 (Figure 10.5). In the pH range 4 to 5, the resonances of the terminal carboxyl and CH_2 shift upfield, evidence of the deprotonation of $AlHcit_2{}^{2-}(aq)$ to form $Alcit_2{}^{3-}(aq)$. Above pH 5, the resonances of carboxyl and CH_2 shift downfield, indicating the dissociation of $Alcit_2{}^{3-}(aq)$ to form $Al(H_{-1}cit)cit^{4-}(aq)$ and $Al(H_{-1}ct)_2{}^{5-}(aq)$. The resonance of the hydroxyl shifts upfield when solution pH is increased above 2. However, this shift was not discussed relative to Al–citrate speciation. The Gregor and Powell model predicts a very different view of Al–citrate speciation, compared to the Öhman and the Motekaitis and Martell models, under conditions of similar cit_{TS}/Al_{TS} (> 5 : 1) (Figure 10.4*a* and *b*). Both the Öhman and the Motekaitis and Martell models predict Al_{TS} speciation to be dominated by 1 : 1 Al/cit complexes throughout the pH range 4 to 8. However, the Gregor and Powell model predicts a greater contribution from $Alcit^{0}(aq)$ (up to pH 4), and the predominance of 1 : 2 complexes (mononuclear dicitrate complexes) when pH > 4.

10.3.1.4. Venturini and Berthon Model The chemical model for Al–citrate speciation developed by Venturini and Berthon (1989) closely mirrors that of Gregor and Powell (1986), except for the inclusion of di- and trinuclear species (Table 10.4). Venturini and Berthon (1989) conducted potentiometric

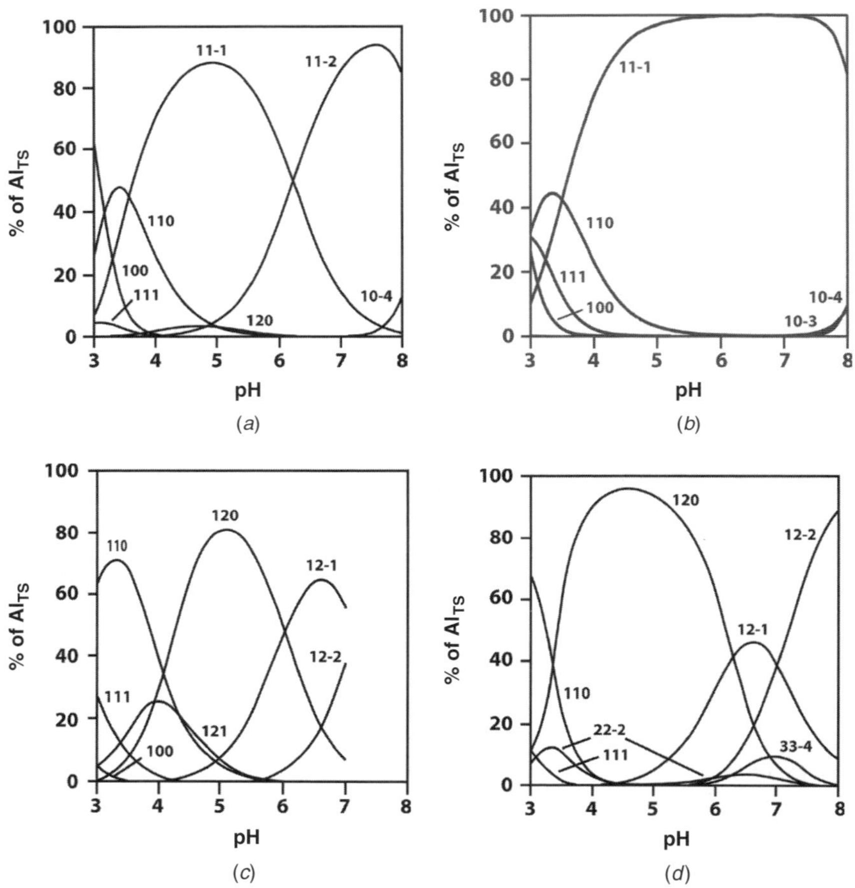

Figure 10.4. The predicted distribution of Al–containing species at 25°C as a function of pH according to various the Al–citrate speciation models (Table 10.4): (*a*) model of Öhman (1988) in 0.6 mol L^{-1} NaCl with $Al_{TS} = 10^{-6}$ mol L^{-1} and $cit_{TS} = 10^{-4}$ mol L^{-1} ($cit_{TS}/Al_{TS} = 100:1$); (*b*) model of Motekaitis and Martell (1984) in 0.01 mol L^{-1} $NaNO_3$ with $Al_{TS} = 10^{-6}$ mol L^{-1} and $cit_{TS} = 10^{-4}$ mol L^{-1} ($cit_{TS}/Al_{TS} = 100:1$); (*c*) model of Gregor and Powell (1986) in 0.1 mol L^{-1} KCl with $Al_{TS} = 4.02 \times 10^{-4}$ mol L^{-1} and $cit_{TS} = 2.55 \times 10^{-3}$ mol L^{-1} ($cit_{TS}/Al_{TS} = 6.34:1$); (*d*) model of Venturini and Berthon (1989) in solutions containing $Al_{TS} = 5 \times 10^{-4}$ mol L^{-1} and $cit_{TS} = 3.152 \times 10^{-2}$ mol L^{-1} ($cit_{TS}/Al_{TS} = 63:1$; undefined ionic strength condition; 37°C); (*e*) model of Essington et al. (2005) in 0.01 mol L^{-1} $NaNO_3$ with $Al_{TS} = 10^{-6}$ mol L^{-1} and $cit_{TS} = 10^{-4}$ mol L^{-1} ($cit_{TS}/Al_{TS} = 100:1$; 20 to 22°C). The species are identified by stoichiometric coefficients according to Eq. (10.7).

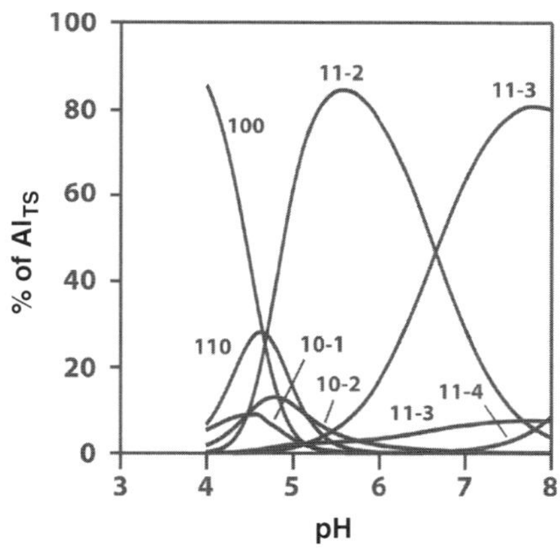

(*e*)

Figure 10.4. *continued*

titrations at 37°C in 0.15 mol L^{-1} NaCl, with cit_{TS}/Al_{TS} ranging from 1 : 2 to 10 : 1 and in the pH range 1.8 to 9. During the data analysis, the log $^{0.15}K_{10-j}$ values for $Al^{3+}(aq)$ hydrolysis were treated as adjustable parameters. As a result, only the $AlOH^{2+}(aq)$ (log $^{0.15}K_{10-1} = -3.17 \pm 0.03$) and $Al(OH)_3{}^0(aq)$ (log $^{0.15}K_{10-3} = -14.32 \pm 0.06$) species were resolved and included in the data interpretations. The authors also employed a 2-min equilibration time, surmising that this reaction period was sufficient to achieve equilibrium. The chemical model that described the Al–citrate titration data consisted of $Alcit^0(aq)$, $AlHcit^+(aq)$, $Alcit_2{}^{3-}(aq)$, $AlOHcit_2{}^{4-}(aq)$, $Al(OH)_2cit_2{}^{5-}(aq)$, $Al_2(OH)_2cit_2{}^{2-}(aq)$, and $Al_3(OH)_4cit_3{}^{4-}(aq)$. Note that these species are denoted by the general formula $M_m(OH)_j cit_n{}^{3m-j-3n}(aq)$. The authors correctly concluded that they could not differentiate between metal hydrolysis [e.g., $(AlOH)cit_2{}^{4-}(aq)$] and citrate hydroxyl dissociation [e.g., $Al(H_{-1}cit)cit^{4-}(aq)$] from the potentiometric measurements. The similarities between the speciation models of Gregor and Powell (1986) and Venturini and Berthon (1989) are evidenced by comparing Figure 10.4*c* and *d*.

10.3.1.5. Harris Model Harris et al. (2003) examined Al–citrate speciation using potentiometric titrations and electrospray ionization mass spectrometry. The titrations were performed in the pH range 2 to 9 at 25°C in 0.10 mol L^{-1} KCl solutions using cit_{TS}/Al_{TS} ratios of 1 : 1 or 2 : 1 and equilibration times of < 1 h. Data analysis included the consideration of citric acid dissociation (Table 10.1) and the formation of monomeric $Al^{3+}(aq)$ hydrolysis products (Baes and Mesmer, 1986). Two chemical models were employed to describe the titration data. One model included $Al_2(H_{-1}cit)_2{}^{2-}(aq)$ (Table 10.4), while the other included $Al_2cit_3{}^{3-}(aq)$ at the expense of $Al_2(H_{-1}cit)_2{}^{2-}(aq)$. Although the authors considered the $Al_2(H_{-1}cit)_2{}^{2-}(aq)$ model to provide a

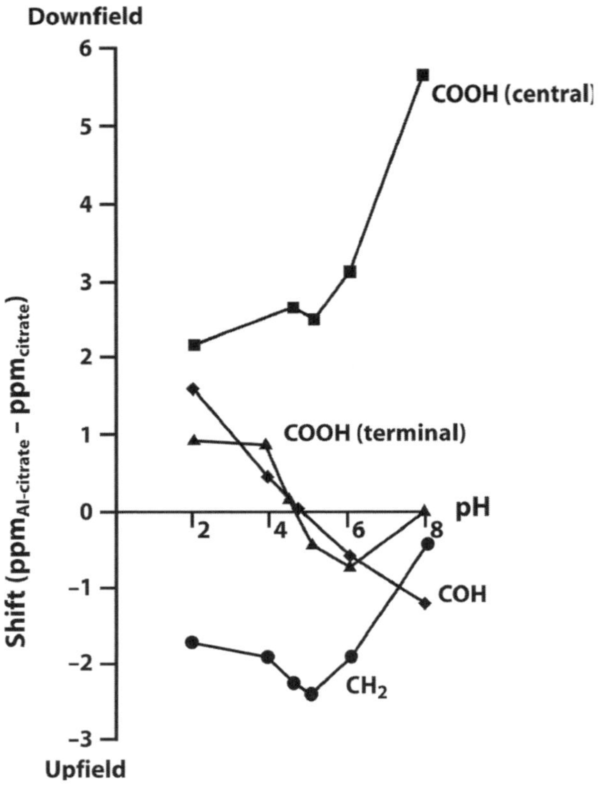

Figure 10.5. Chemical shifts in the ^{13}C NMR resonance lines associated with the structural moieties of the citrate molecule as a function of pH and the presence or absence of Al. Reference citrate resonance lines were obtained using 1 mol L^{-1} cit_{TS} solutions. Aluminum–citrate resonance lines were obtained using 1 mol L^{-1} cit_{TS}+ 0.25 mol L^{-1} Al_{TS} solutions. (Modified from Gregor and Powell, 1986.)

better description of the titration data, the electrospray mass spectrometry results indicate that both $Al_2(H_{-1}cit)_2{}^{2-}(aq)$ and $Al_2cit_3{}^{3-}(aq)$ are present in solution throughout a broad pH range, with the former species being found in greater abundance (Figure 10.6). From their titration data, they predicted that the trinuclear species, $Al_3(OH)_4cit_3{}^{4-}(aq)$, was essentially the sole Al species in solution when pH $>$ 7. In general, the modeled Al–citrate species distribution (predicted from potentiometric titration) is consistent with mass spectrometry determinations. The $Al_3(OH)_4cit_3{}^{4-}(aq)$ species (denoted by the mass spectra of $[Al_3(H_{-1}cit)_3{\cdot}H]^{2-}$ and $[Al_3(H_{-1}cit)_3{\cdot}H_2O,2H]^-$) predominates throughout the pH range 4 to 8, with significantly lesser amounts of $[Al(cit)_2{\cdot}2H]^-$ $[Alcit_2{}^{3-}(aq)$ and $Al(H_{-1}cit)cit^{4-}(aq)]$, $[Al_2(H_{-1}cit)_2{\cdot}H]^-$ $[Al_2(H_{-1}cit)_2{}^{2-}(aq)]$, and $[Alcit{\cdot}Cl]^-$ $[Al(H_{-1}cit)^-(aq)]$. These authors also found that the half-life for the formation of the trinuclear complex was approximately 80 minutes.

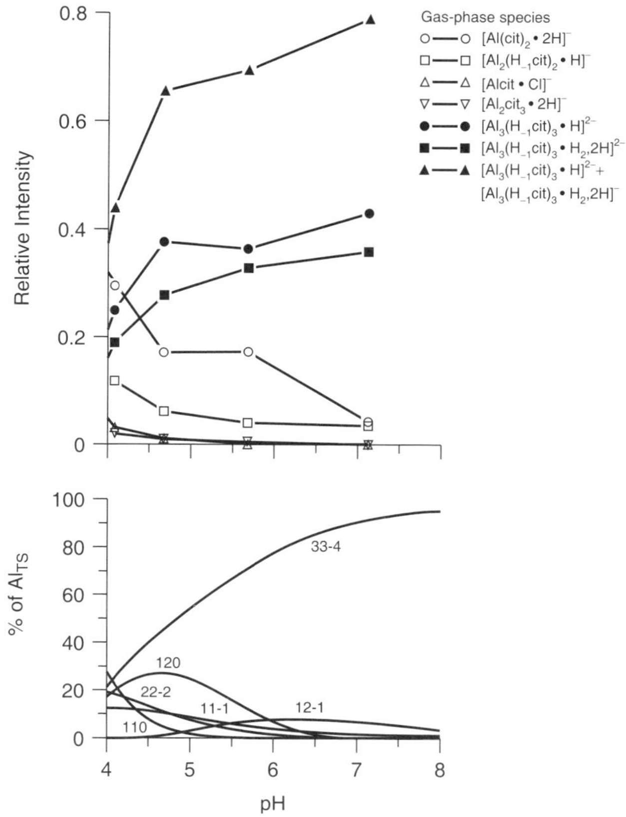

Figure 10.6. Electrospray mass spectrometry results of Harris et al. (2003) and the corresponding predicted distribution of Al–containing species according to the Al–citrate speciation model of Harris et al. (2003) (Table 10.4) as a function of pH in 0.1 mol L^{-1} KCl with $Al_{TS} = 1.0 \times 10^{-3}$ mol L^{-1} and $cit_{TS} = 2.0 \times 10^{-3}$ mol L^{-1} ($cit_{TS}/Al_{TS} = 2:1$; 25°C). The species are identified by stoichiometric coefficients according to Eq. (10.7).

10.3.1.6. Lakatos Model and Related Spectroscopic Studies Lakatos et al. (2001) observed that the time required for Al–citrate systems to reach equilibrium depended on the initial pH. In strongly acidic (pH ≈ 3) and slightly alkaline (pH ≈ 7.5) solutions, equilibrium was achieved within approximately 2 hours. However, when the starting pH was slightly acidic (pH ≈ 5.5), the equilibrium state was not achieved until approximately 30 hours. They attributed the long equilibration times to the slow kinetics involved in the formation of the trinuclear species, $Al_3(OH)_4cit_3{}^{4-}(aq)$. Potentiometric titrations (0.2 mol L^{-1} KCl, 25°C, pH 2.5 to 8.5) were performed as a function of reaction time (1 minute to 30 hours) and cit_{TS}/Al_{TS} ratio (1 : 1 and 3 : 1), and interpreted using their measured citric acid ionization constants (Table 10.1) and the $Al^{3+}(aq)$ hydrolysis constants of Öhman and Sjöberg (1983): $AlOH^{2+}(aq)$ ($\log {}^{0.2}K_{10-1} = -5.49$),

$Al_3(OH)_4^{5+}(aq)$ (log $^{0.2}K_{30-4} = -13.54$), $Al_{13}(OH)_{32}^{7+}(aq)$ (log $^{0.2}K_{13,0,-32} = -97.2$), and $Al(OH)_4^-(aq)$ (log $^{0.2}K_{10-4} = -23.40$). With the exception of $Alcit^0(aq)$ and $AlHcit^+(aq)$ formation, which equilibrate rapidly, the distribution predicted for Al_{TS} is influenced by both reaction time and the cit_{TS}/Al_{TS} ratio (Figure 10.7). In the equimolar cit_{TS}/Al_{TS} systems, the trinuclear species

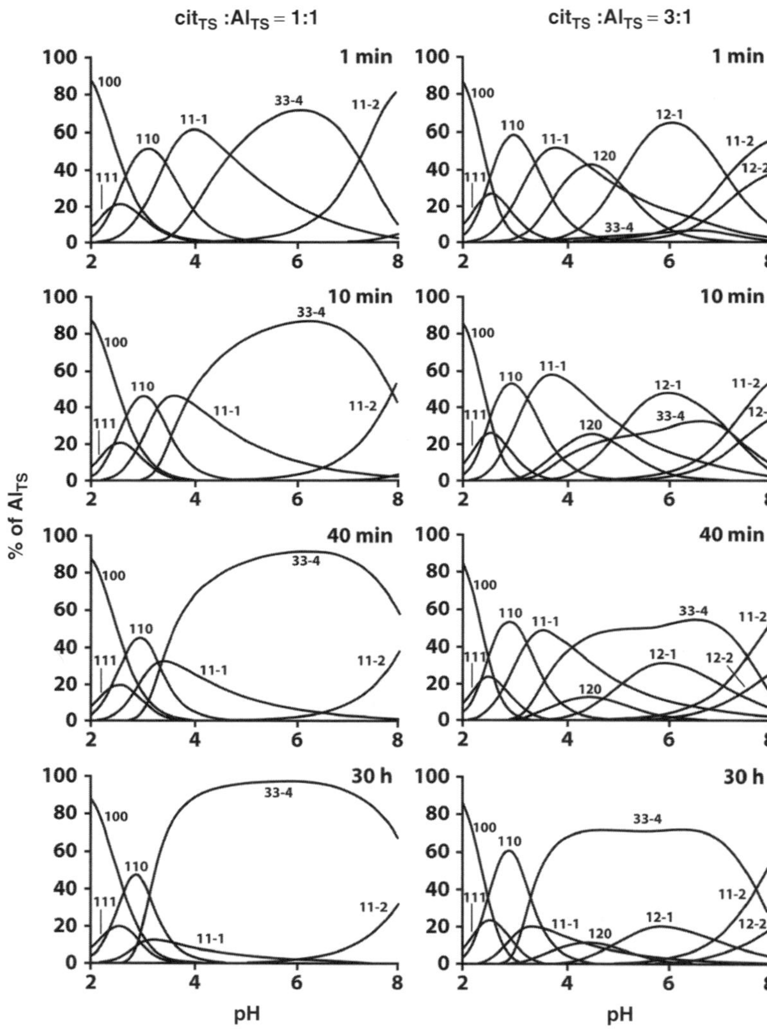

Figure 10.7. Predicted distribution of Al–containing species according to the Al–citrate speciation model of Lakatos et al. (2001) (Table 10.4) as a function of pH and reaction time in 0.2 mol L^{-1} KCl with $Al_{TS} = cit_{TS} = 4.0 \times 10^{-3}$ mol L^{-1} for equimolar cit_{TS}/Al_{TS} systems, and $cit_{TS} = 4.0 \times 10^{-3}$ mol L^{-1} and $Al_{TS} = 1.3 \times 10^{-3}$ mol L^{-1} for $cit_{TS}/Al_{TS} = 3:1$ systems at 25°C. The species are identified by stoichiometric coefficients according to Eq. (10.7). (Modified from Lakatos et al., 2001.)

predominates in the pH range 5 to 7 at a 1-minute reaction time. This predominance region expands with increasing reaction time to encompass the pH > 3 range at equilibrium (30 hours). A similar equilibrium result is observed for the 3 : 1 cit_{TS}/Al_{TS} system, although the mononuclear monocitrate and dicitrate species are fast-forming, then disappear at the expense of the trinuclear complex. These results are nearly identical to those of Öhman (1988) for the equimolar cit_{TS}/Al_{TS} systems (Figure 10.3).

In addition to potentiometric titrations, Lakatos et al. (2001) performed ^{1}H, ^{13}C, and ^{27}Al NMR spectroscopic measurements of solutions having cit_{TS}/Al_{TS} ratios of 1 : 1 and 3 : 1 as a function of pH and reaction times. The spectra were then compared with the titration results to identify the probable sources of the various signals. In pH ~ 2 solution, the ^{1}H NMR spectrum suggested the presence of the mononuclear species $AlHcit^{+}(aq)$ (Figure 10.8). Both $AlHcit^{+}(aq)$ and $Alcit^{0}(aq)$ are resolved from the pH ~ 2.5 spectrum, whereas $Alcit^{0}(aq)$ and $AlH_{-1}cit^{-}(aq)$ are resolved from the pH ~ 3 spectrum. Both the pH 2.5 and 3 ^{1}H NMR spectra indicated fast proton exchange and also contained numerous signals which were interpreted to indicate the formation (oligomerization) of $Al_3(OH)_4cit_3{}^{4-}(aq)$ [denoted as $Al_3(OH)(H_{-1}cit)_3{}^{4-}(aq)$ by the authors]. In pH ≈ 5 (and 7) equilibrium solutions, the ^{1}H, ^{13}C, and ^{27}Al NMR spectra were dominated by signals from the $Al_3(OH)(H_{-1}cit)_3{}^{4-}(aq)$ species. The spectra indicate that the ligand exchange reactions between trinuclear, mononuclear, and free citrate are slow. Additional signals were also observed, particularly at reaction times less than that required for equilibrium (<30 hours). However, the source of these signals could not be identified, although they were attributed to the precursors of $Al_3(OH)(H_{-1}cit)_3{}^{4-}(aq)$.

The NMR spectra interpretations of Lakatos et al. (2001) were similar to those derived from prior studies (Powell and Heath, 1996). In the pH range 7 to 9 and after an 18-hour equilibration period, Feng et al. (1990) identified the $Al_3(OH)(H_{-1}cit)_3{}^{4-}(aq)$ species using ^{1}H, ^{13}C, and ^{27}Al NMR spectroscopy of solutions having a cit_{TS}/Al_{TS} ratio of 1 : 1. Upon acidification to pH < 3, they observed the hydrolysis of $Al_3(OH)(H_{-1}cit)_3{}^{4-}(aq)$ and the subsequent formation of a 1 : 1 Al–citrate complex [either $AlHcit^{+}(aq)$ or $Alcit^{0}(aq)$] and possibly the $Alcit_2{}^{3-}(aq)$ species. Based on the ^{27}Al NMR spectra of Karlik et al. (1983) and Feng et al. (1990), and the structure described by Powell and Heath (1996), Kubicki et al. (1999) performed ab initio computations to support the occurrence of $Al_3(OH)(H_{-1}cit)_3{}^{4-}(aq)$ and the monodentate $AlH_2cit^{2+}(aq)$ species in acidic solutions. And although undifferentiated, they also surmised that the monodentate $AlOH(H_2cit)^{+}(aq)$, bidentate $AlHcit^{+}(aq)$, or the tridentate $Alcit^{0}(aq)$ species could be supported by observed ^{27}Al NMR peaks.

Although generally restricted to solutions with cit_{TS}/Al_{TS} ratios less than 4 : 1, numerous studies have documented the slow oligomerization of Al–citrate complexes to form $Al_3(OH)(H_{-1}cit)_3{}^{4-}(aq)$ (Bodor et al., 2002; Clausén et al., 2005). Using ^{1}H and ^{13}C NMR, Matzaptekakis et al. (1999 and 2001) proposed a reaction scheme for the time-dependent formation of the $Al_3(OH)(H_{-1}cit)_3{}^{4-}(aq)$ oligomer in alkaline solution. The dissolution of $(NH_4)_5Al(H_{-1}cit)_2 \cdot 2H_2O(s)$

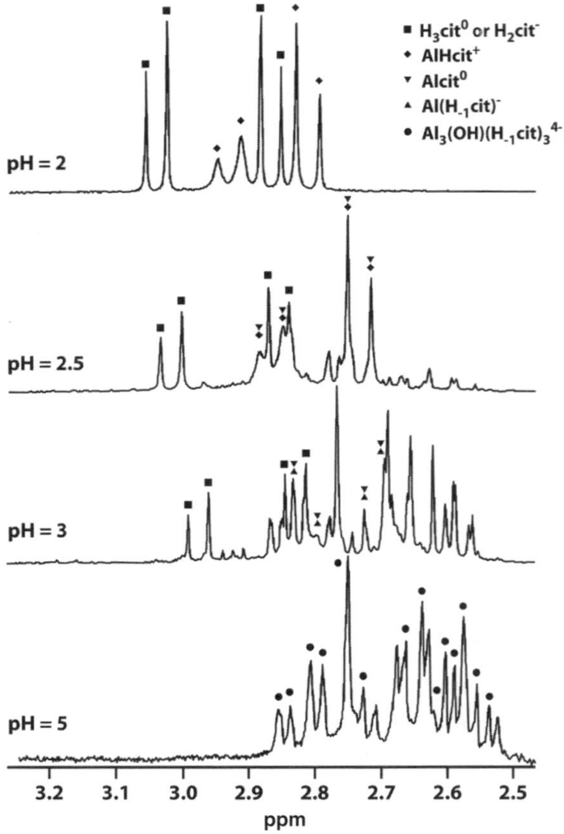

Figure 10.8. Influence of pH on the ^{1}H NMR spectra of Al–citrate solutions. The spectra were obtained at equilibrium (30–h reaction time) with $Al_{TS} = cit_{TS} = 0.04$ mol L^{-1}. (Modified from Lakatos et al., 2001.)

(cit_{TS}/Al_{TS} of 2 : 1) in pH 7.89 solution leads to the formation of $Al_3(OH)(H_{-1}cit)_3{}^{4-}(aq)$ according to the following:

$$Al(H_{-1}cit)_2^{5-}(aq) + H_2O = Al(H_{-1}cit)(OH)^{2-}(aq) + cit^{3-}(aq) \qquad (10.9)$$

$$3Al(H_{-1}cit)(OH)^{2-}(aq) + H_2O = Al_3(H_{-1}cit)_3(OH)^{4-}(aq) + 2OH^-(aq) \qquad (10.10)$$

where an increase in pH (to 8.36) was viewed to support the proposed scheme. Similarly, the dissolution of $(NH_4)_4Al(H_{-1}cit)cit{\cdot}3H_2O(s)$ (cit_{TS}/Al_{TS} of 2 : 1) resulted in solutions that ultimately (after more than 26 hours) contained $cit^{3-}(aq)$, mononuclear monocitrate complexes, and the trinuclear oligomer (Figure 10.9). Upon dissolution of $(NH_4)_4Al_2(H_{-1}cit)cit_2{\cdot}4H_2O(s)$ (cit_{TS}/Al_{TS} of 3 : 2) and using ^{1}H NMR, Dakanali et al. (2003) observed the time-dependent and

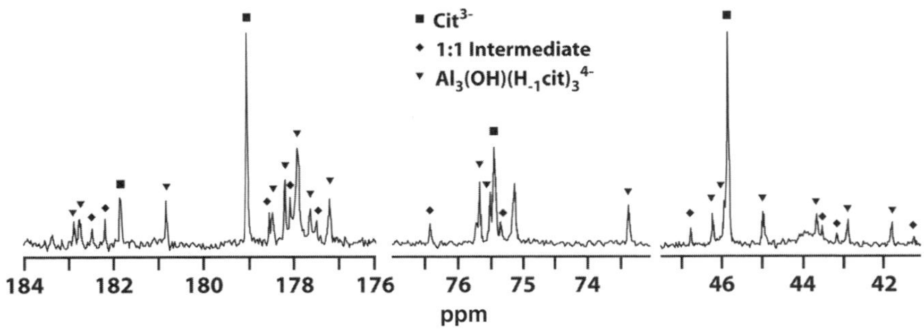

Figure 10.9. ^{13}C NMR spectrum of a 0.07 mol L^{-1} solution of $(NH_4)_4Al(H_{-1}cit)cit \cdot 3H_2O(s)$ in $D_2O(l)$. The spectrum was recorded 26 hours following the dissolution of the Al–citrate salt (the pD of the solution was not defined by the authors). (Modified from Matzapetakis et al., 2001.)

gradual transformation of $Al_2(H_{-1}cit)cit_2{}^{4-}(aq)$ to $cit^{3-}(aq)$, dinuclear dicitrate species (in various states of deprotonation), and mononuclear Al–citrate species [e.g., $Al(H_{-1}cit)_2{}^{5-}(aq)$ and $Al(H_{-1}cit)(OH)^{2-}(aq)$]. The authors concluded that these mono- and dinuclear Al–citrate species would ultimately oligomerize to form the thermodynamically stable $Al_3(OH)(H_{-1}cit)_3{}^{4-}(aq)$ species.

10.3.1.7. Essington Model Like Motekaitis and Martell (1984), Essington et al. (2005) employed a minimalist model, consisting of mononuclear monocitrate species, to describe Al–citrate complexation in the pH range 4 to 8. They examined the equilibrium solubility of gibbsite [$Al(OH)_3(s)$] (12-week equilibrations) in 0.05 and 0.43 mol L^{-1} $NaNO_3$, 0.05 mol L^{-1} Na_3cit, and 0.45 mol L^{-1} $NaNO_3+$ 0.05 mol L^{-1} Na_3cit solutions at ambient temperatures (20 to 22°C). The elevated Al_{TS} concentrations supported by gibbsite in solutions containing citrate (Table 10.2), relative to those in nitrate solutions, were used in conjunction with the mass balance expressions:

$$Al_{TS} = {}^{I}K_{sp.gibbsite}\left([H^+]^3 + {}^{I}K_{10-1}[H^+]^2 + {}^{I}K_{10-2}[H^+] + {}^{I}K_{10-3} + \frac{{}^{I}K_{10-4}}{[H^+]} + \sum_i {}^{I}K_{1n_ij_i}[H^+]^{3+j_i}[cit^{3-}]^{n_i}\right) \quad (10.11)$$

$$cit_{TS} = [cit^{3-}]\left(1 + \frac{[H^+]}{{}^{I}K_{013}} + \frac{[H^+]^2}{{}^{I}K_{012}{}^{I}K_{013}} + \frac{[H^+]^3}{{}^{I}K_{011}{}^{I}K_{012}{}^{I}K_{013}}\right) + \sum_i n_i^I K_{1n_ij_i}[Al^{3+}][H^+]^{3+j_i}[cit^{3-}]^{n_i} \quad (10.12)$$

to establish an Al–citrate complexation model. Data analysis included the consideration of citric acid dissociation (May and Murray, 2000) and the formation of monomeric $Al^{3+}(aq)$ hydrolysis products (Baes and Mesmer, 1986).

Unlike potentiometric titration experiments, the cit_{TS}/Al_{TS} ratio was not controlled, but varied as a function of pH ($\approx$ 5:1 at pH 4 to $\approx$ 50 : 1 at pH > 6). The chemical model derived consisted of the mononuclear species $Alcit^0(aq)$, $AlOH(H_{-1}cit)^{2-}(aq)$, and $Al(OH)_2(H_{-1}cit)^{3-}(aq)$ (Table 10.4). Di- and trinuclear species and mononuclear dicitrate species, previously inferred from titration and spectroscopic studies, could not be resolved. However, based on previous findings (e.g., Gregor and Powell, 1986; Öhman, 1988) and the high cit_{TS}/Al_{TS} ratios in the gibbsite systems, the formation of polynuclear species would have been of little significance. In the pH range 4 to 8, the chemical model proposed by Essington et al. (2005) bears very little resemblance to those generated from potentiometric titrations (Figure 10.4*e*). The $Alcit^0(aq)$ species is the only species common to all Al–citrate chemical models. In the model of Essington et al. (2005), $Alcit^0(aq)$ achieves a maximum expression when the pH is approximately 4.6, whereas in all other models, maximum expression is predicted when the pH is less than 3.5. The $Al(OH)_2cit^{2-}(aq)$ species is predicted to predominate in the pH range 4.7 to 6.7 (Figure 10.4*e*). However, the models of Öhman (1988) and Lakatos et al. (2001) predict $Al(OH)_2cit^{2-}(aq)$ to dominate Al_{TS} when solution pH is greater than approximately 6.5 (Figures 10.4*a* and 10.7, and excluding the trinuclear species). The $Al(OH)_3cit^{3-}(aq)$ species appears only in the model of Essington et al. (2005). It is predicted to predominate in alkaline solutions.

10.3.2. Summary of Aluminum–Citrate Models

The aqueous speciation of Al–citrate systems has been examined in numerous studies that involve differing methodologies and chemical conditions. Also variable across studies are the parameters employed to describe $Al^{3+}(aq)$ hydrolysis and citric acid ionization. Despite such differences, it is surprising that the chemical models of Al–citrate speciation disagree relative to the species included and their significance. In general, the chemical speciation models agree that the $AlHcit^+(aq)$, $Alcit^0(aq)$, and $Al(H_{-1}cit)^-(aq)$ species describe Al–citrate complexation in strongly acidic solutions (pH < 4). Further, the formation constants for each species, as generated from potentiometric titration studies, are relatively consistent across studies. The log K_{111} values for the formation of $AlHcit^+(aq)$, obtained by converting the log ${}^IK_{nmj}$ values in Table 10.4. using a modified Davies equation (Sposito and Mattigod, 1980), range from 12.47 to 12.98, averaging 12.76 ± 0.20. Similarly, the log K_{110} values for the formation of $Alcit^0(aq)$ range from 9.93 to 10.79, averaging 10.33 ± 0.30; and log K_{11-1} values for the formation of $Al(H_{-1}cit)^-(aq)$ range from 5.91 to 6.84, averaging 6.36 ± 0.38.

Unfortunately, Al–citrate speciation in the pH range that encompasses normal soil solutions (pH 4 to 9) is not well established. Actually, it is correct to state that studies have been interpreted in several different ways, leading to several different chemical models. The available information indicates that Al–citrate speciation in the pH range 4 to 9 is a function of the molar cit_{TS}/Al_{TS} ratio and the period of time that Al_{TS} and cit_{TS} are in contact. When cit_{TS}/Al_{TS} is low (< 3:1), fast-forming mononuclear complexes initially dominate Al–citrate speciation. During the approximately 30 hours that is required to achieve equilibrium, these mononuclear species undergo oligomerization to form the polynuclear complexes, most notably $Al_3(OH)(H_{-1}cit)_3^{4-}(aq)$. Indeed, this trinuclear species appears to be a common thread that runs through many of the pH titration-based chemical models, its occurrence having been confirmed by various spectroscopic examinations. However, when cit_{TS}/Al_{TS} is high (>5 : 1), equilibrium is achieved rapidly and the chemical models developed for these systems consist predominantly of mononuclear mono- and dicitrate species. It may be concluded that the aqueous speciation of Al–citrate in soil solutions cannot be predicted with any degree of certainty: (1) it cannot be ascertained which of the chemical models provides a true picture of Al–citrate speciation, and (2) the residency time of citrate in soil may be substantially less than the time required to achieve Al–citrate equilibrium, due to the rapid decomposition of citrate. Based on the latter, the time-dependent formation of polynuclear species in soil solutions can be discounted. However, based on the former, considerable variability in model predictions exists, even among the mononuclear models.

10.3.3. Iron(III)–Citrate Interactions

Iron(III)–citrate speciation has received very little attention relative to that of Al–citrate. In general, the chemical models that describe Fe(III)–citrate speciation in acidic solutions are similar to those that describe Al–citrate. In excess citrate (high $cit_{TS}/Fe(III)_{TS}$) and pH < 4 solutions, Fe(III)–citrate complexation is thought to be well established in terms of $FeHcit^+(aq)$, $Fecit^0(aq)$, and $FeOHcit^-(aq)$ [or $Fe(H_{-1}cit)^-(aq)$] (Martin, 1986; Pierre and Gautier-Luneau, 2000). However, in slightly acidic to slightly alkaline solutions, Fe(III)–citrate speciation is ill defined, a consequence of the exceedingly low solubility of Fe(III) hydroxides. Like the Al–citrate systems, evidence is available to suggest that Fe(III)–citrate speciation may be influenced by the $cit_{TS}/Fe(III)_{TS}$ ratio, where lower values favor the formation of di- and trinuclear species. However, the influence of reaction time on speciation remains unknown. Also, the chemical models that describe Fe(III)–citrate speciation, as derived from potentiometric titration and solubility studies, are generally restricted to mononuclear monocitrate species. Perhaps this divergence from Al–citrate models, which predict the predominance of polynuclear Al–citrate species throughout a broad pH range when cit_{TS}/Al_{TS} is low (< 3 : 1), can be attributed to the high $cit_{TS}/Fe(III)_{TS}$ ratios necessary to establish analytical Fe(III) concentrations (where the formation of polynuclear species is less favorable).

Warner and Weber (1953) predicted the formation of $Fecit^0(aq)$ and $FeOHcit^-(aq)$ ($\log {}^{0.1}K_{110} = 11.38$ and $\log {}^{0.1}K_{11-1} = 9.46$ at 30°C) in pH < 3 solutions containing 0.1 mol L^{-1} $NaNO_3$. Hamm et al. (1954) also performed potentiometric titration experiments in pH < 1 to ≈ 4 solutions at 25°C, using a 1 mol L^{-1} $NaClO_4$ background electrolyte and equimolar $cit_{TS}/Fe(III)_{TS}$. Their chemical model describes Fe(III)–citrate speciation in terms of $FeHcit^+(aq)$, $Fecit^0(aq)$, and $FeOHcit^-(aq)$ ($\log {}^1K_{111} = 11.44$, $\log {}^1K_{110} = 11.85$, and log $^{0.1}K_{11-1}$ = 9.40). Hamm et al. (1954) also reported a $^{1}K_{11-2}$ value for $Fe(OH)_2cit^{2-}(aq)$ formation, but noted a lack of confidence in the value, as their results were confounded by the precipitation of colloidal Fe(III) hydroxide that was observed. Timberlake (1964) examined Fe(III)–citrate speciation in pH < 3 solutions at 20°C, with varied $cit_{TS}/Fe(III)_{TS}$ ratios and in 0.1 mol L^{-1} $NaClO_4$. Timberlake (1964) employed the $Fecit^0(aq)$ (log $^{0.1}K_{110} = 11.40 \pm 0.02$) and $Fe_2(OH)_2cit_2^{2-}(aq)$ (log $^{0.1}K_{22-2} = 21.17 \pm 0.13$) species to describe the pH titration data. The dimer was imposed in the model to conform to the author's hypothesis that polynuclear Fe(III) complexes should form in addition to mononuclear species. Although no other chemical models in the literature employ the dimer, the findings of Shweky et al. (1994) (discussed below) have been used to substantiate its occurrence. Ramamoorthy and Manning (1973) performed potentiometric titrations in 0.1 mol L^{-1} $NaClO_4$ at 25°C in 1 : 1 and 2 : 1 $cit_{TS}/Fe(III)_{TS}$ systems. In the pH range 1.6 to 4, they modeled Fe(III)–citrate speciation to consist of $Fecit^0(aq)$ (log $^{0.1}K_{110} = 10.24 \pm 0.01$) and $Fecit_2^{3-}(aq)$ (log $^{0.1}K_{120} = 15.94 \pm 0.01$).

The authors of several studies have inferred the formation of polymeric hydroxy–Fe(III)–citrate complexes, particularly in alkaline solutions and low $cit_{TS}/Fe(III)_{TS}$ ratios. From pH titration and dialysis data, Spiro et al. (1967a,b) postulated that $FeOHcit^-(aq)$ forms in slightly acidic solutions (pH $\approx$ 6) when $cit_{TS}/Fe(III)_{TS}$ is 1 : 1. However, in alkaline solutions (pH 8 to 9), $FeOHcit^-(aq)$ polymerizes to form hydroxy–Fe(III)–citrate agglomerates that are approximately 7.2 nm in diameter (they are not aqueous species). They also found that the formation of these microparticles was prevented in excess citrate ($cit_{TS}/Fe(III)_{TS}$ > 20 : 1); instead, the formation of $Fe(OH)_2cit_2^{5-}(aq)$ was postulated. Aisen et al. (1978) derived log $^{0.1}K_{12-2} = 3.29$ for the formation of $Fe(OH)_2cit_2^{5-}(aq)$, using the data reported by Spiro et al. (1967b). Similarly, Hegenauer et al. (1979) estimated log $^{0.1}K_{12-2} = 1.48$.

10.3.3.1. Field Model Field et al. (1974) employed a combination of spectrophotometric and potentiometric titration techniques to elucidate Fe(III)–citrate speciation at 25°C in the pH range 1.5 to 3. Titrations were performed in 0.1 mol L^{-1} KNO_3 with a cit_{TS}/Fe_{TS} ratio of 2:1. Data analysis included Fe(III) hydrolysis characterized by log $K_{10-1} = -2.6$ and the citrate dissociation constants shown in Table 10.1. Both the titration and spectrophotometric results were characterized by a chemical model that consisted of the $FeHcit^+(aq)$, $Fecit^0(aq)$, and $FeOHcit^-(aq)$ species (Table 10.5.) Further, the computed log K_{mnj} values obtained by the two methods were in good agreement. On average, log $^{0.1}K_{111} = 12.38 \pm 0.38$, log $^{0.1}K_{110} = 11.19 \pm 0.12$, and log $^{0.1}K_{11-1} = 8.49 \pm 0.64$. The model of Field et al. (1974) is limited to the pH < 4 range. Martin (1986) modified this model to include the $Fecit_2^{3-}(aq)$ species, such that the model could be applied to solutions up to pH 7.4. The formation constant for $Fecit_2^{3-}(aq)$ species was generated by comparison to the Al–citrate complexation models of Öhman and Sjöberg (1983) and Motekaitis and Martell (1984). The distribution of Fe(III)–citrate species, as computed by the modified chemical model of Field et al. (1974), predicts that the $Fecit_2^{3-}(aq)$ species predominates from pH 4 up to approximately pH 6.5 (where goethite precipitates) (Figure 10.10*a*). When the $Fecit_2^{3-}(aq)$ species is not included in the computations, as in the original model of Field et al. (1974), the $FeOHcit^{-}(aq)$ species would predominate throughout a pH bounded by pH 2.8 and the precipitation of goethite, albeit the modeled predictions would be valid only for the pH range 1.5 to 3.

10.3.3.2. Königsberger Model Königsberger et al. (2000) performed potentiometric titrations of 1 mol L^{-1} NaCl solutions at 25°C with $cit_{TS}/Fe(III)_{TS}$ ranging from 1 : 1 to 5 : 1. The study covered the pH range from 1.3 to 5.4. Data analysis included Fe(III) hydrolysis (May and Murray, 2000), and the citrate dissociation constants shown in Table 10.1. They employed a chemical model consisting of $Fecit^{0}(aq)$, $FeOHcit^{-}(aq)$, $Fecit_2^{3-}(aq)$, $FeHcit_2^{2-}(aq)$, and $FeOHcit_2^{4-}(aq)$ to describe their Fe(III)–citrate systems (Table 10.5.) The predicted distribution of Fe(III)–citrate species as a function of pH illustrates that the Königsberger et al. (2000) model predictions are considerable different from those of Field et al. (1974) (Figures 10.10*a* and *b*). In the former, the $FeOHcit^{-}(aq)$ species predominates throughout a broad pH range; whereas in the later, $Fecit_2^{3-}(aq)$ is predicted to predominate.

10.3.3.3. Essington Model Essington et al. (2005) examined the equilibrium solubility of goethite [FeOOH(*s*)] (8-week equilibrations) in 0.05 and 0.43 mol L^{-1} $NaNO_3$, 0.05 mol L^{-1} Na_3cit, and 0.45 mol L^{-1} $NaNO_3$+ 0.05 mol L^{-1} Na_3cit solutions at ambient temperatures (20 to 22°C). The concentrations of $Fe(III)_{TS}$ in goethite suspensions containing citrate were used in conjunction with Eq. (10.12) and the mass balance expression:

$$Fe(III)_{TS} = {}^{I}K_{sp,goethite}\Big([H^+]^3 + {}^{I}K_{10-1}[H^+]^2 + {}^{I}K_{10-2}[H^+] + {}^{I}K_{10-3} + \frac{{}^{I}K_{10-4}}{[H^+]} + \sum_i {}^{I}K_{1nj_i}[H^+]^{3+j_i}[cit^{3-}]^{n_i}\Big) \quad (10.13)$$

to establish an Fe(III)–citrate complexation model. Data analysis included the consideration of citric acid dissociation (May and Murray, 2000) and the formation of monomeric $Fe^{3+}(aq)$ hydrolysis products (Liu and Millero, 1999). Unlike potentiometric titration experiments, the $cit_{TS}/Fe(III)_{TS}$ ratio was not controlled but varied as a function of pH (≈ 500:3 at pH 4 to ≈ 5000 : 1 at pH > 6). The chemical model derived consisted of the mononuclear monocitrate species $Fecit^{0}(aq)$, $Fe(OH)_2cit^{2-}(aq)$, and $Fe(OH)_3cit^{3-}(aq)$ (Table 10.5.) Additional species, such as $Fecit_2^{3-}(aq)$, inferred from the titration and spectroscopic studies of Field et al. (1974) and Königsberger et al. (2000), could not be resolved. In the pH range 4 to 8, the chemical model proposed by Essington et al. (2005) bears very little similarity to those generated from potentiometric titration studies (Figure 10.10*c*). The $Fecit^{0}(aq)$ species is the only species common to the

TABLE 10.5. Chemical Models and the Conditional Formation Constants [log $^{I}K_{mnj}(\pm\sigma)$ at 25°C] That Have Been Developed to Describe the Aqueous Speciation of Fe(III)–Citrate Systems[a]

Aqueous Complex (mnj)[b]	Complexation Model				
	Field et al. (1974)[c], $I = 0.1$	Field et al. (1974)[d], $I = 0.1$	Königsberger et al. (2000)[e], $I = 1$	Essington et al. (2005)[f] $I = 0.15$	Essington et al. (2005)[f] $I = 0.6$
$Fecit^{0}(aq)$	11.21	11.16	9.50	10.96	10.18
(110)	(0.10)	(0.07)	(0.02)	(0.02)	(0.03)
$FeHcit^{+}(aq)$	12.38	12.37	—	—	—
(111)	(0.34)	(0.18)			
$FeOHcit^{-}(aq)$	8.60	8.38	7.31	—	—
(11−1)	(0.64)	(0.64)	(0.02)		
$Fecit^{3}{}_{2}{}^{-}(aq)$	18.2[g]	—	15.30	—	—
(120)			(0.06)		
$Fe(OH)_2cit^{2-}(aq)$	—	—	—	2.60	2.16
(11−2)				(0.01)	(0.01)
$Fe(OH)_3cit^{3-}(aq)$	—	—	—	−4.60	−4.69
(11−3)				(0.03)	(0.03)
$FeHcit^{2}{}_{2}{}^{-}(aq)$	—	—	19.12	—	—
(121)			(0.06)		
$FeOHcit^{4}{}_{2}{}^{-}(aq)$	—	—	10.46	—	—
(12−1)			(0.08)		

[a] Stability constants are for the reaction defined by Eq. (10.6).

[b] The stoichiometric denotation associated with each complex (*mnj*), as defined by Eq. (10.7).

[c] cit_{TS}/Fe_{TS} ratio of 2 : 1 and potentiometric titrations in 0.1 mol L^{-1} KNO_3.

[d] From the spectrophotometric examination of 2 : 1 cit_{TS}/Fe_{TS} in 0.1 mol L^{-1} KNO_3.

[e] cit_{TS}/Fe_{TS} ratios that range from 1 : 1 to 5 : 1, and titration in 1.0 mol L^{-1}.

[f] Fe-citrate speciation in 0.05 mol L^{-1} Na_3cit and 0.45 mol L^{-1} $NaNO_3$+ 0.05 mol L^{-1} Na_3cit solutions in equilibrium with goethite (cit_{TS}/Fe_{TS} ratios ranged from approximately 170 : 1 to 5000 : 1).

[g] The log $^{c}K_{mnj}$ value for the formation of $Fecit^{3}{}_{2}{}^{-}$ is proposed by Martin (1986) as an addendum to the chemical model developed by Field et al. (1974).

Fe(III)–citrate chemical models of Field et al. (1974), Königsberger et al. (2000), and Essington et al. (2005). In the model of Essington et al. (2005), $Fecit^0(aq)$ achieves a maximum expression when the pH is ≈ 3, whereas in all other models, maximum expression is predicted when the pH is between 2 and 2.5. The $Fe(OH)_2cit^{2-}(aq)$ and $Fe(OH)_3cit^{3-}(aq)$ species, which are not components of any other Fe(III)–citrate model, are predicted to predominate at increasingly higher pH values (Figure 10.10*d*).

10.3.3.4. Direct Characterizations of Iron(III)–Citrate Species A small number of studies have led to the direct characterization of Fe(III)–citrate complexes. In general, however, the species identified by direct characterization techniques do not match those found in chemical models based on potentiometric titrations. Shweky et al. (1994) precipitated salts of $Fe_2(OH)_2cit_2{}^{2-}(aq)$ and $Fe_2cit_3{}^{3-}(aq)$ from solutions of unknown character. They postulated that the two species would exist in solutions containing excess citrate. They also concluded that their findings were consistent with the aqueous speciation model proposed by Timberlake (1964), which consisted of $Fecit^0(aq)$ and $Fe_2(OH)_2cit_2{}^{2-}(aq)$. Matzapetakis et al. (1998) precipitated a salt of $Fe(OH)_2cit_2{}^{5-}(aq)$ from a pH ≈ 8 solution, and inferred its existence in the aqueous phase. Gautier-Luneau et al. (2005) examined aqueous Fe(III)–citrate speciation as a function of pH and $cit_{TS}/Fe(III)_{TS}$ using electrospray mass spectrometry. Equilibrium (20-hour) solutions containing high $cit_{TS}/Fe(III)_{TS}$ ratios (10 : 1 and 20 : 1) were generally dominated by mononuclear dicitrate species of the form $[Fecit_2{\cdot}H]^-$ (Figure 10.11). Trinuclear species, detected as $[Fe_3cit_3{\cdot}H]^{2-}$ and $[Fe_3cit_4{\cdot}H_6]^-$, and dinuclear dicitrate species ($[Fe_2cit_2{\cdot}H]^-$), predominated in 4 : 1, 2 : 1, and 1 : 1 $cit_{TS}/Fe(III)_{TS}$ solutions at all pH values and in pH-neutral 10 : 1 $cit_{TS}/Fe(III)_{TS}$ solutions. Mononuclear monocitrate complexes such as those predicted from potentiometric titration and solubility studies [e.g., $Fecit^0(aq)$ and $FeOHcit^-(aq)$, Table 10.5. and Figure 10.10*a*, to *c*] were not detected by electrospray mass spectrometry even in strongly acidic solutions (pH 2.4 and 3.5). This method-specific inconsistency has yet to be resolved.

10.3.4. Summary of Iron(III)–Citrate Models

The chemical models that describe Fe(III)–citrate speciation in aqueous solutions are generally applicable only to strongly acidic conditions. The models that are considered most reliable—Timberlake (1964) for $\log {}^{0.1}K_{110} = 11.40$ and Field et al. (1974) for $\log {}^{0.1}K_{111} = 12.38$, $\log {}^{0.1}K_{110} = 11.19$, and $\log {}^{0.1}K_{11-1} = 8.49$—are restricted to pH < 4 solutions and thus are not applicable to soil solutions. The model of Field et al. (1974) was modified to include an estimated $\log {}^{0.1}K_{120} = 18.2$ (Martin, 1986), purportedly extending the range to pH ≈ 7.4. Recently, Königsberger et al. (2000) extended the range of applicability to pH ≈ 5.4. Their model ignored the $FeHCit^+(aq)$ species and included several mononuclear dicitrate species. These dicitrate species, however, accounted for only a minor percentage of $Fe(III)_{TS}$ in the pH range 3 to ≈ 6.5, which was instead

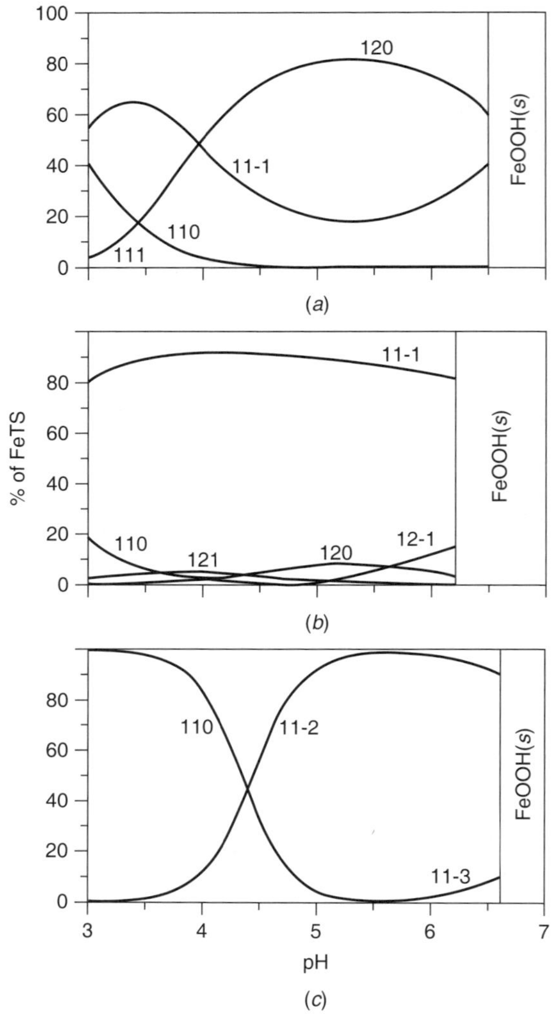

Figure 10.10. Predicted distribution of Fe(III)–containing species as a function of pH in 0.01 mol L^{-1} $NaNO_3$ with $Fe(III)_{TS} = 10^{-6}$ mol L^{-1} and $cit_{TS} = 10^{-3}$ mol L^{-1} [$cit_{TS}/Fe(III)_{TS} = 1000:1$; 25°C] according to the Fe(III)–citrate speciation models (Table 10.5) of (*a*) Field et al. (1974), as modified by Martin (1986); (*b*) Königsberger et al. (2000); and (*c*) Essington et al. (2005). The species are identified by stoichiometric coefficients according to Eq. (10.7).

dominated by $FeOHcit^-(aq)$. Finally, Essington et al. (2005) generated a chemical model from goethite solubility studies that encompassed the pH range 4 to 8. Their model did not include the $FeOHcit^-(aq)$ species, which is common to the pH titration-generated models of Field et al. (1974) and Königsberger et al. (2000).

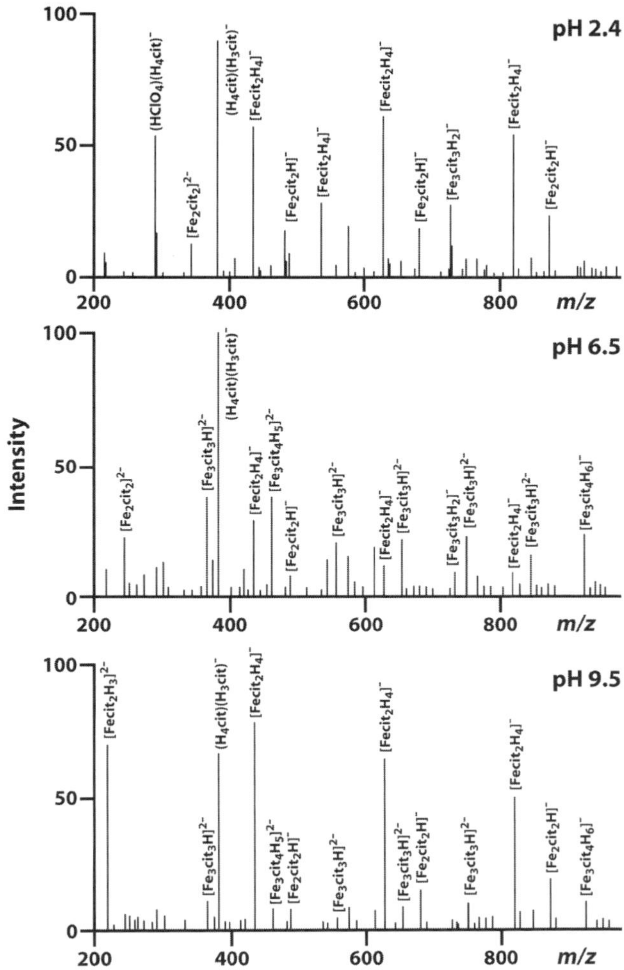

Figure 10.11. Electrospray mass spectra, as a function of pH, of Fe(III)–citrate solutions containing 0.1 mol L^{-1} Fe(III)$_{TS}$ and 1 mol L^{-1} cit$_{TS}$ [cit$_{TS}$/Fe(III)$_{TS}$ = 10]. The spectra were obtained following a 20–hour equilibration period. (Modified from Gautier-Luneau et al., 2005.)

Many of the titration-based chemical models employed to describe Al–citrate speciation were validated using ancillary spectrometry data (Lakatos et al., 2001; Harris et al., 2003). Results from spectroscopic studies of Fe(III)–citrate systems do not support the titration-based models, which are defined by mononuclear monocitrate species. Instead, mononuclear dicitrate, dinuclear, and trinuclear Fe(III)–citrate species are observed. It may be concluded that the aqueous speciation of Fe(III)–citrate in soil solutions cannot be predicted. The current speciation models are not applicable to the pH range of soil solutions, or as in the case of the Essington et al. (2005) model, have not been validated. There is also evidence to suggest that Fe(III)–citrate equilibrium is not instantaneous, even though the residency time of citrate in soil may be substantially less than the time required to achieve equilibrium.

10.4. METAL–CITRATE COMPLEXATION AND MINERAL SOLUBILITY

Despite decades of research, the aqueous speciation of the Al–and Fe(III)–citrate systems remains enigmatic. Although there is general consistency among the species and predicted distributions offered by various chemical models for strongly acidic solutions (pH < 4), there exists no commonality among the model predictions within the pH range of normal soil solutions (pH $\sim$ 4 to 9). Thus, the aqueous chemistry of Al–and Fe(III)–citrate, and the impact predicted for the citrate ligand on the soil chemistry of Al and Fe(III), remain untenable.

As evidence of the impact of selected chemical models on the elucidation and interpretation of a soil chemical process, consider the equilibrium solubility [Al_{TS} and $Fe(III)_{TS}$ concentrations] of two common soil accessory minerals, gibbsite and goethite, in the presence of citrate. For gibbsite, five Al–citrate speciation models were selected for comparison: Motekaitis and Martell (1984), Gregor and Powell (1986), Lakatos et al. (2001), Harris et al. (2003), and Essington et al. (2005) (Table 10.4.) [note that only the chemical model of Essington et al. (2005) is based on gibbsite solubility measurements]. The models of Motekaitis and Martell (1984) and Gregor and Powell (1986) are valid for pH solutions below 7.5 and below 7, while those of Lakatos et al. (2001), Harris et al. (2003), and Essington et al. (2005) are valid for pH solutions below 9. The Al–citrate aqueous speciation and the Al_{TS} concentrations controlled by gibbsite dissolution were computed using GEOCHEM-PC (Parker et al., 1995). The controlling conditions included 0.01 mol L^{-1} $NaNO_3$ background electrolyte and a cit_{TS} value of 10^{-4} mol L^{-1}. The thermodynamic constants for gibbsite solubility, $cit^{3-}(aq)$ protonation, and $Al^{3+}(aq)$ hydrolysis were taken from Essington et al. (2005). Figure 10.12*a* illustrates the concentrations of Al_{TS} supported by gibbsite, as predicted by employing the four Al–citrate chemical models. The Al_{TS} concentrations predicted are model dependent, varying as much as 1.5 orders of magnitude in pH solutions of 6.5 to 8. The predictions offered by the models of Motekaitis and Martell (1984), Gregor and Powell (1986), and Lakatos et al. (2001) are nearly identical in the pH range 4 to $\approx$ 7.3, despite their very different interpretations of Al–citrate aqueous speciation (Figure 10.12*b* to *d*). Further, these three models predict the greatest Al_{TS} concentrations within the pH range 4 to 8, indicating that a relatively large percentage of Al_{TS} is predicted to exist in Al–citrate complexes. The models of Harris et al. (2003) and Essington et al. (2005) predict significantly smaller Al_{TS} concentrations (Figure 10.12*e* and *f*), suggesting that citrate is a relatively less effective chelate. It is also evident that the five models provide very different views of Al–citrate chemistry (Figure 12).

The concentrations of $Fe(III)_{TS}$ in equilibrium with goethite in the pH range 4 to 8 was evaluated using the chemical models of Field et al. (1974) [as modified

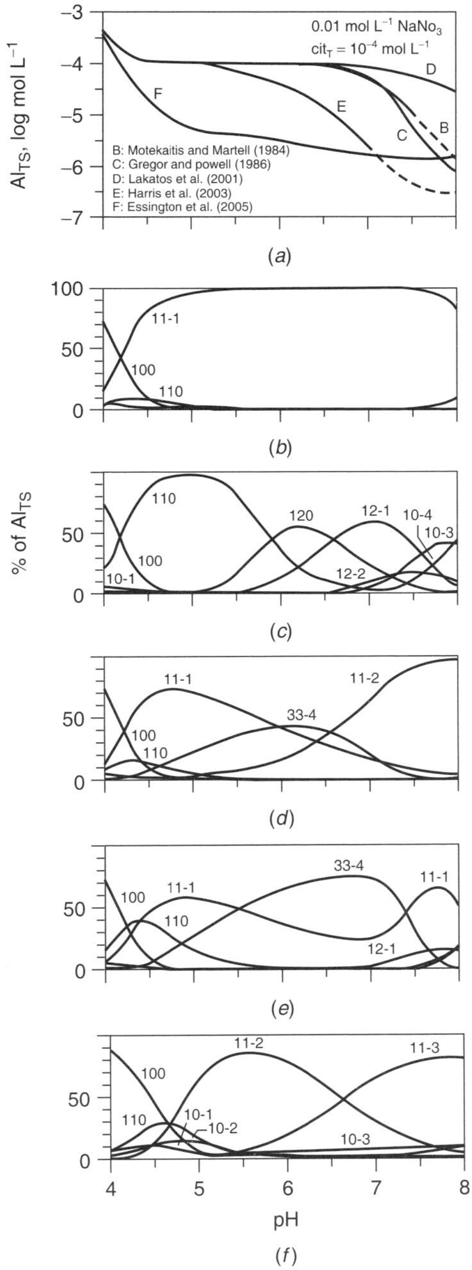

Figure 10.12. (*a*) Predicted total soluble concentration of aluminum (Al_{TS}) supported by gibbsite as a function of pH and the imposed Al–citrate chemical speciation models. The associated distributions of Al–citrate species in the equilibrium solutions are shown for the following models: (*b*) Motekaitis and Martell (1984); (*c*) Gregor and Powell (1986); (*d*) Lakatos et al. (2001); (*e*) Harris et al. (2003); (*f*) Essington et al. (2005). The imposed conditions included 0.01 mol L^{-1} $NaNO_3$, 10^{-4} mol L^{-1} cit_{TS}, 25°C and 1 atm, and log K_{sp} = 8.05 for the reaction $Al(OH)_3(s) + 3H^+(aq) = Al^{3+}(aq) + 3H_2O(l)$.

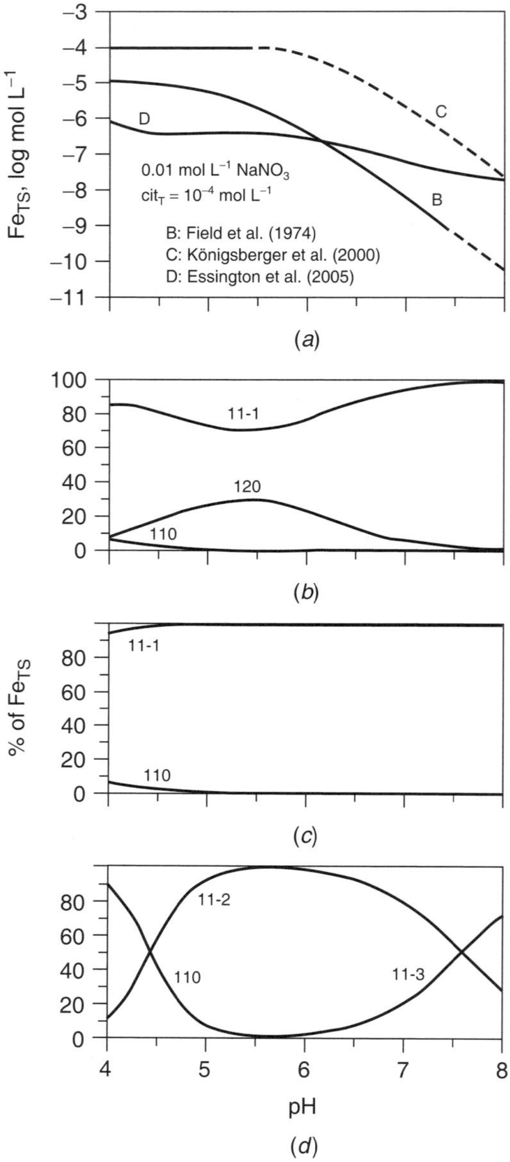

Figure 10.13. (*a*) Total soluble concentration predicted for iron(III) ($Fe(III)_{TS}$) supported by goethite as a function of pH and the imposed Fe(III)–citrate chemical speciation models. The associated distributions of Fe(III)–citrate species in the equilibrium solutions are shown for the following models: (*b*) Field et al. (1974); (*c*) Königsberger et al. (2000); (*d*) Essington et al. (2005). The imposed conditions include 0.01 mol L^{-1} $NaNO_3$, 10^{-4} mol L^{-1} cit_{TS}, 25°C and 1 atm, and log $K_{sp} = -0.02$ for the reaction $FeOOH(s) + 3H^+(aq) = Fe^{3+}(aq) + 3H_2O(l)$.

TABLE 10.6. Aqueous Al–Citrate and Fe(III)–Citrate Species for Which Formation Constants Are Tabulated in the Joint Expert Speciation System (JESS) (May and Murray, 2000) and the National Institute of Standards and Technology (NIST) (Martell et al., 2004) Thermodynamic Database Compilations

	Al–Citrate		Fe(III)–Citrate	
Species[a]	JESS	NIST	JESS	NIST
111	X	X	X	X
110	X	X	X	X
11–1	X	X	X	X
120	X	X	X	
121			X	
122			X	
11–2		X	X	
12–1	X			
12–2	X		X	
130	X			
22–2	X	X	X	X
33–4	X	X		
33–7	X	X		

[a]Species are denoted by the metal, citrate, and proton or hydroxide stoichiometry of each complex (*mnj*), as defined in Eq. (10.7).

by Martin (1986) to include the $Fecit_2^{3-}(aq)$ species], Königsberger et al. (2000), and Essington et al. (2005). The Fe(III)–citrate speciation model of Essington et al. (2005) is valid throughout the pH range 4 to 8, whereas that of Field et al. (1974) is valid up to pH 7.4. The model of Königsberger et al. (2000) is the most restrictive relative to the range of pH, valid only up to pH 5.4. GEOCHEM-PC was again employed to predict $Fe(III)_{TS}$, with the controlling conditions identical to those used above to predict gibbsite solubility. The solubility product constant for goethite was obtained from Lindsay (1979). Of the three models, Königsberger et al. (2000) predicts the greatest $Fe(III)_{TS}$ concentrations, which are one to two orders of magnitude greater than those predict by the other models (Figure 10.13*a*). Greater than 95% of $Fe(III)_{TS}$ is predicted to occur in a single species [$FeOHcit^{-}(aq)$] (Figure 10.13*c*), even though the chemical model of Königsberger et al. (2000) contains the greatest number of Fe(III)–citrate complexes (Table 10.5.) The $Fe(III)_{TS}$ concentrations predicted by the models of Essington et al. (2005) and Field et al. (1974) are approximately within one order of magnitude of each other, with greater $Fe(III)_{TS}$ concentrations predicted by Field et al. (1974) at pH < 6.2. The significant contribution of the $FeOHcit^{-}(aq)$ species predicted by the Field et al. (1974) model is consistent with the predictions of Königsberger et al. (2000), even though the percentages differ (Figure 10.13*a*). The model of Essington et al. (2005), which was derived from goethite solubility studies, does not contain the $FeOHcit^{-}(aq)$ species. Instead, the $Fe(OH)_2cit^{2-}(aq)$ species is predicted to predominate throughout a wide pH range (Figure 10.13*c*).

10.5. CONCLUSIONS

Citrate is a naturally occurring organic ligand that enjoys widespread occurrence in nature. It is a chelate, exuded by plant roots and having the effect of enhancing the phytoavailability of Fe, while detoxifying Al. Despite the relative importance of this ligand in soil systems, and its potential impact on the fate and behavior of Fe(III) and Al in the rhizosphere, the aqueous complexation chemistry of Fe(III)–and Al–citrate in the pH range of normal soil solutions (pH 4 to 9) is ambiguous. There are a number of established difficulties that confound one's ability to generate chemical models of Fe(III)–and Al–citrate speciation. Both Fe(III) and Al form sparingly soluble precipitates in the pH range of soil solutions. Exceedingly high concentrations of citrate are required to maintain analytical Fe(III) and Al concentrations in all but strongly acidic systems ($pH < 3$ to 4). Thus, the protonation and deprotonation of the citrate moieties tends to obscure metal complexation reactions when studied by potentiometric titration (the preferred method of characterizing Fe(III)–and Al–citrate speciation). The aqueous speciation at equilibrium is a function of the initial pH and the citrate/metal molar ratio. Low pH and high citrate/metal molar ratios favor the rapid formation of 1 : 1 metal/citrate complexes, whereas neutral pH and low citrate/metal molar ratios favor the additional formation of 1:2 and 2 : 2 species, which slowly oligomerize to form a thermodynamically stable 3 : 3 metal/citrate species, particularly when the metal is Al.

Numerous potentiometric titration-based chemical models have been proposed, a small number of which has been verified by ^{1}H, ^{13}C, and ^{27}Al NMR and electrospray mass spectroscopic techniques. The models range in character relative to the number and nature of species considered, from simple (a small number of mononuclear monocitrate species) to inordinately complex (several mono- to trinuclear and mono- to tricitrate species). The models are incongruous, in that no two models are composed of the same species, nor do they predict similar Fe(III)–or Al–citrate aqueous speciation. Similarly, the models differ relative to the predicted impact of citrate on the solubility of Fe(III)- or Al-bearing accessory minerals.

The findings in this chapter bring into question the ability to predict the influence of citrate, if not other rhizosphere products, on the soil chemistry of metals. Clearly, additional studies are needed to validate an existing model or to establish an Fe(III)–or Al–citrate aqueous speciation model that may be applied to the soil environment. Equally problematic are thermodynamic data compilations relative to the Fe(III)–or Al–citrate species that are included. Chemical models, as defined by various studies in Tables 10.4 and 10.5, cannot be mixed. They are constructed specifically to describe a given series of chemical systems. Presumably, the number of species, their stoichiometric characteristics, and their formation constants are exactly what is required to model the experimental observations to a high degree of confidence (minimal variance between predicted and observed values). A review of two compilations, the Joint Expert Speciation System (JESS) (May and Murray, 2000) and the National Institute of Standards and Technology (NIST) database (Martell et al., 2004), shows that formation constants are provided for numerous Fe(III)–or Al–citrate species (Table 10.6). For example, the NIST database includes formation constants for the following Al–citrate species: $Alcit^0(aq)$, $AlHcit^+(aq)$, $AlOHcit^-(aq)$, $Al(OH)_2cit^{2-}(aq)$, $Alcit_2{}^{3-}(aq)$, $Al_2(OH)_2cit_2{}^{2-}(aq)$, $Al_3(OH)_4cit_3{}^{4-}(aq)$, and $Al_3(OH)_7cit_3{}^{7-}(aq)$.

As illustrated in Table 10.4., none of the chemical models cited employ all eight species. Thus, the Al–citrate species distribution in a soil solution, predicted using the NIST compilations, would be erroneous, as would any correlations to soil chemical processes and the resulting interpretations. At present, the most appropriate mechanism for predicting the impact of citrate on Al and Fe(III) aqueous speciation is to select a specific chemical model and employ the model unmodified.

REFERENCES

Aisen, P., Leibman, A., and Zweier, J. (1978). Stiochiometric and site characteristics of the binding of iron to human transferrin. *J. Biol. Chem.* **253**, 1930–1937.

Amico, P., Daniele, P. G., Cucinotta, V., Rizzarelli, E., and Sammartano, S. (1979). Equilibrium study of iron(II) and manganese(II) complexes with citrate ion in aqueous solution: relevance to coordination of citrate to the active site of aconitase and to gastrointestinal absorption of some essential metal ions. *Inorg. Chim. Acta* **36**, 1–7.

Arena, G., Cali, R., Grasso, M., Musumeci, S., and Sammartano, S. (1980). The formation of proton and alkali-metal complexes with ligands of biological interest in aqueous solution, I: Potentiomtric and calometric investigations of H^+ and Na^+ complexes with citrate, tartrate and malate. *Thermochim. Acta* **36**, 329–342.

Baes, C. F., Jr., and Mesmer, R. E. (1986). *The Hydrolysis of Cations*, R. E, Krieger, Malabar, FL.

Bertsch, P. M., and Parker, D. R. (1995). Aqueous polynuclear aluminum species. In *The Environmental Chemistry of Aluminum*, 2nd ed., ed. Sposito, G., Lewis Publishers, Boca Raton, FL, 117–168.

Bodor, A., Bányai, I., and Tóth, I. (2002). ^{1}H- and ^{13}C-NMR as tools to study aluminum coordination chemistry: aqueous Al(III)–citrate complexes. *Coord. Chem. Rev.* **228**, 175–186.

Byrne, R. H., and Kester, D. R. (1976a). Solubility of hydrous ferric oxide and iron speciation in seawater. *Mar. Chem.* **4**, 255–274.

Byrne, R. H., and Kester, D. R. (1976b). A potentiometric study of ferric ion complexes in synthetic media and seawater. *Mar. Chem.* **4**, 275–287.

Byrne, R. H., and Kester, D. R. (1978). Ultraviolet spectroscopic study of ferric hydroxide complexation. *J. Sol. Chem.* **7**, 373–383.

Byrne, R. H., and Kester, D. R. (1981). Ultraviolet spectroscopic study of ferric equilibria at high chloride concentrations. *J. Sol. Chem.* **10**, 51–67.

Byrne, R. H., Luo, Y.-R., and Young, R. W. (2000). Iron hydrolysis and solubility revisited: observations and comments on iron hydrolysis characterizations. *Mar. Chem.* **70**, 23–25.

Clausén, M., Öhman, L.-O., and Persson, P. (2005). Spectroscopic studies of aqueous gallium(III) and aluminum(III) citrate complexes. *J. Inorg. Biochem.* **99**, 716–726.

Dakanali, M., Raptopoulou, C. P., Terzis, A., Lakatos, A., Banyai, I., Kiss, T., and Salifoglou, A. (2003). A novel dinuclear species in the aqueous distribution of aluminum in the presence of citrate. *Inorg. Chem.* **42**, 252–254.

Daniele, P. G., Rigano, C., Sammartano, S., and Zelano, V. (1994). Ionic strength dependence of formation constants, XVIII: The hydrolysis of iron(III) in aqueous KNO_3 solution. *Talanta* **41**, 1577–1582.

Dietzel, M., and Böhme, G. (2005). The dissolution rates of gibbsite in the presence of chloride, nitrate, silica, sulfate, and citrate in open and closed systems at 20°C. *Geochim. Cosmochim. Acta* **69**, 1199–1211.

Ekström, L. G., and Olin, A. (1979). Complex-formation between lead-(II) and citrate ions in acid, neutral and weakly alkaline-solution. *Chem. Scr.* **13**, 10–15.

Essington, M. E., Nelson, J. B., and Holden, W. L. (2005). Gibbsite and goethite solubility: the influence of 2-ketogluconate and citrate. *Soil Sci. Soc. Am. J.* **69**, 996–1008.

Feng, T. L., Gurian, P. L., Healy, M. D., and Barron, A. R. (1990). Aluminum citrate: isolation and structural characterization of a stable trinuclear complex. *Inorg. Chem.* **29**, 408–411.

Field, T. D., McCourt, J. L., and McBryde, W. A. E. (1974). Composition and stability of iron and copper citrate complexes in aqueous-solution. *Can. J. Chem.* **52**, 3119–3124.

Gautier-Luneau, I., Merle, C., Phanon, D., Lebrun, C., Biaso, F., Serratrice, G., and Pierre, J.-L. (2005). New trends in the chemistry of iron(III) citrate complexes: correlations between x-ray structure and solution species probed by electrospray mass spectrometry and kinetics of iron uptake from citrate by iron chelators. *Chem. Eur. J.* **11**, 2207–2219.

Geelhoed, J. S., Hiemstra, T., and Van Riemsdijk, W. H. (1998). Competitive interaction between phosphate and citrate on goethite. *Environ. Sci. Technol.* **32**, 2119–2123.

Gorzsás, A., Getty, K., Andersson, I., and Pettersson, L. (2004). Speciation in the aqueous $H^+/H_2VO_4^-/H_2O_2$/citrate system of biomedical interest. *Dalton* **2002**, 2873–2882.

Gräfe, M., Eick, M. J., Grossl, P. R., and Sanders, A. M. (2002). Adsorption of arsenate and arsenite on ferrihydrite in the presence and absence of dissolved organic carbon. *J. Environ. Qual.* **31**, 1115–1123.

Greenaway, F. T. (1986). Aluminum-27 NMR study of aluminum(III) interactions with hydroxyl carboxylic acids. *Inorg. Chim. Acta* **116**, L21–L23.

Gregor, J. E., and Powell, H. K. J. (1986). Aluminum(III)-citrate complexes: a potentiometric and C-13 NMR-study. *Aust. J. Chem.* **39**, 1851–1864.

Hamm, R. E., Shull, C. M., Jr., and Grant, D. M. (1954). Citrate complexes with iron(II) and iron(III). *J. Am. Chem. Soc.* **76**, 2111–2114.

Harris, W. R., Wang, Z., and Hamada, Y. Z. (2003). Competition between transferring and the serum ligands citrate and phosphate for binding aluminum. *Inorg. Chem.* **42**, 3262–3273.

Haynes, R. J., and Mokolobate, M. S. (2001). Amelioration of Al toxicity and P deficiency in acid soils by additions of organic residues: a critical review of the phenomenon and the mechanisms involved. *Nutr. Cycl. Agroecosyst.* **59**, 47–63.

Hegenauer, J., Saltman, P., and George, N. (1979). Iron(III)–phosphoprotein chelates: stoichiometric equilibrium constant for interaction of iron(III) and phosphorylserine residues of phosvitin and casein. *Biochemistry*. **18**, 3865–3879.

Hocking, P. J. (2001). Organic acids exuded from roots in phosphorus uptake and aluminum tolerance of plants in acid soils. *Adv. Agron.* **74**, 63–97.

Huang, P. M., and Violante, A. (1986). Influence of organic acids on crystallization and surface properties of precipitation products of aluminum. In *Interactions of Soil Minerals with Natural Organics and Microbes*, ed. Huang, P. M., and Schnitzer, M., Spec. Publ. 17, Soil Science Society of America, Madison, WI, 159–221.

Inskeep, W. P., and Silvertooth, J. C. (1988). Inhibition of hydroxyapatite precipitation in the presence of fulvic, humic, and tannic acids. *Soil Sci. Soc. Am. J.* **52**, 941–946.

Jardine, P. M., and Zelazny, L. W. (1995). Surface reactions of aqueous aluminum species. In *The Environmental Chemistry of Aluminum* 2nd ed., ed. Sposito, G., Lewis Publishers, Boca Raton, FL, 221–270.

Jones, D. L. (1998). Organic acids in the rhizosphere: a critical review. *Plant Soil* **205**, 25–44.

Jones, D. L., Dennis, P. G., Owen, A. G., and van Hees, P. A. W. (2003). Organic acid behavior in soils: misconceptions and knowledge gaps. *Plant Soil* **248**, 31–41.

Kafkafi, U., Bar-Yosef, B., Rosenberg, R., and Sposito, G. (1988). Phosphorus adsorption by kaolinite and montmorillonite, II. Organic anion competition. *Soil Sci. Soc. Am. J.* **52**, 1585–1589.

Karlik, S. J., Tarien, E., Elgavish, G. A., and Eichhorn, G. L. (1983). Aluminum-27 nuclear magnetic resonance study of aluminum(III) interactions with carboxylate ligands. *Inorg. Chem.* **22**, 525–529.

Kester, D. R., Byrne, R. H., and Liang, Y. J. (1975). Redox reaction and solution complexes of iron in marine systems. In *Marine Chemistry in the Coastal Environment*, ed. Church, T. M. ACS Symp. Ser. 18, American Chemical Society, Washington, DC, 56–79.

Königsberger, L. C., Königsberger, E., May, P. M., and Hefter, G. T. (2000). Complexation of iron(III) and iron(II) by citrate: implications for iron speciation in blood plasma. *J. Inorg. Biochem.* **78**, 175–184.

Kubicki, J. D., Sykes, D., and Apitz, S. E. (1999). Ab initio calculation of aqueous aluminum and aluminum–carboxylate complex energetics and ^{27}Al NMR chemical shifts. *J. Phys. Chem.* **103**, 903–915.

Kuma, K., Nakabayashi, S., Suzuki, Y., and Matsunaga, K. (1992). Dissolution rate and solubility of colloidal hydrous ferric oxide in seawater. *Mar. Chem.* **38**, 133–143.

Kuma, K., Nishioka, J., and Matsunaga, K. (1996). Controls on iron(III) hydroxide solubility in seawater: the influence of pH and natural chelators. *Limnol. Oceanogr.* **41**, 396–407.

Lackovic, K., Angove, M. J., Wells, J. D., and Johnson, B. B. (2004). Modeling the adsorption of Cd(II) onto goethite in the presence of citric acid. *J. Colloid Interface Sci.* **269**, 37–45.

Lakatos, A., Bányai, I., Decock, P., and Kiss, T. (2001). Time-dependent solution speciation of the Al^{III}–citrate system: potentiometric and NMR studies. *Eur. J. Inorg. Chem.* **2001**, 461–469.

Lebron, I. and Suarez, D. L. (1999). Mechanisms and precipitation rate of rhodochrosite at 25 EC as affected by p_{CO2} and organic ligands. *Soil Sci. Soc. Am. J.* **63**, 561–568.

Lindsay, W. L. (1979). *Chemical Equilibria in Soils,* Wiley, New York.

Liu, X., and Millero, F. J. (1999). The solubility of iron hydroxide in sodium chloride solutions. *Geochim. Cosmochim. Acta* **63**, 3487–3497.

Liu, X., and Millero, F. J. (2002). The solubility of iron in seawater. *Mar. Chem.* **77**, 43–54.

Lopez-Quintela, M. A., Knoche, W., and Veith, J. (1984). Kinetics and thermodynamics of complex formation between aluminum(III) and citric acid in aqueous solution. *J. Chem. Soc. Faraday Trans.* **80**, 2313–2321.

Martell, A. E., Smith, R. M., and Motekaitis, R. J. (2004). *NIST Critically Selected Stability Constants of Metal Complexes*, NIST Standard Reference Database 46, Version 8.0, National Institute of Standards and Technology, Gaithersburg, MD.

Martin, R. B. (1986). Citrate binding of Al^{3+} and Fe^{3+}. *J. Inorg. Biochem.* **28**, 181–187.

Matzapetakis, M., Raptopoulou, C. P., Terzis, A., Lakatos, A., Kiss, T., and Salifoglou, A. (1998). Synthesis, structural characterization, and solution behavior of the first mononuclear, aqueous aluminum citrate complex. *Inorg. Chem.* **38**, 618–619.

Matzapetakis, M., Kourgiantakis, M., Dakanali, M., Raptopoulou, C. P., Terzis, A., Lakatos, A., Kiss, T., Bányai, I., Iordanidis, L., Mavromoustakos, T., and Salifoglou, A. (2001). Synthesis, pH-dependent structural characterization, and solution behavior of aqueous aluminum and gallium citrate complexes. *Inorg. Chem.* **40**, 1734–1744.

May, P. M., and Murray, K. (2000). *Joint Expert Speciation System (JESS)*, http://jess.murdoch.edu.au/jess/jess_home.htm (verified Aug. 12, 2004.)

Millero, F. J., Yao, W., and Aicher, J. (1995). The speciation of Fe(II) and Fe(III) in natural waters. *Mar. Chem.* **50**, 21–39.

Motekaitis, R. J., and Martell, A. E. (1984). Complexes of aluminum(III) with hydroxyl carboxylic acids. *Inorg. Chem.* **23**, 18–23.

Nordstrom, D. K., and May, H. M. (1995). Aqueous equilibrium data for mononuclear aluminum species. In *The Environmental Chemistry of Aluminum*, ed. Sposito, G., Lewis Publishers, Boca Raton, FL, 39–80.

Öhman, L.-O. (1988). Equilibrium and structural studies of silicon(IV) and aluminum(III) in aqueous solution, 17: Stable and metastable complexes in the system H^{+}–Al^{3+}–citric acid. *Inorg. Chem.* **27**, 2565–2570.

Öhman, L.-O., and Sjöberg, S. (1983). Equilibrium and structural studies of silicon (IV) and aluminum(III) in aqueous solution, 9: A potentiometric study on mono- and polynuclear aluminum citrates. *J. Chem. Soc. Dalton Trans.* **11**, 2513–2517.

Parker, D. R., Norvell, W. A., and Chaney, R. L. (1995). GEOCHEM-PC: a chemical speciation program for IBM and compatible personal computers. In *Chemical Equilibrium and Reaction Models,* ed. Loeppert, R. H., et al., Spec. Publ. 42, Soil Science Society of America, Madison, WI, 253–269.

Pattnaik, R. K., and Pani, S. (1961). The citrate complex of aluminum. *J. Indian Chem. Soc.* **38**, 379–384.

Pierre, J. L., and Gautier-Luneau, I. (2000). Iron and citric acid: a fuzzy chemistry of ubiquitous biological relevance. *Biometals* **13**, 91–96.

Powell, A. K., and Heath, S. L. (1996). X-ray structural analysis of biologically relevant aluminum (III) complexes. *Coord. Chem. Rev.* **149**, 59–80.

Rajan, K. S., Mainer, S., and Davis, J. M. (1978). Formation and stabilities of the ternary metal chelates of L-3,4-dihydroxyphenyl alanine (L-DOPA) with a number of secondary ligands. *J. Inorg. Nucl. Chem.* **40**, 2089–2099.

Rajan, K. S., Mainer, S., Rajan, N. L., and Davis, J. M. (1981). Studies on the chelation of aluminum for neurobiological application. *J. Inorg. Biochem.* **14**, 339–350.

Ramamoorthy, S., and Manning, P. G. (1973). Equilibrium studies of metal-ion complexes of interest to natural waters, VI: Simple and mixed complexes of Fe(III) involving NTA as primary ligand and a series of oxygen-bonding organic anions as secondary ligands. *J. Inorg. Nucl. Chem.* **35**, 1571–1575.

Shweky, I., Bino, A., Goldberg, D. P., and Lippard, J. (1994). Synthesis, structures, and magnetic properties of two dinuclear iron(III) citrate complexes. *Inorg. Chem.* **33**, 5161–5162.

Smith, R. M., and Martell, A. E. (1976). *Critical Stability Constants*, Vol. 4, *Inorganic Complexes*, Plenum Press, New York.

Spiro, T. G., Pape, L., and Saltman, P. (1967a). The hydrolytic polymerization of ferric citrate, I: The chemistry of the polymer. *J. Am. Chem. Soc.* **89**, 5555–5559.

Spiro, T. G., Bates, G., and Saltman, P. (1967b). The hydrolytic polymerization of ferric citrate, II: The influence of excess citrate. *J. Am. Chem. Soc.* **89**, 5559–5562.

Sposito, G., and Mattigod, S. V. (1980). *GEOCHEM: A Computer Program for the Calculation of Chemical Equilibria in Soil Solution and Other Natural Water Systems,* Kearney Foundation of Soil Science, University of California, Riverside, CA.

Strobel, B. W. (2001). Influence of vegetation on low-molecular-weight carboxylic acids in soil solution: a review. *Geoderma* **99**, 169–198.

Sunda, W., and Huntsman, S. (2003). Effect of pH, light, and temperature on Fe–EDTA celation and Fe hydrolysis in seawater. *Mar. Chem.* **84**, 35–47.

Timberlake, C. F. (1964). Iron–malate and iron–citrate complexes. *J. Chem. Soc.* **1964**, 5078–5085.

van Hees, P. A. W., Lundström, U. S., and Giesler, R. (2000). Low molecular weight organic acids and their Al-complexes in soil solution: composition, distribution and seasonal variation in three podzolized soils. *Geoderma* **94**, 173–200.

Vanni, A., and Gennaro, M. C. (1974). Equilibrium studies of beryllium complexes: interaction of beryllium with citric and tricarbalylic acids. *Ann. Chim.* **64**, 397–408.

Venturini, M., and Berthon, G. (1989). Aluminum speciation studies in biological fluids, 2: Quantitative investigations of aluminum–citrate complexes and appraisal of their potential significance in vivo. *J. Inorg. Biochem.* **37**, 69–90.

Wada, N., Yamashita, K., and Umegaki, T. (1999). Effects of carboxylic acids on calcite formation in the presence of Mg^{2+} ions. *J. Colloid Interface Sci.* **212**, 357–364.

Warner, R. C., and Weber, I. (1953). The cupric and ferric citrate complexes. *J. Am. Chem. Soc.* **75**, 5086–5094.

Wiese, G., and Veith, J. A. (1975). Formation of complexes with citric acid, I: Formation of complexes between citric acid and aluminum. *Z. Naturforsch.* **30b**, 446–453.

Wijnja, H., and Schulthess, C. P. (2000). Interaction of carbonate and organic anions with sulfate and selenate adsorption on an aluminum oxide. *Soil Sci. Soc. Am. J.* **64**, 898–908.

Xu, R. K., and Ji, G. L. (2003). Effects of anions of low molecular weight organic acids on adsorption and desorption of aluminum by and from a kaolinite at different pH. *Soil Sci.* **168**, 39–44.

Yao, H. L. and Yeh, H. H. (1996). Fumarate, maleate, and succinate adsorption on hydrous δ-Al_2O_3, 1: Comparison of the adsorption maxima and their significance. *Langmuir* **12**, 2981–2988.

PART III

SPECIATION, MOBILITY, AND BIOAVAILABILITY OF METALS AND METALLOIDS AND RESTORATION OF CONTAMINATED SOILS

11

CHEMICAL SPECIATION AND BIOAVAILABILITY OF TRACE METALS

G. S. R. Krishnamurti

313-855 West 16th Street, North Vancouver, British Columbia, Canada

R. Naidu

Cooperative Research Centre for Contamination Assessment and Remediation of the Environment, University of South Australia, Mawson Lakes, South Australia

Biophysico-Chemical Processes of Heavy Metals and Metalloids in Soil Environments, Edited by Antonio Violante, Pan Ming Huang, and Geoffrey Michael Gadd

11.1. INTRODUCTION

It is now generally accepted that the distribution, mobility, bioavailability, and toxicity of chemicals depend not simply on their concentrations but critically, on their forms. Changes in environmental conditions, whether natural or anthropogenic, can strongly influence the behavior of both essential and toxic elements by altering the form in which they occur. Some of the more important factors controlling element behavior are pH, redox potential, and the presence of reactive species such as complexing ligands (both organic and inorganic), particle surfaces for adsorption, and colloidal matter. To comprehend and fully predict the environmental chemistry of an element, it would be necessary to characterize in detail the proportions and chemistry of all its various forms under the diverse range of conditions possible in natural systems (Ure and Davidson, 2002).

There is no general definition of *speciation*. The IUPAC has published a document on chemical speciation terminology which recommends the term *chemical species* as the form of an element defined as to isotopic composition, electronic or oxidation state, and/or complex or molecular structure (Templeton et al., 2000). As pointed out by Ure and Davidson (2002), this narrow definition is perfectly appropriate to speciation in solution samples but would exclude most studies on solid materials such as soils, sediments, and geochemical and biological materials from the field of speciation and relegate to the category of fractionation methodology the selective extraction and fractionation procedures relevant to such solid materials. Most current use would include such solid sample techniques in a broader definition of speciation that would encompass the foregoing IUPAC type of narrow definition of species, perhaps best described as *classical speciation*, as well as speciation in which species are described *functionally* or *operationally*. The 1st International Conference on Trace Element Speciation in Biomedical, Nutritional and Environmental Sciences (Michalke and Schrammel, 1999) also concluded that a narrow definition of speciation, on the lines of the IUPAC proposal, was still not attainable for solid samples and found merit in the use of functionally and operationally defined species. The terminology used in this chapter is based on the definition of speciation as recommended by the IUPAC and therefore includes three types of speciation based on species defined functionally, operationally for chemical pools in the solid phase, or as classical speciation for solution species.

1. *Classical (solution) speciation.* Measurement of the precise chemical form of an element as present in the solution phase is the most difficult to achieve since analytical methodology of great selectivity and sensitivity is required. Some success has been achieved by using chromatographic and electrochemical techniques, as well as through chemical modeling. It is also important to distinguish chemical species according to their distribution among soil components such as organic matter and hydrous oxides. This information will be useful for assessing the potential of soil to supply sufficient micronutrients for plant growth or to contain toxic quantities of

trace metals, and for determining amelioration procedures for soils at the risk of causing trace metal contamination of waterways.

2. *Operationally defined speciation.* In operationally defined speciation the physical or chemical fractionation procedure applied to the sample defines the fraction or chemical pools isolated for measurement.
3. *Functionally defined speciation.* Functionally defined species are exemplified by the plant-available species or chemical pools in which the "function" is plant availability. Available forms of trace metal cations are not necessarily associated with one particular chemical species or a specific soil component. Hence, to predict the availability of trace metals, we either have to establish the species involved and develop methods that specifically determine those forms only, or we have to establish an empirical relationship between an accepted diagnostic measure of the metal and plant growth. Both speciation in solution and fractionation of the solid phase to identify the chemical pools can affect plant uptake (phytoavailability) of trace metals and water pollution.

The bioavailability of metal to plants and soil organisms is mainly through the solution phase. However, soil solutions are in intimate contact with the soil solid phase and are therefore influenced by mineral equilibria and exchange and sorption processes that involve soil organic matter and mineral phases, as well as complexation by organic matter in both the solid phase and that dissolved in the soil solution. Le Châtelier's principle of chemical equilibrium would suggest that metal bioavailability in soils is thus controlled by both solution speciation and the metal associated with the solid phase.

The element bioavailability is reported to be a function of at least three parameters (Brümmer, 1986):

1. The total amount of potentially available elements (the quantity factor)
2. The concentration or activity and ionic ratios of elements in the soil solution (the intensity factor)
3. The rate at which elements transfer from the solid to the liquid phase and to plant roots (reaction kinetics)

The fate of toxic metals in soils is influenced by its initial chemical form and the environmental and edaphic conditions, such as pH, redox status, and soil organic matter content. The metal speciation in soils can be discussed broadly using the following approaches:

- Computer-based modeling on geochemical principles, using either the equilibrium constants or Gibbs free-energy values. Both approaches are subject to the conditions of equilibrium and mass balance.
- Soil environmental constraints, based on an understanding of the soil processes and conditions that control the formation and transformation of metal species.

- Chemical analysis using extractants, based on the reactivity of the extractants with the metal under study.

These different approaches, individually, provide an idea of the processes involved in the fate and behavior of toxic metals. However, there are no published studies that have utilized a combination of these approaches to understand the fate of metals in a soil ecosystem. Furthermore, although numerous studies have used chemical extractants to quantify different metal fractions, only a limited few have attempted to characterize metal phytoavailability by correlating soil-extractable metal fractions with plant metal uptake (e.g., Lake et al., 1984; Krishnamurti et al., 1995a).

11.2. CLASSICAL (SOLUTION) SPECIATION

Speciation aims to determine the various physicochemical forms of an element that together make up its total soluble concentration and to understand either the biological or geochemical cycling of the elements. Biological cycling includes bioavailability, bioaccumulation, and toxicity; geochemical cycling involves transport, sorption, and precipitation. No meaningful interpretation of either biological or geochemical cycling can be made without speciation information. Electrochemical techniques can be used to provide speciation information and have been applied to over 30 elements, including trace elements of primary environmental concern, such as copper, zinc, lead, and cadmium. Polarography, anodic stripping voltammetry (ASV), cathodic stripping voltammetry (CSV), and potentiometric stripping analysis (PSA) are a few of the most important techniques that are in use for trace element speciation analysis. Florence (1989) has provided an excellent review of the details of the techniques and their application.

11.2.1. Thermodynamic Models

Calculation of trace element speciation using mathematical models based on fundamental thermodynamic concepts may provide insight into the concentration of species. Application of mathematical models to the elucidation of trace element speciation appears particularly appealing in systems where thermodynamic parameters are known with some certainty. However, application of such geochemical models to complex systems such as natural waters, which contain ill-defined soluble and particulate organic (humic) materials, presents difficulties. Mathematical approaches to equilibrium models have been dealt with in detail by Waite (1989) and Loeppert et al. (1995). While the foundation for general-purpose algorithms for the computation of chemical equilibrium was laid by Brinkley (1947), the first model of chemical speciation in natural water was the ion-association model of Garrels and Thompson (1962). A number of reviews on the principles and applications of the models developed since then are available (Jenne, 1981; Nordstrom et al., 1989; Waite, 1989).

Using basic chemical theory, Sposito (1986) and Morgan (1987) outlined principles for elucidating ion speciation in the natural environment. Computer-based geochemical modeling approaches have been developed specifically for aqueous systems including soil solution (Sposito and Mattigod, 1980; Sposito and Coves, 1988). Two different thermodynamic approaches can be used to calculate metal ion speciation, which includes using either equilibrium constants or Gibbs free-energy values, which are subject to conditions of equilibrium and mass balance. Even though the same approach to thermodynamic calculations can be applied to both inorganic and organic metal complexes, we do not have reliable thermodynamic data for realistic metal–organic complexes. The character of dissolved organic matter in soil solutions differs from environment to environment and contains a chemically diverse group of chemical substances, collectively termed *humic acid* and *fulvic acid*. Relative to humic acid, fulvic acid is soluble and is the predominant secondary synthesis compound in soil solution and other natural waters, due largely to its lower molecular mass and greater total acidity. Indeed, fulvic acids are ubiquitous in natural waters. In addition, fulvic acids are important complexing agents of metal cations, directly affecting metal bioavailability and transport (Essington, 2004).

Computer-based chemical equilibrium models of natural systems, such as soil systems, have undergone a great deal of development in recent years and have become useful tools for studying water quality criteria (Jenne, 1979). The widely used programs—GEOCHEM (Sposito and Mattigod, 1980), used exclusively for soil systems, and MINEQL (Westall et al., 1976; James and Parks, 1978), used to study water bodies—are progeny of the program REDEQL2 (Morel and Morgan, 1972; McDuff and Morel, 1973). Details of the REDEQL2 program have been presented by Legatt (1977) and Ingle et al. (1978). The equilibrium constant approach and Newton–Raphson iteration were used in REDEQL, which minimizes the difference between measured and calculated total component concentrations (Morel and Morgan, 1972). REDEQL has served as a parent for a number of widely used programs, including MINEQL (Westall et al., 1976), MICROQL (Westall, 1979), and GEOCHEM (Sposito and Mattigod, 1980). Some typical applications of GEOCHEM include (Sposito, 1983):

- Prediction of the concentration of inorganic and organic complexes of a metal cation in a soil solution
- Calculation of the concentration of a particular chemical form of a nutrient element in a solution bathing plant roots so as to correlate that form to nutrient uptake
- Prediction of the chemical fate of pollutant metal added to a soil solution of known characteristics
- Estimation of the effect of changing pH, ionic strength, redox potential, water content, or the concentration of some constituent on the solubility of a chemical element of interest in soil solution

GEOCHEM has more than twice as many thermodynamic data as those in REDEQL2. However, it differs from REDEQL2 principally in using thermodynamic data that have been selected especially for soil systems and in containing a method for describing the cation exchange and employing a different subroutine for correcting thermodynamic equilibrium constants for the effect of nonzero ionic strength. MINEQL differs from REDEQL2 in the subroutines that describe solid phases and the adsorption phenomena.

In recent years several computer-based models have been widely available and used successfully [e.g., SOILCHEM (the updated version of GEOCHEM, Sposito and Coves, 1988), HYDRAQL (Papelis et al., 1988), ECOSAT (Keizer, 1991), and MINTEQA2 (Allison et al., 1991)]. Excellent reviews on the theory and use of computerized speciation models have been presented: GEOCHEM-PC (Parker et al., 1995b), SOILCHEM (Sposito and Coves, 1995), and MINTEQA2/PRODEFA2 (Allison and Brown. 1995). Despite the success reported by many investigators, most models still suffer from limited information on metal-dissolved organic matter (DOM) stability constants. Such a constraint can lead to incorrect partitioning of metals in soil solution. An increasing number of studies have focused on modeling the impact of anthropogenic metal inputs, such as sewage sludge (Mattigod and Sposito, 1979), geothermal brines (Sposito et al., 1979), acid rain (Sposito et al., 1980), and coal gasification (Ireland et al., 1982) in soil systems. Elemental speciation was predicted by GEOCHEM. These studies provided valuable insights into the chemistry of speciation of metals in soil systems and a basis for predicting the behavior of the elements in contaminated soils.

The speciation of Cd in soil solutions of soils from northern France, which were contaminated by effluent from a Pb–Zn metallurgical plant, was calculated using the program SOILCHEM (Charlatchka et al., 1997). The database Geodata was complemented with the stability constants of Cd-butyrate and Cd-propionate (Sillen and Martel, 1971) and of Cd-fulvates (Lamy et al., 1994). The free species Cd^{2+} accounted for around 60%, and the inorganic species of $CdCl^{+}$ and $CdHCO_3{}^{+}$ accounted for around 40%. The organic species increased from $< 1\%$ to 20% with increased flooding, with concomitant decrease in the concentration of Cd^{2+} species (Charlatchka et al., 1997). The input data consisted of the measured total concentration of Cd, pH, and the thermodynamic constants from the database. The ionic strength was calculated by the program as well as activity coefficients (Sposito, 1981; Sposito and Coves, 1988). A number of studies have indicated that either free Cd^{2+} ions or Cd complexed with inorganic ligands are the dominant species present in the soil solution of most sludged and arable soils (Emmerich et al., 1982; Hirsch and Banin, 1990). It is generally believed that organic ligands do not have great relevance in the complexation of Cd, in contrast to other metals, such as Cu, Pb, or Zn, in soil solutions (Adriano, 2001).

11.2.1.1. Dissolved Organic Matter–Solution Speciation Few studies report actual measurement of Cd^{2+} speciation in soil solutions. Many authors report Cd^{2+} speciation based on measured total dissolved Cd concentration and computation of free Cd^{2+} using chemical equilibrium models. The accuracy of the

TABLE 11.1. Speciation of Cd in Soil Solution of Kapinnie Soil Using Various log K Values for the Cd–DOM Complexes

	Cd Species in Solution (μg L^{-1})		
log K_{Cd-DOM}	Cd^{2+}	Cd–DOM	Cd–Inorganic Ligand[a]
4.10 (Schnitzer and Hansen, 1970)	418.80 (81.8)[b]	43.50 (8.5)	49.65 (9.7)
5.30 (Almas et al., 2000)	185.85 (36.3)	304.10 (59.4)	22.05 (4.3)
5.80 (Alberts and Giesy, 1983)	81.90 (16.0)	420.50 (82.1)	9.60 (1.9)

Source: Krishnamurti and Naidu (2003).
[a]Cd–inorganic ligand complexes such as $CdCl^+$, $CdSO_4$, and $CdNO_3{}^+$.
[b]Values in parentheses are percent distribution of the species.

speciation depends on the correctness of Cd-complex stability constants (Turner, 1995). For example, the proportion of free Cd (as Cd^{2+}) in solution varies from 16 to 82%, depending on the log K values for Cd–DOM complexes (Table 11.1). The stability constants of Cd-humics are particularly uncertain. The high values reported for pCd^{2+} ranged between 8 and 5, and the proportion of Cd–organic complexes was usually reported as negligible (Hirsch and Banin, 1990; Jopony and Young, 1994; Temminghoff et al., 1995; Candelaria and Chang, 1997; Elzinga et al., 1999). In contrast to other metals, such as Cu, Pb, and Zn, Adriano (2001) stated that "it is plausible that organic ligands do not have great relevance in the complexation of Cd in soil solutions." A number of studies also indicated that free Cd^{2+} ions or Cd complexed with inorganic ligands are the dominant Cd species in soil solution of most dredged and arable soils (Emmerich et al., 1982; Hirsch and Banin, 1990). Nevertheless, the association of Cd with organics in soil solution is not insignificant (Del Castilho and Rix, 1993; Naidu and Harter, 1998). Several investigations have shown that soluble organics increase the Cd concentration in soil solutions (Dunnivant et al., 1992; McBride et al., 1997; Krishnamurti et al., 1997a; Naidu and Harter, 1998).

In recent years the understanding of colloid surfaces and soil constituents has increased tremendously. Surface coordinating functional groups on particulate inorganic and humic materials are viewed as complexant ligands (Stumm, 1992). The surface complex models (SCM) are now finding increased application in the fields of pollutant retention behavior (Zachara et al., 1992), the chemistry of plant nutrient retention (Goldberg and Traina, 1987), and the transport of pollutants by colloids (Goldberg, 1992). The elegance of the surface complexation approach lies in the fact that it can be incorporated into the thermodynamic speciation models used for soluble complexes. The surface complexation models that are commonly in use are the diffuse double-layer model (DDLM) (Huang and Stumm, 1973; Dzombek and Morel, 1990), the constant-capacitance model

(CCM) (Stumm et al., 1970, 1976, 1980; Schindler et al., 1976), the triple-layer model (TLM) (Davis and Leckie, 1978, 1980; Davis et al., 1978; Hayes and Leckie, 1987; Hayes et al., 1988), and the 1 p*K* basic Stern model (Bolt and van Riemsdijk, 1982; Van Riemsdijk et al., 1986, 1987). The application of many of the commonly used computer models in the determination of the speciation in solution phase has been dealt with exhaustively by Lumsdon and Evans (1995).

The organic matter found in natural waters is composed of an extremely complex mixture of compounds, most of which (80%) are not yet identified. Moreover, in the case of thoroughly studied homologous groups, such as fulvic and humic matter, scant attention has been paid to the identification of the nature of binding sites, which are probably more relevant for trace metals than for the major metals (Schnitzer, 1991). Development of methods to better characterize the nature of humic functional groups is in progress, with techniques now used including nuclear magnetic resonance (NMR) (Cook et al., 1996; Sihombing et al., 1996; Cook and Langford, 1998; Lee et al., 1998), pyrolysis–mass spectrometry (Saiz-Jimenez et al., 1993; Van Loon et al., 1993; Schulten, 1996; Peuravuori and Pihlaja, 1997), x-ray absorption spectroscopy (Xia et al., 1997a, b), x-ray analysis near-edge spectroscopy (XANES), and extended x-ray absorption fine structure spectroscopy (EXAFS) (Davies et al., 1997; Morra et al., 1997). However, the amounts and types of different functional groups associated with fulvic acids vary significantly between solutions of different origin, potentially limiting the utility of an experimentally determined binding constant. The type of metal complexation process—very strong inner-sphere and polydentate (type 1a), strong inner-sphere and polydentate (type 1b), and weak outer-sphere and monodentate (type 2)—is evidenced by the magnitude of the conditional stability constants (Table 11.2).

The geochemical equilibrium speciation model MINTEQA2 (Allison et al., 1991; Allison and Brown, 1995; Serkiz et al., 1996) is now widely used to derive metal speciation in soil solutions. Calculations can be performed to take into account varied environmental conditions, such as pH, ionic strength, temperature, and redox conditions. The program includes an extensive database,with thermodynamic data for the soluble complexes, mineral solubilities, gas solubilities, and redox couples. The database includes over 900 dissolved species, including 13 species of trace metal complexes with dissolved organic matter (Allison and Brown, 1995). In this model, dissolved organic matter is treated as a complex material consisting of various types of monoprotic acid sites and was assumed to be normally distributed with respect to their log K values for protons and metals (Dobbs et al., 1989a,b). The database of the metal–DOM complexation constants is based on a study of metal interactions with Suwannee River fulvic acid using lanthanide ion probe spectroscopy (Susetyo et al., 1990, 1991).

Chemical speciation of two acidic stream waters from Scotland was calculated by Lumsdon and Evans (2002), using the MINTEQA2 model. An estimate of the DOM concentration was based on the titratable acidity of 7 mol kg^{-1}, which is typical of the humic substances found in Scottish surface waters (Anderson et al.,

TABLE 11.2. Conditional Association Constants (log ^cK) for the Formation of Metal–Fulvate Complexes

Ion	log ^cK			Reference
	Type 1a	Type 1b	Type 2	
Cu^{2+}	7.47	5.82	4.79	Luster et al. (1996)[a]
	7.34	5.78	4.08	Brown et al. (1999b)[b]
	8.7	—	4.2	Berbel et al. (2001)[c]
Pb^{2+}	6.6	—	—	Christl et al. (2001)[d]
	—	4.59	—	Estevas da Silva et al. (2002)[e]
	8.38	—	3.79	Berbel et al. (2001)[c]
Cd^{2+}	—	4.17	—	Estevas da Silva et al. (2002)[e]
	—	4.42	—	Brown et al. (1999a)[b]
	6.29	—	2.3	Berbel et al. (2001)[c]

Source: Essington (2004).
[a]Ion-exchange method and molecular fluorescence spectroscopy and assumed 1 : 1 metal/fulvate mole ratio in the complex.
[b]Ion-exchange method and assumed 1 : 1 metal/fulvate mole ratio in the complex.
[c]Potentiometric and voltammetric titrations and assumed 1 : 1 metal/fulvate mole ratio in the complex .
[d]Potentiometric titration and assumed 1 : 1 metal/fulvate mole ratio in the complex.
[e]Potentiometric and acid–base conductimetric titrations and assumed 1 : 1 metal/fulvate mole ratio in the complex.

1990). The input data and the results of the calculation are given in Table 11.3. High concentrations of free metals, especially Al, are toxic to aquatic life. In natural systems the effect of DOM can ameliorate the toxic effect of Al by forming metal–organic complexes. The results show that free Al for the Keltry site is about 49% of the total Al, whereas for the Rumster site the free Al is only 6%.

Krishnamurti and Naidu (2002, 2003) used the stability constants of the metal–DOM complexes, as reported by Stevenson and Fitch (1986), in the database of the MINTEQA2 model for speciating metals in soil solutions. The stability constants of metal–organic complexes obtained using the Scatchard plot approach has been the method of choice in most studies on metal binding by organic acids (Fitch and Stevenson, 1983, 1984). The structure of DOM in the soil interstitial water was assumed to resemble that of fulvic acid (FA), and an estimate of DOM concentration was arrived at using the structural model of FA from the Suwannee River (U.S. Geological Survey Staff, 1989). The soil solution speciation of Cu and Zn of a few typical soils of South Australia (Krishnamurti and Naidu, 2002), as calculated from the model, are presented in Table 11.4. The pH of the soil solutions varied widely between 4.55 and 8.41. The Cu–DOM and Zn–DOM complexes were found to be the most dominant species, averaging 97.1% and 68.5%, respectively, in the neutral and mildly acidic soils (pH < 7.0). It was earlier demonstrated that Cu was present almost exclusively in organically complexed form in soil solutions (Stevenson, 1991; Sauvé et al., 1997;

TABLE 11.3. Results of a MINTEQA2 Calculation of the Chemical Speciation of Soluble Metal Complexes in Two Acidic Stream Waters as a Percentage of the Total Component

		Kelty Water	Rumster
	pH:	4.26	4.23
	DOM[a]:	3 mg dm^{-3} (20.7 μmol dm^{-3})	30 mg dm^{-3} (200 μmol dm^{-3})
Component	Species		
Al	Al^{3+}	48.8	6.1
	$AlOH^{2+}$	4.0	0
	AlDOM	44.5	93.4
	$AlSO_4{}^+$	1.6	0
Cu	Cu^{2+}	63.7	9.3
	CuDOM	35.8	90.7
Zn	Zn^{2+}	96.9	71.6
	ZnDOM	2.2	27.9
Fe	Fe^{3+}	0	0
	$Fe(OH)_2{}^+$	60.4	8.2
	$FeOH^{2+}$	6.5	0
	FeDOM	33.0	91.3

Source: Lumsdon and Evans (2002).

[a]The units of DOM in mol dm^{-3} are based on DOC (mg dm^{-3}) and assume a titratable acidity of 7 mol kg^{-1} for the humic substance in surface waters (Anderson et al., 1990).

Van Riemsdijk et al., 1997) and that Zn–DOM complexes were quite significant in soil solutions (Almas et al., 2000). In mildly alkaline soils (pH > 8.0), the Cu $(OH)_2$ species were dominant, averaging 85.3%, while Zn–hydroxy species were significant, averaging 16.5% (Table 11.4). The $Cu(OH)_{2(aq)}$ and $Zn(OH)^+$ species dominate in soil solutions with pH > 7.0 (Sauvé et al., 1997).

The speciation of Cd in soil solutions obtained from soils incubated with 10 mg Cd kg^{-1} soil was conducted using the MINTEQA2 computer model (Krishnamurti and Naidu, 2003). The Cd–DOM complexes were found to be the most dominant species with small amounts (average < 5%) of free Cd^{2+} species (Table 11.4). Recent reports (Almas et al., 2000; Sauvé et al., 2000) have also shown the Cd–organic complexes to be the dominant Cd species in most soil solutions, using differential pulse anodic stripping voltammetry for measuring Cd^{2+} activity.

The study of metal speciation in the soil solution has been encouraged by the free metal ion hypothesis in environmental toxicology (Lund, 1990). This hypothesis states that the toxicity and bioavailability of a metal is related to the activity of the free aquo ion. Although this hypothesis is gaining popularity in studies of soil–plant relations (Parker et al., 1995a), some evidence is now emerging that the free metal ion hypothesis may not be valid in all situations (Tessier and Turner, 1995). Therefore, the role of metal–organic and metal–inorganic complexes in metal uptake merits attention.

TABLE 11.4. Results of a MINTEQA2 Calculation of the Speciation of Soluble Heavy Metal Complexes in Soil Solutions of Selected Soils of South Australia as a Percentage of the Total Component

		Soils with Acidic pH	Soils with Alkaline pH
	No. of soils studied:	7	4
	pH range:	4.55–6.35	8.07–8.41
	DOC range:	89–1629 mg L^{-1}	156–318 mg L^{-1}
	DOM[a] range:	0.31–5.657 mM	0.54–1.10 mM
Component	Species		
Cu	Free Cu^{2+}	2.8	0.2
	CuDOM	97.2	11.7
	CuL	0	88.0[b]
Zn	Free Zn^{2+}	21.8	28.8
	ZnDOM	76.8	54.9
	Zn L	1.3	17.2[b]
Cd	Free Cd^{2+}	5.0	3.4
	CdDOM	94.4	96.2
	Cd-L	0.5	0.3[b]

Source: Krishnamurti and Naidu (2002, 2003).
[a]Units of DOM are based on the structural model of fulvic acid from the Suwannee River (Stevenson, 1994).
[b]The ML species consist of metal–inorganic ligand complexes and M-OH species dominant (>90%) in ML species.

The validity of the predictions of solution speciation offered by geochemical models (GEOCHEM, MINTEQA2, etc.) is a function of three basic factors: the thoroughness of the solution chemical analysis, the robustness of the thermodynamic data file, and the critical nature of the thermodynamic data. The accuracy of the estimate of solution species concentration depends on the stability constants of the metal–ligand complexes (Turner, 1995; Krishnamurti and Naidu, 2003). It should also be noted that the model predictions are valid only for 25°C. Many researchers accept geochemical model predictions without any knowledge of the pitfalls. Utmost caution has to be exercised before giving too much importance to the results obtained for solution speciation using geochemical models.

11.2.2. Solution Speciation: Bioavailability

It is generally recognized that metal ion availability and uptake by plants are controlled by several factors, such as pH, ionic strength, redox potential, solution composition, ionic size, and valence (Frausto da Silva and Williams, 1976). It has been suggested that typically, strongly complexed metals are less toxic than are weakly complexed forms, which, in turn, are less toxic than free ions (Allen et al., 1980; Cabrera et al., 1988). In contrast, elevated Cd concentrations in potato tubers (McLaughlin et al., 1994) and sunflower kernels (Li et al., 1994) were correlated with high Cl concentrations in soils, caused by the salinity of irrigation water. Possible mechanisms offered to explain the synergistic effect of Cl^- on Cd uptake include (1) ion exchange of Na^+ or Ca^{2+} for Cd^{2+} displacing Cd^{2+} from the absorption site; (2) chloro complexation, increasing the diffusion of Cd through soil to plant roots; and (3) increasing Cd concentrations in soil solutions and direct uptake of Cd–Cl complexes by plants (Smolders and McLaughlin, 1996a,b). Norvell et al. (2000) also reported enhanced bioaccumulation of Cd in durum wheat by the presence of Cl^- in soil. On the contrary, Cabrera et al. (1988) reported that barley plants preferred Cd^{2+} to $CdCl^+$ for uptake, whereas Cd complexed with humic acid was not absorbed.

Therefore, elemental speciation is a major factor in controlling the availability and uptake of various nonessential elements by plants (Sposito and Bingham, 1981). Speciation of 10 metals and 13 ligands in saturation extracts of soils was computed by GEOCHEM, and the uptake of Cd by sweet corn grown on these soils subjected to known additions of Cd was studied (Sposito and Bingham, 1981). The results showed that Cd uptake by sweet corn was highly correlated with the concentration of $CdCl^+$ in soil solution and not with Cd^{2+} concentration in soil solution (Figure 11.1). The authors concluded that reduced charge on Cd through complex formation appeared to enhance Cd uptake.

Early studies of heavy metal toxicity were carried out using ionic metal additions to laboratory test organisms, with the conclusions based on acute toxicity tests. The mortality of test organisms was found to correlate strongly with ionic metal concentration. Free Cu ion was found to be toxic to the freshwater alga *Scendesmus quadricada* in the range 10^{-10} to 10^{-12} M (Petersen, 1982) and to the bacterium *Klebsiella aerogenes* in the range 10^{-6} to 10^{-10} M (Zevenhuisen

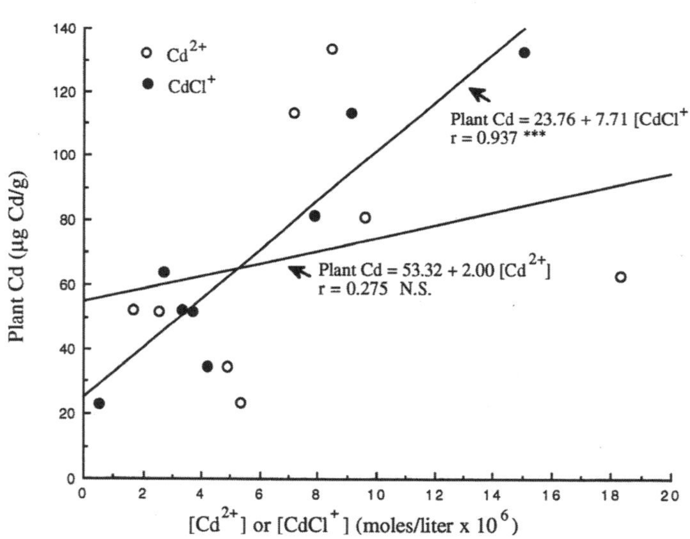

Figure 11.1. Cd uptake by corn shoots against molar concentrations of $CdCl^+$ and Cd^{2+} in saturation extracts of soils. (Redrawn from Sposito and Bingham, 1981; copyright © 1981, Taylor & Francis, Ltd., www.tandf.co.uk/journals.)

et al., 1979). Further, the metal toxicity was also shown to be pH dependent. It was proposed that hydrogen ion competes with free metal for cellular binding sites at low pH, thereby reducing metal toxicity. This theory was supported by the results on Cu toxicity to the green alga *S. quadricada* (Peterson et al., 1984) and on Zn toxicity to the alga *Chlamydomonas variabilis*.

With respect to potentially toxic ions, it appears that the free ion, M^{n+}, and monomeric hydrolysis species are the major toxic chemical forms. Complexing with both inorganic and organic ligands was shown to decrease toxicity (Chen and Stevenson, 1986). In the case of Al, complexes with SO_4^{2-}, F^-, PO_4^{3-}, and organic ligands were reported to be nontoxic to plants grown in nutrient solutions (Ritchie, 1989). The free ion, Cd^{2+}, appears to be the form that is taken up readily by plants, whereas $CdCl^+$ is taken up more slowly, and Cd–humate is not adsorbed (Bingham et al., 1984, 1986; Cabrera et al., 1988). Zinc initially complexed with citrate was reported to be taken up by barley from nutrient solutions (Chairidchai and Ritchie, 1993) and complexation with organic ligands has been found to increase Fe uptake (Chen and Aviad, 1990).

In many instances, when only ionic additions were made, conclusions regarding the toxic species have been based on calculated species distribution using known equilibrium data for metals with other dissolved ions in solution (Morrison et al., 1989). Guy and Kean (1980) showed that the five organic ligands with relatively high Cu stability constants were toxic at a concentration of 10^{-8} M Cu. The toxicity was higher for citric acid (10^{-10} M Cu) and ethylenediamine (10^{-9} M Cu), which have low stability constants with Cu. The organic ligands may transport Cu into the cell via the biological membrane (Luoma, 1983). Organic ligands of high molecular weight in pond water were found to reduce Cd toxicity to the crustacean *Simocephalus serrulatus* but not to the mosquito fish *Gambusia affinis* (Geisy et al., 1977). It is believed that complexation with humic substances reduces metal bioavailability and, in turn, potential toxicity. The results on toxicity of Cu to the crustacean *Daphnia pulex* (Winner, 1985) and of Cd to the alga *Selenastrum capricornutum* Printz (Sedlacek et al., 1983) demonstrate the observation. It is also reported that the root exudates, which contain low-molecular-mass organic (LMMO) acids (Denny and Wilkins, 1987; Cumming and Weinstein, 1990), and the rhizosphere bacteria exudates, which contain sideophores (Raymond et al., 1984), complex with heavy metals and make them unavailable. On the contrary, the bioavailability of Cd complexed with citrate and DOM (dissolved organic matter) to soil algae has been reported (Krishnamurti et al., 2004) (Tables 11.5 and 11.6). The Cd^{2+} ion concentration in soil solutions as determined by ASV compared reasonably well with that computed by the MINTEQA2 model assuming the stability constant of the Cd–DOM complex as 5.80, which suggests a polydentate binding mechanism that contains both inner- and outer-sphere character. Further, increased Cd bioavailability to crops on a sewage sludge–amended soil does not indicate immobilization of Cd under the influence of increased organic matter content (Tichy et al., 1997). On the other hand, it may indicate the availability of Cd–organic complexes formed in soil solution.

TABLE 11.5. Speciation of Cd[a] in Growth Medium With and Without Citrate, and Uptake of Cd by the Algae *Chlorococcum* sp.

Cd	Citric Acid		Cd Species in Solution (μg L^{-1})					Cd Uptake
(μg L^{-1})	(M)	pH	Cd^{2+}	$CdCl^+$	$CdSO_4(aq)$	$CdNO_3^+$	Cd–citrate	(μg L^{-1})
100	0	7.0	90.5	4.8	4.2	0.5	—	75.4
100	0.001	6.5	9.2	0.5	0.4	0.1	89.8	64.7

Source: Data from Krishnamurti et al. (2004).

[a] Speciation of Cd in solution was carried out using the MINTEQA2 computer program.

TABLE 11.6. Influence of Cd and Cd Species on Growth of the Algae *Chlorococcum* sp.[a]

Growth Medium				Soil Pore Water			
	Cd Species ($\mu g\ L^{-1}$)				Cd Species ($\mu g\ L^{-1}$)		
Total Cd ($\mu g\ L^{-1}$)	Cd^{2+}	Cd–DOM	Inhibition of Algal Growth (%)[b]	Total Cd ($\mu g\ L^{-1}$)	Cd^{2+}	Cd–DOM	Inhibition of Algal Growth (%)[b]
100	91	0	−25	115	5	109	11
413	372	2	50	302	31	269	50
500	452	2	54	479	50	427	79

Source: Data from Krishnamurti et al. (2004).

[a]Solution speciation was determined using the MINTEQA2 computer model with an updated database.

[b]Negative number indicates promotion of growth at low concentration of Cd in solution.

Complexation of Cu by dissolved organics has practical implications regarding the bioavailability of Cu in soil. The bioavailability of soluble forms of Cu depends most probably on both the molecular mass of Cu complexes and on the amounts present. Compounds of low molecular mass liberated during decay of plant and animal residues as well as those applied with sewage sludge may greatly enhance the availability of Cu to plants (Kabata-Pendias and Pendias, 2001).

The formation of metal complexes in solution can increase their uptake by increasing the concentration of metals in solution, which, in turn, increases the readily available source of metal or increases its diffusion to the root surface. The extent to which these two mechanisms are important will depend on the ability of soil to release the metal ion from solid-phase sources quickly. It appears that plants can take up metals that may initially have been complexed with ligands, be they organic or inorganic, but it is less clear whether the complex always dissociates at the root surface or the metal is adsorbed by the root as the complex. The heavy metal content of sewage sludge, which usually has been used to improve soil fertility and the resulting effect of the application of sewage sludge on plant metal uptake, is presented in Table 11.7. Tiffin (1977) found the ^{109}Cd concentration in tomato xylem exudates to be from 1.5 to 3.5 μM after treatment with this radionuclide, which suggested that Cd, as other heavy metals, can easily be transported within a plant in the form of metalloorganic complexes.

Chelation by LMMO substances is a major factor in the transport of micronutrient cations such as Cu, Zn, Ni, Fe, and Mn to root surface by mass flow and diffusion. In soil solutions of calcareous soils 40 to 75% of the zinc and 98% of copper have been found in organic complexes (Hodgson et al., 1966; Sanders, 1982, 1983; McGrath et al., 1988). For plants the importance of complexed micronutrients in the soil solution is particularly evident in calcareous soils. This is also indicated by the fact that soil extractions with synthetic chelators such as EDTA and DTPA provide suitable soil tests for estimation of available micronutrients (Sims and Johnson, 1991) (summarized in detail in Section 11.3).

There is often, however, a poor correlation between the concentrations of chelated micronutrients in the soil solution and their uptake by plants. This is

TABLE 11.7. Plant Uptake of Heavy Metals from Sewage Sludge

	Heavy Metal Content (mg kg^{-1})				Plant Uptake of Metals (mg kg^{-1})		
	Sewage Sludge						
		Median					
	Range	U.S.	Europe	Soil	Range	Mean	Maximum
Cu	84–17,000	800	1230	10–80	7–30	20	150
Zn	101–49,000	1700	2780	10–300	21–70	50	300
Cd	1–3410	15	31	0.01–0.7	0.05–0.2	0.1	3

Source: Baker (1990).

TABLE 11.8. Effect of Metal Chelator on the Availability of Micronutrients to Bean

	Content in Bean Leaves (μg g^{-1} dry weight)				
	Zn	Cu	Fe	Mn	Ni
Soil solution culture					
Control	23.4	7.6	124	108	2.0
+ 10^{-3} M DTPA	26.8	18.6	230	136	12.8

Source: Based on Wallace (1980a,b).

because the metal–organic complexes in the soil solution differ in both electric charge (Sims and Patrick, 1978) and size. In general, the uptake of metal cations from metal–organic complexes is lower than that of free cations (free ion activity model) and decreases with the size of the organic ligand, as has been demonstrated for Cu (Jarvis, 1987). In contrast to such findings from experiments in nutrient solutions, chelation of Cu and nickel in soil gives rise not to a decrease but an increase in their content in plants (Table 11.8).

11.3. FUNCTIONALLY DEFINED SPECIATION

Functionally defined species are exemplified by the plant-available species in which the "function" is plant availability. The category "plant-available species" may well include a number of different chemical forms or binding types, all contributing to the available forms assessed. Undoubtedly, selective extraction methods can assess the amount of mobile or potentially mobile species, which in turn may correlate well with plant-available contents in certain environmental or agricultural contexts. The use of chemical extraction for the assessment or prediction of agricultural crop plant content, plant growth, or plant health has been a major topic of study in agricultural laboratories around the world for almost five decades, although the term *speciation* was not used. There exists exhaustive information on soil extractants for these purposes for which plant contents and soil-extractable elements are correlated and the methods validated for various crops and soil types. The extractants developed are not universal reagents, but remain soil and crop specific. Bingham and Bartels (1996) presented a wide overview of trace element soil tests. Most methods are based on the establishment of a significant correlation between the amounts of extractable heavy metal ions and the metal content of plants. Although such comparisons reflect neither soil processes controlling metal phytoavailable fraction nor the mechanism of metal uptake by plants, strong correlations between the amounts of extractable metals and the metal contents of plants have led to acceptance of such methods as indices of metal phytoavailability.

The specific solid-phase fraction that is immediately available to a plant is dependent on the microenvironment conditions of the soil rhizosphere, which modifies the distribution of the heavy metal among various fractions. Broadly, the availability can be visualized as decreasing in the order exchangeable > metal–organic complex and organic-bound > adsorbed > Fe and Mn oxide-bound > silicate mineral lattice-bound.

11.3.1. Exchangeable Species

It is generally reported that the exchangeable species is the most mobile fraction of soil (Sparks, 1983) and hence is expected to be bioavailable. Use of electrolytes such as $CaCl_2$, KNO_3, and NH_4NO_3 as extractants promotes displacement of exchangeable metal ions, which are held by electrostatic attraction to negative sites on soil surfaces. Usually, 1 M solutions are employed as extractants, although more dilute solutions have also been preferred, since they more closely resemble the electrolyte concentrations that can occur in natural systems. Use of NH_4OAc was also advocated as an effective reagent for the extraction of exchangeable phase. However, acetate ion has a special complexing effect, particularly with heavy metals.

11.3.2. Metal–Organic Complexes

By virtue of their strong complexing ability, complexing extractants such as EDTA or DTPA, displace metals from insoluble organic and organometallic complexes. However, EDTA extracts more Cd and Zn than DTPA does (Tlustos et al., 1994). The DTPA reagent was designed to minimize dissolution of carbonate-occluded Cu, Fe, Mn, and Zn (Lindsay and Norvell, 1978). In predicting crop Cd concentrations, a DTPA extractant was found to be superior to Mehlich 3 extractant, which is favored in the southeastern United States (King and Hajjar, 1990), to 0.05 M $CaCl_2$, to M NH_4NO_3, and to 0.05 M Na_2–EDTA extractants (Jackson and Alloway, 1991). Table 11.9 lists some of the extractants used to assess plant-available element contents.

Although many of the procedures listed in Table 11.9 are used for contaminated soils, they will not necessarily measure plant-available contents but rather, the labile or mobile species contents. They can best be indicated as potentially plant-available contents rather than actual contents. The most commonly used extractants are diethylenetriaminepentaacetic acid–triethanolamine (DTPA–TEA), buffered at pH 7.3 (Lindsay and Norvell, 1978) and 0.05 M $CaCl_2$ (Sauerbeck and Styperek, 1984). Some authors also reported no relationship between extractable metals and test plant metal concentrations (e.g., Haq and Miller, 1972; Rappaport et al., 1988). O'Connor (1988) has subsequently identified an entire series of "misuses" of the DTPA test, which probably account for failure of the test,. Two major constraints with the DTPA extractant include the high pH (that may not typify soil pH) and chelation effect of the ligand ion. The chelate-based extractants tend to extract significantly higher amounts of trace elements and thus may not necessarily reflect the plant-available content in soils. To compensate for the high pH of the extractant, O'Connor (1988) suggested inclusion of pH as one of the variables in the correlation studies.

TABLE 11.9. Examples of Extractants Used to Assess Plant-Available Trace Element Contents

Target Species	Extractant	Elements	References
Exchangeable	M NH_4OAc, pH 7	Mo, Ni, Pb, Zn	John (1972), John et al. (1972), Sedberryand Reddy (1976), Haq et al. (1980), Soon and Bates (1982), Sterckeman et al. (1996)
	M NH_4NO_3	Cd, Ni	Symeonides and McRae (1977), DIN (1995), Merkel (1996)
	0.1 M NaNO3	Cd, Zn, Cu, Pb	Haeni and Gupta (1983)
Exchangeable + metal–chloride complex	0.01/0.05/0.1 M $CaCl_2$	Zn, Cd, Pb	Sauerbeck and Styperek (1984), Jackson and Alloway (1991), Smilde et al. (1992), Andrewes et al. (1996), Merkel (1996), McBride et al. (2003)
	M NH_4Cl	Cd	Krishnamurti et al. (1995b, 2000), Gray et al. (1999)
Exchangeable + organic and organometallic complex	M NH_4OAc + 0.01 M EDTA, pH 7	Cu, Mn, Zn	Sterckemen et al. (1996)
	0.5 M NH_4OAc + 0.02/0.05 M EDTA	Cu, Fe, Mn, Zn	Lakanen and Ervio (1971), Davis (1979), Sillenpaa (1982)
		Mo, Pb, Zn, Cd	Sauerbeck and Styperek (1984), Sillanpaa and Jansson (1992)
	0.005 M DTPA + 0.01–0.1 M $CaCl_2$	Cd, Cu, Fe, Mn, Zn, Ni	Street et al. (1977), Lindsay and Norvell (1978), Soltanpour and Schwab (1977), Davis (1979), Sillanpaa (1982), Soon and Bates (1982), Houba et al. (1990), Whitten and Ritchie (1991)

There is, nevertheless, concern among scientists regarding the objective of such data processing, which is rather empirical and does not reflect processes occurring at the soil–plant–root interface.

Comparison of many of the extractants in use indicated that the extractant 0.01 M or 0.05 M $CaCl_2$ provides a measure of mobile heavy metal fraction in soils. This may indicate potential availability of some heavy metals, particularly Cd, Zn, Mn, and Pb, in contaminated soils (Hornburg et al., 1995; Andrewes et al., 1996).

11.3.3. Phytoavailability: Native Soils

Phytoavailability of a metal ion varies with the particular metal, soil properties, and plant species. It is difficult to assess the value of the large number of studies that have reported metal extractants and plant availability of metals on different soils with or without the application of contaminants such as sewage sludges. Comparisons between studies can be virtually impossible because the duration of treatments prior to extraction of the metals was often widely different. Moreover, few studies report metal availability tests, determined using chemical extractions, for native species under real field conditions (e.g., Gough et al., 1980; Krishnamurti et al., 1995a).

Reviewing the bioavailability of heavy metals, Naidu et al. (1997) concluded that soil extractants are often used with little consideration of their mode of action and the chemistry of heavy metals. For instance, 0.01 M or 0.05 M $CaCl_2$ could never be expected to predict bioavailable Cu given that Cu is largely present in soil associated with soluble organic matter. Where researchers have used this extractant, no correlation was found between the extractable Cu and Cu content of plants (e.g., Lee and Zheng, 1994). Similarly, the use of DTPA for acid soils has yielded no correlation with plant metal contents. This is not surprising given that this extractant was originally developed for calcareous soils. There are few guidelines for assessment of the risk of contamination of the food chain with toxic metals, but the best extraction reagents for indicating their soil-to-plant transfer appear to be weak, unbuffered salt solutions (Lebourg et al., 1996).

Few studies report metal availability tests, determined using chemical extractions, for native species under real field conditions (e.g., Gough et al., 1980; Krishnamurti et al., 1995b). In recent studies, a number of commonly used extractants were compared for assessing the plant-available Cd from soil extractions (Krishnamurti et al., 1995b, 2000; Gray et al., 1999). Based on published data on native and uncontaminated soils, it appears that soil extraction with M NH_4Cl provides the best overall prediction of plant-available Cd (Table 11.10).

Use of ion-exchange membranes for the selective extraction of heavy metals from soil suspensions and soil solutions have been developed, especially for Cd, Cr, Ni, and Pb. The extractable metals had better correlation with wheat seedling uptake than chemical extraction with 0.01 M $CaCl_2$ or 0.005 M DTPA (Lee and Zheng, 1994; Liang and Schoenau, 1996).

TABLE 11.10. Concentrations of Cd Extracted from Soils by Various Procedures and Their Correlation with Plant-Available Cd

Extractant	Cd Extracted (μg Cd/kg soil)		Simple Correlation Analysis[a]	
	Range	Mean	r^2	P
Soils of South Australia (Krishnamurti et al., 2000)				
0.01 M $CaCl_2$	10.0–45.0	23.7	0.051	5×10^{-1}
M NH_4NO_3	1.0–34.0	12.1	0.706	1×10^{-3}
0.05 M $CaCl_2$	21.0–103.0	46.3	0.581	6×10^{-3}
M NH_4Cl	22.0–162.0	75.8	**0.862**	9×10^{-5}
0.1 M EDTA	24.0–171.0	102.2	0.433	3×10^{-2}
Soils of Saskatchewan, Canada (Krishnamurti et al., 1995b)				
M NH_4OAc	3.1–74.0	23.9	0.684	3×10^{-3}
ABDTPA	36.0–182.0	105.1	0.594	5×10^{-3}
M NH_4NO_3	0.2–15.9	3.4	0.320	2×10^{-2}
0.05 M $CaCl_2$	0.1–50.3	7.3	0.844	2×10^{-3}
M NH_4Cl	6.3–191.2	60.3	**0.884**	6×10^{-5}
Soils of New Zealand (Gray et al., 1999)				
M NH_4OAc		55.7	0.52	5×10^{-2}
AAAc-EDTA		174.4	0.52	5×10^{-2}
M NH_4NO_3		18.9	0.54	3×10^{-2}
0.05 M $CaCl_2$		58.2	0.51	5×10^{-2}
M NH_4Cl		121.9	**0.59**	1×10^{-2}

[a]Simple correlation between the extractable Cd and Cd concentration in plant shoots and leaves.

11.4. OPERATIONALLY DEFINED SPECIATION

In operationally defined speciation the physical or chemical fractionation procedure applied to the sample defines the fraction isolated for measurement. Selective extraction procedures are used to isolate fractions of elements, such as *exchangeable, reducible, oxidizable*, and *residual*. These fractions are often equated with the metals associated, bound or adsorbed in the exchangeable sites, in the iron–manganese oxyhydroxides, organic matter/sulfides, and silicate phases, respectively. Although this is often a convenient concept, it should be emphasized that the reagents used may not be that selective to regard the fractions as defined by the operational procedures.

The present state of knowledge of solid matter speciation of trace elements is still somewhat unsatisfactory because the appropriate techniques are only operational tools with associated conceptual and practical problems. With respect to estimating bioavailable concentrations, one such conceptual problem is the effect of competition between binding sites on the solid substrate and selective mechanisms of metal translocation by the different organisms involved, a situation that cannot be improved by more sophisticated speciation approaches (Kersten and Förstner, 1989). On the other hand, the usefulness of a differentiated approach to the interactive processes between water/biota and, even operationally defined, solid phase has been evidenced clearly.

11.4.1. Sequential Extraction Schemes

Most of the instrumental techniques available are too sophisticated to be included routinely in trace element speciation studies, although rapid progress is being achieved with analytical proton-induced x-ray emission (PIXE) spectroscopy and surface analysis techniques such as Auger electron spectroscopy (AES) and secondary-ion mass spectrometry (SIMS). The method of sequential extraction is the least sophisticated and most convenient technique available for solid-state speciation assessment. The number of steps required in the extraction scheme depends on the purpose of the study. The schemes of Engler et al. (1977) and Tessier et al. (1979) have been proved adequate for the specific tasks undertaken, ranging from sewage sludge and sludge amendments (e.g., Patsakis et al., 1998; Walter and Cuevas, 1999), to waste material (e.g., Bodog et al., 1996; Flyhammer, 1997), and even to street dust (e.g., Stone and Marsalek, 1996; Lee et al., 1997). A compilation of more than 20 different schemes developed subsequently, with details on experimental parameters, can be had from Kersten and Förstner (1995), and an exhaustive treatment of this technique can be had from Sheppard and Stevenson (1997) and Kersten (2002). A few typical sequential extraction schemes are presented in Tables 11.11 and 11.12. The absence of standardized conditions makes it difficult to compare data obtained from studies in which experimental conditions are significantly different.

A more sophisticated sequential extraction scheme, however, significantly improves the specificity and efficiency of extraction, by a carefully designed combination of various extractants. Not many extraction schemes have used the results obtained from such studies to identify the specific species contributing to bioavailability. The first concerted attempt in this direction was reported by Krishnamurti et al. (1995a). The differentiation of the metal/organic complex–bound metal species, as distinct from the other organically bound species, was the innovation in the selective sequential extraction scheme suggested by Krishnamurti et al. (1995a). The particulate-bound metal species in soils were fractionated as exchangeable carbonates, metal–organic complexes, easily reducible metal oxides, organic material, amorphous mineral colloids, crystalline Fe oxides, and residual. The metal–organic complexes were extracted selectively using 0.1 M sodium pyrophosphate (pH 10) as the extractant in the sequential extraction scheme. Extraction with 0.1 M sodium or potassium pyrophosphate was in use in the extraction schemes proposed earlier (MacLaren and Crawford, 1973; Schalscha et al., 1982; Miller and McFee, 1983) for delineating organically bound trace metals, although not specifically for extracting metal–organic complexes. Manganese oxides may also be partially extracted at pH 10 by this reagent. An improved method for the selective dissolution of manganese oxides involving

TABLE 11.11. Selected Sequential Extraction Schemes Commonly Used for the Fractionation of Particulate-Bound Metal Species (Post-1979)[a]

Species	Förstner et al. (1981)	Shuman (1985)	Krishnamurti et al. (1995a)	Ma and Uren (1998)
Exchangeable	(I) M NH_4OAc, pH 7	(I) M $Mg(NO_3)_2$, pH 7	(I) M $Mg(NO_3)_2$, pH 7	(I) M $MgCl_2$, pH 7
Specifically adsorbed	—	—	—	(II) 1% NaCaHEDTA in M NH_4OAc, pH 8.3
Carbonates	(II) M NaOAc, pH 5	—	(II) M NaOAc, pH 5	(IV) 0.5 M NaOAc, pH 4.74
Easily reducible metals	(III) 0.1 M $NH_2OH{\cdot}HCl$ in 0.01 M HNO_3	(II) 0.1 M $NH_2OH{\cdot}HCl$ in 0.01 M HNO_3	(IV) 0.1 M $NH_2OH{\cdot}HCl$ in 0.01 M HNO_3	(III) 0.2% quinol in M NH_4OAc, pH 7
Oxidizable oxides and sulfides	(V) 30% H_2O_2, pH 2, 85°C; extract with M NH_4OAc	(III) 0.7 M NaOCl, pH 8.5, 85°C	(V) 30% H_2O_2, pH 2, 85°C extract with M $Mg(NO_3)_2$ in 20% HNO_3	(V) 30% H_2O_2, pH 4.74, and extract with 0.5 M NaOAc, pH 4.74
Moderately reducible Fe, Al oxides	(IV) 0.2 M NH_4Ox/HOx, pH 3	—	—	(VI) 0.175 M NH_4Ox/HOx, pH 3.25 (daylight)
Amorphous oxides	—	(IV) 0.2 M NH_4Ox/HOx, pH 3 (dark)	(VI) 0.2 M NH_4Ox/HOx, pH 3 (dark)	—
Crystalline Fe oxides	—	(V) 0.2 M NH_4Ox/HOx, pH 3 in 0.1 M ascorbic acid, 95°C	(VII) 0.2 M NH_4Ox/HOx, pH 3 in 0.1 M ascorbic acid; 95°C	—
Metal–organic complexes	—	—	(III) 0.1 M $Na_4P_2O_7$ (pH 10) (VIII)	—
Residual	(VI) HNO_3	(VI) HF–$HClO_4$ digestion	(VIII) HF–$HClO_4$ digestion	(VII) (total– sum of the extractable)

[a]Roman numerals in parentheses indicate various stages in the sequential extraction scheme.

TABLE 11.12. Selected Sequential Extraction Schemes Commonly Used for the Fractionation of Particulate-Bound Metal Species (Pre-1979)[a]

Species	Gibbs (1973)	Gupta and Chen (1975)	Stover et al. (1976)	Engler et al. (1977)	Tessier et al. (1979)
Exchangeable	(I) M $MgCl_2$	(I) M NH_4OAc	(I) M KNO_3	(I) M NH_4OAc	(I) M $MgCl_2$, pH 7
Specifically adsorbed	—	(II) 0.5 M KF, pH 6.5	—	—	—
Carbonates	—	—	—	—	(II) M NaOAc, pH 5
Easily reducible	—	—	—	(II) $NH_2OH{\cdot}HCl$, pH 2	
Organic	(III) NaOCl, pH 8.5	(III) 0.1 M $Na_4P_2O_7$	(II) 0.1 M $Na_4P_2O_7$	(III) H_2O_2/NH_4OAc	(IV) 3 mL H_2O_2, pH 2, 85°C; extract with 3.2 M NH_4OAc in 20% HNO_3
Fe and Mn oxides	(II) $Na_2S_2O_4$ + Na citrate	(IV) 0.1 M EDTA, pH 6.5	(III) 0.1 M EDTA, pH 6.5	(IV) DCB	(III) $NH_2OH{\cdot}HCl$ in 25% HOAc, 95°C
Residual	(IV) $LiBO_2$, 1000°C	(V) HNO_3	(IV) M HNO_3	(V) HF/HNO_3	(V) HF–$HClO_4$ digestion

[a] Roman numerals in parentheses indicate various stages in the sequential extraction schemes.

nonacidified 0.1 mol L^{-1} hydroxylamine solution at pH 3.6 as an extractant has been proposed (Neaman et al., 2004). However, inclusion of this extractant in selective extraction schemes for delineating trace element species has yet to be explored.

Krishnamurti et al. (1995a) determined the distribution of particulate-bound Cd species in a few typical soils of southern Saskatchewan, Canada, following the schemes of Tessier et al. (1979) and the modified scheme. They found that Cd in these soils was bound predominantly to metal–organic complexes, accounting for on an average of 42% of the total Cd present in the soils (Figure 11.2*b*), whereas on average 43% of Cd in the soils was observed to be in Fe and Mn oxide-bound form following the fractionation scheme of Tessier et al. (1979) (Figure 11.2*a*). The Cd in the surface horizons of the temperate soils (Krishnamurti et al., 1995a) is present predominantly in metal–organic complex-bound form accounting for 31 to 55%, with an average of 42% of the total Cd present in the soils. The metal–organic complex-bound Cd is also generally highest among the particulate-bound Cd species of the tropical soils (Onyatta and Huang, 1999), accounting for 25 to 46%, with an average of 37% of the total Cd in the soils. However, in

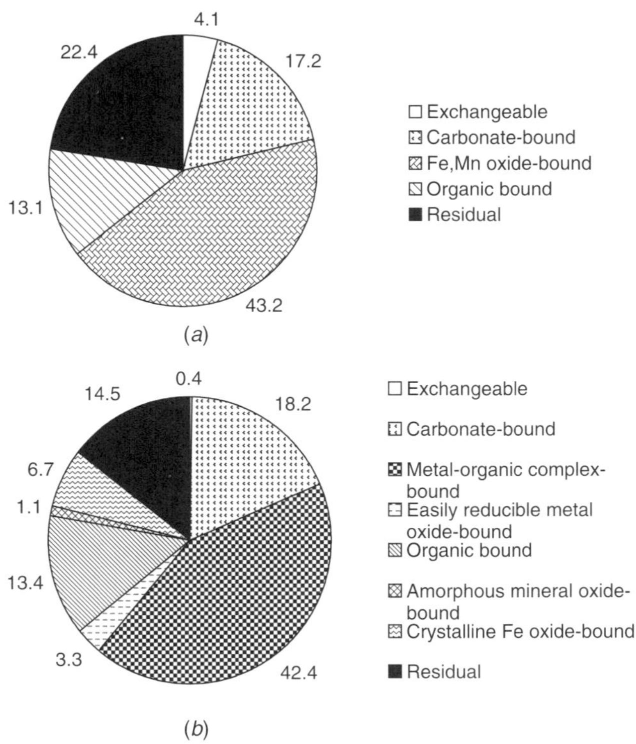

Figure 11.2. Average percent distribution of particulate-bound Cd fractions in selected soils of Saskatchewan, Canada, following the method of (*a*) Tessier et al. (1979) and (*b*) Krishnamurti et al. (1995a). (Redrawn from Krishnamurti et al., 1995a.)

certain tropical soils, residual or crystalline Fe oxide-bound Cd is predominant (Onyatta and Huang, 1999).

Compared with bulk soils, solid-phase speciation of Cd differs substantially in phosphate fertilizer-treated rhizosphere soils (Krishnamurti et al., 1996). The amounts of carbonate-bound Cd and metal–organic complex-bound Cd species of the rhizosphere soils at the 2-week plant growth stage, particularly in soils treated with Idaho phosphate fertilizer, are appreciably higher than those of the corresponding bulk soils. Compared to the corresponding bulk soils, the amount of carbonate-bound Cd species of the rhizosphere soils increased by 15 to 18% in the control soils and by 79 to 92% in soils treated with Idaho phosphate fertilizer, whereas the metal–organic complex-bound Cd species increased by 4 to 7% in the control soils and two- to threefold in soils treated with Idaho phosphate fertilizer. The increase in carbonate-bound Cd species in the rhizosphere soils is attributed to the increased amounts of carbonate, a product of plant respiration, present at the soil–root interface. Xian and Shokohifard (1989) suggested that exudation of H_2CO_3 by roots may help to solubilize metal carbonates and make them more bioavailable. The prolific root growth of the plants, particularly in soils treated with Idaho phosphate fertilizer, might have resulted in the secretion of increased amounts of low-molecular-mass organic acids (LMMOAs) into the rhizosphere, which resulted in high amounts of metal–organic complex-bound Cd species by chelation. Appreciable amounts of LMMOAs were detected in the root exudates of durum wheat, with the actual amount depending on the cultivar (Cieśliński et al., 1998; Szmigielska et al., 1995). The sustaining Cd release from the soils by LMMOAs was shown to have the same trend as that of Cd accumulated in the plant (Krishnamurti et al., 1997a).

11.4.2. Sequential Extraction and Bioavailability

An attempt to identify the specific species of Cd important to assess its phytoavailability was made by Krishnamurti et al. (1995a). The authors have carried out multiple regression analysis between the cadmium availability index (CAI), as measured by ABDTPA-extractable Cd (Soltanpour and Schwab, 1977), and different forms of particulate-bound Cd. The importance of metal–organic complex-bound Cd species in the bioavailability of Cd and the nature of bonding of Cd sites was also worked out in detail using multiple regression analysis and differential FTIR analysis (Krishnamurti et al., 1995a, 1997b). Based on the differential FTIR spectra of the metal–organic complexes extracted by the 0.1 M sodium pyrophosphate extractant used in the speciation scheme, Krishnamurti et al. (1997b) had shown that Cd in the soils was apparently bonded at the COO– of carboxyl and the OH of the phenolic groups and of the Fe, Al, and Mn in the metal–organic complexes Wang et al. (1998) also made an attempt to bring together results of sequential extraction and biotoxicity testing in order to produce information on bioavailability of trace elements.

11.4.2.1. Novel Sequential Extraction Scheme Krishnamurti and Naidu (2000) modified the sequential extraction scheme developed by Krishnamurti et al.

(1995a) by subfractionating the trace element bound to metal–organic complexes as extracted by 0.1 M sodium pyrophosphate. This fraction may contain metal associated with both humic and fulvic acid fractions of soil organic matter, and was fractionated following the method suggested by Schnitzer and Schuppli (1989) as metal–fulvate complexes and metal–humate complexes. The metal solubilized by acid (pH 1.0) in the pyrophosphate extracts is termed *metal–fulvate complex*, and the metal bound strongly to humic acid that resist desorption at pH 1.0 is termed *metal–humate complex*. They have shown that the information obtained from the novel sequential extraction scheme, consisting of nine extraction steps, fractionated the solid components into specific "species" with operationally defined binding mechanisms, such as exchangeable, specifically adsorbed/carbonate-bound, metal–fulvic complex-bound, metal–humic complex-bound, easily reducible metal oxide-bound, organic site-bound, amorphous metal oxide-bound, crystalline Fe oxide-bound, and residual (aluminosilicate lattice-bound). They showed that the extraction scheme could be used to identify the specific species contributing to the bioavailability of the trace element. Details of the extraction scheme are presented in Table 11.13.

The distribution of solid-phase fractions of Cu, Zn, and Cd of a few typical surface soils of South Australia, carried out following the fractionation scheme, is presented in Figures 11.3 and 11.4. The trace elements in these soils are dominantly (on an average 40% of Cu, 52.4% of Zn, and 33.4% of Cd) associated with the alumnosilicate mineral lattices, identified as residual fraction in the scheme, followed by the fraction associated with organic sites (on average, 32.4% of Cu, 28.0% of Zn, and 28.5% of Cd)

The mobility and phytoavailability of trace elements are mainly through the solution phase. However, the plant uptake of an element depends not only on its activity in the solution, but also on the relation that exists between solution ions and solid-phase ions. The transfer of an element between soil and plant does not necessarily go through the solution phase (Kabata Pendias and Pendias, 2001). Trace elements might be phytoavailable directly from the solid-phase, possibly through a diffusion process.

An attempt was made by Krishnamurti and Naidu (2000, 2002) to understand the importance of solid-phase fractions in assessing the phytoavailability of Cu, Zn, and Cd using multiple regression analysis (Table 11.14). Phytoavailable Cu, Zn, and Cd were found to be correlated significantly with fulvic complex Cu ($r = 0.944$, $p < 0.0001$), exchangeable Zn ($r = 0.832$, $p = 0.002$), and fulvic complex Cd ($r = 0.824$, $p = 0.002$). It is observed that fulvic complex Cu could explain 89% of variation in phytoavailable Cu, whereas fuvic complex element together with exchangeable element could explain 79% and 92% of variation in phytoavailable Zn and Cd, respectively. Inclusion of solution element concentration in the regression analysis was not found to improve the predictability of phytoavailable element. The regression analysis indicated that the phytoavailability of Cu, Zn, and Cd in these soils is mainly through solid-phase fractions.

TABLE 11.13. Multistep Selective Sequential Extraction Scheme for Fractionation of Solid Metal Phases

Step	Target Species	Reagent	Shaking Time and Temperature
1	Exchangeable	10 mL of M NH_4NO_3 (pH 7)	4 h at 25°C
2	Specifically adsorbed	25 mL of M CH_3COONa (pH 5)	6 h at 25°C
3	Metal–organic complex-bound[a]	30 mL of 0.1 M $Na_4P_2O_7$ (pH 10)	20 h at 25°C
4	Easily reducible metal oxide-bound	20 mL of 0.1 M $NH_2OH{\cdot}HCl$ in 0.01 M HNO_3	30 min at 25°C
5	Organic-bound	5 mL of 30% H_2O_2 (pH 2), 3 mL of 0.02 M HNO_3	2 h at 85°C
		3 mL of 30% H_2O_2 (pH 2), 1 mL of 0.02 M HNO_3	2 h at 85°C
		Cool, add 10 mL 2 M NH_4NO_3 in 20% HNO_3	30 min at 25°C
6	Amorphous mineral colloid-bound	10 mL of 0.2 M $(NH_4)_2C_2O_4$/0.2 M $H_2C_2O_4$ (pH 3)	4 h at 25°C (dark)
7	Crystalline Fe oxide-bound	25 mL of 0.2 M $(NH_4)_2C_2O_4$/0.2 M $H_2C_2O_4$ (pH 3) in 0.1 M ascorbic acid	30 min at 95°C
8	Residual	Digestion with HF–$HClO_4$	

Source: Krishnamurti and Naidu (2000).

[a] 30 mL of 0.1 M $Na_4P_2O_7$ extract was brought to pH 1.0 with the addition of 6 M HCl, and the suspension was left overnight for the coagulation of humic acid. The suspension was centrifuged at 12,000*g* for 10 min. Metal–fulvate complexes were determined in the supernatant. The residue was solubilized with 0.1 M $Na_4P_2O_7$, and the metal–humate complexes were determined in the solution.

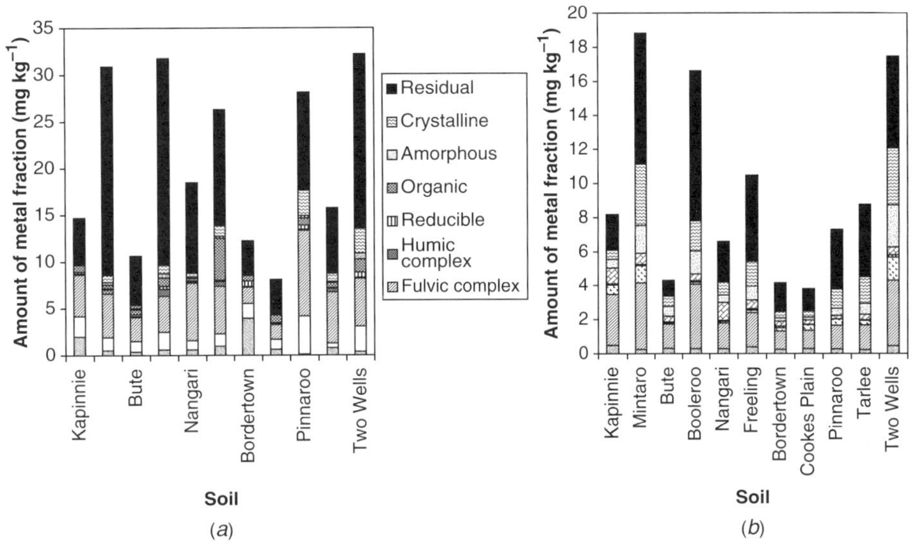

Figure 11.3. Distribution of solid-phase (*a*) zinc and (*b*) copper fractions of selected soils of South Australia using the sequential selective extraction scheme of Krishnamurti and Naidu, 2000; (Reprinted with permission from Krishnamurti and Naidu (2002); copyright © 2002 American Chemical Society.)

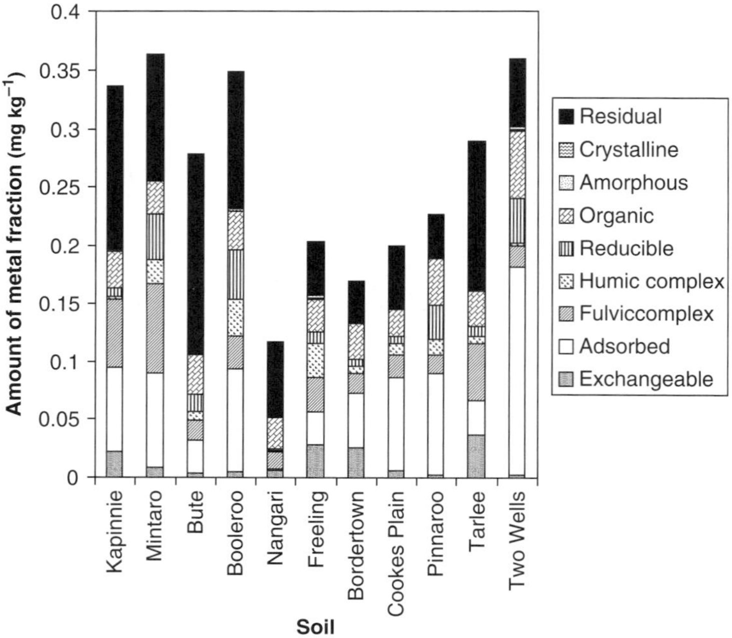

Figure 11.4. Distribution of solid-phase Cd fractions of selected soils of South Australia using the sequential selective extraction scheme of Krishnamurti and Naidu (2000). (Reprinted with permission from Krishnamurti and Naidu, 2000; copyright © 2000 CSIRO, Australia.)

However, more research is needed to achieve an understanding of the mechanisms involved in the mobility and bioavailability at the soil–root interface and to develop rational predictive models for assessing bioavailability.

11.4.3. Nondestructive Analysis

The experimental detection and quantification of surface species on in situ soil particles and other natural colloids is a difficult area of research because of the sample heterogeneity, low surface concentrations, and the necessity to investigate the solid adsorbents in the presence of water. Unambiguous information can be obtained only with in situ surface spectroscopy, such as x-ray photoelectron (XPS), extended x-ray absorption fine structure (EXFAS), x-ray absorption near-edge structure (XANES), inelastic electron tunneling (IETS), and electron energy loss (EELS) spectroscopies. Recent advances in the development of nonevasive, in situ spectroscopic scanned-probe and microscopic techniques have been applied successfully to study mineral particles in aqueous suspensions (Hawthorne, 1988; Hochella and White, 1990).

TABLE 11.14. Data on the Multiple Regression Analysis Between[a] Phytoavailable Metal, Soil Solution Metal Concentration (M_S), and Solid-Phase Metal Fractions[b]

Simple correlation factors

	Fulvic complex	Exchangeable	M_s
Phytoavailable Cu	0.944 (<0.0001)	0.424 (ns)	−0.029 (ns)
Phytoavailable Zn	− 0.669 (0.02)	0.832 (0.002)	0.753 (0.008)
Phytoavailable Cd[†]	0.824 (0.002)	0.735 (0.01)	

Multiple regression analysis

Phytoavailable Cu	= 1.79 ± 1.48 + 5.11 ± 0.59 fulvic Cu (ns) (< 0.0001)	[Eq. 1]	$r^2 = 0.89$ (< 0.0001)
	= 1.02 ± 2.23 + 5.13 ± 0.62 fulvic Cu + 3.89 ± 8.14 Cu_s (ns) (< 0.0001) (ns)		R^2 =0.895 (0.0001)
Phytoavailable Zn	= 20.62 ± 3.54 + 11.04 ± 2.45 exch. Zn (0.003) (0.002)		r^2 =0.69 (0.002)
	= 12.73 ± 5.80 + 7.87 ± 2.96 exch. Zn + 39.33 ± 23.94 Zn_s (0.06) (0.03) (ns)		$R^2 = 0.77$ (0.003)
	= 33.58 ± 7.43 + 8.81 ± 2.45 exch. Zn − 2.37 ± 1.23 fulvic Zn (0.002) (0.007) (0.09)	[Eq. 2]	$R^2 = 0.79$ (0.002)
	= 32.33 ± 6.72 + 8.06 ± 2.25 exch. Zn − 2.43 ± 1.11 fulvic Zn + 26.19 ± 15.52 Zn_s (0.002) (0.009) (0.06) (ns)		$R^2 = 0.85$ (0.003)
Phytoavailable Cd	= 0.03 ± 0.03 + 3.36 ± 0.77 fulvic Cd (ns) (0.002)		$r^2 = 0.68$ (0.002)
	= 0.00 ± 0.02 + 2.65 ± 0.45 fulvic Cd + 3.57 ± 0.76 exch. Cd (ns) (0.0003) (0.002)	[Eq. 3]	$R^2 = 0.92$ (0.0001)

Source: Data from Krishnamurti and Naidu (2002); copyright © 2002 American Chemical Society.

[a]Metal concentration in shoots and leaves of durum wheat plant harvested after 5 weeks of plant growth.

[b]The values in parentheses are the P (level of significance) values.

Among the various spectroscopic methods, the most notable one is synchrotron-based x-ray technique that is revolutionizing soil and environmental chemistry research (Manceau et al., 1992). The most utilized synchrotron-based x-ray technique in soil and environmental chemistry to date has been x-ray absorption spectroscopy. True solid-state speciation can be accomplished only by spectroscopic methods [e.g., using near-edge structure of electron energy loss (ELNES: van Aken et al., 1998) or of x-ray absorption (XAFS: Manceau et al., 1992)], The latter technique has recently been shown to be particularly useful in studies of trace metal speciation in contaminated soils (Manceau et al., 1996; Scheidegger and Sparks, 1996; Peterson et al., 1997) or of early diagenetic reactions in sediments (Helz et al., 1996), even in situ in a sediment core with a spatial resolution at the millimeter level (Tokunaga et al., 1998). The extended x-ray absorption fine structure spectroscopy (EXAFS) provides specific information on the local environment of the absorber, including coordination number, identity, and distances to nearest and sometimes to next-nearest neighboring atoms (Fendorf et al., 1994; Schulze and Bertsch, 1995; Fendorf and Sparks, 1996; Bertsch and Hunter, 1998; Sparks, 2000). However, sampling techniques required for these methods often annihilate or change irreversibly the surface species of interest. Molecular-level information about the mechanisms, orientation, or dynamics of surface species inferred from data obtained by these methods may bear little resemblance to the chemical mechanisms operating in a natural soil–water system (Ritchie and Sposito, 1995). Therefore, improvement of sampling techniques merits close attention in uncovering the in situ metal species at the molecular level.

11.5. CONCLUSIONS AND FUTURE DIRECTIONS

Speciation and bioavailability form the basis of any risk assessment and risk-based management process. The lack of appropriate techniques to estimate these parameters has led to significant difficulty in the setting up of policies on contaminated site management and remediation in many countries. Altough there has been significant research on speciation and bioavailability, research regulators are demanding scientific basis for ion toxicity and bioavailability, including the need for standardized techniques for speciation and in situ bioavailability assessment. Although geochemical models and a selected few wet chemical techniques provide some insight into the nature of ions present in aqueous systems, such as the riverine environment, there is still a great dearth of information on the nature of ionic species, their toxicity, and their availability in soil solutions, which is the medium of soil reactions. The free ion activity model (FIAM) is utilized uniformly in understanding bioavailability as related to the aqueous metal ion species. However, recent reports suggest the importance of metal–organic species in assessing bioavailability (e.g., Krishnamurti et al., 2004). Detailed research is needed to understand the role of metal–organic species, which are accepted as dominant aqueous ionic species in soil solution, and in assessing bioavailability.

Many chemical models have been developed over the years for problems solving chemical speciation. A comprehensive survey of many of the models and detailed analysis of computer modeling in relation to trace element speciation has been by made by Waite (1989), with particular reference to MINEQL, and by Lumsdon and Evans (2002), with particular reference to MITEQA2. Soil solution has significant amounts of DOM, which is poorly quantified and understood. The trace element speciation information obtained using computer modeling with data on complexation constants of metal–DOM complexes has to be substantiated with data on free metal ion species obtained from sophisticated physical techniques.

The lack of information on speciation and bioavailability in the terrestrial environment arises from the lack of both appropriate techniques for in situ chemical speciation and bioavailability in the field and analytical tools for speciation or fractionation of metal species. Many researchers continue to compare published data without paying much attention to the technique used during speciation. There is therefore a need to standardize speciation and techniques for assessing bioavailability that as a minimum all researchers are using the same technique. The Community Bureau of Reference (BCR) of the Commission of European Communities suggested a simple three-stage extraction procedure (Ure et al., 1993) to improve the comparability between sequential extraction techniques. This will allow comparison of data between labs and also among researchers. However, caution has to be exercised in attributing the importance of extracted ionic species and fractions to bioavailability in as much as the species identified by sequential extraction techniques do not per se identify the targeted species exclusively. Utilizing a novel sequential extraction technique, the importance of metal–fulvic species in assessing phytoavailability of Cu, Cd, and Zn has been demonstrated in selected soils (Krishnamurti and Naidu, 2000, 2002). Results reported look promising and need assessment as applicable to soils developed on diverse parent materials under varying climatic environments.

The mechanism of metal and metalloid sorption in soils was postulated some 30 years ago. The interaction of arsenic in soils and sediments at molecular level has been studied using extended x-ray absorption fine structure (EXFAS) and x-ray absorption near-edge structure (XANES) spectroscopies. Much work remains to be conducted to investigate the nature of metal and metalloid species sorbed at varying pH values and ionic strength in highly contaminated sites (see Chapter 5).

REFERENCES

Adriano, D. C. (2001). *Trace Elements in Terrestrial Environments: Biogeochemistry, Bioavailability and Risks of Metals*, 2nd ed., Springer-Verlag, New York.

Alberts, J. J., and Giesy, J. P. (1983). Conditional stability constants of trace metals and naturally occurring humic materials: their application in equilibrium models and verification with field data. In *Aquatic and Terrestrial Humic Material*, ed. Christman, R. F., and Gjessing, E. T., Ann Arbor Science, Ann Arbor, MI, 333–348.

Allen, H. E., Hall, R. H., and Brisbin, T. D. (1980). Metal speciation: effects on aquatic toxicity. *Environ. Sci. Technol.*, **14**, 441–443.

Allison, J. D., and Brown, D. S. (1995). MINTEQA2/PRODEFA2: a geochemical speciation model and interactive preprocessor. In *Chemical Equilibrium and Reaction Models*, ed. Loeppert, R. H., Schwab, A. P., and Goldberg, S., Soil Science Society of America, Madison, WI, 241–252.

Allison, J. D., Brown, D. S., and Novo-Gradac, K. J. (1991). *MINTEQA2/PRODEFA2: A Geochemical Assessment Model for Environmental Systems,* EPA/600/3-91/021, U.S. Environmental Protection Agency, Athens, GA.

Almas, A. R., McBride, M. B., and Singh, B. R. (2000). Solubility and lability of cadmium and zinc in two soils treated with organic matter. *Soil Sci.* **165**, 250–259.

Anderson, H. A., Hepburn, A., Miller, J. D., Stewart, M., Ferrier, R. C., and Walker, T. A. B. (1990). Humic substances of surface waters: litter and soil through flow relationships in two forested ecosystems. *Anal. Chim. Acta* **232**, 3–10.

Andrewes, P., Town, R. M., Hedley, M. J., and Loganathan, P. (1996). Measurement of plant available cadmium in New Zealand soils. *Aust. J. Soil Res.* **34**, 441–452.

Baker, D. E. (1990). Copper. In *Heavy Metals in Soils,* ed. Alloway, B. J., Blackie & Son, Glasgow, Scotland, 151–176.

Berbel, F., Diaz-Cruz, J. M., Arino, C., Esteban, M., Mas, F., Garces, J. L., and Puy, J. (2001). Voltammetric analysis of heterogeneity in metal ion binding by humics. *Environ. Sci. Technol.* **35**, 1097–1102.

Bertsch, P. M., and Hunter, D. B. (1998). Elucidating fundamental mechanisms in soil and environmental chemistry: the role of advanced analytical, spectroscopic, and microscopic methods. In *Future Prospects for Soil Chemistry*, ed. Huang, P. M., Sparks, D. L., and Boyd S. A., Spec. Publ. 55, Soil Science Society of America, Madison, WI, 103–122.

Bingham, F. T., and Bartels, J. M. (eds.) (1996). *Methods of Soil Analysis,* Part 3, *Chemical Methods*, Soil Science Society of America, Madison, WI.

Bingham, F. T., Sposito, G., and Strong, J. E. (1984). The effect of chloride on the availability of cadmium. *J. Environ. Qual.* **13,** 71–74.

Bingham, F. T., Sposito, G., and Strong, J. E. (1986). The effect of sulfate on the availability of cadmium. *Soil Sci.* **141**, 172–175.

Bodog, I., Polyak, K., Csikos-Hartyanyi, Z., and Hlavay, J. (1996). Sequential extraction procedure for the speciation of elements in fly ash samples.*Microchem. J.* **54**, 320–330.

Bolt, G. H., and Van Riemsdijk, W. H. (1982). Ion adsorption on inorganic variable charge constituents. In *Soil Chemistry*, Part B, *Physico-Chemical Models*, 2nd ed., ed. Bolt, G. H., Elsevier, Amsterdam, The Netherlands, 495–502.

Brinkley, S. R., Jr. (1947). Calculation of equilibrium composition of systems of many constituents. *J. Chem. Phys.* **15**, 107–110.

Brown, G. K., MacCarthy, P., and Leenheer, J. A. (1999a). Simultaneous determination of Ca, Cu, Ni, Zn and Cd binding strengths with fulvic acid fractions by Schubert's method. *Anal. Chim. Acta* **402**, 169–181.

Brown, G. K., Cabaniss, C. A., MacCarthy, P., and Leenheer, J. A. (1999b). Cu(II) binding by a pH-fractionated fulvic acid. *Anal. Chim. Acta* **402**, 183–193.

Brümmer, G. W. (1986). Heavy metal species, mobility and availability. In *The Importance of Chemical Speciation in Environmental Processes,* ed. Bernhard, M., Brinkman, F. E., and Sadler P. J., Springer-Verlag, Berlin, 169–192.

Cabrera, D., Young, S. D., and Rowell, D. (1988). The toxicity of cadmium to barley plants as affected by complex formation with humic acid. *Plant Soil* **105**, 195–204.

Candelaria, L. M., and Chang, A. C. (1997). Cadmium activities, solution speciation, and solid-phase distribution of cadmium in cadmium nitrate- and sewage sludge-treated soil systems. *Soil Sci.* **162**, 722–732.

Chairidchai, P., and Ritchie, G. S. P. (1993). The effect of citrate and pH on Zn uptake by wheat. *Agron. J.* **85,** 322–328.

Charlatchka, R., Cambier, P., and Bourgeois, S. (1997). Mobilization of trace metals in contaminated soils under anaerobic conditions. In *Contaminated Soils, Proc. 3rd International Conference on the Biogeochemistry of Trace Elements*, Paris, May 15–19, 1995, ed. R. Prost, INRA, Paris, 159–174.

Chen, Y., and Aviad, T. (1990). Effects of humic substances on plant growth. In *Humic Substances in Soil and Crop Sciences: Selected Readings*, ed.MacCarthy, P., Clapp, C. E., Malcolm, R. L., and Bloom, P. R., American Society of Agronomy, Madison, WI, 161–186.

Chen, Y., and Stevenson, F. J. (1986). Soil organic matter interactions with trace elements. In *The Role of Organic Matter in Modern Agriculture,* ed. Chen, Y., and Avnimelech, Y., Kluwer Academic, Dordrecht, The Netherlands, 73–116.

Christl, I., Milne, C. J., Kinniburgh, D. G., and Ktretzschmar, R. (2001). Relating ion binding by fulvic and humic acids to chemical composition and molecular size, II: Metal binding. *Environ. Sci. Technol.* **35,** 2512–2517.

Cieśliński, G., Van Rees, K. C. J., Szmigielska, A. M., Krishnamurti, G. S. R., and Huang, P. M. (1998). Low-molecular-weight organic acids in rhizosphere soils of durum wheat and their effect on cadmium bioaccumulation. *Plant Soil* **203**, 109–117.

Cook, R. L., and Langford, C. H. (1998). Structural characterization of a fulvic acid and a humic acid using solid-state ramp-CP-MAS ^{13}C nuclear magnetic resonance. *Environ. Sci. Technol.* **32**, 719–725.

Cook, R. L., Langford, C. H., Yamdagni, R., and Preston, C. M. (1996). A modified cross-polarization magic angle spinning ^{13}C NMR procedure for the study of humic materials. *Anal. Chem.* **68**, 979–986.

Cumming, J. R., and Weinstein, L. H. (1990). Aluminum–mycorrhizal interactions in the physiology of pitch pine seedlings. *Plant Soil* **125**, 7–18.

Davies, G., Fataftah, A., Cherkasskiy, A., Ghabbour, E. A., Radwan, A., Jensen, S. A., Kolla, S., Piciolla, M. D., Sein, L. T., Jr., Buermann, W., Balasubramanian, M., Budnick, J., and Xing, B. (1997). Tight metal binding by humic acids and its role in biomineralization. *J. Chem. Soc. Dalton Trans.* **1997**, 4047–4060.

Davis, R. D. (1979). Uptake of Cu, Ni and Zn by crops growing in contaminated soil. *J. Sci. Food Agric.* **30**, 937–947.

Davis, J. A., and Leckie, J. O. (1978). Surface ionization and complexation at the oxide/water interface, II: Surface properties of amorphous iron oxyhydroxide and adsorption of metal ions. *J. Colloid Interface Sci.* **67**, 90–107.

Davis, J. A., and Leckie, J. O. (1980). Surface ionization and complexation at the oxide/water interface, III: Adsorption of anions. *J. Colloid Interface Sci.* **74**, 32–43.

Davis, J. A., James, R. O., and Leckie, J. O. (1978). Surface ionization and complexation at the oxide/water interface, I: Computation of electrical double layer properties in simple electrolytes. *J. Colloid Interface Sci.* **63**, 480–499.

Del Castilho, P., and Rix, I. (1993). Ammonium acetate extraction for soil heavy metal speciation; model aided soil test interpretation. *Int. J. Environ. Anal. Chem.* **51**, 59–61.

Denny, H. J., and Wilkins, D. A. (1987). Zinc tolerance in *Betula* sp., IV: The mechanism of ectomycorrhizal amelioration of zinc toxicity. *New Phytol.* **106**, 545–553.

DIN (Deutsches Institut für Normung) (1995). *Bodenbeschaffenheit: Extraktion von Spurenelemente mit Ammonium-nitratlosung,* Vornorm DIN V19730, Boden-Chemische Bodenuntersuchungsverfahren, Berlin.

Dobbs, J. C., Susetyo, W., Knight, F. E., Castles, M. A., Carriera, L. A., and Azarraga, L. V. (1989a). Characterization of metal binding sites in fulvic acids (a novel approach to metal-humic complex studies) by lanthanide ion probe spectroscopy. *Anal. Chem.* **61**, 483–488.

Dobbs, J. C., Susetyo, W., Carriera, F. E., and Azarraga, L. V. (1989b). Competitive binding of protons and metal ions in humic substances by lanthanide ion probe spectroscopy. *Anal. Chem.* **61**, 1519–1524.

Dunnivant, F. M., Jardine, P. M., Taylor, D. L., and McCarthy, J. F. (1992). Cotransportation of cadmium by hexachlorobiphenyl by dissolved organic carbon through columns containing aquifer material. *Environ. Sci. Technol.* **26**, 360–368.

Dzombek, D. A., and Morel, F. M. M. (1990). *Surface Complexation Modeling: Hydrous Ferric Oxide*, Wiley, New York.

Elzinga, E. J., van Grissen, J. J. M., and Swartjes, F. A. (1999). General purpose Freundlich isotherms for cadmium, copper and zinc in soils. *Eur. J. Soil Sci.* **50**, 139–149.

Emmerich, W. E., Lund, L. J., Page, A. L., and Chang, A. C. (1982). Predicted solution phase forms of heavy metals in sewage sludge-treated soils. *J. Environ. Qual.* **11**, 182–186.

Engler, R. M., Brannon, J. M., and Rose, J. (1977). A practical selective extraction procedure for sediment characterization. In *Chemistry of Marine Sediments*, ed. Yen, T. F., Ann Arbor Science, Ann Arbor, MI, 163–178.

Essington, M. E. (2004). *Soil and Water Chemistry: An Integrative Approach*, CRC Press, Boca Raton, FL.

Estevas da Silva, J. C. G., and Oliveira, C. G. S. (2002). Metal ion complexation properties of fulvic acids extracted from composted sewage sludge as compared to a soil fulvic acid.*Water Res.* **36**, 3404–3409.

Fendorf, S. E., and Sparks, D. L. (1996). X-ray absorption fine structure spectroscopy. In *Methods of Soil Analysis,* Part 3, *Chemical Methods*, ed. Sparks, D. L., Soil Science Society of America, Madison, WI, 377–416.

Fendorf, S. E., Sparks, D. L., Lamble, G. M., and Kelley, M. J. (1994). Application of x-ray absorption fine structure spectroscopy to soils. *Soil Sci. Soc. Am. J.* **58**, 1583–1595.

Fitch, A., and Stevenson, F. J. (1983). Stability constants of metal–organic matter complexes: theoretical aspects and mathematical models. In *The Significance of Trace Elements in Solving Paragenetic Problems and Controversies,* ed. Augustithis, S. S., Theoprastus Publications, Athens, Greece, 645–649.

Fitch, A., and Stevenson, F. J. (1984). Comparison of models for determining stability constants of metal complexes with humic substances. *Soil Sci. Soc. Am. J.* **48**, 1044–1049.

Florence, T. M. (1989). Electrochemical techniques for trace element speciation in waters. In *Trace Element Speciation: Analytical Methods and Problems*, ed. Batley, G. E., CRC Press, Boca Raton, FL, 77–116.

Flyhammer, P. (1997). Estimation of heavy metal transformations in municipal solid waste. *Sci. Total Environ.* **198**, 123–133.

Förstner, U., Calmano, K., Conrad, H., Jaksch, H., Schimkus, C., and Schoer, J. (1981). Chemical speciation of heavy metals in solid waste materials (sewage sludge, mining wastes, dredged materials, polluted sediments) by sequential extraction. *Proc. International Conference on Heavy Metals in the Environment,* WHO/EED, 698–704.

Frausto da Silva, J. J. R., and Williams, R. J. P. (1976). The uptake of elements by biological systems. *Struct. Bond.* **29**, 67–121.

Garrels, R. M., and Thompson, M. E. (1962). A chemical model for seawater at 25°C and one atmosphere total pressure. *Am. J. Sci.* **260**, 57–66.

Geisy, J. P., Leversee, G. J., and Williams, D. R. (1977). Effects of naturally occurring aquatic organic fractions on cadmium toxicity to *Simocephalus serrulatus* (Daphnidae) and *Gambusia affinis* (Poeciliidae). *Water Res.* **11**, 1013–1020.

Gibbs, R. J. (1973). Mechanisms of trace metals transport in rivers. *Science* **180**, 171–173.

Goldberg, S. (1992). Use of surface complexation models in soil chemical systems. *Adv. Agron.* **47**, 223–329.

Goldberg, S., and Traina, S. J. (1987). Chemical modeling of anion competition on oxides using the constance capacitance model-mixed-ligand approach. *Soil Sci. Soc. Am. J.* **51**, 929–932.

Gough, L. P., McNeal, J. M., and Severson, R. C. (1980). Predicting native plant copper, iron, manganese and zinc levels using DTPA and EDTA soil extractants in northern Great Plains. *Soil Sci. Soc. Am. J.* **44**, 1030–1036.

Gray, C. W., McLaren, R. G., Roberts, A. H. C., and Condron, L. M. (1999). Cadmium phytoavailability in some New Zealand soils. *Aust. J. Soil Res.* **37**, 461–477.

Gupta, S. K., and Chen, K. Y. (1975). Partitioning of trace elements in selective chemical fractions of near shore sediments. *Environ. Lett.* **10**, 129–158.

Guy, R. D., and Kean, A. R. (1980). Algae as a chemical speciation monitor, I: A comparison of algal growth and computer calculated speciation. *Water Res.* **14**, 891–899.

Haeni, H., and Gupta, S. (1983). Total and biorelevant heavy metal contents and their usefulness in establishing limiting values in soils. In *Environmental Effects of Organic and Inorganic Contaminants in Sewage Sludge, Proc. Workshop*, Stevenage, Hertfordshire, England, May 25–26, 1982, ed. Davis, R. D., Hucker, G., and L'Hermite, P., Kluwer Academic, Dordrecht, The Netherlands, 121–129.

Haq, A. U., and Miller, M. H. (1972). Prediction of available soil Zn, Cu and Mn using chemical extractants. *Agron. J.* **64**, 779–782.

Haq, A. U., Bates, T. E., and Soon, Y. K. (1980). Comparison of extractants for plant-available zinc, cadmium, nickel, and copper in contaminated soils. *Soil Sci. Soc. Am. J.* **72**, 772–777.

Hawthorne, F. C. (1988). *Spectroscopic Methods in Mineralogy and Geology*, Mineralogical Society of America, Washington, DC.

Hayes, K. F., and Leckie, J. O. (1987). Modeling ionic strength effects on cation adsorption at hydrous oxide/solution interfaces. *J. Colloid Interface Sci.* **115**, 564–572.

Hayes, K. F., Papelis, C., and Leckie J. O. (1988). Modeling ionic strength effects on anion adsorption at hydrous oxide/solution interfaces. *J. Colloid Interface Sci.* **125**, 717–726.

Helz, G. R., Miller, C. V., Charnock, J. M., Mosselmans, J. F. W., Pattrick, R. A. D., Garner, C. D., and Vaughan, D. J. (1996). Mechanism of molybdenum removal from the sea and its concentration in black shales: EXAFS evidence. *Geochim. Cosmochim. Acta* **60**, 3631–3642.

Hirsch, D., and Banin, A. J. (1990). Cadmium speciation in soil solutions: the residual effect of sewage sludge on heavy metal content of tobacco and peanut. *J. Environ. Qual.* **19**, 366–372.

Hochella, M. F., and White, A. F. (eds.) (1990). *Mineral–Water Interface Geochemistry*, Mineralogical Society of America, Washington, DC.

Hodgson, J. F., Lindsay, W. L., and Trierweiler, J. T. (1966). Micronutrient cation complexing in soil solution,. II: Complexing of zinc and copper in displaced solution from calcareous soils. *Soil Sci. Soc. Am. Proc.* **30**, 723–726.

Hornburg, V., Welp, G., and Brümmer, G. W. (1995). Behaviour of heavy metals in soils, 2: Extraction of mobile heavy metals in soils with $CaCl_2$ and NH_4NO_3. *Z. Pflanzenernähr. Bodenkd.* **158**, 137–145.

Houba, V. J. G., Novozamsky, I., Lexmond, T. M., and van der Lee, J. J. (1990). Applicability of 0.01 M $CaCl_2$ as a single extraction solution for the assessment of nutrient status of soils and other diagnostic purposes. *Commun. Soil Sci. Plant Anal.* **21**, 2281–2290.

Huang, C. P., and Stumm, W. (1973). Specific adsorption of cations on hydrous γ-aluminum oxide. *J. Colloid Interface Sci.* **43**, 409–417.

Ingle, S. E., Schuldt, M. D., and Schults, D. W. (1978). *A User's Guide for REDEQL*-EPA, EPA/600/3-78/024, U.S. Environmental Protection Agency, Corvallis, OR.

Ireland, R. R., McConachie, W. A., Stuermer, D. H., Wang, F. T., and Koszykowski, R. F. (1982). The analysis of inorganic constituents in the ground water at an underground coal gassification site. In *Atomic and Nuclear Methods in Fossil Energy Research, Proc. Topical Conference,* American Nuclear Society and American Chemical Society, Dec. 1980, Mayaguez, Puerto Rico, ed. Filby, R. H., Plenum Press, New York, 175–188.

Jackson, A. P., and Alloway, B. J. (1991). The bioavailability of cadmium to lettuce and cabbage in soils previously treated with sewage sludges. *Plant Soil* **132**, 179–186.

James, R. O., and Parks, G. A. (1978). *Application of the Computer Program MINEQL to Solution of Problems in Surface Chemistry*, Tech. Note, Stanford University, Stanford, CA.

Jarvis, J. C. (1987). Copper concentrations in plants and their relationship to soil properties. In *Copper in Soils and Plants*, ed. Loneregan, J. F., Robson, A. D., and Graham, R. D., Academic Press, London, 265–285.

Jenne, E. A. (1979). *Chemical Modeling in Aqueous Systems*, ACS Symp. Ser. 93, American Chemical Society, Washington, DC.

Jenne, E. A. (1981). *Ground Water Modeling: A Review*, PNL-3574, Pacific Northwest Laboratory, Richland, WA.

John, M. K. (1972). Lead availability related to soil properties and extractable lead. *J. Environ. Qual.* **1**, 295–298.

John, M. K., Van Laerhoven, J., and Chauh, H. H. (1972). Factors affecting plant uptake and phytotoxicity of cadmium added to soils. *Environ. Sci. Technol.* **6**, 1005–1009.

Jopony, M., and Young, S. D. (1994). The solid–solution equilibria of lead and cadmium in polluted soils. *Eur. J. Soil Sci.* **45**, 59–70.

Kabata-Pendias, A., and Pendias, H. (2001). *Trace Elements in Soils and Plants*, CRC Press, Boca Raton, FL.

Keizer, M. G. (1991). *ECOSAT: A Computer Program for the Calculation of Speciation in Soil–Water Systems*, Department of Soil Science and Plant Nutrition, Agricultural University, Wageningen, The Netherlands.

Kersten, M. (2002). Speciation of trace metals in sediments. In *Chemical Speciation in the Environment,* 2nd ed., ed. Ure, A. M., and Davidson, C. M., Blackie, Glasgow, Scotland, 301–321.

Kersten, M., and Förstner, U. (1989). Speciation of trace elements in sediments. In *Trace Element Speciation: Analytical Methods and Problems*, ed. Batley, G. E., CRC Press, Boca Raton, FL., 245–317.

Kersten, M., and Förstner, U. (1995). Speciation of trace metals in sediments and combustion wastes. In *Chemical Speciation in the Environment*, ed. Ure, A. M., and Davidson, C. M., Backie, Glasgow, Scotland, 234–275.

King, L. D., and Hajjar, L. M. (1990). The residual effect of sewage sludge on heavy metal content of tobacco and peanut. *J. Environ. Qual.* **19**, 738–748.

Krishnamurti, G. S. R., Cieśliński, G., Huang, P. M., and Van Rees, K. C. J. (1997a). Kinetics of cadmium release from soils as influenced by organic acids: implication in cadmium availability. *J. Environ. Qual.* **26**, 271–277.

Krishnamurti, G. S. R., Huang, P. M., and Van Rees, K. C. J. (1996). Studies on soil rhizosphere: speciation and availability of cadmium. *Chem. Speciat. Bioavailab.* **8**, 23–28.

Krishnamurti, G. S. R., Huang, P. M., Van Rees, K. C. J., Kozak, L. M., and Rostad, H. P. W. (1995a). Speciation of particulate-bound cadmium of soils and its bioavailability. *Analyst* **120**, 659–665.

Krishnamurti, G. S. R., Huang, P. M., Van Rees, K. C. J., Kozak, L. M., and Rostad, H. P. W. (1995b). A new soil test method for the determination of plant available cadmium in soils. *Commun. Soil Sci. Plant Anal.* **26**, 2857–2867.

Krishnamurti, G. S. R., Huang, P. M., Van Rees, K. C. J., Kozak, L. M., and Rostad, H. P. W. (1997b). Differential FTIR study of pyrophosphate extractable material of soils: implication in cadmium-bonding sites and availability. In *Contaminated Soils, Proc. 3rd International Conference on the Biogeochemistry of Trace Elements*, May 15–19, 1995, Paris, ed. Prost, R., INRA, Paris, 012pdf. 1–10.

Krishnamurti, G. S. R., Megharaj, M., and Naidu, R. (2004). Bioavailability of cadmium–organic complexes to soil alga: an exception to the free ion model. *J. Agric. Food Chem.* **52**, 3894–3899.

Krishnamurti, G. S. R., and Naidu, R. (2000). Speciation and phytoavailability of cadmium in selected surface soils of south Australia. *Aust. J. Soil Res.* **38**, 991–1004.

Krishnamurti, G. S. R., and Naidu, R. (2002). Solid–solution speciation and phytoavailability of copper and zinc in soils. *Environ. Sci. Technol.* **36**, 2645–2651.

Krishnamurti, G. S. R., and Naidu, R. (2003). Solid–solution equilibria of cadmium in soils. *Geoderma* **113**, 17–30.

Krishnamurti, G. S. R., Smith, L. H., and Naidu, R. (2000). Method for assessing plant-available cadmium in soils. *Aust. J. Soil Res.* **38**, 823–836.

Lakanen, E., and Ervio, R. (1971). A comparison of eight extractants for the determination of plant available micronutrients in soils. *Acta Agral. Fenn.* **123**, 223–232.

Lake, D. L., Kirk, P. W. W., and Lester, J. N. (1984). Fractionation, characterization, and speciation of heavy metals in sewage sludge and sludge-amended soils: a review. *J. Environ. Qual.* **13**, 175–183.

Lamy, I., Cambier, P., and Bourgeois, B. (1994). Lead and cadmium complexation with soluble organic carbon and speciation in alkaline soil leachates. *Environ. Geochem. Health* **16**, 1–16.

Lebourg, A., Sterckeman, T., Ciesielski, H., and Proix, N. (1996). Suitability of chemical extraction to assess risks of toxicity induced by soil trace metal bioavailability. *Agronomie* **16**, 201–215.

Lee, D.-Y., and Zheng, H.-C. (1994). Simultaneous extraction of soil phytoavailable cadmium, copper, and lead by chelating resin membrane. *Plant Soil* **164**, 19–23

Lee, G. S. H., Wilson, M. A., and Young, B. R. (1998). The application of "WATERGATE" suppression technique for analyzing humic substances by nuclear magnetic resonance. *Org. Geochem.* **28**, 549–559.

Lee, P. K., Touray, J. C., Bailiff, P., and Ildefonse, J. P. (1997). Heavy metal contamination of settling particles in a retention pond along the A-71 motorway in Sologne, France.*Sci. Total Environ.* **201**, 1–15.

Legatt, D. J. (1977). Machine computation of equilibrium concentrations: some practical considerations. *Talanta* **24**, 535–542.

Li, Y.-M., Chaney, R. L., and Schneiter, A. M. (1994). Effect of soil chloride level on cadmium concentration in sunflower kernels. *Plant Soil* **167**, 275–280.

Liang, J., and Schoenau, J. J. (1996). Speciation in metal contaminated soils as revealed by an ion-exchange resin membrane fractionation procedure. *Commun. Soil Sci. Plant Anal.* **27**, 3013–3026.

Lindsay, W. L., and Norvell, W. A. (1978). Development of a DTPA soil test for zinc, iron, manganese, and copper. *Soil Sci. Soc. Am. J.* **42**, 421–428.

Loeppert, R. H., Schwab, A. P., and Goldberg, S. (eds.) (1995). *Chemical Equilibrium and Reaction Models,* Spec. Publ. 42, Soil Science Society of America, Madison, WI.

Lumsdon, D. G., and Evans, L. J. (1995). Predicting chemical speciation and computer simulation. In *Chemical Speciation in the Environment*, ed. Ure, A. M., and Davidson, C. M., Blackie Academic & Professional, London, 86–134.

Lumsdon, D. G., and Evans, L. J. (2002). Predicting chemical speciation and computer simulation. In *Chemical Speciation in the Environment,* ed. Ure, A. M., and Davidson, C. M., Blackwell Science, Oxford, 89–131.

Lund, W. (1990). Speciation analysis: Why and how? *Fresenius'J. Anal. Chem.* **354,** 337–357.

Luoma, S. N. (1983). Bioavailability of trace metals to aquatic organisms: a review. *Sci. Total Environ.* **28**, 3–22.

Luster, J., Lloyd, G., Sposito, G., and Fry, I. V. (1996). Multi-wavelength molecular fluorescence spectrometry for quantitative characterization of copper(II) and aluminum(III) complexation by dissolved organic matter. *Environ. Sci. Technol.* **30**, 1565–1574.

Ma, Y. B., and Uren, N. C. (1998). Transformation of heavy metals added to soil: application of new sequential extraction procedure. *Geoderma* **84**, 157–168.

MacLaren, R. G., and Crawford, D. (1973). Studies on soil copper, 1: The fractionation of Cu in soils. *J. Soil Sci.* **24**, 172–181.

Manceau, A., Charlet, L., Boisett, M.-C., Didier, B. and Spadini, L. (1992). Sorption and speciation of heavy metals on hydrous Fe and Mn oxides: from microscopic to macroscopic. *Appl. Clay Sci.* **7**, 201–223.

Manceau, A., Boisset, M.-C., Sarret, G., Hezemann, J.-L., Mench, M., Cambier, Ph., and Prost, R. (1996). Direct determination of lead speciation in contaminated soils by EXAFS spectroscopy. *Environ. Sci. Technol.* **30**, 1540–1552.

Mattigod, S. V., and Sposito, G. (1979). GEOCHEM: *A Computer Program for the Calculation of Chemical Equilibria in Soil Solutions and Other Water Systems*, Department of Soils and Environmental Sciences, University of California, Riverside, CA.

McBride, M. B., Nibarger, E. A., Richards, B. K., and Steenhius, T. (2003). Trace metal accumulation by red clover grown on sewage sludge amended soils and correlation to Mehlich3 and calcium chloride extractable metals. *Soil Sci.* **168**, 29–38.

McBride, M. B., Sauvé, S., and Hendershot, W. (1997). Solubility control of Cu, Zn, Cd, and Pb in contaminated soils. *Eur. J. Soil Sci.* **48**, 337–346.

McDuff, R. E., and Morel, F. M. M. (1973). *Description and Use of the Chemical Equilibrium Program REDEQL2,* Tech. Rep. EQ-73-02, California Institute of Technology, Pasadena, CA.

McGrath, S. P., Sanders, J. R., and Shalaby, M. H. (1988). The effect of soil organic matter levels on soil solution concentrations and extractabilities of manganese, zinc and copper. *Geoderma* **42**, 177–188.

McLaughlin, M. J., Tiller, K. G., Beech, T. A., and Smart, M. K. (1994). Increased soil salinity causes elevated cadmium concentrations in field-grown potato tubers. *J. Environ. Qual.* **23**, 1013–1018.

Merkel, D. (1996). Cadmium, copper, nickel, lead and zinc contents of wheat grain and soils extracted with $CaCl_2$/DTPA (CAD), $CaCl_2$ and NH_4NO_3. *Agrobiol. Res. Z. Agrobiol. Agrikulturchem. Okol.* **49**, 30–37.

Michalke, B., and Schrammel, P. (1999). Editorial: First International Conference on trace element speciation in biomedical, nutritional and environmental sciences. *Fresenius' J. Anal. Chem.* **363**, 429–430.

Miller, W. P., and McFee, W. W. (1983). Distribution of Cd, Zn, Cu and Pb in soils of industrial north western Indiana. *J. Environ. Qual.* **12**, 29–33.

Morel, F. M. M., and Morgan, J. J. (1972). A numerical method for computing equilibria in aqueous systems. *Environ. Sci. Technol.* **6**, 58–67.

Morgan, J. J. (1987). General affinity concepts, equilibria and kinetics in aqueous metal chemistry. In *Metal Speciation, Separation and Recovery*, ed. Patterson, J. W., and Passino, R., Lewis Publishers, Chelsea, MI, 27–61

Morra, M. J., Fendorf, S. E., and Brown, P. D. (1997). Speciation of sulfur in humic and fulvic acids using x-ray absorption near-edge structure (XANES) spectroscopy. *Geochim. Cosmochim. Acta* **61**, 683–688.

Morrison, G. M. P., Batley, G. E., and Florence, T. M. (1989). Metal speciation and toxicity. *Chem. Br.* **1989**, 791–796.

Naidu, R., and Harter, R. D. (1998). Effect of different organic ligands on cadmium sorption by and extractability from soils. *Soil Sci. Soc. Am. J.* **62**, 644–650.

Naidu, R., Kookana, R. S., Sumner, M. E., Harter, R. D., and Tiller, K. G. (1997). Cadmium sorption and transport in variable charge soils: a review. *J. Environ. Qual.* **26**, 602–617.

Neaman, A., Mouélé, F., Trolard, F., and Bourrié, G. (2004). Improved methods for selective dissolution of Mn oxides: applications for studying trace element associations. *Appl. Geochem.* **19**, 973–979.

Nordstrom, D. K., Plummer, L. N., Wigley, T. M. L., Wolery, T. J., Ball, J. W., Jenne, E. A., Bassett, R. L., Crerar, D. A., Florence, T. M., Fritz, B., Hoffman, M., Holdren, G. R., Lafon, G. M., and Waite, T. D. (1989). Mathematical modeling of trace element speciation. In *Trace Element Speciation: Analytical Methods and Problems*, ed. Batley, G. E., CRC Press, Boca Raton, FL, 117–184.

Norvell, W. A., Wu, J., Hopkins, D. G., and Welch, R. M. (2000). Association of cadmium in durum wheat grain with soil chloride and chelate-extractable soil cadmium. *Soil Sci. Soc. Am. J.* **64,** 2162–2168.

O'Connor, G. A. (1988). Use and misuse of the DTPA soil test. *J. Environ. Qual.* **17**, 715–718.

Onyatta, J. O., and Huang, P. M. (1999). Cadmium speciation and bioavailability index for selected Kenyan soils. *Geoderma* **91**, 87–101.

Papelis, C., Hayes, K. F., and Leckie, J. O. (1988). *HYDRAQL: A Program for the Computation of Chemical Equilibrium Composition of Aqueous Batch Systems Including Surface Complexation Modeling of Ion Adsorption at the Oxide/Solution Interface*, Tech. Rep. 306. Department of Civil Engineering, Stanford University, CA.

Parker, D. R., Chaney, R. L., and Norvell, W. A. (1995a). Chemical equilibrium models: applications to plant nutrition. In *Soil Chemical Equilibrium and Reaction Models*, ed. Loeppert, R. H., Schwab, A. P., and Goldberg, S., Spec. Publ. 42, American Society of Agronomy, Soil Science Society of America, Madison, WI, 163–200.

Parker, D. R., Norvell, W. A., and Chaney, R.L., (1995b). GEOCHEM-PC: a chemical speciation program for IBM and compatible personal computers. In *Soil Chemical Equilibrium and Reaction Models*, ed. Loeppert, R. H., Schwab, A. P., and Goldberg, S., Spec. Publ. 42, American Society of Agronomy, Soil Science Society of America, Madison, WI, 253–270.

Patsakis, N., Savvides, C., Haralambous, K. J., and Loizidou, M. (1998). Chemical partitioning of metals in thermally treated sewage sludge. *Environ. Technol.* **19**, 331–337.

Petersen, R. 1982. Influence of copper and zinc on the growth of a fresh water alga, *Scendesmus quadricada:* the significance of chemical speciation. *Environ. Sci. Technol.* **16**, 443–447.

Peterson, H. G., Healy, F. P., and Wagemann, R. (1984). Metal toxicity to alga: a highly pH-dependent phenomenon. *Can. J. Fish. Aquat. Sci.* **41**, 974–979.

Peterson, M. L., Brown, G. E., Jr., Parks, G. A., and Stein, C. L. (1997). Differential redox and sorption of Cr(III/VI) on natural silicate and oxide minerals: EXAFS and XANES results. *Geochim. Cosmochim. Acta* **61**, 3399–3412.

Peuravuori, J., and Pihlaja, K. (1997). Pyrolysis electron impact mass spectrometry in studying aquatic humic substances. *Anal. Chim. Acta* **350**, 241–247.

Rappaport, B. D., Martens, D. C., Reneau, R. B., and Simpson, T. W. (1988). Metal availability in sludge amended soils with elevated metal levels. *J. Environ. Qual.* **17**, 42–47.

Raymond, K. N., Muller, G., and Mayzanke, B. F. (1984). Complexation of iron by sideophores: a review of their solution and structural chemistry and biological function. *Top. Curr. Chem.* **123**, 49–102.

Ritchie, G. S. P. (1989). The chemical behaviour of aluminium, hydrogen and manganese in acid soils. In *Soil Acidity and Plant Growth,* ed. Robson, A. D., Academic Press, Sydney, NSW, Australia, 1–60.

Ritchie, G. S. P., and Sposito, G. (1995). Speciation in soils. In *Chemical Speciation in the Environment*, ed. Ure, M., and Davidson, C. M., Blackie Academic & Professional, London, 201–233.

Saiz-Jimenez, C., Hermosin, B., and Ortega-Calvo, J. J. (1993). Pyrolysis/methylation: a method for structure elucidation of the chemical nature of aquatic humic substances. *Water Res.* **27**, 1693–1696.

Sanders, J. R. (1982). The effect of pH upon the copper and cupric ion concentrations in soil solution.*J. Soil Sci.* **33**, 679–685.

Sanders, J. R. (1983). The effect of pH on the total and free ionic concentrations of manganese, zinc and cobalt in soil solutions. *J. Soil Sci.* **34**, 315–323.

Sauerbeck, D., and Styperek, P. (1984). Predicting the cadmium availability from different soils by $CaCl_2$ extraction. In *Processing and Use of Sewage Sludge*, ed. L'Hermite, P., and Ott, H., D. Reidel, Dordrecht, The Netherlands, 431–435.

Sauvé, S., McBride, M. B., Norvell, W. A., and Hendershot, W. (1997). Copper solubility and speciation of in situ contaminated soils: effects of copper levels, pH and organic matter. *Water Air Soil Pollut.* **100**, 133–149.

Sauvé, S., Hendershot, W., and Allen, H. E. (2000). Solid-solution partitioning of metals in contaminated soils: dependence of pH, total metal burden and organic matter. *Environ. Sci. Technol.* **34**, 1125–1131.

Schalscha, E. G., Marlaes, M., Vergara, I., and Chang, A. C. (1982). Chemical fractionation of heavy metals in waste-water affected soils. *J. Water Pollut. Control Fed.* **54**, 175–180.

Scheidegger, A. M., and Sparks, D. L. (1996). A critical assessment of sorption–desorption mechanisms at the mineral/water interface. *Soil Sci.* **161,** 813–831.

Schindler, P. W., Furst, R. D., and Wolf, P. U. (1976). Ligand properties of surface silanol groups, 1: Surface complex formation with Fe^{3+}, Cu^{2+}, Cd^{2+} and Pb^{2+}. *J. Colloid Interface Sci.* **55**, 469–475.

Schnitzer, M. (1991). Soil organic matter: the next 75 years. *Soil Sci.* **151**, 41–58.

Schnitzer, M. and Hansen, E.H. (1970). Organo-metallic interactions in soils, 8. An evaluation of methods for the determination of stability constants of metalfulvic acid complexes. *Soil Sci.* **109**, 333–340.

Schnitzer M. and Schuppli, P. (1989). The extraction of organic matter from selected soils and particle size fractions with 0.5 M NaOH and 0.1 M $Na_4P_2O_7$ solutions. *Can J. Soil Sci.* **69**, 253–262.

Schulten, H.-R. (1996). Three-dimensional, molecular structure of humic acids and their interactions with water and dissolved contaminants. *Int. J. Environ. Anal. Chem.* **64**, 47–162.

Schulze, D. G., and Bertsch, P. M. (1995). Synchrotron x-ray techniques in soil, plant and environmental research. *Adv. Agron.* **55**, 1–66.

Sedberry, J. E., and Reddy, C. N. (1976). The distribution of zinc in selected soils in Indiana. *Commun. Soil Sci. Plant Anal.* **7**, 787–795.

Sedlacek, J., Kalquist, T., and Gjessing, E. (1983). Effect of aquatic humus on uptake and toxicity of cadmium to *Selenastrum capricornutum* Printz. In *Aquatic and Terrestrial Humic Materials*, ed . Christman, R. F., and Gjessing, E., Ann Arbor Science, Ann Arbor, MI, 495–516.

Serkiz, S. M., Allison, J. D., Perdue, E. M., Allen, H. E., and Brown, D. S. (1996). Correcting errors in the thermodynamic database for the equilibrium speciation model MINTEQA2. *Water Res.* **30**, 1930–1933.

Sheppard, M. I., and Stephenson, M. (1997). Critical evaluation of selective extraction methods for soils and sediments. In *Contaminated Soils, Proc. 3rd International Conference on the Biogeochemistry of Trace Elements*, Paris, May 15–19, 1995, ed . Prost, R., INRA, Paris, 69–97.

Shuman, L. M. (1985). Fractionation method for soil microelements. *Soil Sci.* **140**, 11–22.

Sihombing, R., Greenwood, P. F., Wilson, M. A., and Hanna, J. V. (1996). Composition of size exclusion fractions of swamp water humic and fulvic acids as measured by solid state NMR and pyrolysis–gas chromatography–mass spectrometry. *Org. Geochem.* **24**, 859–873.

Sillanpaa, M. (1982). *Micronutrients and the Nutrient Status of Soils: A Global Study,* FAO Soils Bull. 48, Food and Agricultural Organization of the United Nations, Rome.

Sillanpaa, M., and Jansson, H. (1992). *Status of Cadmium, Lead, Cobalt and Selenium in Soils and Plants of Thirty Countries*, FAO Soils Bull. 65, Food and Agricultural Organization of the United Nations, Rome.

Sillen, L. G., and Martell, A. E. (1971). *Stability Constants of Metal-Ion Complexes,* Chemical Society, London.

Sims, J. T., and Johnson, G. V. (1991). Micronutrient soil tests. In *Micronutrients in Agriculture,* ed. Mortvedt, J. J., Cox, F. R., Shuman, L. M., and Welch, R. M., SSSA Book Ser. 4, Soil Science Society of America, Madison, WI, 427–476.

Sims, J. L., and Patrick, W. H., Jr. (1978). The distribution of micronutrient cations in soil under conditions of varying redox potential and pH. *Soil Sci. Soc. Am. J.* **42**, 258–262.

Smilde, K. W., van Luit, B., and van Driel, W. (1992). The extraction by soil and absorption by plants of applied zinc and cadmium. *Plant Soil* **143**, 233–238.

Smolders, E., and McLaughlin, M. J. (1996a). Chloride increases Cd uptake in Swiss chard in a resin-buffered nutrient solution. *Soil Sci. Soc. Am. J.* **60**, 1443–1447.

Smolders, E., and McLaughlin, M. J. (1996b). Effect of Cl and Cd uptake by Swiss chard in nutrient solution. *Plant Soil* **179**, 57–64.

Soltanpour, P. N., and Schwab, A. P. (1977). A new soil test for the simultaneous extraction of macro- and micro-nutrients in alkaline soils. *Commun. Soil Sci. Plant Anal.* **8**, 195–207.

Soon, Y. K., and Bates, T. E. (1982). Chemical pools of Cd, Ni and Zn in some polluted soils and some preliminary indications of their availability to plants. *J. Soil Sci.* **33**, 477–488.

Sparks, D. L. (1983). Ion activities: an historical and theoretical overview. *Soil Sci. Soc. Am. J.* **47**, 514–518.

Sparks, D. L. (2000). Soil chemistry: past achievements and new frontiers. In *Soil Science: Accomplishments and Changing Paradigm Towards the 21st Century, Proc. International Symposium on Soil Science*, Bangkok, Thailand, Apr. 17–18, 2000, ed. Kheoruenromme, I., and Theerawong, S., International Society of Soil Science, Bangkok, Thailand.

Sposito, G. (1981). Trace metals in contaminated waters. *Environ. Sci. Technol.* **15**, 396–403.

Sposito, G. (1983). The chemical form of trace metals in soils. In *Applied Environmental Geochemistry*, ed. Thornton, I., Academic Press, London, 123–170.

Sposito, G. (1986). Distribution of potentially hazardous trace metals. In *Metal Ions in Biological Systems*, Vol. 20, ed. Siegel, H., Marcel Dekker, New York.

Sposito, G., and Bingham, F. T. (1981). Computer modeling of trace metal speciation in soil solutions: correlation, with trace metal uptake by higher plants. *J. Plant Nutr.* **3**, 35–49.

Sposito, G., and Coves, J. (1988). *SOILCHEM: A Computer Program for the Calculation of Chemical Speciation in Soils*, Kearney Foundation of Soil Science, University of California, Berkeley, CA.

Sposito, G., and Coves, J. (1995). SOILCHEM on the Macintosh. In *Soil Chemical Equilibrium and Reaction Models*, ed. Loeppert, R. H., Schwab, A. P., and Goldberg, S., SSSA Spec. Publ. 42, American Society of Agronomy, Soil Science Society of America, Madison, WI, 271–288.

Sposito, G., and Mattigod, S. V. (1980). *GEOCHEM: A Computer Program for the Calculation of Chemical Equilibria in Soil Solutions and Other Natural Water Systems,* University of California, Riverside, CA.

Sposito, G., Page, A. L., and Frink, M. E. (1980). *Effects of Acid Precipitation on Soil Leachate Quality: Computer Calculations,* Environmental Research Laboratory, U.S. Environmental Protection Agency, Corvallis, OR.

Sposito, G., Page, A. L., and Mattigod, S. V. (1979). *Trace Metal Speciation in Saline Waters Affected by Geothermal Brines*, Kearney Foundation of Soil Science, University of California, Berkeley, CA.

Sterckeman, T., Gomez, A., and Ciesielski, H. (1996). Soil and waste analysis for environmental risk assessment in France. *Sci. Total Environ.* **178**, 63–69.

Stevenson, F. J. (1991). Organic matter reactions involving micronutrients in soils. In *Micronutrients in Agriculture*, ed. Mortvedt, J. J., Giordano, P. M., and Lindsay, W. L., American Society of Agronomy, Madison, WI, 145–186.

Stevenson, F. J. (1994). *Humus Chemistry,* Wiley, New York.

Stevenson, F. J., and Fitch, A. (1986). Chemistry of complexation of metal ions with soil solution organics. In *Interaction of Soil Minerals with Natural Organics and Microbes*, ed. Huang, P. M., and Schnitzer, M., Spec. Publ. 17, Soil Science Society of America, Madison, WI, 29–76.

Stone, M., and Marsalek, J. (1996). Trace metal composition and speciation in street sediment, Sault Ste Marie, Canada. *Water Air Soil Pollut.* **87**, 149–169.

Stover, R. C., Sommers, L. E., and Silviera, D. J. (1976). Evaluations of metals in waste water sludge. *J. Water Pollut. Control Fed.* **48**, 2165–2175.

Street, J. J., Lindsay, W. L., and Sabey, B. R. (1977). Solubility and plant uptake of cadmium in soils amended with cadmium and sewage sludge. *J. Environ. Qual.* **6**, 72–77.

Stumm, W. (1992). *Chemistry of the Solid–Water Interface: Processes at the Mineral–Water and Particle–Water Interface in Natural Systems*, Wiley, New York.

Stumm, W., Huang, C. P., and Jenkins, S. R. (1970). Specific chemical interactions affecting the stability of dispersed systems. *Croat. Chim. Acta* **42**, 223–231.

Stumm, W., Hohl, H., and Dalang, F. (1976). Interaction of metal ions with hydrous oxide surfaces. *Croat. Chim. Acta* **48,** 491–499.

Stumm, W., Kummertt, R., and Sigg, L. (1980). A ligand exchange model for adsorption of inorganic and organic ligands at hydrous oxide interfaces. *Croat. Chim. Acta* **53**, 291–312.

Susetyo, W., Carrierra, L. A., and Grimm, D. M. (1991). Fluorescence techniques for metal humate interactions. *Fresenius' J. Anal. Chem.* **339**, 624–635.

Susetyo, W., Dobbs, J. C., Carrierra, L. A., Azarraga, L. V., and Grimm, D. M. (1990). Development of a statistical model for metal–humic interactions. *Anal. Chem.* **62**, 1215–1221.

Symeonides, C., and McRae, S. G. (1977). The assessment of plant available cadmium in soils. *J. Environ. Qual.* **6**, 120–123.

Szmigielska, A. M., Van Rees, K. C. J., Cieśliński, G., Huang, P. M., and Knott, D. R. (1995). Determination of low molecular weight dicarboxylic acids in root exudates by gas chromatography. *J. Agric. Food Chem.* **43**, 956–959.

Temminghoff, E. J. M., Van Der Zee, S. E. A. T. M., and De Haan, F. A. M. (1995). Speciation and calcium competition effects on cadmium sorption by sandy soil at various pHs. *Eur. J. Soil Sci.* **46**, 649–655.

Templeton, D. M., Ariese, F., Cornelis, R., Danielson, L.-G., Munau, H., van Leeuwen, H. P., and Lobinski, R. (2000). Guidelines for terms related to chemical speciation and fractionation of elements: definitions, structural aspects and methodological approaches (IUPAC Recommendations 2000). *Pure Appl. Chem.* **72**, 1453–1470.

Tessier, A., Campbell, P. G. C., and Bissom, M. (1979). Sequential extraction procedure for the speciation of particulate trace metals. *Anal. Chem.***51**, 844–851.

Tessier, A., and Turner, D. R.(eds.) (1995). *Metal Speciation and Bioavailability in Aquatic Systems*, IUPAC Series on Analytical and Physical Chemistry of Environmental Systems, Vol. 3, Wiley, Chichester, West Sussex, England.

Tichy, R., Nidyl, V., Kuzel, S., and Kolar, L. (1997). Increased cadmium availability to crops on sewage-sludge amended soils. *Water Air Soil Pollut.* **94**, 361–372.

Tiffin, L. O. (1977). The form and distribution of metals in plants: an overview. In *Hanford Life Sciences Symposium. Proc.,* Symp. Ser., U.S. Department of Energy, Washington, DC, 315.

Tlustos, P., van Dijk, D., Szakova, J., and Parlikova, D. (1994). Cd and Zn release through the use of selected extractants. *Rostl. Vyroba* **40**, 1107–1121.

Tokunaga, T. K., Sutton, S. R., Bajt, S., Nuessle, P., and Shea-McCarthy, G. (1998). Selenium diffusion and reduction at the water–sediment boundary: microXANES spectroscopy of reactive transport. *Environ. Sci. Technol.* **32**, 1092–1098.

Turner, D. R. (1995).Problems in trace metal speciation modeling. In *Metal Speciation and Bioavailability in Aquatic Systems*, ed. Tessier, A., and Turner, D. R., Wiley, New York, 149–203.

Ure, A. M., and Davidson, C. M. (2002). Introduction. In *Chemical Speciation in the Environment*, 2nd edition, ed. Ure, A. M., and Davidson, C. M., Blackwell Science, Oxford, 1–5.

Ure, A., Quevauviller, Ph., Muntau, H., and Griepink, B. (1993). The speciation of heavy metals in soils and sediments: an account of the improvement and harmonization of extraction techniques undertaken under the auspices of Community Bureau of Reference (BCR). *Int. J. Environ. Anal. Chem.* **51**, 135–151.

U.S. Geological Survey Staff (1989). *Humic Substances in Suwannee River, Georgia; Interactions, Properties, and Proposed Structures*, Open File Rep. 87-557, USGS, Denver, CO.

Van Aken, P. A., Liebscher, B., and Styrsa, V. J. (1998). Quantitative determination of iron oxidation states in minerals using Fe $L_{2,3}$-edge electron energy-loss near-edge structure spectroscopy. *Phys. Chem. Miner.* **25,** 323–327.

Van Loon, W. M. G. M., Boon, J. J., and de Groot, B. (1993). Quantitative analysis of sulfonic acid groups in macromolecular lignosulfonic acids and aquatic humic substances by temperature-resolved pyrolysis–mass spectrometry. *Environ. Sci. Technol.* **27**, 2387–2396.

Van Riemsdijk, W. H., Benedetti, M. F., and Hiemstra, T. (1997). Possibilities for predicting mobility of trace elements in soils via modeling. In *Contaminated Soils, Proc. 3rd International Conference on the Biogeochemistry of Trace Elements*, Paris, May 15–19, 1995, ed. Prost, R., INRA, Paris, 147–158.

Van Riemsdijk, W. H., Bolt, G. H., Koopal, L. K., and Blackmeer, J. (1986). Electrolyte adsorption on heterogeneous surfaces: adsorption models. *J. Colloid Interface Sci.* **109**, 219–228.

Van Riemsdijk, W. H., De Witt, J. C. M., Koopal, L. K., and Bolt, G. H. (1987). Metal ion adsorption on heterogeneous surfaces: adsorption models. *J. Colloid Interface Sci.* **116**, 511–522.

Waite, T. D. (1989). Mathematic modeling of trace metal speciation. In *Trace Metal Speciation: Analytical Methods and Problems*, ed. Batley, J. E., CRC Press, Boca Raton, FL, 117–184.

Wallace, A. (1980a). Effect of excess chelating agents on micronutrient concentrations in bush beans grown in solution culture. *J. Plant Nutr.* **2**, 163–171.

Wallace, A. (1980b). Effect of chelating agents on uptake of trace metals when chelating agents are supplied to soil in contrast to when they are applied to solution culture. *J. Plant Nutr.* **2**, 171–175.

Walter, I., and Cuevas, G. (1999). Chemical fractionation of heavy metals in a soil amended with repeated sewage application. *Sci. Total Environ.* **226**, 113–119.

Wang, W. H., Wong, M. H., Leharne, S., and Fisher, B. (1998). Fractionation and biotoxicity of heavy metals in urban dusts collected from Hongkong and London. *Environ. Geochem. Health* **20**, 185–198.

Westall, J. C. (1979). *MICROQL: A Chemical Equilibrium Program in BASIC*, Swiss Federal Institute of Technology (EAWAG), Zurich, Switzerland.

Westall, J. C., Zachary, J. L., and Morel, F. M. M. (1976). *MINEQL: A Computer Program for the Calculation of Chemical Equilibrium Composition of Aqueous Systems*, Tech. Note 18, Ralph M. Parsons Laboratory, Massachusetts Institute of Technology, Cambridge, MA.

Whitten, M. G., and Ritchie, G. S. P. (1991). Calcium chloride extractable cadmium as an estimate of cadmium uptake by subterranean clover. *Aust. J. Soil Res.* **29**, 215–221.

Winner, R. W. (1985). Bioaccumulation and toxicity of copper as affected by interactions between humic acid and water hardness. *Water Res.* **19**, 449–455.

Xia, K., Bleam, W., and Helmke, P. A. (1997a). Studies of the nature of Cu^{2+} and Pb^{2+} binding sites in soil humic substances using x-ray absorption spectroscopy. *Geochim. Cosmochim. Acta* **61**, 2221–2222.

Xia, K., Bleam, W., and Helmke, P. A. (1997b). Studies of the nature of binding sites of first row transition elements bound to aquatic and soil humic substances using x-ray absorption spectroscopy. *Geochim. Cosmochim. Acta* **61**, 2223–2236.

Xian, X., and Shokohifard, G. (1989). Effect of pH on chemical forms and plant availability of cadmium, zinc and lead in polluted soils. *Water Air Soil Pollut.* **45**, 265–273.

Zachara, J. M., Smith, S. C., Resch, C. T., and Cowan, C. E. (1992). Cadmium sorption to soil separates containing layer silicates and iron and aluminum oxides. *Soil Sci. Soc. Am. J.* **56**, 1074–1084.

Zevenhuisen, L. P. T. M., Dolfing, J., Eshuis, E. J., and Scholten-Koersehman, I. J. (1979). Inhibitory effects of copper on bacteria related to the free ion concentration. *Microb. Ecol.* **5**, 139–146.

12

FRACTIONATION AND MOBILITY OF TRACE ELEMENTS IN SOILS AND SEDIMENTS

P. S. FEDOTOV

Vernadsky Institute of Geochemistry and Analytical Chemistry, Russian Academy of Sciences, Moscow, Russia

M. MIRÓ

University of the Balearic Islands, Illes Balears, Spain

Biophysico-Chemical Processes of Heavy Metals and Metalloids in Soil Environments,
Edited by Antonio Violante, Pan Ming Huang, and Geoffrey Michael Gadd

12.1. INTRODUCTION

The accumulation of metal ions and metalloids in different compartments of the biosphere and their possible mobilization under environmentally changing conditions induces a perturbation of both the structure and function of the ecosystem and might cause adverse health effects to biota. Nowadays, it is widely recognized that the impact of anthropogenic metal and metalloid ions on the environment cannot be evaluated by measuring merely the total concentration of individual trace element (TE) species, because the mobility, bioavailability, and consequently, toxicity depend strongly on their chemical forms and type of binding. Thus, procedures for discrimination of different solid-phase associations of TEs are required. The identification and quantification of metals associated with predefined phases or soil compartments is defined as *fractionation analysis* according to the IUPAC recommendation (Templeton et al., 2000). The extractants used in the sequential extraction procedures (SEPs) are intended to simulate conditions whereby metals associated with certain mineralogical phases can be released. For example, changes in the ionic composition affecting adsorption–desorption reactions or the pH decrease in the soil solution may lead to the release of metals retained on a matrix by weak electrostatic interactions or coprecipitated with carbonates (*exchangeable* and *acid-soluble* forms, respectively). Decreasing the redox potential can result in dissolution of oxides, unstable under reducing conditions, and liberation of scavenged metals (*reducible* forms). Changes in oxidizing conditions may cause the degradation of organic matter and release of complexed metals (*oxidizable* form). Finally, the destruction of primary and secondary mineral lattice releases metals retained within the crystal structure, [e.g., due to isomorphous substitution, *residual* form]. It should be taken into consideration that special sequences of extracting reagents are needed for the fractionation of metalloids and nutrient species as well, as a consequence of their anionic nature. In general, sequential extraction provides relevant information on the origin, physicochemical mobility, and transport of TEs. Actually, the nominal "forms" determined by operational fractionation assist in the estimation of the amounts of elements in different reservoirs that can be mobilized under changes in the chemical properties of soil (Davidson et al., 1998).

The fundamental principles of equilibrium-based sequential extraction are detailed in the Section 12.3. Yet, to prevent biased information, special care should be taken on the sampling strategies and further sample handling of the environmental solid prior to use of the partitioning scheme, as discussed in Section 12.2. We also describe the state of the art regarding the development of novel fractionation methods conducted in a flow-through dynamic mode aimed at the investigation of leaching kinetics and elucidation of the fractional distribution of TE in a semi- or fully automated fashion.

12.2. SAMPLING AND SAMPLE PREPARATION

In recent years, problems associated with sampling and preanalytical sample preparation have received increasing attention from the scientific community

(Theocharopoulos et al., 2001; Paschke, 2003). Hence, the aim of an outstanding European project (i.e., the CEEM soil project) was to assess the existing guidelines and the degree and state of standardization in the field of soil sampling in the European countries, and to compare the results when applied to the same test area under comparable conditions by different sampling teams (Wagner et al., 2001). In addition, the European Commission has recently convened a workshop (Langenkamp and Marmo, 2001) to consider the topic of sampling of environmental parameters in soil and sludges.

A clear message regarding this initiative was the identification of a need for a concerted effort to achieve better harmonization of sampling and analysis methods that should possibly relate not merely to a single matrix but, if possible, should have "horizontal" applicability to a range of similar solid matrices (e.g., soil, sludges, biowastes) without compromising precision. The Horizontal Project seeks to fulfil these goals as standardized procedures in the form of CEN (European Standards organization) standards (Gawlik et al., 2004; Lambkin et al., 2004). In this chapter some aspects of sampling and sample preparation that may affect the fractionation and mobility of heavy metals and metalloids in soil environments are discussed briefly. In addition, novel strategies for in situ sampling of soil solution directly with minimum disturbance of natural conditions are addressed.

Sampling is often the major source of uncertainty in analytical results and the limiting step in assuring quality. In fact, the estimated variance due to sampling (S_s) can be expressed as

$$S_{\mathrm{s}} = S_{\mathrm{ps}} + S_{\mathrm{ss}} \tag{12.1}$$

where S_{ps} is the primary (field) sampling variability and S_{ss} is the subsampling variability comprising errors associated to mass reduction procedures. Uncertainty related to field sampling is typically much greater than errors associated with sample preparation, handling, and analytical analysis, as stressed by several authors (Muntau et al., 2001; Lambkin et al., 2004). The larger uncertainty component affecting sampling is the heterogeneity of the matrix. Therefore, for reliable fractionation of trace elements (TEs) in soils, sediments, or sludges, a representative sample is required, thereby adequately reflecting the area under investigation. An appropriate sampling design would thus include (Schleuß and Müller, 2001):

- Identification of all substantial compartments
- Selection of an optimum number of samples and a composite size to ensure a reduced copy of the sampling site
- Assessment of the sampling pattern (i.e., random, systematic, or judgmental)
- Statistical evaluation of distribution parameters of the elements of interest
- Implementation of quality measures for the assurance of data quality
- Incorporation of a special depth function for element distribution, particularly for urban soils

In addition, as reviewed by Hlavay et al., (2004), the seasonal variability is known to influence significantly the results of trace element fractionation in soils. Due to the variation in weather conditions, seasonal patterns of extractable fractions are not necessarily predictable from a few years of observation and may differ from site to site. Hence, it is virtually possible to obtain biased results when sampling is used only once. A distinction should also be made between sampling of natural, agricultural, grassland, forest, or moorland soils where element distribution and speciation can be regarded as being homogeneous, and sampling of industrially contaminated soils where the distribution is heterogeneous not only over the surface area but also with depth. In the latter case, random sampling may be desirable but will often be uneconomical and time consuming. The judgmental approach (Hämmann and Desaules, 2003), where the distribution of sampling sites is derived from the contamination hypothesis, may be needed.

It should also be noted that soil properties may vary considerably on a micro scale of about 1 to 100 mm. Thus, metal solubility and extractability may be affected either directly by micro nonhomogeneity of the total metal content or by simultaneous variation in soil properties (i.e., pH, organic matter, mineral composition, exchangeable cations, and soil texture; Hlavay et al., 2004). The preparation of soil samples generally involves the following steps: (1) drying or rewetting, (2) homogenizing and sieving, (3) storage, and occasionally, (4) grinding (Hlavay et al., 2004). Soil samples intended for fractionation analysis are seldom milled, as this sample pretreatment procedure can appreciably influence the content of soluble pollutants (Houba et al., 1994). Air drying prior to extraction is a standard procedure but leads to increased extractability of metal ions bound to Fe and Mn oxyhydroxides (Wenzel and Blum, 1994). The effect of air drying depends on soil properties and the initial moisture conditions; consequently, no general correlation is made available for the prediction of metal levels under field-moist conditions from the analysis of air-dried samples. Since extraction of field-moist samples cannot be recommended for routine analysis, individual relations on a local or regional scale should be obtained to avoid biased results in the determination of mobile pools of target species. Homogenization is an essential step for performing accurate soil analysis. This sample pretreatment approach may contribute to the creation of novel reaction sites (Beckett, 1989), thereby enhancing the metal readsorption phenomena during fractionation analysis, and the modification of pH of soil solution and carbonate content whenever applied to materials from different horizons. Sieving is also a central step in the preparation of soil samples. In most cases, soil samples are sieved to 2 mm (Theocharopoulos et al., 2001). If more detailed patterns for element distribution are needed, clay (< 2 μm), silt (2 to 50 μm), and sand (50 to 2000 μm) fractions can be separated and subjected to subsequent speciation analysis (Brady and Weil, 1996).

Regarding sediments, special care should be taken during sampling to prevent the alteration of natural biogeochemical processes, thus preserving the chemical speciation of TEs (Kersten and Förstner, 1995). Sampling for pollution mapping should consider the heterogeneity of the deposit by methods such as particle size analysis and geochemical normalization (Hlavay et al., 2004). As summarized by Förstner (2004), several factors might influence the results of the sediment sample analysis, including type of water system, prevailing bottom dynamics, size of the water body, bottom roughness, anthropogenic factors, sediment chemical conditions, sediment physical and biological characteristics, number of samples, type of sampling net, sampling devices, sample handling, and reliability of laboratory analysis. However, no systematic study has yet been made to assess the effect of even half of these factors.

It is generally accepted that fine-grained suspended and bottom-sediment particles below 63 μm accumulate greater concentrations of contaminants (particularly those with low water solubility) than coarse particles (particle size > 63 μm) (Förstner, 2004). The fine-grained particles exhibit higher adsorption and ion-exchange capacity for contaminants than that of the coarse fragments (Horowitz, 1991). Further, sediments support a large part of the benthic community by supplying the food as organic matter associated with the fine-grained particles. The assessment of sediment quality must therefore be carried out on the fine-grained fraction sampled in areas of the water body where permanent accumulation of sediments is taking place (Mudroch and Azcue, 1995).

Freezing is considered to be an acceptable preservation method for wet sediments collected for the determination of inorganic constituents (Förstner, 2004). It has been shown that rapid deep-freezing can best maintain sample integrity, thus enabling reliable evaluation of contaminant concentrations. The lower the deep-freezing temperature the better; a minimum temperature of −80°C was suggested.

Handling operations of sediment samples include drying, sieving, grinding, mixing, and homogenization. Three types of drying may be used to prepare solid samples prior to analysis (Mudroch and MacKnight, 1994; Förstner, 2004):

1. *Air-drying* is rarely used for the preparation of sediments for pollution studies, since it may generate undesirable changes in metal complexation and availability. In some cases, air-drying has been used to avoid losses of components, such as mercury, which are volatile at temperatures above 50 to 60°C.
2. *Oven-drying* of sediments is usually performed on samples collected for the determination of inorganic components, such as major and trace elements. However, oven-drying is not acceptable for either organic or inorganic sediments that contain any volatile or oxidizable component.
3. *Freeze-drying* is the most suitable option for the determination of both organic and inorganic components. The principal advantages of freeze-drying for sediments are the following: (a) chemical changes in temperature-labile components do not occur; (b) loss of volatile constituents and oxidation of targeted phases are entirely circumvented; (c) Most particles of dried sediments remain dispersed; and (d) sterility is maintained.

Generally, due to the instability of polluted sediments, sample handling and storage prior to the fractionation of TEs are somewhat problematic. In particular, changes from reducing to oxidizing conditions, which involve transformations of sulfides and the shift to more acidic conditions, might increase the mobility of metal ions and metalloids (Förstner, 1993). In addition, the physicochemical properties of soils and sediments (pH, cation-exchange capacity, redox potential, organic carbon content, total nitrogen content, particle-size distribution, contents of amorphous and crystalline iron oxides) should be considered carefully for appropriate sampling design and sample handling (Paschke, 2003) as well as for the interpretation of fractionation results.

Environmental researchers have, however, recognized that sampling and sample preparation are in fact the Achilles' heel of metal partitioning studies. Sampling itself is a distortion of the natural environment, which can severely alter the composition of the sample and the original element-fractionation distribution as a consequence of analyte transformations and change of equilibria at the soil–solution interface. Any further steps needed for sample preparation (e.g., homogenization, drying, crushing, grinding) enhance these risks and might call into question whether the appropriate information with respect to metal mobility within soils, sediments, sludges, or solid biowastes, and availability to plants via root uptake, can be received. Noninvasive or in situ measurement techniques (e.g., via remote sensing or implantable sensors) applied directly to the natural sample in its original state would be ideal. However, there are very few parameters (e.g., water content, pH) that can be determined in such a way. For many years, gravity-driven zero-tension lysimeters (Edwards et al., 1993; Lawrence and David, 1996) have been proven effective tools for in situ sampling of soil solution and for assessment of the effect of anthropogenic actions and natural occurrences on metal accessibility from field-moist materials. On-site lysimeter experiments are performed by implementation of a plastic cylindrical bucket (with a capacity ranging from the cubic decimeter to the cubic meter level) into the soil under investigation, which is further connected to a deeper receiver canister via a drainage pipe. Long-term explorations are conducted by periodical analysis of water percolated through the lysimeter soil. To facilitate the free infiltration of soil water, the field lysimeter is composed of a series of screen layers of variable thickness and composition (e.g., quartz gravel and quartz sand) placed in sequence below the overlying soil layer. This methodology has been regarded as the most appropriate approximation for sampling of soil solution under natural conditions. In fact, the traditional paste method is based on off-site collection of soil solution and involves the addition of water to a fresh sample, which, after equilibration, is further subjected to centrifugation or filtration.

The installation of the various lysimeter components needs careful consideration to limit disruption of the sampling area, and hence, to ensure maintenance of drainage conditions close to those in undisturbed soils. Lysimeter analyses average the metal concentration released from the bulk solid sample for long monitoring periods, whereby the resulting temporal and spatial resolution may be insufficient for following fast metal leaching kinetics and processes occurring locally.

To understand the local behavior and mobility of TEs in definite sites of environmental concern such as the rhizosphere (i.e., the plant–root environment) and assess the root-induced changes of soil, *rhizobox systems* have attracted increasing interest (Fitz et al., 2003). They consist of a plastic acrylic container divided into a set of nylon compartments of appropriate mesh size capable of isolating the plant roots from the rhizosphere soil (e.g., the soil in direct contact with the root-only plane) and the bulk sample (Wenzel et al., 2001b). The technique of diffusive gradients in thin films (DGTs) has been presented as a promising tool for predicting element fractions available for plant uptake (Zhang and Davison, 1995; Zhang et al., 2001). It is based on a simple implantable device that accumulates the mobilized elements on a binding reagent (e.g., iron oxide gel) after diffusion through a well-defined hydrogel layer (Fitz et al., 2003). This methodology is founded on kinetic rather than equilibrium principles, thus allowing the measurement of metal fluxes and monitoring of plant–soil interfacial concentrations under dynamic conditions. Although DGT-mediated analyses involve, in most cases, manual treatment of the soil collected (e.g., cutting the sample in root parallel sections, moistening the resulting layers, and effecting off-site entrapment of target species with subsequent digestion of the metal-containing gel), recent trends are focused on the implementation of these diffusion-based samplers into the soil body or the various rhizobox compartments for in situ measurements.

The microsuction cup method, based on the utilization of tension lysimeters, is one of the most popular techniques for gathering interstitial liquids and root exudates from environmental solids (Grossmann and Udluft, 1991; Göttlein et al., 1996). Tension lysimeters are porous cups or plates with a thickness of a few millimeters placed within specific locations at the monitored area for probing the pore soil water periodically, mostly via hand-powered or automatic vacuum pumping systems (Haines et al., 1982). Lysimeters may also be used to monitor any chemical change of contaminants while traveling through the vadose zone. It has, however, been proved that the filter material of the suction cups is a potential source of bias, owing to the partial adsorption of the isolated compounds (Wenzel and Wieshammer, 1995; Wenzel et al., 1997). Nylon membranes rather than ceramic materials are strongly recommended for sampling metal ions, due to their negligible influence on the concentration of dissolved elements (Wenzel et al., 1997). It should also be noted that each time the soil solution is sampled, interstitial water is removed from the sampling site, thus changing the soil moisture and conditions for water and nutrient transport (Vetterlein and Jahn, 2004). As a consequence, the interpretation of kinetic processes at field sites becomes troublesome.

To tackle the aforementioned problem, a novel and interesting concept based on microdialysis sampling has been addressed (Torto et al., 2001; Miró and Frenzel, 2004a). Although flow-through microdialysis is a continuous sample cleanup technique used conventionally in neurochemistry and pharmacokinetic studies, current trends have been directed to expand the scope of this technique to the environmental and agricultural fields (Torto et al., 2002; Miró and Frenzel, 2005). It is based on deploying a dedicated capillary-type probe acting as a buried diffusion-controlled passive sampler for in situ microsampling of low-molecular-mass compounds from soil solution. The major attractive features of microdialysis are the minimal depletion of dialyzable compounds from interstitial soil water, and the miniature dimensions of the flow-through probe that facilitate straightforward implementation into the target site (e.g., rhizosphere soil) with negligible distortion of the natural distribution equilibria. In addition, microdialysis is a unique strategy to detect hot spots in the soil body as a result of the unrvaled spatial resolution (i.e., ca. 200 μm for capillary probes) (Miró and Frenzel, 2004a). It has also attracted particular interest because of the capability of continuous monitoring of the ongoing leaching processes in the outer solid microenvironment in almost real time, as detailed in Section 12.3. *Stimulus–response* (also termed retrodialysis) *approaches* based on the provision of the leaching agent (or alternatively, the target species) via the perfusion liquid should be regarded as appropriate vehicles for the investigation of adsorption–desorption profiles and evaluation of labile metal pools at the sampling area (Miró et al., 2005b). As with in vivo assays, there is a lack of harmonization of on-site calibration methods for estimating the dialysis yields of microanalytical devices after implantation into targeted moist solid samples, thus hindering the accurate comparison of quantitative analytical results reported by various authors.

12.3. FRACTIONATION OF TRACE ELEMENTS INTO OPERATIONALLY DEFINED FORMS

12.3.1. Equilibrium-based sequential extraction procedures

In geological studies, fractionation of TEs into operationally defined forms under the sequential action of given leachants with increasing aggressiveness is a common approach to distinguishing various species of TEs according to their physicochemical mobility and potential bioavailability (Ure and Davidson, 2001; Hlavay et al., 2004). Thus, eight extractable fractions of trace metals can be separated, as shown in Table 12.1. The fractionation of elements existing in anionic forms (As, Se, P) needs obviously specific sequences of leaching reagents, as discussed later.

There are two well-accepted approaches to the classification of operationally defined fractions. The first is based on the specific action of the reagents used. Hence, fractions can be called *acid soluble, easily reducible, easily oxidizable*, and so on. These terms are virtually correct and reflect conditions whereby TEs associated with certain soil (sediment) components can be released. The second approach is based on the mineralogical phases that are supposed to be dissolved under the action of a more-or-less selective reagent. Therefore, fractions can be termed as *elements coprecipitated with carbonates, elements bound to Mn oxides, elements bound to Fe oxides*, and so on. Although these names are definitely useful for the interpretation of fractionation results, in some cases they might lead to misunderstanding because the reagents used are not highly selective and might occasion the shift or overlapping of fractions. Also, for example, the term *coprecipitated with carbonates* should not be used for a noncalcareous soil.

TABLE 12.1. Operationally Defined Fractions and Corresponding Extracted Components

Operationally Defined Fraction	Extracted components (Possible Forms of Binding TEs)	Physicochemical Mobility	Potential Bioavailability	Common Extractants
Water soluble	Free ions	Mobile	Easily available	Deionized water
Exchangeable (A)	Exchangeable ions (metals retained by weak electrostatic interactions)	Mobile	Easily available	$Ca(NO_3)_2$, $Mg(NO_3)_2$, $CaCl_2$, $MgCl_2$, CH_3COONH_4
Acid soluble (B)	Carbonates (for noncalcareous soils: other specifically adsorbed forms of TEs)	Easily mobilizable	Easily available	CH_3COOH, CH_3COONa ($pH \leq 6$)
Easily reducible (C)	Mn oxyhydroxides	Readily mobilizable	Readily available	$NH_2OH \cdot HCl$
Easily oxidizable (D)	Metal–organic complexes	Readily mobilizable	Readily available	$K_4P_2O_7$, $Na_4P_2O_7$, NaClO, NaOH
Moderately reducible (E)	Amorphous Fe (and Al) oxyhydroxides	Poorly mobilizable	Poorly available	Oxalate buffer, $NH_2OH \cdot HCl$
Moderately and poorly oxidizable (F)	Refractory organic compounds and sulfides	Poorly mobilizable	Poorly available	H_2O_2/CH_3COONH_4
Poorly reducible (G)	Crystalline Fe and Al oxyhydroxides	Poorly mobilizable	Poorly available	$(NH_4)_2C_2O_4$/ascorbic acid
Residual	Mineral lattice (metals retained within the crystal structure)	Immobile	Unavailable	HF, HNO_3, $HClO_4$

Batchwise sequential extraction schemes, also termed *end-over-end extraction procedures*, have, however, been criticized, and several researchers (Filgueiras et al., 2002a; Gleyzes et al., 2002) claimed the existence of the following shortcomings: lack of harmonization of procedures, insufficient selectivity of a given reagent for a specific phase, readsorption of released forms, lack of quality control, and poor reproducibility. In fact, as the term implies, the results of operationally defined fractionation are strongly dependent on the experimental conditions selected, (i.e., type of leaching reagents, contact time between solid and liquid phases, mode of shaking, and sample/reagent ratio).

Despite these drawbacks, sequential extraction is still regarded as an essential tool for distinguishing TE fractions of different solubility related to mineralogical phases. By using SEPs we expect to obtain insights into both the principal types of binding of TEs with a soil (sediment) matrix and the factors governing the behavior of TEs (Ladonin, 2002). For example, in calcareous soils, where the formation of poorly soluble compounds appears to be one of the principal mechanisms for TE fixing, determination of the "carbonate" fraction is of paramount importance. In humus-enriched soil horizons, special attention should be paid to TEs bound to organic matter.

In addition, the overall content of TEs in extractable (mobile) forms can be estimated prior to SEP by a single-extraction procedure using electrolyte solutions, such as 1.0 mol L^{-1} HNO_3 or 1.0 mol L^{-1} HCl (Ladonin, 2002). If the contents of TEs obtained by the screening protocol are rather low, it might then be unreasonable to use a sequential extraction method.

12.3.1.1. Leaching of trace metals Sequential extraction schemes for metal partitioning in environmental solid samples have recently been summarized in a comprehensive fundamental review by Filgueiras et al. (2002a) and in an IUPAC technical report by Hlavay et al. (2004). Gleyzes et al. (2002) have reviewed the reagents used in the various schemes, their advantages and shortcomings being pinpointed. In the present chapter, some representative SEPs are described and relevant information on separated fractions is supplied. New trends for development, improvement, and application of SEP are also discussed briefly.

Sequential extraction schemes for TEs were proposed for the first time in the late 1960s (Ladonin, 2002). During the last 30 years, a few dozen different SEPs have been proposed (Filgueiras et al., 2002a; Gleyzes et al., 2002) and several reliable schemes have received widespread acceptance. However, none of these procedures has been found to be absolutely preferable. A few representative examples of SEPs are summarized in Table 12.2. McLaren and Crawford (1973) were the first to develope a scheme capable of discriminating a variety of extractable fractions: exchangeable, weakly bound to specific sites, organically bound, and occluded by oxide material. According to this SEP, the organically bound fraction should be extracted prior to dissolution of the oxide material. A few years later, Tessier et al. (1979) introduced one of the most popular protocols. This scheme also addresses four extractable fractions, however, it employs different reagents, and TEs bound to Mn and Fe oxides are separated prior to the sulfidic/organic fraction.

TABLE 12.2. Representative Sequential Extraction Procedures for the Fractionation of Trace Elements

SEP	Leaching Reagents	Extractable Fractions (Residual Fractions Are Not Included)
McLaren and Crawford (1973)	0.05 mol L^{-1} $CaCl_2$	Exchangeable (A)
	2.5% CH_3COOH	Weakly bound to specific sites (B)
	0.1 mol L^{-1} $K_4P_2O_7$, pH 10	Organically bound (D)
	0.1 mol L^{-1} $H_2C_2O_4$ + 0.175 mol L^{-1} $(NH_4)_2C_2O_4$, pH 3.2	Occluded by Mn, Fe, and Al oxyhydroxides (C + E)
Tessier et al. (1979)	1 mol L^{-1} $MgCl_2$	Exchangeable (A)
	1 mol L^{-1} CH_3COONa, pH 5	Bound to carbonates (B)
	0.04 mol L^{-1} $NH_2OH \cdot HCl$ + 25% CH_3COOH, pH 2, 96°C	Bound to Mn and Fe oxyhydroxides (C + E)
	30% H_2O_2, pH 2 (85 °C); then 3 mol L^{-1} CH_3COONH_4	Bound to organic matter (D + F)
Kersten and Förstner (1986)	1 mol L^{-1} CH_3COONH_4, pH 7	Exchangeable (A)
	1 mol L^{-1} CH_3COOH, pH 5	Bound to carbonates (B)
	0.01 mol L^{-1} $NH_2OH \cdot HCl$ + 0.01 mol L^{-1} HNO_3, pH 2	Bound to Mn oxyhydroxides (C)
	0.1 mol L^{-1} oxalate buffer, pH 3	Bound to Fe and Al oxyhydroxides (E)
	30% H_2O_2, pH 2 (85 °C); then 1 mol L^{-1} CH_3COONH_4 in 6% HNO_3	Sulfidic/organic (D + F)
Zeien and Brümmer (1989)	1 mol L^{-1} NH_4NO_3, pH 7	Exchangeable (A)
	1 mol L^{-1} CH_3COONH_4, pH 6	Specifically sorbed (B)
	0.1 mol L^{-1} $NH_2OH \cdot HCl$ + 1 mol L^{-1} CH_3COONH_4, pH 6	Bound to Mn oxyhydroxides (C)
	0.025 mol L^{-1} NH_4-EDTA, pH 4.6	Organically bound (D)
	0.2 mol L^{-1} $(NH_4)_2C_2O_4$, pH 3.2	Bound to amorphous Fe and Al oxyhydroxides (E)
	0.2 mol L^{-1} $(NH_4)_2C_2O_4$ + 0.1 mol L^{-1} ascorbic acid (boiling)	Bound to crystalline Fe and Al oxyhydroxides (G)

(continued overleaf)

TABLE 12.2. ***(continued)***

SEP	Leaching Reagents	Extractable Fractions (Residual Fractions Are Not Included)
Krishnamurti et al. (1995)	1 mol L^{-1} $Mg(NO_3)_2$, pH 7	Exchangeable (A)
	1 mol L^{-1} CH_3COONH_4, pH 5	Carbonates (B)
	0.1 mol L^{-1} $Na_4P_2O_7$, pH 10	Metal–organic complexes (D)
	0.1 mol L^{-1} $NH_2OH \cdot HCl$ + 0.01 mol L^{-1} HNO_3	Easily reducible metals (C)
	30% H_2O_2, pH 2 (85°C); then 1 mol L^{-1} $Mg(NO_3)_2$ + 20% HNO_3	Oxidizable organic matter, sulfides (F)
	0.2 mol L^{-1} $(NH_4)_2C_2O_4$, pH 3	Amorphous Fe and Al oxyhydroxides (E)
	0.2 mol L^{-1} $(NH_4)_2C_2O_4$, pH 3 + 0.1 mol L^{-1} ascorbic acid; 95°C	Crystalline Fe and Al oxyhydroxides (G)
Ure et al. (1993), Original BCR	0.11 mol L^{-1} CH_3COOH (16 h)	Exchangeable and acid soluble (A + B)
	0.1 mol L^{-1} $NH_2OH \cdot HCl$ + HNO_3 to adjust pH 2 (16 h)	Fe and Mn oxyhydroxides (C + E)
Rauret et al. (1999), Modified BCR	8.8 mol L^{-1} H_2O_2 (2 × 1 h, 85°C); then 1.0 mol L^{-1} CH_3COONH_4 + HNO_3 to adjust pH 2 (16 h)	Sulfidic/organic (D + F)
	0.11 mol L^{-1} CH_3COOH (16 h)	Exchangeable and acid soluble (A + B)
	0.5 mol L^{-1} $NH_2OH \cdot HCl$ + HNO_3 to adjust pH 1.5 (16 h)	Fe and Mn oxyhydroxides (C + E + partially G)
	8.8 mol L^{-1} H_2O_2 (2 × 1 h, 85°C); then 1.0 mol L^{-1} CH_3COONH_4 + HNO_3 to adjust pH 2 (16 h)	Sulfidic/organic (D + F)

Kersten and Förstner's (1986) procedure is based on Tessier's protocol, but it addresses five extractable fractions and allows one to distinguish between Mn and Fe oxides, which are characterized by their different mobility and availability. Zeien and Brümmer (1989) proposed to separate six extractable fractions and to distinguish between Mn oxides, amorphous Fe oxides, and crystalline Fe oxides. Manganese oxides are separated before the organically bound TEs. It should be noted that Krishnamurti's scheme addressing seven extractable fractions is capable of distinguishing between the easily oxidizable (humic/fulvic acid-bound) fractions using pyrophosphate as extractant and poorly oxidizable (sulfidic/organic) fractions using hydrogen peroxide at 85°C (Krishnamurti et al., 1995). Both the sulfide and organically bound fractions are separated prior to the three-step dissolution of oxide material. According to the modified scheme of Krishnamurti and Naidu (2000), TEs associated with fulvic and humic substances can be fractionated additionally. Readers are referred to Chapter 11 for further details on the above-mentioned SEPs.

The lack of harmonization of sequential extraction schemes impedes the worldwide comparison of fractionation results and validation of operational methods. To overcome this limitation, the European Commission through the BCR (Community Bureau of Reference) Programme and its successor [Standards, Measurements, and Testing Programme, (SM&T)] launched a collaborative project (Ure et al., 1993) aimed to (1) design a simplified three-step sequential extraction scheme, (2) test the developed scheme in interlaboratory studies involving expert European laboratories, and (3) certify the extractable TE contents in a sediment reference material. This project has been described extensively in the literature (Quevauviller et al., 1994; Rauret and López-Sánchez, 2001). A detailed description of the conditions during each of the three steps, as well as the preparation of leaching solutions, was given to prevent any difference affecting reproducibility of the procedure. Nevertheless, when the scheme was applied to the fractionation of Cd, Cr, Cu, Ni, Pb, and Zn in a sediment reference material (CRM 601), the contents of Cu and Cr extracted at the second step (reducible fraction) as well as the contents of Cu, Cr, and Zn extractable at the third step (oxidizable fraction) were not certified, due to the lack of reproducibility (Quevauviller et al., 1997; López-Sánchez et al., 1998). To identify the main sources of uncertainties, the influence of several factors, such as pH, temperature, extraction time, and concentration of reagents, was investigated thoroughly (Rauret et al., 1999; Sahuquillo et al., 1999). It was realized that the extraction step attacking reducible forms (step 2) was the main source of irreproducibility. Then a modified three-step SM&T leaching scheme was proposed (Rauret et al., 1999), employing an increased concentration of hydroxylamine hydrochloride (0.5 mol L^{-1} rather than 0.1 mol L^{-1}) and a lower pH value (1.5 rather than 2.0). The extractable contents of Cd, Cr, Cu, Ni, Pb, and Zn in a new sediment reference material (BCR 701) were certified on the basis of a modified SM&T leaching scheme (Pueyo et al., 2001). The suitability of this scheme has also been studied for reference soil samples [CRM 483 (Rauret et al., 2000), SRM 2710 (Sutherland and Tack, 2002; Kubová et al., 2004) and SRM 2711 (Sutherland and Tack, 2002; Kubová et al.,

2004)], road-deposited sediments (Sutherland et al., 2000), and sludge (Pueyo et al., 2003).

It should be noted that the results obtained by the use of different SEPs are often difficult to compare. For example, fraction-specific concentrations, percentages, and relative recoveries for Cu, Pb, and Zn in SRM 2710 and 2711 were used to explore differences between the optimized (modified) SM&T procedure and three other sequential extraction schemes (i.e., the Tessier scheme, the Geological Survey of Canada scheme, and the original SM&T scheme) (Sutherland and Tack, 2003). The experimental results revealed significant differences among the four schemes, even for those closely associated. Thus, the TEs released by the various SEPs in the different fractions must be assessed carefully because different trends may be obtained for the same metal species, soil-phase association, and environmental solid.

For a better understanding of the peculiarities of SEPs, we discuss the most commonly elucidated TE fractions briefly in the following sections. Particular attention is given to the selectivity of leachants toward mineralogical phases.

a. Water-Soluble Fraction This fraction is usually negligible although TEs extracted by water are most mobile and thus potentially most bioavailable (Filgueiras et al., 2002a). The water-soluble species are free ions or ions complexed with soluble organic matter and other constituents. However, the use of water as an extractant is connected with some methodological problems: (1) the organic matter content of the sample considerably influences the interaction between the water and the solid phase, and (2) water as an extractant has no buffering capacity; hence, easily mobilizable major components of soils or sediments strongly affect the results of a long-term leaching of TE. The amount of bioavailable trace metals in soils and sediments often correlates satisfactorily with the sum of the water-soluble fraction and the fraction extracted with a diluted salt solution (Brümmer, 1986). Hence, the water-soluble form of TEs is generally determined together with the exchangeable fraction.

b. Exchangeable Fraction Exchangeable (weakly adsorbed) cations are retained on the solid surface by relatively weak electrostatic interactions or by ion-exchange processes. Hence, changing the ionic composition or lowering the pH of a soil solution could result in the mobilization of exchangeable metals (Salomons and Förstner, 1984; Ure and Davidson, 2001). The choice of a proper reagent is critical for adequate determination of exchangeable TEs in soils and sediments. A potential electrolyte cation should be characterized by high displacement strength and an ionic radius close to that of the target metal, but also by the inability of specific interaction with soil (sediment) components that might lead to the formation of precipitates or complexed compounds (Ladonin, 2002). The latter requirement should also be fulfilled by the extracting electrolyte anion.

In practice, unbuffered salt solutions such as $MgCl_2$, $CaCl_2$, $BaCl_2$, $NaNO_3$, KNO_3, NH_4NO_3, $Mg(NO_3)_2$, $Ca(NO_3)_2$, $Pb(NO_3)_2$, and CH_3COONH_4 are usually employed (Ure, 1996; Kennedy et al., 1997; Hlavay et al., 2004) for the leaching of metals retained on the negative sites of the solid phase by electrostatic interactions. $MgCl_2$ and CH_3COONH_4 at a concentration level of 1.0 mol L^{-1} are the most widely used extractants. However, besides leaching the exchangeable fraction, acetate and chloride salts may form rather stable complexes with a number of TEs. As a result, equilibrium toward the liquid phase is shifted, whereby the extractable amounts are overestimated. Nitrate salts are more advantageous (Krishnamurti et al., 1995; Filgueiras et al., 2002a) because only ion-exchange processes are involved in the reaction. Concerning the nature of the electrolyte cation, Ca^{2+} is preferable over other divalent ions. Sr^{2+} or Ba^{2+} are characterized by higher displacement strength, thereby enhancing release of the exchangeable fraction (Ladonin, 2002). However, the selectivity for exchangeable TEs is generally poorer than that of Ca^{2+}. This can be explained by the possible parallel reactions of complexation and/or precipitation with matrix ingredients. Therefore, calcium nitrate that meets the above-mentioned requirements should be regarded as an attractive reagent for the exchangeable fraction.

The concentration of electrolyte should suffice for quantitative stripping of weakly retained ions, but an excessive surplus is not recommended because the increase in ionic strength might affect the stability of other matrix components of soils or sediments as well as the accuracy of the analytical instrument for final TE quantitation. A concentration range from 0.05 to 1.0 mol L^{-1} is usually acceptable.

It should be noted that the exchangeable fraction can be separated as a single extraction procedure using electrolyte solutions [e.g., NH_4NO_3, $Ca(NO_3)_2$, $CaCl_2$, $NaNO_3$)] for fast assessment of TE availability for plant uptake (Kennedy et al., 1997; Pueyo et al., 2004). Further details can be found in Chapter 11.

c. Acid-Soluble Fraction In general, the acid-soluble fraction contains metals released at a pH close to 5. The most popular leachant is the sodium acetate/acetic acid buffer (pH 5) at a 1.0 mol L^{-1} concentration level (Tessier et al. 1979; Ure, 1996; Gleyzes et al., 2002). The fraction recovered under these experimental conditions addresses metals coprecipitated with carbonates as well as TEs specifically adsorbed to certain sites on the surface of clays, organic matter, and hydrous oxides of iron and manganese. Besides, the lowering of pH from 7 (typically used for separation of the exchangeable fraction) to 5 results in the release of a fraction of weakly adsorbed cations that was not extracted in the preceding step. For soils and sediments with a high carbonate content, an incomplete dissolution of carbonate minerals in the sodium acetate/acetic acid buffer is sometimes observed (Tack and Verloo, 1995). Consequently, the release of metals coprecipitated with carbonate might still occur during separation of the following fractions, resulting in TE overestimation due to phase overlapping.

Nonbuffered acetic acid solutions (0.1 to 1.0 mol L^{-1}) are frequently used to strip the specifically adsorbed fraction. A 0.11 mol L^{-1} acetic acid solution is used in both the original and modified SM&T procedures for the simultaneous extraction of water-soluble, exchangeable, and acid-soluble fractions (Ure et al., 1993; Rauret et al., 1999). The higher the concentration of acetic acid, the higher was the recovery of TEs. For a set of contaminated soils and sediments, the 0.43

to 0.11 mol L^{-1} acetic acid recovery ratio was found to be within the ranges 1 to 1.4, 1.3 to 3.2, 2.1 to 6.4, and 2.1 to 9.4 for Cd, Zn, Cu, and Pb, respectively (Sahuquillo et al., 2003).

In general, TEs in exchangeable and acid-soluble forms are considered to be easily bioavailable (see Table 12.1). The metal–organic complex-bound form is also relatively mobile (Krishnamurti et al., 1995). The reducible and oxidizable forms are relatively stable under standard soil conditions, yet easily reducible and oxidizable fractions may be readily mobilized, as detailed in the ensuing sections.

d. Pyrophosphate Extractable Fraction Metal species bound to organics are more reactive physicochemically than are amorphous inorganic metal oxides. (Bascomb, 1968). Sodium pyrophosphate (0.1 mol L^{-1}, pH 10.0) was shown to be a specific reagent for the removal of Al–and Fe–organic complexes (McKeague, 1967; Bascomb, 1968). The differentiation of metal–organic complexes as distinct from other amorphous metal oxide–bound and other organically bound metal species (Krishnamurti et al., 1995) is a recent innovation in the selective sequential extraction scheme proposed by Krishnamurti et al. (2005). This scheme distinguishes particulate-bound metal species in soils as exchangeable, carbonate-bound, organic complex–bound, easily reducible metal oxide–bound, organic-bound, amorphous mineral colloid–bound, crystalline metal oxide–bound, and residual.

Krishnamurti et al. (1995) determined the distribution of particulate-bound Cd species in selected typical soils of southern Saskatchewan, Canada, following the schemes of Tessier et al. (1979) and the scheme modified by them. They found that Cd in these soils was bound predominantly to organic complexes, accounting for, on average, 40% of the total Cd present in the soils; whereas, on average, 36% of the Cd in the soils was observed to be in the Fe, Mn oxide–bound form, following the fractionation scheme of Tessier et al. (1979). The average percent distribution of solid-bound Cd species in surface horizons of the temperate soils of southern Saskatchewan studied is in the order organic complex–bound > carbonate-bound > residual > organic-bound > crystalline Fe oxide–bound > easily reducible metal oxide–bound > amorphous mineral colloid–bound; that of the tropical soils of Kenya (Onyatta and Huang, 1999) is in the order organic complex–bound > residual > crystalline Fe oxide–bound > organic-bound > amorphous mineral colloid–bound > easily reducible metal oxide–bound > carbonate-bound (not detectable). The fraction of Cd bound to organic complexes in the surface horizons of the temperate soils accounts for 31 to 55%, with an average of 40%, of the total Cd present in the soils. The amount of Cd bound to organic complexes is also generally highest among the particulate-bound Cd species of tropical soils, accounting for 25 to 46%, with an average of 37%, of the total Cd in the soils.

Compared with bulk soils, solid-phase speciation of Cd differs substantially in phosphate fertilizer–treated rhizosphere soils (Krishnamurti et al., 1996). The amounts of Cd associated with organic complexes in rhizosphere soils at the 2-week plant growth stage, particularly in soils treated with Idaho phosphate fertilizer, are appreciably higher than those of the corresponding bulk soils. Compared to the corresponding bulk soils, the amount of organic-complexed Cd in rhizosphere soils increased by 4 to 7% in the control soils and two- to threefold in soils treated with Idaho phosphate fertilizer. The prolific root growth of the plants, particularly in the soils treated with Idaho phosphate fertilizer, might have resulted in the secretion of increased amounts of low-molecular-mass organic acids (LMMOAs) into the rhizosphere, which resulted in high amounts of metal–organic complexes by chelation. Appreciable amounts of LMMOAs were detected in the root exudates of durum wheat, with the actual amount depending on the cultivar (Szmigielska et al., 1995, 1997; Cieśliński et al., 1998). The sustaining Cd release from the soils by LMMOAs was shown to have the same trend as that of Cd accumulated in the plant (Krishnamurti et al., 1997). Therefore, the importance of the organic complex–bound metal species merits attention.

e. Reducible Fractions Hydrous oxides of manganese and iron, present as a coating on mineral surfaces or as fine discrete particles, are well-known sinks of heavy metals in soils, sediments, and sludge. The following mechanisms can be involved in the scavenging of TEs: coprecipitation, adsorption, surface complex formation, and ion exchange (Filgueiras et al., 2002a). The iron and manganese oxides are thermodynamically unstable under anoxic conditions. The reduction of Fe(III) and Mn(IV) results in their dissolution and hence the release of bound TEs. Hydroxylamine, oxalic acid/ammonium oxalate buffer (Tamm's reagent), and sodium citrate/sodium dithionite buffer (Mehra and Jackson's reagent) are commonly exploited to leach TEs bound to manganese and iron oxyhydroxides.

The reducible fraction can be recovered in three steps: easily reducible fraction (manganese oxides), moderately reducible fraction (amorphous iron oxides), and poorly reducible fraction (crystalline iron oxides). A few SEPs address all three fractions (Miller et al., 1986; Zeien and Brümmer, 1989; Krishnamurti et al., 1995). A number of SEPs are only able to differentiate between easily reducible and moderately reducible fractions (called *readily* and *poorly available forms of TE*, respectively) (Shuman, 1983; Salomons and Förstner, 1984; Kersten and Förstner, 1986). Due to the lack of selectivity of reagents, with consequent overlapping of separated fractions, all reducible forms are typically separated together (Ure et al., 1993; Rauret et al., 1999). Despite the fact that the SEPs may be simplified, in this way, simultaneous recovery of readily and poorly available forms of TE seems rather disadvantageous for risk assessment purposes.

It has been reported that use of a 0.01 to 0.1 mol L^{-1} hydroxylamine hydrochloride solution acidified at pH 2 causes the selective dissolution of manganese oxides with minimum extraction of iron oxides (Shuman, 1985; Kersten and Förstner, 1986; Berti and Cunningham, 1997). Nevertheless, the low pH value may lead to a partial release of TE bound to organic matter. The separated reducible fraction can thus be overestimated. An improved method for the selective dissolution of manganese oxides involving a nonacidified 0.1 mol L^{-1} hydroxylamine solution at pH 3.6 as a leachant has been proposed by Neaman et al. (2004). Actually, the nonacidified hydroxylamine solution is not expected to dissolve considerable amounts of iron oxides, organic matter, or

Al-containing minerals. As a result, reliable information on associations of TE with manganese oxides is to be attained. Therefore, this reagent can be regarded as an attractive alternative to common reducing agents of sequential extraction schemes for the fractionation of TEs in soils and sediments.

The dissolution of amorphous and poorly crystallized iron oxyhydroxides (moderately reducible fraction) may be achieved using 0.04 to 0.1 mol L^{-1} hydroxylamine hydrochloride in 25% acetic acid (pH 2) at high temperatures (96 to 100°C) (Tessier et al., 1979, Belzile et al., 1989). However, this procedure was considered to be insufficiently effective for quantitative extraction of iron oxides, especially for materials with a high total content of iron (La Force and Fendorf, 2000; Gleyzes et al., 2001). If the concentration of reductant is increased up to 0.5 to 1.0 mol L^{-1}, a relatively selective leaching of the moderately reducible fraction can be achieved without appreciable stripping of silicates (Cabral and Lefebvre, 1998). According to the modified SM&T scheme (Rauret et al., 1999), manganese oxides, amorphous iron oxides, and poorly crystallized iron oxides are extracted simultaneously by 0.5 mol L^{-1} hydroxylamine at pH 1.5.

Another alternative for the selective dissolution of amorphous and poorly crystallized iron oxides involves chemical attack of the targeted phases with oxalic acid/ammonium oxalate at pH 3.2 (Tamm's reagent). Amorphous and poorly crystalline oxyhydroxides of aluminum are, however, extracted together with iron oxides due to the formation of very stable Al–oxalate chelates (Gleyzes et al., 2002). Tamm's reagent has been used in a number of SEPs for leaching of the moderately reducible fraction at room temperature, and it is considered to be superior to hydroxylamine (Filgueiras et al., 2002a; Davidson et al., 2004). Nevertheless, leaching of TEs bound to organic matter (fulvates and humates) is likely to occur due to the complexing capacity of oxalate (Vodyanitskii and Vostokova, 2004). This reagent is also able to dissolve partially manganese oxides. Hence, it seems reasonable to extract both the easily reducible and oxidizable "organic" fractions prior to the application of Tamm's reagent (Shuman, 1983; Zeien and Brümmer, 1989; Krishnamurti et al., 1995).

The action of Tamm's reagent has been reported to be light sensitive, the ultraviolet catalytic effect being used to expedite the attack of iron oxides (Gleyzes et al., 2002). Yet, it has been proven that the scattered light of luminescent lamps affects the release of iron insignificantly. The treatment in the darkness leads to quite similar results (Vodyanitskii and Vostokova, 2004). On contrary, the temperature is a core parameter for the dissolution of iron oxides. If the temperature increases from 20°C to 26°C, the recovery of iron from sod–podzol soils increases by 14 to 28% (Vodyanitskii and Vostokova, 2004). The redox reactions facilitated by Tamm's reagent should thus be thermostated, a temperature of 25°C often being recommended. The mixture of ascorbic acid and oxalate buffer at high temperature (96 to 100°C) enables crystalline iron oxides to be dissolved successfully (Zeien and Brümmer, 1989; Krishnamurti et al., 1995). The dissolution rate depends on the concentration of ascorbate ions and increases with decreasing reagent pH. Sodium dithionite/sodium citrate buffer (Mehra and

Jackson's reagent) can be used for the extraction of TEs bound to even well-crystallized iron oxides (Harrington et al., 1998), yet one of the main drawbacks of the commercially available dithionite salts is the presence of metal (e.g., zinc) impurities (Gleyzes et al., 2002). As a consequence, a rather tedious purification procedure is frequently required. In addition, a lack of selectivity has been observed by Shuman (1982), due to the facility of attacking silicates.

f. Oxidizable Fractions In sediments and soils, the organic matter comprises mainly complex macromolecular material (humic substances) and other compounds, such as carbohydrates, proteins, peptides, amino acids, fats, waxes, and resins. Under oxidizing conditions, the organic matter tends to be degraded, leading to the release of bound TE. Since organic substances may form stable coordination compounds with divalent metal ions coming from anthropogenic sources, the oxidizable fraction is especially important for contaminated soils, sediments, and sludges, where it might be the dominant TE solid-phase association form. The oxidizable fraction also contains TEs occurring in the environment as sulfide forms.

As already mentioned in Section 12.3.1.1d, a "mild" reagent employed for leaching of metals associated with organic matter is potassium/sodium pyrophosphate at pH 10 (McLaren and Crawford, 1973; Miller et al., 1986; Krishnamurti et al., 1995). This reagent is relatively selective for separation of the easily oxidizable organic fraction (e.g., metals associated with humates and fulvates), but, the extraction efficiency is rather poor (Papp et al., 1991). To increase the solubility of organic soil components, a 0.1 mol L^{-1} $K_4P_2O_7$ + 0.1 mol L^{-1} KOH mixture has been recommended in lieu of 0.1 mol L^{-1} $K_4P_2O_7$ (pH 10) (Ladonin, 2002). It should be noted that manganese oxides might also be extracted at pH 10 to 12. Hence, the easily reducible fraction is commonly isolated prior to use of the pyrophosphate extraction protocol (Miller et al., 1986).

Use of nonoxidizing strong complexing reagents such as salts of ethylenediaminetetraacetic acid (EDTA) have also been attempted for the extraction of readily mobilizable or available metals associated with humates and fulvates (Zeien and Brümmer, 1989; Ladonin, 2002). For proper phase discrimination, exchangeable and acid-soluble fractions should be separated before EDTA treatment. In addition, the fact that EDTA is capable of releasing metals bound to nonsilicate iron compounds, owing to the high stability constant of the Fe-EDTA chelates, should be taken into account.

The use of sodium hypochlorite as an oxidizing reagent in alkaline medium leads to the destruction of both soluble organic components (e.g., humic and fulvic acids) and refractory organic compounds tightly included in the mineral soil phases (Shuman, 1983; Krishnamurti et al., 1995), while minimizing the attack of manganese and iron oxides as well as clay materials. The extraction is generally performed with 0.7 mol L^{-1} NaClO at pH 8.5 and heated at 85 to 100°C (Shuman, 1983; Mikutta et al., 2005).

Hydrogen peroxide is the most popular reagent for separation of the oxidizable fraction (Hlavay et al., 2004). In general, the oxidation of soil matrix components is performed at high temperature for 2 to 3 hours as a compromise between complete dissolution of the organic matter and minimum alteration of silicates (Tessier et al., 1979). Although hydrogen peroxide heated at 85°C is an excellent oxidizing reagent, the oxidation of resistant organics as well as sulfide minerals may be incomplete for high-organic-matter content samples. The combination of hydrogen peroxide and ammonium acetate (adjusted at pH 2), where the addition of the latter after oxidation facilitates the redissolution of metals adsorbed onto the oxidized substrate, has been adopted in most SEPs for leaching of TEs associated with organic matter and sulphides (Tessier et al., 1979; Salomons and Förstner, 1984; Ure et al., 1993; Rauret et al., 1999). Unlike sodium hypochlorite, hydrogen peroxide is able to attack manganese and iron oxides. Hence, this reagent should be used after extraction of the reducible fraction.

12.3.1.2. Leaching of As, Se, and P Special sequences of extracting reagents are needed for the fractionation of nutrients and pollutants as a consequence of their anionic nature. It should be taken into consideration that metalloids such as As and Se could be found in soils and sediments under different oxidation states, each with a particular behavior. Several steps in fractionation analysis are performed under oxidizing or reducing conditions, whereby the original oxidation state can be altered. Although the extraction results might be questioned, they are definitely appropriate to assess the potential bioavailability of metalloids under environmentally changing conditions (Gleyzes et al., 2002).

The adaptability of extraction steps from commonly used SEPs to fractionating As and Se has been tested by Gruebel et al. (1988) taking standard minerals and their mixtures as an example. In particular, two steps were examined: the reductive dissolution of amorphous iron oxides by hydroxylamine, and the oxidation of organic matter by hydrogen peroxide and sodium hypochlorite. It has been shown that during reductive and oxidative dissolution of target elements from a certain mineral phase, readsorption on other mineral phases with subsequent desorption of elements in the next extraction step is a serious limitation on the appropriate fractionation of metalloids.

Nevertheless, since the fractionation of P, Se, and especially As is of particular relevance for environmental studies, different sequences of extracting reagents have been attempted for the partitioning of these elements (i.e., Williams et al., 1976; Hieltjes and Lijklema, 1980; Ruttenberg, 1992; Golterman, 1996; and Ruban et al., 1999, for P; Johnston and Barnard, 1979; Onken and Adriano, 1997; Dauset al., 1998; La Force et al., 2000; Lombi et al., 2000; and Wenzel et al., 2001a, for As; and Zhang and Moore, 1996; Martens and Suarez, 1997; and Ponce de León et al., 2003, for Se). Taking into account the similarities in the chemical properties of As and P, sequential extraction schemes initially developed for P were also applied to the fractionation of As. A number of such schemes designate soluble or exchangeable, Al- and Fe-bound, and Ca-associated extractable fractions using NH_4Cl or NH_4F, NaOH, and H_2SO_4 or HCl, respectively. The principal difference between these schemes and SEPs for metal ions is the evaluation of Fe and Al associated fractions under alkaline conditions after separation of the exchangeable fraction.

Wenzel et al. (2001a) developed and tested a novel SEP for As based on the selection of extraction reagents commonly used for sequential extraction of metals, Se, and P. Two eligibility criteria were adopted: (1) extraction efficiency and (2) specificity for targeted phases. The final sequence includes four extraction steps, as summarized in Table 12.3, and separation of the residual fraction. Despite the inherent limitations of chemical fractionation, the correspondence between applied reagents and separated fractions was supported by selective chemical analysis of the extracts of soils and pure mineral phases, and by energy-dispersive x-ray microanalysis of As in selected soils. The developed SEP has been proved suitable for predicting changes in As lability in various solid phases as a result of soil remediation or alteration of environmental factors. A series of extracting reagents employed for the fractionation of Se have been evaluated for contaminated sediment samples (Ponce de León et al., 2003). The different extraction methodologies assayed involved the sequential use of water, phosphate buffer (pH 7), and either acid solution (e.g., HNO_3 of HCl) or neutral and alkaline solutions (e.g., ammonium acetate, NaOH, or tetramethylammonium hydrochloride). Neutral and alkaline extractions were more efficient than the acidic ones. The most advantageous SEP involved four extraction steps (Table 12.3). The species separated, identified by chromatographic methods, were selenate, selenite, elemental Se, and organically bound Se. It has been also stressed that pyrophosphate extraction, usually utilized to strip TEs bound to humic acids, might also be tailored to the fractionation of Se in soils and sediments for isolation of the easily oxidizable fraction.

Besides the fractionation of metalloids through SEPs, a series of methods has been proposed for the determination of individual species in the various oxidation states (Gong et al., 2002; Kahakachchi et al., 2004). The most popular detectors for metalloid speciation are inductively coupled plasma–mass spectrometry (ICP-MS) and atomic fluorescence spectrometry (AFS), especially after liquid chromatographic separation and hydride formation, which are increasingly replacing atomic absorption spectrometry (AAS). Speciation analysis of pollutants in terrestrial environments is, however, beyond our scope in this chapter.

Phosphorus is not a TE but a major nutrient element. Nevertheless, fractionation of this element is essential for environmental studies, and hence it seems reasonable to highlight here some relevant SEPs. Four different procedures for the fractionation of P in lake sediment samples have been tested in an interlaboratory study in the framework of the SM&T program (Ruban et al., 1999). As a result, a novel scheme based on the Williams protocol (Williams et al., 1976) has been developed aimed at the restoration of lake sediments. The scheme comprises three separate assays: (1) sequential extraction of NaOH-extractable (Fe- and Al-bound) and HCl-extractable (Ca-bound) fractions, (2) sequential extraction of "inorganic" and "organic" phosphorus and (3) single extraction, after calcination, of "concentrated HCl-extractable" (total P) fraction (see Table 12.3 for further details). Further discrimination of specific compounds is made feasible by the use of chromatographic and capillary separation techniques as reviewed by Spivakov et al. (1999).

TABLE 12.3. Some Relevant SEP for the Fractionation of As, Se, and P

Element, SEP	Leaching Reagents	Extractable Fractions (Residual Fractions Are Not Included)
As, Wenzel et al. (2001a)	0.05 mol L^{-1} $(NH_4)_2SO_4$	Nonspecifically sorbed
	0.05 mol L^{-1} $(NH_4)H_2PO_4$	Specifically sorbed
	0.2 mol L^{-1} $(NH_4)_2C_2O_4$, pH 3.25	Amorphous Fe and Al oxides
	0.2 mol L^{-1} $(NH_4)_2C_2O_4$ + 0.1 mol L^{-1} ascorbic acid, pH 3.25, 96°C	Crystalline Fe and Al oxides
Se, Ponce de León et al. (2003)	Water	Soluble (selenites, selenates)
	0.5 mol L^{-1} KH_2PO_4/K_2HPO_4, pH 7	Adsorbed (selenates, organic species)
P, Hieltjes and Lijklema (1980)	5% tetramethylammonium hydrochloride	Non-water-soluble selenites, organic species
	1 mol L^{-1} Na_2SO_3, pH 7	Elemental, remaining extractable
	1.0 M NH_4Cl	Exchangeable
	0.1 M NaOH	Fe- and Al-bound phosphorus
	0.5 M HCl	Ca-bound phosphorus
P[a], Ruban et al. (1999)	I. 1 mol L^{-1} NaOH	Fe- and Al-bound phosphorus
	1 mol L^{-1} HCl	Apatite phosphorus
	II. 1 mol L^{-1} HCl	Inorganic phosphorus
	Calcination at 450°C, then 1 mol L^{-1} HCl	Organic phosphorus
	III. Calcination at 450°C, then 3.5 mol L^{-1} HCl	Concentrated HCl-extractable (total P)

[a] In fact, the extraction scheme recommended under the auspices of the SM&T program for the fractionation of P comprises two different two-step SEPs and one single-extraction procedure. A new portion of sample is required for each stage (I, II, and III).

12.3.1.3. General comments It should be mentioned as a general remark that the kinetics of TE extractions from solid samples are usually slow. For example, the recovery of metals into each of the three extractable fractions according to the SM&T leaching schemes (Ure et al., 1993; Rauret et al., 1999) requires at least 16 hours of contact between solid and liquid phases. The entire procedure therefore lasts more than 2 days. As a consequence, routine SEPs are rather time consuming although several samples can be treated simultaneously. Besides, the readsorption and redistribution of TEs among phases during the extraction may keep fractionation analysis from being performed accurately (Filgueiras et al., 2002a).

Despite the above-mentioned drawbacks, SEPs are used extensively for fractionation and mobility assessment of TEs in soils, sediments, and sludges. Sahuquillo et al. (2003) have formulated the following useful steps and criteria for the selection of most relevant leaching tests needed for risk assessment:

1. Selection of TEs of interest in terms of leachability from soils or sediments.
2. Assessment of the contamination level of TEs and the need for further investigation in terms of soil and sediment leachability. The first step is to search for legislation that includes this type of information (target and/or intervention limits based on soil type and use).
3. Evaluation of the environmental conditions that might affect the behavior of TEs (e.g., drastic changes in pH, redox potential, temperature, and water regime).

The validation and harmonization of SEPs in the framework of EUs SM&T program as well as preparation of the corresponding reference materials, provide an excellent base for the interlaboratory comparison of fractionation data. The three-step SM&T leaching scheme (Rauret et al., 1999) offers the possibility of evaluating the bulk contents of trace metals in acid-soluble, reducible, and oxidizable fractions. Each fraction may contain various species differing in mobility and bioavailability (Table 12.1). If recoveries of TEs into extractable fractions are relatively low, use of the SM&T scheme is justified. However, for highly contaminated samples, further discrimination of leachable forms may be required. In such a way, the development of a reliable SEP addressing exchangeable, acid-soluble (specifically adsorbed), easily reducible (bound to Mn oxides), and easily oxidizable (metal–organic complexes) fractions of TEs is desirable.

Some general trends for the current and future development and application of SEPs have been summarized by Filgueiras et al. (2002a) as follows:

- Preparation of new reference materials with certified extractable contents of elements
- Further harmonization of SEPs through round-robin tests to facilitate the comparability of data

- Design of accelerated schemes based on the application of external energy sources to decrease the overall treatment time
- Development of extraction schemes optimized specifically for the characteristics of the target sample
- Implementation of online procedures for performing sequential extractions
- Application of single-extraction methods that concomitantly enable the simplification of operational protocols
- Development of small-scale extractions to minimize sample and reagent consumption
- Application of SEP for characterization of TE mobility in other environmental samples, such as coal fly ash, solid waste incineration bottom ash, airborne dust, and workroom aerosols
- Use of chemometric approaches for robustness testing of SEP, for finding relationships between soil metal fractions and plant uptake, and for characterization of pollution sources on the basis of fractionation data
- Further investigations of changes in TE mobility in soils subjected to amendment

In addition, further studies should be conducted for the replacement of current leaching reagents of SEPs by more selective ones. In this context, chemometric approaches should be regarded as powerful tools for ascertaining the selectivity of a given extractant through elucidation of the chemical components in the leachates, as demonstrated by Santamaria-Fernández et al. (2006).

12.3.2. Flow-through fractionation procedures

Despite recent efforts toward settling operational conditions for metal and metalloid fractionation assays—in terms of concentration, pH, and temperature for each of the leaching reagents, sample weight/extractant volume ratio, extraction time, shaking protocol, analytical instrumentation, and phase separation method—conventional sequential extraction schemes lack automation and are inherently rather time consuming and laborious. This is the consequence of a number of steps needed for the separation, identification, and quantification of TEs in each fraction. For example, the SM&T recommended protocol lasts more than 50 hours, whereas the operating time of Tessier's scheme approaches 20 hours.

Some attempts to expedite the extraction processes by exploiting microwaves (Campos et al., 1998; Pazos-Capeáns et al., 2005) or ultrasound (Pérez-Cid et al., 1998; Davidson and Delevoye, 2001) as auxiliary energies, or to miniaturize the overall fractionation assays (Filgueiras et al., 2002b) have been described. However, all these batchwise procedures involve steady-state measurements whereby the ecotoxicological relevance of the information provided is rather questionable, since naturally occurring processes are always dynamic. Two additional main problems mentioned earlier are (1) phase overlapping, that is, the possibility of releasing metals associated with different geological phases by a single leaching agent (Whalley and Grant, 1994; Bermond, 2001), and (2) the metal readsorption phenomenon: that is, the fact that TEs released by one extractant could associate with other undissolved solid components or freshly exposed surfaces within the time scale of the extraction step (Gómez-Ariza et al., 1999; Bermond, 2001). Depending on the degree to which readsorption occurs, the contents of trace metals or metalloids bound to a given soil fraction will be underestimated, whereas the metal mobility in subsequent phases will be overestimated.

To alleviate these drawbacks, alternative methodologies relying on the continuous provision of fresh extractant volumes to the solid sample under investigation have been developed, characterized, and contrasted with the classical end-over-end extraction procedures. The fundamental principles of these novel, dynamic (nonequilibrium) strategies, based primarily on the use of continuous-flow analysis (Ruzicka and Hansen, 1988), flow injection analysis (Ruzicka and Hansen, 1988; Trojanowicz, 2000; Miró and Frenzel, 2004b), or sequential injection analysis (Ruzicka and Marshall, 1990; Lenehan et al., 2002), are described in detail below, and their advantageous features and limitations for fractionation explorations are discussed critically.

12.3.2.1. Large columns To mimic field conditions as closely as possible and to obtain data on a realistic time scale, dynamic column extraction experiments have been proposed as a simple means for estimating metal mobility in soil and sediment substrates (Anderson et al., 2000; Wisotzky and Cremer, 2003b) as well as in compost (Song and Greenway, 2004). The packing of long, thick nonmetallic columns with a large amount of material (typically ranging from 500 g to 10 kg) ensures an environmentally representative sample mass. For column leaching experiments, the extractant volume (0.1 to 10 L day^{-1}) is applied atop the column either manually or automatically, and the leachates are collected in various consecutive subfractions. Although in-field lysimeter solutions cannot be reproduced in laboratory extractions consistently and accurately, column leaching protocols have been developed for discriminating the effects of sampling and the storage of samples on the metal fractions mobilized as well as for producing leachates that resemble lysimeter soil solutions (MacDonald et al., 2004a,b).

Column experiments proved to be useful tools for the investigation of (1) the fate and distribution of TEs in soils receiving polluted wastewater or biosolids (Tam and Wong, 1996; Gove et al., 2001) (2) the crucial factors influencing the immobilization of TEs in environmental solids (Camobreco et al., 1996 Gong and Donahoe, 1997), (3) the potential of compost as remediation material for metal-contaminated waste (Song and Greenway, 2004), (4) the addition of sorptive materials in metal-polluted soils aimed at the development of reclamation strategies (Karczewska et al., 1996; Jang et al., 1998), and (5) the efficiency of single or sequential extraction schemes for soil clean-up (Wasay et al., 2001).

Entirely enclosed column systems have also been proposed to reduce contamination, personal errors, and contact of the packed sample with atmospheric oxygen (anoxic conditions), whereby undesired decomposition of reduced mineral phases and oxidation of reduced water constituents are prevented (Wisotzky and

Cremer, 2003a). As opposed to batch tests, this approach warrants well-adjusted amounts of extracting reagents to be used. The immediate consequence of the optimization of the sample/reagent volume ratio is the likelihood of determining the TE content of reactive mineral phases, which could not be detected through standard mineralogical procedures due to the excessive dilution of the resulting extracts (Wisotzky and Cremer, 2003b).

Column leaching is an extremely tedious and time-consuming extraction method, but it contributes to a deeper understanding of hydrogeological processes (Wisotzky and Cremer, 2003a,b) and represents an elegant approach for long-term risk assessment of metal pollution under environmentally simulated events such as acid rainfall (Anderson et al., 2000). However, the progressively tighter packing of the solid material may cause the creation of preferential flow channels or impede free percolation of the extractant. Hence, novel microanalytical column-based flow-through systems equipped with liquid drivers (e.g., peristaltic pumps or syringe pumps) for provision of the extraction reagent by positive (or eventually negative) pressure have been designed and exploited as complementary dynamic partitioning techniques, as described in ensuing sections.

12.3.2.2. Rotating coiled column Rotating coiled columns (RCCs) have traditionally been used for the separation and purification of organic (Ito, 1991) and inorganic (Maryutina et al., 1999) solutes by countercurrent chromatography (support-free liquid–liquid partition chromatography). The technique is based on the retention of one phase (stationary) of a two-phase solvent system in a column without solid support under the action of centrifugal forces while the other liquid phase (mobile) is pumped through continuously. Solutes are subjected to partitioning between the immiscible phases and are eluted in increasing order of their partition coefficients.

It has been demonstrated that RCCs can be used in liquid–solid systems for continuous leaching of trace metals (Fedotov, 2002; Fedotov et al., 2002, Fedotov et al., 2005b) and metalloids (Fedotov et al., 2005a) from contaminated soils, lake sediments, and river sludges. The solid sample (typically, 0.5 g) is retained as a stationary phase, whereas leaching solutions of mineral acids, electrolytes (salts), and complexing agents are provided to the sample successively at a common pumping rate of 1.0 mL min^{-1} using a conventional peristaltic pump as a liquid driver. The fractionation of elements is performed on a planetary centrifuge furnished with a vertical one-layer coiled polytetrafluoroethylene (PTFE) column, as shown in Figure 12.1*a*. During the experimental protocol the spiral column rotates around its axis and, simultaneously, revolves around the central axis of the unit with the aid of a planetary gear. The mobile phases have been selected according to the Kersten and Förstner (1986), McLaren and Crawford (1973), and SM&T (Ure et al., 1993; Rauret et al., 1999) procedures for metal species, and to an optimized four-step protocol for arsenic (Wenzel et al., 2001a). Information on metal and metalloid chemical forms and type of binding supplied by the various SEPs has been critically discussed and compared with that of traditional batchwise protocols. As a consequence of the continuous flowing nature

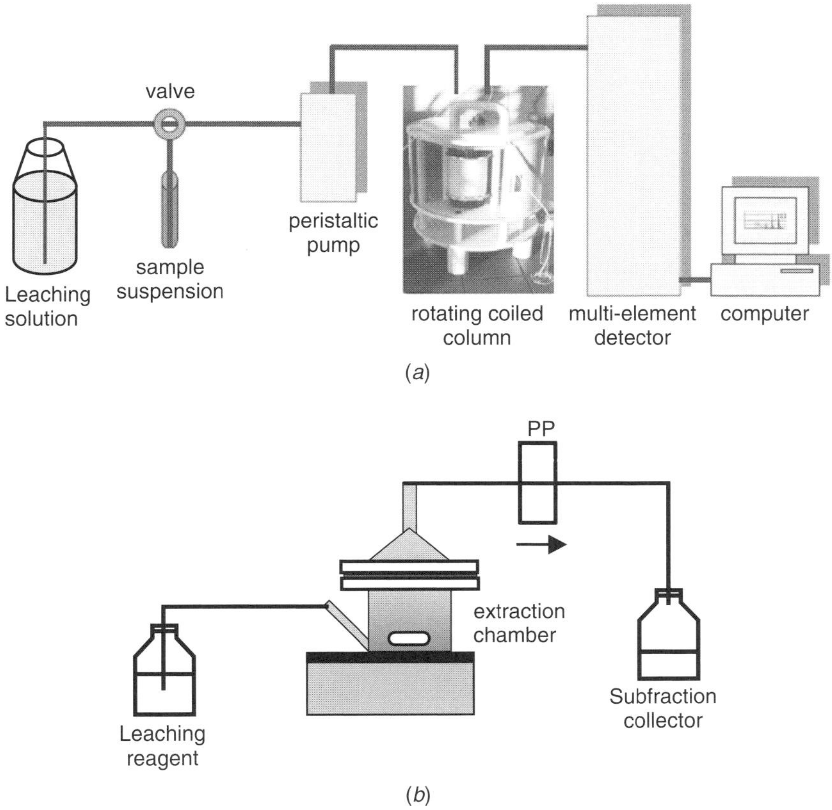

Figure 12.1. Schematic representation of continuous-flow systems for dynamic fractionation: (*a*) rotating coiled column (RCC); (*b*) stirred flow-through cell. [(*b*) From Shiowatana et al., 2001b.]

of the extracting (mobile) phase that makes the contribution of metal refixation negligible, the most mobilizable metal forms (i.e., those provided by the first fractions of the SEP) were released to a larger extent then when equilibrium-based methods were used (Fedotov et al., 2002, 2005b). Despite the slow mobilization kinetics of TEs from solid samples, the use of multistage continuous extraction in RCCs facilitates decrease in contact time between phases down to 1 to 2 hours per fraction. In addition, RCCs have been used to evaluate labile and moderately labile metal forms in different fractions for a variety of environmental solids (Fedotov et al., 2005b).

Further research has been conducted aiming at extending the RCC concept to heterogeneous solids with a lower TE content. Hence, the rotating column extraction process has been scaled up by using large-diameter column tubing and bobbins with variable rotation radii (Fedotov et al., 2003). It has been shown that the greater the rotation-to-revolution radius ratio (the β-value), the better the retention of solid material. By selection of β-values close to 0.8, particulate samples up to 5 g can be completely retained in the coiled PTFE tube (total inner volume 34 mL) with minimal loses during the fractionation procedure.

TEs and major elements released from the stationary phase can also be determined online by coupling RCCs with an inductively coupled plasma–atomic emission spectrometer (ICP-AES) (Schreiber et al., 2005; Fedotov et al., 2006). This novel approach was utilized for preliminary leaching studies using single extractants: water and simulated acid rain at pH 4. For the particular set of samples analysed, water was more effective for the mobilization of both arsenic and metal ions (Schreiber et al., 2005). The hyphenated flow-through analytical system proposed has been applied to the fractionation of TEs in soil, sediment, and sludge samples (Fedotov et al., 2006) using a new sequence of leachants that ensure improved selectivity. Simultaneous recording of the elution profiles of TEs and major elements facilitates the elucidation of TE associations in the separated fractions, and hence the efficiency of extractants for targeted mineralogical phases may be ascertained. The time-resolved dissolution of various organic complexes of metals was observed for sediments. It was also possible to discriminate the iron and aluminum oxyhydroxides released by Tamm's reagent as a consequence of the slight differences in leaching rates. In fact, the combined method looks rather promising for developing effective leaching schemes for risk assessment of pollutants in environmental solids.

12.3.2.3. Stirred-flow cells Continuous-flow systems furnished with closed extraction chambers have recently been presented as an appealing approach to speed up, facilitate, and improve the accuracy of the chemical fractionation of metals in solid samples with minimum risk of sample contamination. Exploited primarily by the research groups of Shiowatana (Shiowatana et al., 2001a,b) in Bangkok and Strobel (Strobel et al., 2001; Sukreeyapongse et al., 2002) in Copenhagen, this continuous extraction concept involves placement of a weighed soil sample (commonly ranging from 0.25 to 5 g) into a purpose-made magnetically stirred reactor through which extractants flow sequentially by either positive or negative pressure via a peristaltic pump for leaching of analyte species from different targeted soil phases. A sketch of a flow manifold devised for that purpose is depicted schematically in Figure 12.1*b*. Different three or four-step SEPs not only for trace metal and metalloid partitioning studies (Shiowatana et al., 2001a,b,c), but also for nutrient [e.g., reactive phosphorus species (Tiyapongpattana et al., 2004)] fractionation have been performed successfully in the stirred-flow cell mode, thus rendering time- and cost-effective procedures.

It has been demonstrated that pH changes during extraction occur only at the very beginning of the sequence, so that the entire protocol can be considered to be effected at the pH value selected with no requirements for pH adjustment during the extraction protocol (Shiowatana et al., 2001b). Capitalizing on this asset, stirred-flow cells have been used to evaluate the bonding strength and potential mobility of metals in soils under proton-induced desorption reactions at a steady-state pH value (Sukreeyapongse et al., 2002). The dependence of both pH and dissolved organic matter (DOC) on the release rates of TEs (e.g., Cu,

Cd) from nonpolluted soils has also been investigated thoroughly, and pH–DOC synergic effects have been identified for some elements under certain scenarios (Strobel et al., 2001). Stirred-flow cells have also been presented as a promising vehicle to assess the metal readsorption phenomenon in the various steps of SEPs, the extent of which was found to be influenced by the element itself, the organic matter content of the sample, the extracting flow rate, and the metal concentration level (Chomchoei et al., 2002).

One of the main constraints of this continuous semiautomated flow approach is the lack of flexibility for reagent manipulation as a consequence of its unidirectional flow nature. Hence, the intimate sample–extractant contact time can only be enhanced by decreasing the flow rate of the extractant. In addition, there is an overlapping of consecutive fractions due to the dead volume of the sample reservoir. It should also be mentioned that the various analytical procedures reported to date cannot be strictly compared, owing to the variable operational conditions (e.g., the delivery rate of the influent solutions ranges from 1.6 mL h^{-1} to 10 mL min^{-1}, which have been chosen according to the timescale of the processes under investigation (Shiowatana et al., 2001a, Sukreeyapongse et al., 2002).

12.3.2.4. Packed microcolumns

a. Flow-Injection Assemblies In recent years the implementation of sample packed microcartridges in the conduits of flow injection (FI) setups has attracted special interest in dynamic leaching studies (Zhi et al., 1996; Miró and Frenzel, 2004b; Morales-Muñoz et al., 2004) as well as metal partitioning and release-rate investigations. Microcolumn leaching experiments are typically conducted by filling a cylindrical microcolumn with an amount of sample at the low-milligram level (typically, 5 to 25 mg) (Beauchemin et al., 2002; Jimoh et al., 2004, 2005; Dong and Yan, 2005) through which different extracting agents are pumped sequentially either in a continuous or discrete fashion for fast assessment of element distribution among different phases. This flow-through microcolumn approach is characterized by the likelihood of accommodating three conceptually different modes for leaching the targeted phases: continuous leaching, static extraction, or recirculation mode, via exploitation of the liquid-handling capabilities of flow injection analysis. The first extraction protocol can be performed by continuous delivery of the leaching agent through the packed material (Beauchemin et al., 2002; Dong and Yan, 2005) (see Figure 12.2*a*) or by discrete plug injection based on inserting a metered volume of reagent via a six-port rotary injection valve into a carrier stream flowing continuously through the microcartridge (Beauchemin et al., 2002), as illustrated in Figure 12.2*b*.)

The second alternative, also termed the stopped-flow approach, relies on stopping the fresh extractant within the packed column in whatever of the above-mentioned configurations aimed at enhancing the reagent–sample contact time (Jimoh et al., 2005). This is also the main goal of the third approach, involving reutilization of the leaching agent in a closed-loop fashion (Jimoh et al., 2004). In this case the leachate is redirected to the sample cartridge and continues to

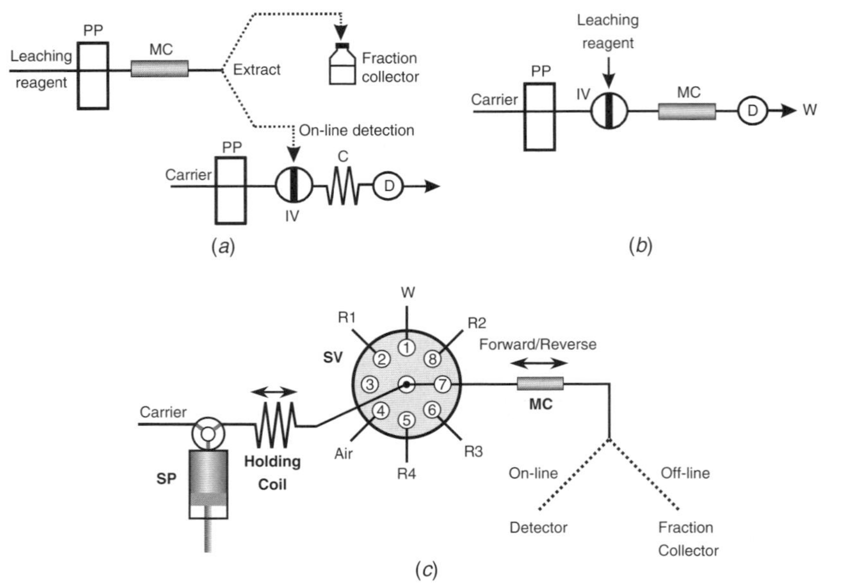

Figure 12.2. Microcolumn flow-through manifolds for fractionation analysis in environmental solids: (*a*) continuous-flow system with either fraction collection or online detection; (*b*) flow-injection setup with discrete injection of extractant; (*c*) sequential injection system as assembled for fully automated sequential or single extractions. MC: microcolumn, C: coil, SP: syringe pump, PP: peristaltic pump, R: reagent, D: detector, W: waste, SV: selection valve, IV: injection valve. (Adapted from Miró et al., 2005a.)

circulate for a preset of time. The concentration change with repetitive cycling is followed by successive collections of minute volumes of extract for further analysis. Moreover, the direction of the leaching agent can be programmed for a forward–backward change at fixed intervals to prevent undesirable compaction of the soil in the extraction column. It should be stressed that the optimum leaching procedure for each solid material, element, and single or sequential extraction protocol is ruled by the lability of the metal form under investigation as well as by the binding strength of the extractant, which determines the extent to which TE redistribution occurs.

The most severe limitations of FI–microcolumn procedures arise from the progressive aging of the flexible tubes of peristaltic pumps, which necessitates frequent recalibration of the liquid driver, and the small inner capacity of the microcartridges, which prohibits sample weights above 100 mg from being used. Therefore, this flow-through fractionation technique is applicable primarily to fast metal pollution screening of highly homogeneous environmental solids.

b. Sequential Injection Assemblies As a result of accurate fluidic control via syringe pumps, a new variant of flow analysis, sequential injection analysis (SIA) (Lenehan et al., 2002; Wang and Hansen, 2004) has opened new avenues in the miniaturization of sample processing and handling of solid samples. Whereas most FI procedures employ continuous, unidirectional pumping of solutions, SIA is based on exploiting programmable, bidirectional discontinuous flow that is coordinated and controlled precisely by a computer. Basically, SI analyzers encompass a multiposition selection valve with a central communication channel that can be made to address individual peripheral ports furnished with the appropriate solutions and/or sample pretreatment modules for executing the unit operations desired (Marshall et al., 2003; Wang and Hansen, 2004). In the present context, SIA should be regarded as a versatile, rugged, and rapid approach to ascertaining the mobility and binding sites of specific elements in complex matrices by the operation of both single and sequential extraction microcolumn protocols at will (Miró and Hansen, 2005). Using traditional extraction solutions, the automated sequence comprises consecutive aspiration of individual extractants from different external ports of a valve into a holding coil, which, via flow reversal, are exposed sequentially to sample contained in a microcartridge attached to another outer position, as in Figure 12.2*c*. Most recently, a new sample container involving dual conical microcolumns with minimum dead volume has been designed (Chomchoei et al., 2004). As opposed to the cylindrical microcolumns assembled in FI manifolds, the optimal hydrodynamics of this custom-built column permit precise handling not only of highly homogeneous but, most notably, also of poorly homogeneous solid materials (e.g., sewage-amended CRM 483 soil) by increasing the sample size (up to 300 mg of soil) with no adverse back-pressure or clogging effects, thereby assuring automatic performance of the metal and metalloid release protocols with minimal operational maintenance (Chomchoei et al., 2005a).

Taking advantage of the versatility of the movements of the syringe pump, four operational modes for handling leaching solutions (i.e., unidirectional flow, bidirectional flow, multi-bidirectional flow, and stopped flow) are implemented easily in a single-SI set-up with no need for manifold reconfiguration (Chomchoei et al., 2005b) as demanded in FI-based assays. Therefore, consecutive sequential extraction steps with diverging kinetics can be implemented in the same setup with no further hindrance since the extraction time can be controlled automatically. It is important to emphasize the existence of a correlation between the extraction time and the extent of the readsorption process, as the longer the extraction time, the greater the influence of this phenomenon. However, it is possible to discern between the two effects by comparing the metal-leachable contents obtained for the stopped-flow and forward–backward approaches for a given extraction time. In fact, the stopped-flow mode provides relevant information on the redistribution process as a consequence of the stagnant nature of the leaching solution, whereas the effect of the contact time may be ascertained via the bi- or multidirectional modes, since the diffusion layer of the extractant solution in contact with the soil matrix (containing the highest concentration of metal species released) is renewed in each extraction step. Table 12.4 provides a selection of relevant analytical use of various flow-through extraction techniques for dynamic metal–metalloid fractionation.

TABLE 12.4. Techniques for Dynamic Metal and Metalloid Fractionation

Technique	Flow System	Analyte(s)	Sample	Detection Principle[a]	Detection Mode	Subfraction Volume	Fractionation Protocol	References
Large columns	Continuous flow	Cu, Cr,Ni, Zn, V, Pb, Cd, Mn	Field-moist soil from a chemical manufacturing site	FAAS/ ETAAS	Off-line	500 mL	Artificial rainwater at pH 4.3 and 3.5	Anderson et al. (2000)
	Continuous flow	Cr, Zn, Cd, Ni, Cu, Pb, As	Mature compost	ICP-MS	Off-line	12 mL	Single extractants: diluted sulfuric acid, oxalate, citrate, and EDTA	Song and Greenway (2004)
	Continuous flow	Cd, Zn, Ni, Cu, Pb	Soils subjected to trace metal emissions from smelting centers	ICP-AES/ ETAAS	Off-line	30 mL	Simulated lysimeter solution (calcium chloride–calcium sulfate mixture)	MacDonald et al. (2004b)
	Continuous flow	Ni, Co, As, Fe	Overburden sediments from lignite mines	ETAAS	Off-line	100 mL	Water soluble, exchangeable, acid soluble and reducible fractions	Wisotzky and Cremer (2003b)
Rotating coiled columns	Continuous flow	Pb, Cu, Zn, As, Mn, Fe	Highly contaminated SRM 2710 Montana soil Highly contaminated sediment from As–Pb mine	ICP-AES	Off-line	10 mL	Five-step Kersten–Förstner, and four-step Mc Laren–Crawford	Fedotov et al. (2002)
	Continuous flow	Cd, Cu, Pb, Zn, Cr, Ni	Highly contaminanted SRM 2710 Montana soil Lake sediments: CRM-601 and BCR-701 River sludges after flood	ICP-AES/ ICPMS	Off-line	10 mL	Three-step SM&T sequential extraction	Fedotov et al. (2005b)

	Continuous flow	As, Mn, Fe	River sludges As-contaminated forest soils	ICP-AES/ ICPMS/ HG-AAS	Off-line	10 mL	Easily exchangeable, surface-bound, amorphous hydrous oxide–bound, and crystalline hydrous oxide–bound fractions	Fedotov et al. (2005a)
	Continuous flow	As, Cu, Pb, Zn, Fe	Soils and sediments influenced by a flood	ICP-AES	Online	Continuous leaching	Single extractants: water and simulated acid rain at pH 4	Schreiber et al. (2005)
	Continuous flow	Cu, Pb, Zn, As, Fe, Al, Mn, Mg	Soil, sediment, and sludge samples	ICP-AES	Online	Continuous leaching	Five-step SEP: exchangeable, specifically sorbed, bound to Mn oxides, bound to humates and fulvates, bound to amorphous and poorly crystalline Fe and Al oxides	Fedotov et al. (2006)
Stirred chamber	Continuous flow	Cu, Zn, Mn, Fe, Cs	Highly contaminated SRM 2710 Montana soil	FAAS	Off-line	20 mL	Three-step SM&T sequential extraction	Shiowatana et al. (2001b)
	Continuous flow	Cd, Cu, Pb	Sewage sludge–amended and nonamended soils	ETAAS	Off-line	2.7 mL	pH gradient from 2.9 to 7.1	Sukreeya-pongse et al. (2002)
	FI and continuous flow	Ca, Fe, Mn, Cu, Zn	Highly contaminated SRM 2710 Montana soil	FAAS	Off-line/ online	10–60 mL (off-line)	Three-step SM&T sequential extraction	Shiowatana et al. (2001a)
	Continuous flow	Cd, Cu	Unpolluted cultivated soil	ETAAS	Off-line	2.8 mL	Solutions with various levels of dissolved organic matter	Strobel et al. (2001c)
	Continuous flow	As	Sediment (SRM 2704) and soils (SRM 2710, SRM 2711, and As-polluted soils)	ETAAS	Off-line	20 mL	Water soluble, surface adsorbed, associated with Fe- and Al-oxo-hydroxides, and carbonate-bound fractions	Shiowatana et al. (2001c)

(*continued overleaf*)

TABLE 12.4. ***(continued)***

Technique	Flow System	Analyte(s)	Sample	Detection Principle[a]	Detection Mode	Subfraction Volume	Fractionation Protocol	References
Micro-column leaching	SI	Cu, Zn, Mn, Fe, Ca	Highly contaminated SRM 2710 Montana soil	FAAS	Off-line	5 mL	Three-step SM&T sequential extraction	Chomchoei et al. (2004), (2005b)
	SI	Cu, Zn, Pb	Nonhomogeneous sewage sludge–amended CRM 483 soil	FAAS	Online	175μL	Exchangeable, EDTA and acetic acid single extraction, and three-step SM&T sequential extraction	Chomchoei et al. (2005a)
	FI	Ni, Cu, Zn, Cd, Pb, Al	SRM 1648 urban air particulars, SRM 1570a spinach leaves, CRM 3616 pyrrhotine ore, and CRM 701 lake sediment	FAAS/ ICP-MS	Online	Loop injection	Three-step acid scheme, acid soluble and reducible fractions	Jimoh et al. (2004), (2005)
	FI and continuous flow	Co, Fe, Ni, Zn, Cd, Cr, Pb, Mn, Cu, Sr,Th, U, Ca	Sandstone	ICP-MS	Online	Continuous leaching or loop injection	Three-step acid scheme	Beauchemin et al. (2002)
	FI	As	Soil materials (GBW 08302 and GBW 07405)	HG-AFS	Online	Continuous leaching	Water soluble, associated with oxohydroxides, and acid soluble phases	Dong and Yan (2005)

[a]FI, flow injection; SI, sequential injection; ICP-AES, inductively coupled plasma–atomic emission spectrometry; FAAS, flame atomic absorption spectrometry, ETAAS: electrothermal atomic absorption spectrometry, ICP-MS: inductively coupled plasma–mass spectrometry, HG-AAS: hydrige generation–atomic absorption spectrometry, HG-AFS: hydrige generation–atomic fluorescence spectrometry.

12.4. INVESTIGATION OF LEACHING KINETICS: COMPLEMENTARY METHOD TO ACCESS THE MOBILITY OF TRACE ELEMENTS IN ENVIRONMENTAL SAMPLES

Classical end-over-end extraction approaches are intended to estimate the environmental impact of metal or metalloid pollution on biota. As stated above, these methods are based on establishment of the partitioning equilibria for target species between the solid phase and the leaching solution. Enhanced information on the mobility and accessibility of elements in environmental solids can be attained by the exploitation of dynamic fractionation systems that offer intrinsic capabilities for continuous monitoring of the metal leaching processes. The resulting extraction profile, the extractogram, which is obtained by a graphical plot of the concentration of released metals versus extraction time or volume of extracting reagent, renders valuable information on the lability of the various metal reservoirs, the chemical associations of the elements within the solid phases, and the maximum pool size of targeted fractions. Extractograms can be regarded as unique tools to test the efficiency of leaching reagents (i.e., the volume of extractant required for quantitative release of a given metal form), thus obtaining sample weight/extractant volume ratios that can be of great interest for further design and harmonization of SEPs.

To obtain detailed information on the leaching kinetics for particular elements attending the various binding forms, batchwise kinetic extractions as well as flow-through-based partitioning techniques are receiving particular attention. The latter may be performed by continuous-flow, flow injection or sequential injection in an off- or online detection fashion, as detailed in the following sections and illustrated in Figures 12.1 and Figure 12.2 via flow injection/sequential injection microcolumn manifolds.

12.4.1. Batchwise kinetic leaching

Kinetic fractionation of TEs in environmental solids is readily performed by minor modification of the experimental conditions for manual end-over-end extraction. The methodology consists of exposing various sample batches to the extracting solution for time intervals ranging from 5 to 15 minutes to 12 to 24 hours. Thus, the classical steps required for metal leaching and analysis of the supernatant solution are repeated at defined times. This protocol has been described to provide more reliable results than collecting aliquots from a single sample at fixed intervals (Fangueiro et al., 2002).

Batchwise kinetic leaching is aimed at understanding both the rate and extent of TE remobilization by changes in environmental scenarios. Moreover, this technique has the potential to overcome the recognized limitation of phase overlapping of several extractants (e.g., the single and sequential extractants of SM&T procedures) whenever the several pools attacked feature different kinetic behavior. Contributions of readsorption phenomena to the thermodynamic leachate results may also be discerned by evaluation of kinetic data (Bermond et al., 1998). Manual fractionation protocols based on kinetic explorations are capable of discerning three operationally defined fractions with potential for risk assessment: the labile (i.e., fast leachable) fraction, which is of environmental concern, the moderately or nonlabile (i.e., slowly leachable) fraction, and the nonextractable (inert) phase (Bermond et al., 1998; Ghestem and Bermond, 1999). In addition to discrimination of the metal content associated with bioavailable phases, kinetic constants may also be calculated, as demonstrated by Gismera et al. (2004) and Bermond and Varrault (2004).

This methodology has been utilized primarily for the investigation of extraction patterns of TEs and major elements using EDTA as a nonselective single extractant (Ghestem and Bermond, 1999; Fangueiro et al., 2002; Manouchehri et al., 2006). EDTA is usually selected in these studies because of its effectiveness in mobilizing metal ions from the main compartments of the solid substrate and fitting of experimental results to the sum of two exponential (first-order kinetics) functions (Bermond et al., 1998). Different models, the diffusion model and two first-order reaction model, have been exploited for ascertaining the rate-limiting step in TE extraction by EDTA (Fangueiro et al., 2005). In this context it should be stressed that the kinetic models have proven suitable for evaluating the availability of TEs (e.g., Cu and Cd) from reducible and oxidizable phases of soil samples (Varrault et al., 2001). The application to soil model compounds (i.e., goethite and humic acids) revealed that elements associated with iron oxyhydroxides are in a rather labile form, whereas those bound to organic compounds are more likely in a moderately labile form. By applying mathematical models to the experimental data of column leaching approaches, it was also possible to differentiate chemical components according to their lability (Song and Greenway, 2004).

Special attention has been paid to the modeling of kinetic curves for other chemical reagents. Thus, common leaching agents from SEP [i.e., ammonium acetate, acetic acid, hydroxylamine, and organic acids (e.g., oxalate and citrate)] have been assessed as independent single extractants for soil and sediment samples and the experimental results contrasted with those of equilibrium-based methods (Bermond et al., 1998; Gismera et al., 2004). It has been concluded that the kinetic approach renders complementary information to steady-state methods in terms of readily available elements in different compartments and offers more accurate insight into the size of mobilizable pools via renewal of the leaching reagent at defined intervals (Krishnamurti et al., 1997). A good correlation is reported for various TEs (i.e., Cu, Zn, Pb, Cd) between the labile phase and equilibrium data for the exchangeable (ammonium acetate) fraction, as expected for fast leaching processes that do not involve matrix dissolution. On the other hand, the labile acetic acid fraction is generally more than two fold lower than the extractable amount in the first leaching step of the SM&T procedure. This information is especially relevant to predicting the immediate consequences of soil acidification due to acid rainfall, mining activities, or anthropogenic spills.

The combination of kinetic single and steady-state sequential extraction methods has proved suitable for monitoring the fate of arsenic in slightly polluted soils through irrigation (Cornu et al., 2004). The EDTA–kinetic procedure confirmed that anthropogenic arsenic is fixed in soils mainly under labile forms and is easily extracted with the chelating agent. Furthermore, the fast leachable fraction was identified with the exchangeable and surface-bound fraction of the sequential extraction, while the slowly released As fraction is identified with that occluded in amorphous oxides and organic matter.

Despite the aforementioned benefits and supplementary information provided, kinetic methods are tedious and time consuming, use large amounts of sample and reagents, and so far have been limited to single extractions. In fact, the analytical procedure requires as many manual extractions as data points in the extractrogram.

12.4.2. Flow-through methods with off-line detection

This operational mode for ascertaining leaching kinetics using the above-mentioned flow through dynamic extraction approaches relies on the consecutive collection of defined volumes of leachates. Subfraction volumes, typically ranging from 3 to 10 mL, offer adequate temporal resolution to follow moderately fast release rates. Off-line detection enables the batchwise application of high dilution factors to the leachates prior to measurement by atomic spectrometric detectors [i.e., flame atomic absorption spectrometry (FAAS), electrothermal atomic absorption spectrometry (ETAAS), hydride generation–atomic fluorescence spectrometry (HG-AFS), ICP-MS and ICP-AES], thereby minimizing potential spectral and nonspectral interferences, which are commonly caused by the high salt content of several extractants in SEPs and the constituents of the soil matrix itself, with no need for matrix matching calibration procedures. Furthermore, sequential multielemental determination is applicable even for spectrometers with single detection capabilities, such as ETAAS and FAAS.

As a consequence of the ability to obtain continuous information on the extraction steps without requiring the construction of any interface between the flow setup and the spectrometer, off-line detection is the preferred modality to investigate leaching kinetics in most of the continuous-flow systems presented above, especially rotating coiled columns and stirred-flow cells (see Table 12.4). As an example, Figure 12.3*a* shows the extractrograms from a highly metal-contaminated material (i.e., SRM 2710 Montana soil), recorded by resorting to a flow-through microcolumn-containing assembly with 5-mL-fraction collections (Chomchoei et al., 2005b). Comparing the leaching patterns of the various metals using the three-step SEPs endorsed by the SM&T program in a unidirectional continuous extraction mode, it is noticeable that the kinetics of metal release from the solid material follow the same trends for all trace metals: Each target analyte is leached mostly in the first subfraction ($> 30\%$ for Cu and $> 40\%$ for Zn and Pb), and the free concentration in solution decreases rapidly in the following extractant zones. According to the extraction patterns of SRM 2710 soil, Zn and Cu are the most labile elements for this reference material, being especially sensitive to acidification processes. Moreover, the target phases are rapidly leachable, since more than 80% of the maximum metal available is obtained in

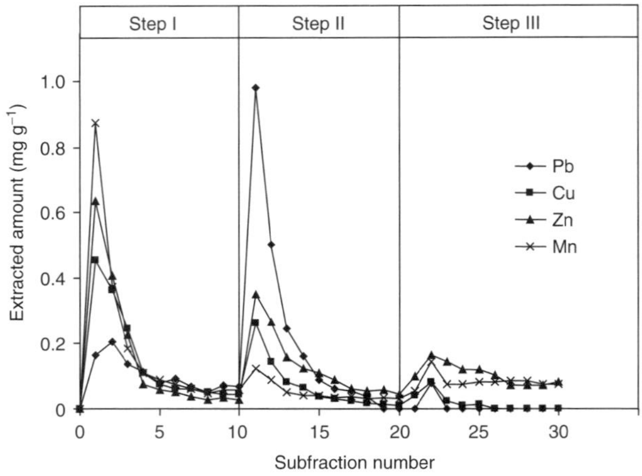

(*a*)

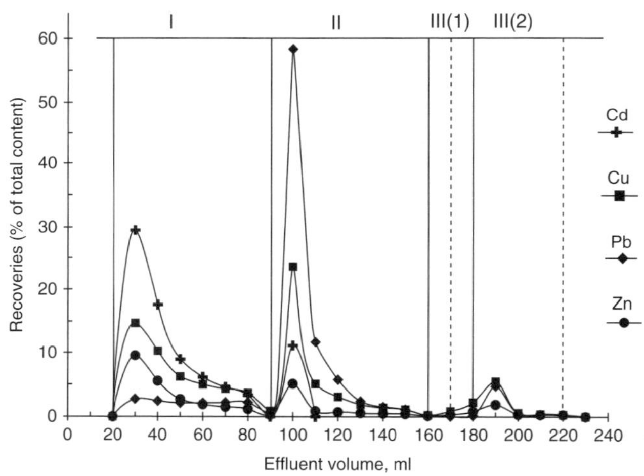

(*b*)

Figure 12.3. Leaching profiles of metals in SRM 2710 soil obtained by the application of extractants according to the original SM&T three-step sequential extraction protocol using (*a*) a flow-through microcolumn-based system (column inner volume, 0.36 mL; sample, 25 mg; extraction flow rate, 3 mL min^{-1}; subfraction volume, 5 mL), and (*b*) a flow-through rotating coiled column-based system (column inner volume, 20 mL; sample, 500 mg; extraction flow rate, 1 mL min^{-1}; subfraction volume, 10 mL). [(a) From Chomchoei et al., 2005b); (*b*) from Fedotov et al., 2005b, by permission of the Royal Society of Chemistry.)

the first four subfractions (i.e., in the first 20 mL of the leaching solution). On the other hand, lead is highly affected by reduction processes since the maximum extractability was obtained in step II of the SEP, shown in Figure 12.3*a*. The leaching profiles for the same sample obtained using RCC (Fedotov et al.,

2005b) are shown for comparison in Figure 12.3*b*. (The "0 mL" value corresponds to the injection of a suspended sample into an RCC. The profiles start at an effluent volume matching the free column volume (i.e., 20 mL). As can be seen, although the experimental conditions are quite different (i.e., the column volume, sample weight, and fraction volume), the resulting curves show the same trends. This proves the relevance of both methods for time-resolved studies of metal leaching in well-homogenized environmental solids.

It should be noted that both the available fractions and leaching profiles depend heavily on the chemical composition and binding sites of the environmental solid. The dynamic extractable amounts of Pb, Cu, and Zn from a sewage-amended soil (CRM 483) for which certified batchwise metal-extractable contents for single extraction procedures (acetic acid and EDTA) are available are shown in Table 12.5. For the CH_3COOH microcolumn-based extraction of Pb and Cu, the total volume of the extractant was fixed at 12 mL to maintain a soil weight/volume ratio identical to that in the classical manual method. For these elements, a steady continuous leaching profile was detected, as illustrated in Figure 12.4*a*, owing to the nonselectivity of the reagent (Whalley and Grant, 1994), which is also capable of releasing various metal pools susceptible to acidification processes (Sahuquillo et al., 2003). On the other hand, Zn was leached rapidly from the sewage-amended soil (see Figure 12.4*b*). The CH_3COOH extraction pattern was monitored until the increase in metal leached in various consecutive subfractions was negligible with respect to the total amount accumulated (Chomchoei et al., 2005a). As shown in Table 12.5, the acetic acid–extractable content of Zn is statistically comparable to the certified value, thereby indicating that the two conceptually different methods yield similar results when applied to weakly bound forms. This also holds true for reagents with a strong capacity for metal mobilization and stabilization, such as EDTA, as illustrated in Figure 12.5. In this context, it should be borne in mind that EDTA is not merely capable of mobilizing carbonate and organic (humate and fulvate) metal forms by competing complexing reactions, but also metals entrapped in amorphous hydrous

TABLE 12.5. Comparison of the Extractable Metal Contents (mg/kg) in CRM 483 obtained with Manual and Dynamic Flow-through Procedures Using Single Extractants

	Acetic Acid ($n = 3$)		EDTA ($n = 3$)	
Element	Dynamic[a]	Certified Value	Dynamic[a]	Certified Value
Pb	1.4 ± 0.1	2.10 ± 0.25	206 ± 9	229 ± 8.0
Cu	19 ± 2	33.5 ± 1.6	227 ± 11	215 ± 11
Zn	637 ± 38	620 ± 24	617 ± 5	612 ± 19

Source: Chomchoei et al. (2005b).

[a] Soil sample, 300 mg; extraction flow rate, 0.6 mL min^{-1}; subfraction volume/total extractant used, 1 mL/50 mL and 2 mL/12 mL (except Zn) for EDTA and CH_3COOH, respectively.

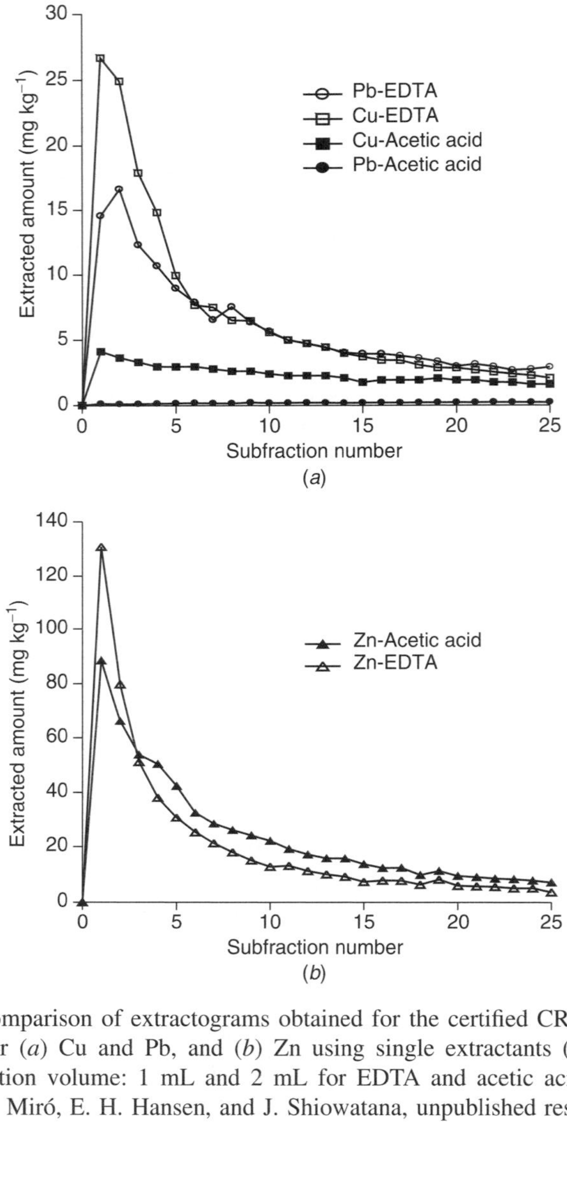

Figure 12.4. Comparison of extractograms obtained for the certified CRM 483 sewage-amended soil for (*a*) Cu and Pb, and (*b*) Zn using single extractants (acetic acid and EDTA). Subfraction volume: 1 mL and 2 mL for EDTA and acetic acid, respectively. (R. Petersen, M. Miró, E. H. Hansen, and J. Shiowatana, unpublished results, 2005.)

oxides of iron because of the high stability constant of the Fe–EDTA chelates. This explains the good agreement in Table 12.5 between the dynamic and batch results for the EDTA-extractable fractions. In contrast, the leachable content under the dynamic mode for less labile forms, such as acid-soluble Pb and Cu, is appreciably lower than that obtained under a steady-state regime. This is a consequence of the inherently longer intimate contact time between soil and solution in the batchwise method. The results provided by both procedures are expected

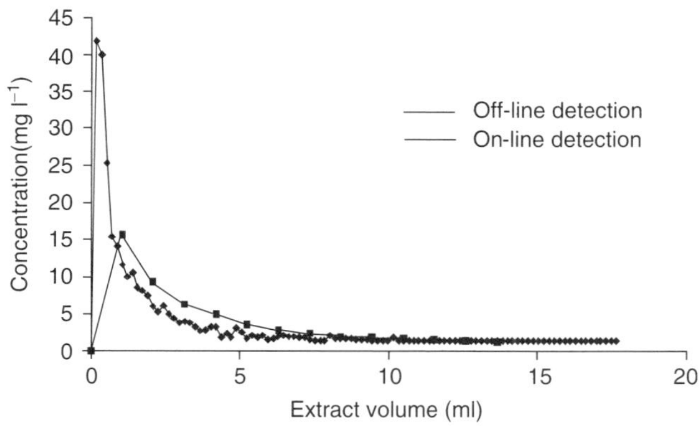

Figure 12.5. Comparison of EDTA extractrograms for Cu as obtained via off-line and online measurements. Subfraction volume: 1 mL and 175 μL for off-line and online detection, respectively. (Reproduced from Chomchoei et al., 2005a; copyright © American Chemical Society.)

to differ considerably whenever the leaching agent attacks soil phases containing reservoirs of slowly accessible elements. Actually, higher metal availability for CRM 483 soil than that estimated from the certified values was reported by increasing the acid extract volume (e.g., > twofold increase for Cu and Pb using 50 mL of extractant) (Chomchoei et al., 2005a) as a result of the continuous shift of the metal distribution equilibria. Therefore, it is possible to conclude that automated fractionation studies are better suited than their traditional counterparts to predicting actual risks associated with soil contamination.

Another outstanding asset of flow-through dynamic fractionation systems is the feasibility of comparing both the peak positions and the profile shapes between elements to identify and elucidate pollutant–soil phase interactions. To gain more detailed information on the leaching kinetics and elemental association, continuous detailed extractograms need to be explored. This can be effected either by collection of smaller volumes of extract per subfraction or as described in Section 12.4.3, by coupling the sample-containing column with an appropriate flow-through detector, thereby allowing online determination of the individual metals.

Although the discussion has been focused mainly on the performance of off-site measurements using column leaching methods, it also holds true for in situ sampling of interstitial soil solution as capitalized on implanted microdialyzers (Miró and Frenzel, 2005). As a result of the continuous flow of the perfusion liquid, microdialysates containing elements that have been leached from the surrounding environment under natural scenarios can be collected readily in small vials prior to detection. It should be borne in mind that the perfusing solution is commonly pumped through the miniaturized dialytic probes at flow rates between 1 and 10 μL min^{-1} to warrant appropriate recovery of dissolved species. In this context, ETAAS determinations are advantageous, as the low volume requirements inherent in the detection principle ensure high temporal resolution for environmentally occurring processes.

12.4.3. Flow-through methods with online detection

Fully automated flow-through methods involving the hyphenation of a microextraction device with a continuously operating atomic spectrometer have received widespread acceptance for single and sequential element partitioning studies (Miró et al., 2005a). Besides preventing contamination risks, the totally enclosed environment reduces auxiliary sample manipulations dramatically, as neither separation nor dilution steps are needed. Very importantly, the hyphenated microanalytical technique ensures highly time-resolved information on the ongoing extraction process. This is especially beneficial for monitoring metal pools weakly bound to soil phases or easily available by the action of aggressive extractants such as strong chelating agents. The extractograms depicted in Figure 12.5 illustrate clearly the different kinetic information rendered by online analysis of minute volumes of leachate (<200 μL), and measurements after manual dilution of 1-mL subfractions when EDTA was used as a single extractant. The averaged concentrations of the off-line procedure are thus inadequate for short-term assessment of the ecological impact of pollutants liberated under the action of soluble ligands.

Success in the performance of online detection procedures resides in the accurate control of the volume and flow rate of the leaching agent through the solid sample. For stirred-flow cell experiments, although the flow rate of the influent solution is compatible with the free uptake rate of the FAAS nebulizer, direct connection with the detector has been reported to be troublesome, owing to the continuous change in the delivery volume occasioned by gradual blockage of the membrane filter with fine particles. To minimize this drawback, periodical calibration of the flow network is recommended by implementation of a rotary injection valve at the inlet of the nebulizer to inject control standards into the extract stream (Shiowatana et al., 2001a). The flow rate of the extracting reagents in an RCC mounted onto a planetary centrifuge drum is close to that of the pneumatic nebulizer of ICP–MS or ICP–AES detectors, yet direct coupling is subjected to the same demands as those for stirred-flow cells. Microcolumn-based partitioning protocols accommodated in flow injection or sequential injection networks have currently received a broad appeal for online monitoring of the leaching kinetics of metal and metalloid species. The robustness of the liquid driver of the latter flow approach allows minute well-defined volumes of extract to be handled with great precision (Chomchoei et al., 2005a). With respect to detection systems, much effort have been dedicated to exploitation of the simultaneous multielemental detection capabilities of ICP–AES and ICP–MS. In the latter, monitoring several isotopes of a given analyte whose isotopic composition varies in the environment may be of significant relevance for discriminating between sources of this species. However, the salt and carbon content of the leachates, the contribution of isobaric interferences, and the in-instrument formation of polyatomic ions may jeopardize the accuracy of measurements of certain isotopes (e.g., Fe) by quadrupole ICP–MS (Beauchemin et al., 2002). Spectral interferences can, however, be reduced to a great extent by using to high-resolution mass analyzers such as ICP–sector field–MS or multicollector ICP–MS or by applying derivatization reactions in a dynamic reaction cell prior to ICP–MS detection. Further minimization of interferences is readily achieved by interfacing the flow-injection network with the detector via a rotary valve, by which small leachate volumes are introduced into a carrier stream supplied to the nebulizer (Jimoh et al., 2004) (see Figure 12.2). The resulting high wash-to-sample exposure time allows complex matrices to be analyzed with no risk of solid deposition and etching of the MS interface. Similar strategies are obviously applicable to FAAS or AFS to reduce the influence of interfering matrix components.

Similar to FI systems, in-line coupling of a sequential injection network with suitable detectors renders a detailed insight into leaching kinetics. As a consequence of the discontinuous flow nature of SIA, the in-line hyphenation of microcolumn assays with ETAAS is relatively simple, because of the discrete, noncontinuous operation of the detector. On the other hand, interfacing the miniaturized flow assembly with continuously operating detectors such as FAAS, HG-AFS, ICP-AES and ICP-MS requires the incorporation of conventional sample injectors (e.g., solenoid valves, rotary valves) for discrete introduction into the detection instrument of metered leachate volumes.

The moderate dilution factors attainable with fully automated flow injection or sequential injection systems may impose severe difficulties for the online development of accurate partitioning studies in environmental solids of relative complexity. The contribution to the analytical results of multiplicative (nonspectral) matrix interferences has been solved elegantly by the use of in-line standard addition methods (Dong and Yan, 2005), although at the expense of a threefold decrease in sampling frequency.

12.5. CONCLUSIONS AND FUTURE DIRECTIONS

Fractionation analysis has been established as a routine tool for attaining valuable information about the mode of occurrence, bioavailability, mobility, and ecotoxicological significance of metal ions in solid environmental samples. This is commonly done by exploiting sequential extraction procedures with given leaching solutions under operationally defined conditions, as reviewed in this chapter. The ability to elucidate soil-phase associations for trace elements is discussed in detail, as is their adaptability for testing the accessibility of anionic species such as metalloid compounds from targeted soil–sediment compartments.

Special emphasis has been paid to relevant aspects for appropriate sampling and sample preparation which might affect the quality of fractionation results. Novel techniques for in situ sampling of soil solution directly from the solid substrate with minimum disturbance of the sampling site have also been addressed. The performance of traditional batchwise partitioning methods based on establishing equilibrium between solid and liquid phases has been compared critically with that of novel flow-through dynamic fractionation approaches that utilize steady renewal of the extractant phase. The outstanding advantages of these recent trends in solid–liquid partitioning can be summarized as follows: (1) accurate monitoring of the ongoing leaching of trace and major elements; (2) simplification of the overall procedure (from days to hours), (3) miniaturization and automation of the extraction procedure, (4) minimization of risks of sample contamination and straightforward manipulation of leaching agents, and (5) evaluation of the efficiency and selectivity of extractants as well as the maximum pool of available fractions. Further research is needed for harmonization and establishment of quality control tools for dynamic extraction procedures, including traceability issues. Actually, current efforts are focused on the generalization of similarities and discrepancies between batchwise and kinetic leaching methods for the various operationally defined fractions using endorsed sequential extraction schemes and quality control materials.

REFERENCES

Anderson, P., Davidson, C. M., Duncan, A. L., Littlejohn, D., Ure, A. M., and Garden, M. L. (2000). Column leaching and sorption experiments to assess the mobility of potentially toxic elements in industrially contaminated land. *J. Environ. Monit.* **2**, 234–239.

Bascomb, C. L. (1968). Distribution of pyrophosphate-extractable iron and organic carbon in soils of various groups. *J. Soil Sci.* **19**, 251–258.

Beauchemin, D., Kyser, K., and Chipley, D. (2002). Inductively coupled plasma mass spectrometry with on-line leaching: a method to assess the mobility and fractionation of elements. *Anal. Chem.* **74**, 3924–3928.

Beckett, P. H. T. (1989). The use of extractants in studies on trace metals, in soils, sewage sludge and sludge-treated soils. *Adv. Soil. Sci.* **9**, 143–176.

Belzile, N., Lecomte, P., and Tessier A. (1989). Readsorption of trace elements during partial chemical extractions of bottom sediments. *Environ. Sci. Technol.* **23**, 1015–1020.

Bermond, A. (2001). Limits of sequential extraction procedures re-examined with emphasis on the role of H^+ ion reactivity. *Anal. Chim. Acta* **445**, 79–88.

Bermond, A., and Varrault, G. (2004). Application of a kinetic fractionation of trace elements (Cd, Cu & Pb) in unpolluted soil samples. *Environ. Technol.* **25**, 293–300.

Bermond, A., Yousfi, I., and Ghestem, J. P. (1998). Kinetic approach to the chemical speciation of trace metals in soils. *Analyst* **123**, 785–789.

Berti, W. R., and Cunningham, S. D. (1997). In-place inactivation of Pb in Pb-contaminated soils. *Environ. Sci. Technol.* **31**, 1359–1364.

Brady, N. C., and Weil, R. R. (1996). *The Nature and Properties of Soils*. 11th ed., Prentice Hall, Upper Saddle River, NJ.

Brümmer, G. W. (1986). Heavy metal species, mobility and availability. In *The Importance of Chemical Speciation in Environmental Processes,* ed. Bernhard, M., Brinkman, F. E., and Sadler P. J., Springer-Verleg, Berlin, 169–192.

Cabral, A. R., and Lefebvre, G. (1998). Use of sequential extraction in the study of heavy metal retention by silty soils. *Water Air Soil Pollut.* **102**, 329–344.

Camobreco, V. J., Richards, B. K., Steenhuis, T. S., Peverly, J. H., and McBride, M. B. (1996). Movement of heavy metals through undisturbed and homogenized soil columns. *Soil Sci.* **161**, 740–750.

Campos, E., Barahona, E., Lachica, M., and Mingorance, M. D. (1998). A study of the analytical parameters important for the sequential extraction procedure using microwave heating for Pb, Zn and Cu in calcareous soils. *Anal. Chim. Acta* **369**, 235–243.

Chomchoei, R., Shiowatana, J., and Pongsakul, P. (2002). Continuous-flow system for reduction of metal readsorption during sequential extraction of soil. *Anal. Chim. Acta* **472**, 147–159.

Chomchoei, R., Hansen, E. H., and Shiowatana, J. (2004). Utilizing a sequential injection system furnished with an extraction microcolumn as a novel approach for executing sequential extractions of metal species in solid samples. *Anal. Chim. Acta* **526**, 177–184.

Chomchoei, R., Miró, M., Hansen, E. H., and Shiowatana, J. (2005a). Automated sequential injection-microcolumn approach with on-line flame atomic absorption spectrometric detection for implementing metal fractionation schemes of homogeneous and nonhomogeneous solid samples of environmental interest. *Anal. Chem.* **77**, 2720–2726.

Chomchoei, R., Miró, M., Hansen, E. H., and Shiowatana, J. (2005b). Sequential injection system incorporating a micro-extraction column for automatic fractionation of metal ions in solid samples: comparison of the extraction profiles when employing uni-, bi-, and multi-bidirectional flow plus stopped flow sequential extraction modes. *Anal. Chim. Acta*. **536**, 183–190.

Cieśliński, G., Van Rees, K. C. J., Szmigielska, A. M., Krishnamurti, G. S. R., and Huang, P. M. (1998). Low-molecular-weight organic acids in rhizosphere soils of durum wheat and their effect on cadmium bioaccumulation. *Plant Soils* **203**, 109–117.

Cornu, S., Montagne, D., and Conil, P. (2004). Comparison of sequential and kinetic extractions for As speciation in contaminated soils. *C. R. Geosci.* **336**, 1007–1015.

Daus, B., Weiss, H., and Wennrich, R. (1998). Arsenic speciation in iron hydroxide precipitates. *Talanta* **46**, 867–873.

Davidson, C. M., and Delevoye, G. (2001). Effect of ultrasonic agitation on the release of copper, iron, manganese and zinc from soil and sediment using the BCR three-stage sequential extraction. *J. Environ. Monit.* **3**, 398–403.

Davidson, C. M., Duncan, A. L., Littlejohn, D., Ure, A. M., and Garden, L. M. (1998). A critical evaluation of the three-stage BCR sequential extraction procedure to assess the potential mobility and toxicity of heavy metals in industrially-contaminated land. *Anal. Chim. Acta* **363**, 45–55.

Davidson, C. M., Hursthouse, A. S., Tognarelli, D. M., Ure, A. M., and Urquhart, G. J. (2004). Should acid ammonium oxalate replace hydroxylammonium chloride in step 2 of the revised BCR sequential extraction protocol for soil and sediment? *Anal. Chim. Acta* **508**, 193–199.

Dong, L.-M., and Yan, X.-P. (2005). On-line coupling of flow injection sequential extraction to hydride generation atomic fluorescence spectrometry for fractionation of arsenic in soils. *Talanta* **65**, 627–631.

Edwards, W. M., Shipitalo, M. J. and Owens, L. B. (1993). Gas, water and solute transport in soils containing macropores: a review of methodology. *Geoderma* **57**, 31–49.

Fangueiro, D., Bermond, A., Santos, E., Carapuça, H., and Duarte, A. (2002). Heavy metal mobility assessment in sediments based on a kinetic approach of the EDTA extraction: search for optimal experimental conditions. *Anal. Chim. Acta* **459**, 245–256.

Fangueiro, D., Bermond, A., Santos, E., Carapuça, H., and Duarte, A. (2005). Kinetic approach to heavy metal mobilization assessment in sediments: choose of kinetic equations and models to achieve maximum information. *Talanta* **66**, 844–857.

Fedotov, P. S. (2002). Untraditional applications of countercurrent chromatography. *J. Liq. Chromatogr. Relat. Technol.*, **25**, 2065–2078.

Fedotov, P. S., Zavarzina, A. G., Spivakov, B. Ya., Wennrich, R., Mattusch, J., Titze, K de P. C., and Demin, V. V. (2002). Accelerated fractionation of heavy metals in contaminated soils and sediments using rotating coiled columns. *J. Environ. Monit.* **4**, 318–324.

Fedotov, P. S., Sutherland, I. A., Wood, P., and Spivakov, B. Ya. (2003). Retention of solids in rotating coiled columns: the effect of beta value and tubing material. *J. Liq. Chromatogr. Relat. Technol.*, **26**, 1641–1649.

Fedotov, P. S., Fitz, W. J., Wennrich, R., Morgenstern, P., and Wenzel, W. W. (2005a). Fractionation of arsenic in soil and sludge samples: continuous-flow extraction using rotating coiled columns versus batch sequential extraction. *Anal. Chim. Acta* **538**, 93–98.

Fedotov, P. S., Wennrich, R., Stärk, H.-J., and Spivakov, B. Ya. (2005b). Continuous-flow fractionation of trace metals in environmental solids using rotating coiled columns: some kinetic aspects and applicability of three-step BCR leaching schemes. *J. Environ. Monit.* **7**, 22–28.

Fedotov, P. S., Savonina, E. Yu., Wennrich, R., and Spivakov, B. Ya. (2006). A hyphenated flow-through analytical system for the study on the mobility and fractionation of trace and major elements in environmental solid samples. *Analyst* **131**, 509–515.

Filgueiras, A. V., Lavilla, I., and Bendicho, C. (2002a). Chemical sequential extraction for metal partitioning in environmental solid samples. *J. Environ. Monit.* **4**, 823–857.

Filgueiras, V., Lavilla, I., and Bendicho, C. (2002b). Comparison of the standard SM&T sequential extraction method with small-scale ultrasound-assisted single extractions for metal partitioning in sediments. *Anal. Bioanal. Chem.* **374**,103–108.

Fitz, W. J., Wenzel, W. W., Zhang, H., Nurmi, J., Stipek, K., Fischerova, Z., Schweiger, P., Köllensperger, G., Ma, L. Q. and Stingeder, G. (2003). Rhizosphere characteristics of the arsenic hyperaccumulator *Pteris vittata* L. and monitoring of phytoremoval efficiency. *Environ. Sci. Technol.* **37**, 5008–5014.

Förstner, U. (1993). Metal speciation: general concepts and applications. *Int. J. Environ. Anal. Chem.* **51**, 5–23.

Förstner, U. (2004). Traceability of sediment analysis. *Trends Anal. Chem.* **23**, 217–236.

Gawlik, B. M., van der Sloot, H. A., Ulberth, F., Nortcliff, S., Simonart, T., Cooper, B., Leschber, R., Andersen, K., and Wichman, K. (2004). On the development of horizontal CEN standards supporting the implementation of EU directives on sludge, soil and biowaste-project HORIZONTAL. *Trends Anal. Chem.* **23**, 686–703.

Ghestem, J. P., and Bermond, A. (1999). Feasibility study of a fractionation of trace elements in soil samples based on kinetics. *Environ. Technol.* **20**, 1119–1128.

Gismera, M. J., Lacal, J., Da Silva, P., García, R., Sevilla, M. T., and Procopio, J. R. (2004). Study of metal fractionation in river sediments: a comparison between kinetic and sequential extraction procedures. *Environ. Pollut.* **127**, 175–182.

Gleyzes, C., Tellier, S., Sabrier, R., and Astruc, M. (2001). Arsenic characterisation in industrial soils by chemical extractions. *Environ. Technol.* **22**, 27–38.

Gleyzes, C., Tellier, S., and Astruc, M. (2002). Fractionation studies of trace elements in contaminated soils and sediments: a review of sequential extraction procedures. *Trends Anal. Chem.* **21**, 451–466.

Golterman, H. L. (1996). Fractionation of sediment phosphate with chelating compounds: a simplification, and comparison with other methods. *Hydrobiology* **335**, 87–95.

Gómez-Ariza, J. L., Giráldez, I., Sánchez-Rodas, D., and Morales E. (1999). Metal readsorption and redistribution during the analytical fractionation of trace elements in oxic estuarine sediments. *Anal. Chim. Acta* **399**, 295–307.

Gong, C. R., and Donahoe, R. J. (1997). An experimental study of heavy metal attenuation and mobility in sandy loam soils. *Appl. Geochem.* **12**, 243–254.

Gong, Z.-L., Lu, X.-F., Ma, M.-S., Watt, C., and Le, X. C. (2002). Arsenic speciation analysis. *Talanta* **58**, 77–96.

Göttlein, A., Hell, U., and Blasek, R. (1996). A system for microscale tensiometry and lysimetry. *Geoderma* **69**, 147–156.

Gove, L., Cooke, C. M., Nicholson, F. A. and Beck, A. J. (2001). Movement of water and heavy metals (Zn, Cu, Pb and Ni) through sand and sandy loam amended with biosolids under steady-state hydrological conditions. *Bioresour. Technol.* **78**, 171–179.

Grossmann, J., and Udluft, P. (1991). The extraction of soil water by suction-cup method: a review. *J. Soil Sci.* **42**, 83–93.

Gruebel, K. A., Davis, J. A., and Leckie, J. O. (1988). The feasibility of using sequential extraction techniques for arsenic and selenium in soils and sediments. *Soil Sci. Soc. Am. J.* **52**, 390–397.

Haines, B. L., Waide, J. B., and Todd, R. L. (1982). Soil solution nutrient concentrations sampled with tension and zero-tension lysimeters: report of discrepancies. *Soil Sci. Soc. Am. J.* **46**, 658–661.

Hämmann, M., and Desaules, A. (2003). *Sampling and Sample Pre-treatment in Soil Pollutant Monitoring*, Swiss Agency for the Environment, Forests and Landscape (SAEFL), Berne, Switzerland.

Harrington, J. M., LaForce, M. J., Rember, W. C., Fendorf, S. E., and Rosenzweig, R. F. (1998). Phase associations and mobilization of iron and trace elements in Coeur d'Alene Lake, Idaho. *Environ. Sci. Technol.* **32**, 650–656.

Hieltjes, A. H. M., and Lijklema, L. (1980). Fractionation of inorganic phosphates in calcareous sediments. *J. Environ. Qual.* **3**, 405–407.

Hlavay, J., Prohaska, T., Weisz, M., Wenzel, W.W., and Stingeder, G. J. (2004). Determination of trace elements bound to soils and sediment fractions. IUPAC Tech. Rep. *Pure Appl. Chem.* **76**, 415–442.

Horowitz, A. (1991). *A Primer on Sediment–Trace Element Chemistry*, 2nd ed., Lewis Publishers, Chelsea, MI.

Houba, V. J. G., Novozamsky, I. and van der Lee, J. J. (1994). Aspects of pre-treatment of soils for inorganic chemical analysis. *Quim. Anal.* **13**, S94–S99.

Ito, Y. (1991). Recent advances in countercurrent chromatography. *J. Chromatogr.* **538**, 3–25.

Jang, A., Choi, Y. S. and Kim I. S. (1998). Batch and column tests for the development of an immobilization technology for toxic heavy metals in contaminated soils of closed mines. *Water Sci. Technol.* **37**, 81–88.

Jimoh, M., Frenzel, W., Müller, V., Stephanowitz, H., and Hoffmann, E. (2004). Development of a hyphenated microanalytical system for the investigation of leaching kinetics of heavy metals in environmental samples. *Anal. Chem.* **76**, 1197–1203.

Jimoh, M., Frenzel, W., and Müller, V. (2005). Microanalytical flow-through method for assessment of the bioavailability of toxic metals in environmental samples. *Anal. Bioanal. Chem.* **381**, 438–444.

Johnston, S. E., and Barnard, W. M. (1979). Comparative effectiveness of fourteen solutions for extracting arsenic from four western New York soils. *Soil Sci. Soc. Am. J.* **43**, 304–308.

Kahakachchi, C., Uden, P. C., and Tyson, J. F. (2004). Extraction of arsenic species from spiked soils and standard reference materials. *Analyst* **129**, 714–718.

Karczewska, A., Chodak, T., and Kaszubkiewicz, J. (1996). The suitability of brown coal as a sorbent for heavy metals in polluted soils. *Appl. Geochem.* **11**, 343–346.

Kennedy, V. H., Sanchez, A. L., Oughton, D. H., and Rowland, A. P. (1997). Use of single and sequential chemical extractants to assess radionuclide and heavy metal availability from soils for root uptake. *Analyst* **122**, 89R–100R.

Kersten, M., and Förstner, U. (1986). Chemical fractionation of heavy metals in anoxic estuarine and coastal sediments. *Water Sci. Technol.* **18**, 121–130.

Kersten, M., and Förstner, U. (1995). Speciation of trace metals in sediments and combustion waste. In *Chemical Speciation in the Environment*, ed. Ure, A. M., and Davidson, C. M., Blackie Academic, London.

Krishnamurti, G. S. R., and Naidu, R. (2000). Speciation and phytoavailability of cadmium in selected surface soils of South Australia. *Aust. J. Soil Res.* **38**, 991–1004.

Krishnamurti, G. S. R., Huang, P. M., Van Rees, K. C. J., Kozak, L. M., and Rostad, H. P. W. (1995). Speciation of particulate-bound cadmium of soils and its bioavailability. *Analyst* **120**, 659–665.

Krishnamurti, G. S. R., Huang, P. M., and Van Rees, K. C. J. (1996). Studies on soil rhizosphere: speciation and availability of cadmium. *Chem. Spec. Bioavailab.* **8**, 23–28.

Krishnamurti, G. S. R., Cieśliński, G., Huang, P. M., and Van Rees, K. C. J. (1997). Kinetics of cadmium release from soils as influenced by organic acids: implication in cadmium availability. *J. Environ. Qual.* **26**, 271–277.

Krishnamurti, G. S. R., McArthur, D. F. E., Wang, M. K., Kozak, L. M., and Huang, P. M., (2005). Biogeochemistry of soil cadmium and the impact on terrestrial food chain contamination. In *Biogeochemistry of Trace Elements in the Rhizosphere*, ed. Huang, P. M., and Gobran, G., Elsevier, Amsterdam, The Netherlands, pp. 197–257.

Kubová, J., Stresko, V., Bujdos, M., Matús, P., and Medved, J. (2004). Fractionation of various elements in CRMs and in polluted soils. *Anal. Bioanal. Chem.* **379**, 108–114.

Ladonin, D. V. (2002). Heavy metal compounds in soils: problems and methods of study. *Eurasian Soil Sci.* **35**, 605–613.

La Force, M. J., and Fendorf, S. (2000). Solid-phase iron characterization during common selective sequential extractions. *Soil Sci. Soc. Am. J.* **64**, 1608–1615.

La Force, M. J., Hansel, C. M., and Fendorf, S. (2000). Arsenic speciation, seasonal transformations, and co-distribution with iron in a mine waste–influenced palustrine emergent wetland. *Environ. Sci. Technol.* **34**, 3937–3943.

Lambkin, D., Nortcliff, S., and White, T. (2004). The importance of precision in sampling sludges, biowastes and treated soils in a regulatory framework. *Trends Anal. Chem.* **23**, 704–715.

Langenkamp, H., and Marmo, L. (eds.) (2001). *Workshop on Harmonisation of Sampling and Analysis Methods for Heavy Metals, Organic Pollutants and Pathogens in Soil and Sludge: Summary and Conclusions*, Feb. 8–9, 2001, Stresa, Italy, European Commission, EUR 198909 EN.

Lawrence, G. B. and David, M. B. (1996). Chemical evaluation of soil-solution in acid forest soils. *Soil Sci.,* **161**, 298–313.

Lenehan, C. E., Barnett, N. W., and Lewis, S. W. (2002). Sequential injection analysis. *Analyst* **127**, 997–1020.

Lombi, E., Sletten, R. S., and Wenzel, W. W. (2000). Sequentially extracted arsenic from different size fractions of contaminated soils. *Water Air and Soil Pollut.* **124**, 319–332.

López-Sánchez, J. F., Sahuquillo, A., Fiedler, H. D., Rubio, R., Rauret, G., Muntau, H., and Quevauviller, Ph. (1998). CRM 601: a stable material for its extractable content of heavy metals. *Analyst* **123**, 1675–1677.

MacDonald, J. D., Bélanger, N., and Hendershot W. H. (2004a). Column leaching using dry soil to estimate solid-solution partitioning observed in zero-tension lysimeters, 1: Method development. *Soil Sediment Contam.* **13**, 361–374.

MacDonald, J. D., Bélanger, N., and Hendershot, W. H. (2004b). Column leaching using dry soil to estimate solid-solution partitioning observed in zero-tension lysimeters, 2: Trace metals. *Soil Sediment Contam.* **13**, 375–390.

Manouchehri, N., Besancon, S., and Bermond, A. (2006). Major and trace metal extraction from soil by EDTA: equilibrium and kinetic studies. *Anal. Chim. Acta* **559**, 105–112.

Marshall, G., Wolcott, D., and Olson, D. (2003). Zone fluidics in flow analysis: potentialities and applications. *Anal. Chim. Acta* **499**, 29–40.

Martens, D. A., and Suarez, D. L. (1997). Selenium speciation of soil/sediment determined with sequential extractions and hydride-generation atomic absorption spectrophotometry. *Environ. Sci. Technol.* **31**, 133–139.

Maryutina, T. A., Fedotov, P. S., and Spivakov, B. Ya. (1999). Application of countercurrent chromatography in inorganic analysis. In *Countercurrent Chromatography*, ed. Menet, J.-M., Thiébaut, D., Chromatographic Science Series, Vol. 82, Marcel Dekker, New York, **82**, 171–221.

McKeague, J. A. (1967). An evaluation of 0.1 M pyrophosphate and pyrophosphate-dithionite in comparison with oxalate as extractants of the accumulation products in Podzols and some other soils. *Can. J. Soil Sci.* **47**, 95–102.

McLaren, R. G., and Crawford, D. (1973). Studies on soil copper., 1: The fractionation of copper in soils. *J. Soil Sci.* **24**, 172–179.

Mikutta, R., Kleber, M., Kaiser, K., and Jahn, R. (2005). Review: Organic matter removal from soils using hydrogen peroxide, sodium hypochlorite, and disodium peroxodisulfate. *Soil Sci. Soc. Am. J.* **69**, 120–135.

Miller W. P., Martens, D. C., and Zelazny, L. W. (1986). Effect of sequence on extraction of trace metals from soils. *Soil Sci. Soc. Am. J.* **50**, 598–601.

Miró, M., and Frenzel, W. (2004a). Implantable flow-through capillary-type microdialyzers for continuous in situ monitoring of environmentally relevant parameters. *Anal. Chem.* **76**, 5974–5981.

Miró, M., and Frenzel, W. (2004b). What flow injection has to offer to the environmental analytical field. *Microchim. Acta* **148**, 1–20.

Miró, M., and Frenzel, W. (2005). The potential of microdialysis as an automatic sample-processing technique for environmental research. *Trends Anal. Chem.* **24**, 324–333.

Miró, M., and Hansen, E. H. (2005). Miniaturisation and automation of metal fractionation schemes applied to environmental solid samples by sequential injection microcolumn extraction procedures. *Anal. Bional. Chem.* **382**, 878–880.

Miró, M., Hansen, E. H., Chomchoei, R., and Frenzel, W. (2005a). Dynamic flow-through approaches for metal fractionation in environmentally relevant solid samples. *Trends Anal. Chem.* **24**, 759–771.

Miró, M., Jimoh, M., and Frenzel, W. (2005b). A novel dynamic approach for automatic microsampling and continuous monitoring of metal ion release from soils exploiting a dedicated flow-through microdialyser. *Anal. Bioanal. Chem.* **382**, 396–404.

Morales-Muñoz, S., Luque-García, J. L., and Luque de Castro, M. D. (2004). A continuous approach for the determination of Cr(VI) in sediment and soil based on the coupling of microwave-assisted water extraction, preconcentration, derivatization and photometric detection. *Anal. Chim. Acta* **515**, 343–348.

Mudroch, A., and Azcue, J. M. (1995). *Manual of Aquatic Sediment Sampling*, Lewis Publishers, Boca Raton, FL.

Mudroch, A., and MacKnight, S. D. (1994). *Techniques for Aquatic Sediment Sampling*, 2nd ed., Lewis Publishers, Boca Raton, FL.

Muntau, H., Rehnert, A., Desaules, A., Wagner, G., Theocharopoulos, S., and Quevauviller, Ph. (2001). Analytical aspects of the CEEM soil project. *Sci. Total Environ.* **264**, 27–49.

Neaman, A., Mouélé, F., Trolard, F., and Bourrié, G. (2004). Improved methods for selective dissolution of Mn oxides: applications for studying trace element associations. *Appl. Geochem.* **19**, 973–979.

Onken, B. M., and Adriano, D. C. (1997). Arsenic availability in soil with time under saturated and subsaturated conditions. *Soil Sci. Soc. Am. J.* **61**, 746–752.

Onyatta, J. O., and Huang, P. M. (1999). Cadmium speciation and bioavailability index for selected Kenyan soils. *Geoderma* **91**, 87–101.

Papp, C. S. E., Filipek, L. H., and Smith, K. S. (1991). Selectivity and effectiveness of extractants used to release metals associated with organic matter. *Appl. Geochem.* **6**, 349–353.

Paschke, A. (2003). Consideration of the physicochemical properties of sample matrices: an important step in sampling and sample preparation. *Trends Anal. Chem.* **22**, 78–89.

Pazos-Capeáns, P., Barciela-Alonso, M. C., Bermejo-Barrera, A., and Bermejo-Barrera, P. (2005). Chromium available fractions in Arousa sediments using a modified microwave BCR protocol based on microwave assisted extraction. *Talanta* **65**, 678–685.

Pérez-Cid, B., Lavilla, I., and Bendicho, C. (1998). Speeding up of a three-stage sequential extraction method for metal speciation using focused ultrasound. *Anal. Chim. Acta* **360**, 35–41.

Ponce de León, C. A., de Nicola, K., Montes Bayón, M., and Caruso, J. A. (2003). Sequential extraction of selenium in soils from Stewart Lake: total selenium and speciation measurements with ICP-MS detection. *J. Environ. Monit.* **5**, 435–440.

Pueyo, M., Rauret, G., Lück, D., Yli-Halla, M., Muntau, H., Quevauviller, Ph., and López-Sánchez, J. F. (2001). Certification of the extractable contents of Cd, Cr, Cu, Ni, Pb and Zn in a freshwater sediment following a collaboratively tested and optimised three-step sequential extraction procedure. *J. Environ. Monit.* **3**, 243–250.

Pueyo, M., Sastre, J., Hernández, E., Vidal, M., López-Sánchez, J. F., and Rauret, G. (2003). Prediction of trace element mobility in contaminated soils by sequential extraction. *J. Environ. Qual.* **32**, 2054–2066.

Pueyo, M., López-Sánchez, J. F., and Rauret, G. (2004). Assessment of $CaCl_2$, $NaNO_3$ and NH_4NO_3 extraction procedures for the study of Cd, Cu, Pb and Zn extractability in contaminated soils. *Anal. Chim. Acta* **504**, 217–226.

Quevauviller, Ph., Rauret, G., Muntau, H., Ure, A. M., Rubio, R., López-Sánchez, J. F., Fiedler, H. D., and Griepink, B. (1994). Evaluation of a sequential extraction procedure for the determination of extractable trace metal in sediments and soils. *Fresenius' J. Anal. Chem.* **349**, 808–814.

Quevauviller, Ph., Rauret, G., López-Sánchez, J. F., Rubio, R., Ure A., and Muntau, H. (1997). Certification of trace metal extractable contents in a sediment reference material (CRM 601) following a three-step sequential extraction procedure. *Sci. Total Environ.* **205**, 223–234.

Rauret, G., and López-Sánchez, J. F. (2001). New sediment and soil CRMs for extractable trace metal content. *Int. J. Environ. Anal. Chem.* **79**, 81–95.

Rauret, G., López-Sánchez, J. F., Sahuquillo, A., Rubio, R., Davidson, C., Ure, A., and Quevauviller, Ph. (1999). Improvement of the BCR three step sequential extraction procedure prior to the certification of new sediment and soil reference materials. *J. Environ. Monit.* **1**, 57–61.

Rauret, G., López-Sánchez, J. F., Sahuquillo, A., Barahona, E., Lachica, M., Ure, A., Davidson, C. M., Gomez, A., Lück, D., Bacon, J., Yli-Halla, M., Muntau H., and Quevauviller, Ph. (2000). Application of a modified BCR sequential extraction (three-step) procedure for the determination of extractable trace metal contents in a sewage sludge amended soil reference material (CRM 483), complemented by a three-year stability study of acetic acid and EDTA extractable metal content. *J. Environ. Monit.* **2**, 228–233.

Ruban, V., López-Sánchez, J. F., Pardo, P., Rauret, G., Muntau, H., and Quevauviller, Ph. (1999). Selection and evaluation of sequential extraction procedures for the determination of phosphorus forms in lake sediment. *J. Environ. Monit.* **1**, 51–56.

Ruttenberg, K. C. (1992). Development of a sequential extraction method for different forms of phosphorus in marine sediments. *Limnol. Oceanogr.* **7**, 1460–1482.

Ruzicka, J., and Hansen, E. H. (1988). *Flow Injection Analysis*, 2nd ed., Wiley-Interscience, New York.

Ruzicka, J., and Marshall, G. D. (1990). Sequential injection: a new concept for chemical sensors, process analysis and laboratory assays. *Anal. Chim. Acta* **237**, 329–343.

Sahuquillo, A., López-Sánchez, J. F., Rubio, R., Rauret, G., Thomas, R. P., Davidson, C. M., and Ure, A. M. (1999). Use of a certified reference material for extractable trace metals to assess sources of uncertainty in the BCR three-stage sequential extraction procedure. *Anal. Chim. Acta* **382**, 317–327.

Sahuquillo, A., Rigol, A., and Rauret, G. (2003). Overview of the use of leaching/extraction tests for risk assessment of trace metals in contaminated soils and sediments. *Trends Anal. Chem.* **22**, 152–159.

Salomons, W., and Förstner, U. (1984). *Metals in the Hydrocycle,* Springer-Verlag, Berlin.

Santamaria-Fernández, Cave, M. R., and Hill, S. (2006). Trace metal distribution in the Arosa Estuary: the application of a recently developed sequential extraction procedure for metal partitioning. *Anal. Chim. Acta* **557**, 344–352.

Schleuß, U., and Müller, F. (2001). Requirements for soil sampling in the context of ecosystem research. *Sci. Total Environ.* **264**, 193–197.

Schreiber, M., Otto, M., Fedotov, P. S., and Wennrich, R. (2005). Dynamic studies on the mobility of trace elements in soil and sediment samples influenced by dumping of residues of the flood in the Mulde River region in 2002. *Chemosphere* **61**, 107–115.

Shiowatana, J., Tantidanai, N., Nookabkaew, S., and Nacapricha, D. (2001a). A flow system for the determination of metal speciation in soil by sequential extraction. *Environ. Int.* **26**, 381–387.

Shiowatana, J., Tantidanai, N., Nookabkaew, S., and Nacapricha, D. (2001b). A novel continuous-flow sequential extraction procedure for metal speciation in solids. *J. Environ. Qual.* **30**, 1195–1205.

Shiowatana, J., McLaren, R. G., Chanmekha, N., and Samphao, A. (2001c). Fractionation of arsenic in soil by a continuous-flow sequential extraction method. *J. Environ. Qual.* **30**, 1940–1949.

Shuman, L. M. (1982). Separating soil iron- and manganese-oxide fractions for microelement analysis. *Soil Sci. Soc. Am. J.* **46**, 1099–1102.

Shuman, L. M. (1983). Sodium hypochlorite methods for extracting microelements associated with soil organic matter. *Soil Sci. Soc. Am. J.* **47**, 656–660.

Shuman, L. M. (1985). Fractionation method for soil microelements. *Soil Sci.* **140**, 11–22.

Song, Q. J., and Greenway, G. M. (2004). A study of the elemental leachability and retention capability of compost. *J. Environ. Monit.* **6**, 31–37.

Spivakov, B. Ya., Maryutina, T. A., and Muntau, H. (1999). Phosphorus speciation in water and sediments. *Pure Appl. Chem.* **71,** 2161–2176.

Strobel, B. W., Hansen, H. C. B., Borggaard, O. K., Andersen, M. K., and Raulund-Rasmussen, K. (2001). Cadmium and copper release kinetics in relation to afforestation of cultivated soil. *Geochim. Cosmochim. Acta* **65**, 1233–1242.

Sukreeyapongse, O., Holm, P. E., Strobel, B. W., Panichsakpatana, S., Magid, J., and Hansen, H. C. B. (2002). pH-dependent release of cadmium, copper, and lead from natural and sludge-amended soils. *J. Environ. Qual.* **31**, 1901–1909.

Sutherland, R. A., and Tack, F. M. G. (2002). Determination of Al, Cu, Fe, Mn, Pb and Zn in certified reference materials using the optimized BCR sequential extraction procedure. *Anal. Chim. Acta* **454**, 249–257.

Sutherland, R. A., and Tack, F. M. G. (2003). Fractionation of Cu, Pb and Zn in certified reference soils SRM 2710 and SRM 2711 using the optimized BCR sequential extraction procedure. *Adv. Environ. Res.* **8**, 37–50.

Sutherland, R. A., Tack, F. M. G., Tolosa, C. A., and Verloo, M. G. (2000). Operationally defined metal fractions in road deposited sediment, Honolulu, Hawaii. *J. Environ. Qual.* **29**, 1431–1439.

Szmigielska, A. M., Van Rees, K. C. J., Cieśliński, G., Huang, P. M., and Knott, D. R. (1995). Determination of low molecular weight dicarboxylic acids in root exudates by gas chromatography. *J. Agric. Food Chem.* **43**, 956–959.

Szmigielska, A. M., Van Rees, K. C. J., Cieśliński, G., and Huang, P. M. (1997). Comparison of liquid and gas chromatography for analysis of low molecular weight organic acids in rhizosphere soil. *Commun. Soil Sci. Plant Anal.* **28,** 99–111.

Tack, F. M. G., and Verloo, M. G. (1995). Chemical speciation and fractionation on soil and sediment heavy metal analysis: a review. *Int. J. Environ. Anal. Chem.* **59**, 225–238.

Tam, N. F. Y., and Wong, Y. S. (1996). Retention and distribution of heavy metals in mangrove soils receiving wastewater. *Environ. Pollut.* **94**, 283–291.

Templeton, D. M., Ariese, F., Cornelis, R., Danielson, L.-G., Munau, H., van Leeuwen, H. P., and Lobinski, R. (2000). Guidelines for terms related to chemical speciation and fractionation of elements. definitions, structural aspects and methodological approaches (IUPAC Recommendations 2000). *Pure Appl. Chem.* **72**, 1453–1470.

Tessier, A., Campbell, P. G. C., and Bisson, M. (1979). Sequential extraction procedure for the speciation of particulate trace metals. *Anal. Chem.* **51**, 844–851.

Theocharopoulos, S. P., Wagner, G., Sprengart, J., Mohr, M.-E., Desaules, A., Muntau, H., Christou, M., and Quevauviller, Ph. (2001). European soil sampling guidelines for soil pollution studies. *Sci. Total Environ.* **264**, 51–62.

Tiyapongpattana, W., Pongsakul, P., Shiowatana, J., and Nacapricha, D. (2004). Sequential extraction of phosphorus in soil and sediment using a continuous-flow system. *Talanta* **62**, 765–771.

Torto, N., Mwatseteza, J., and Laurell, T. (2001). Microdialysis sampling challenges and new frontiers. *LC-GC Eur.* **14**, 536–546.

Torto, N., Mwatseteza, J., and Sawula, G. (2002). A study of microdialysis sampling of metal ions. *Anal. Chim. Acta* **456**, 253–261.

Trojanowicz, M. (2000). *Flow Injection Analysis: Instrumentation and Applications*, World Scientific, Singapore.

Ure, A. M. (1996). Single extraction schemes for soil analysis and related applications. *Sci. Total Environ.* **178**, 3–10.

Ure, A. M., and Davidson, C. M. (2001). *Chemical Speciation in the Environment*, Blackie, Glasgow, Scotland.

Ure, A., Quevauviller, Ph., Muntau, H., and Griepink, B. (1993). The speciation of heavy metals in soils and sediments: an account of the improvement and harmonization of extraction techniques undertaken under the auspices of Community Bureau of Reference (BCR). *Int. J. Environ. Anal. Chem.* **51**, 135–139.

Varrault, G., Fabre, Y., and Bermond, A. (2001). Kinetic fractionation of copper and cadmium from soil model compounds. *Can J. Anal. Sci. Spectrosc.* **46**, 101–107.

Vetterlein, D., and Jahn, R. (2004). Gradients in soil solution composition between bulk soil and rhizosphere: in situ measurement with changing soil water content. *Plant Soil* **258**, 307–327.

Vodyanitskii, Yu. N., and Vostokova, T. A. (2004). The effect of extraction conditions on iron transfer to Tamm's extract. *Eurasian Soil Sci.* **37**, 60–68.

Wagner, G., Mohr, M.-E., Sprengart, J., Desaules, A., Muntau, H., Theocharopoulos, S., and Quevauviller, Ph. (2001). Objectives, concept and design of the CEEM soil project. *Sci. Total Environ.* **264**, 3–15.

Wang, J.-H., and Hansen, E. H. (2004). The three generations of flow injection analysis. *Anal. Lett.* **37**, 345–359.

Wasay, S. A., Barrington, S., and Tokunaga, S. (2001). Organic acids for the in situ remediation of soils polluted by heavy metals: soil flushing in columns. *Water Air Soil Pollut.* **127**, 301–314.

Wenzel, W. W., and Blum, W. E. H. (1994). Assessment of metal mobility in soil: methodological problems. In *Metal Speciation and Contamination of Soil,* ed. Allen, H. E., and Huang, C. P., Lewis Publishers, Boca Raton, FL, Chap. 9.

Wenzel, W. W., and Wieshammer, G. (1995). Suction cup materials and their potential to bias trace metal analyses of soil solutions: a review. *Int. J. Environ. Anal. Chem.* **59**, 277–290.

Wenzel, W. W., Sletten, R. S., Brandstetter, A., Wieshammer, G., and Stingeder, G. (1997). Adsorption of trace metals by tension lysimeters: nylon membrane vs. porous ceramic cup. *J. Environ. Qual.* **26**, 1430–1434.

Wenzel, W. W., Kirchbaumer, N., Prohaska, T., Stingeder, G., Lombi, E., and Adriano, D. C. (2001a). Arsenic fractionation in soils using an improved sequential extraction procedure. *Anal. Chim. Acta* **436**, 309–323.

Wenzel, W. W., Wieshammer, G., Fitz, W. J., and Puschenreiter, M. (2001b). Novel rhizobox design to assess rhizosphere characteristics at high spatial resolution. *Plant Soil* **237**, 37–45.

Whalley, C., and Grant, A. (1994). Assessment of the phase selectivity of the European Community Bureau of Reference (BCR) sequential extraction procedure for metals in sediment. *Anal. Chim. Acta* **291**, 287–295.

Williams, J. D. H., Jaquet, J. M., and Thomas, R. L. (1976). Forms of phosphorus in the surficial sediments of Lake Erie. *J. Fish. Res. Board Can.* **33**, 413–429.

Wisotzky, F., and Cremer, N. (2003a). Organic acids for the in situ remediation of soils polluted by heavy metals: soil flushing in columns. *Environ. Geol.* **44**, 799–804.

Wisotzky, F., and Cremer, N. (2003b). Sequential extraction procedure in columns, 2: Application of a new method. *Environ. Geol.* **44**, 805–810.

Zeien, H., and Brümmer, G. W. (1989). Chemische extraktionen zur bestimmung der bindungsformen von schwermetallen in böden. *Mitt. Dtsch. Bodenkundl. Ges.* **59**, 505–510.

Zhang, H., and Davison, W. (1995). Performance characteristics of diffusion gradients in thin films for the in situ measurement of trace metals in aqueous solution. *Anal. Chem.* **67**, 3391–4000.

Zhang, Y.-Q., and Moore, J. N. (1996). Selenium fractionation and speciation in a wetland system . *Environ. Sci. Technol.* **30**, 2613–2619.

Zhang, H., Zhao, F. J., Sun, B., Davison, W., and McGrath, S. P. (2001). A new method to measure effective soil solution concentration predicts copper availability to plants. *Environ. Sci. Technol.* **35**, 2602–2607.

Zhi, Z.-L., Ríos, A., and Valcárcel, M. (1996). Direct processing and analysis of solid and other complex samples with automatic flow injection systems. *Crit. Rev. Anal. Chem.* **26**, 239–260.

13

SOURCES AND MOBILITY OF METALLIC RADIONUCLIDES IN SOIL SYSTEMS

S. Staunton

INRA–UMR Biogéochimie du Sol et de la Rhizosphère, Montpellier, France

C.-S. Haudin

Andra, Châtenay-Malabry, France

G. Wang

Fujian Agriculture and Forestry University, Fuzhou, Fujian, People's Republic of China

G. Shaw

University of Nottingham, Nottingham, England

Biophysico-Chemical Processes of Heavy Metals and Metalloids in Soil Environments,
Edited by Antonio Violante, Pan Ming Huang, and Geoffrey Michael Gadd

13.1. INTRODUCTION

Radioactive isotopes of some metals and metalloids are present naturally or as a result of human activity in soil systems. Their mobility and their availability to living sources are almost identical to that of their stable analogs and similar techniques and concepts are necessary to investigate both radioactive and nonradioactive metals. However, there are important differences that warrant separate treatment of radioactive metals in this book. First, in addition to the biological effects they share with other isotopes of the same element, they may have adverse effects, due to the radiation they emit. Second, in addition to the complex role of soil in determining the mobility of metals and the exposure of living organisms in soil systems, the physical interaction of radiation with soil also determines the external exposure of living organisms to potentially damaging ionizing radiation. Third, there is little overlap between the inventory of metals studied because of their nutrient or potentially toxic properties and those studied in radioecology. There is often, therefore, a dearth of information on the metals of interest to radioecology. Finally, the threshold concentration at which they can be detected or at which adverse biological effects of radiation may ensue also differs from the concentration range of interest for nonradioactive metals.

The aim of this chapter is to illustrate the similarities and differences between the behavior of metallic radionuclides and other metals in soil systems. We summarize some basic principles to show the hazards associated with radionuclides, give a brief inventory of the metals that may exist in soil systems, and give a concise description of the behavior of some of the more common and most widely studied radioactive metals.

13.2. BASIC PHYSICAL AND BIOLOGICAL CONCEPTS

A comprehensive description of the nature of radioactivity and its interaction with matter is beyond the scope of this chapter, and such information can easily be found elsewhere (Shleien et al., 1998; CPEP, 2003; Tykva, 2004). Here we give only a brief sketch of the basic principles. Radioactivity is a natural phenomenon, discovered in 1896 by Henri Becquerel. The nuclei of some atoms are unstable and decay spontaneously, emitting ionizing radiation to attain a more energetically favorable state. Radioisotopes are characterized by the nature of the radiation they emit and its energy and their half-life: the time required to decay to half the initial number and hence for the radiation to decrease by a factor of 2. The SI unit of radioactivity is the becquerel (Bq), defined as 1 disintegration per second. Radiation may consist of particles or be electromagnetic. The physical nature of the radiation and its energy determine how it interacts with matter. *Alpha particles* consist of helium nuclei and have both charge (+2) and mass (4 atomic units). They interact strongly with matter, are easily stopped (e.g., by a sheet of paper or the epidermal layer of skin) but are consequently strongly ionizing and cause considerable damage when in direct contact with biological systems. Only heavier radioisotopes emit alpha particles, and most are actinides, but these are not considered in this chapter. *Beta radiation* consists of a stream of electrons (β^-) or positrons (β^+) formed by the disintegration of a neutron or a proton within the nucleus. This radiation is thus charged with some mass. It is less ionizing than alpha radiation and the physical barrier required to stop it depends on the energy of the radiation, which is itself a function of the isotope. For example, β radiation of ^{137}Cs has an energy of 0.512 and 1.17 MeV and requires 2.1 mm of glass to be stopped, whereas the radiation of ^{99}Tc is less energetic, 0.3 MeV, and is stopped completed by 0.3 mm of glass. *Gamma radiation* may be emitted alone or in association with β radiation. It is purely electromagnetic and is never completely stopped by interaction with matter. The thickness of a medium required to attenuate γ radiation depends on its density and the energy of the radiation. For example, the γ radiation of ^{137}Cs (0.662 MeV) is attenuated by a factor of 10 by 2.4 cm of lead or 7.2 cm of steel, whereas the same attenuation is provided by less than 0.1 cm of the same materials for the radiation produced by ^{210}Pb (0.013 and 0.047 MeV).

The biological consequences of radiation depend on the dose received. The *adsorbed* or *physical dose* is the amount of energy transferred to matter, with the unit gray (Gy), J kg^{-1}. The *biological dose*, or *dose equivalent*, with the unit sievert (Sv), takes into account a quality factor depending on the biological damage that a type of radiation induces. For humans very large doses are rapidly fatal, with mortality depending on the dose. Such large exposures are associated with accidents and are not usually encountered in the environment. Lower doses (below 1 Sv) lead to increased risk of cancer, with the probability of cancer depending on the dose. There is ongoing debate as to the effect of very low doses, and in the absence of concrete data it is generally assumed that there is a linear relation between cancer fatalities and dose. At very low doses, similar to those usually encountered in the natural environment, there is no clear link between the risk of cancer and the dose received, but the principle of precaution prevails. It is therefore necessary to reduce the exposure of living organisms to radiation as much as possible . There are two possible forms of exposure: internal, after assimilation of a radioactive source by a living organism, often by way of the food chain, in common with other forms of pollution, and external, caused by the penetrating nature of ionizing radiation. The latter has important consequences for the remediation of pollutant lands, since agricultural and leisure activities can lead to exposure. It is important to consider the exposure to radiation due to anthropogenic activity with respect to natural exposure. For example, cosmic radiation accounts for about 0.25 mSv or 10% of the radiation dose received, and exposure to telluric emanations, another 1.3 mSv. This may be twice as great in areas of high radon concentrations, such as Cornwall in the United Kingdom and Brittany in France. In comparison, medical exposure adds approximately 0.3 mSv in developed countries. The dose received by an average worker in the nuclear industry is about 1.1 mSv, whereas the maximum allowable dose is 20 mSv, and the average dose caused by fallout from the Chernobyl accident is estimated to have been between 0.006 and 0.5 mSv in various countries of western Europe.

13.3. SOURCES OF RADIOACTIVE METALS IN SOIL

Radioisotopes of some metals are naturally present in soils. The most common is potassium-40, with a natural abundance of 0.012% and a half-life of 1.28×10^9 years. Since potassium is an essential nutrient, the average ^{40}K content of the human body is 4000 Bq. Similarly, another alkali earth metal, rubidium, has a natural radioisotope, ^{87}Rb, with a half-life of 4.75×10^{10} years and accounts for about 600 Bq in the human body. Other natural radionuclides, such as ^{238}U and ^{232}Th, whose distribution and relative abundance may be altered by mining and other human activity, may contribute significantly to radiation dose but are not considered in this chapter.

Fallout from atmospheric nuclear weapons testing in the 1950s and 1960s introduced considerable amounts of radiocesium (^{137}Cs, half-life 30.6 years, and ^{134}Cs, half-life 2.34 years) into soils and waters in the northern hemisphere. Nearly 50 years later, traces of the shorter-lived isotope have almost disappeared. Accidents and discharges from nuclear treatment plants have also introduced various amounts of these and other radioisotopes. The most notorious such incident was the accident that took place near Chernobyl in 1986. A cocktail of radioisotopes, mostly metals, were released into the environment. The heavier elements, released in particulate form, were deposited in a zone close to the reactor. Other elements, Cs, Sr, and I, traveled farther and were deposited with rainfall over western and northern Europe. Estimates of deposition are variable, largely because of the heterogeneous nature of precipitation, leading to sampling bias. In addition, some of the contaminated areas, such as Cumbria in the United Kingdom, had already been contaminated by releases after the Sellafield accident and by other discharges (Baker and Cawse, 1990). Although the isotopic signature of the various releases can be used to calculate sources of the measured contamination, this leads inevitably to uncertainties.

Another source of radionuclides in the environment, one that is likely to increase in the future, arises from the storage of nuclear waste. Nuclear waste is classified according to the activity and half-life of the isotopes it contains. The highly active, often alpha-emitting waste and longest-lived isotopes, must be stored securely for very long periods. No country has yet decided on a definitive repository for such waste. Unless research makes transmutation a feasible option in the future, and since deep-sea disposal has been ruled out, it seems likely that storage will be in deep underground repositories. Confinement conditions of the waste and the thick geological barrier (clay, granite, or salt), along with radioactive decay, will be designed to limit the movement of radioisotopes to the biosphere. However, various scenarios that could lead to contamination must be envisaged and the associated risk assessed. Furthermore, the much larger volumes of lower-level-activity waste are destined for surface or shallow land storage, and until definitive repositories have been identified for high-level waste, storage conditions may lead to release into the environment. Apart from the actinides, which are outside the scope of this chapter, a large number of radioisotopes have been identified in nuclear waste. These include ^{41}Ca, ^{59,63}Ni, ^{79}Se, ^{90}Sr/^{90}Y, ^{93}Zr/^{93m}Nb, ^{93}Mo/^{93m}Nb, ^{94}Nb, ^{99}Tc, ^{108m}Ag, ^{107}Pd, ^{113m}Cd, ^{125}Sb/^{125m}Te, ^{126}Sn, ^{126}Sb, ^{135,137}Cs, ^{205}Pb, and various actinides (uranium, neptunium, americium, and plutonium). They originate not only from the nuclear industry, but also from other industries and from medicine and research (Amiro, 1993; Ouzounian, 1996).

The chemical and biological properties of these radioisotopes differ very little from those of their stable isotopes. To a very good approximation, all the isotopes of a given element behave identically. Isotope effects are almost always of biological origin and involve the formation and breaking of covalent bonds whose energy depends on the respective masses of the atoms involved. Until very recently the isotope effect was considered only for light elements such as H and C, but isotope discrimination, even over short periods has been reported more often in recent years (Maréchal et al., 1999; Anbar et al., 2000; Maréchal and Albarède, 2002; Zhu et al., 2002; Beard et al., 2003; Weiss et al., 2005) probabily because accurate mass spectroscopic techniques permit measurements that are very hitherto impossible.

13.4. FACTORS AFFECTING THE MOBILITY AND BIOAVAILABILITY OF RADIO NUCLIDES IN SOIL

13.4.1. Adsorption–Desorption

As for other metals in soil, the *mobility* of radionuclides is limited by their adsorption on the mineral and organic constituents of soil. To a good first approximation, adsorbed metals are immobilized. Their mobility thus depends on the fraction that is in solution and the ease with which it can be desorbed: namely, its *lability*. There is some evidence that adsorbed cations retain some mobility, particularly when they are present in the diffuse double layer of charged clay particles and not associated with soil solids via covalent bonds (Staunton, 1990; Ochs et al., 2003). In compact clays such as those in natural or engineered clay barriers, this residual mobility may dominate (Bourg et al., 2003), but in soils its effect is difficult to measure and is often ignored. There is no evidence that anions retain any mobility after adsorption on soil surfaces. Adsorption is often quantified using the *distribution coefficient*, K_d, defined as the ratio of concentrations in the adsorbed and solution phases. This has the advantage of simplicity, but it is assumed that adsorption is reversible and that equilibrium has been attained. Adsorption is not always completely and instantaneously reversible (Comans et al., 1991; Guivarch et al., 1999; de Koning and Comans, 2004), particularly when adsorption involves inner rather than outer-sphere complexes, with cations remaining in the diffuse double layer. This is an important limitation to the K_d concept and is often overlooked. There is no consensus regarding the reason for irreversibility of adsorption. When a cation or anion is not easily desorbed from a surface, it is said to be *fixed*. In some cases, especially that of cesium, fixation appears to increase with time (Noordijk et al., 1992; Absalom et al., 1995, 1999). Some authors consider the mechanism of fixation to be the slow diffusion toward internal sites (Tamura and Jacobs, 1960). More recently, McKinley et al. (2004) and Liu et al. (2003) have found direct evidence using microscopy and x-ray microprobe analysis of such interlayer diffusion for Cs adsorbed on mica. Other mechanisms of fixation or attenuation include, as for nonradioactive trace elements, incorporation within biological systems and consequently, soil organic matter, and coprecipitation with sparingly soluble soil minerals. In many cases the evidence for fixation is not conclusive, since great care must be taken to perform measurements of adsorption and desorption under exactly the same conditions, to avoid confusing irreversibility with the fact that adsorption depends on experimental conditions such as reaction time, ionic strength, pH, and solution composition, including the amount of dissolved organic matter, that may be influenced by the soil/solution ratio or by the repeated washing of soil. Furthermore decreasing bioavailability may be interpreted as being due to fixation on soil minerals, but could be due to changing spatial distribution of the radioisotope in soil after vertical movement down a soil profile or slow penetration into soil aggregates.

As for their stable analogs, the adsorption of radionuclides may depend on experimental conditions. Any variation results from the chemistry of the element, not form its radioactive nature. For example, the adsorption of the alkali metal cesium is largely independent of pH, whereas pH dependence is to be expected for oxyanions (Chapter 5, this volume) such as selenite and selenate and for transition metals such as nickel, silver, and palladium.

Adsorption of metals on soil is often concentration dependent. Many studies of the fate of metals in soil present and interpret adsorption isotherms. Since the relevant range of concentrations for radionuclides is often orders of magnitude less than that of the stable elements, care must be taken in extrapolating data and trends to radionuclides. For example, heavy metal–polluted soil can contain several percentages by weight of the metal (although the distinction between soil and spoil may become difficult to make), whereas soils in Belarus and the Ukraine that are heavily contaminated with ^{137}Cs following the accident at Chernobyl may contain more than 555 kBq m^{-2} (AEN, 1987; Savechenko, 1995), equivalent to about 10^{-9} mol kg^{-1}. When a radionuclide is added to a soil, the soil may already contain a much greater amount of the stable isotopes, so much of the immediate reaction would be isotopic dilution.

13.4.2. Uptake and Assimilation by Biological Systems

The general principles of metal uptake by biota from soil systems are treated in chapters 7 to 10 of the book. The same rules govern radionuclide uptake by biological systems. Uptake and assimilation depend on the chemical and biological properties of the element. The rate-limiting step for the uptake of many radionuclides, those strongly immobilized by soil constituents, is the movement to the interface between the soil solution and the biological membrane. If absorption is more rapid than the movement of the solute to the interface, uptake becomes diffusion limited. The adsorption properties of the soil are therefore determinant.

Since the presence of most radionuclides in soil is related to human activity, they are not distributed homogeneously. Radionuclides introduced into the environment by atmospheric fallout are deposited on the surface of soil and often do not penetrate very deeply for years after deposition. Deeply rooting plants would therefore not be exposed to large concentrations and so are essentially protected. In contrast, if the source of contamination is underground, such as the contamination of groundwaters from leaking storage facilities as is widely documented for the Hanford site in the United States (Brady et al., 2002) and the potential risk following deep storage of high-level waste, deeply rooting plants would be exposed first. They might even contribute to the redistribution of radionuclides with soil depth if elements are transported up the root system and exuded or secreted at shallower depths.

When a radioactive solute comes into contact with a biological membrane the initial reaction may be adsorption. When the naturally occurring element is present before the arrival of the radioisotope, isotopic equilibration accounts for much of the removal of the radioisotope from solution. It is difficult to distinguish experimentally between adsorption and absorption, and most studies do not attempt to do so. *Adsorption* is a purely physical reaction that takes place at a surface, where as *absorption* is a biological process, so is determined by the biological role of the element and the capacity of the living organism to obtain its necessary supply of nutrients and to deal with potentially toxic elements. The extent to which living organisms discriminate against radioisotopes therefore depends on their biological role and their similarity to other elements. For example, Cs has no known biological role, but its chemical similarity to K means that it is taken up by similar mechanisms. K–Cs discrimination in plants seems to depend on the relative and absolute concentration of both elements and the potassium status of the plant. Technetium, all of whose isotopes are radioactive, has no beneficial biological role and can become toxic if it is present in sufficiently large concentrations to disrupt the nitrogen cycle. Other radioisotopes that are or could be present as soil contaminants are toxic, including Cd, Pb, and Se.

The fact that radioisotopes decay to produce ionizing radiation does not influence their biological role. Radioisotopes are taken up and assimilated in the same way as are stable isotopes. Enzymes may be sufficiently sensitive to discriminate slightly between isotopes of the same element, leading to changes in relative isotopic abundance, but the effect is small (Maréchal et al., 1999; Anbar et al., 2000; Maréchal and Albarède, 2002; Zhu et al., 2002; Beard et al., 2003; Weiss et al.,

2005). To a large extent, the redistribution of a radioisotope in a plant or other living system will be the same as that of the stable isotopes when these exist. At quite high levels of radioactivity, genetic damage may be induced, but foodstuffs are considered unfit for consumption far below these levels. Therefore, since plants and other living organisms do not suffer side effects from the accumulation of radioisotopes, they are under no pressure to develop strategies to limit this accumulation. Some of strategies employed by living organisms to deal with elevated concentrations of potentially toxic elements (complexation, precipitation) do not reduce accumulation and may even lead to hyperaccumulation of complexed or precipitated species. These are not useful strategies to reduce either radiation damage (rare) or, more important, contamination of the food chain. When the same element has been studied as a chemical pollutant and in radioecology, care must be taken in transferring findings. Often, the concentrations involved are very different. Biological systems adapt and react to the pressure of a chemical pollutant beyond a threshold level. These changes will not be initiated in the presence of a chemically tiny level of radioactive pollution. For example, selenium may be present in large concentrations, and microorganisms have been identified that precipitate elementary Se to avoid toxicity (Garbisu et al., 1996; Losi and Frankenberger, 1997; Roux et al., 2001). The accumulation of radio selenium would never be expected to be as high in the absence of stable Se.

A vast literature exists on soil-to-plant transfer factors for various plants and soils, especially for the most often studied radionuclides (Cs, Sr, U, Tc, ...). Comprehensive surveys of data are given in books by Coughtrey and Thorne (1983a,b) and Coughtrey et al. (1983). Databanks have been compiled, especially those of the International Atomic Energy Agency and the International Union of Radioecology (IUR, 1989; IAEA, 1994). Usually, the data are for the readily sampled aboveground part of the plant, and sometimes edible organs are selectively sampled. Rather limited information exists for other living systems in soil. The role of soil microorganisms in determining the fate of radionuclides has been reviewed by Gadd (1996), Keith-Roach and Livens (2002), and Deneux-Mustin et al. (2003). It seems unlikely that soil microorganisms would be an important sink for strongly adsorbed radionuclides, because adsorption on soil constituents would swamp their effect. However, despite the experimental difficulties that hamper measurements, it is possible that a purely abiotic approach to immobilization in soil overlooks processes that may be important for less strongly adsorbed and more readily bioavailable elements, such as Tc and Se.

Biological systems do not act simply as passive sinks for metals. They also modify their chemical environment and consequently the chemical speciation and mobility of nutrient and potentially toxic elements. Their influence on the mobility of radionuclides is less widely studied than that of other metals, but is certainly as important. The relevant factors include pH, redox potential, concentration of complexing ligands in solution, and mineral weathering. The relative importance depends on the soil and on the chemistry of the element considered, and no generalities can be made. It is not surprising that elements such as Se and Tc that exist in various oxidation states they sensitive to redox conditions. For other elements the effect may be indirect; for example, ammonium produced under reducing conditions may inhibit Cs adsorption. Other possible indirect effects would be the redox-dependent dissolution of soil constituents that contribute to the adsorption properties of soil. There is no pH dependence of the adsorption of Cs in mineral soil (except under extreme conditions when rapid weathering can take place, whereas adsorption of most transition metals (Ni, Ag, Pd, Zr) and oxyanions (TcO_3^-, SeO_4^{2-}, and SeO_3^{2-}) varies over a wide range of pH. Cs adsorption is highly selective on some clay minerals, notably illites and this selectivity depends strongly on weathering and the presence of organomineral coatings (Maes et al., 1999; Staunton et al., 2002). Such effects have not been reported for other common radionuclides in soil. Finally, the solubility of transition metals is increased by complex-forming ligands in solution, but there is no equivalent mechanism for alkali metals (Cs and Rb) or metalloid anions (Tc and Se).

13.4.3. Remediation Strategies

As for all pollution, the best strategy is to avoid contamination as much as possible. However, when a soil is contaminated, various approaches are possible to limit impact (Zhu and Shaw, 2000). These are broadly similar to those employed or envisaged for metallic pollution: phytostabilization, phytoextraction, choice of low-uptake or nonfoodcrops, soil amendment to reduce or enhance uptake, and semiindustrial agricultural practices, such as covering or diluting contaminated soil with clean soil. These approaches are treated in much greater detail by Gräfe and Naidu in Chapter 14. There are, however, important differences between radionuclides and other metal pollutants. First, in themselves they are unlikely to have adverse effects on crop production or soil biological activity except in exceptionally contaminated areas. Second, the impact of land use on exposure to radioactivity must be considered. The various factors taken into account would include contamination of surface waters by erosion and runoff, transport and inhalation of dust, consumption of soil by animals and humans (poorly washed vegetables, dust inhalation, and direct consumption by children), and external exposure of agricultural workers and other land users when soil is contaminated by γ-emitters. Third, risk assessment takes into account the cumulative dose to a population and not simply to individuals, because the risk of cancers for a population is proportional to total exposure. This approach is applied less systematically for other forms of pollution. An important consequence is that diluting contaminated foodstuffs to bring the level of radioactivity below a threshold level is not an appropriate strategy.

13.5. SOURCES AND FATE OF SOME RADIONUCLIDES IN SOIL

Table 13.1 shows the environmentally significant radioisotopes of metals and metalloids considered in this chapter. Figure 13.1 shows some of the fluxes of radioactive nuclides in the environment.

TABLE 13.1. Environmentally Significant Radioisotopes of Metals and Metalloids

Element	Atomic Number	Decay Products, Radioactive Emissions, and Half-Lives
Cobalt	27	$^{60}Co \rightarrow {}^{60}Ni(\beta, \gamma)$, 5.27 years
Selenium	34	$^{79}Se \rightarrow {}^{79}Br(\beta)$
		Estimates of half-life range from 2.8×10^5 to 1.13×10^6 years
		$^{75}Se \rightarrow {}^{75}As(\gamma)$, 119.8 days
Strontium	38	$^{90}Sr \rightarrow {}^{90}Y(\beta)$, 28.8 years
Zirconium	40	$^{95}Zr \rightarrow {}^{95m,95}Nb(\beta, \gamma)$, 64 days
Niobium	41	$^{95}Nb \rightarrow {}^{95}Mo(\beta, \gamma)$, 35 days
Technetium	43	$^{99}Tc \rightarrow {}^{99}Ru(\beta)$, 2.11×10^5 years
Ruthenium	44	$^{103}Ru \rightarrow {}^{103}Rh(\beta, \gamma)$, 39.3 days
		$^{106}Ru \rightarrow {}^{106}Rh(\beta)$, 373.6 days $\rightarrow {}^{106}Pd$ (β, γ) 29.8 s
Silver	47	$^{110m}Ag \rightarrow {}^{110}Cd(\beta, \gamma)$, 249.8 days
Antimony	51	$^{125}Sb \rightarrow {}^{125}Te(\beta, \gamma)$, 2.76 years
Cesium	55	$^{134}Cs \rightarrow {}^{134}Ba(\beta, \gamma)$, 2.06 years
		$^{135}Cs \rightarrow {}^{135}Ba(\beta)$, 2.3×10^6 years
		$^{137}Cs \rightarrow {}^{137}Ba(\beta, \gamma)$, 30.1 years

13.5.1. Cobalt (Atomic No. 27)

Cobalt has only one naturally occurring stable isotope, ^{59}Co (abundance 100%), although it has 22 radionuclides, including ^{56}Co, ^{57}Co, ^{58}Co, and ^{60}Co. Of these radionuclides, however, only ^{60}Co, with a physical half-life of 5.27 years, is of environmental significance. Cobalt-60 is produced by neutron activation of stable cobalt and decays by beta emission, accompanied by the emission of two relatively high-energy gamma rays at 1173 and 1332 keV. Ultimately, ^{60}Co decays to nonradioactive nickel (^{60}Ni). Because of its convenient gamma-ray emissions ^{60}Co is very commonly used in industrial applications such as gamma-ray tomography, radiotherapy, and food and instrument sterilization. It is also produced as a contaminant in irradiated steels in nuclear facilities. Cobalt-60 enters the environment in various waste streams from both laboratories and nuclear facilities. Its dynamics in terrestrial and aquatic systems has been reviewed by Coughtrey and Thorne (1983b) and Frissel et al. (1990).

Stable Co is relatively abundant in soils, with concentrations ranging from 0.05 to 65 mg kg^{-1}. Bowen (1979) quotes a median soil concentration for stable Co of 8 mg kg^{-1}. Smith (1994) quotes a range of stable cobalt concentrations in soils from 0.05 to 300 mg kg^{-1} with an average range of 10 to 15 mg kg^{-1}. Co is an essential element for plants across the entire range of taxa from fungi and blue-green algae up to the higher plants, although it plays a more important role in microorganisms. In vascular plants stable Co concentrations range from 0.005 mg kg^{-1} in woody angiosperms to 4.6 mg kg^{-1}in herbaceous vegetables (Bowen,

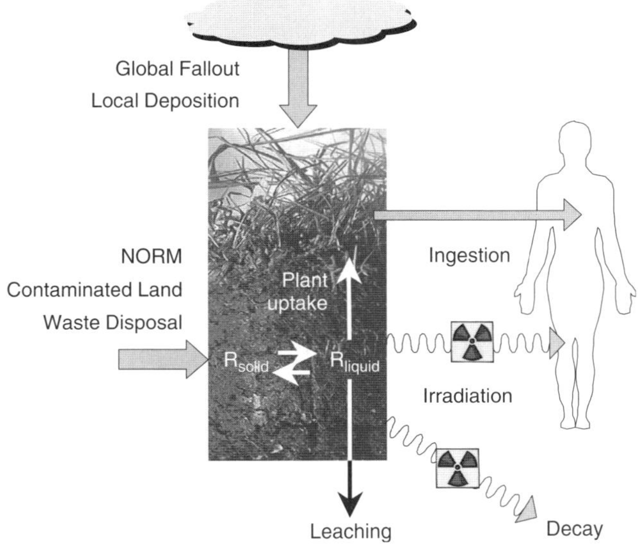

Figure 13.1. Soils contain naturally occurring radioactive metals (NORM) such as ^{40}K, ^{226}Ra and uranium isotopes. In addition, global and local deposition due to nuclear weapons testing and accidents such as Chernobyl have contributed radionuclides such as ^{137}Cs and ^{90}Sr. Significant contaminated land legacies remain around the world from nuclear fuel cycle activities and, in future, radioactive waste disposal is likely to increase the inventory of radionuclides in soils. Once contaminated, sorption-desorption reactions between solid and liquid phases in the soil and microbial immobilisation are critical in controlling plant uptake and, eventually, contamination of the human food chain. However, unlike non-radioactive metals, direct exposure to gamma and beta emitting radionuclides from contaminated soils can occur even in the absence of biological uptake, via irradiation. Accumulation of radionuclides in soils is balanced by losses due to leaching and radioactive decay, though the latter is dependent on the physical half lives of individual radionuclides.

1979). Several "indicator" and "accumulator" species of plant for Co have been identified (Brooks, 1972; Brooks et al., 1977).

Co exists in either the 2+ or 3+ valence states and forms complexes readily with organic matter. Acetic acid and EDTA have commonly been used to extract cobalt from soils and yield approximately 70% of "labile" cobalt. Soil drainage and redox potential play a major role in controlling the bioavailability of Co in soils, with wet soils at low redox potential yielding more acetic acid–extractable cobalt than dry soils at higher redox potentials (Mitchell et al., 1957). This effect may be due to the tendency of cobalt to adsorb to Mn oxides in soils, the solubility of which are strongly affected by redox potential. As residence time in the soil increases, cobalt may gradually become more strongly fixed as it replaces Mn in the crystal lattice. However, Shaw et al. (1990) extracted ^{60}Co from silty loam and silty clay soils after a two-year residence time in lysimeters and found that

70 to 90% was still extractable with 1 M $CaCl_2$, while approximately 10 to 30% was extractable with a reducing agent (hydroxylamine hydrochloride), indicating probable association with Fe and Mn sesquioxides.

The ranges of solid–liquid distribution coefficients reported for ^{60}Co in different soil types are rather similar. In general, K_d values from 0.2 to 20,000 L kg^{-1} are representative of sand, loam, clay, and organic soils. Sheppard and Thibault (1990) quote median K_d values of 60, 1300, 540, and 990 L kg^{-1} for each of these soil types, respectively. Soil–plant transfer factors for numerous crop species have been published by the International Union of Radioecologists (IUR, 1989). These values range from 0.037 for cereal grain to 1.1 for alfalfa fodder. For most of the crop types represented in the IUR (1989) database, a range of soil–plant transfer factors from 0.01 to 0.1 seems most applicable.

The committed effective dose per unit intake following ingestion of ^{60}Co is 3.4×10^{-9} Sv Bq^{-1} for adults > 17 years of age (IAEA, 1996). As a result of its relatively high energy gamma-ray emissions and its relatively high uptake by crop plants, ^{60}Co is a potentially significant contributor to both ingestion and external doses when released to the soil.

13.5.2. Selenium (Atomic No. 34)

Selenium, although not *sensu stricto* a metalloid, has some metallic properties, and because of the environmental importance of some of its radioisotopes, it is considered in this chapter. Twenty-six isotopes of selenium exist, of which six stable isotopes occur naturally and two radioisotopes are of interest: the fission product, ^{79}Se, with a half-life of 1.1×10^6 years (this is a recently revised estimate; the half-life was previously thought to be 6.5×10^4 years) and ^{75}Se, with a half-life 119 days, often used as a tracer. Various risk assessment evaluations have identified ^{79}Se as presenting a risk of entering the biosphere after waste disposal. The stable element is studied widely because it is both an essential nutrient for animals and bacteria (there is not yet evidence of an essential role for plants) (Combs and Combs, 1986) and potentially toxic in some naturally rich environments and in polluted ecosystems (Rosenfeld and Beath, 1964). The Se content of soils is around 0.4 mg kg^{-1}. Below 0.2 mg kg^{-1}, soils are considered to be deficient, whereas above 5 mg kg^{-1} they are classed as seleniferous (Coughtrey et al., 1983). A very narrow range of concentrations separates toxicity and deficiency (Lakin, 1972). Like sulfur, which precedes it in the group VI of the periodic table, selenium is often found in the −2 state, as selenides associated with sulfides in metal ores (Lumbroso, 1977). Elemental selenium as well as two oxyanions, selenite (SeO_3^{2-}) and selenate (SeO_4^{2-}), may also be present in soils and geological formations. Much of the research on Se concerns either polluted or naturally rich systems, with an emphasis on toxicity and how living species tolerate large concentrations, or conversely, Se deficiency and the relation between the Se content of species and the onset and severity of symptoms (Tan et al., 2002; Tinggi, 2003; Wu, 2004). Although these studies cannot necessarily be extrapolated to situations where the presence of radio-Se gives rise to concern but the level of stable Se is low, they are reviewed briefly below.

The speciation of the element depends strongly on redox conditions (Masscheleyn et al., 1991; Jayaweera and Biggar, 1996) and microbial activity (Neal, 1995). Under aerated conditions, selenide and elementary selenium are slowly oxidized to selenate and selenite. The salts of both selenite and selenate are very soluble, except in strongly acidic soils, where $MnSeO_3$ may be sparingly soluble (Elrashidi et al., 1987). The major species in high-redox -solution conditions ($pH + pe > 15$) is SeO_4^{2-}, whereas SeO_3^{2-} and $HSeO_3^-$ dominate in intermediate conditions ($pH + pe$ 7.5 –15), with HSe^- at lower redox values and H_2Se present only at very low pH. There is some indication that the selenium oxidation state may be influenced by relatively localized changes in redox conditions in the rhizosphere (Blaylock and James, 1994; Wu et al., 1997). Selenate behaves much like sulfate in soil and is not strongly adsorbed (Neal and Sposito, 1989), whereas selenite is adsorbed, with adsorption decreasing with increasing pH and in the presence of competing anions, such as phosphate (Neal et al., 1987a,b). Of the two oxyanions, selenite is less mobile; since it is more strongly adsorbed, it is thought that selenate is not directly toxic but must first be reduced to selenite (Turner et al., 1998).

Biomethylation is an essential feature of Se dynamics in soil (Zieve and Peterson, 1981). Biomethylation by microorganisms of organic and mineral Se occurs under both aerobic and anaerobic conditions (Doran and Alexander, 1977). It is assumed that fungi are the predominant organisms methylating in aerobic conditions, and conversely, that bacteria dominate when conditions are anaerobic (Doran, 1982), although in the latter case there is some suggestion that this conclusion is partially due to artifacts in the isolation of bacteria (Dungan and Frankenberger, 1999). The remediation of Se-rich soil via the volatilization of Se has been studied extensively in relation to the enrichments made, mainly of carbohydrates (Karlson and Frankenberger, 1988; Frankenberger and Karlson, 1989). However, the volatile forms, mostly dimethyselenide and dimethyldiselenide (Chasteen, 1998), may react with soil components (Zieve and Peterson, 1985), especially organic matter, and soil microorganisms (Dungan et al., 2002) and hence be reimmobilized by further transformations (Guo et al., 1999; Martens and Suarez, 1999). Under anaerobic conditions, methyl-Se is degraded by methanogen microorganisms and sulfato-reducing bacteria (Oremland and Zehr, 1986). Flooded conditions also decrease the flux of volatile molecules, increasing their residence time in soil and hence the probability that they may be chemically or microbially immobilized before they can be released from the soil (Zhang and Frankenberger, 1999, 2000, 2002; Guo et al., 2000; Zhang et al., 2000).

Soil microorganisms may also oxidize elementary Se (Sarathchandra and Watkinson, 1981; Dowdle and Oremland, 1998) to selenite and selenate. However, microbial reduction of selenite and selenate to elementary selenium is thought to be more important and is certainly investigated more than is microbial oxidation. Both bacteria and fungi (Bautista and Alexander, 1972) play roles in selenium reduction, but the former have been studied more extensively (White et al., 1995). Bacteria may reduce Se under both aerobic (Lortie et al., 1992; Garbisu et al., 1995; Dungan et al., 2003) and anaerobic conditions (Oremland et al., 1994; Blum et al., 1998). Under anaerobic conditions, selenite and selenate may be used by bacteria as terminal electron acceptors for respiration (Stolz and Oremland, 1999). It has been shown that sulfato-respiring bacteria can also use selenate as an alternative acceptor (Zehr and Oremland, 1987) but to a lesser extent, implying the involvement of other enzymatic systems for Se respiration. Interferences with the NO_3^- respiration have been reported frequently, and reductases are well characterized for only a few bacteria, such as *Thauera selenatis* (Macy et al., 1993), *Sulfurospirillum barnesii* (Oremland et al., 1994, 1999), and *Enterobacter cloacae* (Losi and Frankenberger, 1998). Elementary Se can accumulate in large quantities in cells (Garbisu et al., 1996; Bebien et al., 2001; Roux et al., 2001) or can be expelled in the medium (Losi and Frankenberger, 1997).

Plants also vary with respect to their uptake of Se and their capacity to accommodate large, potentially toxic concentrations (Rosenfeld and Beath, 1964; Sors et al., 2005). Wetland plants have been reported to volatilize Se (Pilon-Smits et al., 1999; Lin and Terry, 2003; Wu, 2004), and factors influencing volatilization have been reviewed by Terry and Zayed (1994). As stated above, it has been suggested that changes in redox conditions in the rhizosphere of some plants may influence Se speciation in soil. Some species accumulate Se in Se-rich soils, but not to the extent reported for hyperaccumulators of toxic metals such as Cd and Ni (Banuelos et al., 1997; Prasad and Freitas, 2003; Srivastava et al., 2005). There are few recent studies of Se uptake from soils with normal or deficient levels of Se, as would be the probable situation in the case of radio-Se pollution, but the data have been reviewed comprehensively by Coughtrey et al. (1983) and Ihnat (1989).

13.5.3. Strontium (Atomic No. 38)

Strontium is an alkaline earth metal. Whose four naturally occurring isotopes are stable (^{88}Sr, ^{86}Sr, ^{87}Sr, and ^{84}Sr). Short-lived radioisotopes are used in medicine and as tracers (^{82}Sr, ^{85}Sr, ^{89}Sr, and ^{90}Sr), but the last of these, and the longest-lived (29 years), has been introduced into the environment to a significant extent. Sources of ^{90}Sr in the environment are atmospheric weapons testing, emissions from reactors or reprocessing plants, and fallout from the Chernobyl accident. Strontium is a chemical analog of calcium and has no known biological role. Biological systems discriminate poory between Ca and Sr. The chemistry of Sr in soils is much simpler than that of Cs. Not only is its coordination chemistry simple, but its adsorption on soil constituents, both organic and mineral, is largely nonselective. No significant aging or fixation processes have been reported. This means that Sr can be recycled continuously in soil systems, and the radioecological risk does not diminish with time.

The chemical similarity between Sr and the essential nutrient, Ca, and the resulting poor biological discrimination (Nisbet and Woodman, 2000; Morgan et al., 2001; Sysoeva et al., 2005) between them is the key to determination of the radiological impact of the contaminant. Like Ca, Sr can be taken up directly by plants following foliar interception (Carini and Lombi, 1997; Madoz-Escande et al., 2005). However, the major route for its entry into the food chain is via soil–root uptake. Calcium is a major cation in many soils, both as exchangeable cation adsorbed on the negatively charged surfaces of soil components, and also, associated with sparingly soluble carbonates. Strontium substitutes for Ca both in carbonates and on the exchange complex made up from clay minerals, organic matter, and oxides of iron, manganese, and aluminum. Its mobility in soil depends strongly on its cation-exchange capacity (Ohnuki, 1994) and hence texture, mineralogy, and organic matter content. Unlike transition metals, Ca and Sr do not form strong coordination complexes with organic ligands. They are therefore rather indifferent to soluble organic matter, which may solubilize transition metals, but both adsorb to solid-phase organic matter as a result of its cation-exchange capacity. The role of organic matter in immobilizing Sr is important in very organic soils, including forest soils and sandy soils, when the mineral content is low. Strontium is less strongly retained by organic matter than is Ca (Ehlken and Kirchner, 2002), the reverse of the trend observed for exchange on clay minerals (Liszewski et al., 1998). This is the case for many of the soils in eastern Europe that were severely contaminated by fallout from the Chernobyl accident. The soil-to-plant transfer factor has been reported to decrease with increasing organic matter content (Lakanen and Paasikallio, 1970; Mistry et al., 1973; van Bergeijk et al., 1992; Entry et al., 2001). When Sr is associated with organic matter, the lighter fractions of organic matter, including fulvic acids, appear to be most active (Agapkina et al., 1995; Lee and Lee, 2000).

Soil pH has an important effect on Sr mobility and biological uptake. First, adsorption on soil constituents is pH dependent, and for a large number of soils, adsorption tends to be greater in high-pH calcareous soils (van Bergeijk et al., 1992; Ohnuki, 1994; Chen and Hayes, 1999; Sahai et al., 2000; Twining et al., 2004). There may also be some precipitation of $SrCO_3$ on goethite (Sahai et al., 2000). Absorption on biological membranes is also pH dependent, since the cationic exchange capacity of roots and other membranes is due to weak acid functional groups (Meychik and Yermakov, 1999).

The reactions of Sr with soil are largely instantaneous and reversible, although Coughtrey and Thorne (1983a) cite a range of nonexchangeable Sr in various soils between 3 and 33%. The adsorption mechanism of Sr on most surfaces is largely physical adsorption, with the formation of weak outer-sphere complexes (Axe et al., 1998, 2000; Chen and Hayes, 1999; Trivedi and Axe, 1999; Sahai et al., 2000). This adsorption is easily reversible (Ohnuki and Kozai, 1994). Although Trivedi and Axe (1999) found appreciable diffusion of Sr to internal sites of porous oxides that would lead to fixation, there is no strong evidence of decreasing availability of Sr with time due to fixation of slow ongoing adsorption (Noordijk et al., 1992; Rigol et al., 1999; Cross et al., 2002), or aging. There is some evidence of specific adsorption on hydrous oxides or iron and aluminum (Kinniburgh et al., 1975) and the formation of inner-sphere complexes on hematite (Karasyova et al., 1999). Wang and Staunton (2005) reported that an apparent decrease in the desorbability of Sr over a period of four months in flooded soil was due to changing soil conditions and not to fixation.

13.5.4. Zirconium (Atomic No. 40)

Zirconium has five stable isotopes with different natural abundances,—^{90}Zr (51.5%), ^{91}Zr (11.2%), ^{92}Zr (17.1%), ^{94}Zr (17.4%), and ^{96}Zr (2.8%)—and eight radioactive isotopes, including ^{86}Zr, ^{88}Zr, ^{89}Zr, ^{93}Zr, ^{95}Zr, and ^{97}Zr. Of the radioactive isotopes, only ^{95}Zr is significant in terms of environmental contamination. It is produced by both fission reactions and by neutron activation of ^{94}Zr and has a physical half-life of 64.02 days (LBNL Isotopes Project, 2004), decaying to ^{95}Nb and ^{95m}Nb. Zirconium-95 has entered the environment as a result of nuclear weapons detonations, radioactive waste discharges, and also during the Chernobyl accident, when significant contamination of the local environment occurred due to the scattering of irradiated reactor fuel fragments ("hot particles"). It was estimated by Borovoi et al. (1988, cited by Konoplev et al., 1993) that the Chernobyl reactor fuel contained 3.3×10^{10} Bq g^{-1} of ^{95}Zr. Zirconium-95 decays by beta emission with a maximum beta energy (E_{max}) of 880 keV. As a result of its beta decay, it also emits gamma rays with energies of 724 and 757 keV. In the case of the Chernobyl accident, these gamma rays contributed to the extremely high gamma dose rates observed in the region of the Chernobyl reactor in the weeks following the accident in 1986. These dose rates, up to a maximum of 5 milligray h^{-1} in May 1986, were sufficient to result in the mortality of some tree species, notably pine, in the worst-affected areas (Smith and Beresford, 2005).

The natural crustal concentrations of stable (nonradioactive) zirconium are surprisingly high. Bowen (1979) quoted Zr concentrations in soils ranging from 400 to 2000 mg kg^{-1}, with a median concentration (based on literature values) of 400 mg kg^{-1}. Concentrations in plant tissues are significantly lower, with values ranging from 0.2 to 5.0 mg kg^{-1} across a variety of photosynthetic plants and fungi (Bowen, 1966).

Zirconium is polyvalent, but only the 4+ valence state is observed in aqueous environments. In common with other polyvalent elements, it is adsorbed rapidly and relatively intensely by soils and sediments, rendering it relatively poorly available for biological absorption in the terrestrial environment. Solid–liquid distribution coefficients quoted for ^{95}Zr by Sheppard and Thibault (1990) range from 600 L kg^{-1} for sandy soil to 7300 L kg^{-1} for organic soil.

Soil–plant transfer factors for ^{95}Zr are rare. An "expected value" of 0.001 was quoted by Ng et al. (1982), although Coughtrey and Thorne (1983a) reviewed literature evidence indicating that although zirconium was likely to be strongly excluded from plants, a range of soil–plant transfer factors from 0.001 to 1.0 is possible. However, specific transfer factors quoted from various sources by these authors are for values of 0.006 to 0.002 for the grass *Bromus rubens*, 0.0013 to 0.004 for peas, and 0.012 for wheat straw.

The committed effective dose per unit intake following ingestion of ^{95}Zr is 9.5×10^{-10} Sv Bq^{-1} for adults > 17 years of age (IAEA, 1996). However, due to its relatively short physical half-life, the major environmental impact of ^{95}Zr is likely to be short-term external irradiation.

13.5.5. Niobium (Atomic No. 41)

The name *niobium* was adopted by IUPAC in 1949 in place of the original name, *columbium* (Cb), designated at the time of the discovery of element 41 in 1801, although *columbium* is still used in mining and metallurgy. Niobium has only one stable isotope (^{93}Nb) but 11 radioactive isotopes, of which one, ^{95}Nb, is of environmental significance. ^{95}Nb has a half-life of 34.98 days (LBNL Isotopes Project, 2004). It is formed by nuclear fission, neutron activation, and by the beta decay of ^{95}Zr. Niobium-95 decays by beta emission (E_{max} 165 keV) to ^{95}Mo with two accompanying gamma rays with energies of 753 and 768 keV. As for ^{95}Zr, the major sources of ^{95}Nb within the environment have been nuclear weapons tests, radioactive waste, and the Chernobyl accident. Borovoi et al. (1988, cited by Konoplev et al., 1993) calculated that the specific activity of ^{95}Nb within the Chernobyl reactor fuel at the time of the accident was 3.4×10^{10} Bq g^{-1}. Just as ^{95}Zr contributed to the high gamma dose rates in the immediate vicinity of the Chernobyl reactor, so did ^{95}Nb. Despite its shorter physical half-life, ^{95}Nb activities in the Chernobyl exclusion zone were sustained by ingrowth from decaying ^{95}Zr.

Environmental concentrations of stable (nonradioactive) niobium are lower than those of zirconium, with a likely range of 6 to 300 mg kg^{-1} in soils and concentrations on the order of 0.3 mg kg^{-1} in woody plants (Bowen, 1979). Niobium is chemically and physically similar to tantalum, with four valence states of 2+, 3+, 4+, and 5+. As a polyvalent element it would be expected that sorption of niobium to soils and sediments would be strong. However, the range of solid–liquid distribution coefficients for soils available in the literature suggests a lower degree of sorption and retention than for ^{95}Zr. Sheppard and Thibault (1990) quote K_d values for ^{95}Nb ranging from 160 L kg^{-1} for sandy soils to 2000 L kg^{-1} for organic soils. These K_d values are still sufficiently high to limit the potential biological uptake of ^{95}Nb.

Soil–plant transfer factors for ^{95}Nb are extremely rare in the literature. The most recently available appear to be from Gerzabek et al. (1993), who quoted values of 0.017 for bean pods and 0.05 for oil seed rape. Lisk (1972) proposed a transfer factor of 0.11, and Coughtrey and Thorne (1983a) suggested a transfer factor of 0.023 based on a literature review of stable niobium concentrations in soils and plant tissues. Thus, ^{95}Nb appears to have a slightly stronger tendency than ^{95}Zr for accumulation in plants, which corresponds with its comparatively lower tendency for sorption within the soil. The committed effective dose per unit intake following ingestion of ^{95}Nb is 5.8×10^{-10} Sv Bq^{-1} for adults $>$ 17 years of age (IAEA, 1996). As for ^{95}Zr, however, its short half-life suggests that the major environmental impact of ^{95}Nb is likely to be short-term external irradiation.

13.5.6. Technetium (Atomic No. 43)

Technetium is a transition metal that can exist in all oxidation states between +7 and −1. All of its isotopes are radioactive. One of these, ^{99}Tc, exists naturally as a result of the spontaneous fission of ^{238}U. However, the major sources of Tc in the environment are artificial and the isotope that causes most concern is ^{99}Tc, because of its abundance and its long half-life (4.2×10^6 years). The sources of ^{99}Tc in the environment have been reviewed by Desmet and Myttenaere (1986), Schulte and Scoppa (1987), and more recently by Bennett and Willey (2003). In general, Tc is released in volatile or aqueous forms. In either case, in contact with aerated soil the prevalent form will be the soluble pertechnetate anion TcO_4^-. This anion has little interaction with soil components and so is assumed to be very mobile in soil. Furthermore, it is readily assimilated by plants and microorganisms. Its uptake from solution is reduced by anions, including NO_3^-, $H_2PO_4^-$, SO_4^{2-}, and MoO_4^{2-} (Cataldo et al., 1983; Echevarria et al., 1998). Symptoms of chemical toxicity are frequently reported. Tc is often associated with proteins in plants and is probably substituted for S in thio-containing ligands. The dominant oxidation state of Tc in aqueous solution is +7, although pertechnetate is a weak oxidant. It is stable over a wide range of pH and concentration. Technetium(VII) is known to be reduced to Tc(IV) by organic matter and by various reducing agents, including some commonly used in chemical extraction procedures for soil (hydroxylamine and ascorbic acid). Aqueous forms of Tc(IV) may be reoxidized to Tc(VII) to varying degrees by H_2O_2, HNO_3, MnO_4^-, Cl_2, or air.

The chemistry of Tc in soil is more complex than in aqueous solution. In aerated conditions, the pertechnetate form is thought to dominant. This anion, like nitrate, is adsorbed only to a very limited extent by soil constituents and so is expected to be very mobile. Indeed, the distribution coefficient, K_d is usually found to be close to zero (Wildung et al., 1974; Bird and Schwartz, 1997; Routson et al., 1977; Sheppard et al., 1982, 1990; Bachhuber et al., 1986). Furthermore, Tc added as pertechnetate to soils shows considerable vertical movement (Sheppard and Thibault, 1991; Denys et al., 2003) and is readily taken up by plant roots (Murphy and Johnson, 1993; Echevarria et al., 1997). However, even under oxidizing conditions, Tc is not as soluble and as mobile as would be expected for a simple monovalent anion. For example, Schimmack et al. (1993) showed that ^{95m}Tc was less vertically mobile in an organic forest soil than was bromide. Wildung et al. (1974) found that sorption, although small, was correlated directly with the organic matter content of soil and inversely with pH. This indicates both the role of organic matter and possibly that of Fe and Al oxides. Sheppard et al. (1990) also report that sorption increases with increasing organic matter content of soils. Recent studies have shown that Tc(VII) is associated with organic matter, notably humic acids, by oxidation–reduction reactions (Rössler et al., 2000; Geraedts et al., 2002; Maes et al., 2003). Strictly speaking, these are not adsorption reactions, but it may be difficult experimentally to distinguish adsorption and reaction, since the net result is a removal of Tc from solution. However, even soluble forms of Tc may be associated with low-molecular-mass organic compounds, and these complexes are less available for uptake by roots than the aqueous pertechnetate anion (Yanagisawa et al., 1997).

The immobilization of Tc under reducing conditions is even more marked. About 98% of Tc was lost from solution when soils were incubated in sealed bottles (Landa et al., 1977). Sheppard et al. (1990) found much larger K_d values for anaerobic soils than for aerobic soils. In contrast, about 40% of Tc was found to remain soluble in sediments of an anoxic fjord (Keith-Roach and Roos, 2004). Yanagishawa and Muramatsu (1995a,b) reported that Tc(VII) was reduced to insoluble forms of Tc(IV) with time under reducing conditions but less so under aerated conditions. Technetium was immobilized in two soils over a six-month period under anaerobic conditions, but not when the same soils were kept aerated (Tagami and Uchida (1998). Tagami and Uchida (1999) report that the solubility of Tc gradually decreased in a waterlogged soil over a 52-day period and that little Tc was remobilized when the soil was air-dried and submitted to aerobic conditions. Microbial activity certainly contributes to this immobilization, because of its influence on redox conditions (Tagami and Uchida 1996), but these authors found no conclusive evidence of a direct effect whereby Tc is immobilized as a result of microbial absorption. More recently, Ashworth and Shaw (2005) reported that Tc applied to the base of soil columns tended to remain near the water-saturated base where conditions were reducing. They concluded that pertechnetate was reduced to sparingly soluble Tc(IV), making migration up the column very slow. Few studies attempt to identify the sparingly soluble forms of Tc in soil. Keith-Roach et al. (2003) used a chemical sequential extraction scheme to fractionate Tc. They found different fractionations for the three sediments studied. In one case, the major fraction was attributed to carbonates; however, the authors overlooked the fact that sulfide may also be solubilized by the acidic acetate extraction (Wallmann et al., 1993). Abdelouas et al. (2005) concluded that most insoluble Tc in organic matter–rich soils after anoxic incubation was associated with organic matter (60 to 66% being extracted by hydrogen peroxide) with a considerable proportion (23 to 31%) being associated with Fe-oxyhydroxides (extracted by hydroxylamine). X-ray absorption spectroscopy has confirmed that the insoluble form of Tc formed in anoxic sediments is hydrous TcO_2 (Burke et al., 2005).

The precise role of microorganisms in Tc reduction has been studied more extensively in bacterial culture or in solution than in soil. Biosorption of pertechnetate by cyanobacteria is not energy dependent and easily reversible, with no reduction of Tc (Garnham et al., 1993). However, other bacteria, with known metal-reducing capacity, have been shown to reduce Tc(VII) to Tc(IV) and occasionally, Tc(V) (Lloyd and Macaskie, 1996). Dissimilatory reduction of metals in anoxic environments is well documented (Lovley, 1993) and has been studied increasingly in the last decade with particular attention paid to iron, manganese, chromium, uranium, arsenic, and selenium. Many examples are given in chapter 7 to 10 of the book. There are fewer published examples of Tc reduction. Nevertheless, various sulfur-reducing bacteria, including *Shewanella putrefaciens*, *Geobacter metallireducens*, *G sulfurreducens*, and *Desulfovibrio dessulfuricans*, are reported to reduce Tc(VII) under anaerobic conditions (Lloyd and Macaskie, 1996; Lloyd et al., 1997, 1998, 1999, 2000, 2002; Abdelouas et al., 2002, 2005; Burke et al., 2005). Under anaerobic conditions, *Escherichia coli* (Lloyd et al., 1997, 1999) and *Clostridia* (Francis et al., 2002), which are strict anaerobic fermentative bacteria, have also been reported to cause Tc reduction and precipitation. Gram-negative bacteria may even transform Tc to insoluble forms under oxidizing conditions (Ishii and Uchida, 2005). There is no evidence that aerobic and facultative denitrifying bacteria cause Tc reduction (Abdelouas et al., 2002, 2005). In some studies two distinct mechanisms of reduction have been distinguished. There can be both direct enzymatic reduction and indirect mechanisms catalyzed by electron shuttling compounds, including humic analogs, sometimes followed by precipitation on freshly formed biogenic minerals (Lloyd et al., 2000, 2002; Abdelouas et al., 2002). Various electron donors are used in these reactions, the most efficient being formate and pyruvate (Lloyd et al., 1997; Burke et al., 2005). Reduction appears to take place during active Fe(III) reduction and to be complete before sulfate reduction (Burke et al., 2005). There is no evidence to date of microbial reduction to volatile methylated forms of Tc, as is reported for Se and various metals.

13.5.7. Ruthenium (Atomic No. 44)

Ruthenium has seven stable isotopes (^{96}Ru, ^{98}Ru, ^{99}Ru, ^{100}Ru, ^{101}Ru, ^{102}Ru, and ^{104}Ru), of which the most abundant is ^{102}Ru (~32%). It also has five radioactive isotopes (^{94}Ru, ^{97}Ru, ^{103}Ru, ^{105}Ru, and ^{106}Ru), of which ^{103}Ru and ^{106}Ru are of significance as environmental contaminants. ^{103}Ru is formed by nuclear fission and neutron activation of ^{102}Ru, while ^{106}Ru is formed solely by nuclear fission. ^{103}Ru and ^{106}Ru have half-lives of 39.26 and 373.6 days, respectively, and have both been introduced into the environment following releases from nuclear weapons tests as well as accidental and planned releases from nuclear facilities. Total deposited activities of ^{106}Ru determined within the Chernobyl 30-km exclusion zone ranged from 0.1 to 1.02 MBq m^{-2} (Wang, 1996), the exact value depending on the location. Krouglov et al. (1998) reported total deposited ^{106}Ru activities of 1.2 to 9.2 MBq m^{-2} in the same area. Following the Chernobyl accident, most of the ruthenium released was probably associated with irradiated reactor fuel fragments ("hot particles") deposited within a few tens of kilometers of the reactor. However, Ronneau et al. (1995) have reported that up to 100% of ruthenium in a uranium oxide matrix can be volatilized at the high temperatures reached during the reactor fire. ^{106}Ru detected in soils in western Europe and Scandinavia in the early weeks following the Chernobyl accident must have been transported through the atmosphere in aerosol form. The initial physicochemical form of ^{106}Ru deposited to soil surfaces in the near- and far-field zones was, therefore, quite different.

Data on natural concentrations of stable ruthenium in environmental materials are limited, although then appear to be very rare, with values of 0.001 mg kg^{-1} being reported for igneous rocks (Bowen, 1966). Published measurements of stable ruthenium in plant tissues are also rare, although Bowen (1966) suggested values of 0.005 mg kg^{-1} in angiosperms, some five times higher than those in igneous rocks.

Ruthenium has a complex chemistry with up to seven valence states, although the 2+, 3+, and 4+ valences are encountered most commonly. Essington et al.

(1963) determined that under acidic soil conditions, the 3+ and 4+ oxidation states were dominant. Schultz (1965) suggested that ruthenium in nitrosyl complexes in soil is likely to be more mobile than in noncomplexed forms, although Jones (1960) found that nitrosyl ruthenium can interact with the surface layer of ferric hydroxide coatings on marine sediment particles to form insoluble complexes. In soils of northwestern England, however, Livens and Baxter (1988) found that ^{106}Ru was rather weakly associated with the oxide phase. In soils, ruthenium appears to have a particularly high affinity for organic matter, and Livens and Baxter (1988) found 30% of ^{106}Ru in a gley soil associated with organic matter despite the fact that the soil had a total organic carbon content of only 3%. Andersson and Roed (1994) found that almost 70% of ^{106}Ru derived from Chernobyl fallout in Swedish forest soil was associated with organic matter, the remainder not being extractable even with 7 M HNO_3. Despite this clear tendency to sorb to organic matter, however, ^{106}Ru had migrated down the soil profile significantly faster than both ^{137}Cs and ^{134}Cs.

As expected from the studies cited above, solid–liquid distribution coefficients for ^{103}Ru and ^{106}Ru are highest in organic soils. Sheppard and Thibault (1990) have suggested a median K_d of 6600 L kg^{-1} for organic soils, with a possible range of 3600 to 120,000 L kg^{-1}. This contrasts with median K_d values of 55, 990, and 400 L $kg^{-1}$for sand, loam, and clay soils, respectively. Only a limited number of soil–plant transfer factors have been published for ^{103}Ru and ^{106}Ru, the most extensive source being the soil–plant transfer factor database of the International Union of Radioecologists (IUR, 1989), although even this contains only 17 experimental values for a small number of crop types. These values span a range from 0.005 to 1.45, with a median value of 0.231. Ng et al. (1982) recommended a soil–plant transfer factor of 0.04 for ruthenium isotopes. On the basis of a variety of studies in the 1960s and 1970s, Coughtrey and Thorne (1983a) suggested transfer factors of 0.3 for nitrosyl ruthenium complexes and from 0.01 to 0.1 for noncomplexed forms of ruthenium.

Recently, Willey and Fawcett (2006) augmented the literature data on ruthenium uptake by plants with their own experimental data to produce a data set of comparative $^{103/106}Ru$ concentrations in 116 flowering plant species. Statistical analysis demonstrated that there was almost a 2500-fold difference in the degree to which different plant taxonomic groups took up ruthenium, with 36% of the variance within the data set being accounted for by taxonomic differences. The committed effective doses per unit intake following ingestion of ^{103}Ru and ^{106}Ru are 7.3×10^{-10} and 7.0×10^{-9} Sv Bq^{-1}, respectively, for adults older than 17 years of age (IAEA, 1996).

13.5.8. Silver (Atomic No. 47)

Silver occurs naturally in the form of two stable isotopes, ^{107}Ag (51.4%) and ^{109}Ag (48.6%). Of its 12 radionuclides, ^{108m}Ag, ^{110m}Ag, and ^{111}Ag may be emitted as a result of the nuclear fuel cycle. However, ^{110m}Ag has received the most attention as an environmental contaminant since it was emitted during the

Chernobyl accident of 1986 and was detected in ecosystems up to 2000 km west of the reactor (Beresford, 1989a). It was found to concentrate in animal livers (Beresford, 1989b) and in mushroom species (Battiston et al., 1989) in western Europe. ^{110m}Ag has a half-life of 249.8 days (LBNL Isotopes Project, 2004). It is produced by neutron activation of ^{109}Ag and decays by beta emission to ^{110}Cd, emitting in the process a range of accompanying gamma rays from 450 to 1560 keV.

Natural concentrations of stable silver in soils range from 0.01 to 8 mg kg^{-1}, according to Bowen (1979). Silver is the most reactive of the noble metals, with three cations: Ag^{+}, Ag^{2+}, and Ag^{3+}. Chemical equilibrium calculations by Lindsay (1974), however, have indicated that only $Ag^{0}$and Ag^{+} are likely to be significant in soils. Redox potential and pH are critical controlling factors of the behavior of silver in soils, and under reducing conditions, precipitation of Ag as AgS is likely (Lepp et al., 1995). Lepp et al. (1995) have reviewed literature which suggests that silver is immobilized in the upper soil layers by reaction with soil organic matter. According to these authors, "organically bound soil Ag may represent a significant proportion of the total Ag component and play an important part in controlling the cycling, mobility and behaviour of Ag in soils." More recently, Handl et al. (2000) studied the long-term (ca. five year) behavior of a stable silver tracer in a soil in an orthi-humic-gleysol (6.8% organic carbon) in Germany. Downward movement of Ag from the soil surface was very slow, and a residence half-time of 33 years was estimated for the surface 5-cm soil layer. This indicates that ^{110m}Ag would be subject to almost complete physical decay before any significant migration and leaching in the soil occurred.

Solid–liquid distributions quoted for Ag by Sheppard and Thibault (1990) generally support the observation that Ag is of limited mobility in soils, especially those with significant organic carbon. The best estimated K_d values for sand, loam, clay, and organic soils are 90, 120, 180, and 15,000 L kg^{-1}, respectively. The range of K_d estimates for organic soils is 2400 to 89,000 L kg^{-1}.

Handl et al. (2000) reported measured transfer factors for silver in pasture grass ranging from 0.009 to 0.065. This range is slightly higher than the rather limited collection of soil–plant transfer factors quoted by IUR (1989) of 0.00027 for lettuce, 0.0008 for tomato, and 0.0013 for radish. Ng et al. (1982) quoted a transfer factor value of 0.15 for Ag in unspecified plant material. The committed effective dose per unit intake following ingestion of ^{110m}Ag is 2.8×10^{-9} Sv Bq^{-1} for adults older than 17 years of age (IAEA, 1996).

13.5.9. Antimony (Atomic No. 51)

Two stable isotopes of antimony occur naturally, ^{121}Sb (57.2%) and ^{123}Sb (42.8%). Antimony has 20 radionuclides, but of principal environmental concern is ^{125}Sb, which has a half-life of 2.76 years. Antimony-125 is a fission product that is also produced by neutron activation of ^{124}Sn (tin) to produce ^{125m}Sn, which decays with a half-life of 9.7 minutes to ^{125}Sb and subsequently to ^{125}Te. Antimony-125 decays by beta emission and also produces a range of gamma rays, chief among which are at 427 keV (30.6%) and 598 keV (18.6%).

Antimony is a metalloid with chemical similarities to arsenic, although it is less abundant than As in the natural environment. Bowen (1979) reported a range of Sb concentrations in soils of 0.2 to 10 mg kg^{-1} with a median value of 1.0 mg kg^{-1}. Coughtrey et al. (1983) reported an overall range of Sb concentrations in soils of 0.05 to 260 mg kg^{-1}. Coughtrey et al. (1983) note that the world average Sb concentration in soils (1.0 mg kg^{-1}) is a factor of 5 greater than the world average crustal concentration (0.2 mg kg^{-1}); thus, Sb is usually concentrated in soil relative to parent materials. Anthropogenic enhancement of Sb concentrations in the environment are usually related to industrial activities such as mining and metal working, the application of soil amendments such as superphosphate fertilizers and soil, and the burning of coal, which can contain up to 20 mg kg^{-1} (Lepp et al., 1995).

As in the case of arsenic, antimony can exist in the 3+ (antimonite) and 5+ (antimonate) oxidation states, but simple Sb^{3+} and Sb^{5+} cations are not likely to exist in solution. Peterson and Girling (1981) suggested that antimony probably exists in soils primarily in the form of antimonate (SbO_3^-), and Fuentes et al. (2003) have recently confirmed that the most common species extracted from soil is Sb(V). Bodek et al. (1988) have suggested that hydrolyzed forms of Sb probably dominate in solution. In oxic soils at pH > 3.0, $Sb(OH)_6^-$ probably dominates, whereas under low redox conditions, $Sb(OH)_3$ is likely to be dominant. Kabata-Pendias and Pendias (1984) classed antimony as being moderately mobile in soils. This classification has been confirmed by D'Souza and Mistry (1980), who found that ^{125}Sb applied to a soil surface under tropical conditions became distributed down to 10 cm depth within 33 months. This observation is broadly consistent with published K_d values for Sb within "generic" soil classes. These range from 45 L kg^{-1} in sands to a maximum of 540 L kg^{-1} in organic soils (Sheppard and Thibault, 1990). Nakamura et al. (2005) have recently reported a mean K_d of 62 L kg^{-1} for ^{124}Sb in Japanese agricultural soils, although the maximum K_d they observed was 2065 L kg^{-1}. Sorption of Sb onto the common soil mineral hematite is reported to be strongly pH dependent (Ambe et al., 1986), with sorption increasing strongly as pH decreases below 7.0, in a fashion analogous to the sorption of arsenate species. Nakamaru et al. (2005) have recently confirmed this pH dependency in 110 Japanese agricultural soils. For two of the soils, an Andosol and a Fluvisol, they studied the effect of pH adjustment and report sorption increased logarithmically as pH decreased below 7.0. Sb sorption on soils also decreased with increasing phosphate concentration.

Antimony concentrations in plant tissues are reported to be in the range 0.0001 to 0.2 mg kg^{-1} dry weight (Bowen, 1979). Given a global average soil concentration of 1 mg kg^{-1}, this implies that soil–plant transfer factors for Sb are likely to be considerably less than 1. Coughtrey et al. (1983) surveyed published (non-radioactive) Sb concentrations in a number of different plant types and concluded that on the basis of a "normal" Sb concentration in soil of 1 mg kg^{-1}, typical soil–plant transfer factors for Sb were 0.1 for natural vegetation; 0.05 for pasture grass, herbaceous vegetables, grain, and cereals; and 0.005 for legumes, root vegetables, and garden fruits. The IUR (1989) reported "recommended" transfer factors, based on experiments with ^{125}Sb, which are substantially lower than the values suggested by Coughtrey et al. (1983): 5.6×10^{-4} for radish (root), 1.3×10^{-4} for lettuce, and 8.0×10^{-5} for tomato. However, the IUR (1989) database does contain a small number of rather high transfer factors for Sb: 0.4 for bean (seed), 1.5 for radish (root), and 5.8 for lettuce. Lux et al. (1995) reported transfer factors of 0.001 to 0.01 for ^{125}Sb in vegetation of the Chernobyl 30 km zone and concluded that accumulation by plants of radionuclides such as ^{125}Sb was not significant. In summary, there appears to be considerable uncertainty over the most appropriate transfer factor for antimony, which is perhaps exacerbated by differences in the degree of plant uptake of stable Sb and ^{125}Sb, which are likely to have been present in the soil for very different time periods. The committed effective dose per unit intake following ingestion of ^{125}Sb is 1.1×10^{-9} Sv Bq^{-1} for adults older than 17 years of age (IAEA, 1996).

13.5.10. Cesium (Atomic No. 55)

Cesium is an alkali metal with a very simple solution chemistry. The natural abundance of the stable isotope (^{133}Cs) is 100%, and three artificial radioisotopes may be present in the environment. Two are already present at various concentrations in many soil systems: ^{134}Cs with a half-life of 2.06 years and ^{137}Cs with a half-life of 30.1 years. Both were important constituents of fallout from atmospheric weapons testing in the 1950s and 1960s, have been released more locally around some nuclear processing plants and waste storage facilities, and accounted for much of the radioactivity released during the accident at Chernobyl in 1986. The longer-lived isotope, ^{135}Cs (2.3×10^6 years), is an important component of long-lived radioactive waste, so waste management strategies must consider the possibility that it could enter the environment, although it has not been detected to date. Cesium has no known biological role, at least at the concentrations usually encountered. At very large concentrations (millimolar) it is known to be toxic (Avery, 1995; Hampton et al., 2004) and is used in research to block potassium channels in biological membranes. Despite the capacity of biological systems to discriminate between the two elements, Cs is supposed to behave much like K. The chemistry of Cs in soil is dominated by its complex adsorption properties. Cs is strongly, and some claim largely irreversibly, immobilized on the mineral surfaces of soil. Organic matter does not adsorb Cs selectively, and its role in Cs adsorption is usually largely indirect.

There are various sources of radiocesium in the environment. The input from atmospheric weapons testing in the mid-twentieth century leads to the most widespread and homogeneous contamination of soil and water. The accumulated contribution of ^{137}Cs in Europe from this source is estimated to be about 1000 Bq m^{-2}. The fairly uniform deposition of ^{137}Cs from fallout following the nuclear weapons testing has led to the widespread use of this isotope as a tracer of erosion (Ritchie and McHenry, 1990; Agudo, 1998). The accident that occurred at the Chernobyl nuclear plant in April 1986 led to more localized contamination. The events that led to this accident and the consequences have been widely reported (AEN, 1987; Savchenko, 1995; Aarkrog et al., 1999; IAEA, 2001). For example, in Ukraine more than 8.4 Mha of agricultural land was contaminated to a level of 3.7 kBq m^{-2}, or more. In France, the fallout due to this accident varied from less than 1000 Bq m^{-2} to 10,000 to 12,000 Bq m^{-2} in eastern France and up to 37,000 Bq m^{-2} in some areas, such as the Jura and Alp mountains and Corsica, depending largely on rainfall in early May of that year. Similar variability has been observed for other affected countries in western, central, and eastern Europe. Due to rainfall patterns and the direction of the radioactive cloud after the accident, the OECD countries most affected were the Scandinavian countries, mountainous areas including the Alps, and parts of the British Isles. The heterogeneous pattern of fallout from this accident has complicated the use of ^{137}Cs as a tracer for erosion in Europe.

The chemistry of Cs in soils is dominated by its adsorption properties (Livens and Loveland, 1988; Smolders et al., 1997; Guivarch et al., 1999; Staunton et al., 2002). Soil adsorption properties depend to a large extent on the mineralogical composition of the soils. This is because clay minerals selectively adsorb Cs, particularly at trace concentrations, as is always the case for radiocesium contamination, in preference to other cations. The minerals with the greatest affinity for trace Cs are illites, or hydrous mica. For example, the limiting Cs–Ca Gaines–Thomas selectivity coefficient for illite has been reported to be about 10^{10}, whereas the Cs–Na selectivity coefficient is three orders of magnitude less (Staunton and Roubaud, 1997). On montmorillonite, the equivalent selectivity coefficients are about 10^5 and 10^2. The corresponding K_d values are about 10^5 on illite and 10^3 on montmorillonite. The strong affinity of micaeous minerals is attributed to the presence of a small proportion of frayed edge sites. When Cs is exchanged at these sites, partial layer collapse follows, with the loss of hydration water of Cs (Sawhney, 1972). The decrease in interlayer spacing of an expanding clay when the exchangeable cation is replaced by large amounts of stable Cs can be measured using x-ray diffraction. Frayed edges are visible using microscopy, so Cs adsorption can be related to particle shape and density of frayed edges (Rajec et al., 1999) and stable Cs identified in edge sites and even in interlayer sites on mica (McKinley et al., 2004). However, few studies attempt to follow layer collapse in the presence of trace amounts of radiocesium. Until recently, analytical limitations made such studies impossible. Liu et al. (2003) have used x-ray microprobe analysis to demonstrate that ^{137}Cs is accumulated at the edges of mica and within internal channels, implying adsorbed state diffusion.

The K_d values for soils may be as large as that reported for pure illite, and can be much lower than the value reported for montmorillonite, in sandy or highly organic soils (Cornell, 1993; Wauters et al., 1996; Smolders et al., 1997). The Cs-adsorption properties of many soils are dominated by the presence of trace amounts of illites (Cremers et al., 1988). It has been argued that illite is so important that almost undetectable quantities can determine Cs adsorption on soils, making it impossible to verify this claim using regression analysis (Hird et al., 1995). This may be the case for many soils of temperate climates where illites are usually present, albeit in varying amounts, but is unlikely to be the case in soil from the tropics. The degree of weathering certainly influences the affinity of soil minerals for Cs, but this has not been studied extensively. It would appear that soil clays, which are more strongly weathered than mineral clays, have a greater affinity for Cs. For example, Staunton and Levacic (1999) report K_d values of Cs on the clay-sized fraction of various soils that were as large as would be expected for illites, irrespective of the dominant clay mineralogy determined by x-ray diffraction. Dumat et al. (1997) compared the adsorption of Cs on clay extracted from a subsoil and that extracted from the equivalent topsoil after chemical removal of organic matter, and they attributed the greater adsorption of Cs on topsoil samples to the greater weathering in topsoils due to climatic variations and biological activity. Maes et al. (1998) report a complex effect of weathering on the Cs fixation demonstrated by comparing samples from different horizons of a soil. The nature of the weathering and the consequent effect on Cs fixation properties was found to depend on organic matter content. In mineral layers, Al polymers blocked weathered sites and prevented layer collapse, so the sites were not strongly selective for Cs. In contrast, in organic layers, Al was complexed by organic ligands, so layers could collapse when weakly hydrated Cs was adsorbed. The same group demonstrated that weathering induced chemically under controlled conditions modified the Cs adsorption and fixation properties with respect to Cs (Maes et al., 1999). Liu et al. (2003) have shown that secondary minerals precipitated on micaeous particles also adsorbed ^{137}Cs strongly. These minerals could be removed using acidified ammonium oxalate, suggesting that they included iron oxides and aluminosilicates.

Given that adsorption of Cs on organic matter is nonselective, no effect of soil organic matter content is expected on Cs adsorption. Data collected in situ suggests, however, an inverse correlation between soil organic matter content and Cs immobilization (van Bergeijk, 1992; Rigol et al., 2002). This has been attributed to the absence of clay minerals in such soils, however, as outlined above, the presence of small amounts of illites should provide a large affinity. It is now accepted that organic matter may have an indirect effect on Cs adsorption on clay minerals (Staunton et al., 2002). The addition of organic macromolecules to clay minerals leads to a decrease in the adsorption of Cs . This has been reported for humic substances (Staunton and Roubaud, 1997; Dumat and Staunton, 1999), water-extractable soil organic matter (Staunton et al., 2002), soil polysaccharides, and a protein (Dumat et al., 2000). The effect is greater for illite than for montmorillonite and is most marked at trace concentrations of Cs than at larger concentrations. These observations suggest that the adsorption of the polymer near the frayed edge sites of illite is responsible for the effect. The same conclusion has been drawn using the opposite approach: the progressive removal of soil organic matter using chemical (Dumat et al., 1997) or physical methods (Staunton et al., 2000). Physical removal of soil organic matter using cold oxygen plasma had less effect than chemical removal since it removes the organic matter on physically accessible surfaces, which may not play such an important role in physically impeding Cs adsorption and layer collapse.

In addition to the importance of soil mineralogy and organic matter, the composition of solution and the exchange complex are important. Since the adsorption of radiocesium is an exchange process, the nature of the other competing cations cannot be overlooked. The same exchange sites that have a strong affinity for Cs also adsorb similar cations, such as potassium and ammonium, selectively. Thus, when a large proportion of the exchange complex is saturated with potassium, the soil has less affinity for Cs; additionally, potassium in solution competes effectively with Cs for adsorption, thereby reducing adsorption. These considerations have important implications for seasonal variation in Cs mobility. For example, during plant growth the rhizosphere may become depleted in potassium. Up to a certain degree of depletion, the exchange complex replenishes the solution, and the charge of high-affinity sites remains compensated by potassium. However, when depletion is severe, both the soil solution and the exchange complex are depleted in K, and this should enhance Cs adsorption (Smolders et al., 1997). Conversely, under anoxic conditions, such as in water-saturated soils, ammonium production and Fe-oxide dissolution can lead to the mobilization of adsorbed Cs (Evans et al., 1983; Comans et al., 1989). In contrast to the effect of similar cations, pH has no direct effect on Cs mobility in temperate-zone soils and in sediments containing illite. There may be statistically significant correlations between soil pH and Cs adsorption (Sanchez et al., 2002), but these are not thought to be causal and would arise from other factors, such as organic matter content. There may be a long-term effect of pH on mineral weathering, which could in turn influence Cs adsorption, but no such data exist.

The adsorption of Cs on soils and clays is strongly concentration dependent. Thus, data obtained using large concentrations of Cs, especially stable Cs, cannot be extrapolated to concentrations usually encountered in the environment. The concentration dependence arises because of the heterogeneity of the adsorbing surfaces, which contain sites of varying affinity for Cs (Brouwer et al., 1983). At trace concentrations of Cs, only the high-affinity frayed-edge, sites take part in adsorption. As the Cs concentration is increased, the proportion of low-affinity or regular exchange sites increases, and the overall affinity of the surface, often assessed by the K_d value decreases. Many authors fit stepwise relations between the selectivity coefficient and the concentration from which they then obtain affinity coefficients and relative abundances for each of the two or three classes of adsorption sites they had defined (Brouwer et al., 1983; Poinssot et al., 1999). However, the accurate calculation of such a number of parameters (four or six) would require more experimental points than these studies employ. A more cautious estimation of surface heterogeneity is given by the value of the gradient of the log-transformed Freundlich adsorption. This gradient is unity for linear adsorption isotherms, and the smaller the gradient, the greater the concentration dependence and the more heterogeneous the surface with respect to Cs adsorption (Sposito, 1984). Staunton and co-workers (Staunton, 1994; Staunton and Roubaud, 1997; Staunton and Levacic, 1999) have reported values of about 0.5 to 0.6 for soils and soil clays, with little difference between soils of contrasting mineralogy, whereas a marked contrast between reference clays was observed (0.5 for illite and 0.8 for montmorillonite).

Another much studied feature of Cs dynamics in soil is its fixation with time, or aging. Field studies of Cs bioavailability have showed a decrease over the years following contamination (van Bergeijk, 1992). This has been attributed to the slow transformation of adsorbed Cs to less readily exchangeable forms, possibly due to slow solid-state diffusion to internal interlayer clay sites, or from external to internal frayed-edge sites in illite (Comans et al., 1991). However, experimental and modeling studies have not reproduced this phenomenon to an extent that would explain field observations. Adsorption is to a large extent rapid, as would be expected for cation exchange, with only a very small proportion of ongoing reaction. Desorbability is more difficult to assess, and many studies do not use identical conditions for adsorption and desorption, and so confuse strong adsorption with irreversible adsorption. Guivarch et al. (1999) found that desorption K_d values for ^{137}Cs were between two and six times greater than adsorption K_d values, depending on stable Cs concentration, corresponding to between 50 and 80% of adsorbed Cs being fixed (i.e., not readily desorbed). However, they made no attempt to follow changes in fixation with time. More recently, de Koning and Comans (2004) showed that between 44 and 91% of Cs adsorption on illite is not immediately reversible, depending on the saturating cation. With little indication of initial fixation, it is difficult to establish a trend with time. Although some degree of time-dependent fixation may exist, the aging phenomenon is certainly also due to changes in Cs distribution in aggregated soil after nonhomogeneous deposition at the surface. This involves penetration into soil aggregates and slow movement down the soil profile. Modeling studies have illustrated how closely root uptake depends on the relative profiles of roots and contaminant with soil depth (Darrah and Staunton, 2000).

Plant roots take up cesium readily because of its similarity with the essential nutrient, potassium. Plants differ in their accumulation of cesium (Broadley and Willey, 1997; Lasat et al., 1998; Broadley et al., 1999a,b). This may be due to differences in rooting pattern, root uptake, or translocation within the plant, but the cause is not clearly identified. Cesium is mobile within plants and has similar trends in the translocation within plants to K (Broadley and Willey, 1997; Zhu and Smolders, 2000). Rather little is known about differences between species in their capacity to discriminate between Cs and K. There appears to be a threshold below which Cs uptake by roots increases with decreasing K concentration. The value of this threshold has been reported to be 1 mmol L^{-1}, 250 μmol L^{-1}, or 20 μmol L^{-1} (Shaw and Bell, 1989, 1991; Shaw et al., 1992; Buysse et al., 1996; Smolders et al., 1996; Zhu, 2001). This inhibition may be due in part to increased efflux of absorbed Cs (Zhu et al., 1999). Model prediction of Cs uptake often overpredict uptake and are improved if account is take of soil–solution potassium concentration below 1 mM (Smolders et al., 1997; Absolom et al., 1999; Roca-Jove and Vallejo-Calzada, 2000), Cs uptake is found to be enhanced by potassium starvation (Jones et al., 1998; Broadley et al., 1999b; Zhu et al., 2000), but the effect is short-lived (Willey and Martin, 1997; Staunton et al., 2003).

Although most plants in natural and agricultural systems are infected with mycorrhiza, which contribute to their water and nutrient supply, very little is known about the effect of mycorrhizal infection on the uptake and translocation of cesium (Drissner et al., 1998). Rosen et al. (2005) report that arbuscular mycorrhiza enhanced Cs uptake by leek, but had no effect on rye grass. They could not affirm whether the absence of an observed effect for rye grass was due to poor infection or to its greater root density. In contrast, Riesen and Brunner (1996) found that ectomycorrhizal infection reduced Cs uptake. Declerck et al. (2003) demonstrated that mycorrhiza can take up and translocate radiocesium, and possibly accumulate it also. Later the same group showed that radiocesium uptake is lower for mycorrhizal hyphae than for roots (de Boulois et al., 2005). They noted that intraradical fungal structures allowed some local accumulation of Cs, thus reducing its translocation within roots. In contrast, Clint and Dighton (1992) found that mycorrhizal infection led to an increased transfer of Cs to shoots, despite a lowering of overall uptake from soil. Clearly, the effect of mycorrhizal infection on the uptake and translocation of radiocesium from soil is variable and depends on both the soil and the fungus–host partnership. More work is needed on this subject.

Rather more data exist on the uptake of Cs by fruiting bodies of fungi, particularly edible fungi. It is well known that the consumption by animals and humans of fungi grown on contaminated soil is an important pathway for its entry into the food chain (Barnett et al., 1999; Baeza et al., 2005) and that Cs in mushroom is easily digestible (Baeza and Guillen, 2004). The soil–fungus transfer of Cs is highly variable, and no clear trend as to what factors determine the differences has been established (Gillet and Crout, 2000). Inedible fungi also accumulate Cs, and this may influence not only the speciation of Cs, but also led to its redistribution in soil (Shand et al., 1995; Anderson et al., 1997).

There is little evidence of a direct or indirect role or soil bacteria on the dynamics of Cs. Russell et al. (2004) recently reported that bacterial sulfate reduction decreased the adsorption of Cs on arid and tropical soils, but proposed no mechanism. It is unlikely that the accumulation of Cs makes soil microorganisms an important pool for immobilized Cs in mineral soils (because adsorption on soil clays would be much greater). However, this may not be the case in organic soils (Sanchez et al., 2000).

REFERENCES

Aarkrog, A., Dahlgaard, H., and Nielsen, S. P. (1999). Marine radioactivity in the arctic: a retrospect of environmental studies in Greenland waters with emphasis on transport of ^{90}Sr and ^{137}Cs with the East Greenland Current. *Sci. Total Environ.* **237–238**, 143–151.

Abdelouas, A., Fattahi, M., Grambow, B., Vichot, K., and Gautier, E. (2002). Precipitation of technetium by subsurface sulfate-reducing bacteria. *Radiochim. Acta* **80**, 773–777.

Abdelouas, A., Grambow, B., Fattahi, M., Andres, Y., and Leclerc-Cessac, E. (2005). Microbial reduction of Tc-99 in organic matter–rich soils. *Sci. Total Environ.* **336**, 255–268.

Absalom, J. P., Young, S. D. and Crout, N. M. J. (1995). Radiocaesium fixation dynamics: measurement in 6 Cumbrian soils. *Eur. J. Soil Sci.* **46**, 461–469.

Absalom, J. P., Young, S. D., Crout, N. M. J., Nisbet, A. F., Woodman, R. F. M., Smolders, E., and Gillet, A. G. (1999). Predicting soil to plant transfer of radiocesium using soil characteristics. *Environ. Sci. Technol.* **33**, 1218–1223.

AEN (1987). *Les incidences radiologiques de l'accident de Tchernobyl dans les pays de l'OCED*, Agence pour l'Energie Nucléaire–Organisation de Coopération et de Développement Economiques, Paris.

Agapkina, G. I., Tikhomirov, F. A., Shcheglov, A. I., Kracke, W., and Bunzl, K. (1995). Association of Chernobyl-derived $^{239+240}Pu$, ^{241}Am, ^{90}Sr and ^{137}Cs with organic matter in the soil solution. *J. Environ. Radioact.* **29**, 257–269.

Agudo, E. G. (1998). Global distribution of ^{137}Cs inputs for soil erosion and sedimentation studies. In *Use of ^{137}Cs in the Study of Soil Erosion and Sedimentation*, IAEA–TEC DOC-1028, International Atomic Energy Agency, Vienna, Austria, 117–121.

Ambe, F., Ambe, S., Okada, T., and Sekizawa, H. (1986). In situ Mössbauer studies of metal oxide–aqueous solution interfaces with adsorbed cobalt-57 and antimony-119 ions. In *Geochemical Processes at Mineral Surfaces*, ed. Davis, J. A., and Hayes, K. F., Symp. Ser. Vol. 323, American Chemical Society, Washington, DC, 403–424.

Amiro, B. D. (1993). Protection of the environment from nuclear fuel waste radionuclides: a framework using environmental increments. *Sci. Total Environ.* **128**, 157–189.

Anbar, A. D., Roe, J. E., Barling, J., and Nealson, K. H. (2000). Nonbiological fractionation of iron isotopes. *Science* **288**, 126–128.

Anderson, P., Davidson, C. M., Littlejohn, D., Ure, A. M., Shand, C. A., and Cheshire, M. V. (1997). The translocation of caesium and silver by fungi in some Scottish soils. *Commun. Soil Sci. Plant Anal.* **28**, 635–650.

Andersson, K., and Roed, J. (1994). The behaviour of Chernobyl ^{137}Cs, ^{134}Cs and ^{106}Ru in undisturbed soil: implications for external radiation. *J. Environ. Radioact.* **22**, 183–196.

Ashworth, D. J., and Shaw, G. (2005). Soil migration and plant uptake of technetium from a fluctuating water table. *J. Environ. Radioact.* **81**, 155–171.

Avery, S. V. (1995). Cesium accumulation by microorganisms: uptake mechanisms, cation competition, compartmentalization and toxicity. *J. Ind. Microbiol.* **14**, 76–84.

Axe, L., Bunker, G. B., Anderson, P. R., and Tyson, T.A. (1998). An XAFS analysis of strontium at the hydrous ferric oxide surface. *J. Colloid Interface Sci.* **199**, 44–52.

Axe, L., Tyson, T., Trivedi, P., and Morrison, T. (2000). Local structure analysis of strontium sorption to hydrous manganese oxide. *J. Colloid Interface Sci.* **224**, 408–416.

Bachhuber, H., Bunzl, K., and Schimmack, W. (1986). Spatial variability of the distribution coefficients of Cs-137, Zn-65, Sr-85, Co-57, Cd-109, Ce-141, Ru-103, Tc-95m and I-131 in a cultivated soil. *Nucl. Technol.* **72**, 359–371.

Baeza, A., and Guillen, F. J. (2004). Dose due to mushroom ingestion in Spain. *Radiat. Prot. Dosim.* **111**, 97–100.

Baeza, A., Guillen, J., and Bernedo, J. M. (2005). Soil-fungi transfer coefficients: importance of the location of mycelium in soil and of the differential availability of radionuclides in soil fractions. *J. Environ. Radioact.* **81**, 89–106.

Baker, S. J., and Cawse, P. A. (1990). *A Survey of Radioactive Caesium in Soils of Cumbria and North Lancashire: Comparison of Accumulation Pre- and Post-Chernobyl*, UK Atomic Energy Report AEA-EE-0047.

Banuelos, G. S., Ajwa, H. A., Mackey, B., Wu, L., Cook, C., Akohoue, S., and Zambruzuski, S. (1997). Evaluation of different plant species used for phytoremediation of high soil selenium. *J. Environ. Qual.* **26**, 639–646.

Barnett, C. L., Beresford, N. A., Self, P. L., Howard, B. J., Frankland, J. C., Fulker, M. J., Dodd, B. A., and Marriott, J. V. R. (1999). Radiocaesium activity concentrations in the fruit-bodies of macrofungi in Great Britain and an assessment of dietary intake habits. *Sci. Total Environ.* **231**, 67–83.

Battiston, G. A., Degetto, S., Gerbasi, R., and Sbrignadello, G. (1989). Radioactivity in mushrooms in northeast Italy following the Chernobyl accident. *J. Environ. Radioact.* **9**, 53–60.

Bautista, E. M., and Alexander, M. (1972). Reduction of inorganic compounds by soil microorganisms. *Soil Sci. Soc. Am. Proc.* **36**, 918–920.

Beard, B. L., Johnson, C. M., Skulan, J. L., Nealson, K. H., Cox, L., and Sun, H. (2003). Application of Fe isotopes to tracing the geochemical and biological cycling of Fe. *Chem. Geol.* **195**, 87–117.

Bebien, M., Chauvin, J. P., Adriano, J. M., Grosse, S., and Verméglio, A. (2001). Effect of selenite on growth and protein synthesis in the phototrophic bacterium *Rhodobacter sphaeroides*. *Appl. Envir. Microbiol.* **61**, 4440–4447.

Bennett, R., and Willey, N. (2003). Soil availability, plant uptake and soil to plant transfer of Tc-99: a review. *J. Environ. Radioact.* **65**, 215–231.

Beresford, N. A. (1989a). Field observations of ^{110m}Ag, originating from the Chernobyl accident, in west Cumbrian vegetation and soil samples. *J. Radiol. Prot.* **9**, 281–283.

Beresford, N. A. (1989b). The transfer of Ag-110m to sheep tissues. *Sci. Total Environ.* **85**, 81–90.

Bird, G. A., and Schwartz, W. J. (1997). Partition coefficients (K_d values) for Tc in shield lake sediments under oxidizing and reducing conditions. *Water Res.* **31**, 1673–1678.

Blaylock, M. J., and James, B. R. (1994). Redox transformations and plant uptake of selenium resulting from root–soil interactions. *Plant Soil* **158**, 1–12.

Blum, J. S., Bondi, A. B., Buzzelli, J., Stolz, J. F., and Oremland, R. S. (1998). *Bacillus arsenicoselenatis* sp. nov., and *Bacillus selenitireducens*, sp. nov.: two haloalkalophiles from Mono Lake, California, which respire oxyanions of selenium and arsenic. *Arch. Microbiol.* **171**, 19–30.

Bodek, I., Lyman, W. J., Reehl, R. F., and Rosenblatt, D. H. (1988). *Environmental Inorganic Chemistry Properties, Processes, and Estimation Methods*, Pergamon Press, New York, 7.1-1 to 7.1-5.

Bourg, I.,C., Bourg, A. C. M., and Sposito, G. (2003). Modeling diffusion and adsorption in compacted bentonite: a critical review. *J. Contam. Hydrol.* **61**, 293–302.

Bowen, H. J. M. (1966). *Trace Elements in Biochemistry*, Academic Press, London.

Bowen, H. J. M. (1979). *Environmental Chemistry of the Elements*, Academic Press, London.

Brady, P. V., Jové-Colon, C. F., Carr, G., and Huang, F. (2002). Soil radionuclide plumes. In *Geochemistry of Soil Radionuclides*, Soil Science Society of America, Spec. Publ. 59, Madison, WI, 165–190.

Broadley, M. R., and Willey, N. J. (1997). Differences in root uptake of radiocaesium by 30 plant taxa. *Environ. Pollut.* **97**, 11–15.

Broadley, M. R., Willey, N. J., and Mead, A. (1999a). A method to assess taxonomic variation in shoot caesium concentration among flowering plants. *Environ. Pollut.* **106**, 341–349.

Broadley, M. R., Willey, N. J., Philippidis, C., and Dennis, E. R. (1999b). A comparison of caesium uptake kinetics in eight species of grass. *J. Environ. Radioact.* **46**, 225–236.

Brooks, R. R. (1972). *Geobotany and Biogeochemistry in Mineral Exploration*, Harper & Row, New York.

Brooks, R. R., McCleave, J. A., and Malaisse, F. (1977). Copper and cobalt in African species of *Crotalaria* L. *Proc. R. Soc.* ***B*** **197**, 231–236.

Brouwer, E., Baeyens, B., Maes, A., and Cremers, A. (1983). Cesium and rubidium ion equilibria in illite clay. *J. Phys. Chem.* **87**, 1213–1219.

Burke, I. T., Boothman, C., Mortimer, R. J. G., Lloyd, J. R., Livens, F. R., and Morris, K. (2005). Effects of progressive anoxia on the solubility of technetium in sediments. *Environ. Sci. Technol.* **39**, 4109–4116.

Buysse, J., van den Brande, K., and Merckx, R. (1996). Genotypic differences in the uptake and distribution of radiocaesium in plants. *Plant Soil* **178**, 265–171.

Carini, F., and Lombi, E. (1997). Foliar and soil uptake of Cs-134 and Sr-85 by grape vines. *Sci. Total Environ.* **207**, 157–164.

Cataldo, D. A., Wildung, R. E., and Garland, T. R. (1983). Root absorption and transport behavior of technetium in soybean. *Plant Physiol.* **73**, 849–852.

Chasteen, T. G. (1998). Volatile chemical species of selenium, In *Environmental Chemistry of Selenium*, ed. Frankenberger, W.T., Jr., and Engberg, R. A., Marcel Dekker, New York, 589–612.

Chen, C.-C., and Hayes, K. F. (1999). X-ray absorption spectroscopy investigation of aqueous Co(II) and Sr(II) sorption at clay–water interfaces. *Geochim. Cosmochim. Acta* **19–20**, 3205–3215.

Clint, G. M., and Dighton, J. (1992). Uptake and accumulation of radiocesium by mycorrhizal and non mycorrhizal heather plants. *New Phytol.* **121**, 555–561.

Comans, R. N. J., Middelburg, J. J., Zonderhuis, J., Woiyyiez, J. R. W., de Lange, G. J., van Das, H. A., and Weijden, C. H. (1989). Mobilization of radiocaesium in pore water of lake sediments. *Nature* **339**, 367–369.

Comans, R. N. J., Haller, M., and de Preter, P. (1991). Sorption of cesium on illite: non-equilibrium behaviour and reversibility. *Geochim. Cosmochim. Acta* **55**, 433–440.

Combs, G. F., Jr., and Combs, S. B. (1986). *The Role of Selenium in Nutrition*, Academic Press, London.

Cornell, R. M. (1993). Adsorption of cesium on minerals: a review. *J. Radioanal. Nucl. Chem.* **171**, 483–500.

Coughtrey, P. J., and Thorne, M. C. (1983a). *Radionuclide Distribution and Transport in Terrestrial and Aquatic Ecosystems: A Critical Review of Data*, Vol. 1, A.A. Balkema, Rotterdam, The Netherlands.

Coughtrey, P. J., and Thorne, M. C. (1983b). *Radionuclide Distribution and Transport in Terrestrial and Aquatic Ecosystems: A Critical Review of Data*, Vol. 2, A.A. Balkema, Rotterdam, The Netherlands.

Coughtrey, P. J., Jackson, D., and Thorne, M. C. (1983). R*adionuclide Distribution and Transport in Terrestrial and Aquatic Ecosystems: A Critical Review of Data*, Vol. 3, A.A. Balkema, Rotterdam, The Netherlands.

CPEP (Contemporary Physics Education Project) (2003). Radiation in the environment. In *Nuclear Science: A guide to the Nuclear Science Wall Chart*, *www.CREPweb.org*, 9 pp.

Cremers, A., Elsen, A., de Preter, P., and Maes, A. (1988). Quantitative analysis of radioacesium retention in soils. *Nature* **335**, 247–249.

Cross, M. A., Smith, J. T., Saxèn, R., and Timms, D. (2002). An analysis of the environmental mobility of radiostrontium from weapons testing and Chernobyl in Finnish river catchments. *J. Environ. Radioact* **60**, 149–163.

Darrah, P. R., and Staunton, S. (2000). A mathematical model of root uptake incorporating root turnover, distribution within the plant and recycling of absorbed species.*Eur. J. Soil Sci.* **51**, 643–653.

de Boulois, H. D., Delvaux, B., and Declerck, S. (2005). Effects of arbuscular mycorrhizal fungi on the root uptake and translocation of radiocaesium. *Environ. Pollut.* **134**, 515-524.

Declerck, S., Dupre de Boulois, H., Bivort, C., and Delvaux, B. (2003). Extraradical mycelium of the arbuscular mycorrhizal fungus *Glomus lamellosum* can take up, accumulate and translocate radiocaesium under root–organ culture conditions. *Environ. Microbiol.* **5**, 510–516.

de Koning, A., and Comans, R. N. J. (2004). Reversibility of radiocesium sorption on illite. *Geochim. Cosmochim. Acta* **68**, 2815–2823.

Deneux-Mustin, S., Roussel-Debet, S., Mustin, Ch., Munier-Lamy, C., Berthelin, J., Garnier-Laplace, J., and Leyval, C. (2003). *Mobilité et transfert racinaire des éléments en traces : influence des micro-organismes du sol*, Editions Tec & Doc, Lavoisier, Paris.

Denys, S., Echevarria, G., Florentin, L., Leclerc-Cessac, E., and Morel, J.-L. (2003). Availability of 99Tc in undisturbed soil cores. *J. Environ. Radioact.* **70**, 115–126.

Desmet, G., and Myttenaere, C. (1986). *Technetium in the Environment*, Elsevier Applied Science, London.

Doran, J. W. (1982). Microorganisms and the biological cycling of selenium. *Adv. Microb. Ecol.* **6**, 1–32.

Doran, J. W., and Alexander, M. (1977). Microbial formation of volatile selenium compounds in soil. *Soil Sci. Soc. Am. J.* **41**, 70–73.

Dowdle, P. R., and Oremland, R. S. (1998). Microbial oxidation of elemental selenium in soil slurries and bacterial cultures. *Environ. Sci. Technol.* **32**, 3749–3755.

Drissner, J., Bürmann, W., Enslin, F., Heider, R., Klemt, E., Miller, R., Schick, G., and Zibold, G. (1998). Availability of caesium radionuclides to plants: classification of soils and role of mycrorrhiza. *J. Environ. Radioact.* **41**, 19–32.

D'Souza, T. J., and Mistry, K. B. (1980). Behaviour of gamma-emitting fission products ^{106}Ru, ^{125}Sb, ^{134}Cs and ^{144}Ce deposited on pastures in tropical environs. *J. Nucl. Agric. Biol.* **9**, 50–53.

Dumat, C., and Staunton, S. (1999). Reduced adsorption of caesium on clay minerals caused by various humic substances. *J. Environ. Radioact.* **46**, 187–200.

Dumat, C., Cheshire, M. V., Fraser, A., Shand, C., and Staunton, S. (1997). The effect of removal of soil organic matter and iron on the adsorption of radiocaesium. *Eur. J. Soil Sci.* **48**, 675–683.

Dumat, C., Quiquampoix, H., and Staunton, S. (2000). Adsorption of caesium by synthetic clay–organic matter complexes: effect of the nature of organic polymers.*Environ. Sci. Technol.* **34**, 2985–2989.

Dungan, R. S., and Frankenberger, W. T., Jr. (1999). Microbial transformations of selenium and the bioremediation of seleniferous environments. *Bioremediat.* **3**, 171–188.

Dungan, R. S., Yates, S. R., and Frankenberger, W. T., Jr. (2002). Volatilization and degradation of soil-applied dimethylselenide. *J. Environ. Qual.* **31**, 2045–2050.

Dungan, R. S., Yates, S. R. and Frankenberger, W. T., Jr. (2003). Transformations of selenate and selenite by *Stenotrophomonas maltophilia* isolated from a seleniferous agricultural drainage pond sediment. *Environ. Microbiol.* **5**, 287–295.

Echevarria, G., Vong, P. C., Leclerc-Cessac, E., and Morel, J. L. (1997). Bioavailability of technetium-99 as affected by plant species and growth, application form, and soil incubation. *J. Environ. Qual.* **26**, 947–956.

Echevarria, G., Vong, P. C., and Morel, J. L. (1998). Effect of NO_3^- on the fate of $^{99}TcO_4^-$ in the soil–plant system, *J. Environ. Radioact.* **38**, 163–171.

Ehlken, S., and Kirchner, G. (2002). Environmental processes affecting plant root uptake of radioactive trace elements and variability of transfer factor data: a review. *J. Environ. Radioact.* **58**, 97–112.

Elrashidi, M. A., Adriano, D. C., Workman, S. M., and Lindsay, W. L. (1987). Chemical equilibria of selenium in soils: a theoretical development. *Soil Sci.* **144**, 141–152.

Entry, J. A., Watrud, L. S., and Reeves, M. (2001). Influence of organic amendments on the accumulation of ^{137}Cs and ^{90}Sr from contaminated soil by three grass species. *Water Air Soil Pollut.* **126**, 385–398.

Essington, E., Nishita, H., and Wallace, A. (1963). Effect of chelating agents on the uptake of Y-91, Ce-144, Ru-106, and Pr-147 by beans grown in a calcareous soil. *Soil Sci.* **95**, 331–337.

Evans, C. H., Alberts, J. J., and Clark, R. A. (1983). Reversible ion-exchange fixation of caesium-137 leading to mobilization from reservoir sediments. *Geochim. Cosmochim. Acta* **47**, 1041–1049.

Francis, A. J., Dodge, C. J., and Meinken, G. E. (2002). Biotransformation of pertechnate by *Clostridia*. *Radiochim. Acta* **90**, 791–797.

Frankenberger, W. T., Jr., and Karlson, U. (1989). Environmental factors affecting microbial production of dimethylselenide in a selenium-contaminated sediment. *Soil Sci. Soc. Am. J.* **53**, 1435–1442.

Frissel, M. J., Noordijk, H., and Van Bergeijk, K. E. (1990). The impact of extreme environmental conditions, as occurring in natural ecosystems, on the soil-to-plant transfer of radionuclides. In, *Transfer of Radionuclides in Natural and Semi-natural Environments*, ed. Desmet, G., Nassimbeni, P., and Belli, M., Elsevier, London, pp. 40–47.

Fuentes, E., Pinochet, H., De Gregori, I., and Potin-Gautier, M. (2003). Redox speciation analysis of antimony in soil extracts by hydride generation atomic fluorescence spectrometry. *Spectrochim. Acta* B **58**, 1279–1289.

Gadd, G. M. (1996). Influence of microorganisms on the environmental fate of radionuclides. *Endeavour* **20**, 150–156.

Garbisu, C., Gonzalez, S., Yang, W. H., Yee, B. C., Carlson, D. L., Yee, A., Smith, N. R., Otero, R., Buchanan, B. B., and Leighton, T. (1995). Physiological mechanisms regulating the conversion of selenite to elemental selenium by *Bacillus subtilis*. *Biofactors* **5**, 29–37.

Garbisu, C., Ishii, T., Leighton, T., and Buchanan, B. B. (1996). Bacterial reduction of selenite to elemental selenium. *Chem. Geol.*132, 199–204.

Garnham, G. W., Codd, G. A., and Gadd, G. M. (1993). Accumulation of technetium by cyanobacteria. *J. Appl. Phycol.* **5**, 307–315.

Geraedts, K., Bruggeman, C., Maes, A., Van Loon, L. R., Rossberg, A., and Reich, T. (2002). Evidence for the existence of Tc(IV): humic substance species by x-ray absorption near-edge spectroscopy. *Radiochim. Acta* **90**, 879–884.

Gerzabek, M. H., Horak, O., Mohamad, S., and Muck, K. (1993). *Gefässversuche zum Radionuklidtransfer Boden-Planze, Fusionreaktor–Erste Wand*, Rep. OEFZS A-2531, Österreichisches Forschungszentrum Seibersdorf, 22–31.

Gillett, A. G., and Crout, N. M. J. (2000). A review of Cs-137 transfer to fungi and consequences for modelling Environmental transfer. *J. Environ. Radioact.* **48**, 95–121.

Guivarch, A., Hinsinger, P., and Staunton, S. (1999). Root uptake and distribution of radiocaesium from contaminated soils and the enhancement of Cs adsorption in the rhizosphere. *Plant Soil* **211**, 131–138.

Guo, L., Frankenberger, W. T. and Jury, W. A. (1999). Adsorption and degradation of dimethyl selenide in soil. *Environ. Sci. Technol.* **33**, 2934–2938.

Guo, L., Jury, W. A. and Frankenberger, W.T., Jr. (2000). Characterizing kinetics of sequential selenium transformation in soil. *J. Environ. Qual.* **29**, 1041–1048.

Hampton, C. R., Bowen, H. C., Broadley, M. R., Hammond, J. P., Mead, A., Payne, K. A., Pritchard, J. and White, P. J. (2004). Cesium toxicity in *Arabidopsis*. *Plant Physiol.* **136**, 3824–3837.

Handl, J., Kallweit, E., Henning, M., and Szwec, L. (2000). On the long-term behaviour of ^{110m}Ag in the soil–plant system and its transfer from feed to pig. *J. Environ. Radioact.* **48**, 159–170.

Hird, A. B., Rimmer, D. L., and Livens, F. R. (1995). Total caesium-fixing potentials of acid organic soils. *J. Environ. Radioact.* **26**, 103–118.

IAEA (1994). *Handbook of Parameter Values for the Prediction of Radionuclide Transfer in Temperate Environments*, Tech. Rep. Seri. 364, International Atomic Energy Agency, Vienna, Austria.

IAEA (1996). *Present and Future Environmental Impact of the Chernobyl Accident*, TECDOC-1240, International Atomic Energy Agency, Vienna, Austria.

IAEA (2001). *International Basic Safety Standards for Protection Against Ionizing Radiation and for the Safety of Radioactive Sources,* Saf. Seri. 115, International Atomic Energy Agency, Vienna, Austria.

Ihnat, M. (1989). *Occurence and Distribution of Selenium,* CRC Press, Boca Raton, FL.

Ishii, N., and Uchida, S. (2005). Gram-negative bacteria responsible for insoluble technetium formation and the fate of insoluble Tc in the water column above flooded paddy soil. *Chemosphere* **60**, 157–163.

IUR (1989). *VIth Report of the Working Group Soil-to-Plant Transfer Factors,* report of the working group meeting in Guttannen, Grimsellpass, Switzerland, May 24–26, 1989, ed. Frissel, M. J., RIVM, Bilthoven, The Netherlands.

Jayaweera, G. R. and Biggar, J. W. (1996). Role of redox potential in chemical transformations of selenium in soils. *Soil Sci. Soc. Am. J.***60**, 1056–1063.

Jones, R. F. (1960). The accumulation of nitrosyl ruthenium by fine particles and marine organisms. *Limnol. Oceanogr.* **5**, 312–325.

Jones, D. R., Eason, W. R., and Dighton, J. (1998). Investigation of spatial and temporal patterns of Cs-137 partitioning in *Eriophorum vaginatum* L. in relation to its nutrient retrieval and storage strategy. *J. Environ. Radioact.* **40**, 271–288.

Kabata-Pendias, A., and Pendias, H. (1984). *Trace Elements in Soils and Plants*, CRC Press, Boca Raton, FL.

Karasyova, O. N., Ivanova, L. I., Lakshtanov, L. Z., and Lovgren, L. (1999). Strontium sorption on hematite at elevated temperatures. *J. Colloid Interface Sci.* **220**, 419–428.

Karlson, U., and Frankenberger, W. T., Jr. (1988). Determination of gaseous selenium-75 evolved from soil. *Soil Sci. Soc. Am. J.* **52**, 678–681.

Keith-Roach, M. J., and Livens F. R. (2002). *Interactions of Microorganisams with Radionuclides*, Elsevier, Amsterdam, The Netherlands.

Keith-Roach, M. J., and Roos, P. (2004). Redox-dependent behaviour of technetium-99 entering a permanently stratified anoxic fjord (Framvaren fjord, Norway). *Estuar. Coast. Shelf Sci.* **60**, 151–161.

Keith-Roach, M. J., Morris, K., and Dahlgaard, H. (2003). An investigation into technetium binding in sediments. *Mari. Chem.* **81**, 149–162.

Kinniburgh, D. G., Syers, J. K., and Jackson, M. L. (1975). Specific adsorption of trace amounts of calcium and strontium by hydrous oxides of iron and aluminium. *Soil Sci. Soc. Am. J.* **39**, 464–470.

Konoplev, A. V., Viktorova, N. V., Virchenko, E. P., Popov, V. E., Bulgakov, A. A., and Desmet, G. M. (1993). Influence of agricultural countermeasures on the ratio of different chemical forms of radionuclides in soil and soil solution. *Sci. Total Environ.* **137**, 147–162.

Krouglov, S. V., Kurinov, A. D., and Alexakhin, R. M. (1998). Chemical fractionation of ^{90}Sr, ^{106}Ru, ^{137}Cs and ^{144}Ce in Chernobyl-contaminated soils: an evolution in the course of time. *J. Environ. Radioact.* **38**, 59–76.

Lakanen, E., and Paasikallio, A. (1970). The effects of soil fractors on the uptake of radiostrontium by plants. *Ann. Agric. Fenn.* **9**, 133–138.

Lakin, H. K. (1972). Selenium accumulation in soils and its absorption by plants and animals. *Geol. Soc. Am. Bull.* **83**,181–190.

Landa, E. R., Thorvig, L. H., and Gast, R. G. (1977). Effect of selective dissolution, electrolytes, aeration, and sterilization on technetium-99 sorption by soils, *J. Environ. Qual.* **6**, 181–187.

Lasat, M. M., Furhmann, M., Ebbs, S. D., Cornish, J. E., and Kochian, L. V. (1998). Phytoremediation of a radiocesium-contaminated soil: evaluation of cesium-137 bioaccumulation in the shoots of three plant species. *J. Environ. Qual.* **27**, 165–169.

LBNL Isotopes Project (2004). *WWW Table of Radioactive Isotopes*, Lund University, Lund, Sweden. http://ie.lbl.gov/toi/.

Lee, M. H., and Lee, C. W. (2000). Association of fallout-derived 137Cs, 90Sr and 239,240Pu with natural organic substances in soils. *J. Environ. Radioact.* **47**, 253–262.

Lepp, N. W., Edwards, R., and Jones, K. C. (1995). Other less abundant elements of potential environmental significance. In *Heavy Metals in Soils*, 2nd ed., ed. Alloway, B. J., Blackie Academic and Professional, London 306–352.

Lin, Z. Q., and Terry, N. (2003). Selenium removal by constructed wetlands: quantitative importance of biological volatilization in the treatment of selenium-laden agricultural drainage water. *Environ. Sci. Technol.* **37**, 606–615.

Lindsay, W. L. (1974). *Chemical Equilibria in Soils*, Wiley, New York.

Lisk, D. J. (1972). Trace metals in soils, plants and animals. *Adv. Agron.* **24**, 267–325.

Liszewski, M. J., Bunde, R. L., Hemming, C., Rosentreter, J., and Welhan, J. (1998). The use of synthesized aqueous solutions for determining strontium sorption isotherms. *J. Contam. Hydrol.* **29**, 93–108.

Liu, C. X., Zachara, J. M., Smith, S. C., McKinley, J. P., and Ainsworth, C. C. (2003). Desorption kinetics of radiocesium from subsurface sediments at Hanford site, USA. *Geochim. Cosmochim. Acta* **67**, 2893–2912.

Livens, F. R., and Baxter, M. S. (1988). Chemical associations of artificial radionuclides in Cumbrian soils. *J. Environ. Radioact.* **7**, 75–86.

Livens, F. R., and Loveland, P. J. (1988). The influence of soil properties on the environmental mobility of caesium in Cumbria. *Soil Use Manage* **4**, 69–75.

Lloyd, J. R., and Macaskie, L. E. (1996). A novel PhosphorImager-based technique for monitioring the microbial reduction of technetium. *Appl. Environ. Microbiol.* **62**, 578–582.

Lloyd, J. R., Cole, J. A., and Macaskie, L. E. (1997). Reduction and removal of heptavalent technetium from solution by *Escherichia coli*. *J. Bacteriol.* **179**, 2014–2021.

Lloyd, J. R., Nolting, H.-F., Solé, V. A., Bosecker, K. and Macaskie, L. E. (1998). Technetium reduction and precipitation by sulfate-reducing bacteria. *Geomicrobiol J.* **15**, 45–58.

Lloyd, J. R., Thomas, G. H., Finlay, J. A., Cole, J. A., and Macaskie, L. E. (1999). Microbial reduction of technetium by *Escherichia coli* and *Desulfovibric desulfuricans:* enhancement via the use of high-activity strains and effet of process parameters. *Biotechnol. Bioeng.* **66**, 122–130.

Lloyd, J. R., Sole, V. A., Van Praagh, C. V. G. and Lovley, D. R. (2000). Direct and Fe(II)-mediated reduction of technetium by Fe(III)-reducing bacteria. *Appl. Environ. Microbiol.* **66**, 3743–3749.

Lloyd, J. R., Chesnes, J., Glasauer, S., Bunker, D. L., Livens, F. R., and Lovley, D. R. (2002). reduction of actinides and fission products by Fe(III)-reducing bacteria. *Geomicrobiol J.* **19**, 103–120.

Lortie, L., Gould, W. D., Rajan, S., McCready, R. G. L., and Cheng, K. J. (1992). Reduction of selenate and selenite to elemental selenium by *Pseudomonas stutzeri* isolate. *Appl. Environ. Microbiol.* **58**, 4042–4044.

Losi, M. E., and Frankenberger, W. T., Jr. (1997). Reduction of selenium oxyanions by *Enterobacter cloacae* SLD1a-1: isolation and growth of the bacterium and its expulsion of selenium particles. *Appl. Environ. Microbiol.* **63**, 3079–3084.

Losi, M. E., and Frankenberger W. T., Jr. (1998). Reduction of selenium oxyanions by *Enterobacter cloacas* strain SLD I a-1. In *Environmental Chemistry of Selenium,* ed. Frankenberger, W. T., Jr., and Engberg, R. A., Marcel Dekker, New York, 515–544.

Lovley, D. R. (1993). Dissimilatory metal reduction. *Annu. Rev. Microbiol.* **47**, 263–290.

Lumbroso, H. (1977). *Compléments au nouveau traité de chimie minérale. Tone 7; Sélénium*, Masson, Paris.

Lux, D., Kammerer, L., Ruhm, W., and Wirth, E. (1995). Cycling of Pu, Sr, Cs, and other longliving radionuclides in forest ecosystems of the 30km zone around Chernobyl. *Sci. Total Environ.* **173–174**, 375–384.

Macy, J. M., Rech, S., Auling, G., Dorsch, E., Stackebrandt, E., and Sly, L. I. (1993). *Thauera selenatis* gen. nov., sp. nov., a member of the beta subclass of Proteobacteria with a novel type of anaerobic respiration. *Int. J. Syst. Bacteriol.* **43**, 135–142.

Madoz-Escande, C, Garcia-Sanchez, L., Bonhomme, T., and Morello, M. (2005). Influence of rainfall characteristics on elimination of aerosols of cesium, strontium, barium and tellurium deposited on grassland. *J. Environ. Radioact.* **84**, 1–20.

Maes, E., Delvaux, B., and Thiry, Y. (1998). Fixation of radiocaesium in an acid brown forest soil. *Eur. J. Soil Sci.* **49**, 133–140.

Maes, E., Vielvoye, L., Stone, W., and Delvaux, B. (1999). Fixation of radiocaesium traces in a weathered sequence mica → vermiculite → hydroxy interlayered vermiculite. *Eur. J. Soil Sci.* **50**, 107–115.

Maes, A., Bruggeman, C., Geraedts, K., and Vancluysen, J. (2003). Quantification of the interaction of Tc with dissolved boom clay humic substances. *Environ. Sci. Technol.* **37**, 747–753.

Maréchal, C. N., and Albarède, F. (2002). Ion-exchange fractionation of copper and zinc isotopes. *Geochim. Cosmochim. Acta* **66**, 1499–1509.

Maréchal, C. N., Télouk, P., and Albarède, F. (1999). Precise analysis of copper and zinc isotopic compositions by plasma-source mass spectrometry. *Chem. Geol.* **156**, 251–273.

Martens, D. A., and Suarez, D. L. (1999). Transformations of volatile methylated selenium in soil. *Soil Biol. Biochem.* **31**, 1355–1361.

Masscheleyn, P. H., DeLaune, R. D., and Patrick, W. H., Jr. (1991). Arsenic and selenium chemistry as affected by sediment redox potential and pH. *J. Environ. Qual.* **20**, 522–527.

McKinley, J. P., Zachara, J. M., Heald, S. M., Dohnalkova, A., Newville, M. G., and Sutton, S. R. (2004). Microscale distribution of cesium sorbed to biotite and muscovite. *Environ. Sci. Technol.* **38**, 1017–1023.

Meychik, N. R., and Yermakov, I. P. (1999). A new approach to the investigation on the tonogenic groups of root cell walls. *Plant Soil* **217**, 257–264.

Mistry, K. B., Bhujbal, B. M., and D'Souza, T. J. (1973). *Influence of Agronomic Practices on Uptake of Fission Products by Crops from Soils of Regions Adjoining Nuclear Industry*, STI-PUB-345. IAEA, Vienna, Austria, 303–320.

Mitchell, R. L., Reith, J. W. S., and Johnson, I. M. (1957). Trace-element uptake in relation to soil content. *J. Sci. Food Agric.* **8**, 51.

Morgan, J. E., Richards, S. P. G., and Morgan, A. J. (2001). Stable strontium accumulation by earthworms: a paradigm for radiostrontium interactions with its cationic analogue, calcium. *Environ. Toxicol. Chem.* **20**, 1236–1243.

Murphy, C. E., and Johnson, T. L. (1993). Vegetative uptake of Tc-99 from buried, solidified, low-level radioactive waste. *J. Environ. Qual.* **22**, 793–799.

Nakamaru, Y., Tagami, K., and Uchida, S. (2006). Antimony mobility in Japanese agricultural soils and the factors affecting antimony sorption behaviour. *Environ. Pollut.* **141**, 321–326.

Neal, R. H. (1995). Selenium. In *Heavy Metals in Soils,* 2nd ed., ed. Alloway, B. J., Blackie Academic Professional, London, 260–283.

Neal, R. H., and Sposito, G. (1989). Selenate adsorption on alluvial soils. *Soil Sci. Soc. Am. J.* **53**, 70–74.

Neal, R. H., Sposito, G., Holtzclaw, K. M., and Traina, S. J. (1987a). Selenite adsorption on alluvial soils, I: Soil composition and pH effects. *Soil Sci. Soc. Am. J.* **51**, 1161–1165.

Neal, R. H., Sposito, G., Holtzclaw, K. M. and Traina, S. J. (1987b). Selenite adsorption on alluvial soils, II: Solution composition effects. *Soil Sci. Soc. Am. J.***51**, 1165–1169.

Ng, Y. C., Colsher, C. S., and Thomson, S. E. (1982). *Soil-to-Plant Concentration Factors for Radiological Assessments*, Rep. NUREG/CR-2975 UCID-19463, Lawrence Livermore National Laboratory, Berkelay, CA.

Nisbet, A. F., and Woodman, R. F. M. (2000). Soil-to-plant transfer factors for radiocesium and radiostrontium in agricultural systems. *Health Phys.* **78**, 279–288.

Noordijk, H., van Bergeijk, K. E., Lembrechts, J., and Frissel, M. J. (1992). Impact of ageing and weather conditions on soil-to-plant transfer of radiocesium and radiostrontium. *J. Environ. Radioact.* **15**, 277–286.

Ochs, M., Lothenbach, B., Shibata, M., Sato, H., and Yui, M. (2003). Sensitivity analysis of radionuclide migration in compacted bentonite: a mechanistic model approach. *J. Contam. Hydrol.* **61**, 313–328.

Ohnuki, T. (1994). Sorption characteristics of strontium on sandy soils and their components. *Radiochim. Acta* **64**, 237–245.

Ohnuki, T., and Kozai, N. (1994). Sorption characteristics of radioactive cesium and strontium on smectite. *Radiochim. Acta* **66**, 327–331.

Oremland, R. S., and Zehr, J. P. (1986). Formation of methane and carbon dioxide from dimethylselenide in anoxic sediments and by a methanogenic bacterium. *Appl. Environ. Microbiol.***52**, 1031–1036.

Oremland, R. S., Blum, J. S., Culbertson, C. W., Visscher, P. T., Miller, L. G., Dowdle, P., and Strohmaier, F. E. (1994). Isolation, growth, and metabolism of an obligately anaerobic, selenate-respiring bacterium, strain SES-3. *Appl. Environ. Microbiol.* **60**, 3011–3019.

Oremland, R. S., Blum, J. S., Bindi, A. B., Dowdle, P. R., Herbel, M., and Stolz, J. F. (1999). Simultaneous reduction of nitrate and selenate by cell suspensions of selenium-respiring bacteria. *Appl. Environ. Microbiol.* **65**, 4385–4392.

Ouzounian, G. (1996). Les déchets radioactifs en France: situation et perspectives. *Rev. Gen. Nucl.* **4**, 71–80.

Peterson, P. J., and Girling, C. A. (1981). Other trace metals. In *Effect of Heavy Metal Pollution on Plants*, Vol. I, ed. Lepp, N. W., Applied Science Publishers, London, 213–278.

Pilon-Smits, E. A. H., de Souza, M. P., Hong, G., Amini, A., Bravo, R. C., Payabyab, S. T., and Terry, N. (1999). Selenium volatilization and accumulation by twenty aquatic plant species. *J. Environ. Qual.* **28**, 1011–1018.

Poinssot, C., Baeyens, B., and Bradbury, M. H. (1999). Experimental and modelling studies of caesium sorption on illite. *Geochim. Cosmochim. Acta* **63**, 3217–3227.

Prasad, M. N. V., and Freitas, H. M. D. (2003). Metal hyperaccumulation in plants: biodiversity prospecting for phytoremediation technology. *Electron. J. Biotechnol.* **6**, 285–321.

Rajec, P., Sucha, V., Eberl, D. D., Srodon, J., and Elsass, F. (1999). Effect of illite particle shape on cesium sorption. *Clays Clay Miner.* **47**, 755–760.

Riesen, T. K., and Brunner, I. (1996). Effect of ectomycorrhizae and ammonium on Cs-134 and Sr-85 uptake into *Picea abies* seedlings. *Environ. Pollut.* **93**, 1–8.

Rigol, A., Roig, M., Vidal, M., and Rauret, G (1999). Sequential extractions for the study of radiocesium and radiostrontium dynamics in mineral and organic soils from western Europe and Chernobyl areas. *Environ. Sci. Technol.* **33**, 887–895.

Rigol, A., Vidal, M., and Rauret, G. (2002). An overview of the effect of organic matter on soil-radiocaesim interaction: implications in root uptake. *J. Environ. Radioact.* **58**, 191–216.

Ritchie, J. C., and McHenry, J. R. (1990). Application of radioactive fallout cesium-137 for measuring soil erosion and sediment accumulation rates and patterns: a review. *J. Environ. Qual.* **19**, 215–233.

Roca-Jove, M. C., and Vallejo-Calzada, V. R. (2000). Prediction radiocesium root uptake based on potassium parameters: a mechanistic approach. *Plant Soil* **222**, 35–49.

Ronneau, C., Cara, J., and Rimski-Korsakov, A. (1995). Oxidation-enhanced emission of ruthenium from nuclear fuel. *J. Environ. Radioact.* **26**, 63–70.

Rosen, K., Zhong, W. L., and Martensson, A. (2005). Arbuscular mycorrhizal fungi mediated uptake of Cs-137 in leek and ryegrass. *Sci. Total Environ.* **338**, 283–290.

Rosenfeld, I., and Beath, O. A. (1964). *Selenium Geobotany, Biochemistry, Toxicity and Nutrition*, Academic Press, London.

Rössler, D., Franke, K., and Suss, R. (2000). Synthesis and chromatographic characterization of [Tc-99m]technetium–humic acid species. *Radiochim. Acta* **88**, 95–100.

Routson, R. C., Hansen, G., and Robinson, A. V. (1977). ^{241}Am, ^{237}Np, and ^{99}Tc sorption on two United States subsoils from different weathering intensity areas. *Health Phys.* **33**, 311–317.

Roux, M., Sarret, G., Pignot-Paintrand, I., Fontecave, M., and Coves, J. (2001). Mobilization of selenite by *Ralstonia metallidurans* CH34. *Appl. Environ. Microbiol.* **67**, 769–773.

Russell, R. A., Holden, P. J., Payne, T. E., and McOrist, G. D. (2004). The effect of sulfate-reducing bacteria on adsorption of ^{137}Cs by soils from arid and tropical regions. *J. Environ. Radioact.* **74**, 151–158.

Sahai, N., Carroll, S. A., Roberts, S., and O'Day, P. A. (2000). X-ray absorption spectroscopy of strontium(II) coordination, II: Sorption and precipitation at kaolinite, amorphous silica, and goethite surfaces. *J. Colloid Interface Sci.* **222**, 198–212.

Sanchez, A. L., Parekh, N. R., Dodd, B. A., and Ineson, P. (2000). Microbial component of radiocaesium retention in highly organic soils. *Soil Biol. Biochem.* **32**, 2091–2094.

Sanchez, A. L., Smolders, E., van en Brande, K., Merckx, R., Wright, S. M., and Naylor, C. (2002). Predictions of in situ solid/liquid distribution of radiocaesium in soils. *J. Environ. Radioact.* **63**, 35–47.

Sarathchandra, S. U., and Watkinson, J. H. (1981). Oxidation of elemental selenium to selenite by *Bacillus megaterium*. *Science* **211**, 600–601.

Savechenko, V. K. (1995). *The Ecology of the Chernobyl Catastrophe: Scientific Outlines of an International Programme of Collaborative Research,* Parthenon Publishing Groups, Paris.

Sawhney, B. L. (1972). Selective sorption and fixation of cations by clay minerals: a review. *Clays Clay Miner.* **20**, 93–100.

Schimmack, W., Bunzl, K., and Klotz, D. (1993). Infiltration of radionuclides with high mobility (Br-82, and ^{95m}Tc) into a forest soil: effect of the irrigation intensity. *Sci. Total Environ.* **138**, 169–181.

Schulte, E. H., and Scoppa, P. (1987). Sources and behavior of technetium in the environment. *Sci. Total Environ.* **64**, 163–179.

Schultz, R. K. (1965). Soil chemistry of radionuclides. *Health Phys.* **11**, 1317–1324.

Shand, C. A., Cheshire, M. V., Smith, S., Campbell, C. D., Anderson, P., Davidson, C. M., Littlejohn, D., and Jamieson, N. (1995). Radiocesium in an organic soil and the effect of treatment with the fungicide Captan. *Plant Soil* **170**, 315–322.

Shaw, G., and Bell, J. N. B. (1989). The kinetics of cesium absorption by roots of winter wheat and the possible consequences for the derivation of soil-to-plant transfer factors for radiocesium.*J. Environ. Radioact.* **10**, 213–232.

Shaw, G., and Bell, J. N. B. (1991). Competitive effects of potassium and ammonium on cesium uptake kinetics in wheat. *J. Environ. Radioact.* **13**, 283–296.

Shaw, G., Bell, J. N. B., Minski, M. J., and Nair, S. (1990). *Experimental Studies of Radionuclide Uptake and Retention by Crops Under UK Field Conditions*, final report of research contract undertaken by the Imperial College Reactor Centre (Silwood Park, Ascot) for Berkeley Nuclear Laboratories (CEGB, Berkeley, CA), Gloucestershire, England.

Shaw, G., Hewamanna, R., Lillywhite, J., and Bell, J. N. B. (1992). Radiocaesium uptake and translocation in wheat with references to the transfer factor concept and ion competition effects. *J Environ. Radioact.* **16**, 167–180.

Sheppard, M. I., and Thibault, D. H. (1990). Default soil solid/liquid partition coefficients, K_ds, for four major soil types: a compendium. *Health Phys.* **59**, 471–482.

Sheppard, M. I., and Thibault, D. H. (1991). A 4-year mobility study of selected trace-elements and heavy metals. *J. Environ. Radioact.* **20**, 101–114.

Sheppard, M. I., Reid, J. A. K., Thibault, D. H., Chen, J. D., and Vandergraaf, T. T. (1982). Technetium uptake by and yield of Swiss chard grown in peat and sand. In *Environmental Migration of Long-Lived Radionuclides*, IAEA, Vienna, Austria, 359–364.

Sheppard, S. C., Sheppard, M. I., and Evenden, W. G. (1990). A novel method used to examine variation in Tc sorption among 34 soils, aerated and anoxic. *J. Environ. Qual.* **11**, 215–233.

Shleien, B., Slaback, L. A., Jr. and Birky, B. K., (1998). *Handbook of Health Physics and Radiological Health*, 3rd ed., Williams & Wilkins, Baltimore, MD.

Smith, K. A. (1994). Manganese and cobalt. In *Heavy Metals in Soils*, 2nd ed., ed. Alloway, B. J., Blackie Academic & Professional, London, 224–244.

Smith, J. T., and Beresford, N. A. (2005). *Chernobyl: Catastrophe and Consequences*, Springer/Praxis, Chichester, West Sussex, England.

Smolders, E., Kiebooms, L., Buysse, J., and Merckx, R. (1996). ^{137}Cs uptake in spring wheat (*Tritium aestivum* L. cv. Tonic). at varying K supply, I: A potted soil experiment. *Plant Soil* **181**, 211–220.

Smolders, E., van den Brande, K., and Merckx, R. (1997). Concentrations of ^{137}Cs and K in soil solution predict the plant availability of ^{137}Cs in soils. *Environ. Sci. Technol.* **31**, 3432–3438.

Sors, T. G., Ellis, D. R., Na, G. N., Lahner, B., Lee, S., Leustek, T., Pickering, I. J., and Salt, D. E. (2005). Analysis of sulfur and selenium assimilation in *Astragalus* plants with varying capacities to accumulate selenium. *Plant J.* **42**,785–797.

Sposito, G. (1984). *The Surface Chemistry of Soils*, Oxford University Press, New York.

Srivastava, M., Ma, L. Q., and Cotruvo, J. A. (2005). Uptake and distribution of selenium in different fern species. *Int. J. Phytoremediation* **7**, 33–42.

Staunton, S. (1990). A comparison of the surface impedance factors of isotopic Na, Ca, Rb and Cs in various soils. *J. Soil Sci.* **41**, 643–653.

Staunton, S. (1994). Adsorption of radiocaesium on various soils: interpretation and consequences of the effects of soil: solution ratio and solution composition on the distribution coefficient. *Eur. J. Soil Sci.* **45**, 409–418.

Staunton, S., and Levacic, P. (1999). Cs adsorption on the clay sized fraction of various soils: effect of organic matter destruction and charge compensating cation. *J. Environ. Radioact.* **45**, 161–172.

Staunton, S., and Roubaud, M. (1997). Adsorption of radiocaesium on montmorillonite and illite: effect of charge compensating cation, ionic strength, concentration of potassium, caesium and fulvic acid. *Clays Clay Miner.* **45**, 251–260.

Staunton, S., Malfere, J. L., and D'Acqui, L. P. (2000). Effect of various techniques of organic matter removal on adsorption of caesium by soils. *Proc. 10th International Conference IHSS*, Vol. 2, 565–568.

Staunton, S., Dumat, C., and Zsolnay, A. (2002). Possible role of soil organic matter in the availability of radiocaesium. *J. Environ. Radioact.* **58**, 163–173.

Staunton, S., Hinsinger, P., Guivarch, A., and Brechignac, F. (2003). Root uptake and translocation of radiocaesium from agricultural soils by various plant species. *Plant Soil* **254**, 443–455.

Stolz, J. F., and Oremland, R. S. (1999). Bacterial respiration of arsenic and selenium. *FEMS Microbiol. Rev.* **23**, 615–627.

Sysoeva, A. A., Konopleva, I. V., and Sanzharova, N. I. (2005). Bioavailability of radiostrontium in soil: experimental study and modelling. *J. Environ. Radioact.* **81**, 269–282.

Tagami, K., and Uchida, S. (1996). Microbial role in immobilization of technetium in soil under waterlogged conditions. *Chemosphere* **33**, 217–225.

Tagami, K., and Uchida, S. (1998). Aging effect on bioavailability of Mn, Co, Zn and Tc in Japanese agricultural soils under waterlogged conditions. *Geoderma* **84**, 3–13.

Tagami, K., and Uchida, S. (1999). Chemical transformation of technetium in soil during the change of soil water conditions. *Chemosphere* **38**, 963–971.

Tamura, T., and Jacobs D. G., (1960). Structural implications in cesium sorption. *Health Phys.* **2**, 391–398.

Tan, J. A., Zhu, W., Wang, W., Li, R., Hou, S., Wang, D., and Yang, L. (2002). Selenium in soil and endemic diseases in China. *Sci. Total Environ.* **284**, 227–235.

Terry, N., and Zayed, A. M. (1994). Selenium volatilization by plants. In *Selenium in the Environment*. ed. Frankenberger, W.T., Jr., and Benson, S., Marcel Dekker, New York, 343–367.

Tinggi, U. (2003). Essentiality and toxicity of selenium and its status in Australia: a review. *Toxicol. Lett.* **137**, 103–110.

Trivedi, P., and Axe, L. (1999). A comparison of strontium sorption to hydrous aluminum, iron, and manganese oxides. *J. Colloid Interface Sci.* **218**, 554–563.

Turner, R. J., Weiner, J. H., and Taylor, D. E. (1998). Selenium metabolism in *Escherichia coli. Biometals* **11**, 223–227.

Twining, J. R., Payne, T. E., and Itakura, T. (2004). Soil–water distribution coefficients and plant transfer factors for ^{134}Cs, ^{85}Sr and ^{65}Zn under field conditions in tropical Australia. *J. Environ. Radioact.* **71**, 71–87.

Tykva, R. (2004). Basic terms of radioactivity. In *Man-made and Natural Radioactivity in Environmental Pollution and Radiochronology*, ed. Tykva, R., and Berg, D., Kluwer Academic, Doredrecht, The Netherlands, 1–11.

Van Bergeijk, K. E., Noordijk, H., Lembrechts, J., and Frissel, M. J. (1992). Influence of pH, soil type and soil organic matter content on soil-to-plant transfer of radiocesium and strontium as analyzed by a nonparametric method. *J. Environ. Radioact.* **15**, 265–276.

Wallmann, K., Kersten, M., Gruber, J., and Förstner, U. (1993). Artifacts in the determination of trace-metal binding forms in anoxic sediments by sequential extraction. *Int. J. Environ. Anal. Chem.* **51**, 187–200.

Wang, X. (1996). Physico-chemical behaviour of radionuclides in forest soils of the evacuated region of the Chernobyl nuclear power plant. Ph.D. dissertation, University of London (Imperial College), London.

Wang, G., and Staunton, S. (2005). Evolution of Sr distribution coefficient as a function of time, incubation conditions and measurement technique. *J. Environ. Radioact.* **81**, 173–185.

Wauters, J., Elsen, A., Cremers, A., Konoplev, A. V., Bulgakov, A. A., and Comans, R. N. J. (1996). Prediction of solid/liquid distribution coefficients of radiocesium in soils and sediments, 1: A simplified procedure for the solid phase characterisation. *Appl. Geochem.* **11**, 589–594.

Weiss, D. J., Mason, T. F. D., Zhao, F. J., Kirk, G. J. D., Coles, B. J., and Horstwood, M. S. A. (2005). Isotopic discrimination of zinc in higher plants. *New Phytol.* **165**, 703–710.

White, C., Wilkinson, S. C., and Gadd, G. M. (1995). The role of microorganisms in biosorption of toxic metals and radionuclides. *Int. Biodeterior. Biodegrad.* **35**, 17–40.

Wildung, R. E., Poutson, R. C., Serne, R. J., and Garland, T. R. (1974). *Pertechnetate, Iodide and Methyl Iodide Retention by Surface Soils*, USERDA Rep. BNWL-1950, Part 2, National Technical Information Service, Springfield, VA, 37–41.

Willey, N. J., and Fawcett, K. (2006). Inter-taxa differences in root uptake of $^{103/106}Ru$ by plants. *J. Environ. Radioact.* **86,** 227–240.

Willey, N. J., and Martin, M. H. (1997). A comparison of stable caesium uptake by six grass species of contrasting growth strategy. *Environ. Pollut.* **95**, 311–317.

Wu, L. (2004). Review of 15 years of research on ecotoxicology and remediation of land contaminated by agricultural drainage sediment rich in selenium. *Ecotoxicol Environ. Saf.* **57**, 257–269.

Wu, L., Enberg, A. W. and Guo, X. (1997). Effects of elevated selenium and salinity concentration in root zone on selenium and salt secretion in saltgrass (*Distichlis spicata* L.). *Ecotoxicol. Environ. Saf.* **37**, 251–258.

Yanagishawa, K., and Muramatsu, Y. (1995a). Transfer of technetium from soils to paddy and upland rice. *J. Radiat. Res.* **36**, 171–178.

Yanagishawa, K., and Muramatsu, Y. (1995b). Transfer of technetium in the soil–rice plant system. *J. Radioanal. Nucl. Chem.* **197**, 203–310.

Yanagisawa, K., Muramatsu, Y., and Ban-Nai, T. (1997). Behaviour of technetium in paddy soils. *J. Radioanal. Nucl. Chem.* **226**, 221–223.

Zehr, J., and Oremland, R. S. (1987). Reduction of selenate to selenide by sulfate-respiring bacteria: experiments with cell suspensions and estuarine sediments. *Appl. Environ. Microbiol.* **53**, 1365–1369.

Zhang, Y. C., and Frankenberger, W. T., Jr. (1999). Effects of soil moisture, depth and organic amendments on selenium volatilization. *J. Environ. Qual.* **28**, 1321–1326.

Zhang, Y. C., and Frankenberger, W. T., Jr. (2000). Formation of dimethylselenonium. *Environ. Sci. Technol.* **34**, 776–783.

Zhang, Y. Q., and Frankenberger, W. T., Jr. (2002). Fate of dimethyldiselenide in soil. *J. Environ. Qual.* **31**, 1124–1128.

Zhang, Y. C., Frankenberger, W. T. Jr., and Moore, J. N. (2000). Effect of soil moisture on dimethylselenide transport and transformation to non-volatile selenium. *Environ. Sci. Technol.* **33**, 3415–3420.

Zhu, Y. G. (2001). Effect of external potassium (K) supply on the uptake of ^{137}Cs by spring wheat (*Triticum aestivum* cv. Tonic): a large scale hydroponic study. *J. Environ. Radioact.***55**, 303–314.

Zhu, Y. G., and Shaw, G. (2000). Soil contamination with radionuclides and potential remediation. *Chemosphere* **41**, 121–128.

Zhu, Y. G., and Smolders, E. (2000). Plant uptake of radiocaesium: a review of mechanisms, regulation and application. *J. Exp. Bot.* **51**, 1635–1645.

Zhu, Y. G., Shaw, G., Nisbet, A. F., and Wilkins, B. T. (1999). Effects of external potassium supply on compartmentation and flux characteristics of radiocaesium in intact spring wheat roots. *Ann. Bot.* **84**, 639–644.

Zhu, Y. G., Shaw, G., Nisbet, A. F., and Wilkins, B. T. (2000). Effect of potassium starvation on the uptake of radiocaesium by spring wheat (*Triticum aestivum* cv. Tonic). *Plant Soil* **220**, 27–34.

Zhu, X. K., Guo, Y., Williams, R. J. P., O'Nions, R. K., Matthews, A., Belshaw, N. S., Canters, G. W., Waal, E. C., de Weser, U., Burgess, B. K., and Salvato, B. (2002). Mass fractionation processes of transition metal isotopes. *Earth Planet. Sci. Lett.* **200**, 47–62.

Zieve, R., and Peterson, P. J. (1981). Factors influencing the volatilization of selenium from soil. *Sci. Total Environ.* **19**, 277–284.

Zieve, R., and Peterson, P. J. (1985). Sorption of dimethylselenide by soils. *Soil Biol. Biochem.* **17**, 105–107.

14

REMEDIATION OF METAL-CONTAMINATED SOILS: AN OVERVIEW

M. GRÄFE

The University of Sydney, Sydney, New South Wales, Australia

R. NAIDU

Cooperative Research Centre for Contamination Assessment and Remediation of the Environment, Mawson Lakes, South Australia, Australia

Biophysico-Chemical Processes of Heavy Metals and Metalloids in Soil Environments, Edited by Antonio Violante, Pan Ming Huang, and Geoffrey Michael Gadd

14.1. INTRODUCTION

Contamination of soils by metals and metalloids (Cr, Fe, Co, Ni, Cu, Zn, As, Se, Cd, Pb, U, etc.) through anthropogenic and geogenic activities is a widespread and serious problem confronting scientists, regulators, and societies throughout the world. Unlike organic contaminants, metals and metalloids are nondegradable and tend to accumulate in soils (Naidu et al., 1999) because of their association with organic matter and mineral surfaces. The disappearance of metal(loid)s from soils usually indicates that they have leached into lower strata and may possibly be released into groundwater or surface/subsurface runoff. For this reason, metal(loid)-contaminated sites that have not been remediated or made safe often result in surface and groundwater contamination.

The origins of metal(loid) contamination are classified into point and nonpoint sources. *Point sources* of metal(loid) pollution are mostly anthropogenic and can usually be traced to a specific industrial site, such as a tannery, a timber treatment plant, a septic tank, wastewater treatment plants, or the smokestack of an ore smelter. Geogenic point sources of pollution which are the natural counterparts to a smokestack include hydrothermal vents and volcanos (Pierzynski et al., 1994; Sparks, 2003), although neither of these are responsible for the geogenic As found in groundwaters in Bengal delta (Naidu et al., 2006). Anthropogenic and geogenic *non-point-source* pollution cannot easily be traced to a single site or source, but usually stems from large areas. Anthropogenic non-point sources include agricultural, forestry and mining activities, golf courses, and other broad-scale activities that receive and release contaminants. Land in the vicinity of metal ore mines may be naturally enriched in certain metals and constitutes a geogenic non-point source of metal pollution (Pierzynski et al., 1994; Sparks, 2003).

For a long time, little or no attention was paid to metal contamination until the consequences became apparent for soil quality and for plant and human health. Some researchers draw a distinction between contamination and pollution in order to assess and distinguish between metal-affected sites: *Contaminated sites* contain contaminant concentrations above naturally occurring levels but *without consequences* to living organisms, whereas *polluted sites* contain contaminant levels high enough to *cause negative effects* on the environment and organisms (Pierzynski et al., 1994). This distinction is problematic because it does not distinguish between acute and chronic negative effects on the health of organisms and the environment. Whereas many *polluted sites* cause both acute and chronic effects, the effects from *contaminated sites* can often only be assessed over many years, and often it is difficult to trace negative health effects back accurately to a particular contaminated site. Nevertheless, the negative effects caused by metallic pollutants have become undeniable: reduced agricultural productivity, contaminated food sources, and noninhabitable areas are just a few of the soil degradation effects, while virtually most forms of mammalian cancer, respiratory diseases, organ failures, and retardation of the intellect (caused by metal poisoning) can be in part linked to metal(loid)s transferred through the food chain to mammalian recipients (Brams et al., 1989; Davydova, 1999; Hall, 2002). Given the potential adverse impact of pollutants on environment and human health, there has been a significant move toward developing strategies to manage contaminated sites. The initial focus on remediation was Europe and the United States, where the commercial value of contaminated land and the liability from risks associated with contamination drove this industry. However, in the last 10 years there has been a global interest in the remediation of contaminated sites, with communities and legislators demanding cleanup of such sites, especially where contaminants pose potential risk to the end user. Current approaches to the management of metal(loid)-contaminated sites include (1) in situ treatments based on metal(loid) bioavailability and risks to the environment and human health, (2) ex situ treatments, and (3) excavation and transport to a prescribed landfill site. Of these strategies, excavation and transport to prescribed landfill sites have been most common in Australia until recently, when regulators demanded a more sophisticated approach to on-site management of such contamination, including an assessment of contaminant bioavailability. In this chapter, following an initial discussion on bioavailability, we present an overview of strategies currently used to remediate metal(loid)-contaminated soils.

14.2. BIOAVAILABILITY

In an excellent review, Pierzynski et al. (1994) note that the management of contamination can be split into three categories: reducing inorganic concentration to an acceptable level; isolation of contamination to prevent any further reaction of the soil with the environment; and reducing the biological availability (i.e., bioavailability) of inorganic contaminants. The underlying basis for each of these management options is the bioavailability of contaminants and that remediation technologies act by minimizing the bioavailability fraction.

The bioavailability of a metal(loid) is controlled by its physicochemical speciation [e.g., $Cr(OH)_3$ vs. $HCrO_4^-$] in the soil environment and the biological (microbes, plant roots), physical (point of zero charge, soil moisture content), and chemical factors (pH, ionic strength, redox potential) interacting with metal(loid)s in soils (McBride, 1994; Sparks, 2003; Krishnamurti and Naidu, Chapter 11, this volume). It is the bioavailability fraction of inorganic contaminants that pose a toxicological or an environmental risk (Hinwood, 1991). Total concentrations of metal(loid) pollutants are therefore poor indicators of metal(loid) bioavailability and the resulting health risk to plants, animals, and humans (Pierzynski et al., 1994). Thus, metal(loid)s dissolved in the soil solution are more bioavailability than those sequestered in the soil's solid phase. Among the solid-phase bound species, long-range van der Waals forces, electrostatic attraction, hydrogen bonding, and other outer-sphere mechanisms tend to hold metal(loid)s less strongly when compared with those metal(loid)s that have formed ionic and covalent bonds at soil constituent surfaces. It has been shown that metal(loid)s which are sequestered in organic matrices such as humic and fulvic acids have less labile bonds but are potentially more mobile, due to the nature of the organic matrix (Stevenson, 1994). Metal(loid)s that form two-dimensional surface complexes have faster desorption kinetics than metal(loid) contaminants that are sequestered in three-dimensional precipitates (Stumm, 1992). The association between metal(loid) bioavailability and metal(loid) speciation was based initially on indirect desorption and sequential extraction data of contaminated soils, but in the last 15 years has been manifested through the use of bulk-scale and (more recently) microspectroscopic techniques (Manceau et al., 2002; Sparks, 2003). Much research has been and is focused on factors (pH, ionic strength, time, temperature, competitive ions, etc.) that govern the transition from soluble to solid-phase bound metal(loid)s and on mechanisms (ligand exchange, surface precipitation, inter- and intraparticle diffusion, etc.) that promote the most stable solid-phase configuration.

In recent years, research into plant- and microbially based siderophores has shed new light on the concept of metal(loid) bioavailability as certain plants and microbes excrete specialized organic acids that preferentially scavenge micronutrients such as iron, copper, and zinc (Cornell and Schwertamann, 1996). The discussion surrounding siderophores suggests that physicochemical properties of the metal(loid) contaminated environment [e.g., the presence of soluble (organic) ligand species] play an important role in the bioavailability or bioaccessible fraction of metal(loid)s (Hofte et al., 1993; Hu and Boyer, 1997; Neubauer and Furrer, 1999; Dhungana and Crumbliss, 2005; Siebner-Freibach et al., 2005). Given that the toxicity of metal(loid)s is governed by its bioavailability, there is now an increasing tendency toward the use of chemicals to convert the forms of the contaminants from those of greater hazard (i.e., high water solubility, high mobility, high bioavailability) to those of lesser hazard (low bioavailability) as a means of managing contaminated soils. The term *inplace inactivation* (U.S. EPA, 1996) has been coined to describe this process of chemically and physically inactivating contaminants, both in soil and in other materials found at the Earth's surface. Other names for this strategy are *phytostabilization, agronomic stabilization, in situ stabilization*, and *phytorestoration*. The risk reduction is provided by chemical processes in the soil so that the soil can remain in place. Chemicals and materials that appear to be most promising for in-place inactivation include phosphates, mineral fertilizers, iron oxyhydroxides, other minerals, biosolids, and limestone. Conversion of Pb to less toxic forms has been demonstrated in soils amended with safe additives using common agricultural techniques (Rabinowitz, 1993; Bertie and Cunningham, 1994; Mench et al., 1994; Ruby et al., 1994). In this strategy, no actual reduction in pollutant concentration occurs. Other examples of metal(loid) immobilization are presented in Section 14.3.2.3.

14.3. REMEDIATION

Under the regulatory regimes of most developed countries, the act of remediating a polluted site is one of many mandated steps involved in the long process of dealing with contamination. The framework used for assessing metal(loid) and other contaminated sites in Australia is based on strategies used in the United States and Europe, although Australian guidelines do not include remediation and/or management of site contamination. In the United States Superfund program [Comprehensive Environmental Response, Compensation, and Liability Act (CERCLA), page 21], for example, there are five steps, starting with docket notification, preliminary inspection, site assessment, documentation package, and adopting a site into the National Priority List (NPL) (Page, 1997). In the United States and in European Union member nations, the initial phase works toward the acceptance of a polluted site into a state or federal registry (e.g., the NPL). This phase includes several levels of on- and off-site assessment and defines the

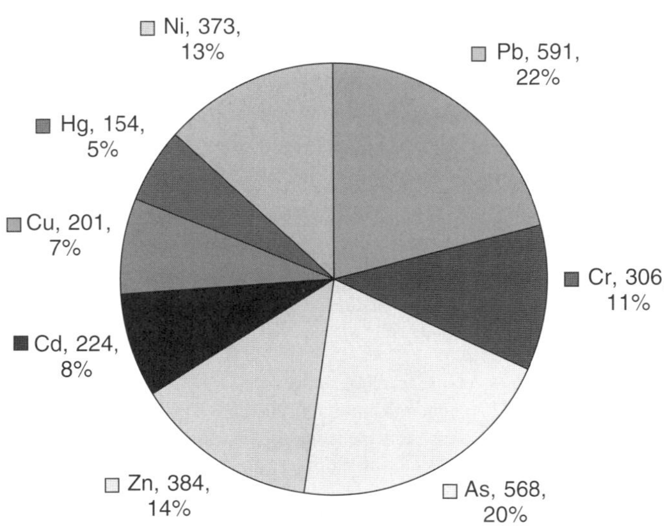

Figure 14.1. Most commonly encountered metals at Superfund (CERCLA) sites in the United States. The numbers represent the sites, and "%" indicates the percentage of sites relative to the total Superfund sites assessed. (Data from Ma and Kingscott, 1997; Green and Hoffnagle, 2004.)

degree of urgency. The urgency level of a site is usually based on a risk versus hazard assessment in which the type of contaminant, the extent of contamination, the concentrations of contaminants (trigger values), and human health risks associated with the contaminants and their concentrations are assessed (Page, 1997; Prokop et al., 2000). These criteria enable the assessment, for instance, to differentiate between the toxicity of mercury and zinc and are quantified in trigger values (or concentrations) of a particular contaminant that may not be exceeded in soils, sediment, or water for fear of (toxic) overexposure in living organisms. Trigger values for metal(loid)s in soils are often based on total elemental concentrations [mg metal(loid) kg^{-1} soil]. Alternatively, trigger values may be determined in leaching tests [e.g., toxicity characteristic leaching procedure (TCLP), extraction procedure toxicity test (EP-TOX)] of metal(loid)-affected soils. Figure 14.1 displays the most commonly encountered metal(loid)s at Superfund sites in the United States. In relation to this figure, Table 14.1 lists TCLP, MCL, MCLG, and RSDC values for these and other metal(loid)s.

TABLE 14.1. Frequently Encountered Metals at Contaminated Sites and Their TCLP[a], MCL[b], MCLG[c], and RDSC[d] Values

Metal	TCLP (mg L^{-1})	MCL (μg L^{-1})	MCLG (μg L^{-1})	RDSC (mg kg^{-1})	No. Sites Encountered[e]
Ag	5	100	N.A.[f]	39	N.A.
As	5	10	0	23	568
Ba	100	2,000	2,000	5,500	N.A.
Be	N.A.	4	4	160	N.A.
Cd	1	5	5	39	224
Cr	1	100	100	230	306
Cu	N.A.	1,300	1300	N.A.	201
Hg	0.2	0.2	2	23	154
Pb	5	15	0.6	N.A.	591
Sb	N.A.	6	50	31	N.A.
Se	1	50	0.5	39	N.A.
Th	N.A.	2	0.5	N.A.	N.A.
Zn	N.A.	1,100	N.A.	23,000	384

Source: Data from Ma and Kingscott (1997), U.S. EPA (2000, 2004a), Townsend et al. (2003).

[a]TCLP toxicity characteristic leaching procedure.

[b]MCL, highest level of a contaminant that is allowed in drinking water. MCLs are set as close to MCLGs as feasible using the best available treatment technology and taking cost into consideration. MCLs are enforceable standards in the United States.

[c]MCLG, level of a contaminant in drinking water below which there is no known or expected risk to health. MCLGs allow for a margin of safety and are nonenforceable public health goals in the United States.

[d]RDSC, reference dose screen concentration in soils.

[e]Superfund sites only.

[f]N.A., not available.

Many decisions concerning remediation are based on risk assessments. Scientific considerations coexist with a myriad of co- and interdependent socioeconomic factors such as political and community concerns about pollution, risk reduction, the timeliness and feasibility of remediation, the availability of suitable remediation technologies, future land uses, and foremost, the *cost* of remediation. Figure 14.2 shows the cost span of different remediation technologies. The scientific discourse about contamination focuses primarily on assessment of the hazards (toxicology and ecotoxicology) and the type of remediation process (e.g., environmental chemistry, physics, biology, geology, engineering), but does not detail the socioeconomic implications arising from contamination. The choice of remediation technologies (see next section) is therefore not only governed by its ability to remove metal(loid)s or reduce metal(loid) bioavailability (see below) of a contaminant, but is also evaluated for its performance in a socioeconomic context: for example, phytoremediation works well and is cheap, but is also (too) slow. Landfilling is relatively quick, but disposal costs are very high. Figure 14.3 summarizes the remedial actions for source control that were applied at Superfund sites in the United States between 1982 and 2002. The design and functionality of the remediation technologies, many of which are discussed below, reflect the socioeconomic context by addressing issues of on-site versus off-site treatments,

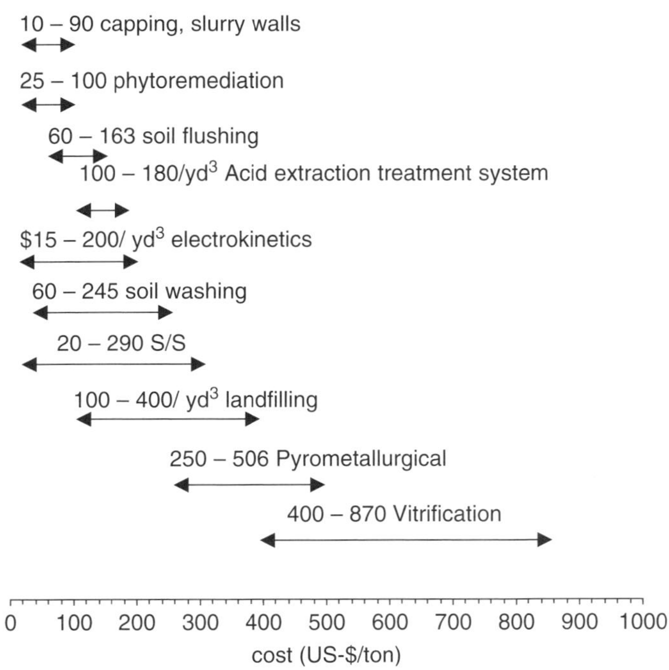

Figure 14.2. Average cost spans of remediation technologies. (Data from Paff et al., 1994; Ma and Kingscott, 1997; Sogorka et al., 1997c; U.S. EPA, 1997; Green and Hoffnagle, 2004.)

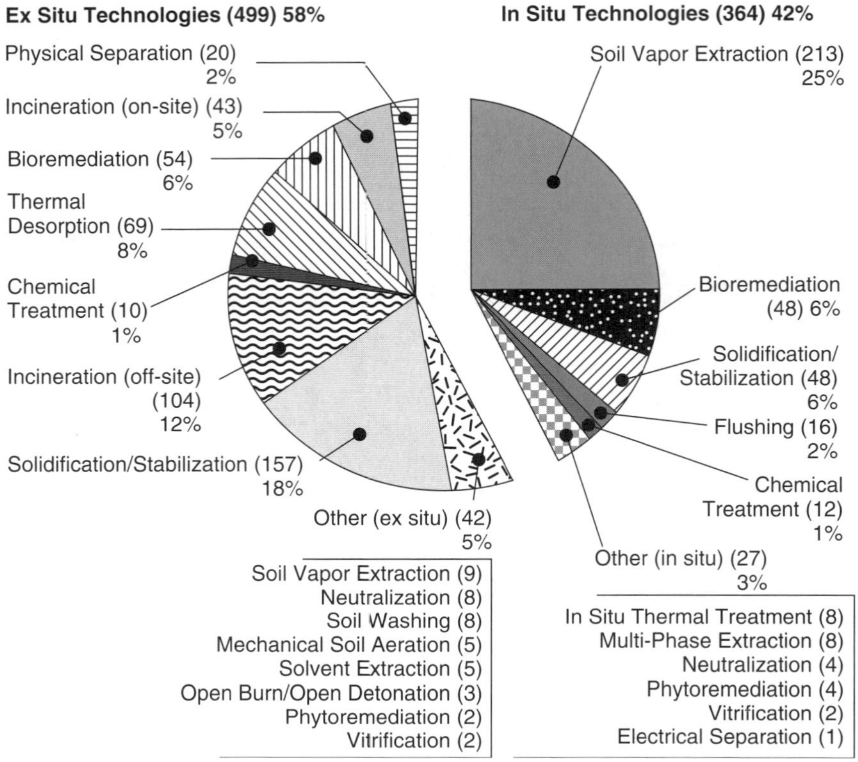

Figure 14.3. Remedial actions undertaken at Superfund (CERCLA) sites (1982–2002) for source control. The numbers in parentheses reflect the number of sites where remediation has been conducted, and "%" is the proportion relative to the total remediated. (From U.S. EPA, 2004b.)

contaminant stabilization versus contaminant removal, acute versus long-term contaminant reduction, and the value of remediation.

14.4. REMEDIATION TECHNOLOGIES

Metal(loid)-contaminated soil can be remediated by chemical, physical, or biological techniques (or a combination of all three). The technology applied is often specific to the metal(loid) contaminant to be removed and the site characteristics and may be further classified into in situ and ex situ categories. In situ *remediation* takes place on site and does not require excavation of the contaminated soil, limiting the exposure pathways to other organisms, including humans. Many in situ techniques aim to stabilize the metal(loid) fraction in the soil via specific soil amendments (e.g., lime), vitrification, electrokinetics, solidification (thermal treatments) or passive remediation methods. Such techniques do not reduce the total metal(loid) content but act by minimizing the bioavailability fraction as discussed in Section 14.2. Other in situ remediation methods aim to remove the metal(loid)s from the soil by leaching or washing procedures using surfactants and collecting effluent downstream from the remediation site. Bioremediation refers generally to all remedial activity involving organisms such as microbes and plants. Phytoextraction and phytostabilization are two in situ methods in which plants are used either to actively take up metal(loid)s from the soil (phytoextraction) or to stabilize otherwise barren sites against soil erosion (phytostabilization). Microbial remediation of most metal(loid)s offers a limited scope. Arsenic, mercury, and selenium may be volatilized and removed from soils by microbe-mediated reactions (e.g., methylation of arsenite); however, the majority of metal(loid)s do not exist as gases in the environment. Unlike most of these metal(loid)s, hexavalent chromium has been reduced to less toxic trivalent forms by chromium-reducing microbes. However, this technique has rarely been used in the field.

Ex situ remediation techniques require the excavation of polluted soil for subsequent treatment or disposal. Ex situ treatments can be broadly classified into extraction versus stabilization treatments that will render the polluted soil less harmful and suitable for deposition in a landfill or backfill. Soil washing is an example of an ex situ extraction technique in which the treated soil can either be returned to its original site (backfill) or be land filled, depending on the success of the cleanup stage. Asphalt incorporation, thermal treatment, and encapsulation are ex situ stabilization techniques in which the metal(loid)-contaminated soil is either incorporated (e.g., asphalt) or contained (encapsulation) by secondary materials that are subsequently land filled. Thermal treatments involve the incineration of the metal(loid)-polluted soil and the conversion of the pollutants into their metallic (zero-valent) states. In the following section we present an overview of the various technologies based on their mechanism of action.

14.4.1. Biological Technologies

14.4.1.1. Bioremediation *Bioremediation* refers to the attenuation of metal(loid)-polluted sites by (natural) processes using organisms such as microbes or plants. Both stabilization and removal mechanisms exist within bioremediation. Unlike hydrocarbon-contaminated soils where bioremediation has been extremely successful, the use of microbes to remediate metal(loid)-contaminated soils has been limited to those elements that can be converted into gaseous forms (e.g., arsenic, mercury) or transformed into less toxic forms, an example being reduction of toxic hexavalent chromium to nontoxic trivalent chromium. The microbial interaction with metal(loid)s is partially dependent on whether the microbes are prokaryotic or eukaryotic. Both types can bind metal(loid)s at their surface; however, only prokaryotic microbes have the ability to oxidize [Mn(II), Fe(II), Co(II), Cu(I), AsO_3^{2-}, Se^0, or SeO_3^{2-}], reduce [Mn(IV), Fe(III), Co(III), CrO_4^{2-}, $HAsO_4^{2-}$, SeO_4^{2-}, or SeO_3^{2-}], or volatilize (Hg, As, Se) metal(loid)s (Ehrlich, 1997). Many of the oxidation–reduction reactions of microbes promote the stability of the metal(loid)s: for example, by precipitating them as sulfides, carbonates, or phosphates. Volatilization of certain metal(loid)s by prokaryotic microbes is the only known removal mechanism of metal(loid)s from soils by microbial action. Greater emphasis on the role of microbes with metal(loid)s is given elsewhere in this book (see Chapter 2).

14.4.1.2. Phytoremediation *Phytoremediation* is a term that describes the use of green plants to remove, immobilize, or render harmless inorganic metal(loid)s and organic pollutants (Raskin et al., 1997). Conceptually, this remediation approach treats a living plant as a solar-driven pump capable of extracting and/or concentrating particular pollutants from soil, sediment, and aqueous environments. Phytoremediation of metal(loid)s is recognized increasingly as a cost-effective "green" technology that enjoys high public support (Whiting et al., 2004). A number of subcategories exist which are listed and defined in Table 14.2, which includes the metal(loid)s and the number of sites treated with phytoremediation techniques in the United States.

Chaney et al. (1997) identified three characteristics of hyperaccumulator plants.

1. They are tolerant of high levels of the toxic element in the root and shoot cells by compartmentalizing the metal(loid) in vacuoles and/or by chelation of the metal(loid) by specific ligands.
2. Hyperaccumulators have the ability to translocate the element from roots to shoots rapidly, possibly through chelation with malate, citrate, and/or histidine.
3. They rapidly take up the element at levels that occur in the soil solution. Hyperaccumulators often require higher solution and leaf concentrations of the element to grow normally (e.g., *T. caerulescens)* and, in some cases, toxic elements may even enhance the growth of the plant (e.g., *Pteris. vittata* L.) (Ma et al., 2001).

The evolutionary advantage to a plant of accumulating high concentrations of a toxic element in its leaves and stems may be to reduce herb ivory by chewing insects and animals and attack by bacterial and fungal diseases, as has been observed in certain Ni and Zn hyperaccumulators (Chaney et al., 1997). Although a favorable attribute of any hyperaccumulator is a high biomass, overall hypertolerance or hyperaccumulation is strategically more important than high biomass, because the amount of element removed by a small hyperaccumulator is still greater than that of a larger non-hyperaccumulator. Because the aboveground tissues are usually harvested, the volume and mass relationship of waste is important from a value point of view of the eventual biomass ash: one with 20 to 40% metal(loid) content is a valuable ore, but biomass ash containing only 2 to 4% metal(loid) is a hazardous waste that needs to be disposed of (Chaney et al., 1997).

Many metal(loid)-contaminated sites require the establishment of a plant cover (phytostabilization) to minimize wind and water erosion of topsoil and to

TABLE 14.2. Phytoremediation[a] and Its Categories

Term	Definition	Species[b]	No. of Sites[c]
Phytoextraction	Removal and storage of metals from the contaminated media into the plant; contaminants applied to Ag, As, Cs, Cd, Co, Hg, Na, Pb, Zn	Hybrid willow, Eastern gamma grass, hybrid poplar, golden weeping willow, hybrid prairie cascade willow, laurel-leaved willow, *Koshia scoparia*, *Pteris vittata, mayil, parkeril*, *Brassica juncea*, *Brassica* sp., tall fescue, *Sinapis alba*, *Helianthus* sp., *Helianthus an.*, *Ricinus communis*, *Zea mays*, *Salix viminalis,Pisum sativum,* ryegrass, sunflower, vetiver grass, eucalyptus, tamarisk, bulrush, sedges, cattail, arrowhead, *Sesbiana* sp. birdsfoot trefoil, Ruebans Canada bluegrass, Streeker redtop, alpine pennycress, bladder campion, *Phaesolus vulgaris*	31
Phytostabilization	Isolation and containment of metals within the soil or sediment through the prevention of erosion and leaching; contaminants applied to As, Cd, Co, Cr, Cu, Hg, Ni, Pb, Zn	Forb and subshrub species, grass, hybrid poplar, western wheat grass, loblolly pine, eucalyptus, willows, Bermuda grass, rye grass, white clover, tall fescue, bulrush, sedges, cattail, arrowhead, birdsfoot trefoil, Ruebans Canada bluegrass, Streeker redtop, alpine pennycress, bladder campion, soybeans, radish, oats, alyssum	15
Phytovolatilization	Uptake and transpiration of contaminants form the media through the plant tissue and into the atmosphere, applicable primarily to organics; potential for As, Se, Hg, and other volatilizable contaminants	Eucalyptus, willows, eastern cottonwood	2
Rhizodegradation	Degradation of contaminants by microorganisms in the soil zone that surrounds and is influenced by the roots of plants (i.e., the rhizosphere)		2
Rhizo- and phytofiltration	Precipitation and/or concentration of contaminants from contaminated effluents by seedlings, roots, and dead plant materials		2

Source: Definitions and data from Raskin et al. (1997), Green and Hoffnagle (2004).

[a]The use of vegetation and its associated microorganisms, enzymes, and water consumption to contain, extract, or degrade contaminants from soil and groundwater.

[b]Some species may not apply to the remediation of metals but to co-contaminating organic contaminants. No differentiation is given by source. Where available, the scientific notation of the plant species are provided.

[c]Number of sites that a particular form of phytoremediation is applied to. Multiple metals usually appear per site.

minimize excessive leaching of metal(loid)s to the groundwater. Under many circumstances, soil amendments that reduce the bioavailability of metal(loid)s are necessary to establish a successful plant cover. Traditional amendments are lime, red mud, or clay materials that either increase the pH or are capable of sorbing large concentrations of soil solution metal(loid)s. Kukier and Chaney (2001) observed that liming was the most effective treatment for reducing Ni phytotoxicity in wheat, red beet, and oats grown in acidic (pH $\sim$ 5.5) muck and silt loam soils. By comparison, hydrous ferric oxide (HFO) was not as effective as limestone at reducing Ni phytotoxicity, which emphasizes the importance of pH. Petrisor et al. (2004) showed that the addition of bacteria to mine spoil soils resulted in more effective plant cover establishment than did fertilization with conventional inorganic fertilizers. In all cases, plants were able to establish on highly contaminated phosphogypsum and sulfide tailings that were artificially inoculated with *Bacillus megaterium* and *Azotobacter chroococcum* and that had received no additional fertilizers or amendments such as dolomite. A key factor in plant establishment, however, was pH. Low pH (< 3) retarded plant growth significantly, but could not eliminate it in the presence of the bacteria. Christie et al. (2004) reported that arbuscular mycorrhizal (AM) fungi protect their host plants under conditions of soil metal(loid) contamination, specifically Zn. The mechanisms responsible for the protection are the metal(loid) tolerance of the AM fungi, the effects of balanced mineral nutrition specifically with respect to P, the factors controlling Zn uptake and partitioning in AM fungi plants, and the immobilization of metal(loid)s in the roots of AM fungi plants. The success of biological amendments to metal(loid)-contaminated sites for phytoremediation underscores the importance of reestablishing an ecologically balanced environment consisting of micro and macro flora and fauna. These approaches are gaining increased regulatory and community approval.

A significant limitation to the uptake of metal(loid)s into the roots and/or shoots of plants is posed by the stability of the soil–metal(loid) complex, which reduces the phytoavailable metal(loid) fractions. Chelating agents such as ethylenediaminetetraacetic acid (EDTA) and small aliphatic acids (oxalate, citrate, malate, etc.) are effective desorbing and dissolution agents that significantly increase soil solution metal(loid) concentrations by forming soluble metal–ligand complexes (Blaylock et al., 1997, Huang et al., 1997, 1998; Wu et al., 1999; Chen, et al., 2004; Lai and Chen, 2004). Blaylock et al. (1997) studied the effect of six different chelators on Pb, Cd, Cu, Ni, and Zn uptake in *Brassica juncea* and showed that the choice of chelator was metal(loid) specific. For example, Pb uptake and translocation into the shoot of *B. juncea* was enhanced more by EDTA than by other chelators: cyclohexylenedinitrilotetraacetic acid (CDTA), citric acid, diethylenetriaminepentaacetic acid (DTPA), and ethylene glycol-bis(2-aminoethyl)-*N,N,N'*, and *N'*-tetraacetic acid (EGTA), but EGTA was more effective in promoting uptake and translocation of Cd in *B. juncea*. Studies by Huang and co-workers corroborate EDTA's efficacy in promoting Pb uptake in *Zea mays* and *Pisum sativum* compared to other chelators, such as 2-hydroxyethylethylenediaminetriacetic acid (HEDTA), DTPA, EGTA, or ethylenediamine di(*o*-hydroxyphynyl) acetic acid (EDDHA) (Huang et al., 1997). Wu and co-workers showed that EDTA promotes the transpiration rate of plants subjected to Pb-complexed EDTA and literally drives Pb-EDTA xylem sap out of the roots and into the shoots of *Z. mays* (Wu et al., 1999). More recently, Lim et al. (2004) demonstrated that the increased bioavailability of metal(loid)s in the presence of chelators such as EDTA can be further enhanced if an electric current is applied. The accumulation of Pb in the shoots of *B. juncea* increased two- to fourfold with electric current and EDTA applied compared to the use of EDTA alone. Huang and co-workers showed that uranium (U) phytoremediation by plants of the genus *Brassica* was enhanced significantly when citrate as citric acid (pH 2) was applied to U-contaminated soils (Huang et al., 1997). The significant difference in U uptake by *Brassica* plants in the presence of citrate suggested that U hyperaccumulators need to be assessed in the presence of the soil amendment. In comparison to malic, acetic, nitric, and sulfuric acids as well as bicarbonate and water (control), citric acid promoted the uptake and translocation of U into the shoots of *B. juncea* and *B. chinensis* significantly more (Huang et al., 1998).

In 2001, Ma and co-workers reported that the Chinese break fern (*P. vittata*) accumulated weight percentages of As in its fronds (Ma et al., 2001). Since then additional fern species (*Pityrogramma calomelanos, Pteris cretica, Pteris longifolia, Pteris umbrosa*) have been identified as hyperaccumulators of As (Visoottiviseth et al., 2002; Zhao et al., 2002; Wongkongkatep et al., 2003; Zhang et al., 2004a,b). These ferns are capable of extracting as much as 22,000 mg kg^{-1} As from artificially contaminated soils and translocating the majority of the As into the fronds of the fern (Lai and Chen, 2004; Ma et al., 2001). Work by Cao et al. (2004) has shown that *P. vittata* L. reacts to As by increasing its enzymatic and nonenzymatic antioxidants to detoxify As. The former is more important at low As exposure ($\leq$ 20 mg As kg^{-1} soil), whereas the latter is more important at concentrations above this threshold. A minor role in the detoxification mechanism is also ascribed to thiols in the fronds of the fern (Zhang et al., 2004b). Webb et al. (2003) observed with x-ray absorption spectroscopy (XAS) that As in the fronds was present in the form of As(III) coordinated dominantly by oxygen atoms. Only at concentrations of 1 wt% (dry weight) did the authors observe that As was coordinated to a significant degree by sulfur in addition to oxygen.

14.4.2. Chemical Technologies

14.4.2.1. Electrokinetics *Electrokinetics* (also known as *electrokinetic remediation*) employs low levels of direct current across a network of cathodes and anodes in a contaminated soil to induce a voltage gradient (Acar and Ashawabekh, 1993; Sogorka et al., 1997a). Electrokinetic techniques were explored as early as the 1930s and were used to concentrate metal(loid)s and to dewater soils and sludges (Sogorka et al., 1997a). During the 1970s the Soviet Union began using electrokinetics to concentrate metal(loid)s and to explore for minerals in deep

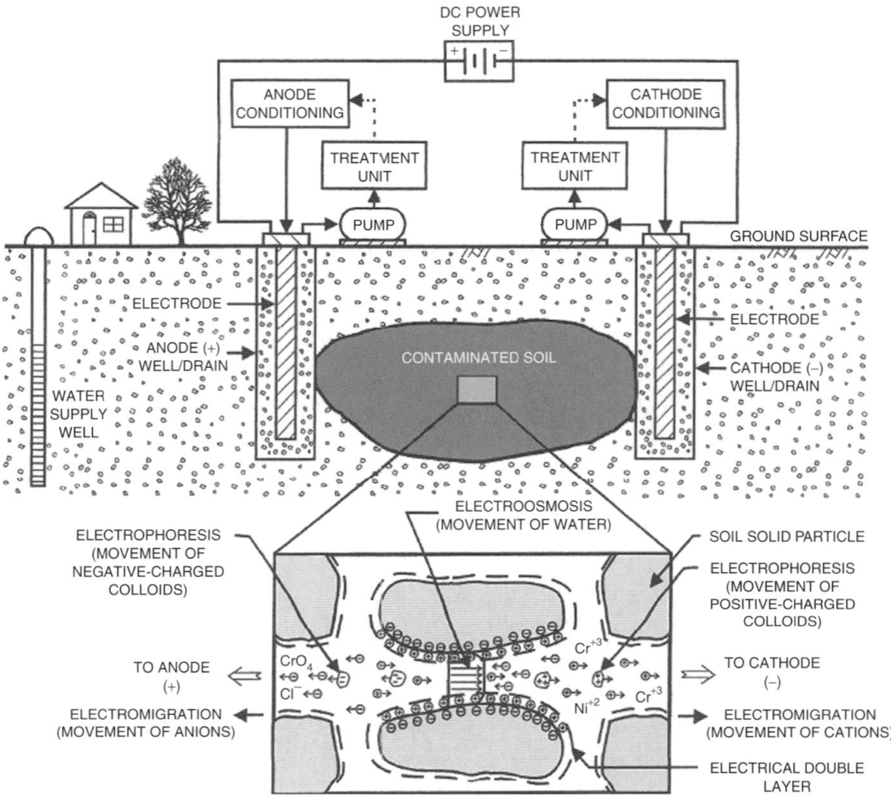

Figure 14.4. Electrokinetic remediation. (From Reddy et al., 2003a.)

soils; by the 1980s, other researchers had realized the potential of electrokinetics to remediate soils contaminated with inorganic and polar organic contaminants (Chilingar et al., 1997).

Electrokinetic remediation involves the installation of trenches or wells that surround the contaminated area. Electrodes (cathodes and anodes) are inserted in these trenches or wells and a low direct current is applied, on the order of mA cm^{-2} (Acar and Ashawabekh, 1993; Reddy and Chinthamreddy, 2004) (Figure 14.4). The voltage gradient generated by the current leads to three transport phenomena in soils: (1) migration of ions toward the oppositely charged electrode (electromigration), (2) bulk water flow (electroosmosis), and (3) movement of charged particles toward one of the electrodes (electrophoresis). In an unenhanced electrokinetic remediation scheme (only H_2O is added at the electrodes), the electrolysis of water will occur at the anode and cathode ends.

$$2H_2O \rightarrow O_2(g) \uparrow + 4H^+ + 4e^- \qquad \text{anode reaction} \qquad (14.1)$$

$$4H_2O + 4e^- \rightarrow 2H_2(g) \uparrow + 4OH^- \qquad \text{cathode reaction} \qquad (14.2)$$

The anode is therefore the source of an acid front, and the cathode is the source of a basic front. As a consequence of the electric potential, excess ions begin to migrate toward the oppositely charged electrode: cations toward the cathode and anions toward the anode. The rate of ion migration is dependent on the ion's diffusion coefficient and ionic mobility, which in turn are dependent on the rate of association and dissociation with the solvating molecules. Consequently, H^+ and OH^- ions have higher diffusion coefficients and greater ionic mobility in water than do metal(loid)s and other anions. The diffusion coefficient and ionic mobility of H^+ ions is greater than that of OH^- ions, allowing H^+ ions to dominate a system under electrokinetic remediation. The transport of H^+ and OH^- ions from anodes and cathodes significantly affects the physicochemical properties of the soil mass and the metal(loid) contaminants associated with them. Desorption and dissolution reactions promote an excess of ions in the soil solution, which begin to migrate toward the oppositely charged electrode. Electrokinetic migration of contaminants is a key advantage over other in situ techniques, such as soil flushing, as the technique is independent of pore size and permeability (Manna et al., 2003). The momentum of the ion migration is transferred to molecules of the solvating medium via viscous forces promoting electroosmosis. The bulk flow of water is co-dependent on the average point of zero charge (PZC) of the soil. Soils that are net negatively charged promote water flow toward the cathode and those that are net positively charged promote water flow toward the anode. The PZC properties of soils depend on their mineralogy and organic matter, but also on the migration of the acid and the base fronts. The migration of the base front toward the acid front is detrimental to the collection of metal(loid) cations at the cathode because they are prematurely precipitated as metal hydroxides and other insoluble phases. These metal-hydroxide precipitates may also cause clogging of soil micropores and reduce ion migration. Therefore, enhanced electrokinetic remediation techniques involve introducing weak acids (e.g., acetic acid) near the cathode to neutralize the base front and/or depolarize the cathodes (Acar et al., 1994).

The success of electrokinetic remediation relies heavily on knowledge of the physicochemical properties of the contaminated soil environment and the speciation of the contaminants. The mineralogical composition of the soil and the presence of anomalies in the subsurface, such as submerged foundations, rubble, pipes, and large amounts of Fe and other substances, can all affect soil properties such as the PZC and subsequently, electroosmotic flow (Sogorka et al., 1997a). Soils that have greater pH-dependent charge characteristics, such as iron and aluminum oxide–rich soils, will exhibit charge dependencies that are affected by the migration of the acid and base fronts. A soil that is dominantly negatively charged (e.g., rich in 2 : 1-type clay minerals) will accumulate water at the cathodic end, while a dominantly positive charged soil (e.g., rich in Feoxides) will induce a reverse electroosmotic condition and promote water accumulation at the anode. Manna et al. (2003) showed that Cr(VI) migration in kaolinite was essentially a tug-of-war between electromigrating Cr(VI) ions toward the anode and electroosmotic transfer of dissolved Cr(VI) in water toward the cathode. The speciation of the contaminant is equally important, as changes in oxidation state may change the chemistry of the contaminant, as is the case for Cr(VI)/Cr(III) or As(V)/As(III). For example, electrokinetic remediation of Cr(VI) in glacial till under unenhanced conditions using tap water at the anode and cathode promoted the migration of Cr(VI) toward the anode, where a portion of the Cr(VI) was reduced to Cr(III). This reduction was most probably caused by OM and the production of electrons at the anode, due to the electrolysis of H_2O (Reddy and Chinthamreddy, 2004).

Enhancement of the electrokinetic remediation process involves the introduction of specific solutions at either the cathodic or anodic end or both. These solutions can have any number of designated effects: for example, acid neutralization of the base front or halide complexation of the soft metal cations. Iodide additions to a mercury (Hg)-contaminated soil promote the electromigration of Hg toward the anode, indicating the dominance of anionic mercury–iodide complexes in solution (Suer and Allard, 2003). However, halide complexation does not work under all circumstances. For example, Cl^- does not promote the formation of soluble anionic mercury–chloride complexes, and iodide achieves a 97% removal of Hg from kaolinite, but iodide was significantly less efficient for Hg removal from glacial till, probably due to mercury–OM complexation (Reddy et al., 2003b). These and many other reported case studies show the intricate and complex nature of electrokinetic remediation and the expert knowledge that one must possess to devise an effective electrokinetic remediation scheme (Reddy and Patrick, 1977; Yeung et al., 1996; Reddy et al., 1997, 2001, 2003a,b; Reddy and Chinthamreddy, 1999; 2003a,b, 2004; 2003c; Kim and Kim, 2001; Kim et al., 2002a,b; Gent et al., 2004).

The extraction and removal of metal(loid)s at the electrodes is accomplished by electrodeposition, precipitation, or ion exchange using resins. The cost of electrokinetic remediation is directly dependent on the hydraulic and chemical properties of the contaminated soils and the appropriateness of the enhancement fluids at the cathode or anode, as these factors govern the energy consumption and the length of treatment. Prices range from \$60 to 225 per yd^3 (not including indirect and accessory costs, such as excavation, permits, and residue treatments) (Sogorka et al., 1997a).

14.4.2.2. In Situ Soil Flushing In situ soil flushing uses water or specific chemical solutions (acids, bases, chelating agents) to flush metal(loid)s from soils toward wells where they are pumped to the surface for further treatment (Mulligan et al., 2001c; U.S. EPA 2001a,c; Wasay et al., 2001). This method supports and improves pump-and-treat cleanup methods and is most successful in soils that have impermeable subhorizons (clayey layers) underlying permeable surface epipedons. The flushing fluids are surface applied or injected into the soil via surface flooding, surface sprinklers, leach fields, vertical or horizontal injection wells, and basin or trench infiltrations systems (Evans et al., 1997). A thorough understanding of the actual or at least potential solid-phase speciation of the metal(loid) contaminants in the soil as well as soil factors governing metal(loid) speciation [pH, soil type, CEC, OM content, metal(loid) concentration, residence time] are paramount in devising effective soil flushing fluids. Many flushing fluids will lower soil pH, promote metal(loid) chelation, or displace toxic metal(loid)s from the exchange complex of clays with nontoxic metal(loid)s (Mulligan et al., 2001c). Vacuum extraction methods in the vadose zone and pump-and-treat methods in saturated zones are applied to remove the flushing fluids once they reach their target location. Recovered groundwater and flushing fluids usually have to be treated in order to be discharged into a common waste stream, returned to the groundwater, or reused for another flush cycle (Mulligan et al., 2001b,c; Di Palma et al., 2003a,b).

Wasay et al. (2001) compared the leaching and removal performance of aliphatic acids (citrate, tartrate, and oxalate) with that of EDTA and DTPA in loamy soils ranging from sandy to sandy clay loam in texture. Sequential extraction studies indicated that the metal(loid)s were mainly sequestered in Fe–Mn and/or organic matter fractions. EDTA and DTPA removed heavy and trace metal(loid)s (Cr, Mn, Hg, Pb, Cd, Cu, Zn) more effectively than the aliphatic acids; however, EDTA and DTPA extracted large concentrations of beneficial macronutrients (Ca, Mg, Fe), thus leaving the soils infertile. Tawinteung et al. (2005) compared the leaching and removal performance of EDTA, M HNO_3, and 0.2 M NH_4–citrate for Pb-contaminated soils. Sequential extraction studies indicated that more than 90% of the Pb was either in the exchangeable, carbonate, or Fe–Mn-oxide-bound fractions. The results showed a similar efficiency between EDTA and M HNO_3, which were both greater than the NH_4–citrate treatment, but the acid treatment affected the physicochemical properties of the treated soils negatively. Lee et al. (2004) determined that As bound mainly by Fe-oxides in mine tailings was most effectively leached from the tailings at low flow rates and EDTA solutions above pH 8, suggesting that Fe-oxide solubility, the PZC, and the solubility of As–Fe-oxide complexes determine the efficiency of As removal. Groudev et al. (2001) showed how sulfuric acid–doped water and local soil flora promoted the leaching of radioactive elements (U, Th, Ra) and heavy metal(loid)s (Cu, Zn, Cd) from a heavily contaminated A-horizon into the underlying B-horizon, which contained sulfate-reducing bacteria. The activity of the indigenous sulfate-reducing bacteria was enhanced by injecting water solutions containing organic compounds and NH_4–phosphate through bore holes into the B-horizon. Subsequently this, promoted the biocatalyzed formation of metal sulfides.

14.4.2.3. Monitored Natural Attenuation Monitored natural attenuation (MNA) depends on natural processes to attenuate contaminants in groundwater that are regularly monitored or screened to ensure that the contaminants do not leave the site (U.S. EPA, 2001d; Brady et al., 2003). Although MNA is not commonly used to describe the attenuation of metal(loid)s in soils, there have been references to the effect of aging on the attenuation of metal(loid)s in soils (Vig et al., 2003) and is now being seen increasingly as a natural phenomenon affecting metal(loid) bioavailability. Natural attenuation processes relevant for metal(loid) contamination of soils and sediments include dispersion, dilution, sorption, precipitation, volatilization, and/or chemical and biochemical stabilization. Parameters that influence solid-phase partitioning of metal(loid)s are the nature and concentration of the metal(loid)s, soil pH and redox potential, ionic strength, soil mineralogy, and the presence of organic matter and other ligands. Natural attenuation of metal(loid)s operates over long periods of time by gradually lowering the bioavailability (hysteresis) of metal(loid)s through specific sorption mechanisms mediated by soil constituent surfaces and soil microfauna. For example, the formation of Ni-, Zn-, and Co–Al layer double-hydroxides is a natural attenuation process in which trace metal(loid)s are converted into stable precipitated matrices in the presence of 2 : 1 and 1 : 1 clay minerals (Scheidegger et al., 1998; Roberts et al., 1999; Scheinost, 1999, 2000; Juillot et al., 2002; Manceau et al., 2002). Because of the long-term processes involved in natural attenuation, few research data have been published on the subject. Matthews and Effler (2003) recorded a 40% decrease over 14 years in Cl^- and total NH_3 loading in Ninemile Creek and Lake Onondaga (both New York State) from upstream waste beds of a soda ash manufacturer. This reduction was attributed to natural attenuation processes, as no related management interventions are known to have ever been implemented since closure of the soda ash plant. Kaplan et al. (2002) investigated an Hg-contaminated wetland with the aim of providing guidance for the selection of an appropriate in situ remediation technology. Sequential extraction experiments showed that less than 8% of total solid-phase Hg was in an exchangeable form, whereas the remainder was associated with Fe-oxides or presumably as a sulfide precipitate. Based on the high stability of the Hg solid phases and the sensitive ecological nature of the wetland, the study concluded that a monitored natural attenuation approach would be most appropriate.

Brady et al. (2003) proposed that a site-specific screening approach followed by sequential extraction measurements could be an appropriate method to determine if a contaminated site is suited for MNA. In the first step, the site's chemical and hydrological conditions are assessed from standard site characterization data. The MNAtoolbox (www.sandia.gov/eesector/gs/gc/na/mnahome.html) provides a rapid initial assessment of the chemical and hydrologic conditions for MNA. The site-specific assessment for natural attenuation is then evaluated using sequential extraction experiments that enhance the site's characterization, determine the bioavailability, fate, and transport of the metal(loid)(s) in question, and assess natural attenuation against other remediation options (such as phytoremediation).

14.4.2.4. Solidification and Stabilization Solidification and stabilization (S/S) constitute a group of in situ and ex situ remediation technologies that either inhibit or retard the release of metal(loid) contaminants from polluted soils. Two essentially different approaches are used. Within solidification, polluted soil is immobilized by mixing it with cementing agents, while stabilization aims at converting soluble, mobile, and toxic metal(loid)s into less mobile and more inert forms [e.g., $Cr^{VI}O_6(aq)$–$Cr^{III}(OH)_3(s)$ reduction and precipitation] (Sogorka et al., 1997b; U.S. EPA, 1997, 2001f). As discussed in Section 14.2, these techniques do not lower the actual concentration of metal(loid) contaminants in the soil; rather, they reduce the risk of harm to target organisms (humans, animals, plants, microbes, etc.) by lowering the contaminants' potential bioavailability.

A common strategy employed for cement-based S/S is to identify the metal(loid) of greatest concern in the contaminated soil and then to arrange for the presence of appropriate counter ions in the stabilized product, which will minimize the solubility of the toxic species (Miller et al., 2000). In ex situ techniques, using cement-based s/s, contaminated materials are mixed with an appropriate ratio of cement, water, and other additives. Portland-type cements and pozzolanic materials are either used by themselves or as a mixture. Portland-type cement consists of calcium silicates, aluminates, alluminoferrites, and sulfates. Pozzolans are small spheroidal particles formed during the combustion of coal (fly ash) and are also found in lime and cement kilns. When mixed with water, pozzolans containing large amounts of silica have cementlike properties (U.S. EPA, 1997). Cement-based S/S reduces the mobility of metal(loid) contaminants by precipitating them as less soluble metalhydroxides, carbonates, silicates, or sulfates, by substitution of the metal(loid)s into mineral structures, by sorption reactions with available surfaces, and by physical encapsulation.

In-situ cement-based S/S consists of two steps: mixing and off-gas treatment. The greatest challenge of in situ cement-based S/S is complete and uniform mixing of the cement with the contaminated soil. Three approaches have been used in the past: (1) vertical auger mixing, (2) in-place mixing of cements with waste by conventional draglines, backhoes, or clamshell buckets, and (3) injection grouting, which forces the cements into the subsurface, where it permeates the contaminated soil (U.S. EPA, 1997).

Polymer encapsulation is an ex situ S/S technique involving the application of thermoplastic resins such as bitumen, polyethylene and other polyelfins, paraffins, waxes, and sulfur-based cements, as opposed to cements and pozzolans. Polymer encapsulation has been used primarily to immobilize low-level radioactive wastes and those waste types that are difficult to immobilize in cement, such as Cl- and SO_4-based salts. Bitumen (asphalt) is the least expensive and (hence) used most often. Thermoplastic encapsulation heats and mixes the contaminated soil with the resin at 130 to 230°C in an extrusion machine. Organic pollutants and water boil off during the extrusion and are collected for treatment or disposal. The final product, a stiff yet plastic resin, is then discharged into a drum or other container and land-filled (U.S. EPA, 1997).

Proper hydration and setting of the cementing agents will promote rigidity, increased density, lower permeability, and an increased resistance to environmentally aggressive conditions. The proper hydration of the cement is therefore paramount for its continued safe in situ disposal or in landfills. Many wastes containing As, Zn, or Cr have the potential to compromise the proper hydration and setting of the cement particles (Hills and Pollard, 1997; Leist et al., 2000).

Polettini et al. (2004) investigated cement-based S/S on a mixed Cr(III)-, Cu(II)-, and Pb(II)- contaminated soil and showed that stabilization of the mixed metal(loid)-contaminated soil was not compromized by the mixed contaminant status, which may have been related to similar immobilization mechanisms for all three metals in the cement. Alpaslan and Yukselen (2002) compared the efficiency of activated carbon, clay, zeolite, sand, lime, and cement for immobilizing Pb from spiked soils against TCLP-leachable Pb. Several ratios of additive to spiked soil were investigated. The most effective additives were activated carbon, lime, and cement. However, the volume of activated carbon required was economically and volumetrically unfeasible. Lime reduced the TCLP-leachable Pb fraction by 88% for a lime/soil ratio of 1 : 21 and a reaction pH of 12.6, presumably through the formation of $Pb(OH)_2(s)$ species. A soil/cement ratio of 1 : 15 and a reaction pH of 8.3 immobilized 99% of Pb in the cement–soil mixture.

Several researchers (Akhter et al., 2000; Miller et al., 2000) have investigated the suitability of various salt amendments to type I Portland cement to stabilize up to four sandy soils previously contaminated with As(III)-based herbicide. These investigators structured their experiments from small laboratory scales (10 to ~3 kg) to field scale, exploring initially a large number of treatment possibilities. Their work showed that additions of ferrous salts performed more reliably and consistently than ferric salts and were more cost-effective than Mn^{2+} and Ba^{2+} salts. Pretreatment of the contaminated soil with $FeSO_4$ for 5 to 7 days in the laboratory and the field scale showed significant reductions in TCLP-leachable As, however, with a further decrease in leachable As following the addition of Portland type I cement. In part this could be ascribed to the proper formation of a normally hydrated cement matrix ascertained through analyses by XRD, SEM, and solid-state NMR of the cemented-waste matrices. Jing et al. (2003) investigated the potential of using one part Portland cement and three parts concrete sand for the stabilization of Fe-oxide groundwater sludges contaminated with arsenate. Using FTIR and XAFS spectroscopies and triple-layer modeling, the authors show the formation of calcium arsenate precipitates in the cemented waste, which become increasingly soluble under the influence of carbonate ions (Na_2CO_3), which compete for Ca^{2+} to form $CaCO_3(s)$. Moon et al. (2004) studied As immobilization in lime-treated soils and observed the formation of calcium arsenates and calcium arsenites via XRD, depending on the initial redox state of As (III vs. V). When a lime–arsenate–kaolinite mixture was investigated, however, the reactions did not result in the formation of pozzolanic reaction products.

14.4.2.5. Permeable Reactive Barriers Permeable reactive barriers deal with groundwater and not soil contamination and so are outside the scope of this chapter. Nevertheless, we wish to make brief mention of this technology for its contribution to protecting and remediating metal contamination in ground and surface waters. Permeable reactive barriers (PRBs) for metal contaminants have found their greatest application in acid mine drainage (AMD) and metallurgy-affected sites to prevent the pollution of the environment by highly acidic and metal-contaminated groundwater plumes. There are two main design configurations for PRBs in groundwater: (1) the funnel and gate system, which uses impermeable curtain walls that direct groundwater flow to treatment zones located between the ends of the curtain walls, and (2) continuous barriers that use trenches filled with reactive media across the entire width of the target plume (Starr and Cherry, 1994; Amos and Younger, 2003). Three modes of action are incorporated in PRBs in combination or by themselves in order to detoxify the plume: (1) an acid-neutralizing mechanism, (2) a mechanism for removing the (heavy) metals and metalloids, and (3) a redox mechanism to convert sulfates to sulfides (Vilensky et al., 2002; Waybrant et al., 2002; Ahn et al., 2003; Amos and Younger 2003; Gibert et al., 2003, 2004; Komnitsas et al., 2004). Many of the soil chemical and colloid physics concepts incorporated into in situ stabilization, soil washing, electrokinetics, and so on, find application in this metal remediation technology.

14.4.2.6. Soil Washing Soil washing is an on-site, but ex situ process in which metal contaminants are separated from the soil by a washing procedure using chemically amended and unamended water following particle-size separation (Mann, 1999; Peters, 1999) (Figure 14.5). When no chemicals are used in the washing fluids, soil washing is strictly a physical process promoting the separation of fine and coarser material; this concentrates the contaminants by reducing the overall soil volume (Petruzzelli et al., 2004). Ultrasonic treatment of larger but coated particles may enhance the removal of fines, crusting sand, and oversized materials (Newman et al., 1997). Contaminants are preferentially partitioned into the finer particle sizes; therefore, particle-size separation of contaminated soil results in a significant concentration of metals in the fine soil fraction (Semer and Reddy, 1996; Peters, 1999; Williford et al., 1999, 2002). The contaminant source may indicate the size fractions in which most of the metal contaminant can be expected to reside. For example, Williford et al. (1999) have shown that Pb concentrations in particles 600 to 2000 μm in size were greater than in particles less than 63 μm in size, in soils from shooting ranges.

Soils with high silt and clay content may not be suitable for soil washing because the volume reduction is not great enough and the metal–soil interaction is so strong that it would require multiple (hence expensive) wash steps (Semer and Reddy, 1996), making the process quite expensive (Tuin and Tels, 1990). Similar conclusions were reached by other researchers (Pagilla and Canter, 1999). Another problem with very fine textured soils is the difficulty of water–solid separation of the slurries. Agglomeration techniques using coagulants have shown some success in overcoming this problem (Majid and Argue, 2001).

Knowledge of the solid-phase speciation of metals in polluted soils and the soil's mineralogy are paramount for devising an appropriate solution for soil washing. A simple aqueous solution may not be sufficient for effective removal of metals (Semer and Reddy, 1996; Davis and Hotha, 1998; Pagilla and Canter, 1999; Tawinteung et al., 2005). Davis and Hotha (1998) demonstrated that the removal efficiency of acidic EDTA solutions was directly dependent on soil mineralogy and the solid-phase speciation of Pb, as these factors determined Pb dissolution kinetics and Pb readsorption to the soil matrix. Similar findings for Cr(VI) desorption from sandy, loamy, and clayey soils were reported (Pagilla and Canter, 1999). Elliott and Herzig (1999) showed that M EDTA solutions were

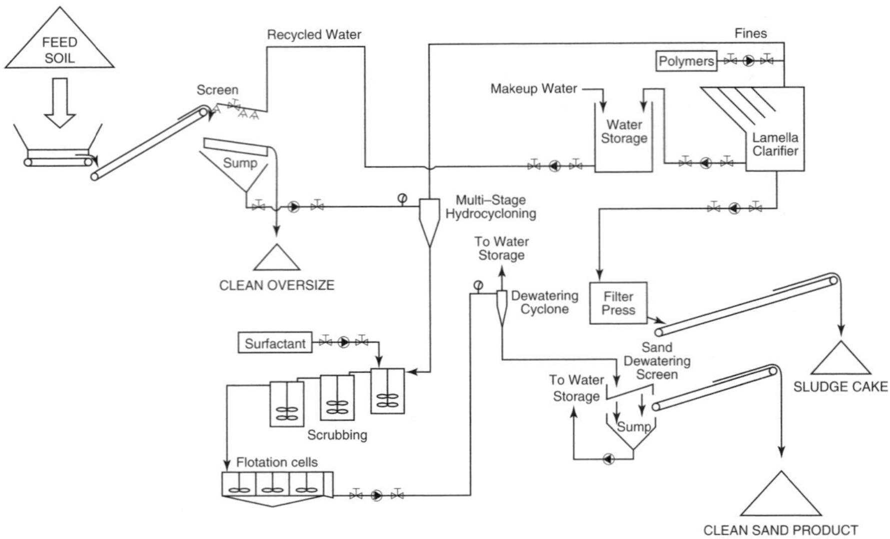

Figure 14.5. Soil washing treatment process. (From Mann, 1999.)

less efficient than M oxalate solutions, as only the latter were able to dissolve the Fe-oxide components with which the contaminants (Zn or Pb) were bound.

Chelators such as EDTA, nitrilotriacetic acid (NTA), 1,2-aminocyclohexane *N,N,N'*,N7-tetraacetic (DCyTA), and ethylene glycol-bis(2-aminoethyl)-*N,N, N',N'*-tetraacetic acid (EGTA) have been studied extensively and are well summarized (Peters, 1999). Chelator concentration and reaction pH influence metal complexation and the success of removal from soils. Sun et al. (2001) observed that batch extraction methods result in 1 : 1 molar extraction ratios of EDTA/metal (Pb, Cd, Zn, Cu) and reveal which metal is more or less soluble in EDTA solutions. Column leaching studies, however, relate the elution patterns and recalcitrance of the metals to desorption and dissolution by EDTA. There is concern over the detrimental effects on soil quality from using chelators because of their biotoxicity, persistence in soil environment, and their removal of beneficial micro- and macronutrients, which leave the washed soil infertile for revegetation when it is backfilled.

Acids and bases have been used as desorption and dissolution agents for metals and oxyanions (van Benschoten et al., 1997; Tawinteung et al., 2005). In Australia, the first full-scale soil washing decontamination project was conducted at a former paint manufacturing facility in Sydney, NSW. The site comprised 7 ha of land contaminated by industrial activity over a period of 70 years. Contamination was mainly by Pb (predominantly as chromate and carbonate), arising from the production of Pb on the site and the use of Pb as a paint pigment. Pilot studies identified the soil fines fraction (approximately 20%) as containing the majority of contaminants and have shown that by using mineral processing the contaminants can be concentrated in a smaller fraction, approximately 20% of the total material requiring remediation (ICI, 1995). Although the technology was used to remediate the site, its application to other sites proved difficult given the cost and excessive corrosive impact on soils, rendering the clean soil ineffective for reuse or backfill. Similar to the Australia study, when HCl was used as a desorbing agent for seven Pb-contaminated soils, the pH of the washing fluid affected the extent, but not the rate, of Pb desorption. The buffering capacity of the contaminated soil counteracts the proton-promoted desorption and dissolution of Pb, reducing the efficiency of acidic washing fluids (Davis and Hotha, 1998). Desorption and dissolution of AsO_4, CrO_4, and other tetrahedral contaminants is enhanced if the wash solution contains a tetrahedral removal agent (e.g., SO_4 or PO_4) (Tokunaga and Hakuta, 2002).

Mixed wastes require mixed washing fluids, but this can mar the efficient removal of one or more contaminants (Semer and Reddy, 1996). For example, HCl was more effective than sulfuric acid in removing heavy metals from a contaminated soil, but H_2SO_4 was more efficient than HCl in removing pesticides and volatile organic halocarbons (VOHs) from the soil, especially in combination with isopropyl alcohol. Abumaizar and Smith (1999) showed that a mixture of 0.01 M Na_2–EDTA and 0.1 M $Na_2S_2O_5$ was more effective in removing Pb, Zn, Cd, and Cr from artificially contaminated soil than the individual wash solutions applied by themselves.

Whereas TCLP and other leach tests may return favorable performance results, bioassays of untreated and treated soils may be better indictors of the success of the remedial action (Chang et al., 1997). Lead-contaminated soils washed with a proprietary acid (no further information on washing fluid is available) and neutralized with lime produced washed soils that affected earthworms, seed germination and root growth more adversely than did the unremediated soils. A rinse step (5 L H_2O kg $soil^{-1}$) simulating two months of rainfall reduced the biotoxicity of the remediated soil. The toxic response of earthworms and seeds was not related to the Pb concentration in the soil but could be attributed to the high salt content in the soil following the acid wash and lime neutralization steps. The osmolality of remediated, but unrinsed, soils was eight times greater than the osmolality of unremediated soil and the remediated but rinsed soil.

Chelant extractors become attractive when they are biodegradable or easily reusable. The ability to recycle the wash fluids is an important cost-saving advantage for this remediation technology. Kim and Ong (1999) showed that lead-contaminated EDTA wastewater can be recycled by introducing Fe(III) at low pH to substitute for EDTA-complexed Pb ions. The Pb ions were subsequently exposed to PO_4 or SO_4 ions, which precipitated readily with Pb. The pH of the Fe(III)–EDTA solution was then increased with NaOH, which promoted the Fe(III)–Na exchange on the EDTA complex and subsequent precipitation of Fe(III) as $Fe(OH)_3$. Wasay et al. (1999) demonstrated the efficiency ($\geq$ 96%) of granular activated carbon (GAC) for removing metals (Cd, Cu, Mn, Pb, and Zn) at optimized pH from leachate used for soil washing. This study raises an interesting point: The pH-dependent sorption behavior of the metals does not always occur as expected (increased sorption at higher pH), which suggests that the metals may have sorbed to the surface of GAC as chelate–metal complexes.

Plant and microbially derived biosurfactants such as surfactin, saponin, rhamnolipids, elastine-like polypeptides, and others have gained increased attention as an alternative to conventional chelants such as EDTA, NTA, or aliphatic acids (Mulligan et al., 1999; Hong et al., 2002; Kos and Lestan, 2003; Neilson et al., 2003; Kos and Lestan, 2004; Prabhukumar et al., 2004; Tandy et al., 2004). Shin et al. (2000) showed that Triton X was an efficient synthetic surfactant for the removal of Cd(II)–iodo and Cu(II)–and Zn(II)–thiocyanato complexes. Shin et al. (2005) showed that increasing the hydrophyllic, nonionic polyethylene oxide chain length on Triton X surfactants on Cd(II) desorption by I^- ligands was detrimental to the removal efficiency of I^-. The authors concluded that the Cd–I^- metal–ligand complex was preferentially stabilized by the micelle's hydrophobic octylphenyl moieties rather than by the hydrophilic functional groups. Unlike synthetic surfactants, biosurfactants have the advantage of being biodegradable, they can be introduced into the contaminated medium by microbes or plant roots, and they are efficient surfactants that lower the surface tension by as much as 27 to 72 mN m^{-1} at minimal concentrations (0.005%) (Mulligan et al., 1999).

Saponin, a triterpene glycoside extracted from the bark of quillaja, was used to extract Cd and Zn from three different types of contaminated soil (Hong et al., 2002). Saponin removed 90 to 100% of Cd and 85 to 95% of Zn sequestered in the exchangeable and carbonated fractions. An increase in leachate pH promoted the release of Cd and Zn from saponin and subsequent precipitation of the metals as their respective hydroxide phases, leaving saponin available for reuse. Specific attention has been paid to the stereoisomers of ethylenediaminedisuccinic acid ([S,S]-EDDS) (Kos and Lestan, 2003, 2004; Prabhukumar et al., 2004; Tandy et al., 2004). Kos and Lestan (2003) showed that although EDTA was more efficient in the desorption of Pb from contaminated soil, EDDS-complexed Pb ions could be more readily sequestered in a permeable reactive barrier consisting of different materials (peat, apatite, vermiculite, hydrogel, and combinations thereof). The permeable reactive barriers promoted the release of Pb from EDDS, Pb sequestration by various ingredients (e.g., pyromorphite formation from apatite), and the degradation of EDDS by microbes in the permeable reactive barrier. Although soil washing technology has found some application in the United States and Europe, the high costs of extractant chemicals and the technology involved (Clijsters and Vangronsveld, 1994) are two of the reasons that this technology is not as popular as in situ capping or excavation.

14.4.3. Physical Technologies

14.4.3.1. In situ Capping In situ capping (cover) involves placing a multilayered cover over a contaminated area to shield the contaminated material physically and chemically from water and wind erosion and to prevent direct contact of the contaminated material with animals and humans (Yland and Wachem,

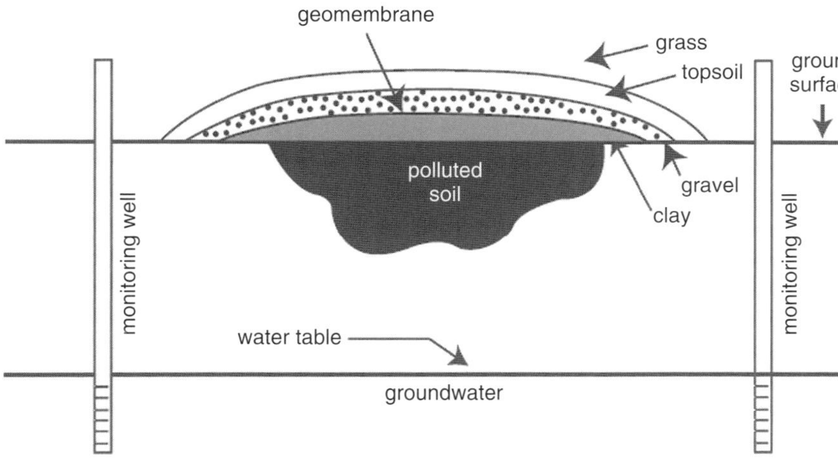

Figure 14.6. Schematic of a typical cap applied to a small area of polluted soil. (From U.S. EPA, 2001a.)

1988; Ellis, 1992; U.S. EPA, 2001a) (Figure 14.6). In situ capping needs only standard earth-moving equipment and the experience that is already available in the construction industry. As a consequence, capping is a popular option (Yland and Wachem, 1988) in Australia and most other countries worldwide. The cover consists of several layers of sand, silt, rock, or geotextile fabric which help prevent water infiltration into the contaminated material. Directly in contact with the contaminants is a clay layer that is covered by a geomembrane. Above this, an optional odor-preventing layer may exist. The top layer consists of soil planted with grasses and other plants, followed by a layer of gravel and pipes to promote drainage.

Although much research still needs to be conducted to define the thickness of caps, which depends on a variety of environmental factors, it is normally from $\frac{1}{2}$ to 2 m thick. Langley and El Saadi (1991) suggested that a $\frac{1}{2} - m$ depth of clean soil is an adequate barrier for residential developments, as this is unlikely to be penetrated by normal gardening activities. In Australia's dry climate, plant root exploration depth, particularly under backyard-gardening activities, is often determined by the depth of domestic watering, about 30 cm (Calder, 1993, although in temperate cooler climates, this is likely to be deeper.

Clean surface covers have been used extensively at Port Pirie in South Australia to reduce exposure to soils contaminated with Pb as a result of the operation of a smelter on the outskirts of the town (Body et al., 1988). Soils in private gardens with greater than 2 500 mg Pb kg^{-1} soil were subject to exposure minimization, either by the installation of clean surface cover or by the removal of contaminated material for disposal.

If upward migration of contaminants (due to hydrogeological conditions) through the surface barrier is considered likely, a surface barrier can be engineered. For instance, impermeable clays or high-density polyethylene (HDPE) liners can be installed immediately below the clean soil cap (Jefferis, 1992). Alternatively, if the downward movement of contaminants is a concern, covers known as capillary break layers, consisting of impermeable layers to prevent infiltration of rain and surface water, may be installed (Cairney, 1992). Most contaminated sites are not uniformly contaminated and have "hot spots" such as the localized spent oxide dumps at gas work sites (ICRCL, 1986). The remainder of the site is relatively uncontaminated or even clean. In instances where the area covered by hot spots is small, the most suitable option may be to excavate the small volume of soil and bury it elsewhere on site in bunkers. Bunkers can be lined with impermeable materials such as clay or HDPE liners to prevent contaminant migration (Dennis, 1993).

One of the major disadvantages of surface barriers is that subsurface flow (i.e., lateral movement of soil water) through the contaminated material cannot be prevented. It is likely that soil contaminants transported in subsurface flow can cause off-site contamination. Obviously, clean cover barrier methods are not appropriate when gaseous or volatile vapor-phase contaminants, such as low-molecular-weight organic and inorganic compounds, are present, as they will not prevent the migration of contaminants into the atmosphere. When on-site containment technologies are applied, soil contaminants remain beneath the barrier in an unchanged form, and one of the main concerns is that the barrier may fail and expose site occupants to hazardous contaminated material. Cracking of clays during drought periods, penetration by plant roots, and breaches of the barrier due to trenching, pipe laying, and so on (Langley and El Saadi, 1991), may all lead to exposure. Failure of barriers to resist forces of expansion and contraction, land shift, and changing groundwater pressure will also limit their life (Ellis, 1992).

The chemistry of soil contaminants is also an important factor determining the success of a barrier layer. Chemical attack on soil surface covers by low-pH soils, and the deflocculation of clays by oils and tars, may also cause failure of the integrity of the barrier, leading to upward migration of the contaminants (Muller-Kirchenbauer et al., 1988). To minimize the risk of exposure due to barrier failure, Tiller (1992) suggests that surface barriers can be used in conjunction with a less sensitive land use that poses the least risk to public health if the barrier fails.

14.4.3.2. Excavation Excavation of contaminated soil is often necessary when small volumes of soil are highly polluted and require either immediate disposal to a landfill or cleaning and backfilling (McNeill and Waring, 1992; U.S. EPA, 2001e). Backhoes, bulldozers, and front-end loaders remove the soil from the ground and place it in tarps or containers to shield workers and others nearby. Excavation has been a cost-effective tool for small quantities of polluted soil for which other technologies are either not suitable or are too expensive.

At present, off-site burial of contaminated soil is used extensively in Australia (Smith, 1993), being applied to over 40% of sites remediated in Victoria between 1988 and 1993 (Shoebridge, 1993). There are a number of reasons why containment in a landfill is currently the most common option in Australia. Containment is well accepted by the remediation industry and its customers, and is perceived to represent the least long-term liability (Shoebridge, 1993). Landfill has also been the subject of significant research concerned with monitoring and predicting contaminant behavior (Gera, 1988; Valocchi, 1988; Barber et al., 1992) and has provided suitable design criteria (Brown, 1993) to model and predict contaminant behavior. For example, in lightly populated countries such as Australia, the availability of suitable landfill sites tends to be higher than in more densely populated northern hemisphere countries. Shoebridge (1993) noted that there were many sites in the Melbourne metropolitan area that were appropriate for the disposal of low-level contaminated soil.

A comparison of remediation in Australia and the United States was made by Shoebridge (1993). Whereas more than 40% of sites in Victoria were remediated by off-site burial, only 7% of U.S. EPA superfund sites were subject to this form of disposal. The difference is due mainly to the strict U.S. regulatory controls, which have forced the cost of landfill up, making it noncompetitive compared to other remediation technologies (DuGuay, 1993). At present off-site containment of contaminated soils in Australia is cost-competitive compared to other soil remediation technologies (Shoebridge, 1993). However, Wheeler (1994) noted that current landfill charges do not accurately reflect the true costs and recommended that charges to the customer should increase by 60 to 90%. Elliott et al. (1989) suggested that off-site burial could not be considered an ideal option, due to the costs involved in removing large volumes of material to landfill. ANZECC/NHMRC (1992) notes that if no appropriate treatment technology is available, contaminated material could be stored *temporarily* in specially designed landfills. The soil could then be recovered for treatment at a later date, once technologies become available or cost competitive.

14.4.3.3. Soil Mixing Soil mixing essentially reduces the hazards associated with metal contaminants through dilution to levels below which exposure is not considered a risk. This can be achieved either by importing clean soil and mixing it with the contaminated soil, or redistributing clean material already on site (Musgrove, 1991). This technique has been used in a number of contaminated sites in Australia, including a former chemical factory site in New South Wales (Finnecy and Johnstone, 1988). Another dilution technique relies on deep ploughing, during which the vertical mixing of surface-contaminated soils with less contaminated subsoils not surprisingly reduces the surface contamination (Thompson-Eagle and Frankenburger, 1992). Vertical mixing of contaminated soils on former agricultural land with low-level agrochemical contamination was proposed in 1995 as a soil remediation strategy, particularly for sites being redeveloped for housing (NSW EPA, 1995). Whether surface contaminants are diluted with clean materials or diluted with subsoils, the total contaminant loading of the site remains the same. In addition, the potential for off-site contaminant transport may even be increased as the contaminant is moved to soil horizons where physicochemical interactions and adsorption are potentially reduced.

14.4.4. Thermal Technologies

14.4.4.1. Incineration Incineration of metal-contaminated media has the advantage of reducing the volume of waste to a minimum prior to land filling. To avoid their volatilization, care must be taken that the incineration temperature does not exceed the boiling points of the metal contaminants (U.S. EPA, 2002). Incineration can be used, for example, to reduce the metals' volume and hence concentrate them in hyperaccumulating plants (see Section 14.3.1.2). The ash of incinerated biomass containing 20 to 40% metals is considered a valuable ore, whereas similar ash containing only 2 to 4% metals is not feasible commercially and would have to be disposed of in a landfill (Chaney et al., 1997).

14.4.4.2. Vitrification *Vitrification* stems from the Latin words *vitrum* and *facere:* meaning "glass" and "to make", respectively. It is an environmental remediation tool that uses Joule resistant heating with graphite electrodes or plasma torches to melt a core of soil in order to allow it to cool into a glasslike material (U.S. EPA, 1997; Mayne et al., 2000; Mulligan et al., 2001a). The glass block is a chemically durable, leach-resistant mass that usually consists of amorphous glassy material resembling volcanic obsidian or partially crystalline igneous rock similar to granite or basalt (Mayne et al., 2000). Vitrification can be carried out both ex situ and in situ. Ex situ vitrification applies heat or a direct electric energy to waste products in a melter. The combustion of organic wastes contributes to heat generation and can reduce energy costs. Size reductions may be required to achieve acceptable throughputs and an even melt. This appears to be less of a problem when the heat is transferred from a plasma torch (U.S. EPA, 1997).

In situ vitrification (ISV) employs an array of electrodes that are inserted vertically into the ground. When current is passed through the electrodes, a starter path consisting of flaked granite, sand, quartz, or other glass-forming material conducts the current between the electrodes until the starter path begins to melt. The melting starter path transfers heat into the surrounding soil, which as it begins to melt, becomes conductive as well. The melt begins to grow outward and downward as the current is gradually increased. Depending on the heat source and the conditions (e.g., moisture content) in the melting soil body, temperatures can reach from 1000 to 7000°C (U.S. EPA, 1997; Mayne et al., 2000; Mulligan et al., 2001b).

Certain site requirements are necessary to prepare the contaminated area for vitrification. Measures include site preparation for the equipment, which may include clearing vegetation and any overburden. Regardless of whether the application is performed in or ex situ, backfill will be required, because the vitreous mass is smaller than the original waste volume.

Vitrification has many limitations, from both a regulatory and a scientific point of view. The large energy costs are disproportionate to the volume of waste that is treated. The U.S. EPA estimates the cost per ton of waste at $400 to 870, which compares with more conventional solidification and stabilization treatment costs of $60 to 290 per ton of waste (U.S. EPA, 1997). Special care must be taken during ISV that volatile organic compounds and volatilizable metals such As, Pb, Cd, and Hg are trapped by an appropriate gas hood spanning the treated area. While combustible organics provide an advantage in ex situ vitrification, during ISV, combustible organic compounds (such as coal residues) can cause underground fires that may overwhelm the capacity of the gas hood. Waste materials that are too wet are unsuitable for vitrification, due to the increased energy required to evaporate the moisture. Waste matrices that contain only a small amount of glass-forming materials will require pretreatment with suitable amendments. Other restrictions exist through underground anomalies such as buried structures and voids that interfere with magmafication of the soil. On sloping land, additional problems arise from melts flowing downhill (U.S. EPA, 1997).

Despite these qualifications, vitrification has some advantages. The vitreous mass is chemically inert and has durable physical properties that lock up contaminants effectively. This is a key advantage when dealing with radioactive contaminants (White et al., 1996). Under some circumstances, the glassy mass can be reused, and some of the energy costs may be recuperated. If the vitrified mass is to be land-filled, the reduction in waste volume is an added cost-saving benefit, in addition to the very inert condition of the glassy material.

14.5. CONCLUSION AND FUTURE DIRECTIONS

Depending on the severity of contamination, heavy metals could be extremely toxic to both the environment and human health. Widespread metal contamination of terrestrial and aquatic environments from both geogenic and anthropogenic sources has been reported in many countries. Risk management of metal-contaminated soil and aquatic ecosystems is an important issue and a great challenge. Its success is necessary to promote sustainable environmental health and to minimize the adverse impact on humans. Unlike organic contaminants, metals do not biodegrade in the environment and have a tendency to accumulate in the genetic horizon unless conditions favor vertical migration into subsurface soils.

A number of physical, chemical, and biological technologies have been developed to remediate metal-contaminated soil. However, apart from the soil washing process, the other techniques are not yet fully developed, and where these have been tried, the duration of the remediation process has been long. Indeed, data on chemical remediation have been obtained using short-term studies, thus causing some concern for the application of outcomes in the field unless specific long-term monitoring is established for bioavailability assessment with time. Soil washing processes are expensive and require expensive excavation and off-site treatment. Dilution of contaminated soils has also been used to minimize the concentration. This process does not, however, remove metals from the contaminated soils. Where immobilization is used with soil amendments, there is concern that changes in soil environmental conditions (e.g., redox conditions) may influence the bioavailability of contaminants bound to soil colloid particles. Bioremediation, using biological wastes and/or microbial strains, offers another avenue for remediation. However, as in the case of physical and chemical technologies, most research involving bioremediation has been demonstrated only in the laboratory, and these studies have focused on metalloids rather than on heavy metals such as Cu, Cd, Pb, and Zn. One of the most attractive techniques being tested involves electrokinetic remediation of contaminated soils. Although this technology has been demonstrated in the lab and at selected field sites, much work still needs to be conducted prior to full-scale in situ remediation. Major challenges for the remediation of metal-contaminated soils include:

- Development of analytical technology for subsurface assessment
- Development of in situ remediation technology that is cheap, effective, and rapid
- Development of an integrated physical, chemical, and biological remediation process
- Remediation of subsurface metal-contaminated soils
- Development of technology for remediation of mixtures (i.e., site contaminated with metals, metalloids, and organics)
- Fundamental process modeling and verification
- Developing genetically engineered microorganisms and genetically modified plants to detoxify metals in contaminated soil
- Enhancement of natural remediation processes

Brown et al. (1998) commented that similar to organic contamination, metal-contaminated soils will go through three stages of evolution: (1) a gross removal process, (2) an active in situ treatment, and (3) a risk-based closure and natural attenuation. They also comment that very little time or money has been spent on developing in situ approaches for metal contamination. It is nine years since Brown et al., published their work, but we have yet to progress to active in situ remediation of metal-contaminated soils, and only a few innovative technologies are currently available for remediation of metal-contaminated soils (U.S. EPA, 1997).

REFERENCES

Abumaizar, R. J., and Smith, E. H. (1999). Heavy metal contaminants removal by soil washing. *J. Hazard. Mater.* **70**(1–2), 71–86.

Acar, Y. B., and Ashawabekh, A. N. (1993). Principles of electrokinetic remediation. *Environ. Sci. Technol.* **27**(13), 2638–2647.

Acar, Y. B., Hamed, J. T., Ashawabekh, A. N., and Gale, R. J.(1994). Removal of cadmium(II) from saturated kaolinite by the application of electrical-current. *Geotechnique* **44**(2), 239–254.

Ahn, J. S., Chon, C. M., Moon, H. S., and Kim, K. W. (2003). Arsenic removal using steel manufacturing by-products as permeable reactive materials in mine tailing containment systems. *Water Res.* **37**(10), 2478–2488.

Akhter, H., Cartledge, F. K., Miller, J., and McLearn, M. (2000). Treatment of arsenic-contaminated soils, I: Soil characterization. *J. Environ. Eng.* **126**(11), 999–1003.

Alpaslan, B., and Yukselen, M. A. (2002). Remediation of lead contaminated soils by stabilization/solidification. *Water Air Soil Pollut.* **133**(1–4), 253–263.

Amos, P. W., and Younger, P. L. (2003). Substrate characterisation for a subsurface reactive barrier to treat colliery spoil leachate. *Water Res.* **37**(1), 108–120.

ANZECC/NHMRC (1992). *Australian and New Zealand Guidelines for the assessment and Management of Contaminated Sites*, Australian and New Zealand Environment and Conservation Council and National Health and Medical Research Council, Canberra, ACT, Australia, 159 p.p., Jan.

Barber, C., Briegel, D. J., Power, T. R.,and Hosking, J. K.(1992). Pollution of groundwater by organic compounds leached from domestic solid wastes: a case study from Morley, Western Australia. In *Groundwater Contamination and Analysis at Hazardous Waste Sites*, ed. Lesage, S., and Jackson, R. E., Marcel Dekker, New York, 357–380.

Bertie, W. R., and Cunningham, S. D. (1994). Remediating soil with green plants. In *Trace Substances: Environment and Health*, ed. Cothern, C. R., Science Reviews, *Northwood, Middlesert, England*, 43–51

Blaylock, M. J., Salt, D. E., Dushenkov, S., Zakharova, O., Gussman, C., Kapulnik, Y., Ensley, B. D., and Raskin, I. (1997). Enhanced accumulation of Pb in Indian mustard by soil-applied chelating agents. *Environ. Sci. Technol.* **31**(3), 860–865.

Body, P. E., Inglis, G. R., and Mulcahy, D. E., (1988). *Lead Contamination in Port Pirie South Australia: A Review of Environmental Lead Which Could Give Rise to an Increase in Childrens Blood Levels*, Rep. 101, South Australian Department of Environment and Planning Adelaide, SA, Australia.

Brady, W. D., Eick, M. J., Grossl, P. R., and Brady, P. V. (2003). A site-specific approach for the evaluation of natural attenuation at metals-impacted sites. *Soil Sediment. Contam.* **12**(4), 541–564.

Brams, E., Anthony, W., and Witherspoon, L. (1989). Biological monitoring of an agricultural food chain: soil cadmium and lead in ruminant tissues. *J. Environ. Qual.* **18**, 317.

Brown, R. V. (1993). The latest trends and developments in waste management in the UK. *Waste Dispos. Water Manag. Aust.* **93**, 9–23.

Brown, R. A., Leahy, M. C., and Pyrih, R. Z., (1998). In Situ *Remediation of Metals Comes of Age*, Wiley, New York, 91–98.

Cairney, T. (1992). Theory and practice of clean cover reclamation. In *Contaminated Land Treatment Technologies*, ed. Rees, J. F., Society of the Chemical Industry, Elsevier Applied Science, London., 97–112.

Calder, I. C. (1993). After remediation how much matters? In *The Health Risk Assessment and Management of Contaminated Sites,* ed. Langley, A., and van Alpen, M., *Proc. 2nd National Workshop on the Health Risk Assessment and Management of Contaminated Sites*, Canberra, ACT, Australia, Apr. 5–7, 1993, South Australian Health Commission, Adelaide, SA, Australia, 65–68.

Cao, X. D., Ma, L. Q., and Tu, C. (2004). Antioxidative responses to arsenic in the arsenic-hyper accumulator Chinese brake fern (*Pteris vittata* L.). *Environ. Pollut.* **128**(3), 317–325.

Chaney, R. L., Malik, M., Li, Y. M., Brown, S. L., Brewer, E. P., Angle, J. S., and Baker, A. J. M. (1997). Phytoremediation of soil metals. *Curr. Opin. Biotechnol* **8**(3), 279–284.

Chang, L. W., Meier, J. R., and Smith, M. K. (1997). Application of plant and earthworm bioassays to evaluate remediation of a lead-contaminated soil. *Arch. Environ. Contam. Toxicol.* **32**(2), 166–171.

Chen, Y. H., Shen, Z. G., and Li, X. D. (2004). The use of vetiver grass (Vetiveria zizanioides) in the phytoremediation of soils contaminated with heavy metals. *Appl. Geochem.* **19**(10), 1553–1565.

Chilingar, G. V., Loo, W. W., Khilyuk, L. F., and Katz, S. A. (1997). Electrobioremediation of soils contaminated with hydrocarbons and metals: progress Report. *Energy Sources* **19**(2), 129–146.

Christie, P., Li, X. L., and Chen, B. D. (2004). Arbuscular mycorrhiza can depress translocation of zinc to shoots of host plants in soils moderately polluted with zinc. *Plant Soil* **261**(1–2), 209–217.

Clijsters, H., and Vangronsveld, J. (1994). Reclamation of contaminated soils by in situ immobilisation of heavy metals. In *Workshop, 1992, Institut National Agronomique,* Paris, ed. Avril, C., and Impens, R., Cadmium Industry Environment, Brussels, Belgium, 57–62.

Cornell, R. M., and Schwertamann, U. (1996). *The Iron Oxides*, VCH, Weinheim, Germany.

Davis, A. P., and Hotha, B. V. (1998). Washing of various lead compounds from a contaminated soil column. *J. Environ. Eng.* **124**(11), 1066–1075.

Davydova, S. L. (1999). Dangerous lead pollution in Russia. *Russ. J. Org. Chem.* **35**(1), 160–163.

Dennis, D. (1993). Columbus Georgia MGP site remediation: case study. *Proc. Management of Manufactured Gas Plant Sites Technology Seminar*, Orlando FL, Mar. 23–24, 1993, Remediation Technologies, Arlington, VA, 283–288.

Dhungana, S., and Crumbliss, A. L. (2005). Coordination chemistry and redox processes in siderophore-mediated iron transport. *Geomicro. J.* **22**(3–4), 87–98.

Di Palma, L., Ferrantelli, P., Merli, C., and Biancifiori, F. (2003a). Recovery of EDTA and metal precipitation from soil flushing. *J. Hazard. Mater.* **103**(1–2), 153–168.

Di Palma, L., Ferrantelli, P., Merli, C., and Petrucci, E. (2003b). Treatment of the solution extracted from metal contaminated soils by reverse osmosis and chemical precipitation. *Ann. Chim.* **93**(12), 1005–1011.

DuGuay, T. (1993). Reclaim metals to clean up soils. *Soils*, Mar., 26–33.

Ehrlich, H. L. (1997). Microbes and metals. *Appl. Microbiol. Biotechnol.* **48**(6), 687–692.

Elliott, H. A., and Herzig, L. M. (1999). Oxalate extraction of Pb and Zn from polluted soils: solubility limitations. *J. Soil Contam.* **8**(1), 105–116.

Elliott, H. A., Linn, J. H. and Shields, G. A.(1989). Role of Fe in extractive decontamination of Pb polluted soils. *Hazard Waste Hazard Mater.* 6, 223–228.

Ellis, B. (1992). On site and in situ treatment of contaminated sites. In *Contaminated Land Treatment Technologies*, ed. Rees, J. F., Society of the Chemical Industry, Elsevier Applied Science, London, 30–46.

Evans, M., Ma, C., and Kingscott, J. (1997). *Recent developments for In Situ Treatment of Metal Contaminated Soils*, U.S. Environmental Protection Agency, Washington, DC, 1–64.

Finnecy, E. E., and Johnstone, E. T. C. (1988). Remedial measures proposed for the rehabilitation of a dioxin contaminated site in New South Wales, Australia. In *Contaminated Soil '88*, ed. Wolf, K., van den Brink, W. J., and Colon, F. J., *Proc. 2nd International TNO/BMFT Conference on Contaminated Soil*, Hamburg, Germany, Apr. 11–15, 1988, Kluwer Academic, Dordrecht, The Netherlands, 1083–1085.

Gent, D. B., Bricka, R. M., Ashawabekh, A. N., Larson, S. L., Fabian, G., and Granade,S. (2004). Bench and field scale evaluation of chromium and cadmium extraction by electrokinetics. *J. Hazard. Mater.* **110**(1–3), 53–62.

Gera, F. (1988). Modelling long term impacts of land disposal of hazardous waste. *In Land Disposal of Hazardous Waste: Engineering and Environmental Issues*, ed. Gronow, J. R., Schofield, A. N., and Jain, R. K., Ellis Horwood, Chichester, West Susset, England, 103–112.

Gibert, O., de Pablo, J., Cortina, J. L., and Ayora, C. (2003). Evaluation of municipal compost/limestone/iron mixtures as filling material for permeable reactive barriers for in-situ acid mine drainage treatment. *J. Chem. Technol. Biotechnol.* **78**(5), 489–496.

Gibert, O., de Pablo, J., Cortina, J. L. and Ayora, C. (2004). Chemical characterisation of natural organic substrates for biological mitigation of acid mine drainage. *Water Res.* **38**(19), 4186–4196.

Green, C., and Hoffnagle, A. (2004). *Phytoremediation Field Studies Database for Chlorinated Solvents, Pesticides, Explosives and Metals*, U.S. Environmental Protection Agency, Washington, DC.

Groudev, S. N., Spasova, I. I., and Georgiev, P. S. (2001). In situ bioremediation of soils contaminated with radioactive elements and toxic heavy metals. *Int. J. Miner. Process.* **62**, 301–308.

Hall, A. H. (2002). Chronic arsenic poisoning. *Toxicol. Lett.* **128**(1–3), 69–72.

Hills, C. D., and Pollard, S. J. T. (1997). The influence of interference effects on the mechanical, micro-structural and fixation characteristics of cement-solidified hazardous waste forms. *J. Hazard. Mater.* **52**(2–3), 171–191.

Hinwood, A.L. (1991). Principles for setting soil criteria approaches to soil guidelines-Environment. In *The Health Risk Assessment and Management of Contaminated Sites*, ed. El Saadi, O., and Langley, A., *Proc. National Workshop on the Health Risk Assessment and Management of Contaminated Sites.* Aug. 5–7, 1991, South Australian Health Commission, Adelaide, SA, Australia, 87–92.

Hofte, M., Buysens, S., Koedam, N., and Cornelis, P. (1993). Zinc Affects siderophore-mediated high affinity iron uptake systems in the rhizosphere *Pseudomonasaeruginosa. Biometals* **6**(2), 85–91.

Hong, K. J., Tokunaga, S., and Kajiuchi, T. (2002). Evaluation of remediation process with plant-derived biosurfactant for recovery of heavy metals from contaminated soils. *Chemosphere* **49**(4), 379–387.

Hu, X., and Boyer, G. L. (1997). Accumulation of copper and cadmium by bacteria through siderophore mediation. *Abstr. Pap. Am. Chem. Soc.* **213**, 82.

Huang, J. W. W., Chen, J. J., Berti, W. R., and Cunningham, S. D. (1997). Phytoremediation of lead-contaminated soils: role of synthetic chelates in lead phytoextraction. *Environ. Sci. Technol.* **31**(3), 800–805.

Huang, J. W. W., Blaylock, M. J., Kapulnik, Y., and Ensley, B. D. (1998). Phytoremediation of uranium contaminated soils: role of organic acids in triggering uranium hyperaccumulation in plants. *Environ. Sci. Technol.* **32**(13), 2004–2008.

ICI, (1995). *Dulux Cabarita Project Turning Full Circle*, Tech. Rep., ICI, Dulux, Australia.

ICRCL, (1986). *Notes on the Redevelopment of Gaswork Sites*, 5th ed., Interdepartmental Committee for the Redevelopment of Contaminated Land, Report 18/79, Department of the Environment, London.

Jefferis, S. A. (1992). Remedial barriers and containment. In *Contaminated Land Treatment Technologies*, ed. Rees, J. F., Society of the Chemical Industry, Elsevier Applied Science, London, 58–82.

Jing, C. Y., Korfiatis, G. P. and Meng, X. G. (2003). Immobilization mechanisms of arsenate in iron hydroxide sludge stabilized with cement. *Environ. Sci. Technol.* **37**(21), 5050–5056.

Juillot, F., Morin, G., Ponthieu, M., Benedetti, M., Trainor, T., Kinniburgh, D., Kretzschmar, R., Brown, G. E., Ildefonse, P., and Calas, G. (2002). Zinc trapping in layered double hydroxydes (LDHs) and phyllosilicates in contaminated soils from Western Europe. *Geochim. Cosmochim. Acta* **66**(15A), A374–A374.

Kaplan, D. I., Knox, A. S., and Myers, J. (2002). Mercury geochemistry in wetland and its implications for in situ remediation. *J. Environ. Eng.* **128**(8), 723–732.

Kim, S. O. and Kim, K. W. (2001). Monitoring of electrokinetic removal of heavy metals in tailing soils using sequential extraction analysis. *J. Hazard. Mater.* **85**(3), 195–211.

Kim, C. S., and Ong, S. K. (1999). Recycling of lead-contaminated EDTA wastewater. *J. Hazard. Mater.* **69**(3), 273–286.

Kim, S. O., Kim, K. W., and Stüben, D. (2002a). Evaluation of electrokinetic removal of heavy metals from tailing soils. *J. Environ. Eng.* **128**(8) 705–715.

Kim, S. O., Moon, S. H., Kim, K. W., and Yun, S. T. (2002b). Pilot scale study of the ex-situ electrokinetic removal of heavy metals from municipal wastewater sludges. *Water. Res.* **36**, 4765–4774.

Komnitsas, K., Bartzas, G., and Paspaliaris, I. (2004). Efficiency of limestone and red mud barriers: laboratory column studies.*Miner. Eng.* **17**(2), 183–194.

Kos, B., and Lestan, D. (2003). Induced phytoextraction/soil washing of lead using biodegradable chelate and permeable barriers. *Environ. Sci. Technol.* **37**(3), 624–629.

Kos, B., and Lestan, D. (2004). Soil washing of Pb, Zn and Cd using biodegradable chelator and permeable barriers and induced phytoextraction by *Cannabis sativa. Plant Soil* **263**(1–2), 43–51.

Kukier, U., and Chaney, R. L. (2001). Amelioration of nickel phytotoxicity in muck and mineral soils. *J. Environ. Qual.* **30**(6), 1949–1960.

Lai, H. Y., and Chen, Z. S. (2004). Effects of EDTA on solubility of cadmium, zinc, and lead and their uptake by rainbow pink and vetiver grass. *Chemosphere* **55**(3), 421–430.

Langley, A. and El Saadi, O. (1991). Protocol for the health risk assessment and management of contaminated Sites. *Summary of a National Workshop on the Health Risk Assessment and management of Contaminated Sites,* South Australian Health Commission, Aug. 5–7 1991, Adelaide, SA, Australia.

Lee, S. W., Kim, J. Y., Lee, J. U., Ko, I. W., and Kim, K. W. (2004). Removal of arsenic in tailings by soil flushing and the remediation process monitoring. *Environ. Geochem. Health* **26**, 403–409.

Leist, M., Casey, R. J., and Caridi, D. (2000). The management of arsenic wastes: problems and prospects. *J. Hazard. Mater.* **76**(1–2), 125–138.

Lim, J. M., Salido, A. L., and Butcher, D. J. (2004). Phytoremediation of lead using Indian mustard (*Brassica juncea*) with EDTA and electrodics. *Microchem. J.* **76**, 3–9.

Ma, C., and Kingscott, J. (1997). *Recent Developments for In Situ Treatment of Metal Contaminated Soils.* U.S. Environmental Production Agency, Washington, DC, 1–48.

Ma, L. Q., Komar, K. M., Tu, C., Zhang, W. H., Cai, Y., and Kennelley, E. D. (2001). A fern that hyper-accumulates arsenic: a hardy, versatile, fast-growing plant helps to remove arsenic from contaminated soils. *Nature* **409**(6820), 579.

Majid, A., and Argue, S. (2001). Remediation of heavy metal contaminated solid wastes using agglomeration techniques. *Miner. Eng.* **14**(11), 1513–1525.

Manceau, A., Marcus, M. A., and Tamura, N. (2002). Quantitative speciation of heavy metals in soils and sediments by synchrotron x-ray techniques. In *Applications of Synchrotron Radiation in Low-Temperature Geochemistry and Environmental Science*, Vol, 49, ed. Fenter, P., and Sturchio, N. C., Mineralogical Society of America, Washington, DC, 341–428.

Mann, M. J. (1999). Full scale and pilot-scale soil washing. *J. Hazard. Mater.* **66**(1–2), 119–136.

Manna, M., Sanjay, K., and Shekhar, R. (2003). Electrochemical cleaning of soil contaminated with dichromate lixiviant. *Int. J. Miner. Process.* **72**, 401–406.

Matthews, D. A., and Effler, S. W. (2003). Decreases in pollutant loading from residual soda ash production waste. *Water Air Soil Pollut.***146**(1–4), 55–73.

Mayne, P. W., Burns, S. E., and Circeo, L. J. (2000). Plasma magmavication of soils by nontransferred arc. *J. Geotech. Geoenviron. Eng.* **126**(5), 387–396.

McBride, M. B. (1994). *Environmental Chemistry of Soils*, Oxford University Press, New York.

McNeill, K. R., and Waring, S. (1992). Vitrification of contaminated soils. In *Contaminated Land Treatment Technologies*, ed. Rees, J. F., Society of Chemical Industry, Elsevier Applied Science, London, 143–159.

Mench, M. J., Didier, V. L., Loffler, M., Gomez, A., and Masson, P. (1994). A mimicked in-situ remediation study of metal-contaminated soils, with emphasis on cadmium and lead. *J. Environ. Qual.* **23**, 58–63.

Miller, J., Akhter, H., Cartledge, F. K., and McLearn, M. (2000). Treatment of arsenic-contaminated soils, II: Treatability study and remediation. *J. Env. Eng.* **126**(11), 1004–1012.

Moon, D. H., Dermatas, D. and Menounou, N. (2004). Arsenic immobilization by calcium–arsenic precipitates in lime treated soils. *Sci. Total Environ.* **330**(1–3), 171–185.

Muller-Kirchenbauer, D., Friedrich, H., Gremmel, W. D., Markwardt, W., and Rogner, J. (1988). New results and aspects in the area of seal wall research. In *Contaminated Soil '88*, ed. Wolf, K., van den Brink, W. J., and Colon, F. J., *Proc. 2nd International TNO/BMFT Conference on Contaminated Soil*, Apr. 11–15, 1988, Hamburg, Germany, Kluwer Academic, Dordrecht, The Netherlands, 549–560.

Mulligan, C. N., Yong, R. N., Gibbs, B. F., James, S. and Bennett, H. P. J. (1999). Metal removal from contaminated soil and sediments by the biosurfactant surfactin. *Environ. Sci. Technol.* **33**(21), 3812–3820.

Mulligan, C. N., Yong, R. N., and Gibbs, B. F. (2001a). An evaluation of technologies for the heavy metal remediation of dredged sediments.*J. Hazard. Mater.* **85**(1–2), 145–163.

Mulligan, C. N., Yong, R. N., and Gibbs, B. F. (2001b). Remediation technologies for metal-contaminated soils and groundwater: an evaluation. *Eng. Geol.* **60**(1–4), 193–207.

Mulligan, C. N., Yong, R. N., and Gibbs, B. F. (2001c). Surfactant-enhanced remediation of contaminated soil: a review. *Eng. Geol.* **60**(1–4), 371–380.

Musgrove, S. (1991). An assessment of the efficiency of remedial treatment for metal polluted soil. In *Land Reclamation: An End to Dereliction,* ed. Davies, M. C. R., Proc International *proc. international Conference on Land Reclamation*, University of Wales, Apr. 1991, Elsevier Science, Barking, Essex, England, 248–255.

Naidu, R., Smith, E., Megharaj, M., Smith, L. H., Smith, J., Sreedaran, R., Churchman, G. J., Kookana, R. S., Juhasz, A., Gates, W., Oliver, D. P., Ragusa, S., Taylor, G., and Peter, P. (1999). Options for remediating metal contaminated soils: an overview. In *Proc. Contaminated Site Remediation Conference: Challenges Posed by Urban and Industrial Contaminants,* Fremantle, WA, Australia, ed. Johnson, C., 483–496.

Naidu, R., Smith, E., Owens, G., Bhattacharya, P., and Nadebaum, P. (eds.) (2006). *Managing Arsenic in the Environment: From Soil to Human Health.* CSIRO Publishing, Collingwood, VIC, Australia, 670 pp.

Neilson, J. W., Artiola, J. F., and Maier, R. M. (2003). Characterization of lead removal from contaminated soils by nontoxic soil-washing agents. *J. Environ. Qual.* **32**(3), 899–908.

Neubauer, U., and Furrer, G. (1999). The use of voltammetry for sorption studies of heavy metals on mineral surfaces in the presence of the siderophore desferrioxamine B. *Anal. Chim. Acta* **392**(2–3), 159–173.

Newman, A. P., Lorimer, J. P., Mason, T. J., and Hutt, K. R. (1997). An investigation into the ultrasonic treatment of polluted solids. *Ultrason. Sonochem.* **4**(2), 153–156.

NSW EPA (1995). *Guidelines for the Vertical Mixing of Soil on Former Broad-Acre Agricultural Land*, New South Wales Environmental Protection Authority, Chatswood NSW, Australia.

Paff, S. W., Bosilivich, B., and Kardos, N. J. (1994). *Acid Extraction Treatment System for Treatment of Metal Contaminated Soils*, Centre for Hazardous Materials Research, Pittsburgh, PA, 1–51.

Page, G. W. (1997). *Contaminated Sites and Environmental Cleanup: International Approaches to Prevention, Remediation, and Reuse,* Academic Press, San Diego, CA.

Pagilla, K. R., and Canter, L. W. (1999). Laboratory studies on remediation of chromium-contaminated soils. *J. Environ. Eng.* **125**(3), 243–248.

Peters, R. W. (1999). Chelant extraction of heavy metals from contaminated soils. *J. Hazard. Mater.* **66**(1–2), 151–210.

Petrisor, I. G., Dobrota, S., Komnitsas, K., Lazar, I., Kuperberg, J. M. and Serban, M. (2004). Artificial inoculation: perspectives in tailings phytostabilization. *Int. J. Phytoremediat.* **6**(1), 1–15.

Petruzzelli, G., Barbafieri, M., Saponaro, B. L. S., Milani, A., and Pedron, F. (2004). Bench scale evaluation of soil washing for heavy metal contaminated soil at a former manufactured gas plant site. Environ. Contam. Toxicol. **73**, 38–44.

Pierzynski, G. M., Sims, J. T., and Vance, G. F. (1994). *Soils and Environmental Quality*, CRC Press, Boca Raton, FL.

Polettini, A., Pomi, R., and Valente, M. (2004). Remediation of a heavy metal–contaminated soil by means of agglomeration. *J. Environ. Sci. Health A Toxic/Hazard. Subst. Environ. Eng.* **39**(4), 999–1010.

Prabhukumar, G., Matsumoto, M., Mulchandani, A., and Chen, W. (2004). Cadmium removal from contaminated soil by tunable biopolymers. *Environ. Sci. Technol.* **38**(11), 3148–3152.

Prokop, G., Schamann, M., and Edelgaard, I. (2000). *Management of Contaminated Sites in Western Europe*, *European Environment Agency*, Copenhagen, Denmark, 1–171.

Rabinowitz, M. B. (1993). Modifying soil lead bioavailability by phosphate addition. *Bull. Environ. Contam.* **51**, 438–444.

Raskin, I., Smith, R. D., and Salt, D. E. (1997). Phytoremediation of metals: using plants to remove pollutants from the environment. *Curr. Opin. Biotechnol.* **8**(2), 221–226.

Reddy, K. R., and Chinthamreddy, S. (1999). Electrokinetic remediation of heavy metal-contaminated soils under reducing environments. *Waste Manag.* **19**(4), 269–282.

Reddy, K. R., and Chinthamreddy, S. (2003a). Effects of initial form of chromium on electrokinetic remediation in clays. *Adv. Environ. Res.* **7**(2), 353–365.

Reddy, K. R., and Chinthamreddy, S. (2003b). Sequentially enhanced electrokinetic remediation of heavy metals in low buffering clayey soils. *J. Geotechnol. Geoenviron. Eng.* **129**(3), 263–277.

Reddy, K. R., and Chinthamreddy, S. (2004). Enhanced electrokinetic remediation of heavy metals in glacial till using different electrolyte solutions. *J. Environ. Eng.* **130**(4), 442–455.

Reddy, C. N., and Patrick, J. W. H. (1977). Effect of redox potential and pH on the uptake of Cd and Pb by rice plants. *J. Environ. Qual.* **6**, 259–262.

Reddy, K. R., Parupudi, U. S., Devulapalli, S. N., and Xu, C. Y. (1997). Effects of soil composition on the removal of chromium by electrokinetics. *J. Hazard. Mater.* **55**(1–3), 135–158.

Reddy, K. R., Xu, C. Y., and Chinthamreddy, S. (2001). Assessment of electrokinetic removal of heavy metals from soils by sequential extraction. *J. Hazard. Mater.* **84**(2–3), 279–296.

Reddy, K. R., Chaparro, C., and Saichek, R. E. (2003a). Iodide-enhanced electrokinetic remediation of mercury-contaminated soils. *J. Environ. Eng.* **129**(12), 1137–1148.

Reddy, K. R., Chaparro, C., and Saichek, R. E. (2003b). Removal of mercury from clay soils using electrokinetics. *J. Environ. Sci. Health A* **38**(2), 307–338.

Reddy, K. R., Chinthamreddy, S., Saichek, R. E., and Cutright, T. J. (2003c). Nutrient amendment for the bioremediation of a chromium-contaminated soil be electrokinetics. *Energy Sources* **25**(9), 931–943.

Roberts, D. R., Scheidegger, A. M., and Sparks, D. L. (1999). Kinetics of mixed Ni–Al precipitate formation on a soil clay fraction. *Environ. Sci. Technol.* **33**, 3749–3754.

Ruby, M. V., Davis, A., and Nicholson, A. (1994). In situ formation of lead phosphates in soils as a method to immobilize lead. *Environ. Sci. Technol.* **28**, 646–654.

Scheidegger, A. M., Strawn, D. G., Lamble, G. M. and Sparks, D. L. (1998). The kinetics of mixed Ni–Al hydroxide formation on clay and aluminum oxide minerals: a time-resolved XAFS study. *Geochim. Cosmochim. Acta* **62**, 2233–2245.

Scheinost, A. C., and Sparks, D. L. (2000). Formation of layered single-and double-metal hydroxide precipitates at the mineral/water interface: a multiple-scattering XAFS analysis. *J. Colloid Interface. Sci.* **223**, 167–178.

Scheinost, A. C., Ford, R. G., and Sparks, D. L. (1999). The role of Al in the formation of secondary Ni precipitates on pyrophyllite, gibbsite, talc, and amorphous silica: a DRS Study. *Geochim. Cosmochim. Acta* **63**, 3193–3203.

Semer, R., and Reddy, K. R. (1996). Evaluation of soil washing process to remove mixed contaminants from a sandy loam. *J. Hazard. Mater.* **45**(1), 45–57.

Shin, M., Umebayashi, Y., Kanzaki, R., and Ishiguro, S. I. (2000). Formation of copper(II) thiocyanato and cadmium(II) iodo complexes in micelles of nonionic surfactants with varying polyethylene chain length. *J. Colloid Interface Sci.* **225**(1), 112–118.

Shin, M., Barrington, S. F., Marshall, W. D., and Kim, J. W. (2005). Effect of surfactant alkyl chain length on soil cadmium desorption using surfactant/ligand systems. *Chemosphere* **58**(6), 735–742.

Shoebridge, L. (1993). A review of soil remediation practice in Australia. In *The Health Risk Assessment and Management of Contaminated Sites,* ed. Langley, A., and van Alpen, M. *Proc. 2nd National Workshop on the Health Risk Assessment and Management of Contaminated Sites*, Canberra, ACT, Australia, South Australian Health Commission, Apr 1993, 5–7, Adelaide, SA, Australia, 223–227.

Siebner-Freibach, H., Yariv, S., Lapides, Y., Hadar, Y. and Chen, Y. N. (2005). Thermo-FTIR spectroscopic study of the siderophore ferrioxamine B: spectral analysis and stereochemical implications of iron chelation, pH, and temperature. *J. Agric. Food Chem.* **53**(9), 3434–3443.

Smith, B. (1993). Remediation update finding the remedy. *Waste Manag. Environ.* **4**, 24–30.

Sogorka, B., Berti, B., Gabert, H., Gombert, D., Goswami, D., Jensen, R. H., Mueller, R., Parker, R., Sogorka, D., and Tool, A. (1997a). *Emerging Technol-ogies for the Remediation of Metals in Soils: Electrokinetics*, Interstate Technology and Regulatory Cooperation, Trenton, NJ, 1–42.

Sogorka, B., Berti, B., Gabert, H., Gombert, D., Goswami, D., Jensen, R. H., Mueller, R., Parker, R., Sogorka, D., and Tool, A. (1997b). *Emerging Technologies for the Remediation of Metals in Soils: In situ Stabilization/Inplace Inactivation, Interstate Technology and Regulatory Cooperation*, Trenton, NJ, 1–31.

Sogorka, B., Berti, B., Gabert, H., Gombert, D., Goswami, D., Jensen, R. H., Mueller, R., Parker, R., Sogorka, D., and Tool, A. (1997c). *Emerging Technologies for the Remediation of Metals in Soils:* Phytoremediation, Interstate Technology and Regulatory Cooperation, Trenton, NJ, 1–15.

Sparks, D. L. (2003). *Environmental Soil Chemistry*, Academic Press, San Diego, CA.

Starr, R. C., and Cherry, J. A. (1994). In-situ remediation of contaminated ground water: the funnel-and-gate system.*Ground Water* **32**(3), 465–476.

Stevenson, F. J. (1994). *Humus Chemistry: Genesis, Composition, Reactions*, Wiley, New York.

Stumm, W. (1992). *Chemistry of the Solid–Water Interface*, Wiley, New York.

Suer, P., and Allard, B. (2003). Mercury transport and speciation during elctrokinetic soil remediation. *Water Air Soil Pollut* **143**, 99–109.

Sun, B., Zhao, F. J., Lombi, E., and McGrath, S. P. (2001). Leaching of heavy metals from contaminated soils using EDTA. *Environ. Pollut.* **113**(2), 111–120.

Tandy, S., Bossart, K., Mueller, R., Ritschel, J., Hauser, L., Schulin, R., and Nowack, B. (2004). Extraction of heavy metals from soils using biodegradable chelating agents. *Environ. Sci. Technol.* **38**(3), 937–944.

Tawinteung, N., Parkpian, P., DeLaune, R. D., and Jugsujinda, A. (2005). Evaluation of extraction procedures for removing lead from contaminated soil. *J. Environ. Sci. Health A* **40**, 384–407.

Thompson-Eagle, E. T., and Frankenburger, W. T., Jr. (1992). Bioremediation of soils contaminated with selenium. In*Advances in Soil Science,* Vol. 17, ed. R Lal R., and Stewart, B. A., Springer-Verlag, New York, 261–310.

Tiller, K. G. (1992). Urban soil contamination in Australia. *Aust. J. Soil Res.* **30**, 937–957.

Tokunaga, S., and Hakuta, T. (2002). Acid washing and stabilization of an artificial arsenic-contaminated soil.*Chemosphere* **46**(1), 31–38.

Townsend, T., Jang, Y. C., and Tolaymat, T. (2003). *A Guide to the Use of Leaching Tests in Solid Waste Management Decision Making,* Gainesville, Department of Environmental Engineering Sciences, University of Florida**,** FL, 1–31.

Tuin, B. J. W., and Tels, M. (1990). Extraction kinetics of six heavy metals from contaminated clay soils. *Environ. Technol.* **11**, 541–554.

U. S. EPA (1996). INERT Soil–Metals Action Team, Research and Development, *Solid Waste and Emergency Response,* 542/F-96/101D, Washington DC.

U.S. EPA (1997). *Technology Alternatives for the Remediation of Soils Contaminated with As, Cd, Cr, Hg, and Pb.* Office of Response, Washington DC, U.S. EPA Eng. Bull. 1–20.

U.S. EPA (2000). *In Situ Treatment of Soil and Groundwater Contaminated with Chromium: Technical Resource Guide*, U.s. Environmental Protection Agency Washington, DC, 1–84.

U.S. EPA.(2001a). *A Citizen's Guide to Capping, The Citizen's Guide Series,* Office of Solid Waste and Emergency Response, U.S. Environmental Protection Agency, Washington, DC.

U.S. EPA (2001b). *A Citizen's Guide to Fracturing, The Citizen's Guide Series*, Office of Solid Waste and Emergency Response, U.S. Environmental Protection Agency, Washington, DC.

U.S. EPA (2001c). *A Citizen's Guide to In Situ Flushing, The Citizen's Guide Series*, Office of Solid Waste and Emergency Response, U.S. Environmental Protection Agency, Washington, DC.

U.S. EPA (2001d). *A Citizen's Guide to Monitored Natural Attenuation, The Citizen's Guide Series*, Office of Solid Waste and Emergency Response, U.S. Environmental Protection Agency, Washington, DC.

U.S. EPA (2001e). *A Citizen's Guide to Soil Excavation. The Citizen's Guide Series*, Office of Solid Waste and Emergency Response, U.S. Environmental Protection Agency, Washington, DC.

U.S. EPA (2001f). *A Citizen's Guide to Solidification/stabilization, The Citizen's Guide Series*, Office of Solid Waste and Emergency Response, U.S. Environmental Protection Agency Washington, DC.

U.S. EPA (2002). *A Citizen's Guide to Incineration, The citizen's guide series*, Office of Solid Waste and Emergency Response, U.S. Environmental Protection Agency, Washington, DC.

U.S. EPA (2004a). *Hazard Ranking System: Hazardous Substance Benchmarks*. U.S. Environmental Protection Agency, Washington, DC.

U.S. EPA.(2004b). *Treatment Technologies for Site Clean-up*: *Annual status report* 11th ed., Office of Solid Waste and Emergency Response, U.S Environmental Protection Agency Washington, DC, 1–228.

Valocchi, A. J. (1988). Mathematical modelling of the transport of pollutants from hazardous waste landfills. In *Land Disposal of Hazardous Waste: Engineering and Environmental Issues*. ed. Gronow, J. R., Schofield, A. N. and Jain, R. K., Ellis Horwood, Chichester, West Sussex, England, 153–164.

van Benschoten, J. E., Matsumoto, M. R., and Young, W. H. (1997). Evaluation and analysis of soil washing for seven lead-contaminated soils. *J. Environ. Eng.* **123**(3), 217–224.

Vig, K., Megharaj, M., Sethunathan, N., and Naidu, R. (2003). Bioavailability and toxicity of cadmium to microorganisms and their activities in soil: a review. *Adv. Environ. Res.* **8**, 121–135.

Vilensky, M. Y., Berkowitz, B., and Warshawsky, A. (2002). In situ remediation of groundwater contaminated by heavy and transition metal ions by selective ion-exchange methods.*Environ. Sci. Technol.* **36**(8), 1851–1855.

Visoottiviseth, P., Francesconi, K. and Sridokchan, W. (2002). The potential of Thai indigeneous plant species for the phytoremediation of arsenic contaminated land. *Environ. Pollut.* **118**, 453–461.

Wasay, S. A., Barrington, S., and Tokunaga, S. (1999). Efficiency of GAC for treatment of leachate from soil washing process. *Water Air Soil Pollut.* **116**(3–4), 449–460.

Wasay, S. A., Barrington, S., and Tokunaga, S. (2001). Organic acids for the in situ remediation of soils polluted by heavy metals: soil flushing in columns. *Water Air Soil Pollut.* **127**(1–4), 301–314.

Waybrant, K. R., Ptacek, C. J., and Blowes, D. W. (2002). Treatment of mine drainage using permeable reactive barriers: column experiments. *Environ. Sci. Technol.* **36**(6), 1349–1356.

Webb, S. M., Gaillard, J. F., Ma, L. Q., and Tu, C. (2003). XAS speciation of arsenic in a hyper-accumulating fern. *Environ. Sci. Technol.* **37**(4), 754–760.

Wheeler, M. (1994). Site remediation: an Australian perspective on best practice. In *Environmental Management Industry Association of Australia,* Yearbook 1994, ed. Cole, J., Executive Media Waterloo, NSW, Australia 156–160.

White, T. P., Toland, R., Jackson, J. A., and Kloeber, J. M. (1996). Simulation and optimization of a new waste remediation process. *Omega Int. J. Manag. Sci.* **24**(6), 705–714.

Whiting, S. N., Reeves, R. D., Richards, D., Johnson, M. S., Cooke, J. A., Malaisse, F., Paton, A., Smith, J. A. C., Angle, J. S., Chaney, R. L., Ginocchio, R., Jaffre, T., Johns, R., McIntyre, T., Purvis, O. W., Salt, D. E., Chat, H., Zhao, F. J., and Baker, A. J. M. (2004). Research priorities for conservation of metallophyte biodiversity and their potential for restoration and site remediation. *Restor. Ecol.* **12**(1), 106–116.

Williford, C. W., Li, Z., Wang, Z. M., and Bricka, R. M. (1999). Vertical column hydroclassification of metal-contaminated soils. *J. Hazard. Mater.* **66**(1–2), 15–30.

Williford, C. W., Bricka, R. M., and Foster, C. C. (2002). Reduction of suspended solids following hydroclassification of metal-contaminated soils. *J. Hazard. Mater.* **92**(1), 63–75.

Wongkongkatep, J., Fukushi, K., Parkpian, P., De Laune, R. D., and Jugsujinda, A. (2003). Arsenic uptake by native fern species in Thailand: effect of chelating agents on hyperaccumulation of arsenic by *Pityrogramma calomelanos*. *J. Environ. Sci. Health, A Toxic/Hazard. Substances Environ Eng.* **38**(12), 2773–2784.

Wu, J., Hsu, F. C., and Cunningham, S. D. (1999). Chelate-assisted Pb phytoextraction: Pb availability, uptake, and translocation constraints. *Environ. Sci. Technol.* **33**(11), 1898–1904.

Yeung, A. T., Hsu, C. N., and Menon, R. M. (1996). EDTA enhanced electrokinetic extraction of lead. *J. Geotech. Eng.* **122**(8), 666–673.

Yland, M. W. F., and Wachem, E. G. (1988). Soil covering systems as remedial action in contaminated housing areas in the Netherlands. In *Contaminated Soil '88*. ed. Wolf, K., van den Brink, W. J., and Colon, F. J., *Proc 2nd International TNO/BMFT Conference on Contaminated Soil*, Apr. Hamburg, Germany, 11–15, 1988, Kluwer Academic, Dordrecht, The Netherlands, 597–599.

Zhang, W. H., Cai, Y., Downum, K. R., and Ma, L. Q. (2004a). Arsenic complexes in the arsenic hyperaccumulator *Pteris vittata* (Chinese brake fern). *J. Chromatogr. A* **1043**(2), 249–254.

Zhang, W. H., Cai, Y., Downum, K. R., and Ma, L. Q. (2004b). Thiol synthesis and arsenic hyperaccumulation in *Pteris vittata* (Chinese brake fern). *Environ. Pollut.* **131**(3), 337–345.

Zhao, F. H., Dunham, S. J., and McGrath, S. P. (2002). Arsenic hyperaccumulation by different fern species. *New Phytol.* **156**, 27–31.

15

PHOSPHATE-INDUCED LEAD IMMOBILIZATION IN CONTAMINATED SOILS: MECHANISMS, ASSESSMENT, AND FIELD APPLICATIONS

R. Melamed
CETEM/MCT–Center for Mineral Technology, Rio de Janeiro, Brazil
L. Q. Ma
University of Florida, Gainesville, Florida

Biophysico-Chemical Processes of Heavy Metals and Metalloids in Soil Environments,
Edited by Antonio Violante, Pan Ming Huang, and Geoffrey Michael Gadd

15.1. INTRODUCTION

Many anthropogenic activities generate wastes and tailings containing toxic elements that require appropriate management to protect the environment adequately. An effective environmental management demands an understanding of various processes involved within the substrate itself, as well as the interactions of the substrate with the surroundings, and possible pathways of toxic elements to the atmosphere, to the soil, to water resources, and to the groundwater. Contamination of ground and surface waters by heavy metals from previously abandoned disposal sites and some currently operating sites has caused much environmental concern. Heavy metals are toxic to animals, humans, and aquatic life. The ubiquitous nature of heavy metals, their toxicity even in trace quantities, and their tendency to bioaccumulate in the food chain have led to stricter environmental regulations in heavy metal discharges. Therefore, it is necessary to develop schemes to remove heavy metals from both wastewaters and landfill leachates and to remediate metal-contaminated sites (Ma et al., 1993).

Among the remediation technologies available for soils contaminated with heavy metals, in situ immobilization techniques are of particular interest because of their cost-effectiveness. The use of phosphorus (P) amendments has been identified as a potentially efficient in situ remediation technology. These amendments are available in various forms, and they are environmentally friendly and simple to use. Numerous studies have been conducted to demonstrate the mechanisms and effectiveness of heavy metal immobilization in soils using various P sources (Ma et al., 1993; Chen et al., 1997a; Singh et al., 2001; Mavropoulos et al., 2002).

Phosphate has been shown to remove Cd and Zn from aqueous solutions (Chen et al., 1997b), to reduce Hg mobility (Melamed and Bôas, 2000), to immobilize Pb effectively in contaminated soils (Ma et al., 1993, 1994a,b, 1995; Ma and Rao, 1999) and scrubber residues (Geysen et al., 2004) and ash (Ioannidis and Zouboulis, 2003), to reduce Pb uptake by plant in soils (Laperche et al., 1997), to mitigate acid mine drainage, and to reduce the transport of heavy metals in soils (Georgopoulou et al., 1995; Evangelou, 1996).

Important mechanisms involved in phosphate-induced immobilization of heavy metals include chemisorption and chemical precipitation. This is affected by physicochemical properties of the system, such as pH, solubility products, competitive species, chemical speciation, and complex formation. The distinction between different mechanisms is important because different mechanisms have different kinetics and efficiencies in metal immobilization. In addition, because of the uniqueness of each site, an understanding of these mechanisms serves as a basis to design an appropriate strategy when applying P amendments to contaminated sites.

In this chapter we review current knowledge on the mechanisms and assessments of P-induced Pb immobilization in soils and water. Field application of P amendments in a Pb-contaminated soil is also discussed. The information gathered here should be useful for field application of this in situ remediation technology.

15.2. PHOSPHORUS-INDUCED LEAD IMMOBILIZATION IN SOILS AND WATER

15.2.1. Mechanisms

Several mechanisms have been proposed for P-induced Pb immobilization in aqueous solutions and contaminated soils (Ma et al., 1993). They include sorption (Xu and Schwartz, 1994; Melamed et al., 2000), cation substitution (Suzuki et al., 1984; Takeuchi and Arai, 1990), precipitation (Ma et al., 1993; Melamed et al., 2000; Sugiyama et al., 2003; Mavropoulos et al., 2005; Raicevic et al., 2005), or a combination of the three (Ma et al., 1993; Mavropoulos et al., 2002; Ioannidis and Zouboulis, 2003).

Phosphate-induced Pb immobilization is achieved primarily through precipitation of pyromorphite [$Pb_{10}(PO_4)_6X_2$; where X = halide or hydroxide] (Ma et al., 1993; Laperche et al., 1996). Depending on the sources of P supplied to the system, different types of pyromorphite minerals can form: hydroxypyromorphite [$Pb_{10}(PO_4)_6(OH)_2$] when using hydroxyapatite [HA; $Ca_{10}(PO_4)_6(OH)_2$], chloropyromorphite [$Pb_{10}(PO_4)_6Cl_2$] when using HA and Cl^-, and carbonated fluorpyromorphite [$Pb_{10}(PO_4)_{6-x}(CO_3)_xF_2$] when using phosphate rock [PR; $Ca_{10}(PO_4)_{6-x}(CO_3)_xF_2$] (Ma et al., 1995). Among the three pyromorphite minerals, chloropyromorphite has the lowest solubility (Chen et al., 1997b). When it is difficult to differentiate the type of pyromorphite minerals formed in a given system, the generic term *pyromorphite* or *pyromorphite-like mineral* has been used (Ma et al., 1995). Table 15.1 shows some relevant phosphate minerals in soil systems, including Pb and Ca phosphates, and their dissolution reactions and solubility products.

Chloropyromorphite is a common constituent of lead-bearing solid phase in lead-mining areas of the UK, and it was considered the end product of lead weathering (Cotter-Howells, 1994). The presence of lead phosphate in metal-contaminated soils was also confirmed by Ruby et al. (1994). Furthermore,

TABLE 15.1. Dissolution Reactions and Solubility Products (K_{sp}) of Selected Phosphate Minerals

Minerals	Reactions	Log K_{sp}
Ca phosphates		
Hydroxyapatite	$Ca_5(PO_4)_3OH + 7H^+ = 5Ca^{2+} + 3H_2PO_4^- + H_2O$	14.46
Fluorapatite	$Ca_5(PO_4)_3F + 6H^+ = 5Ca^{2+} + 3H_2PO_4^{3-} + F^-$	−0.21
$Ca(H_2PO_4)_2H_2O$	$Ca(H_2PO_4)_2H_2O = Ca^{2+} + 2H_2PO_4^- + H_2O$	−1.15
Monetite	$CaHPO_4 + H^+ = Ca^{2+} + H_2PO_4^-$	0.30
Brushite	$CaHPO_4 \cdot 2H_2O + H^+ = Ca^{2+} + H_2PO_4^- + 2H_2O$	0.63
Pb phosphates		
Hydroxypyromorphite	$Pb_5(PO_4)_3OH + 7H^+ = 5Pb^{2+} + 3H_2PO_4^- + H_2O$	−4.14
Chloropyromorphite	$Pb_5(PO_4)_3Cl + 6H^+ = 5Pb^{2+} + 3H_2PO_4^{3-} + Cl^-$	−25.05
Fluorapyromorphite	$Pb_5(PO_4)_3F + 6H^+ = 5Pb^{2+} + 3H_2PO_4^- + F^-$	−12.98
$Pb_4O(PO_4)_2$	$Pb_4O(PO_4)_2 + 6H^+ = 4Pb^{2+} + 2H_2PO_4^- + H_2O$	2.24
$Pb_3(PO_4)_2$	$Pb_3(PO_4)_2 + 4H^+ = 3Pb^{2+} + 2H_2PO_4^-$	−5.26
Pb minerals		
Cerussite	$PbCO_3 + 2H^+ = Pb^{2+} + CO_2 + H_2O$	4.65
Anglesite	$PbSO_4 + 2H^+ = Pb^{2+} + SO_4^{2-}$	−7.79
$Pb(OH)_2$	$Pb(OH)_2 + 2H^+ = Pb^{2+} + 2H_2O$	8.16
PbO	$PbO + 2H^- = Pb^{2+} + H_2O$	12.72
Galena	$PbS = Pb^{2+} + S^{2-}$	−27.51

Source: Adapted from Lindsay (1979).

hydroxypyromorphite was identified in Pb-contaminated soils after reaction with HA (Ma et al., 1993; Laperche et al., 1996). Ma et al. (1993) demonstrated that Pb reacts rapidly with HA to form hydroxypyromorphite, via a dissolution–precipitation mechanism, and as such, P amendment solubility dictates its efficiency in Pb immobilization. The fact that P was able to transform various Pb species, including cerrusite ($PbCO_3$), anglesite ($PbSO_4$), galena (PbS), and goethite-adsorbed Pb, into pyromorphite supports the precipitation mechanism (Ruby et al., 1994; Zhang and Ryan, 1999a,b). Other studies also demonstrated the precipitation of hydroxypyromorphite following HA dissolution (Zhang et al., 1998). Ma et al. (1993) described the dissolution–precipitation mechanisms with the following reactions:

$$Ca_5(PO_4)_3OH + 6H^+ \xrightleftharpoons{\text{dissolution}} 5Ca^{2+} + 3H_2PO_4^- + OH^- \quad (15.1)$$

$$5Pb^{2+} + 3H_2PO_4^- + OH^- \xrightleftharpoons{\text{precipitation}} Pb_5(PO_4)_3OH + 6H^+ \quad (15.2)$$

In addition to the chemical precipitation mechanism, studies conducted by Suzuki et al. (1984) showed that the immobilization of Pb by HA occurs via replacement of Ca by Pb in an HA crystal lattice to form a PbCa–HA solid phase, $Pb_xCa_{(10-x)}(PO_4)_6(OH)_2$. Since the reaction of HA with aqueous Pb to form hydroxypyromorphite is instantaneous (Ma et al., 1993), the slow kinetics of cation exchange between Pb and Ca would preclude such a possibility. While studying the reaction kinetics of HA with aqueous Pb, Mavropoulos et al. (2002) observed formation of PbCa–HA during the first 6 hours of reaction. However, after 24 hours of reaction, PbCa–HA was completely transformed to a more stable Pb–HA phase [$Pb_{10}(PO_4)_6(OH)_2$]. It is possible that the formation of PbCa–HA occurred through precipitation, whereas conversion of PbCa–HA to Pb–HA occurred through cation exchange between Pb and Ca. However, the exact mechanisms are still unclear.

Further, Mavropoulos et al. (2002) have suggested that metal complexation at the HA surface may have contributed to initial Pb immobilization and pH drop from 5.0 to about 4.5 after adding HA to $Pb(NO_3)_2$ solution. However, we believe that at the initial stages of Pb immobilization, surface complexation of Pb is very unlikely in their system, due to three factors: (1) the HA surface is positively charged in that pH range, which is well below the point of zero charge of HA ($PZC_{HA} = 7.5$) (Sposito, 1984); (2) at pH 4.5 to 5.0, Pb hydrolysis in solution is minimum, favoring surface complexation (Hayes and Katz, 1996); and (3) sorption is known to have a relatively slower kinetics (Sposito, 1984).

This initial pH drop was also reported by other authors (Ma et al., 1993; Zhang et al., 1998). They demonstrated that dissolution of HA, a relatively slow process [Eq. (15.1)], causes solution pH to increase. On the contrary, the precipitation of pyromorphite is instantaneous and causes the pH to decrease sharply [Eq. (15.2)]. Thus, the initial pH drop in the system of Mavropoulos et al. (2002) occurred mainly because, initially, the precipitation of pyromorphite (instantaneous) is more rapid than the dissolution of HA (slow and pH dependent), resulting in a decrease in solution pH. In their case, dissolution of HA is the rate-limiting step. As P is exhausted in solution, the precipitation of pyromorphite stops, and hence there is a gradual increase in solution pH. Only at later stages of Pb immobilization, as the system pH approaches the PZC_{HA} and the first hydrolysis constant of Pb, could surface complexation be contributing to Pb immobilization.

Surface complexation was actually shown to play an important role in Pb immobilization by phosphate rock (PR) at high pH. Melamed et al. (2000) concluded that at pH 8.7, surface complexation is responsible for the relatively slow but efficient Pb immobilization, whereas at pH 3.7, chemical precipitation is responsible for instantaneous Pb immobilization. At pH 5.7, PR was less effective in Pb immobilization. In this case, chemical precipitation occurs instantaneously, contributing to 80% of total Pb immobilization, whereas surface complexation contributed to 20% of total Pb immobilization during the period from 6 to 24 hours reaction time.

15.2.2. System pH

Solution pH is a master parameter in controlling the chemistry of a system. When applying P amendments, the system pH is of fundamental importance because of its role in controlling the solubility of minerals. The system pH also plays an important role in the reactivity of metals at the solid–solution interface. In the case of Pb reaction with PR, the system pH is affected by the mechanisms of Pb immobilization, which in turn affects the solubility of PR. Due to the complexity involving the effect of immobilization mechanism on system pH and the system pH on PR solubility, it becomes difficult to isolate and interpret the various changes in solution chemistry, as these processes occur simultaneously.

Many studies monitored the changes in pH during the reaction of Pb with P materials (Xu and Schwartz, 1994; Melamed et al., 2000, 2003; Mavropoulos et al., 2002). Xu and Schwartz (1994) observed that addition of HA to a Pb solution caused an initial pH drop. They attributed this initial drop in pH to the dominance of P–Pb precipitation [Eq. (15.2)], which delivers H^+, over HA dissolution [Eq. (15.1)], which consumes H^+.

A reduction in system pH enhances the solubility of PR, making the precipitation of pyromorphite minerals possible. However, the sorption of Pb decreases sharply as the system pH decreased, producing a sigmoidal function, usually referred to as an adsorption edge, which reflects the affinity of a metal species for a mineral surface (Sposito, 1984). The ability of Pb to form inner-sphere surface complexes is related to the ability of a species in solution to form hydroxides. In fact, it has been shown that surface affinity of metal cations for Fe-oxide and Fe-hydroxide surfaces agrees with their hydrolysis values (Hayes and Katz, 1996). An analogy between solution complexation and surface complexation is represented in the following reactions (Hayes and Katz, 1996):

$$Pb^{2+} + H_2O = Pb(OH)^+ + H^+ \tag{15.3}$$

$$Pb^{2+} + \text{S-OH} = \text{S-OPb}^+ + H^+ \tag{15.4}$$

$$Pb(OH)^+ + \text{S-OH} = \text{S-OPb(OH)} + H^+ \tag{15.5}$$

where S- represents a surface site.

In the system of Mavropoulos et al. (2002), the addition of HA to a solution containing 581 mg L^{-1} Pb caused the pH to drop from 5 to 4.5. The drop in pH was not as sharp when they increased the initial Pb concentration to 917 mg L^{-1}. With time, a progressive increase in system pH was observed, reaching pH levels of 6.2 and 6.8 for initial Pb concentrations of 581 and 917 mg L^{-1}, respectively. We have attributed the initial pH drop to faster precipitation of pyromorphite (decreasing solution pH) than to dissolution of HA (increasing solution pH) [Eqs. (15.1) and (15.2)].

Another approach to investigating the mechanisms of Pb immobilization by PR employed an automatic titration-batch reaction (Melamed et al., 2000). In this case, PR was dissolved previously at a given pH. During the process, the lag between the pH drop and the adjustment by the titrator to the equilibrated pH could have contributed to a degree of nonequilibrium dissolution. At all pH values studied, as soon as Pb was added, the system pH dropped. The drop in pH was approximately 0.7, 2, and 3 units, at equilibrium pH values of 3.7,

5.7, and 8.7, respectively. At pH 8.7, a more pronounced decrease in system pH was accompanied by a relatively long span in pH adjustment. At pH 3.7, PR was dissolved and the precipitation of Pb with P delivered protons to the solution [Eq. (15.2)] (Xu and Schwartz, 1994). At pH 8.7, with relatively low PR dissolution, Pb consumption was attributed to the dominance of the sorption mechanism (Melamed et al., 2000), reflected by the exchange of Pb with surface protons [Eqs. (15.3) to (15.5)] (Hayes and Katz, 1996).

15.2.3. Kinetics

Melamed et al. (2000) studied the reaction kinetics of Pb immobilization by PR at constant pH. They demonstrated that at pH 3.7, approximately 95% of the total Pb added was immobilized in 5 minutes, whereas at pH 8.7, 24 hours was needed to immobilize approximately the same amount of Pb. At pH 5.7, the efficiency of Pb immobilization by PR was decreased by more than 50% compared to the other two pH values studied. Regardless of the equilibrium pH, the P concentration in solution decreased within the first 5 minutes and then stabilized, concomitant with the decrease in Pb from solution. This is consistent with chemical precipitation of Pb, as shown in other studies (Ma et al., 1995; Zhang et al., 1997; Ryan et al., 2001). However, at pH 8.7, with relatively low P solution concentration, the precipitation reaction is relatively slow. The amounts of Pb being immobilized during the first 6 hours were much greater than that during the next 18 hours, which is typical of fast reaction of Pb sorption. The point of zero charge for PR is approximately 7.5, so at pH 8.7, PR surface is negatively charged and thus Pb was probably removed from the system through sorption. Regardless of the mechanism involved, as reflected by the difference in reaction kinetics, Pb immobilization efficiency is high at low and high pH. At pH 5.7, PR had relatively low efficiency, reflected by low chemical precipitation and low sorption (Hayes and Katz, 1996).

15.2.4. The Role of Calcium

Melamed et al. (2000) examined the solubility of a PR at different pH values. The solution molar ratio of Ca/P was found to be 4.0, 2.6, 2.0, and 1.7 at pH 8.7, 5.7, 3.7, and 3.0, respectively. This means that as the pH decreased from 8.7 to 3.0, the Ca/P ratio in solution decreased from 4.0 to 1.7, approaching the Ca/P molar ratio of PR at equilibrium, reflecting the incongruent nature of PR dissolution with respect to Ca and P. In other words, proportionally more Ca than P was dissolved at higher pH. When aqueous Pb is added to each of those pH equilibrated systems, P solution concentration becomes negligible instantaneously, due to the formation of pyromorphite, while the Ca solution concentration varies with time, decreasing and increasing as time progresses.

In a system where a P mineral (e.g., PR or HA) reacts with aqueous Pb, both Ca and P concentrations are controlled mainly by the dissolution of the P mineral [Eq. (15.1)]. This is because the precipitation of pyromorphite from solution is instantaneous (Ma et al., 1993). This is consistent with the observations of Ma et al. (1993), who examined the reactions of HA with 50 mL of solution containing 500 mg L^{-1} Pb up to 24 hours. As expected, with reaction time increasing from 4 hours to 24 hours, Pb concentrations decreased (5.5 to 4.8 μg L^{-1}), whereas Ca and P concentrations, as well as solution pH, increased. The minimum reduction in Pb concentrations from 4 to 24 hours indicates that the precipitation of pyromorphite was essentially complete within 4 hours.

Assuming that all the Pb removed from solution was precipitated as hydroxypyromorphite (from 500 mg L^{-1} to 4.8 to 5.5 μg L^{-1}), approximately 0.12 mmol of hydroxypyromorphite forms. Thus, much of the P dissolved from HA was precipitated as hydroxypyromorphite, resulting in considerably low P concentrations in solution.

Mavropoulos et al. (2002) found that dissolving HA into 581 and 1770 mg L^{-1} Pb solution caused the molar ratio $Ca_{solution}/Pb_{uptake}$ to decrease initially from approximately 1.0 to 0.82 and from 0.92 to 0.65, reflecting a decrease in Ca solution concentration and increased Pb immobilization. This behavior was attributed to slow HA dissolution and the fact that surface complexation was involved in Pb immobilization. However, the same study also showed that an intermediary phase of CaPbHA was formed. Thus, it is possible that the absence of Ca in solution was due not only to slow HA dissolution but also to formation of an intermediary CaPbHA phase. This is contrary to surface complexation of Pb. This formation of an intermediary phase is also supported by Melamed et al. (2000), who found, at pH 3.7, a decrease in Ca solution concentration (from 3.9 to 3.2 mM) concomitant with the decreases in P and Pb solution concentrations. For the alkaline pH studied (8.7), Ca solution concentration actually increased (from 0.08 mM to 0.16 mM), parallel with Pb immobilization. This was attributed to the substitution of Ca for Pb at the PR crystal lattice, as suggested by Suzuki et al. (1984). This is concomitant with Pb sorption in a system characterized by the high energy of Pb sorption, and relatively low Ca and high Pb solution concentrations.

15.2.5. Effect of the Electrolyte System

Although P amendments were found to transform anglesite ($PbSO_4$) into lead phosphate (Ruby et al., 1994), studies conducted by Melamed et al. (2000) demonstrated that PR had little effect on Pb stabilization in an acid mine drainage environment. This simulation was conducted in a batch reactor at a constant pH of 3 in 0.1 M $Na_2SO_4 + H_2SO_4$. In this case, no association between added Pb and previously dissolved P was found. Instead, Pb immobilization was accompanied by a decrease in solution sulfate concentrations. Scanning electron microscopy (SEM) and respective energy-dispersive x-ray analysis (EDX) confirmed the formation of crystals that do not contain P. These results were also confirmed by the speciation program MINTEQA2, which predicted the preferential formation of anglesite. The formation of anglesite in the simulation experiment was accompanied by a minimum decrease in solution pH.

15.3. ASSESSING THE EFFECTIVENESS OF PHOSPHORUS-INDUCED LEAD IMMOBILIZATION IN SOILS

Numerous studies have assessed the effectiveness of P-induced Pb immobilization in soils (Ma et al., 1993, 1995; Ma and Rao, 1999). Various forms of P, including phosphoric acid, phosphate salts (Na_3PO_4 and Na_2HPO_4), fertilizers ($CaHPO_4$ and triple superphosphate), and minerals [hydroxyapatite–$Ca_5(PO_4)_3OH$ and PR–$Ca_5(PO_4, CO_3)_3F$] have been used (Rabinowitz, 1993; Cotter-Howells and Caporn, 1996; Yang et al., 2001; Cao et al., 2002). In addition to P compounds, various P wastes have been tested for Pb immobilization in soils and aqueous solutions, including phosphate basic slag, phosphatic clay, and bonemeal (calcium phosphate) (Mench et al., 1994; Hodson et al., 2001; Singh et al., 2001). Phosphate has been proven to be effective in immobilizing Pb in soils. For example, P can effectively immobilize aqueous Pb in the presence of various anions: NO_3^-, Cl^-, F^-, SO_4^{2-}, or CO_3^{2-} (Ma et al., 1994a), as well as in the presence of various cations: Zn^{2+}, Cd^{2+}, Ni^{2+}, Cu^{2+}, Fe^{2+}, and Al^{3+} (Ma et al., 1994b).

Various methods have been used to evaluate the effectiveness of phosphate-induced Pb immobilization, including chemical (single or sequential extraction), biological (plant uptake, microbial, and physiologically based bioavailability tests), and mineralogical (x-ray diffraction and scanning electron microscopy) methods (Geebelen et al., 2003). It has been observed that P-induced Pb immobilization that is shown to be effective with one method may not be effective when assessed using a different method. This was observed in a study of Brown et al. (2004), where the 1% P–H_3PO_4 and compost treatments resulted in a decrease of 26% in rat tissue Pb concentration compared with the control soil, whereas the 2.5% Fe + 1% P–TSP (triple superphosphate) showed a 39% decrease. The 1% P–H_3PO_4 treatment caused the greatest reduction in in vitro extractable Pb from field samples (pH 2.2), with a measured reduction of 66%, while the compost treatment had a 39% reduction and the 2.5% Fe + 1% P–TSP treatment, a 50% reduction. The in vitro extraction (pH 1.5) run on field samples showed no reduction in the compost or Fe treatments. The 1% P–H_3PO_4 treatment was the most effective at reducing plant Pb.

Although the methods may seem inconsistent, different methods may serve different purposes. In this section we attempt to provide an overview of various methods that have been used to assess the effectiveness of phosphate-induced Pb immobilization in various systems.

15.3.1. Water-Soluble and Leachable Lead

The effectiveness of P in decreasing water-soluble Pb in aqueous solution as well as in soils has been well demonstrated (Ma et al., 1993, 1995; Ma and Rao, 1999). Consistent with the mechanism of dissolution of P minerals and subsequent or concurrent precipitation of pyromorphite minerals, the effectiveness of such a decrease depends on the solubility of the P mineral used. For example, at a rate of 4 g HA L^{-1}, HA effectively decreased aqueous Pb from 500 mg L^{-1}

to 19.7 μg L^{-1} after 0.5 hour of reaction (Ma et al., 1993). On the other hand, at a rate of 1 g PR L^{-1}, PR decreased aqueous Pb from 5 mg L^{-1} to as low as 9.4 μg L^{-1} after 1 hour of reaction (Ma et al., 1993). It is known that HA is much more soluble than PR, therefore being more effective in decreasing aqueous Pb from solution. In the same study, Ma et al. (1993) also demonstrated the effectiveness of P in decreasing aqueous Pb in a contaminated soil. At a rate of 17 g HA L^{-1} and water/soil ratio of 5, HA reduced aqueous Pb concentration from 2273 μg L^{-1} to 220 μg L^{-1} after 5 hours of reaction. In a separate study, Ma and Rao (1999) demonstrated the effectiveness of PR in reducing aqueous Pb in 13 Pb-contaminated soils. However, such effectiveness was also affected by soil properties and the extent of soil contamination. At a rate of 25 g PR L^{-1} and a water/soil ratio of 10, PR decreased aqueous Pb concentrations to below 15 μg L^{-1} in all but three soils after 2 hours of reaction.

Leachable Pb is a better indicator of Pb transport into groundwater. Whereas water-soluble Pb is often determined in batch experiments, leachable Pb is estimated through column experiments. Ma et al. (1995) demonstrated that PR was effective in reducing leachable Pb in a contaminated soil. At a rate of 1 g of PR per 10 g of soil, the PR from Occidental Chemical reduced leachable Pb from 297 μg L^{-1} to 111 μg L^{-1} when PR was mixed with the soil, and from 38 μg L^{-1} to 1.92 μg L^{-1} when PR was placed at the bottom as a layer. Basta and McGowen (2004) conducted a similar experiment, evaluating the effectiveness of PR and diammonium phosphate in reducing subsurface transport of Pb in a smelter-contaminated soil. They reported that placing PR as a layer was more effective in reducing Pb transport (99.9% reduction) than mixing PR with the soil (<27% reduction). Compared to PR, the most soluble form of P, diammonium phosphate, was more effective.

Considering that the same amount of PR was placed in both cases, the higher efficiency of a bottom layer of PR on the immobilization of Pb, than that obtained by mixing PR with the soil might be associated with the stoichiometry or the relative mass balance of P and Pb. The higher concentration of P at the bottom layer will effectively immobilize Pb, inhibiting its transport farther down the soil profile, whereas if the P is distributed over the soil profile, its concentration might not be sufficient to prevent transport of Pb from the lower part of the treated soil. Nevertheless, if the treatment with PR results in a concentration of dissolved P much higher than the concentration of Pb, the two methods should give similar responses.

15.3.2. Sequential and Single Extraction

Aqueous Pb concentrations are critical for evaluating the effectiveness of P-induced Pb immobilization, since it is directly related to Pb bioavailability in contaminated soils (Ma and Rao, 1997). However, geochemical forms of Pb in soil affect its solubility, therefore directly influencing its bioavailability. Thus, assessing environmental impacts of Pb in soils by its aqueous concentrations alone is incomplete. In addition, knowledge of Pb distribution in different geochemical forms will provide more detailed information on the effectiveness of Pb immobilization by P.

Sequential extraction using various leaching agents is commonly used to assess Pb bioavailability in soils. The extraction scheme is based on operationally defined fractions (Tessier et al., 1979). By nature of the fractionation procedure, Pb that is associated with the residual fraction is less bioavailable than is the nonresidual fraction, which is the sum of the Pb associated with the other five fractions (i.e., water soluble, exchangeable, carbonate, Fe-Mn oxides, and organic fraction). Although the conversion of Pb from the nonresidual to the residual fraction may occur during the fractionation process, this procedure has been shown to be very useful to assess Pb immobilization. Using the fractionation method, Ma and Rao (1997) demonstrated that PR was effective in converting Pb associated with the nonresidual to the residual fraction in all eight contaminated soils. The Pb associated with the residual fraction before PR treatment ranged from 4 to 21%, which increased to 18 to 96% after PR treatment. This is consistent with the fact that the extractants used for nonresidual Pb are ineffective in extracting Pb from P–Pb minerals, including hydroxypyromorphite, $Pb_3(PO_4)_2$, and $PbHPO_4$ (Ma and Rao, 1997). These results indirectly support the hypothesis that P-induced Pb immobilization occurred mainly through precipitation of Pb as P–Pb minerals.

Geebelen et al. (2003) used a different fractionation scheme, which separates metals into three fractions (i.e. acid extractable, reducible, and oxidizable) to evaluate the effectiveness of P-induced Pb immobilization in 10 contaminated soils. After the soils were amended with PR for four months, PR significantly decreased the acid-extractable Pb in seven soils, and the reducible and oxidizable Pb in only two soils.

The effects of PR on Pb availability measured using a single chemical extraction have also been assessed (Ma et al., 1997). While PR was effective in reducing $Ca(NO_3)_2$- and HOAc-extractable Pb from the eight Pb-contaminated soils they studied, it had little effect on EDTA/NH_4OAc-extractable Pb. A study conducted by Geebelen et al. (2003) showed that PR was effective in decreasing $CaCl_2$-extractable Pb in all soils. Boisson et al. (1999) showed that HA added at a rate of 0.5% was effective in reducing $Ca(NO_3)_2$-extractable Pb from 18.3 mg L^{-1} to <1 mg L^{-1}. This has been attributed partially to an increase in pH from 5.7 to 6.9 as a result of HA addition, since dissolution of HA consumes H^+ (Ma et al, 1993).

15.3.3. Bioavailability

Metal bioavailability is an important parameter in determining the effectiveness of metal-contaminated soils in remediation. Assessment of metal bioavailability varies from humans to plants to microbes, and as such, various methods have been developed for a specific purpose. In this section we provide an overview of various methods used to determine the effectiveness of P-induced Pb immobilization. Reduction of Pb bioavailability by PR was demonstrated by feeding-trial and plant-uptake studies (Laperche et al., 1996). Hydroxyapatite and PR effectively reduced blood Pb in rats from 3.7 mg L^{-1} to < 0.9 mg L^{-1}. The bioavailability of hydroxypyromorphite in pigs was only 5% of the reference material, Pb acetate.

15.3.3.1. Physiologically Based Extraction Test Although fractionation data are useful to assess the changes in Pb bioavailability in soils before and after P treatment, they are insufficient for evaluation of Pb bioavailability to humans. However, it is almost impossible to determine the bioavailability of metals to human. In addition, animal-based bioavailability tests are time consuming and expensive. To evaluate Pb bioavailability in soils effectively, Ruby et al. (1996) developed a physiologically based extraction test (PBET), which has been widely used to assess the effect of P on Pb bioavailability in soils (Geebelen et al., 2003). It is an in vitro extraction to mimic human gastrointestinal-tract parameters to predict metal bioavailability in the stomach. The PBET-extractable Pb from contaminated soils is well correlated with data generated from Pb feeding studies using rats (Ruby et al., 1996).

Geebelen et al. (2003) determined the PBET-extractable Pb in 10 contaminated soils after they were amended with PR for four months. Among the three amendments tested (lime, ash, and PR), PR was the only amendment capable of reducing PBET-extractable Pb. However, such a reduction occurred in only three soils. The fact that PR is effective in reducing both acid- and PBET-extractable Pb is attributed to increased P concentrations during extraction at low pH in both cases. In the case of PBET, Pb complexation to blycine might have reduced the amount of Pb available to react with P (Geebelen et al., 2003). Using a column experiment, Yang et al. (2002) examined the effectiveness of H_3PO_4 in reducing PBET–Pb in a smelter-contaminated soil containing 2670 mg Pb kg^{-1}. At an application rate of 10 g P kg^{-1}, P reduced PBET–Pb by 16 to 50%, indicating incomplete transformation of soil Pb to nonbioavailable form, due to incomplete dissolution of Pb solid in the soil. In a separate study, Yang et al. (2001) examined the effects of incubation time and P application rates on PBET–Pb in a soil containing 4360 mg Pb kg^{-1} that received 1.25 to 10 g P kg^{-1}. The higher P application rate (10 g P kg^{-1}) and the longer incubation time (70 days) were the most effective in lowering PBET-Pb from 1789 mg L^{-1} to 564 mg kg^{-1}. It should be noted that the effects of incubation were not as obvious as the P application rate (i.e., the longer incubation time didnot necessarily result in lower PBET–Pb).

15.3.3.2. Plant Uptake In addition to metal bioavailability to humans, metal bioavailability to plants is an important parameter for evaluation of a remediation technology. Numerous studies have shown that P is effective in reducing Pb uptake by various plants. For example, P effectively reduced Pb uptake by tobacco from 36 mg kg^{-1} to 16 mg kg^{-1} (Mench, 1994). However, P was not effective in reducing Pb uptake by ryegrass in the same soil, although it was effective for other two soils where Pb uptake was reduced from 16.2 mg kg^{-1} to 12.3 mg kg^{-1} and from 4.9 mg kg^{-1} to 1.4 mg kg^{-1}. Laperche et al. (1996) showed that

Pb concentrations in the shoots of sudangrass decreased as the quantity of PR added increased, whereas the Pb contents in the roots increased. The increased Pb contents in the roots may have been due to the precipitation of pyromorphite on the root surfaces, which was identified using SEM and XRD. However, Boisson et al. (1999) showed that HA was effective in reducing the Pb concentrations in both the leaves and roots of maize, with greater HA addition resulting in lower plant Pb uptake. For example, Pb content in the roots was reduced from 403 mg kg^{-1} to 99.7 mg kg^{-1} after receiving 0.5% HA. On the other hand, Geebelen et al. (2003) showed that PR was ineffective in reducing the Pb uptake by maize in 10 contaminated soils after they were amended with PR for four months. This is consistent with the fact that PR is relatively less effective in mitigating plant phytotoxicity in these soils.

Zhu et al. (2004) evaluated the effectiveness of treating a Pb-contaminated alkaline soil with hydroxyapatite (HA), phosphate rock (PR), water-soluble P fertilizer (single superphosphate, SSP), and a combination of HA with SSP by measuring the bioavailability of Pb to *Brassica campetris* L. var. *communis* and *Brassica oleracea* L. var. *acephala*. The results indicated that the Pb concentrations in both the shoots and roots of the two vegetable plants decreased with increasing quantities of added P compounds. In addition, the HA treatment was most effective at 5 g of P kg^{-1} compared with other treatments, where the Pb concentrations in shoots and roots decreased by an average of 60% compared to the control. However, the SSP treatment had little effect on the Pb concentrations in plant tissues.

Clearly, the effectiveness of P-induced Pb immobilization in soils depends on soil and plant type as well as P source. One form of P may not be applicable to all soils and plants. Soil pH, for instance, determines the solubility of P in a soil, which is important for precipitating available Pb. Soil mineralogy and pH both influence the kinetics of P and Pb transport, which affects P–Pb association, and therefore Pb bioavailability in a soil.

15.3.3.3. Bioassay Using Worms In addition to plants and humans, the impact of heavy metals on other soil organisms needs to be addressed. Earthworms, one of the important soil organisms, are considered excellent organisms for assessing metal bioavailability in soils. Maenpaa et al. (2002) evaluated the effectiveness of P in reducing Pb bioavailability in soils using an earthworm bioassay. They tested two forms of P [i.e., KH_2PO_4 and $Ca(H_2PO_4)_2 \cdot H_2O$], at 600 and 5000 mg P kg^{-1} soil (soil Pb $=$ 1150 mg kg^{-1}). In the control soil, the bioconcentration factor (Pb concentration ratio in earthworm to soil) was 0.24. After 13 days of reaction, it was reduced to 0.05 to 0.19, with both forms of P being equally effective.

Ownby et al. (2005) used four forms of phosphorus to decrease the bioavailability of Pb to the earthworm *Eisenia fetida* in two soils. Triple superphosphate (TSP) at 5 g P kg^{-1} decreased Pb bioavailability in both soils. Phosphate, applied as PR, monocalcium phosphate, and tricalcium phosphate at 5 g P kg^{-1} was not effective in either soil. To optimize phosphorus amendments, additional phosphorus (up to 15 g P kg^{-1}) and lower pH were used in a series of tests.

The combination of lowering the pH to below 6.0 and increasing phosphorus concentrations caused complete mortality in all TSP-amended soils and partial mortality in the highest PR-amended soils, emphasizing the need to optimize amendment quantity and pH when using this remediation technology.

15.3.3.4. Bioassay Using Microbes Lead bioavailability to microbes is also important, since microbes play an important role in maintaining soil quality. The effects of bonemeal on Pb toxicity to microbes were examined by Hodson et al. (2001). They used the common bacteria *Escherichia coli* and *Pseudomonas fluorescens* as test organisms, both of which are bioluminescent. Addition of bonemeal to a soil at a rate of 1 : 100 was ineffective in reducing Pb toxicity to bacteria, whereas at a rate of 1 : 50 it was slightly effective. These results indicated that bonemeal treatment reduced Pb toxicity to bacteria but didn't render it fully nontoxic. Geebelen et al. (2003) examined the effect of PR on Pb toxicity to microbes using a Pb biosensor (BIOMET). Their results showed that PR addition resulted in a slight or no reduction in BIOMET–Pb. Although P was effective in reducing Pb bioavailability to plants and earthworms, it was less effective in reducing Pb toxicity to soil bacteria.

15.3.4. Toxicity Characteristic Leaching Procedure

For remediation purposes it is important to determine Pb concentrations in soils using the toxicity characteristic leaching procedure (TCLP) (U.S. EPA, 1992). Soils that have TCLP–Pb concentrations below 5 mg L^{-1} are considered nonhazardous and can be disposed of in landfills. Although the procedure has little scientific value, it has been used widely for evaluating soil contamination and remediation.

Wang et al. (2001) compared the effectiveness of three P sources (i.e., $CaHPO_4$, Na_2HPO_4, and $MgHPO_4$) in reducing TCLP–Pb in a soil heavily contaminated with Pb and with a pH of 8. The TCLP–Pb in the soil was as high as 956 mg L^{-1} in the surface soil, which was reduced to 3.94 and 1.46 mg L^{-1} after treatment with 10% $CaHPO_4$ for 7 and 14 days, respectively. $CaHPO_4$ was more effective than the other two P sources. The higher solubility of $CaHPO_4$ than of $MgHPO_4$ may have contributed to its greater effectiveness in reducing TCLP–Pb. Although Na_2HPO_4 is more soluble than $CaHPO_4$, addition of high concentrations of Na may have caused more soil dispersion, resulting in higher levels of TCLP–Pb.

15.3.5. Mineralogical Characterization

At Earth's surface, Pb phosphates, including pyromorphites, are considerably less soluble than other Pb compounds, including oxides, hydroxides, carbonates, and sulfates. Therefore, lead phosphates have been considered to be the most stable environmental Pb form under a wide range of conditions (Ruby et al., 1994).

Cotter-Howells and Thornton (1991) reported the presence of substantial amounts of chloropyromorphite (>50% of the total soil Pb) in highly contaminated garden soils in the UK. The amounts of chloropyromorphite increased with increasing Pb residence time as well as total Pb concentrations in a soil, implying that chloropyromorphite acts as a reservoir for lead in the soil. They attributed the low bioavailability of Pb in these soils to the presence of chloropyromorphite. Ruby et al. (1994) also demonstrated the weathering of galena (PbS) to insoluble Pb phosphates in soils historically used for shipping P ore concentrate.

Cotter-Howells and Caporn (1996) reported the formation of chloropyromorphite after amending a mine waste–contaminated soil with Na_2HPO_4. Based on SEM and XRD analysis, Singh et al. (2001) demonstrated the formation of fluoropyromorphite during a Pb immobilization study using phosphatic clay. They attributed the formation of fluoropyromorphite as the dominant mechanism for Pb removal from the aqueous solution. In a greenhouse study, examining the effectiveness of PR and HA in reducing plant Pb uptake, Laperche et al. (1996) observed the formation of chloropyromorphite on the root surface of sudangrass. It is possible that the rhizosphere may have provided favorable conditions (i.e., increased local concentrations of P and/or Pb) for precipitation of pyromorphite, which was observed by Mench et al. (1994). Crannell et al. (2000) observed the formation of both chloropyromorphite and hydroxypyromorphite after treating combustion ash with soluble P. Apparently, the type of pyromorphite formed depended on the sources of P and associated anions, such as OH^-, Cl^-, and F^-. When HA is supplied as the P source, hydroxypyromorphite usually forms. However, in the presence of Cl^- or F^-, chloropyromorphite or fluoropyromorphite forms instead of hydroxypyromorphite (Ma at al., 1994a). When PR is supplied as the P source, carbonated fluoropyromorphite forms (Ma and Rao, 1999). In addition, solid solutions of Ca–Pb phosphates also form as end products [i.e., $Pb_xCa_{10-x}(PO_4)_6X_2$, where $x = 1$ to 10 and X = halides or hydroxides] (Ruby et al., 1994).

Formation of Pb phosphates in laboratory and greenhouse experiments as well as in field observations clearly demonstrates the stability of Pb phosphates in soils. Such data also provide a sound basis for using P to effectively immobilize Pb in soils.

15.4. FIELD APPLICATION OF PHOSPHORUS AMENDMENTS: CASE STUDY

Although much knowledge regarding the mechanisms involved in the in situ immobilization of Pb using P amendments has been acquired, to our knowledge, there has been only limited implementation of this technology in the field. A known field demonstration of this technology, at a site heavily contaminated with Pb and other heavy metals from battery-recycling wastes was reported by Cao et al. (2002) and Melamed et al. (2003).

15.4.1. Strategy and Management

In this section we emphasize that the in situ remediation of a specific contaminated site requires preliminary laboratory studies to assess the mechanisms involved in P applications, the type and rate of the P amendment, and the appropriate application management. The in situ remediation example shown here (Cao et al., 2003) was conducted in a site dominated by a disturbed sandy, slightly alkaline pH and with a high organic matter content. Total Pb concentrations varied greatly at the site, reaching levels of 30,000 mg kg^{-1} at a depth of 10 to 20 cm. The main form of Pb was $PbCO_3$ (cerussite), with TCLP–Pb above the threshold level of 5 mg L^{-1} (U.S. EPA, 1992). Phosphate application sources and rates to be applied to this contaminated soil were optimized using laboratory experiments (Cao et al., 2003). Phosphate amendments were screened with respect to their efficiencies in metal immobilization, which were measured using TCLP and analysis of leachates from batch experiments and soil columns. Sources of P evaluated included soluble KH_2PO_4, $(NH_4)_3PO_4$, H_3PO_4 and $Ca(H_2PO_4)_2 \cdot H_2O$, phosphatic clay, and PR.

Results of these initial screening tests revealed that application of KH_2PO_4 enhanced heavy metals solution concentration, which was correlated with increased organic matter solubility. On the other hand, the P mixture $Ca(H_2PO_4)_2 \cdot H_2O + CaCl_2 + H_3PO_4$ displayed high immobilization efficiency while maintaining a low level of dissolved organic carbon (DOC) in the leachate. The results also demonstrated that application of PR and phosphatic clay alone remained effective for metal immobilization only at the low concentration range, due to the limitation of the high P application rate required. Also, these materials may act as reactive sinks, reducing the effectiveness of metal precipitation. Phosphoric acid was the most effective in reducing TCLP–Pb to below regulatory levels, which was attributed to its effectiveness in enhancing the solubility of metastable bioavailable cerussite ($PbCO_3$) and allowing precipitation of the more stable geochemical phase pyromorphite. The drawbacks of using soluble H_3PO_4 included a relatively large decrease in leachate pH, resulting in increased mobility of heavy metals. We highlight below some of the findings (Cao et al., 2002, 2003; Melamed et al., 2003) on the effects of three P amendments mixed to the surface horizon at this site: T0, control; T1, H_3PO_4; T2, $H_3PO_4 + Ca(H_2PO_4)_2 \cdot H_2O$, and T3, H_3PO_4 + 5% PR.

15.4.2. Effect of Phosphorus Amendments on Soil pH

Application of P amendments affected the pH of the sandy soil with a low buffering capacity at the contaminated site (Melamed et al., 2003). At the surface soil, T1, H_3PO_4, promoted the highest decrease in soil pH, while T3, H_3PO_4 + 5% PR, promoted the least decrease. The range of pH values at the surface during the sampling period varied from 6.5 to 7.0 in T0 to 4.4 to 5.0 in T1, 5.0 to 5.5 in T2, and 5.5 to 5.7 in T3. These data indicate that phosphoric acid should be applied with caution to avoid enhanced mobility of nonimmobilized Pb and other heavy metals (e.g., Zn, Cd, and Cu) (Cao et al., 2003).

The variations in pH at lower depths were attributed to various factors in addition to the associations with P treatment sources at the top. This included redox conditions associated with fluctuations in the water table, lower levels of CO_2 compared to those on the surface, and possibly the effects of sorption–desorption of metals during leaching.

15.4.3. Chemical Fractionation

As mentioned previously, although the geochemical phases in sequential extraction schemes are operationally defined by the reagents used, it is of general consensus that water-soluble, exchangeable, carbonate-bound, Fe–Mn oxides, and organic-bound phases are more bioavailable than is the residual phase. Ma et al. (1997) and Melamed et al. (2003) applied a sequential extraction scheme (Tessier et al., 1979) to assess the effects of phosphate on Pb immobilization in contaminated soils. The data indicate that in T0 (the control plot), Pb was associated primarily with the carbonate fraction in the entire soil profile (50 to 70%). At the surface, all P amendments were able to modify the partitioning of Pb from the nonresidual or potentially available phase to the residual or unavailable phase. In this respect, the major transformation was a decrease in the carbonate-bound Pb up to about 40%, while the residual phase was increased up to 60%. This is consistent with the designed strategy of dissolving cerussite with phosphoric acid, allowing for the precipitation of geochemically stable pyromorphite, according to the following reactions (Ma et al., 1995; Cao et al., 2003):

$$PbCO_3 + 2H_3PO_4 = Pb^{2+} + 2H_2PO_4{}^- + HCO_3{}^- + H^+ \quad (15.6)$$

$$5Pb^{2+} + 3H_2PO_4{}^- + HCO_3{}^- = Pb_5(PO_4)_3CO_3 + 7H^+ \quad (15.7)$$

Upon P applications, Pb associated with the Fe–Mn oxide fraction also decreased by 10% at the surface. This reduction of Pb in the Fe–Mn oxide fraction may be attributed to the desorption of Pb from oxide surfaces as a consequence of reduced system pH (Hayes and Katz, 1996). The reduction in pH was due mainly to phosphoric acid additions and to the precipitation of P with soluble Pb and/or to direct pyromorphite formation from oxide surface–adsorbed Pb (Zhang et al., 1997).

The efficiency of P treatments to immobilize Pb generally decreased with soil depth, as reflected by the progressively smaller reduction in the residual phase. However, compared to the control, this phase was still be increased by up to 20% at a soil depth of 30 to 40 cm. However, at the site, the levels of Pb were much higher at a 20 to 30 cm depth in all plots. Thus, the consumption of P at the upper layer seemed to have suppressed the availability of P for continuous efficiency on Pb immobilization at the lower layer.

Although metals associated with the Fe–Mn oxide fraction are generally considered as bioavailable, we emphasize that this phase plays a significant role in the attenuation of metal leaching via surface complex formation (Hayes and Katz, 1996). As sorption becomes important at lower depths, the system pH becomes critical. The leaching of metals increases as the system pH is decreased, producing a sigmoidal function (Sposito, 1984, 2004).

15.4.4. Stable-Phase Formation

SEM images and respective EDX of the control (T0) at a soil depth of 0 to 10 cm confirm the presence of cerussite (Melamed et al., 2003). All P amendments were eventually able to produce chloropyomorphite. Coupled with the formation of chloropyromorphite, H_3PO_4 promoted the dissolution of cerrusite ($PbCO_3$), the main form of Pb at the site.

SEM–EDX data (Melamed et al., 2003) detected the presence of pyromorphite at the 0 to 10-cm soil depth of T3 treatment and the association of Pb with P. The elemental dot map also showed the association of Ca with P, coherent with addition of PR. Even at the 30 to 40-cm soil layer of T3 treatment, the EDX revealed the presence of pyromorphyte after 217 days of P application.

15.4.5. Toxicity Characteristic Leaching Procedure

As mentioned in Section 15.4.1, total Pb concentrations varied greatly at the site, reaching levels of 30 g kg^{-1} at a depth of 10 to 20 cm. Also, P concentrations are variable at each P treatment because different sources of P were tested. Nevertheless, the data in Table 15.2 address the efficiency of the treatments to reduce TCLP–Pb even with the Pb variability at the site.

In the control plot (T0) where no P amendment was applied, TCLP-extractable Pb concentrations at the surface soil (0 to 10 cm) far exceeded 5 mg L^{-1} (Melamed et al., 2003). Similar to the distribution of total Pb, the highest concentration of TCLP-extractable Pb was observed at 10 to 20 cm. Even at a depth of 20 to 30 cm, TCLP–Pb still exceeded the critical level. The results showed that P amendments were effective in reducing the TCLP–Pb to below the critical level in the surface soil. These results are of great significance with respect to disposal of the soil, because they show that P amendments can amend the soil to a material that would be considered nonhazardous. Although P treatments did reduce TCLP–Pb in the subsurface soils (10 to 20 cm), TCLP-extractable Pb concentrations in T1 and T3 treatments were still higher than the critical level, except in the T2 treatment. Table 15.2 summarizes all the data for a contaminated site at Jacksonville, Florida.

15.5. CONCLUSIONS

In this chapter we demonstrate the efficiency of P on Pb immobilization and discuss various mechanisms involved in the processes. Mechanisms include sorption, cation substitution, and chemical precipitation, which may occur simultaneously and/or sequentially. The complexity of the kinetics in various systems was discussed, and the data shown encourage further research to identify the reactions

TABLE 15.2. Total Lead and Phosphorus Concentrations, pH, Residual Pb, and TCLP–Pb in the Soil[a]

Soil Depth	Pb ($g\ kg^{-1}$)				P ($g\ kg^{-1}$)				pH				Residual Pb[b] (%)				TCLP–Pb[c] ($mg\ L^{-1}$)			
(cm)	T0	T1	T2	T3	T0	T1	T2	T3	T0	T1	T2	T3	T0	T1	T2	T3	T0	T1	T2	T3
0–10	12.0	6.0	16.0	4.0	1.4	4.0	5.9	13.5	6.7	4.4	5.4	5.6	28	55	70	60	80	<5	<5	<5
10–20	32.0	17.0	22.0	12.5	2.0	3.0	3.2	4.5	6.6	5.5	5.4	6.4	26	40	50	25	200	30	<5	60
20–30	7.0	7.0	4.0	4.0	0.8	1.7	1.7	1.7	6.5	5.5	5.3	6.5	10	25	30	25	60	20	<5	20
30–40	0.5	3.5	2.5	0.1	0.0	1.0	1.0	0.8	6.2	5.3	5.5	6.4	5	20	10	5	<5	<5	<5	<5
40–50	0.1	1.5	1.2	0.0	0.1	0.5	0.6	0.4	5.0	5.0	4.9	5.7	5	10	5	10	<5	<5	<5	<5
50–60	0.0	0.0	0.2	0.0	0.1	0.5	0.6	0.5	4.6	5.0	4.5	5.0	10	10	5	10	<5	<5	<5	<5

Source: Adapted from Melamed et al. (2003).

[a]Following 270 days of phosphorus application at the Jacksonville site.

[b]Residual Pb is obtained from the fractionation procedure.

[c]TCLP–Pb is based on the toxicity characteristic leaching procedure.

occurring in these systems. For example, the role of Ca in Pb immobilization by P amendments is at least intriguing. Methods to assess the effectiveness of P-induced Pb immobilization are based on extractants, plants bioassays, and mineralogical characterization. Usually, a combination of these methods is recommended to assess the effectiveness of P amendment. The selection and strategy of P amendment application is site specific and depends primarily on the nature and extent of the contamination and on the characteristics of the soil at the site. Thus, the type and rate of P amendment and the appropriate application management to be utilized require careful scrutiny. As such, preliminary laboratory studies to assess the mechanisms involved in P applications, their efficiencies, and ultimately the consequences on transport and bioavailability, are of fundamental importance.

The use of phosphoric acid may decrease soil pH, especially in low-buffering sandy soils, and consequently, may cause leaching of heavy metals. Thus, low pH and other heavy metal leaching may be potential drawbacks of its indiscriminate utilization. A mixture of H_3PO_4 and calcium phosphate or rock phosphate had excellent efficiencies, and both treatments had less impact on soil pH. At the specific site described in this chapter, these P amendments were efficient in transforming bioavailable Pb into a nonbioavailable form. Effective remediation technology entails minimizing both leaching and bioavailability of Pb in a soil.

REFERENCES

Basta, N. T., and McGowen, S. L. (2004). Evaluation of chemical immobilization treatments for reducing heavy metal transport in a smelter-contaminated soil. *Environ. Pollut.* **127**, 73–82.

Boisson, J., Ruttens, A., Mench, M., and Vangronsveld, J. (1999). Evaluation of hydroxyapatite as a metal immobilizing soil additive for the remediation of polluted soils, 1: Influence of hydroxyapatite on metal exchangeability in soil, plant growth and plant metal accumulation. *Environ. Pollut.* **104**, 225–233.

Brown, S., Chaney, R., Hallfrisch, J., Ryan, J. A., and Berti, W. R. (2004). In situ soil treatments to reduce the phyto- and bioavailability of lead, zinc, and cadmium. *J. Environ. Qual.* **33**, 522–531.

Cao, X., Ma, L. Q., Chen, M., Singh, S. P., and Harris, W. G. (2002). Impacts of phosphate amendments on lead biogeochemistry in a contaminated site. *Environ. Sci. Technol.* **36**, 5296–5304.

Cao, X., Ma, L. Q., Singh, S. P., Chen, M., Harris, W. G., and Kizza, P. (2003). *Field Demonstration of Metal Immobilization in Contaminated Soils Using Phosphate Amendments*, final report to Florida Institute of Phosphate Research, Bartow, FL. Publ. 01-148-194.

Chen, X., Wright, J. V., Conca, L., and Ferrung, L. M. (1997a). Evaluation of heavy metal remediation using mineral apatite. *Water Air Soil Pollut.* **98**, 57–78.

Chen, X., Wright, J. V., Conca, J. L., and Peurrung, L. M. (1997b). Effects of pH on heavy metals sorption on mineral apatite. *Environ. Sci. Technol.* **31**, 624–631.

Cotter-Howells, J. (1994). Lead phosphate formation in soils. *Environ. Geochem. Health*, **16**, 86.

Cotter-Howells, J., and Caporn, S. (1996). Remediation of contaminated land by formation of heavy metal phosphates. *Appl. Geochem.* **11**, 335–342.

Cotter-Howells, J., and Thornton, I. (1991). Sources and pathways of environmental lead to children in a Derbyshire mining village. *Environ. Geochem. Health* **13**, 127–135.

Crannell, B. S., Eighmy, T. T., Krzanowski, J. E., Eusden, J. D., Jr., Shaw, E. L., and Francis, C. A. (2000). Heavy metal stabilization in municipal solid waste combustion bottom ash using soluble phosphate. *Waste Manag.* **20**, 135–148.

Evangelou, V. P. (1996). Pyrite oxidation inhibition in coal waste by PO_4 and H_2O_2 pH-buffered pretreatment. *Int. J. Surf. Mining Reclam. Environ.* **10**, 135–142.

Geebelen, W., Adriano, D., and van der, L. D. (2003). Selected bioavailability assays to test the efficacy of amendment-induced immobilization of lead in soils. *Plant Soil* **249**, 217–228.

Georgopoulou, Z. J., Fytas, K., Soto, H., and Evangelou, V. P. (1995). *Pyrrhotite Coating to Prevent Oxidation*, Mining and the Environment, Sudbury, ON, Canada.

Geysen, D., Imbrechts, K., Vandecasteele, C., Jaspers, M., and Wauters, G. (2004). Immobilization of lead and zinc in scrubber residues from MSW combustion using soluble phosphates. *Waste Manag.* **24**, 471–481.

Hayes, K. F., and Katz, L. E. (1996). Application of x-ray absorption spectroscopy for surface complexation modeling of metal ion sorption. In *Physics and Chemistry of Mineral Surfaces*, ed. Brady, P. V., CRC Press, Boca Raton, FL.

Hodson, M. E., Valsami-Jones, E., Cotter-Howells, J. D., and Dubbin, W. E. (2001). Effect of bone meal (calcium phosphate) amendments on metal release from contaminated soils: a leaching column study. *Environ. Pollut.* **112**, 233–243.

Ioannidis, T., and Zouboulis, A. (2003). Detoxification of a highly toxic lead-loaded industrial solid waste by stabilization using apatites. *J. Hazard. Materi.* **97**, 173–191.

Laperche, V., Traina, S. J., Gaddam, P., and Logan, T. J. (1996). Chemical and mineralogical characterization of Pb in a contaminated soil: reactions with synthetic apatite. *Environ. Sci. Technol.* **30**, 3321–3326.

Laperche, V., Logan, T. J., Gaddam, P., and Traina, S. J. (1997). Effect of apatite amendments on plant uptake of lead from contaminated soil. *Environ. Sci. Technol.* **31**, 2745–2753.

Lindsay, W. L. (1979). *Chemical Equilibria in Soils*, Wiley, New York.

Ma, L. Q., and Rao, G. N. (1997). Effects of phosphate rock on sequential chemical extraction of lead in contaminated soils. *J. Environ. Qual.* **26**, 788–794.

Ma, L. Q., and Rao, G. N. (1999). Aqueous Pb reduction in Pb-contaminated soils by Florida phosphate rocks. *Water Air Soil Pollut.* **110**, 1–16.

Ma, Q. Y., Traina, S. J., Logan, T. J., and Ryan, J. A. (1993). In situ Pb immobilization by apatite. *Environ. Sci. Technol.* **27**:1803–1810.

Ma, Q. Y., Logan, T. J., Traina, S. J., and Ryan, J. A. (1994a). Effects of NO_3^-, Cl^-, F^-, SO_4^{2-}, and CO_3^{2-} on Pb immobilization by hydroxyapatite. *Environ. Sci. Technol.* **28**, 408–418.

Ma, Q. Y., Traina, S. J., Logan, T. J., and Ryan, J. A. (1994b). Effects of Al, Cd, Cu, Fe(II), Ni, and Zn on Pb immobilization by hydroxyapatite. *Environ. Sci. Technol.* **28**, 1219–1228.

Ma, L. Q., Logan, T. J., and Traina, S. J. (1995). Lead immobilization from aqueous solutions and contaminated soils using phosphate rocks. *Environ. Sci. Technol.* **29**, 1118–1126.

Ma, L. Q., Choate, A. L., and Rao, G. N. (1997). Effects of incubation and phosphate rock on Pb extractability and speciation in contaminated soils. *J. Environ. Qual.* **26**, 801–807.

Maenpaa, K. A., Kukkonen, J. V. K., and Lydy, M. J. (2002). Remediation of heavy metal–contaminated soils using phosphorus: evaluation of bioavailability using an earthworm bioassay. *Arch. Environ. Contam. Toxicol.* **43**, 389–398.

Mavropoulos, E., Rossi, A. M., Costa, A. M., Perez, C. A. C., Moreira, J. C., and Saldanha, M. (2002). Studies on the mechanisms of lead immobilization by hydroxyapatite. *Environ. Sci. Technol.* **36**, 1625–1629.

Mavropoulos, E., Rocha, N. C. C., Moreira, J. C., Bertolino, L. C., and Rossi, A. M. (2005). Pb^{2+}, Cu^{2+} and Cd^{2+} ions uptake by Brazilian phosphate rocks. *J. Braz. Chem. Soc.* **16**, 62–68.

Melamed, R., and Bôas, V. R. C. (2000). Application of physico-chemical amendments for the counteraction of mercury pollution. *Environ. Sci. Technol.* **261**, 203–209.

Melamed, R., Self, P., and Smart, R. (2000). Kinetics and spectroscopy of Pb immobilization on rock phosphate at constant pH. In *Environmental Issues and Management of Waste in Energy and Mineral Production, ld,* Singhal, R. K., and Mehrotra, A. K., A. A. Balkema, Rotterdam, The Netherlands, 301–306.

Melamed, R., Ma, L. Q., Cao, X., Chen, M., and Singh, S. P. (2003). Field assessment of Pb immobilization in a contaminated soil after phosphate application. *Environ. Sci. Technol.* **305**, 117–127.

Mench, M. (1994). Evaluation of metal mobility, plant availability and immobilization by chemical agents in a limed-silty soil. *Environ. Pollut.* **86**, 279–286.

Mench, M. J., Didier, V. L., Loffler, M., Gomez, A., and Massion, P. (1994). A mimicked in-situ remediation study of metal-contaminated soils with emphasis on cadmium and lead *J. Environ. Qual.* **23**, 58–63.

Ownby, D. R., Galvan, K. A., and Lydy, M. J. (2005). Lead and zinc bioavailability to *Eisenia fetida* after phosphorus amendment to repository soils. *Environ. Pollut.* **136**, 315–321.

Rabinowitz, M. B. (1993). Modifying soil lead bioavailability by phosphate addition. *Bull. Environ. Contam. Toxicol.* **51**, 438–444.

Raicevic, S., Kaludjerovic-Radoicic, T., and Zouboulis, A. I. (2005). In situ stabilization of toxic metals in polluted soils using phosphates: theoretical prediction and experimental verification. *J. Hazard. Mater.* **117**, 41–53.

Ruby, M. V., Davis, A., and Nicholson, A. (1994). In situ formation of lead phosphates in soils as a method to immobilize lead. *Environ. Sci. Technol.* **28**, 646–654.

Ruby, M. V., Davis, A., Schoof, R., and Eberle, S. (1996). Estimation of lead and arsenic bioavailability using a physiologically based extraction test. *Environ. Sci. Technol.* **30**, 422–430.

Ryan, J. A., Zhang, P., Hesterberg, D., Chou, J., and Sayers, D. E. (2001). Formation of chloropyromorphite in a lead-contaminated soil amended with hydroxyapatite. *Environ. Sci. Technol.* **35**, 3798–3803.

Singh, S. P., Ma, L. Q., and Harris, W. G. (2001). Heavy metal interactions with phosphatic clay: sorption and desorption behavior. *J. Environ. Qual.* **30**, 1961–1968.

Sposito, G. (1984). *The Surface Chemistry of Soils*, Oxford University Press, New York.

Sposito, G. (2004). *The Surface Chemistry of Natural Particles*, Oxford University Press, Oxford.

Sugiyama, S., Ichii, T., Fujisawa, M., and Kawashiro, K. (2003). Heavy metal immobilization in aqueous solution using calcium phosphate and calcium hydrogen phosphates. *J. Colloid Interface Sci.* **259**, 408–410.

Suzuki, T., Ishigaki, K., and Miyake, M. (1984). Synthetic hydroxyapatites as inorganic cation exchangers. *J. Chem. Soc. Farady Trans.* **80**, 3157–3165.

Takeuchi, Y., and Arai, H. (1990). Removal of coexisting Pb^{2+}, Cu^{2+}, and Cd^{2+} ion from water by addition of hydroxyapatite powder. *J. Chem. Eng. Jpn.* **23**, 75–80.

Tessier, A., Campbell, P. G. C., and Bisson, M. (1979). Sequential extraction procedure for the speciation of particulate trace metals. *Anal. Chem.* **51**, 844–850.

U.S. EPA (1992). Toxicity characteristic leaching procedure (Method 1311). In *Test Methods for Evaluating Solid Waste, Physical/Chemical Methods*, Office of Solid Waste, U.S. Environmental Protection Agency, Washington, DC.

Wang, Y. M., Chen, T. C., Yeh, K. J., and Shue, M. F. (2001). Stabilization of an elevated heavy metal contaminated site. *J. Hazard. Mater.* **88**, 63–74.

Xu, Y., and Schwartz, F. W. (1994). Lead immobilization by hydroxyapatite in aqueous solutions. *J. Contam. Hydrol.* **15**, 187–206.

Yang, J., Mosby, D., and Casteel, S. (2001). Lead immobilization using phosphoric acid in a smelter-contaminated urban soil. *Environ. Sci. Technol.* **35**, 3553–3559.

Yang, J., Mosby, D., and Casteel, S. (2002). In vitro lead bioaccessibility and phosphate leaching as affected by surface application of phosphoric acid in lead-contaminated soil. *Arch. Environ. Contam. Toxicol.* **43**, 399–405.

Zhang, P., and Ryan, J. A. (1999a). Formation of chloropyromorphite from galena (PbS) in the presence of hydroxyapatite. *Environ. Sci. Technol.* **33**, 618–624.

Zhang, P., and Ryan, J. A. (1999b). Transformation of Pb(II) from cerrusite to chloropyromorphite in the presence of hydroxyapatite under varying conditions of pH. *Environ. Sci. Technol.* **33**, 625–630.

Zhang, P., Ryan, J. A., and Bryndzia, L. T. (1997). Pyromorphite formation from goethite adsorbed lead. *Environ. Sci. Technol.* **31**, 2673–2678.

Zhang, P., Ryan, J. A., and Yang, J. (1998). In vitro soil Pb solubility in the presence of hydroxyapatite. *Environ. Sci. Technol.* **32**, 2763–2768.

Zhu, Y. G., Chen, S. B., and Yang, J. C. (2004). Effects of soil amendments on lead uptake by two vegetable crops from a lead-contaminated soil from Anhui, China. *Environ. Int.* **30**, 351–356.

INDEX

Biophysico-Chemical Processes of Heavy Metals and Metalloids in Soil Environments,
Edited by Antonio Violante, Pan Ming Huang, and Geoffrey Michael Gadd